MECHANICAL AND ELECTRICAL SYSTEMS IN BUILDINGS

Fourth Edition

Richard R. Janis

M. Arch., P.E.
Affiliate Associate Professor
School of Engineering and School of Architecture
Washington University

William K. Y. Tao

M.S., D.Sc., P.E.
Affiliate Professor
School of Engineering and School of Architecture
Washington University

PEARSON

Prentice Hall

Upper Saddle River, New Jersey
Columbus, Ohio

Library of Congress Cataloging-in-Publication Data

Janis, Richard R.
   Mechanical and electrical systems in buildings/Richard R. Janis, William K. Y. Tao.—4th ed.
      p.   cm.
   Includes bibliographical references and index.
   ISBN 978-0-13-513013-1
   1. Buildings—Mechanical equipment. 2. Buildings—Electric equipment I. Tao, William K. Y. II. Title.
   TH6010.T36 2008
   696—dc22   20008000327

Vice President and Executive Publisher: Vernon R. Anthony
Acquisitions Editor: Eric Krassow
Editorial Assistant: Sonya Kottcamp
Project Manager: Maren L. Miller
Production Coordination: S4Carlisle Publishing Services
Design Coordinator: Diane Y. Ernsberger
Cover Designer: Aaron Dixon
Operations Specialist: Laura Weaver
Director of Marketing: David Gesell
Marketing Manager: Derril Trakalo
Marketing Coordinator: Alicia Dysert

This book was set in Galliard by S4Carlisle Publishing Services. It was printed and bound by Edwards Brothers. The cover was printed by Phoenix Color Book Group.

Pearson Education Ltd.
Pearson Education Singapore Pte. Ltd.
Pearson Education Canada, Ltd.
Pearson Education—Japan

Pearson Education Australia Pty. Limited
Pearson Education North Asia Ltd.
Pearson Educación de Mexico, S. A. de C.V.
Pearson Education Malaysia Pte. Ltd.

10 9 8 7 6 5 4
ISBN-13: 978-0-13-513013-1
ISBN-10:   0-13-513013-1

PREFACE

THIS BOOK ON MECHANICAL AND ELECTRICAL SYSTEMS COVERS FIVE MAJOR DISCIPLINES: HVAC, plumbing and fire protection, electrical power and telecommunications, illumination, and noise and vibration control.

Coauthors Richard R. Janis and William K.Y. Tao have both taught university courses on mechanical and electrical systems for more than 30 years while working as consulting engineers. Their various courses have emphasized the role of engineers in the building process as well as the theories and technologies of system design. In 1989 they finished the first edition of this text in response to the need for a text and reference up to date with current practice, emphasizing the *Why?* and the *How?* as well as the *What?*

The topics covered in this book are in a state of continuous advancement, triggering the need for substantial updating every few years. This fourth edition incorporates new developments in all the major disciplines and reinforces the relationship of mechanical and electrical systems design in the overall context of the built environment.

The U.S. building industry is approaching general acceptance of sustainable design principles, which were part of the authors' practice long before the concept was popular. *Sustainable design* means that engineers must interact with architects, owners, and facility managers in a team effort to provide high-quality, productive environments for people while considering the impact of their design on the environment. This book is a text and reference for all four parties to the building process.

This book is intended both as a textbook and as a reference book for students and professionals interested in building mechanical and electrical systems. The book is organized as follows:

Chapter 1 covers topics that are relevant for all the mechanical and electrical systems covered in subsequent chapters. This chapter will describe the following concepts:

- Basics of energy required to understand mechanical and electrical systems
- How mechanical and electrical systems affect the design of buildings
- Sustainable design principles
- Basic commissioning
- Economics of building operations
- Tools for evaluating options by economics and quality

Chapters 2 through 17 describe the various mechanical and electrical systems that are used in buildings. These chapters contain the basics of science by which the systems operate, descriptions of the many system and equipment options that can be selected, guidelines for which might be appropriate in given applications, and exercises to reinforce the lessons.

Chapter 18 covers noise control in buildings with an emphasis on preventing acoustical problems with mechanical and electrical systems.

Chapter 19 is entitled "Architectural Accommodation and Coordination of Mechanical and Electrical Systems." This chapter is written for readers who are involved in the planning, design, and construction to understand early in the design process what spaces are required for mechanical and electrical systems, how to allocate area, where best to locate systems and equipment, and what construction details are important to make systems work as intended. The chapter covers topics that in the experience of the author can become problematic become if they are not addressed and resolved early in the design.

The authors are indebted to reviewers, students, coprofessionals, technical associations, and leading product manufacturers (listed separately in the Acknowledgments) for their cooperation in providing data, illustrations, and insights.

Richard R. Janis
William Tao

Online Instructor's Resources

To access supplementary materials online, instructors need to request an instructor access code. Go to  **www.pearsonhighered.com/irc**, where you can register for an instructor access code. Within 48 hours after registering you will receive a confirming e-mail including an instructor access code. Once you have received your code, go to the site and log on for full instructions on downloading the materials you wish to use.

Acknowledgments

CONTRIBUTING AUTHORS

We wish to acknowledge several contributing authors who have taken on the responsibility of reviewing and updating a few of the specialty chapters:

Chapter 12: **Communication, Life Safety, and Security Systems**
Steve Brohammer, RCDD, and Timothy D. Ruiz, RCDD
William Tao & Associates, Inc.
St. Louis, MO

Chapter 15: **Lighting Equipment and Systems**
Davis Krailo, LC
Technical Manager—Engineering
Sylvania
Danvers, MA

Chapter 18: **Noise and Vibrations in Mechanical and Electrical Systems**
J. T. Weissenburger, ScD, P.E.
President, Engineering Dynamics
International
St. Louis, MO

ORGANIZATIONS

Special thanks go to the following organizations for providing valuable design data:

ASHRAE   American Society for Heating Refrigeration and Air Conditioning Engineers
IESNA   Illuminating Engineering Society of North America
ASPE   American Society for Plumbing Engineers
NCAC   National Council of Acoustical Consultants
NEC   National Electrical Code
NFPA   National Fire Protection Association
TIA/EIA   Telecommunication Industry Association
NSPC   National Standard Plumbing Code/ National Association of Plumbing-Heating-Cooling Contractors

PRODUCT MANUFACTURERS

We gratefully acknowledge the product manufacturers that provided valuable illustrations and photos:

AAF International, Louisville, KY
Advance Transformer Co., Rosemont, IL
Airtherm Mfg. Co., St. Louis, MO
Alfa-Laval Gen. Corp., Richmond, VA
Allen-Bradley Co., Milwaukee, WI
American Insulated Wire Corp., Pawtucket, RI
AMTROL, Inc., West Warwick, RI
Ansul Incorporated, Marinette, WI
ASHRAE, Atlanta, GA
ASPE, Westlake, CA
Atlas/Soundolier, St. Louis, MO
Aurora Pump, Aurora, IL
Baltimore Aircoil Co., Baltimore, MD
Belden Wire & Cable, Richmond, IN
Ber-Tek, Inc., New Holland, PA
Brasch Mfg. Co., Inc. Maryland Heights, MO
Bryan Steam Corp, Peru, IN
Bussmann, St. Louis, MO
Carnes Company, Inc., Verona, WI
Cerberus Pyrotron, Cedar Knolls, NJ
Challenger Electrical Equipment Corp., Malvern, PA
Chromalox, Pittsburgh, PA
Cleaver Brooks, Milwaukee, WI
Culligan International Co., Northbrook, IL
Cutler-Hammer/Westinghouse, Pittsburgh, PA
Dukane Corp., Saint Charles, IL
Edwards System Technologies, Farmington, CT
Eljer Plumbing, Plano, TX
Elkhart Mfg. Co., Elkhart, IN
Envirovac Inc., Rockford, IL
Fiberstars, Fremont, CA
Figgie Fire Protection System, Charlottesville, VA
Flexonics, Inc., New Braunfels, TX
Fusion Lighting, Inc., Rockville, MD
General Electric, Cleveland, OH
Grinnell Corp., Exeter, NH
Halo/Cooper Ind., Elk Grove Village, IL
IESNA, New York, NY

Industrial Acoustics Company, Inc., Bronx, NY
ITT Bell & Gossett, Morton Grove, IL
Joy Technologies, New Philadelphia, OH
Kidde-Fenwal, Inc., Ashland, MA
Kohler Company, Kohler, WI
Lennox Industries, Richardson, TX
Lightolier, Fall River, MA
MagneTek Drives & Systems, New Berlin, WI
Marley Cooling Towers, Mission, KS
Mason Industries, Hauppauge, NY
Master Publishers, Inc., Richardson, TX
McQuay International, Minneapolis, MN
Minolta Corp., Ramsey, NJ
Mitel Corporation, Canada K2K 1X3
Mueller Company, Springfield, MO
NFPA, Quincy, MA
NSPC, Falls Church, VA
OSRAM/SYLVANIA, Danvers, MA
Palmer Instruments, Inc., Asheville, NC
Panasonic Co., Secaucus, NJ
PASO Sound Systems, Pelham, NY
Pelli & Associates, New Haven, CT
Peerless Lighting Corp., Berkeley, CA
Philips Lighting Co., Somerset, NJ
Reliable Sprinkler Co., Mt. Vernon, NY
Sams Publishing, Indianapolis, IN
Sloan Valve Co., Franklin Park, IL
SMACNA, Chantilly, VA
Southwire Co., Carrollton, GA
Spirax Sarco, Allenton, PA
Sporlan Valve Co., Washington, MO
Square D Co., Lexington, KY

Star Sprinkler Corp., Milwaukee, WI
State Industries, Ashland City, TN
Sterling Heating Equipment, Westfield, MA
Tate Access Floors, Jessup, MD
Thomas Industries, Tupelo, MS
Titus, Richardson, TX
TIR Systems LTD, Burnaby, BC
Trane Company, La Crosse, WI
United McGill Corp., Groveport, OH
Van-Packer Co., Buda, IL
Victaulic Co. of America, Easton, PA
Viking Corporation, Hastings, MI
Walker Systems, Inc., Parkersburg, WV
Watts Regulator Co., N. Andover, MA
Weil McLain, Michigan City, IN
H.E. Williams, Inc., Carthage, MO
Wiremold Company, West Hartford, CT
York International, York, PA
Zurn Industries, Erie, PA

REVIEWERS

We wish to thank the following reviewers for their helpful comments and suggestions:

Leonard R. Bachman, University of Houston
John A. Bryant, Texas A & M University
Ruchi Choudhary, Georgia Institute of Technology
Jere C. Hamilton, University of Wyoming
Michael Horman, Penn State University
Richard M. Kelso, University of Tennessee
Thomas M. Korman, University of Nevada, Las Vegas

CONTENTS

CHAPTER 19

ARCHITECTURAL ACCOMMODATION AND
COORDINATION OF MECHANICAL AND
ELECTRICAL SYSTEMS   593

APPENDIX A

GLOSSARY OF TERMS, ACRONYMS, AND
ABBREVIATIONS   **629**

APPENDIX B

GLOSSARY OF TECHNICAL
ORGANIZATIONS   **643**

APPENDIX C

UNITS AND CONVERSION OF
QUANTITIES   **645**

INDEX   **649**

1

INTRODUCTION TO MECHANICAL AND ELECTRICAL SYSTEMS: ENERGY, SUSTAINABILITY, AND ECONOMICS

1.1 BASICS OF ENERGY

Mechanical and electrical (M/E) systems use and convert energy and move fluids to make buildings habitable and functional. A basic understanding of fluid flow, energy forms, and conversion factors is essential to the study of M/E systems and in understanding discussions of M/E systems in design and construction. Energy forms applicable to building systems include thermal energy, electricity, mechanical energy, and chemically stored energy (fuels).

The words "energy" and "power" are often used interchangeably, but there is an important distinction between the two. Energy is a quantity, such as heat; power is the rate at which the quantity is transferred or used. Table 1–1 shows the forms of energy and power, their units of measure, and conversion factors.

Thermal energy is measured in British thermal units, or Btus. A Btu is the amount of heat required to raise 1 lb of water 1°F. Stated another way, if we heat 1 lb of water (about 1 pint) 1°, the water will have absorbed 1 Btu. If we heat a pound of water 2°, we will need 2 Btu. Pound for pound, water will absorb much more heat than most other materials for a given temperature rise. For example, a Btu will raise 1 lb of water 1°. Only 0.156 Btu will be necessary to raise 1 lb of concrete 1°. If we normalize the heat-absorbing capacity of water at 1.0, the heat capacity (C) of concrete will be 0.156. These relationships can be combined into the following equation:

$$Q = M \times C \times TD \qquad (1-1)$$

where

Q = heat absorbed (or released) (Btu)
M = is mass (lb)

C = heat capacity (often called "specific heat") (Btu lb.°F)
TD = temperature increase or decrease, °F

The quantity C, heat capacity or specific heat, is listed for many common materials in Table 1–2.

Example 1.1 A 10'-by-10' concrete floor is 8" thick. If the floor is warmed by the sun to 80° during the day and cools to 70° overnight, how much heat is stored and released by the floor on a daily basis?

The specific heat of concrete is 0.156 Btu per lb °F. The density of concrete is approximately 144 lb/cu ft. Heat storage is calculated below:

$$Q = M \times C \times TD$$
$$= 144 \times (10 \times 10 \times 8/12) \times 0.156 \times (80 - 70)$$
$$= 15,000 \text{ Btu}$$

The *rate* of energy flow is "power." The unit of power for thermal energy will be Btus per hour, abbreviated Btuh. This unit is used in quantifying the amount of heating gained or lost by a structure (load) and the amount of heating or cooling capacity required by equipment to offset the heat or load.

For all forms of energy the following equation will apply, but units will depend on energy form:

Power = Energy/time
or
Energy = Power × time   (1–2)

Example 1.2 In the previous example, the heat was released from the concrete slab during a night setback period from 10 P.M. to 6 A.M. What was the average

TABLE 1–1
Forms and units of energy and power

Energy Form	Unit of Measure		Conversion to Btu
	Energy	Power	
Heat	British thermal unit (Btu)	British thermal unit per hour (Btuh)	1.00
Electric	Watt-hour (W)	Watt (W)	3.41
	Kilowatt-hour (kWh)	Kilowatt (kW)	3,413
Mechanical	Horsepower-hour (hp-hr)	Horsepower (hp)	2,545

capacity of the slab over this period to assist in heating the building?

The amount of heat is 15,000 Btu. It was released over an 8-hour period, therefore the average capacity was

Power = Energy/time

   = 15,000 Btu/8 hours = 1875 Btuh

Electric power is measured in watts (W) or kilowatts (1000 W). These are power units. If power is applied over time, energy is the product:

Electric energy (kilowatt-hours, or kWh)

   = Electric power (kilowatts, or kW)

   × time (hours)   (1–3)

Example 1.3   A 100-W light is on for 10 hours per day. How much energy will the light use in a year's time?

Energy = power × time

Energy (kWh) = 100 Watts × 10 hours per day

   × 365 days per year

   = 36,500 watt-hours, or 36.5 kWh

Electrical energy can be converted to mechanical energy in a motor, to light in a lamp, or to heat in a resistance heater. All of the electrical energy used in a heater becomes heat. In a motor, the majority of the electrical energy becomes mechanical power, measured in horsepower. A small portion of the electrical energy is lost as heat.

TABLE 1–2
Heat capacities of common materials

Material	Density lb/cu ft	Heat Capacity Btu/°F lb
Water	62.4	1.0
Wood	45	0.57
Foam insulation	1.5	0.38
Air	.075	0.24
Concrete	144	0.156
Steel	489	0.12

Eventually, even the mechanical energy degrades into heat. In a lamp, a portion of the electrical energy becomes light, and a portion becomes heat. Eventually, virtually all the light is absorbed by the surface and becomes heat.

Example 1.4   An electric motor running a large copier draws 1.6 kW. How much heat is produced in the space as a result of the copier's operation?

From Table 1–1 we find the conversion factor from electric to heat energy or power:

Heat power (Btuh)

   = Electric power (kW) × 3413 Btuh/kWh

   = 1.6 × 3413 = 5460 Btuh

1.2 FUELS

Fuels are burned to produce thermal energy, which can be used to heat buildings or run engines to produce mechanical energy. The mechanical energy can be used to operate machinery, vehicles, or to produce electricity in a generator. Fuels commonly associated with building systems include natural gas (primarily methane), propane (LPG), oil (various grades), and coal (for very large applications). The thermal energy produced by burning various fuels is shown in Table 1–3.

If fuel is burned in a boiler or furnace, a small portion of the heat is lost through the flue stack and through radiated heat losses from the equipment. The net heat available for use will generally be 60 to 95 percent of the total. This percentage is the "efficiency" of the boiler or furnace.

Example 1.5   A 75% efficient boiler is required to produce 800,000 Btuh to offset a heating load. If the boiler uses natural gas, what will the input rate be in cu ft per hour?

From Table 1–3, each cubic foot of gas has a heating value of 1000 Btu. At 75% efficiency, each cubic foot will produce a net heating value of .75 × 1000, or 750 Btu. To produce 800,000 Btuh, we will need 800,000 Btuh/750 Btu per cu ft, or 1067 cu ft per hour.

TABLE 1–3
Heating values of various fuels

Fuel	Unit of Measure[1]	Nominal Heating Value/Unit, Btu (kJ)	Combustion Efficiency, %
Natural gas	cu ft	1000 (1055)	70–85
LP (propane gas)	gal	93,000 (98,000)	70–85
No. 1 oil (diesel)	gal	138,000 (146,000)	75–80
No. 5 oil (heavy)	gal	145,000 (153,000)	72–82
No. 6 oil (bunker C)	gal	153,000 (161,000)	75–80
Soft coal (bituminous)	lb	13,000 (14,000) 13,700 (14,800)	75–85
Hard coal (anthracite)	lb	12,500 (13,500) 13,200 (14,300)	75–85
Electrical resistance	kWh	3413 (3600)	100[2]
Electric heat pump	kWh	5100 (5,400)	150–300[3]

[1]1 gallon = 3.78 liters; 1 cubic foot = 28.32 liters; 1 pound = 0.454 kilogram.
[2]Electrical to heat energy conversion efficiency.
[3]Denotes coefficient of performance (COP) percent or 1.5–3.0 per unit.

1.3 FLUID FLOW AND PRESSURE IN MECHANICAL SYSTEMS

Mechanical systems use the flow of air, water, and steam to transfer energy. Airflow is measured in cu ft per minute, abbreviated CFM. Air pressures in heating and air conditioning systems are very low, and measuring in the familiar unit of psi (pound per sq in) would result in numbers too small to be used conveniently. Pressure is measured in "inches of water column" as explained in Figure 1–1.

Water flow is measured in gallons per minute, abbreviated GPM. Pressures are measured in two units: psig and "ft of head." Head, measured in feet, is equal to the pressure at the bottom of a column of water. For instance, a dam that has a headwall holding water 100 ft in depth has a pressure of 100 ft of head at the bottom. One psig is equal to 2.31 ft of head.

Steam is measured in lb and flow is measured in lb per hour.

1.4 HOW MECHANICAL AND ELECTRICAL SYSTEMS AFFECT BUILDING DESIGN

Before modern heating, air conditioning, and illumination systems, building dimensions were limited due to the need to access windows for light and natural ventilation (see Figure 1–2). Floors were typically 60 feet or less in depth, or included light wells. Windows were operable and needed to be large and tall enough to allow deep light penetration. Ceiling heights were high to promote stratification of summer heat and to allow the use of operable transom windows over doors for ventilation of interior spaces. Features shown in Figure 1–3 are now found only in historic buildings. These buildings were much more responsive to the exterior environment than later modern buildings, which rely on artificial light and air conditioning.

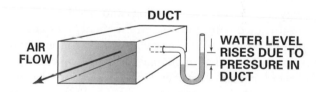

Air pressure measured in "inches of water column."

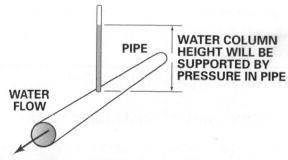

Water pressure measured in "feet of head."

■ **FIGURE 1–1**
Measuring pressure of air and water in HVAC systems.

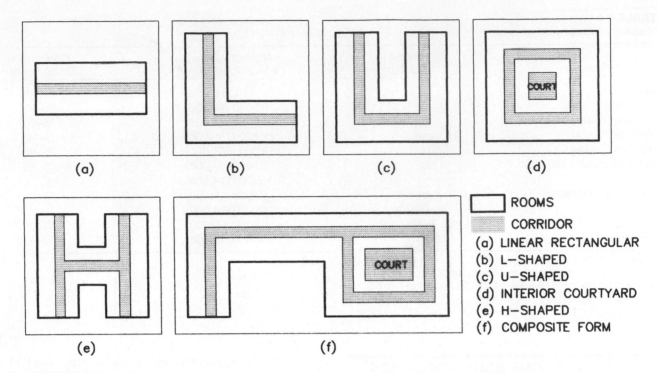

■ **FIGURE 1-2**
Common building geometry prior to development of modern M/E systems.

Air conditioning and good artificial lighting gave architects the flexibility to design larger floors, and good elevators made high-rise construction possible. These developments occurred when energy to operate buildings was inexpensive by today's standards, and there was very little concern about fossil fuel depletion, dependence on foreign oil, or environmental impact of energy use. As a result, buildings and building systems were designed with little regard for energy efficiency.

In the early 1970s, the political and economic context of building design changed with the oil embargo, increased energy costs, and the realization that we needed to take care of the environment. These concerns gradually coalesced into sustainable design in the early 1990s, which embodies a renewed interest in daylighting, natural ventilation systems, and energy-efficient mechanical and electrical systems.

1.5 SUSTAINABLE DESIGN

1.5.1 Overview of Sustainability

Sustainability is a concept that applies not only to buildings, but also to industry, agriculture, transportation, and all other aspects of societal activity. "Sustainable" can be defined simply as *having an overall beneficial effect on productivity, health, resources, economics, and*

the environment. Sustainable design acknowledges responsibility for future as well as current outcomes. Sustainable design decisions are made for their impact not only at the building level, but also at the community and global level. Utility, comfort, energy conservation, environmental impact, and appropriate use of technology are basic criteria for mechanical/electrical systems in a sustainable design process.

The LEED™ rating system, created by the United States Green Building Council (USGBC), is an excellent way to quantify the sustainability of buildings. (see Figure 1–4.)

1.5.2 Design Interactions

Achieving sustainable building solutions requires that many parties work closely together with an understanding of the interactions among building systems and processes. For example, energy usage is affected by architectural form, building materials, lighting, appliances, HVAC systems, and even by access to public transportation. There are many participants in the design process who take responsibility for these issues (e.g., architect, lighting designer, owner, consulting engineers, contractors, suppliers, and others). Too often, each participant makes decisions independent of the others, and opportunities are lost by not understanding the interactions between design factors.

■ **FIGURE 1–3**
Prior to air conditioning, buildings were equipped with features to take advantage of natural ventilation, such as operable sash, louvered shutters (a), and transom lights (b).

■ **FIGURE 1–4**
Nidus Center for Scientific Enterprise, among the first LEED™ Certified buildings, uses principles of sustainable design to conserve resources, reduce environmental impact and produce a healthy productive place to work. LEED™ (Leadership in Energy & Environmental Design) is a "green" building rating system administered by the U.S. Green Building Council. (Courtesy of WTA.)

Decisions made by each member of the team will affect systems in which others are also affected. For instance, an architect might design larger windows, which could increase the size of heating and air-conditioning equipment. Or, the lighting designer might design more light fixtures, which would increase the size of air-conditioning equipment.

Interactions also affect health and productivity of building occupants. Daylight and outdoor views, for example, enhance occupants' sense of well-being in buildings, and the effect on performance in the workplace is obvious, though difficult to quantify. Likewise, effective HVAC systems contribute to good indoor air quality to the benefit of occupants' health.

1.5.3 Environmental Impact of Buildings and Building Systems

A building's impact goes beyond the site boundary through contribution, both directly and indirectly, to fossil fuel depletion, acid rain, ozone depletion, global warming, water pollution, natural resource depletion, urban congestion, and growth of landfills. Sustainable design must also go beyond site boundaries to consider how well buildings work to minimize negative environmental impact or even benefit the environment.

Buildings contribute to disruption of storm-water flow, ground erosion, fouling of natural water, light pollution, and ultimately, the growth of landfills from disposal of building materials as construction waste and, ultimately, demolition. These impacts can be mitigated by good design, and there is potential for well-planned buildings to have zero impact or even improve the environment.

Buildings account for about 30 percent of overall energy usage in the United States and over 60 percent of electrical usage. This represents not only a depletion of energy resources, but also affects the environment by emissions through combustion of fossil fuels, both on the building site and remotely at power-generating stations. Pollution from energy consumption is quantified in Table 1–4.

TABLE 1–4
Air pollutants produced from energy conversion

Energy Converted or Consumed	Air Pollutants Produced, g (lb)		
	CO_2	SO_2	NO_x
1 gallon of fuel oil by combustion[a]	10,500 (23.1)	45.0 (0.10)	18.3 (0.04)
1 gallon of gasoline by automobiles[b]	8500 (18.8)	37.0 (0.08)	15.0 (0.03)
1 pound of coal by combustion[c]	1090 (2.4)	9.0 (0.02)	4.4 (0.01)
1 therm of natural gas by combustion[d]	6350 (14.0)	Nil (–)	24.0 (0.05)
1 kWh of electric energy generated by oil[e]	860 (1.9)	3.7 (0.008)	1.5 (0.003)
1 kWh of electric energy generated by gas[e]	635 (1.4)	Nil (–)	2.4 (0.005)
1 kWh of electric energy generated by coal[e]	1090 (2.4)	9.0 (0.02)	4.4 (0.01)

[a]Calculated by using fuel oil containing 85 percent carbon and 12 percent hydrogen and 7.4 lb/gal.
[b]Calculated by using gasoline mixture of C_8H_{18} and (C_nH_{2n+2}) having 84 percent carbon and 15 percent hydrogen, and 6.1 lb/gal.
[c]Calculated by using bituminous coal containing 65 percent carbon and 3.8 percent sulfur.
[d]Calculated by using mixture of methane (CH_4) and ethane (C_2H_6) and 100,000 Btu/therm.
[e]Data from Green Light Program, Environmental Protection Agency.

1.5.4 Water Conservation

Conserving water is a goal of sustainable design. As with most elements of sustainable design, there are economic benefits. Saving water in buildings will reduce the cost of constructing, improving, and maintaining water and sewer infrastructure. EPact (Energy Policy Act of 1992) became effective in 1996 to mandate that manufacturers produce conventional fixtures that flow less water. Sensor controls have also become commonplace in building design. Currently, designers are using alternative products on a limited basis, which use even less water, such as waterless urinals, rainwater collection, and composting toilets. These measures will require acceptance by owners and code officials before widespread usage.

1.5.5 Energy Conservation

Sustainable design concepts for energy conservation can be quite different from the mindset that engineers and architects have employed in the not too distant past. During the late 1970s and early 1980s the building industry reacted to concerns for rising energy costs and the realization that fossil fuels are limited. Designers responded with a simplistic approach to saving energy by reducing lighting levels, lowering ventilation rates, and operating systems by methods that reduced comfort. In addition, very expensive technologies were often used to save energy with no basis for economic justification. Solar collectors, often installed during the "Energy Crisis" of the 1970s, were generally uneconomical and a short-lived engineering fad (see Figure 1–5). The resources applied to inappropriate solutions would certainly have been better spent on energy solutions that improve overall building performance, have good economics, and could be

maintained over the long term such as the heat recovery wheel shown in Figure 1–6.

Energy conservation solutions and renewable energy should be evaluated and applied within economic constraints and criteria. This can be done in a limited way by applying economic criteria such as "payback" analysis. More advanced methods including discounted cash flow analysis are also described in this chapter.

1.6 INDOOR ENVIRONMENTAL QUALITY

1.6.1 Components of IEQ

In addition to environmental benefits and resource conservation, sustainable design enhances health, well-being, and productivity of building occupants. These benefits are achieved by several goals of sustainable design:

- Healthful indoor air quality
- Thermal comfort and individual control
- Good lighting
- Connection with the outdoors

Combined, these factors contribute to "Indoor Environmental Quality" (IEQ), a term used in the LEED™ rating system, described above.

1.6.2 Indoor Air Quality

Indoor air pollution is preventable by good architectural detailing as shown in Figure 1–7, effective mechanical systems, and proper maintenance. Indoor air

■ **FIGURE 1–5**
Solar Collector fad after the 1970's oil embargo was a well-intentioned, but poorly conceived solution in sustainable terms, which must consider economic realities.

■ **FIGURE 1–6**
Heat recovery wheel allows higher ventilation rates without sacrificing economy of operation, exemplifying sustainable design within economic realities. (Courtesy of WTA.)

pollution in typical buildings, such as offices, comes from finish materials, cleaning products, furniture, and fumes from equipment. In addition, biopollutants such as mold can result if humidity is not properly controlled or there are moisture problems in building assemblies and systems.

Interior chemical pollution and odors can be diluted to acceptable levels with ventilation by liberal quantities of relatively purer outdoor air. Selecting furnishings, interior finish materials, and cleaning products to be nonpolluting will allow lower ventilation rates and save energy. Using local exhaust over offensive equipment is also effective at preventing chemicals from entering the larger occupied space.

Building occupants themselves are also sources of pollution. They consume oxygen and emit carbon

■ **FIGURE 1–7**
Air intakes for Monsanto Research Center in St. Louis are placed high to avoid street level air pollution as a measure to improve indoor air quality.

dioxide and body odors. ASHRAE Standard 62-99, entitled "Ventilation for Acceptable Indoor Air Quality," specifies the amount of outside air needed to cover various levels of occupancy. The amount of outside air required in buildings is based on the nature and density of occupancy.

Condensation in roofs or walls can also be a problem. Venting, insulation, and vapor barriers can avoid condensation if properly applied, and HVAC design for proper dehumidification is essential. Interior surfaces of HVAC systems can also harbor dust, odors, bacteria, and mold. Filters are porous and microorganisms can breed if they are not changed frequently. Acoustical duct liner is also porous, and should be avoided or treated with biocide. Condensate pans in air-conditioning systems are continually moist during hot weather, and should drain properly.

Indoor air quality has been correlated with employee productivity. Measures cited above are reported to render 23 to 76 percent reductions in the incidence of acute respiratory illnesses. Measured data is also available on the relationship between "sick building syndrome (SBS)" symptoms and worker performance. Workers who reported any SBS symptoms took 7 percent longer to respond in a computerized neurobehavioral test. In another test, workers with symptoms had a 30 percent higher error rate. Considering the number of office workers with two or more frequent SBS symptoms yields a 3 percent average decrease in performance.

One study was performed to determine the effects of ventilation rate on absenteeism. Buildings were classified as moderate ventilation (25 CFM/occ) or high ventilation (50 CFM/occ). Absence rate was 35 percent lower in high-ventilation buildings. The moderate ventilation rate cited in the study is higher than rates prescribed by ASHRAE Standard 62, indicating potential for improvement in current design practices.

1.6.3 Thermal Comfort

In ASHRAE Standard 55, entitled "Thermal Environmental Comfort Conditions for Human Occupancy," comfort involves factors including temperature, air velocity, and humidity. In general, the standard asserts that these quantities must be maintained within reasonable levels and not allowed to change rapidly.

Basically, the standard identifies conditions of temperature and humidity that 80 percent of research subjects will find acceptable. The obvious corollary is that 20 percent will be uncomfortable. This implies that there will be a greater likelihood of satisfying everyone if individual temperature controls are provided. Space heaters and thermostat tampering demonstrate the desire for individual control.

Air temperature has been documented to affect worker performance. Small differences in temperature have been reported to have 2 to 20 percent performance impact in tasks such as typewriting, learning performance, reading speed, multiplication speed, and word memory.

1.6.4 Individual Control

HVAC systems can be designed that offer opportunities for individuals to control their local thermal environment. This simple notion is generally ignored in typical institutional buildings designed with the goal of providing uniform temperature control.

Allowing greater personal control of indoor environments, and allowing temperatures to fluctuate with outdoor conditions could improve perceived comfort and reduce energy consumption. Individuals will tolerate a wider range of thermal conditions if they have control over their environment, such as operable windows or the ability to adjust airflow. The effect of individual control on productivity has also been documented. Providing $\pm 5°F$ of individual temperature control has been claimed to increase work performance by 3 to 7 percent.

Individual control is not practical with many HVAC systems. There are, however, several practical options for giving control to individuals. A few furniture manufacturers can integrate "task cooling" into their workstations, which allows the occupant to adjust the quantity and direction of airflow. (See Figure 1–8.) Ironically, table fans used before the advent of air conditioning are similar in concept.

Other options are to deliver air through floor registers, which allows the nearest occupant to adjust the airflow, or to use operable windows, which is appropriate in some climates and/or some seasons. One analysis revealed that occupants of buildings using central HVAC systems were much more sensitive to temperature variation than occupants of buildings that have operable windows. Having control results in higher perceived comfort. People might even be invigorated by the variability of temperature in naturally ventilated buildings. (See Figure 1–9.)

Integrating operable windows with conventional HVAC control systems is a challenge. Typical systems, for instance, might place multiple rooms on the same thermostat. If the occupant with the thermostat opens the window, control will be lost for the other spaces. Other potential problems include possible freezing from cold air through windows left open, security, and infiltration of pollen and dust. Despite potential problems, operable windows are highly desired by building occupants and well worth the effort to work out problems.

The use of operable insulation to reduce heat losses and gains during unoccupied periods would be an effective way to negate the energy penalty of large glazing areas, but cost, operation, and maintenance are problematic.

To date, there is no effective equivalent of the shutter in modern commercial or institutional buildings.

1.6.5 Superior Lighting Systems

Lighting affects quality of space as well as energy consumption. Uniform illumination by recessed fluorescent fixtures is the most common lighting solution for work spaces, and often results in glare, shadows, and reflections in computer screens. Indirect or semi-indirect lighting is an alternative, which produces better visibility of tasks at lower levels of illumination. Indirect lighting is theoretically less efficient than direct lighting due to light lost in reflection from room surfaces. However, indirect light is uniform, eliminates glare, results in less shadows, and can be designed at lower light levels to produce a better environment at lower energy cost. (See Figure 1–10.)

Daylight

Sustainable lighting strategies generally include daylighting. The challenge in using daylight is to control the glare, avoid thermal discomfort, and minimize HVAC loads. Energy interactions must be considered carefully. While one would expect higher air-conditioning loads due to extra window or skylight area, the extra load may be more than offset by reducing the heat gain from artificial lighting.

No one would question that an attractive, visually interesting environment contributes to occupant satisfaction and higher levels of productivity. Having an outdoor view or a source of natural light is desirable. (See Figure 1–11.) The best publicized study on the effects of daylight and view was performed by The Pacific Gas and Electric Company. The following is quoted from their executive summary:

> Controlling for all other influences, we found that students with the most day lighting in their classrooms progressed 20% faster on math tests and 26% on reading tests in one year than those with the least. Similarly, students in classrooms with the largest window areas were found to progress 15% faster in math and 23% faster in reading than those with the least.
>
> And students that had a well-designed skylight in their room, one that diffused the daylight throughout the room and which allowed teachers to control the amount of daylight entering the room, also improved 19–20% faster than those students without a skylight. We also found another window-related effect, in that students in classrooms where windows could be opened were found to progress 7–8% faster than those in rooms with fixed windows. This occurred regardless of whether the classroom also had air conditioning. These effects were all observed with 99% statistical certainty.

■ **FIGURE 1–8**
Personal cooling outlet (left) gives individual personnel control of climate at workstation.

■ **FIGURE 1–9**
Variations in environment are well tolerated when people have a choice; these shoppers prefer an open-air market to the modern climate-controlled grocery store.

■ **FIGURE 1–10**
Indirect lighting in this research laboratory is not only comfortable for occupants, but also illuminates building services in the exposed ceiling, resulting in better maintenance and safety.

■ **FIGURE 1–11**
Light well in this classroom building allows daylight to the interior and gives occupants a sense of outdoor weather and time of day.

■ **FIGURE 1–12**
A simple window at the end of this laboratory corridor provides daylight and view. Lights are rarely turned on during the day in this space.

1.6.6 Connection with Outdoors

Daylight, views outside, natural ventilation, and temperature variation are ways to give building occupants a sense of connection with the outdoors. Occupants feel better and perform better when they have a sense of time of day and outside weather. These connections need not be exaggerated by using large windows, large skylights, or large ventilation openings. Effective placement is more critical in achieving success as shown in Figure 1–12.

1.7 COMMISSIONING

1.7.1 Scope of Commissioning

Commissioning is an essential feature of sustainable design. It is a prerequisite for LEED™ certification and highly recommended for any new building. Commissioning can generally be defined as the process of proving that systems will operate as intended and implementing adjustments necessary to achieve that goal. Typically, the commissioning process would include the following steps.

1. Review system criteria, including design temperatures.
2. Review and assure that design (load calculations, equipment selections) is able to achieve criteria.
3. Review plans and specifications for consistency with design.
4. Observe construction to assure that equipment and systems are installed per plans and specifications.
5. Verify that contractor has performed prefunctional checkout of systems and equipment (e.g., proper wiring connections, clean filters, etc.).
6. Measure system component performance, review test results.
7. Verify control sequences (e.g., thermostat call for cooling starts compressor).
8. Document that these procedures have been performed along with their outcome.
9. Make sure that appropriate Owner's staff are trained in operation of the systems.
10. Verify that operating manuals are turned over to the Owner.

Most of the commissioning scope can be performed by the design and construction team; however, the tasks involving review of design are generally done by a third-party commissioning agent.

1.7.2 Benefits of Commissioning

Making sure that systems operate properly will produce better comfort and save energy. In addition, commissioning reduces the need for warranty work and callbacks to adjust systems during the first year. The commissioning report and associated documentation also provides a baseline of performance for tracking the condition of systems and equipment over the life of the building. Commissioning also aids in organizing maintenance materials (manuals and training) for ongoing use by the building's operations staff.

1.7.3 Range of Applications

The scope of commissioning will depend on how simple or complicated the systems are and on the relative importance of proper system operation. An abbreviated commissioning process might be quite satisfactory for a small commercial building with simple heating and cooling equipment. If system performance is critical, the commissioning process will be extensive. Examples of buildings requiring high emphasis on commissioning include museums, data centers, and correctional facilities.

Museums require that systems operate reliably to produce a precision environment with respect to temperature and humidity. Tight control is needed to prevent damage to valuable artifacts. In most climates, systems have extra components and controls for humidification and dehumidification. Systems must be demonstrated to operate properly before valuable artifacts are moved into the building and placed at risk if systems do not operate properly.

Many enterprises rely on continuous operation of data centers for business-critical and safety-critical functions such as market transactions, air traffic control, and reservations. Systems are designed with redundancy in the event of failure and must transfer load to backup equipment without interruption of service. Commissioning is essential to test failure modes as well as normal operations.

Correctional facilities may have simple HVAC systems, but they are located in facilities that have limited access for correcting systems problems once the facility is put in service. For this reason, a rigorous commissioning process is necessary to make sure the systems are complete and to minimize callbacks. Other systems such as security and alarm require extensive commissioning due to the critical nature of their performance and their complexity in comparison with similar systems for other buildings.

Buildings with less critical functions can generally suffice with the typical start-up and checkout procedures used by conscientious contractors based on manufacturers' recommendations for particular pieces of equipment. For many simple buildings, ongoing maintenance is outsourced, and there is no need for the owner to receive training and operating and maintenance documentation.

1.7.4 Checklists and Forms

Forms are used in commissioning to assist field personnel through the checkout procedure and to record and sign off on results. In most instances, the equipment manufacturers' start-up procedures and forms will be satisfactory with minor modifications for use in commissioning of individual equipment items. Commissioning at the system level (as opposed to individual equipment checkout) requires procedures and checklists customized for the particular system. Control sequences in the specifications or from the control subcontractor's shop drawing submittals are generally the basis for producing system commissioning procedures and forms. Websites of various commissioning organizations and equipment vendors are good sources of standard forms that can be customized for particular projects.

1.8 EVALUATING DESIGN OPTIONS

1.8.1 Subjective Viewpoints

System quality cannot be assessed without defining criteria. Criteria will vary depending on viewpoint. For instance, a contractor might assess a design solely on the basis of ease of construction, whereas a CFO might consider cost most important, and the Director of Physical Plant might look more closely at maintenance issues. The purpose of the building must also be considered. A developer-built speculative office building might be designed with nondurable, low-cost materials, meet budget, and be economically feasible; whereas a corporate headquarters office building might command a higher level of finish. Building life expectations are also important. For instance, a building for a 5-year research program need not be equipped with 20-year life systems; whereas a long-term, institutional building might be designed for 50+-year systems.

1.8.2 Qualitative versus Quantitative Analysis

The result of sustainable design is buildings that are healthy and pleasant and that minimize negative impacts on the local and global environment. Achieving this goal with the best solution requires that many factors be considered in the process. Some factors can be quantified economically and some can only be judged qualitatively based on relative importance.

1.8.3 Decision Matrix Method

The decision matrix is a method for evaluating criteria difficult to quantify in economic terms. The decision matrix can be used to supplement life-cycle cost analyses and weigh options qualitatively as well as quantitatively.

Decision matrix forces the decision makers to assess what is important to them for a successful outcome.

A sample decision matrix analysis is shown in Table 1–5. The analysis includes ranking on quantitative economic factors and on qualitative factors for which no precise economic quantification is available. The key feature of the matrix method is inclusion of weighting factors which allow quantitative inclusion of real, albeit qualitative, criteria in a comprehensive comparison among alternatives.

1.8.4 Economic Evaluation

The basis for making good business decisions is economics, and two methods are commonly used to evaluate options. They are *simple payback period* and *life-cycle cost analysis*. The fallacy of payback analysis is that it considers only the initial cost of implementing the idea and recurring savings in energy. The cost of maintenance and financing are neglected, and the interactions with other building systems are generally ignored. Life-cycle cost analysis includes maintenance and financing cost, but still leaves out many important parameters. Virtually never do these analyses include the effects of interactions with other building systems, and, more important, the impact on productivity and comfort. These are considered "soft costs," beyond the realm of defensible quantification.

1. *Payback analysis.* Despite its limitations, payback analysis is often used to evaluate and compare options. Given options with different initial cost and different operating costs, the simple payback period can be calculated to determine which of the options will recoup initial cost most quickly. Simple payback period is calculated by the following equation:

    Payback Period = Extra Cost/Savings

    where payback period = the time required for savings between two options to equal the difference in cost
    Extra cost = the difference in initial cost between the two options
    Savings = the annual difference in operating cost, generally including utilities and maintenance

    ### Example 1.1
    What is the payback period for an energy-saving device that costs $20,000 to install, lasts 5 years, saves $6,000 per year in utilities (current rates), and requires on average $500 per year for maintenance and repairs?
    Answer:
    Divide $20,000 by net annual savings, which are $6,000 utilities, less $500 maintenance, or $5,500.

Payback Period = $20,000/($6,000 − $500)
          = 3.6 years

Example 1.2
What is the payback period for an alternative energy-saving device that costs $30,000 to install, lasts 20 years, saves $6,000 per year in utilities (current rates), and requires on average $500 per year for maintenance and repairs?
Answer:
Divide $30,000 by net annual savings, which are $6,000 utilities, less $500 maintenance, or $5,500.

Payback Period = $30,000/($5,000 − $500)
          = 5.5 years

Results of payback analysis can be deceiving. A low-investment, quick-payback option (Example 1.1) might appear superior to a higher investment with higher savings and a longer payback (Example 1.2). Over the economic life of the option, the higher investment could produce superior results.

2. *Life-cycle cost analysis.* Economics of alternative decisions are best demonstrated by life-cycle cost analysis. Life-cycle cost analysis is an appropriate tool to compare options on the basis of economics, but is often performed solely on the basis of those criteria that are easiest to document. Cost of construction, financing, maintenance, and utilities can be estimated and have a reportable impact on balance sheets and income statements. Life-cycle cost analysis can be more effective if owners, engineers, and architects are willing and convinced to include cash flows attributable to effects on productivity in the workplace. Good indoor environmental quality results in real economic benefit from productivity. These benefits should be included in life-cycle cost analysis.

    Life-cycle cost analysis uses the discounted cash flow method to compare options by analyzing their annual expenses over the economic life of the building investment. This method of analysis is much more accurate than simple payback analysis, which favors quick-payback solutions and ignores the value of longer-term savings. Mathematics of life-cycle cost analysis includes calculation of escalation, amortization, and net present value. Equations can be found in texts on engineering economics.

    Life-cycle cost analysis is performed by listing the cash flows associated with an option over a given period defined as the life cycle. Initial cost can be accounted for as a single outlay at year zero. Cost could also be treated as payments to amortize a loan or as yearly depreciation of the asset for accounting

TABLE 1–5
Decision matrix method

A. How a Corporate Owner Might Think About His Options for HVAC of an Office Building

Criteria	Weight	VAV/Reheat Score	VAV/Reheat Weighted	VAC/Convectors Score	VAC/Convectors Weighted	VAV/Dual Duct Score	VAV/Dual Duct Weighted	Multizone Score	Multizone Weighted	VAV/FTU Score	VAV/FTU Weighted	Fancoils Score	Fancoils Weighted
Comfort	8	5	30	8	42	5	30	5	30	8	48	7	42
Flexibility	6	10	60	7	42	8	48	1	4	8	48	7	42
Initial cost	3	10	30	8	24	6	18	4	12	7	21	6	18
Energy consumption	6	7	42	8	48	7	42	7	42	9	54	9	54
Ease of maintenance	6	7	42	8	48	9	54	10	60	6	36	5	30
Longevity	6	9	54	7	42	9	54	9	54	6	36	5	30
Acoustics	5	8	40	8	40	8	40	8	40	5	25	5	25
Total score			**299**		**308**		**296**		**252**		**284**		**255**
% score (normalized)			97%		100%		96%		82%		92%		85%
Grade			A		A+		B		F		B		C

B. How a Developer Might Think About His Options for HVAC of an Office Building

Criteria	Weight	VAV/Reheat Score	VAV/Reheat Weighted	VAC/Convectors Score	VAC/Convectors Weighted	VAV/Dual Duct Score	VAV/Dual Duct Weighted	Multizone Score	Multizone Weighted	VAV/FTU Score	VAV/FTU Weighted	Fancoils Score	Fancoils Weighted
Comfort	3	5	15	8	21	5	15	5	15	8	24	7	21
Flexibility	3	9	30	7	21	8	24	1	4	8	24	7	21
Initial cost	10	9	100	8	80	6	60	4	40	7	70	6	60
Energy consumption	2	7	14	8	16	7	14	7	14	9	18	9	18
Ease of maintenance	2	7	14	8	16	9	18	10	20	6	12	5	10
Longevity	2	9	18	7	14	9	18	9	18	6	12	5	10
Acoustics	5	8	40	8	40	8	40	8	40	5	25	5	25
Total score			**213**		**211**		**189**		**151**		**185**		**165**
% score (normalized)			100%		97%		87%		69%		85%		76%
Grade			A+		A−		B		D		B		C

This process involves several steps: 1. Define important criteria. 2. Score options 1 to 10 on these criteria. 3. Assign weight according to perceived importance of each criteria. 4. Multiply weight × score for weighted score. 5. Add weighted scores for total score. 6. Normalize scores as percent of score of highest ranking option. 7. Grade "on curve." Note that different constituencies may have various opinions on weighting of criteria, exemplified here by an owner (A) and a developer (B).

and tax purposes. The single outlay method is used for simplicity to illustrate the process.

During the first year and subsequent years of the life cycle, there will be expenses for utilities, maintenance, and repairs. Utility costs will probably escalate at a higher rate than costs for maintenance and repairs. One might assume, for instance, that utility costs will escalate at 5 percent per year, while maintenance and repair costs will escalate at only 4 percent per year.

All future cash flows must be brought back to current value using a discount rate. The discount rate represents the cost of money or foregone return on investment. Using the "foregone return" logic, a dollar invested today at, say, 6 percent investment return will be worth more in the future. Its value will be escalated by 6 percent per year. Conversely, a dollar in the future is worth less today, its value being "de-escalated" or discounted by 6 percent per year.

The following examples illustrate life-cycle cost for the same energy-saving device described in the prior section on payback.

Example 1.3

What is the life-cycle cost for an energy-saving device that costs $20,000 to install, lasts 5 years, saves $6,000 per year in utilities (current rates), and requires on average $500 per year for maintenance and repairs? Assume that energy cost will escalate at 5 percent per year and that maintenance/repair cost will escalate at 3 percent per year. Assume also that the Owner expects a 15 percent rate of return for investment.

Life-Cycle Cost Analysis: $20,000 Energy-Saving Device, 5-Year Life, 3.6-Year Payback

Life cycle of investment (years)	5
Installation cost	20,000
First-year energy saving (utility rates first year)	6,000
Annual maintenance/repair cost (first-year value)	500
Energy escalation rate	5%
Repair and maintenance escalation rate	3%
Discount rate	15%

		Cash Flows in Year of Occurrence			
Year	*Install Cost*	*Energy Saving*	*Repair and Maintenance*	*Total Annual Cash Flow*	*Present Value Total Annual*
0	(20,000)	0	0	(20,000)	(20,000)
1	0	6,000	(500)	5,500	4,783
2	0	6,300	(515)	5,785	4,374
3	0	6,615	(530)	6,085	4,001
4	0	6,946	(546)	6,399	3,659
5	0	7,293	(563)	6,730	3,346
Net Present Value of Life-Cycle Cash Flows ($)					163

Example 1.4

What is the life-cycle cost of an alternative energy-saving device that costs $30,000 to install, lasts 10 years, saves $6,000 per year in utilities (current rates), and requires on average $500 per year for maintenance and repairs? Assume that energy cost will escalate at 5 percent per year and that maintenance/repair cost will escalate at 3 percent per year. Assume also that the Owner expects a 15 percent rate of return for investment.

Note that the option in Example 1.3 has a shorter payback than the one in Example 1.4, however, the life-cycle savings of the option in Example 1.4 are superior.

Life-Cycle Cost Analysis: $30,000 Energy-Saving Device, 10-Year Life, 5.5-Year Payback

Life cycle of investment (years)	10
Installation cost	30,000
First-year energy saving (utility rates first year)	6,000
Annual maintenance/repair cost (first-year value)	500
Energy escalation rate	5%
Repair and maintenance escalation rate	3%
Discount rate	15%

Year	*Cash Flows in Year of Occurrence*				
	Install Cost	*Energy Saving*	*Repair and Maintenance*	*Total Annual Cash Flow*	*Present Value Total Annual*
0	(30,000)	0	0	(30,000)	(30,000)
1	0	6,000	(500)	5,500	4,783
2	0	6,300	(515)	5,785	4,374
3	0	6,615	(530)	6,085	4,001
4	0	6,946	(546)	6,399	3,659
5	0	7,293	(563)	6,730	3,346
6	0	7,658	(580)	7,078	3,060
7	0	8,041	(597)	7,444	2,798
8	0	8,443	(615)	7,828	2,559
9	0	8,865	(633)	8,231	2,340
10	0	9,308	(652)	8,656	2,140
Net Present Value of Life-Cycle Cash Flows ($)					3,059

The implication of Example 1.4 is that installing the hypothetical energy-saving device will have a net present value of $3,059. Spending the $30,000 will generate savings, which will recoup the investment with profit, and there will be additional savings worth $3,059 today. In essence, we could afford to spend $30,000 plus $3,059 to generate the 10-year savings and meet our expectation of profit. This life-cycle approach can be used to determine how much we can afford to spend for saving energy and for improving productivity in the workplace.

1.9 ECONOMICS OF OWNING AND OPERATING BUILDINGS

1.9.1 Energy Usage in Perspective

This section examines the components of building operating cost and the relative importance of energy in a comprehensive view. An office building is used for example. Office buildings in temperate climates use energy for HVAC, water heating, lighting, office appliances, and vertical transportation. The table below shows how a typical low-rise building might use energy for these functions.

Energy Use	*Percent*
HVAC	53%
Hot water	1%
Lighting	28%
Appliances	18%

Generally, a Midwest office building will experience energy bills of $1.50 to $2.00 per year for gas and electric.

To place the cost of energy in perspective, here is an example of operating costs for a typical office building in a temperate climate:

Facility Expenses, Typical Office Building

Expense Component	*Expense*	*$ Per Yr*
Investment return	59%	8.35
Repairs and maintenance	5%	0.73
Preventative maintenance	4%	0.60
Janitorial	7%	1.01
Site maintenance	0%	0.06
Gas	**1%**	**0.19**
Electric	**13%**	**1.84**
Water	1%	0.09
Sewer	0%	0.05
Environmental	1%	0.13
Life-safety	1%	0.11
Security	5%	0.73
Space planning	2%	0.29
Total facilities expense	**100%**	**14.18**

Energy at $2.00 per sq ft per year represents less than 15 percent of the overall cost of owning and operating a facility. This value might be higher or lower depending on climate and utility rates, but the general relationship will be fairly constant. An ambitious program to reduce energy consumption by, say, 30 percent would net approximately $0.60 per sq ft per year, or about 4 percent of facilities expenses.

1.9.2 Energy Cost versus Employee Cost

A typical office building might have an occupancy density of 200 gross sq ft per workstation including circulation, toilets, lobby, etc. For such a building, the expense of energy would be $400 per year, per employee.

Energy costs can be compared with other expenses on a per-person basis. Auto expenses for a typical 15-mile-per-day work commute, for instance, cost approximately $1,300 per year per employee using current IRS guidelines—over three times the energy cost at the office! Urban parking is generally $1,500 to $3,000 per year. Building design will not directly affect these costs, but location of buildings will.

Energy cost can also be compared with the expense of salaries and fringe benefits for personnel.

Analysis of Workplace Occupant Cost

Occupancy density	200 sq ft per employee
Salary (example)	$50,000 per year
Fringes @ 30%	$15,000 per year
Employee cost	$65,000 per year
Employee cost	$325/sq ft/yr

1.9.3 Appropriate Solutions for Energy Conservation

Relative to energy cost at $2.00 per sq ft per year, employee cost is a very large number! Energy conservation is a key issue in sustainable design, but considering the purpose of buildings, saving energy at the expense of occupant well-being and performance is not advisable. Here are a few arithmetic scenarios to make the point.

Suppose that a facility manager changes thermostat setpoints and reduces lighting levels to effect a 20 percent reduction in utility bills. Savings would be about $0.40 per sq ft per year. Suppose further that these reductions in building service quality reduced employee productivity by 1 percent. The loss in productivity would be 1 percent of $325 per sq ft per year, or $3.25. Obviously, the energy savings, though very impressive, would not be a good idea considering the harmful side effects on productivity.

Conversely, suppose that a facility manager installs better temperature controls and increases lighting levels at the expense of a 20 percent increase in utility bills. Extra cost would be about $0.40 per sq ft per year. Suppose further that these improvements in building service quality increased employee productivity by 1 percent. The gain in productivity would be 1 percent

of $325 per sq ft per year, or $3.25. Obviously, the extra cost for energy, though considerable, would be a great investment considering the beneficial side effects on productivity.

These examples are not intended to discourage energy conservation, but rather to point out that the application of resources should be balanced to benefit all concerned, and energy conservation is just one of many benefits that can be achieved through thoughtful building design and operations.

QUESTIONS

1.1 If the lighting load for a 20,000-sq-ft building is estimated at 2W/sq ft, what will be the resulting heat generated by lighting?

1.2 If the lighting load were increased, what would be the effect on other building systems?

1.3 How much CO_2 will be liberated to the atmosphere in a year's time due to lighting operation in the building in Question 1.1?

1.4 How much heat (Btus) will be stored in a 100-sq-ft concrete wall 1 ft thick if it is warmed from 75°F to 85°F by exposure to sunlight?

1.5 What is the value of the heat in Question 1.4 compared with gas at $1.00 per therms burned in a boiler at 85 percent efficiency?

1.6 How does "sustainable" design differ from energy-effective design?

1.7 What factors should the architect and engineer consider to produce a high-performance environment for building occupants?

1.8 What is the relationship between building codes and sustainable design?

1.9 How does saving energy help to protect the environment?

1.10 What is the role of maintainability in sustainable buildings?

1.11 How could building site selection affect the environment?

1.12 What factors should interior designers consider in terms of indoor air quality? Architects? HVAC engineers? Design teams?

1.13 What design features would you suggest to allow personal climate control in a single-story residence? A high-rise office building? A classroom building?

1.14 What sustainable design issue should architects consider in deciding window materials and locations?

1.15 What is the difference between qualitative and quantitative factors in an analysis? How might we deal with each?

1.16 Prepare a decision matrix to decide between operable windows and fixed windows in an office building. Fill out the matrix as if you were an occupant, a maintenance staffer, the building owner.

1.17 Will a developer use a higher or lower discount rate than a building owner? Why?

1.18 An energy conservation option has a first cost of $25,000. It requires $2,000 per year maintenance and saves $5,000 per year in utilities. What is the simple payback period for the option?

1.19 The system in Question 1.18 will last 15 years with no salvage value. What is the 15-year life-cycle cost assuming energy cost escalation of 4 percent annually, maintenance cost escalation of 2 percent annually, and a 5 percent discount rate? What if the discount rate is 15 percent?

1.20 Assume the option in Question 1.18 is installed in a building with 200 occupants, average personnel cost of $60,000 per year. If the device interferes with temperature control, resulting in a 2 percent decrease in productivity, what would the simple payback be?

1.21 What would the payback be if the option in Question 1.18 improved temperature control and resulted in a 2 percent increase in productivity?

1.22 Calculate the life-cycle costs for the two cases (2 percent decrease, 2 percent increase in productivity) using data from Question 1.20–1.21 for a 5 percent discount rate and 15 percent discount rate.

REFERENCES

LEED Green Building Rating System (Version 2.0). U.S. Green Building Council, 2001.

DeGarmo, E. Paul, *Engineering Economy* (4th ed.), New York, NY: The Macmillan Company, 1967.

ASHRAE Standard 62-99, *Ventilation for Acceptable Indoor Air Quality.*

Milton, D. K., P. M. Glencross, and M. D. Walters, "Risk of Sick Leave Associated with Outdoor Air Supply Rate, Humidification, and Occupant Complaints," *Indoor Air*, 2000.

Fisk, W. J., "How IEQ Affects Health, Productivity." *ASHRAE Journal*, May, 2002.

Nunes, F., R. Menzies, R. M. Tamblyn, E. Boehm, R. Letz, "The Effect of Varying Level of Outside Air Supply on Neurobehavioral Performance Function during a Study of Sick Building Syndrome," Proc. Indoor Air 1993, The 6th International Conference on Indoor Air Quality and Climate, Helsinki. *Indoor Air*, 1: 53–58, 1993.

Fisk, W. J., "How IEQ Affects Health, Productivity." *ASHRAE Journal*, May, 2002.

ASHRAE Standard 55, *Thermal Environmental Comfort Conditions for Human Occupancy.*

Wyon, D. P., "Indoor Environmental Effects on Productivity," *IAQ '96 Paths to Better Building Environments*, ASHRAE.

Brager, Gail S., and Richard de Dear, "A Standard for Natural Ventilation." *ASHRAE Journal*, October, 2000.

Milne, G. R., "The Energy Implications of a Climate-Based Indoor Air Temperature Standard." In *Standards for Thermal Comfort*, Nicol, Humphreys, Sykes, and Roaf (eds.), E and FN Spon, London, pp. 182–189.

The Pacific Gas and Electric Company, "Daylighting in Schools, An Investigation into the Relationship Between Daylighting and Human Performance," *Daylighting Initiative*, August, 1999.

HVAC FUNDAMENTALS

2

THIS CHAPTER CONTAINS BASIC INFORMATION REGARDing comfort, properties of air, load estimation, and determining the proper flow of heat transfer fluids to satisfy loads. This information is a prerequisite for understanding how architecture affects the size of HVAC systems and how HVAC systems operate to control the environment.

The next five chapters describe, in turn, the following concepts, subsystems, and equipment used in HVAC systems:

- HVAC delivery
- Cooling production
- Heating production
- Air handling
- Piping systems

Figure 2–1 shows typical equipment used in an HVAC system for a large building. Illustrated are the subsystems described in the five chapters that follow.

2.1 ENVIRONMENTAL COMFORT

2.1.1 Comfort for Occupants

The temperature of a space is not the only factor affecting a person's comfort. Even if the temperature is within an acceptable range, the space may seem warm if the humidity is too high, the airflow is too low, or warmth is being radiated to the occupants. Conversely, a space may seem cool if the humidity is low, the space is drafty, or warmth is being radiated from the occupants to cold surfaces. Comfort for building occupants is affected by a number of environmental variables, including the following:

- Temperature
- Airflow
- Humidity
- Radiation

Indoor air quality is another aspect of comfort. In air of good quality, sufficient oxygen is present and objectionable impurities such as dust, pollen, odors, and hazardous materials are absent.

Different conditions may be deemed comfortable, depending on the type of activity that goes on in a space. Appropriate conditions for an office would be too warm for a gymnasium and too dry and cool for a natatorium. Expectations must also be considered: Saunas are hot on purpose, and a wide variety of conditions are commonly tolerated in factories. The physical condition of the occupants, including their age and health, also affects their comfort. Even the seasons affect comfort: Warmer environments are tolerated during the summer and cooler environments in winter, because of clothing and acclimatization.

Economics and concerns about energy conservation are also considered in defining comfort. People will be satisfied with less comfort when faced with a worthy cause or a mandate based on sound business practice.

2.1.2 Temperature and Humidity

Both temperature and humidity affect our sense of comfort. Figure 2–2 shows the acceptable range of each for persons wearing typical summer and winter clothing during sedentary activities. The lower comfort limit in cold weather is 68°F at about 30 percent relative humidity (RH), and the upper limit in hot weather is 79°F at about 55 percent RH. HVAC systems are generally designed to maintain temperature and RH within a tighter range than is indicated in the figure.

An interior design temperature of about 75°F is considered comfortable by most people in general-use spaces, as shown in Figure 2–2. During the summer, a slightly higher temperature may be appropriate because of light clothing and acclimatization to warm weather; this should be considered in designing air-conditioning systems. Conversely, slightly cooler temperatures are acceptable and can be considered in the design of

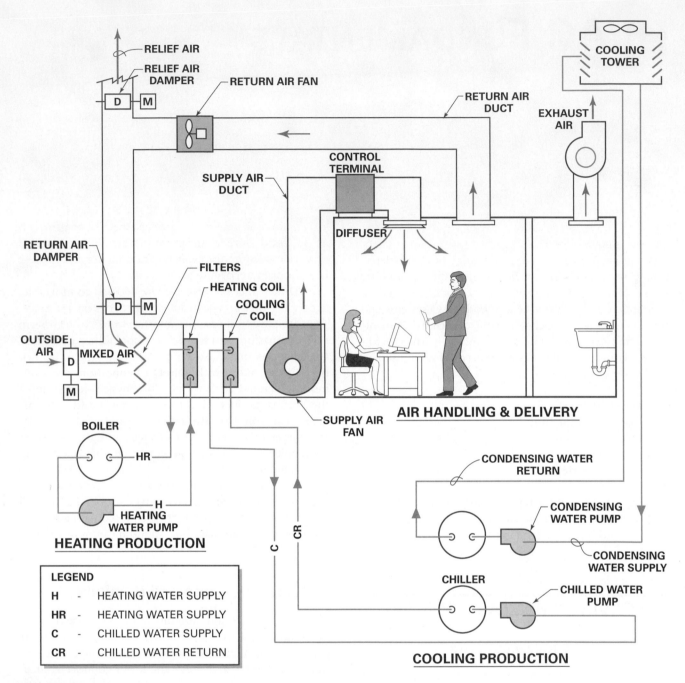

■ FIGURE 2–1

Components of a large HVAC system. (Based on hot–chilled water system.)

heating systems. Most air-conditioning systems are designed to maintain a summer temperature of 72°F–78°F. During winter, heavier clothing and ac-climatization to cold weather result in a recommended design temperature of 68°F–72°F for heating systems. These interior design temperatures will be appropriate for the majority of buildings.

Humidity in excess of 60 percent is considered high in general-use spaces. High humidity not only is uncomfortable but also can result in indoor air-quality problems due to mold growth. Humidity lower than 25–30 percent can result in uncomfortable drying of breathing passages and problems with electronic equipment due to static electricity.

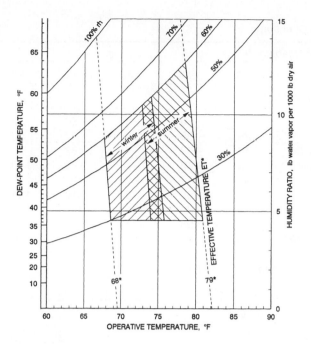

FIGURE 2-2
Standard effective temperature and comfort zones.
(Courtesy: Reprinted by permission from ASHRAE
(www.ashrae.org.))

2.1.3 Airflow

Systems must be designed for adequate airflow to prevent complaints of "stuffiness" or drafts. The measure of airflow is velocity. Space air velocities less than 10 feet per minute will be stuffy; those more than 50 feet per minute will seem drafty.

2.1.4 Air Quality

Systems must provide sufficient amounts of clean air to keep oxygen at an acceptable level and to dilute contaminants generated within occupied spaces. Air should be reasonably free of dust, and spaces free of odors or other pollutants that may be hazardous or objectionable. These conditions are generally achieved through the use of filters and by the introduction of outside air into the system at rates specified in Table 2–9.

2.1.5 Radiant Effects

Even if the temperature, humidity, and airflow in a space are acceptable, the space may be uncomfortable owing to radiant effects from cold windows or walls. Systems must therefore compensate for these effects with radiant heat or higher temperatures. Similarly, cooler temperatures or

higher air velocities will be needed to offset the effects of warm surfaces. Downdrafts from cold surfaces are also uncomfortable and can be offset by proper placement of heating devices.

2.1.6 Special Considerations

Buildings such as museums, computer rooms, and laboratories have special requirements for temperature, humidity, airflow, and air quality. In some instances these requirements are consistent with the comfort of the occupants, but in others they are at odds with comfort.

Interior environmental criteria are often based on specifications for equipment used within an occupied space. Computer rooms, for example, are often drafty and cool to suit the environmental requirements of the computing equipment. This will be an uncomfortable environment for operators of the computers, and special provisions may be desirable to provide better conditions in certain areas of the room. Similarly, materials stored in a warehouse may tolerate cold or hot temperature, but the warehouse employees need a refuge of human comfort.

Economics and expectations of comfort also affect design criteria. Despite the fact that warehouses and factories are occupied by people, it is deemed unnecessary to maintain these buildings at the same interior temperatures as an office building or hospital. The need for energy conservation may also temper expectations of comfort.

Ventilation rates for indoor air quality are also subject to special considerations. Historically, ventilation standards have varied depending on concerns for energy conservation and health. The values shown in Table 2–9 are much higher than those recommended during the energy crisis of the late 1970s.

2.1.7 Wind Chill Factor (WCF)

Both ambient air temperature and wind conditions affect discomfort associated with cold outdoor environment. Siple and Passel in 1945 introduced an empirical formula known as the wind chill index (WCI) to express the combined effect of wind velocity and air (dry-bulb) temperature on the heat loss of a cylindrical body. This formula was later adopted and modified by meteorologists in weather reports to express the severity of cold environment as the equivalent wind chill temperature (EWCT), commonly known as the wind chill factor (WCF). Table 2–1 gives the calculated WCF at various dry-bulb temperatures and wind velocities in conventional and SI units. For temperatures and wind velocities not listed in the tables, linear interpolation may be used. Wind velocity greater than 40 mph (70 km/h) has little added chilling effect. See problems, page 74.

TABLE 2–1
Wind chill factor (WCF) of cold environments (in conventional and SI units)

Wind Speed, mph	Actual Thermometer Reading, °F											
	50	40	30	20	10	0	−10	−20	−30	−40	−50	−60
	Equivalent Chill Temperature, °F											
0	50	40	30	20	10	0	−10	−20	−30	−40	−50	−60
5	48	37	27	16	6	−5	−15	−26	−36	−47	−57	−68
10	40	28	16	3	−9	−21	−34	−46	−58	−71	−83	−95
15	36	22	9	−5	−18	−32	−45	−59	−72	−86	−99	−113
20	32	18	4	−11	−25	−39	−53	−68	−82	−96	−110	−125
25	30	15	0	−15	−30	−44	−59	−74	−89	−104	−119	−134
30	28	13	−3	−18	−33	−48	−64	−79	−94	−110	−125	−140
35	27	11	−4	−20	−36	−51	−67	−83	−98	−114	−129	−145
40	26	10	−6	−22	−38	−53	−69	−85	−101	−117	−133	−148

Wind Speed, km/h	Actual Thermometer Reading, °C												
	10	5	0	−5	−10	−15	−20	−25	−30	−35	−40	−45	−50
	Equivalent Chill Temperature, °C												
Calm	10	5	0	−5	−10	−15	−20	−25	−30	−35	−40	−45	−50
10	8	2	−3	−9	−14	−20	−25	−31	−37	−42	−48	−53	−59
20	3	−3	−10	−16	−23	−29	−35	−42	−48	−55	−61	−68	−74
30	1	−6	−13	−20	−27	−34	−42	−49	−56	−63	−70	−77	−84
40	−1	−8	−16	−23	−31	−38	−46	−53	−60	−68	−75	−83	−90
50	−2	−10	−18	−25	−33	−41	−48	−56	−64	−71	−79	−87	−94
60	−3	−11	−19	−27	−35	−42	−50	−58	−66	−74	−82	−90	−97
70[b]	−4	−12	−20	−28	−35	−43	−51	−59	−67	−75	−83	−91	−99

Little danger: In less than 5 hours, with dry skin. Maximum danger from false sense of security (WCI less than 1400).	**Increasing danger:** Danger of freezing exposed flesh within 1 minute (WCI between 1400 and 2000).	**Great danger:** Flesh may freeze within 30 seconds (WCI greater than 2000).

Source: Reprinted by permission from ASHRAE (www.ashrae.org).

2.2 PROPERTIES OF AIR–WATER MIXTURES

The design of environmental control systems relies on an understanding of the properties of air, including temperature and humidity. These properties affect loads on buildings, and HVAC systems are used to alter the properties of air and produce comfort.

2.2.1 Psychrometry

Psychrometry is the study of properties of air–water mixtures. A psychrometric chart is a convenient source for data on the properties of such mixtures. Figure 2–3 shows how important properties are presented on a psychrometric chart. Figure 2–4 is a complete chart that can be used in analysis of processes associated with HVAC.

2.2.2 Absolute and Relative Humidity

Two basic properties of air–water mixtures are temperature and humidity. The humidity of the air can be expressed in two ways: absolute and relative. *Absolute humidity,* also known as the *humidity ratio* (W), is the amount of water in the air and is measured in grains or pounds of water per pound of dry air. A grain is equivalent to $1/7000$ of a pound. This unit is preferred owing to the very small amount of water present in air. *Relative humidity* (RH) is the ratio of the actual water content to the maximum possible moisture content at a

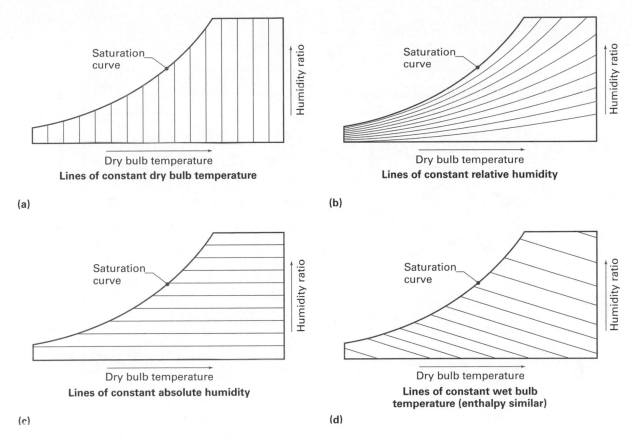

■ FIGURE 2–3

Lines representing major properties of air–water mixture on the ASHRAE psychrometric chart. (a) Vertical lines: constant dry-bulb (DB) temperature. (b) Curved lines: constant relative humidity (RH). (c) Horizontal lines: constant humidity ratio (*W*), also commonly referred to as *absolute humidity*. (d) Sloped lines: constant wet-bulb (WB) temperature; lines with same slope: constant enthalpy (*H*).

given temperature, expressed as a percent. If the air is currently holding all the moisture possible, the relative humidity is 100 percent, and the air is termed *saturated*.

2.2.3 Effect of Temperature on Humidity

The moisture-holding capacity of the air depends on the air temperature. Warm air can hold more moisture than cold air. For this reason, the same absolute humidity results in different relative humidities at different temperatures. The psychrometric chart illustrates the relationship of temperature, absolute humidity, and relative humidity.

2.2.4 Wet-Bulb Temperature

If a wet sock is placed over the bulb of a conventional thermometer, a lower temperature will be recorded owing to evaporative cooling. The drier the air, the more effective will be the evaporative cooling, and the lower will be the temperature measured. If the air is saturated, then there will be no evaporation and the wet-bulb thermometer will measure the same temperature as a dry-bulb thermometer. The temperature and humidity of the air can be determined by measuring both dry-bulb and wet-bulb temperatures. A combination of wet- and dry-bulb temperature represents a discrete point on the psychrometric chart.

2.2.5 Sensible, Latent, and Total Heat

Air contains thermal energy in two forms: sensible heat and latent heat. Water vapor, or humidity, in the air contains the water's latent heat of vaporization (approximately 1000 Btu/lb of water). Temperature is a measure of sensible heat, while water vapor content is a measure of latent heat. Total heat—the sum of sensible and latent heat—is *enthalpy*, symbolized by the Greek

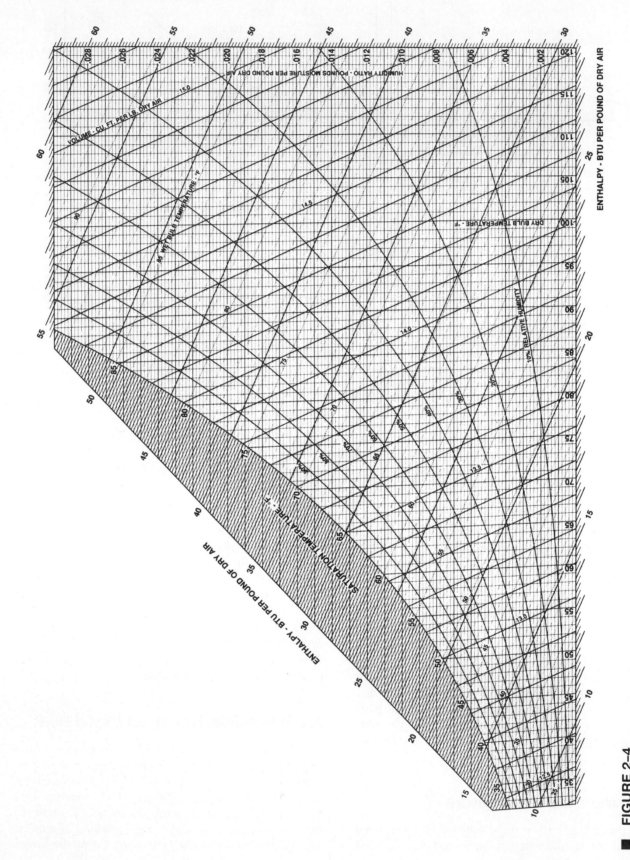

■ **FIGURE 2-4**
Psychrometric chart. Reprinted by permission from ASHRAE (www.ashrae.org).

letter eta, or by *H*. Enthalpy is expressed in units of Btu/lb of dry air. High temperature or high humidity constitutes high energy.

On the psychrometric chart, horizontal movement is associated with sensible heat change (no change in absolute humidity), and vertical movement is associated with latent heat change (no change in temperature). Moving upward or to the right indicates a higher energy level; moving downward or to the left indicates a lower energy level. Lines of constant enthalpy slope upward and to the left at approximately the same slope as lines of constant wet-bulb temperature. This is no coincidence, for wet-bulb temperature is a good measure of total energy.

Often, changes in air conditions result in changes in both humidity and temperature. The net change in energy level, or enthalpy, can be determined by plotting the initial and final conditions on the psychrometric chart, as shown in Figure 2–5.

2.2.6 Sensible Heating and Cooling

Sensible heating (cooling) occurs when the temperature of an air–water mixture is raised (lowered) but the absolute moisture content remains the same. Sensible heating or cooling occurs as air in spaces is warmed or cooled by building loads that do not change the moisture content of the air. Sensible heating or cooling is also performed by systems to compensate for loads. For instance, room air may be cooled by an outside wall during cold winter weather. To compensate, a heater at the base of the wall may warm the air. Sensible heating or cooling is represented by a horizontal movement along the psychrometric chart.

2.2.7 Processes Involving Latent Heat

Heating and cooling represent a change of sensible heat; humidification and dehumidification represent a change of latent heat. The amount of moisture liberated or absorbed by air is measured by its initial and final absolute humidities.

Air can be humidified either by adding dry steam to it or by evaporating moisture into it. If dry steam is added, the air will have a higher energy level, taking on the latent heat of the steam. (There will also be a slight increase in temperature owing to the sensible heat of the steam, but the effect is small and generally ignored in practice.) On the psychrometric chart, this process is represented by a vertical movement.

If water is evaporated into air, the air will cool, but the final energy level of the air does not change. The heat required to vaporize the water cools the air. The sensible heat loss equals the latent heat gain, resulting in constant enthalpy. This process is called *adiabatic saturation*. (No energy is added or removed.) Evaporative humidification is accompanied by evaporative cooling and is represented on the psychrometric chart by an upward movement along a line of constant enthalpy (approximately parallel to a line of constant wet-bulb temperature).

Cooling is a method for dehumidifying air. If moist air is cooled to the saturation curve, further cooling will not only reduce temperature but also remove moisture. The temperature at which moisture begins to condense is termed the *dew point*. Liquid moisture removed from the air by this process is termed *condensate*. The air that results from the process is both cooler and less humid than it was initially.

Air also can be dehumidified by absorption. Some substances are *hygroscopic*, meaning that they absorb moisture. Hygroscopic substances, or desiccants, such as silica gel and lithium bromide are used in certain applications to absorb moisture from the air. As moisture condenses in the desiccant its latent heat is liberated, heating the desiccant and the air. Absorption is represented by a downward movement on the psychrometric chart along a line of constant dry-bulb temperature.

2.2.8 Examples

1. Air at 70°F DB and 75% RH is heated to 84°F. What is the RH of the air at this higher temperature?

    Ans. In Figure 2–4, locate the air at the initial condition (70°F DB and 75% RH), and follow the horizontal line to the right until it meets the 84°F DB line (vertical). The heated air is now at 47% RH.

2. Air at 90°F DB and 70% RH is cooled to 75°F. What is the relative humidity?

    Ans. In Figure 2–4, from the intersecting point of 90°F DB and 70% RH, draw a line to the left. This line meets the saturation curve at 79°F, which is the dew point temperature of the air. The air is then cooled further, following the saturation curve until it stops at 75°F. Between 79°F and 75°F, the air is saturated, and moisture condenses out of it. The RH of the air is now 100%.

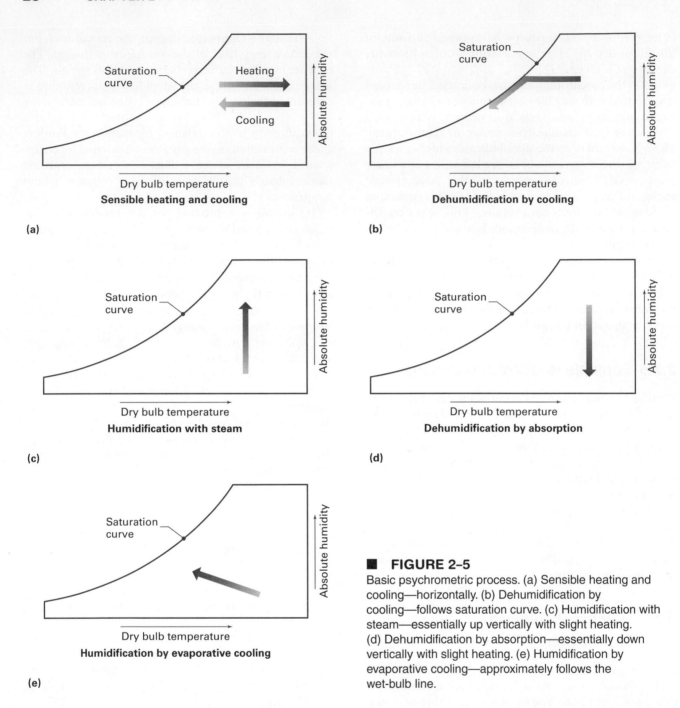

(a) Sensible heating and cooling

(b) Dehumidification by cooling

(c) Humidification with steam

(d) Dehumidification by absorption

(e) Humidification by evaporative cooling

■ **FIGURE 2–5**
Basic psychrometric process. (a) Sensible heating and cooling—horizontally. (b) Dehumidification by cooling—follows saturation curve. (c) Humidification with steam—essentially up vertically with slight heating. (d) Dehumidification by absorption—essentially down vertically with slight heating. (e) Humidification by evaporative cooling—approximately follows the wet-bulb line.

2.3 ENERGY TRANSPORT IN HVAC SYSTEMS

2.3.1 Heat Transport by Fluid Flow

Figure 2–1 shows that HVAC systems use fluids to transport heat and cold to satisfy loads and maintain comfort. Such fluids include air, water, steam, and refrigerant. Equations are developed in this section that can be used to determine heat transport based on the flow rate and the initial and final conditions of the fluid. These equations can also be used in equipment design to specify flow rates or conditions, based on requirements for heat transport.

Heat is measured in *British thermal units,* or Btus. A Btu is the amount of heat required to raise the temperature of 1 lb of water 1°F. The rate of heat flow is measured in Btus per hour, or Btuh. Fluids are used to transport heat in HVAC systems.

2.3.2 Heat Transport by Sensible Heating and Cooling

The natural property of a fluid that affects heat transfer is called *specific heat;* this is the amount of energy (Btu) required to raise the temperature of 1 lb of a substance 1°F. The specific heat of water is 1.0; that of air is 0.24. The heat liberated from a quantity of fluid is equal to the specific heat of the fluid multiplied by the number of pounds of the fluid and the temperature change between the initial and final states of the fluid; that is,

$$q = m \times C \times TD \qquad (2\text{--}1)$$

where q = heat energy (Btu)
      m = mass (lb)
      C = specific heat (Btu/(lb·°F))
      TD = temperature difference, °F (final minus initial temperature)

HVAC equipment loads, equipment capacity, and output are expressed as quantities per unit time, or *rates:*

$$Q = M \times C \times TD \qquad (2\text{--}2)$$

where Q = heat flow (Btu/hr, or Btuh)
      M = mass flow (lbm/hr)
      C = specific heat (Btu/(lb·°F))
      TD = temperature difference, °F

Heat Transfer in Water

Mass flow is quantified in units of gallons per minute (GPM). Knowing that 1 gal of water has a mass of 8.35 lb and that there are 60 minutes in 1 hour, the following equation can be derived:

$$Q = 500 \times GPM \times TD \qquad (2\text{--}3)$$

where Q = heat flow (Btu/hr, or Btuh)
     GPM = water flow (gal/min)
       TD = temperature difference, °F

Heat Transfer in Air

Mass flow is quantified in units of cubic feet per minute, or CFM. An equation for heat transfer in air can be derived given that the density of air at standard pressure is 0.076 lb/ft³ and that the specific heat of air is 0.24 Btu/lb·°F:

$$Q = 1.1 \times CFM \times TD \qquad (2\text{--}4)$$

where CFM = airflow (cu ft/min)

2.3.3 Heat Transport by Fluid Phase Change

Heat Transfer in Steam

Heat is liberated from steam by a change of phase from vapor to liquid. One pound of steam liberates approximately 1000 Btu as it condenses. Conversely, a boiler must produce 1000 Btu to boil 1 lb of water. For steam, heat flow is approximated by the equation

$$Q = 1000 \times SFR \qquad (2\text{--}5)$$

where SFR = steam flow rate (lb/hr)

Heat Transfer in Refrigerants

Refrigerants absorb heat by changing phase from liquid to gas. The heat absorbed is equal to the latent heat of vaporization, measured in Btu per pound, times the refrigerant flow rate, measured in pounds per hour. There are many types of refrigerants, and each has its own distinct latent heat of vaporization.

2.3.4 Selecting Fluid Flow Rates for HVAC Systems

HVAC systems and subsystems are designed to satisfy heat loads by using heat transport fluids. Fans, pumps, boilers, and distribution elements are sized according to flow requirements, which must be determined by the HVAC designer. The first step is to estimate building heat loads. Methods for estimating loads are presented later in this chapter. Once they are estimated, the HVAC designer must decide on the proper combination of flow and conditions for fluids used to transfer heat and thereby compensate for loads. Initial and final conditions are generally selected on the basis of accepted general practice found to achieve satisfactory results. Equations (2 1) through (2–5) can be used to calculate flow, given the heat transfer requirement and the initial and final conditions of the fluid.

Water Flow

For devices using hot water for heating, a supply temperature of 160°F might be chosen and the load equipment selected to allow a 20° drop in water temperature, resulting in a 140° return temperature. Once this decision is made, the required water flow rate can be calculated as

$$GPM = Q/(500 \times TD) \qquad (2\text{--}6)$$

Similarly, chilled water can be used for cooling. Chilled-water supply temperatures between 40° and 50°F are common for building HVAC applications, and systems are designed for water temperature rises ranging from 10° to 15°F. The same equation applies for determining the required chilled-water flow rate.

Airflow

For systems using warm air for heating, supply temperatures between 105° and 140°F will be appropriate to maintain a space at, say, 75°F. Given a space temperature and a selected supply temperature, the required airflow rate can be calculated from

$$\text{CFM} = Q/(1.1 \times \text{TD}) \tag{2-7}$$

Systems using chilled air for cooling generally have supply air temperatures between 50° and 60°F. Equation (2-7) can be used to determine the airflow rate required to satisfy the sensible portion of cooling loads. Once the airflow rate is determined, the humidity can be determined from a psychrometric chart.

Steam Flow

The rate of steam flow required to satisfy a given heating load is determined by the equation

$$\text{SFR} = Q/1000 \tag{2-8}$$

where SFR = steam flow rate (lb/hr)
Q = heat flow (Btuh)
1000 = heat (Btu) liberated by condensation of 1 lb of steam

Refrigerant Flow

The rate of refrigerant flow for cooling is determined by dividing the cooling load by the latent heat of vaporization.

The foregoing concepts and equations are used to estimate theoretical fluid flow rates required to meet a given load. The resulting estimates are the basis for sizing the piping and duct systems, along with the pumps and fans required to transport heating and cooling.

2.4 HVAC LOAD ESTIMATION

2.4.1 Nature of HVAC Loads

The design of HVAC systems starts with an estimation of the loads the system must satisfy. Heating loads represent how much heat is lost and therefore must be made up by the system. Cooling loads represent how much heat is gained and must be removed. Humidity must also be considered. Internal and external moisture gains and losses may need to be counteracted in maintaining proper humidity levels. Estimates of these loads involve both sensible and latent heat transfer and conversion into and within the building.

2.4.2 Methods for Estimating Loads

Many methods are available for calculating heating and cooling loads for buildings. All, however, should be considered estimates, the precision of which depends on how the method accommodates the nonuniform qualities of building assemblies and contents and the non-steady-state nature of building loads. Heat transfer in building systems is a dynamic process, with ever-changing loads from outside and within the building.

2.4.3 Accuracy and Precision

Oversized HVAC equipment operates at lower efficiency than properly sized equipment. The results are higher operating cost and higher initial cost for installation. Accuracy in load calculations and proper equipment sizing are part of the sustainable design process, and have a significant effect of initial and operating costs.

The architect and the HVAC engineer generally work to the same deadline for completion of design. This necessitates that the engineer make many assumptions regarding the construction prior to the architect's documents being finalized. These assumptions are prone to be conservative. In addition, engineers generally apply a safety factor to account for poor construction. Both of these causes of oversizing could be eliminated with proper scheduling of the design/construction process and better commissioning and inspection to reduce the risk of poor-quality construction, which is a major reason why engineers apply safety factors in sizing equipment.

Accepted methods for calculating heating and cooling loads are documented in the ASHRAE *Handbook of Fundamentals,* which is revised periodically. ASHRAE methods have become increasingly precise and have become the basis for algorithms in computer programs. In the following discussion we describe the basic principles of calculating heating and cooling loads using one of the earlier, simpler methods that can be performed with a pencil, calculator, and tables. In most cases, the results will be conservative. Greater precision can be achieved by computer analysis using more complex methods, however, safety factors and allowances for unknown developments still need to be applied in sizing HVAC systems and components.

2.4.4 Critical Conditions for Design

The designer must select an appropriate set of conditions for the load calculation. Relevant conditions include the outside weather, solar effects, the inside temperature and humidity, the status of building operations, and many other factors.

For heating, the critical design condition occurs during cold weather, at a time when there is little or no heating assistance from radiant solar energy or internal heat gains from lights, appliances, or people. The selection of an appropriately cold outside air temperature for design is an important decision.

For cooling load calculations, the critical design condition is the peak coincident occurrence of heat, humidity, solar effects, and internal heat gains from equipment, lights, and people. The position of the sun varies by season and through the day, as does the weather. Building operations also vary. Sometimes, several estimates must be performed for different times to determine the highest combination of individual loads.

2.4.5 Temperature Criteria

The inside temperature and humidity are set by criteria based on expectations of comfort, as discussed earlier.

Outside weather conditions affect heating and air-conditioning loads from introduction of outside air for ventilation, infiltration (leakage) of air and conduction of heat through the building envelope. Historic extremes of temperature and humidity are the basis for design load

calculations here. Statistical data compiled for locations throughout the world are used by HVAC designers.

Table 2–2 shows design temperatures for selected locations throughout the United States and abroad. The table includes data not only on weather but also on latitude, longitude, and relative wind severity, as well as other useful information. Latitude is important in estimating solar loads. Wind conditions are important for judging the severity of infiltration through a building envelope.

The outside temperature criteria used to calculate loads depends on the nature of the building. If it is essential that the system be capable of always meeting demand, the designer might assume the coldest recorded temperature. Buildings are seldom designed according to that criterion, however.

A few buildings are designed to meet an outside temperature criterion corresponding to a median of extremes. For heating systems, this is the mean of the coldest recorded temperatures. The median value has as many annual extremes above as below it. This is a stringent criterion that may be appropriate when system performance is critical, during the coldest temperatures. For instance, performance is important in a hospital because of the condition of the patients. Also, the occupants are present during the early morning hours, when the lowest outside temperatures generally occur.

For most buildings, criteria need not be so stringent. If the inside temperature of an office building falls a few degrees lower than the intent of the design, no great harm results. In addition, the schedule of office

TABLE 2–2
Climatic conditions in the United States

Col. 1	Col. 2		Col. 3		Col. 4	Col. 5		Col. 6			Col. 7	Col. 8		
						Winter,[b] °F				Summer,[c] °F				
State and Station[a]	Lat.		Long.		Elev.	Design Dry Bulb		Design Dry Bulb and Mean Coincident Wet Bulb			Mean Daily	Design Wet Bulb		
	°	′	°	′	Feet	99%	97.5%	1%	2.5%	5%	Range	1%	2.5%	5%
Alabama														
Birmingham AP	33	34	86	45	620	17	21	96/75	94/75	92/74	21	78	77	76
Mobile Co	30	40	88	15	211	25	29	95/77	93/77	91/76	16	80	79	78
California														
Los Angeles AP	33	56	118	24	97	41	43	83/68	80/68	77/67	15	70	69	68
San Francisco AP	37	37	122	23	8	35	38	62/64	77/63	73/62	20	65	64	62
Missouri														
Kansas City AP	39	07	94	35	791	2	6	99/75	96/74	93/74	20	78	77	76
St. Louis AP	38	45	90	23	535	2	6	97/75	94/75	91/74	21	78	77	76

(For convenience, only a portion of Table 2–2 is shown here. The full table is located at the end of the chapter.)

Source: Reprinted by permission from ASHRAE (www.ashrae.org).

occupancy is such that the lowest outside temperature is not coincident with the presence of many people in the building. For most buildings, outside design temperatures are selected on the basis of a percent concept.

For heating load calculations, the percent concept assumes a heating season of December through February, which equals 2160 hours per year. Design conditions are tabulated according to percentages that imply that, statistically, the weather will be at or above the listed condition only for the specified percentage of 2160 hours per year. For example, for a 97.5 percent value listed as 6°F, statistically, the temperature can be expected to be below 6°F for 2.5 percent of 2160 hours, or approximately 50 hours per year. Data are listed for criteria of 99 percent and 97.5 percent. The former criterion is more stringent than the latter.

Cooling load criteria are similar, but they use different percentage values, including 1 percent, 2.5 percent, and 5 percent. If a 5 percent design criterion is used, we can expect that, on average, 5 percent of the summer will be warmer than anticipated by the load calculation, and similarly for 2.5 percent and 1 percent. Summer is defined as June through September, a total of 2928 hours per year.

Extremes of hot weather are expressed as the dry-bulb temperature and the mean coincident wet-bulb temperature. The wet-bulb temperature is also listed in Table 2–2 according to percent occurrence. This information is provided for specifying equipment for which performance is sensitive to the wet-bulb temperature.

Such equipment includes cooling towers, evaporative condensers, and evaporative coolers.

2.5 CALCULATING HEATING LOADS

The first step in the design of HVAC systems is to calculate loads. Heat losses include conducted loads through building envelope elements (walls, windows, roof, etc.) and outside air loads from leakage (infiltration and exfiltration) and ventilation air. See Figure 2–6. The amount of ventilation air required depends on the type of occupancy and the number of occupants contemplated. See Table 2–9 for ASHRAE-recommended occupant density and ventilation air quantity.

2.5.1 Manual versus Computer Calculations

Building load calculations are done almost exclusively by using computer programs. Many good programs are available. Most use data, algorithms, and methods developed by ASHRAE. Manual load calculations should be considered only for preliminary design or simple buildings, therefore, the simplified methods presented here tend to be appropriately conservative.

■ **FIGURE 2–6**
Components of building heating load.

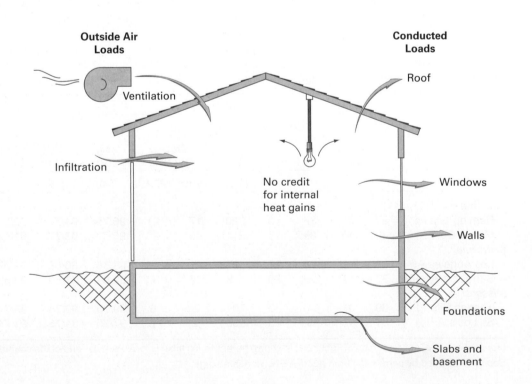

2.5.2 Conduction

Heat transfer by conduction is proportional to the temperature difference between the warm and cold sides of the building envelope element. Inside temperature will depend on design criteria for either personal comfort or manufacturing process, whichever is applicable, and weather conditions—such as outside air temperature, wind velocity, and humidity—selected for design.

Conduction is proportional to the difference between the outside and inside temperatures. It is also proportional to the area through which the heat is transferred; that is, twice the wall begets twice the heat transfer. Conduction also depends on the insulating quality of the wall, which is measured by resistance to heat transfer: the R-value. Envelope elements are made up of several layers, and heat must flow through each layer in sequence. The insulating value of a total assembly is the sum of the R-values of each component. The higher the resistance, the lower is the heat transfer. Heat transfer, temperature difference, area, and resistance are related by the equation

$$Q = A \times TD/R \qquad (2-9)$$

where Q = heat transfer (Btuh)
A = area of assembly (sq ft)
R = resistance (hr·ft²·°F/Btu)

The tendency of an assembly to conduct heat is called the U-factor, or U, and is mathematically the reciprocal of R. Thus, heat transfer is inversely proportional

to R and directly proportional to U. Substituting U for R in the preceding equation, we have

$$Q = U \times A \times TD \qquad (2-10)$$

where U = U-factor (Btu/hr·ft²·°F)

This equation is used in calculating heating load to estimate conduction heat loss through a wall, roof, or window. For calculating air-conditioning load, different equations must be used that consider the effects of the heat of the sun as well as the outside air temperature.

2.5.3 Estimating *U*-Factors for Building Assemblies

Resistance to heat flow and the U-factor for a wall or roof can be calculated using thermal properties for the elements that make up the assembly. The sum of the resistances of individual layers of the assembly will be the total resistance for the assembly.

Thermal properties of commonly used building materials are listed in Table 2–3. For some materials, resistance is tabulated by the inch. For example, concrete has a resistance of 0.08 per inch. Thus, 8 in. of concrete has a thermal resistance of 8 × 0.08, or 0.64.

For certain commonly used modules, resistances are tabulated for the module. An 8-in. concrete masonry unit has a resistance of 1.11. Some insulating materials are specified according to their R-values, such as an R-13 fiberglass batt.

TABLE 2–3
Thermal properties of typical building and insulating materials

Description	Density lb/ft³	Conductivity, λ Btu-in/h·ft²·°F	Conductance, C Btu/h·ft²·°F	Thickness per inch, 1/λ h·ft²·°F/Btu	For thickness listed, 1/C h·ft²·°F/Btu
Building Board					
Boards, Panels, Subflooring, Sheathing					
Woodboard Panel Products					
Asbestos-cement board	120	4.0	—	0.25	—
Asbestos-cement board ....0.125 in.	120	—	33.00	—	0.03
Asbestos-cement board ....0.25 in.	120	—	16.50	—	0.06
Gypsum or plaster board ....0.375 in.	50	—	3.10	—	0.32
Gypsum or plaster board ....0.5 in.	50	—	2.22	—	0.45
Gypsum or plaster board ....0.625 in.	50	—	1.78	—	0.56
Plywood (Douglas fir)	34	0.80	—	1.25	—
Plywood (Douglas fir) ....0.25 in.	34	—	3.20	—	0.31
Plywood (Douglas fir) ....0.375 in.	34	—	2.13	—	0.47
Plywood (Douglas fir) ....0.5 in.	34	—	1.60	—	0.62
Plywood (Douglas fir) ....0.625 in.	34	—	1.29	—	0.77

(For convenience, only a portion of Table 2–3 is shown here. The full table is located at the end of the chapter.)
Source: Reprinted by permission from ASHRAE (www.ashrae.org).

Airspaces contained in building assemblies offer significant resistance to heat flow. Table 2–4 can be used to estimate this resistance. Note that the resistance of an airspace depends not only on the thickness of the space and orientation of the heat flow but also on the emissivity of surfaces facing the airspace. Emissivity is a measure of a surface's ability to reflect and absorb radiant heat.

A film of air clings to any surface and has a resistance to heat flow that depends on the thickness of the film. In still air, the film will be thick. If wind is present, its thickness will be less. The direction of heat flow (up, down, or horizontal) will also affect the resistance of the air film. Resistances of air films are listed in Table 2–5.

Some assemblies have a nonuniform construction, such as stud walls with insulation between framing. The

U-factor of the assembly will be the average based on areas of the different constructions.

Figures 2–7 through 2–10 illustrate the procedure for calculating the total assembly resistance and U-factor.

2.5.4 Infiltration

The HVAC system must have sufficient capacity to heat or cool air infiltration at windows and entries and through loose construction. Infiltration loads will depend on the amount of outside air leakage and the difference in conditions between the outside and inside air. For heating load calculations, the following equation, derived earlier in the chapter (Eq. 2–4), is used:

$$Q = 1.1 \times \text{CFM} \times \text{TD} \qquad (2\text{–}11)$$

If the space must be humidified, the humidification load can be determined by comparing the absolute humidity of the outside air and the desired space condition. Absolute humidity (W) is expressed in terms of grains or pounds of water per pound of dry air (1 pound = 7000 grains). The humidification load is determined approximately using either of the following equations:

$$Q = 4500 \times \text{CFM} \times (W_{\text{room}} - W_{\text{oa}}) \quad (2\text{–}12\text{a})$$

where Q = humidification energy (Btuh)
    W_{room} = desired room humidity ratio (pounds moisture/pound dry air)
    W_{oa} = outside air humidity ratio (pounds moisture/pound dry air)

$$Q = 0.6 \times \text{CFM} \times (W_{\text{room}} - W_{\text{oa}}) \quad (2\text{–}12\text{b})$$

where Q = humidification energy (Btuh)
    W_{room} = desired room humidity ratio (grains moisture/pound dry air)
    W_{oa} = outside air humidity ratio (grains moisture/pound dry air)

If steam is being used to humidify the space, the steam load can be calculated using Eq. 2–8, or

$$\text{Steam flow} = Q/1000$$

Tables 2–6 and 2–7 offer guidelines for estimating the amount of air leakage through construction, in units of cubic feet per hour (cfh) per square foot of wall surface and linear feet of crack for fenestration. Values are listed for various constructions and pressure differences between the inside and outside. A detailed estimate can be prepared assuming the type of construction areas and length of crack for sources of building leakage. This procedure is known as the *crack method* of estimating infiltration.

TABLE 2–4
Thermal resistance of air spaces (h·sq ft °F/Btu)

Position of Airspace	Direction of Heat Flow	Thickness of Airspace, in.			
		0.5	0.75	1.5	3.5
Horizontal	Up	0.91	0.93	0.97	1.03
45° Slope	Up	1.02	1.00	1.04	1.06
Vertical	Horizontal	1.13	1.18	1.12	1.14
45° Slope	Down	1.15	1.26	1.27	1.27
Horizontal	Down	1.15	1.30	1.49	1.62

Values are listed for 0° mean temperature, 20° temperature difference.

Source: Reprinted by permission from ASHRAE (www.ashrae.org).

TABLE 2–5
Surface conductances (Btu/h·sq ft °F) and resistance (h·sq ft °F/Btu) for air

Position of Surface	Direction of Heat Flow	Surface Emittance Nonreflective	
		h	R
Still Air			
Horizontal	Upward	1.63	0.61
Sloping—45°	Upward	1.60	0.62
Vertical	Horizontal	1.46	0.68
Sloping—45°	Downward	1.32	0.76
Horizontal	Downward	1.08	0.92
Moving Air		h	R
(Any position)			
15 mph wind (for winter)	Any	6.00	0.17
7.5 mph wind (for summer)	Any	4.00	0.25

Source: Reprinted by permission from ASHRAE (www.ashrae.org).

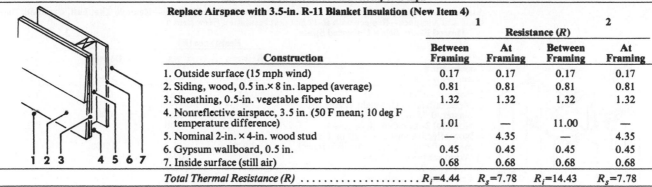

These coefficients are expressed in Btu per (hour) (square foot) (degree Fahrenheit difference in temperature between the air on the two sides), and are based on an outside wind velocity of 15 mph

	Replace Airspace with 3.5-in. R-11 Blanket Insulation (New Item 4)				
		1		2	
		Resistance (R)			
	Construction	**Between Framing**	**At Framing**	**Between Framing**	**At Framing**
	1. Outside surface (15 mph wind)	0.17	0.17	0.17	0.17
	2. Siding, wood, 0.5 in.× 8 in. lapped (average)	0.81	0.81	0.81	0.81
	3. Sheathing, 0.5-in. vegetable fiber board	1.32	1.32	1.32	1.32
	4. Nonreflective airspace, 3.5 in. (50 F mean; 10 deg F temperature difference)	1.01	—	11.00	—
	5. Nominal 2-in. × 4-in. wood stud	—	4.35	—	4.35
	6. Gypsum wallboard, 0.5 in.	0.45	0.45	0.45	0.45
	7. Inside surface (still air)	0.68	0.68	0.68	0.68
	Total Thermal Resistance (R)	R_i=4.44	R_s=7.78	R_i=14.43	R_s=7.78

Construction No. 1: $U_i = 1/4.44 = 0.225$; $U_s = 1/7.81 = 0.128$. With 15% framing (typical of 2-in. × 4-in. studs @ 16-in. o.c.), $U_{av} = 0.8 (0.225) + 0.15 (0.128) = 0.199$ (See Eq 9)

Construction No. 2: $U_i = 1/14.43 = 0.069$; $U_s = 0.128$. With framing unchanged, $U_{av} = 0.8(0.069) + 0.2(0.128) = 0.081$

■ **FIGURE 2–7**

Sample calculation of *U*-factor for frame walls.

Coefficients are expressed in Btu per (hour) (square foot) (degree Fahrenheit difference in temperature between the air on the two sides), and are based on an outside wind velocity of 15 mph

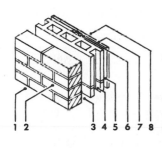

	Replace Furring Strips and Gypsum Wallboard with 0.625-in. Plaster (Sand Aggregate) Applied Directly to Concrete Block-Fill 2.5-in. Airspace with Vermiculite Insulation, 7-8.2 lb/ft³ (New Items 3 and 7)			
		1		2
		Resistance (R)		
	Construction	**Between Furring**	**At Furring**	
	1. Outside surface (15 mph wind)	0.17	0.17	0.17
	2. Common brick, 4 in.	0.80	0.80	0.80
	3. Nonreflective air space, 2.5 in. (30 F mean; 10 deg F temperature difference)	1.10*	1.10*	5.32**
	4. Concrete block, three-oval core, stone and gravel aggregate, 4 in.	0.71	0.71	0.71
	5. Nonreflective airspace 0.75 in. (50 F mean; 10 deg F temperature difference)	1.01	—	—
	6. Nominal 1-in. × 3-in. vertical furring	—	0.94	—
	7. Gypsum wallboard, 0.5 in.	0.45	0.45	0.11
	8. Inside surface (still air)	0.68	0.68	0.68
	Total Thermal Resistance (R)	$R_i = 4.92$	$R_s = 4.85$	$R_i = R_s = 7.79$

Construction No. 1: $U_i = 1/4.92 = 0.203$; $U_s = 1/4.85 = 0.206$. With 20% framing (typical of 1-in. × 3-in. vertical furring on masonry @16-in. (o.c.), $U_{av} = 0.8(0.203) + 0.2(0.206) = 0.204$

Construction No. 2: $U_i = U_s = U_{av} = 1.79 = 0.128$

■ **FIGURE 2–8**

Sample calculation of *U*-factor for masonry cavity walls.

Most designers prefer to use a simpler procedure called the *air change method*. Airflow into a space can be measured in air changes per hour. An air change is equal to the volume of the space. For estimating infiltration, air change rates are assumed on the basis of one's judgment regarding the tightness of the space. Air change rates estimated for design will generally vary from $\frac{1}{4}$ air change per hour for perimeter rooms of a tight building to 2 air changes per hour for a loosely constructed building. Exceptional buildings might warrant higher or lower estimates. Guidelines for estimating air change rates in residential construction are given in Table 2–8.

2.5.5 Ventilation

Outside air is introduced by the HVAC system to dilute building air contaminants and to make up for exhaust. During cold weather, the air must be heated to the temperature of the space. The ventilation load is calculated by the same equations used for infiltration.

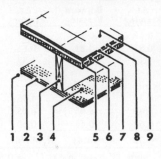

Coefficients are expressed in Btu per (hour) (square foot) (degree Fahrenheit difference between the air on the two sides), and are based on still air (no wind) on both sides

Assume Unheated Attic Space above Heated Room with Heat Flow Up—Remove Tile, Felt, Plywood, Subfloor and Airspace—Replace with R-19 Blanket Insulation (New Item 4)

Heated Room Below Unheated Space				
Construction (Heat Flow Up)	**1** Resistance (R)		**2**	
	Between Floor Joists	At Floor Joists	Between Floor Joists	At Floor Joists
1. Bottom surface (still air)	0.61	0.61	0.61	0.61
2. Metal lath and lightweight aggregate, plaster, 0.75 in.	0.47	0.47	0.47	0.47
3. Nominal 2-in. × 8-in. floor joist	—	9.06	—	9.06
4. Nonreflective airspace, 7.25-in. (50 F mean; 10 deg F temperature difference)	0.93*	—	19.00	—
5. Wood subfloor, 0.75 in.	0.94	0.94	—	—
6. Plywood, 0.625 in.	0.77	0.77	—	—
7. Felt building membrane	0.06	0.06	—	—
8. Tile	0.05	0.05	—	—
9. Top surface (still air)	0.61	0.61	0.61	0.61
Total Thermal Resistance (R)	R_i= 4.44	R_s= 12.57	R_i= 20.69	R_s=10.75

Construction No. 1: U_i= 1/4.45= 0.225; U_s= 1/12.58= 0.079. With 10% framing (typical of 2-in. joists @ 16-in. o.c.), U_{av} = 0.9 (0.225) + 0.1 (0.079)= 0.210

Construction No. 2: U_i = 1/20.69 = 0.048; U_s = 1/10.75 = 0.093. With framing unchanged, U_{av} = 0.9 (0.048) + 0.1 (0.093) = 0.053

■ **FIGURE 2–9**

Sample calculation of U-factor for frame construction ceiling and floor.

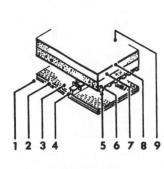

These coefficients are expressed in Btu per (hour) (square foot) (degree Fahrenheit difference in temperature between the air on the two sides), and are based upon an outside wind velocity of 15 mph

Add Rigid Roof Deck Insulation, C = 0.24 (R = 1/C = 4.17) (New Item 7)

Construction (Heat Flow Up)	1	2
1. Inside surface (still air)	0.61	0.61
2. Metal lath and lightweight aggregate plaster, 0.75 in.	0.47	0.47
3. Nonreflective airspace, greater than 3.5 in. (50 F mean; 10 deg F temperature difference)	0.93*	0.93*
4. Metal ceiling suspension system with metal hanger rods	0**	0**
5. Corrugated metal deck	0	0
6. Concrete slab, lightweight aggregate, 2 in. (30 lb/ft^3)	2.22	2.22
7. Rigid roof deck insulation (none)	—	4.17
8. Built-up roofing, 0.375 in.	0.33	0.33
9. Outside surface (15 mph wind)	0.17	0.17
Total Thermal Resistance (R)	4.73	8.90

Construction No. 1: U_{av} = 1/4.73 = 0.211
Construction No. 2: U_{av} = 1/8.90= 0.112

■ **FIGURE 2–10**

Sample calculation of U-factor for flat masonry roofs with built-up roofing, with and without suspended ceiling (winter conditions, upward flow).

Determining the proper amount of outside air requires analyses of the building's exhaust systems and fresh air requirements for occupancy and a consideration of excess air to pressurize the building slightly and prevent undue infiltration.

Minimum values for outside air are specified by code to maintain acceptable indoor air quality. Model codes generally incorporate recommendations of ASHRAE Standard 62. Code requirements for outside air have varied significantly over the years, owing to varying concerns about energy conservation and attention to indoor air quality. Table 2–9 shows the consensus standard for outside air rates used in most current codes. Rates are specified according to usage of space.

TABLE 2–6
Air leakage through walls (cfh per sq ft)

	Pressure Difference, in. Water			
Type of Wall	0.05	0.10	0.20	0.30
Brick Wall: 8.5 in.				
Plain	5	9	16	24
Plastered Two coats on brick	0.05	0.08	0.14	0.2
Brick Wall: 13 in.	5	8	14	20
Plain plastered; two coats on brick	0.01	0.04	0.05	0.09

Plastered, furring, lath;
  two coats gypsum
  plaster
	0.03	0.24	0.46	0.66

Frame Wall:				

Bevel siding painted or
  cedar singles,
  sheathing, building
  paper, wood lath: three
  coats gypsum plaster
	0.09	0.15	0.22	0.29

Source: Reprinted by permission from ASHRAE (www.ashrae.org).

TABLE 2–7
Infiltration through double-hung wood windows (cfh per foot of crack)

	Pressure Difference, in. Water		
Type of Window	0.10	0.20	0.30
A. Wood double-hung window (Locked)			
1. Nonweatherstripped, loose fit or weatherstripped, loose fit	77	122	150
2. Nonweatherstripped, average fit	27	43	57
3. Weatherstripped, average fit	14	23	30

**B. Frame–wall leakage
  (Leakage is that passing between
  the frame of a wood double-hung
  window and the wall)**			

1. Around frame in masonry wall, not caulked	17	26	34
2. Around frame in masonry wall, caulked	3	5	6
3. Around frame in wood frame wall	13	21	29

Source: Reprinted by permission from ASHRAE (www.ashrae.org).

2.5.6 Miscellaneous Loads

In addition to conduction, infiltration, and ventilation, heating loads should take into account miscellaneous factors such as losses through walls below grade and slabs on grade. Guidelines are shown in Tables 2–10 and 2–11.

TABLE 2–8
Infiltration air changes per hour occurring under average conditions in residences

Kind of room	Single Glass, No Weatherstrip	Storm Sash or Weatherstripped
No windows or exterior doors	0.5	0.3
Windows or exterior doors on one side	1	0.7
Windows or exterior doors on two sides	1.5	1
Windows or exterior doors on three sides	2	1.3
Entrance halls	2	1.3

Source: Reprinted by permission from ASHRAE (www.ashrae.org).

TABLE 2–9
Outdoor air requirements for ventilation

	Estimated Maximum Occupants per 1000 sq ft	Outdoor Requirements		
Application		CFM/ person	CFM/ sq ft	CFM/ room
Dry Cleaners, Laundries[a]				
Commercial laundry	10	25		
Commercial dry cleaner	30	30		
Storage, pick up	30	35		
Coin-operated laundries	20	15		
Coin-operated dry cleaner	20	15		

[a]Dry cleaning processes may require more air.
(For convenience, only a portion of Table 2–9 is shown here. The full table is located at the end of the chapter.)

Source: Reprinted by permission from ASHRAE (www.ashrae.org).

2.5.7 Heating Load Problem 2–1

This sample problem demonstrates methods for calculating heating and humidification loads for a small office building, defined in Figure 2–11. The criteria and physical properties of the building are as follows:

- Design conditions—indoor (72°F, 30% RH), outdoor (8°F, near 0% RH)
- Infiltration—2 air changes per hour
- Ventilation—500 CFM

TABLE 2–10
Below-grade heat losses for basement walls and floors

Ground Water Temperature	Basement Floor Loss,[a] Btu/sq ft	Below-Grade Wall Loss,[b] Btu/sq ft
40	3.0	6.0
50	2.0	4.0
60	1.0	2.0

[a]Based on basement temperature of 70°F and U of 0.10.
[b]Assumed twice basement floor loss.
Source: Reprinted by permission from ASHRAE (www.ashrae.org).

TABLE 2–11
Heat loss of concrete floors at or near grade
(Btuh/ft slab edge)

Outdoor Design Temperature, °F	Unheated (Heated) Slab; Resistance of Insulation		
	5.0	3.3	2.5
5 to −5	23 (31)	35 (47)	46 (62)
−5 to −15	26 (36)	39 (54)	53 (72)
−15 to −25	29 (40)	44 (60)	59 (80)

Source: Reprinted by permission from ASHRAE (www.ashrae.org).

2.6 CALCULATING COOLING LOADS

The first step in sizing air-conditioning equipment is to calculate loads. Heat gains include conduction, solar effects, outside air loads, and internal heat loads, as illustrated in Figure 2–12.

2.6.1 Conduction through Walls and Roofs

In the estimation of cooling loads, a simple temperature difference between inside and outside air will not account for solar heat. The outside surface of a wall or roof may be much warmer than the surrounding air, owing to solar effects. Accordingly, conduction through walls and roofs is estimated by equations using a total equivalent temperature difference (TETD) that includes solar effects as well as temperature difference. The value of TETD will vary with the orientation, time of day, absorption property of the surface, and thermal mass of the building assembly, which affects the timing of heat entry to the interior. Values for TETD are shown

in Table 2–12 for roofs and Table 2–13 for walls. Wall construction is defined in Table 2–14. (*Note:* These tables are given in their entirety at the end of the chapter.)

For walls and roofs, the cooling load is calculated by using the equation

$$Q = U \times A \times TETD \qquad (2\text{–}13)$$

where TETD = total equivalent temperature difference (°F)

2.6.2 Conducted and Solar Heat through Glazing

Solar effects must also be considered in estimating heat gains through windows and skylights. These heat loads are considered in two parts—simple conduction and solar transmission—as shown in Figure 2–13. Equation 2–14 applies:

$$Q = U \times A \times TD + SC \times A \times SHGF \quad (2\text{–}14)$$

where SHGF = solar heat gain factor (Btuh/sq ft)
SC = shading coefficient (dimensionless)

The *solar heat gain factor* (SHGF) represents the amount of solar heat that will enter a clear single-pane window at a given time of year and time of day, facing the specified orientation. Values of SHGF are shown in Table 2–15 for 40° north latitude. Tables for other latitudes are included in the ASHRAE *Handbook of Fundamentals.* The tables also include values for the azimuth and altitude of the sun, which are important in accounting for the geometry of shading from fins, overhangs, and adjacent structures.

The *shading coefficient* (SC) is a property of the glazing material and accessories such as blinds or draperies. SC is the ratio of solar heat admitted in comparison with what is admitted by clear single-strength glass, which has a shading coefficient of 1.0. Thus, a glazing with 0.5 SC allows only half as much solar heat into the space as clear single-strength glass does. Shading coefficients as low as 0.12 can be achieved with heavily reflective films. Table 2–16 gives shading coefficients for commonly used glazing materials.

2.6.3 Infiltration and Ventilation

Generally, the amount of air infiltration is much lower during hot weather than during cold weather. This is because winds are milder, and lower temperature differentials cause less of a chimney effect. Accordingly, air change rates should be estimated lower for summer than for winter.

North

East

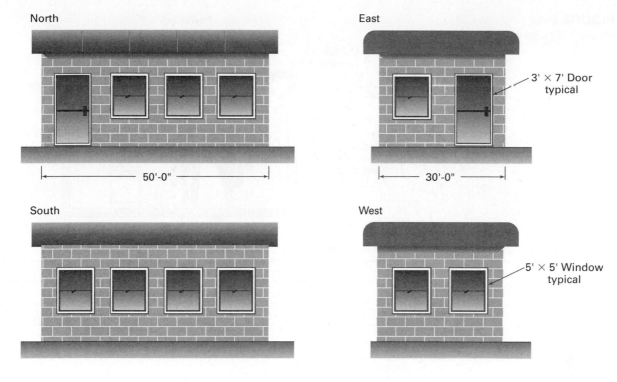

3' × 7' Door typical

50'-0"

30'-0"

South

West

5' × 5' Window typical

Elevations

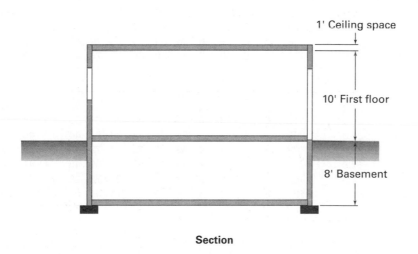

1' Ceiling space

10' First floor

8' Basement

Section

■ **FIGURE 2–11**
Elevations and section of the building, Problem 2–1. *U*-factors are as follows: walls, 0.062; roof, 0.16; windows, 0.58; and doors, 0.64.

Infiltration loads have two components: sensible and latent. The equations governing such loads are identical to those cited earlier for heating load calculations:

$$Q_{sensible} = 1.1 \times \text{CFM} \times \text{TD} \qquad (2\text{–}15)$$

$$Q_{latent} = 4500 \times \text{CFM} \times (W_{final} - W_{initial}) \quad (2\text{–}16)$$

Total heat, termed *enthalpy (H)*, is the sum of sensible and latent heat. The total heat of air at various conditions of temperature and humidity can be taken from a psychrometric chart or tables, and the following equation can be used to determine energy flow:

$$Q = 4.5 \times \text{CFM} \times \Delta H \qquad (2\text{–}17)$$

where ΔH = change in enthalpy (Btu/lb)

■ **FIGURE 2–12**

Components of building cooling loads.

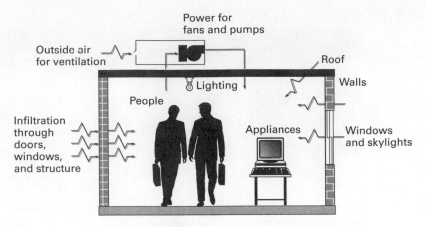

TABLE 2–12

Total equivalent temperature differentials for calculating heat gain through flat roofs

Description of Roof Construction	U-value Wt, lb per sq ft	Btu/(hr) (ft²)(°F)	Sun Time																		
			A.M.						P.M.												
			8		10		12		2		4		6		8		10		12	
			D	L	D	L	D	L	D	L	D	L	D	L	D	L	D	L	D	L
Light Construction Roofs—Exposed to Sun																					
1″ Insulation + steel siding	7.4	0.213	28	11	65	31	90	48	95	53	78	45	43	27	8	6	1	1	−3	−3
2″ Insulation + steel siding	7.8	0.125	24	8	61	29	88	46	96	53	81	46	48	30	10	8	2	2	−3	−3
1″ Insulation + 1″ wood	8.4	0.206	12	2	47	21	77	39	92	50	86	48	61	36	25	16	7	5	0	−1
2″ Insulation + 1″ wood	8.5	0.122	8	0	41	18	72	36	90	48	88	40	65	38	30	19	9	7	1	0
1″ Insulation + 2.5″ wood	12.7	0.193	2	−2	23	8	48	23	70	36	79	42	71	40	50	29	29	17	15	9
2″ Insulation + 2.5″ wood	13.1	0.117	1	−2	19	6	43	20	65	33	76	41	72	40	53	31	33	20	18	11
Medium Construction Roofs—Exposed to Sun																					
1″ Insulation + 4″ wood	17.3	0.183	5	0	14	5	31	14	49	24	62	32	65	35	56	31	41	24	29	17
2″ Insulation + 4″ wood	17.8	0.113	6	1	13	4	28	12	45	22	58	30	63	34	56	31	43	25	32	18
1″ Insulation + 2″ h.w. concrete	28.3	0.206	4	−1	27	11	54	26	74	39	81	44	70	40	45	27	24	15	12	7
2″ Insulation + 2″ h.w. concrete	28.8	0.122	2	−2	23	9	49	23	70	36	79	43	71	40	49	29	28	17	15	9
4″ l.w. concrete	17.8	0.213	1	−3	28	11	59	28	82	43	88	48	74	42	44	27	19	12	6	4
6″ l.w. concrete	24.5	0.157	−2	−4	9	2	31	13	55	27	72	38	76	41	64	36	42	25	25	15
8″ l.w. concrete	31.2	0.125	6	2	6	1	16	6	32	14	49	24	61	32	63	34	55	31	41	24
Heavy Construction Roofs—Exposed to Sun																					
1″ Insulation + 4″ h.w. concrete	51.6	0.199	7	1	17	6	33	15	50	25	61	32	63	34	53	30	40	23	28	16
2″ Insulation + 4″ h.w. concrete	52.1	1.120	7	2	15	6	30	13	46	23	58	30	61	33	54	30	41	23	31	17
1″ Insulation + 6″ h.w. concrete	75.0	0.193	13	6	17	7	26	12	38	18	48	25	53	28	51	27	43	24	35	19
2″ Insulation + 6″ h.w. concrete	75.4	0.117	15	7	17	7	25	11	36	17	46	23	51	27	50	27	43	24	36	20

1. *Application.* These values may be used for all normal air-conditioning estimates, usually without correction in latitude 0° to 50° north or south when the load is calculated for the hottest weather.

2. *Corrections.* The values in the table were calculated for an inside temperature of 75°F and an outdoor maximum temperature of 95°F with an outdoor daily range of 21°F. The table remains approximately correct for other outdoor maximums (93–102°F) and other outdoor daily ranges (16–34°F), provided that the outdoor daily average temperature remains approximately 85°F.

3. *Attics or other spaces between the roof and ceiling.* If the ceiling is insulated and a fan is used for positive ventilation in the space between the ceiling and roof, the total temperature differential for calculating the room load may be decreased by 25 percent. If the attic space contains a return duct or other air plenum, care should be taken in determining the portion of the heat gain that reaches the ceiling.

4. *Light Colors.* Credit should not be taken for light-colored roofs, except where the permanence of light color is established by experience, as in rural areas or where there is little smoke.

Note: h.w. = heavy weight

l.w. = light weight

Calculations for Problem 2–1

Load Components			*Load (Btuh)*

Roof

$Q = U \times A \times \text{TD}$
$Q = 0.16 \times (30 \times 50) \times (72 - 8)$ 15,360

Walls[#]

$Q = U \times A \times \text{TD}$

North	$Q = 0.062 \times (11 \times 50) \times (72 - 8)$	2182
South	$Q = 0.062 \times (11 \times 50) \times (72 - 8)$	2182
East	$Q = 0.062 \times (11 \times 30) \times (72 - 8)$	1309
West	$Q = 0.062 \times (11 \times 30) \times (72 - 8)$	1309

Doors

$Q = U \times A \times \text{TD}$

North	$Q = 0.64 \times (3 \times 7) \times (72 - 8)$	860
East	$Q = 0.64 \times (3 \times 7) \times (72 - 8)$	860

Windows

$Q = U \times A \times \text{TD}$

North	$Q = .58 \times (3 \times 5 \times 5) \times (72 - 8)$	2784
South	$Q = .58 \times (4 \times 5 \times 5) \times (72 - 8)$	3712
East	$Q = .58 \times (1 \times 5 \times 5) \times (72 - 8)$	928
West	$Q = .58 \times (2 \times 5 \times 5) \times (72 - 8)$	1856

Basement floor

$Q = \text{Btuh/ft}^2 \times \text{area}$
$Q = 3 \times (30 \times 50)$ 4500

Basement walls

$Q = \text{Btuh/ft}^2 \times \text{area}$
$Q = 6.0 \times (8 \times (30 + 50 + 30 + 50))$ 7680

Infiltration, sensible only

$Q = 1.1 \times \text{CFM} \times \text{TD}$
$\text{CFM} = (\text{air exchanges per hour} \times \text{volume})/60 \text{ minutes per hour}$
$Q = 1.1 \times ((2 \times (30' \times 50' \times 10'))/60) \times (72 - 8)$ 35,200

Ventilation, sensible only

$Q = 1.1 \times \text{CFM} \times \text{TD}$
$\text{CFM} = 500$
$Q = 1.1 \times 500 \times (72 - 8)$ 35,200

Total Heat Loss = 115,922 Btuh

Humidification (Optional)

Infiltration air

$Q = 4500 \times \text{CFM} \times (W_{\text{room}} - W_{\text{oa}})$
$\text{CFM} = (\text{air exchanges per hour} \times \text{volume})/60 \text{ minutes per hour}$
$Q = 4500 \times ((2 \times (30' \times 50' \times 10'))/60) \times (0.005 - 0)$ 12,240

Ventilation air

$Q = 4500 \times \text{CFM} \times (W_{\text{room}} - W_{\text{oa}})$
$\text{CFM} = (\text{air exchanges per hour} \times \text{volume})/60 \text{ minutes per hour}$
$Q = 4500 \times 500 \times (0.005 - 0)$ 12,240

Total Humidification of Outside Air = 22,500 Btuh

[#]For simplicity, areas of doors and windows are not deducted from wall area calculations in this example. Adjustment is appropriate if doors and window areas are significant in comparison with the respective wall areas.

TABLE 2–13
Total equivalent temperature differentials for calculating heat gain through sunlit walls

North Latitude — Sun Time — Exterior color of wall—D = dark, L = light

Groups A, B, C (A.M.: 8, 10, 12 | P.M.: 2, 4, 6, 8, 10, 12)

Wall Facing	8 D	8 L	10 D	10 L	12 D	12 L	2 D	2 L	4 D	4 L	6 D	6 L	8 D	8 L	10 D	10 L	12 D	12 L
Group A																		
NE	27	16	31	18	26	17	24	17	24	18	23	17	20	15	17	13	15	11
E	32	18	41	24	37	22	29	21	28	20	26	19	23	16	20	14	18	13
SE	25	15	36	21	38	23	33	21	28	20	26	18	22	16	19	14	18	12
S	14	9	20	13	28	18	31	22	31	21	25	18	20	15	17	13	15	11
SW	17	11	20	13	24	16	34	22	42	27	41	26	28	19	20	14	18	12
W	17	11	20	13	24	16	30	20	42	27	48	30	33	22	24	16	19	13
NW	14	9	17	11	21	14	23	17	31	21	38	25	28	19	23	16	16	11
N	14	9	15	10	17	12	20	15	21	16	21	16	18	14	14	11	12	9
Group B																		
NE	12	7	27	14	31	17	30	19	31	21	30	22	27	20	20	15	16	13
E	14	8	34	18	45	24	43	25	39	25	35	24	30	22	23	18	17	14
SE	9	5	25	13	39	21	44	26	41	26	37	25	31	23	22	17	17	14
S	4	3	7	4	18	11	32	19	41	26	39	27	33	24	24	18	18	15
SW	5	3	7	4	11	7	23	15	41	26	54	34	51	33	38	25	26	19
W	6	4	7	4	11	7	18	12	26	18	34	23	55	34	43	28	30	20
NW	5	3	6	4	11	7	17	12	26	18	27	19	41	27	36	24	25	18
N	6	4	9	5	12	8	18	12	22	16	21	16	25	19	21	16	16	14
Group C																		
NE	9	6	19	10	26	15	28	17	29	18	29	20	28	20	24	19	20	16
E	10	7	22	12	36	19	40	23	39	23	36	24	33	23	28	20	22	17
SE	8	6	16	9	29	16	38	21	39	24	37	24	34	23	28	21	23	17
S	7	5	7	4	12	7	21	14	32	20	36	24	34	24	28	21	23	17
SW	9	6	8	5	10	6	16	10	28	18	42	26	48	30	42	28	33	22
W	10	7	9	5	10	6	14	9	24	16	40	25	52	32	47	30	37	24
NW	8	6	8	5	9	6	13	9	19	14	30	20	40	27	38	26	30	21
N	7	5	8	5	10	7	14	9	18	13	22	16	25	19	23	18	19	16

Groups G, H, I (A.M.: 8, 10, 12, 2 | P.M.: 4, 6, 8, 10, 12)

Wall Facing	8 D	8 L	10 D	10 L	12 D	12 L	2 D	2 L	4 D	4 L	6 D	6 L	8 D	8 L	10 D	10 L	12 D	12 L
Group G																		
NE	11	9	15	10	20	12	24	14	26	16	26	17	26	18	23	16	24	17
E	13	9	17	11	26	15	32	18	34	20	34	21	31	22	29	20	29	19
SE	13	9	14	9	21	12	28	16	33	19	34	21	30	22	30	20	28	19
S	12	9	10	7	11	8	16	10	23	15	29	18	30	21	29	20	26	19
SW	16	11	13	9	13	9	14	10	23	15	30	19	37	25	39	25	35	23
W	18	12	15	10	14	9	14	9	27	18	38	25	42	29	42	30	38	25
NW	14	10	12	8	12	8	13	9	21	15	29	18	33	23	33	22	31	21
N	10	8	10	7	10	7	12	8	18	13	20	15	18	15	18	14	16	14
Group H																		
NE	15	11	16	11	18	12	20	13	22	14	24	15	25	16	24	15	24	17
E	18	13	18	12	22	14	26	16	29	17	30	19	31	20	29	20	29	19
SE	18	13	17	12	19	12	23	14	27	16	29	18	30	19	28	18	28	18
S	16	12	14	10	17	11	18	12	16	12	23	15	25	17	26	18	26	18
SW	22	14	19	12	18	12	18	12	12	11	23	15	29	21	32	21	32	22
W	23	15	20	13	18	12	18	12	17	11	22	15	29	22	33	23	34	22
NW	19	13	17	11	15	10	15	11	15	10	18	12	23	15	26	18	27	19
N	13	10	12	9	11	9	12	9	11	9	13	10	17	13	18	14	18	14
Group I																		
NE	16	11	18	12	20	13	22	14	23	15	24	16	24	16	23	16	22	16
E	19	13	21	14	25	16	29	17	30	18	30	19	28	19	28	19	26	18
SE	19	13	19	13	22	14	26	16	28	18	29	18	29	18	28	18	26	18
S	16	12	15	11	16	11	18	12	21	14	25	16	31	20	32	20	23	16
SW	20	14	19	13	18	12	18	12	22	14	27	17	31	20	33	21	30	20
W	22	14	21	14	20	13	20	13	22	14	26	17	31	21	32	21	32	21
NW	18	12	16	11	17	11	18	12	18	12	21	14	25	17	27	18	26	18
N	13	10	12	9	13	9	13	10	15	11	16	12	18	13	18	14	14	14

Table — Equivalent Temperature Differentials for Sunlit Walls (Groups D, E, F and J, K, L)

Sun Time — A.M. (8, 10, 12) and P.M. (2, 4, 6, 8, 10, 12); each hour split into D = dark / L = light.
Exterior color of wall—D = dark, L = light

Groups D, E, F

North Latitude Wall Facing	8 D	8 L	10 D	10 L	12 D	12 L	2 D	2 L	4 D	4 L	6 D	6 L	8 D	8 L	10 D	10 L	12 D	12 L
Group D																		
NE	8	5	19	10	28	15	29	17	30	19	30	21	28	21	24	19	23	16
E	9	6	23	12	38	20	42	24	40	24	37	24	33	23	27	20	21	17
SE	7	5	16	9	30	16	40	22	41	25	38	25	34	24	28	21	22	17
S	5	4	6	4	12	7	23	14	34	21	25	16	35	24	29	21	23	17
SW	8	5	7	4	9	6	16	10	19	12	44	28	51	32	49	28	33	22
W	8	6	7	5	9	6	14	9	16	14	42	27	55	34	37	31	37	25
NW	7	5	7	4	9	6	13	9	14	11	31	21	42	28	40	27	31	21
N	6	4	8	5	10	6	14	10	19	14	23	17	25	19	24	19	19	16
Group E																		
NE	10	6	23	12	30	16	30	18	30	20	30	21	28	21	23	18	18	14
E	11	6	28	15	42	22	43	24	39	24	36	24	32	23	25	19	19	15
SE	8	5	20	11	35	19	42	24	41	25	38	24	33	23	26	20	20	16
S	4	3	6	4	15	9	28	17	38	24	39	26	34	24	27	20	20	16
SW	6	4	7	4	10	6	19	12	35	22	49	31	52	33	41	27	30	21
W	7	5	7	4	10	6	16	11	30	20	48	31	57	36	47	30	34	23
NW	6	4	6	4	10	6	15	10	23	16	36	24	45	30	38	26	28	20
N	6	4	8	5	11	7	16	11	21	15	24	18	26	20	23	18	18	15
Group F																		
NE	9	7	14	9	21	12	25	15	27	17	29	19	28	20	26	19	23	17
E	10	8	17	10	28	15	35	19	37	22	37	23	35	23	31	22	26	19
SE	10	7	13	8	22	12	31	17	36	21	37	23	35	23	32	22	27	19
S	9	7	7	5	10	6	17	10	26	16	32	20	33	22	31	22	27	19
SW	12	9	10	6	9	6	13	8	22	14	33	21	42	27	43	27	37	25
W	14	9	11	7	10	7	12	8	19	12	31	20	43	27	46	29	41	27
NW	12	8	9	6	9	6	11	8	16	11	24	16	33	22	36	24	33	23
N	8	7	6	6	8	6	12	8	15	11	19	14	22	17	23	18	21	17

Groups J, K, L

North Latitude Wall Facing	8 D	8 L	10 D	10 L	12 D	12 L	2 D	2 L	4 D	4 L	6 D	6 L	8 D	8 L	10 D	10 L	12 D	12 L
Group J																		
NE	18	13	17	12	18	12	19	13	21	14	22	15	23	15	23	16	23	16
E	22	15	20	14	21	14	24	15	26	16	28	17	29	18	29	19	20	19
SE	21	15	20	14	20	13	21	14	24	15	26	16	628	17	28	18	28	18
S	19	14	17	12	16	11	17	12	20	13	21	14	22	15	24	16	24	16
SW	24	16	22	15	20	13	19	13	21	14	1	14	24	16	24	16	30	19
W	26	18	24	16	22	14	20	13	20	13	21	14	24	16	28	18	31	20
NW	21	15	19	13	18	12	17	11	17	12	19	13	19	13	22	15	25	17
N	15	11	14	11	13	10	13	10	13	9	14	10	15	11	17	12	17	13
Group K																		
NE	19	14	19	13	20	13	20	13	21	14	21	14	22	14	22	15	22	15
E	23	16	22	15	24	16	26	16	27	16	27	17	27	16	28	18	27	18
SE	23	15	22	15	22	14	24	15	25	15	26	16	26	15	27	17	27	17
S	20	14	21	13	23	14	22	15	20	13	20	13	21	14	22	15	23	15
SW	25	16	23	15	22	15	21	14	21	14	22	14	24	16	24	16	28	18
W	26	17	24	16	23	15	22	14	21	14	21	14	24	15	27	17	28	18
NW	21	15	20	14	19	13	18	13	18	12	19	13	20	14	22	15	23	16
N	15	11	14	11	14	10	14	10	14	10	14	11	15	11	16	12	16	12
Group L																		
NE	18	13	18	13	19	13	20	13	21	14	22	15	23	15	23	16	22	15
E	22	15	22	14	23	15	25	16	27	17	28	18	28	18	28	18	27	18
SE	21	14	21	14	22	14	23	15	24	16	27	17	27	17	27	17	26	18
S	19	13	19	13	22	14	21	14	20	13	21	14	23	15	23	16	23	16
SW	23	15	22	14	21	14	22	14	23	15	24	15	26	17	26	17	28	18
W	25	16	23	15	22	14	21	14	22	15	24	15	26	17	29	19	30	19
NW	20	14	20	14	19	13	18	12	18	12	19	13	21	15	23	16	24	16
N	14	11	14	11	13	10	13	10	14	10	15	11	16	12	17	13	17	13

1. *Application.* These values may be used for all normal air-conditioning estimates, usually without correction when the load is calculated for the hottest weather.

2. *Corrections.* The values in the table were calculated for an inside temperature of 75°F and an outdoor maximum temperature of 95°F with an outdoor daily range of 21°F. The table remains approximately correct for other outdoor maximums (93–102°F) and other outdoor daily ranges (16–34°F), provided that the outdoor daily average temperature remains approximately 85°F.

3. *Color or exterior surface of wall.* Use temperature differentials for light walls only when the permanence of the light wall is established by experience. For cream colors, use the values for light walls. For medium colors, interpolate halfway between the dark and light values. Medium colors are medium blue, medium green, bright red, light brown, unpainted wood, natural color concrete, etc. Dark blue, red, brown, green, etc., are considered dark colors.

43

TABLE 2–14
Description of several representative wall constructions

Group	Components	Wt. lb per sq ft	U-Value
A	1″ stucco + 4″ l.w. concrete block + air space	29	0.27
	1″ stucco + air space + 2″ insulation	16	0.11
B	1″ stucco + 4″ common brick	56	0.39
	1″ stucco + 4″ h.w. concrete	63	0.48
C	4″ face brick + 4″ l.w. concrete block + 1″ insulation	63	0.16
	1″ stucco + 4″ h.w. concrete + 2″ insulation	63	0.11
D	1″ stucco + 8″ l.w. concrete block + 1″ insulation	41	0.14
	1″ stucco + 2″ insulation + 4″ h.w. concrete block	37	0.11
E	4″ face brick + 4″ l.w. concrete block	62	0.33
	1″ stucco + 8″ h.w. concrete block	57	0.35
F	4″ face brick + 4″ common brick	90	0.36
	4″ face brick + 2″ insulation + 4″ l.w. concrete block	63	0.10
G	1″ stucco + 8″ clay tile + 1″ insulation	63	0.14
	1″ stucco + 2″ insulation + 4″ common brick	56	0.11
H	4″ face brick + 8″ clay tile + 1″ insulation	96	0.14
	4″ face brick + 8″ common brick	130	0.28
I	1″ stucco + 8″ clay tile + air space	63	0.21
	4″ face brick + air space+ 4″ h.w. concrete block	70	0.28
J	face brick + 8″ common brick + 1″ insulation	130	0.15
	4″ face brick + 2″ insulation + 8″ clay tile	97	0.09
K	4″ face brick + air space + 8″ clay tile	96	0.20
	4″ face brick + 2″ insulation + 8″ common brick	130	0.10
L	4″ face brick + 8″ clay tile + air space	96	0.20
	4″ face brick + air space + 4″ common brick	90	0.27
M	4″ face brick + air space + 8″ common brick	130	0.22
	4″ face brick + air space + 12″ h.w. concrete	190	0.25

■ **FIGURE 2–13**
Heat gain through glazing.

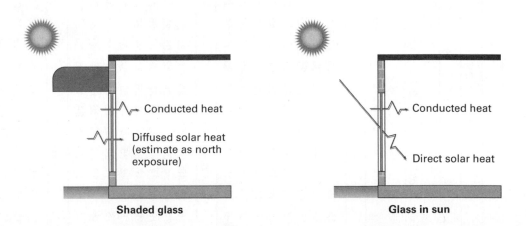

Shaded glass   Glass in sun

Outside air introduced for ventilation by the air-conditioning equipment will result in sensible and latent loads calculated according to the same equations. Recommended minimum ventilation rates are shown in Table 2–9. When determining ventilation requirements, no credit can be taken for infiltration, which will vary depending on wind and temperature, and how well the building is sealed.

2.6.4 Internal Heat Gains

Heat is generated inside buildings by lights, appliances, and people. People liberate both sensible and latent heat. Latent heat results from exhaled moisture and evaporation of perspiration. Loads will depend on the level of activity, as shown in Table 2–17.

TABLE 2–15
Solar position and intensity and solar heat gain factors for 40° north latitude

Date	Solar Time, A.M.	Altitude	Azimuth	Direct Normal Irradiation, Btuh/sq ft	N	NE	E	SE	S	SW	W	NW	Horizontal	Solar Time, P.M.
Jan 21	8	8.1	55.3	141	5	17	111	133	75	5	5	5	13	4
	9	16.8	44.0	238	11	12	154	224	160	13	11	11	54	3
	10	23.8	30.9	274	16	16	123	241	213	51	16	16	96	2
	11	28.4	16.0	289	18	18	61	222	244	118	18	18	123	1
	12	30.0	0.0	293	19	19	20	179	254	179	20	19	133	12
	Half-day Totals				59	68	449	903	815	271	59	59	353	
Feb 21	7	4.3	72.1	55	1	22	50	47	13	1	1	1	3	5
	8	14.8	61.6	219	10	50	183	199	94	10	10	10	43	4
	9	24.3	49.7	271	16	22	186	245	157	17	16	16	98	3
	10	32.1	35.4	293	20	21	142	247	203	38	20	20	143	2
	Half-day Totals				81	144	634	1035	813	250	81	81	546	
Mar 21	7	11.4	80.2	171	8	93	163	135	21	8	8	8	26	5
	8	22.5	69.6	250	15	91	218	211	73	15	15	15	85	4
	9	32.8	57.3	281	21	46	203	236	128	21	21	21	143	3
	10	41.6	41.9	297	25	26	153	229	171	28	25	25	186	2
	11	47.7	22.6	304	28	28	78	198	197	77	28	28	213	1
	12	50.0	0.0	306	28	28	30	145	206	14	30	28	223	12
	Half-day Totals				112	310	849	1100	692	218	112	112	764	
Apr 21	6	7.4	98.9	89	11	72	88	52	5	4	4	4	11	6
	7	18.9	89.5	207	16	141	201	143	16	14	14	14	61	5
	8	30.3	79.3	253	22	128	225	189	41	21	21	21	124	4
	9	41.3	67.2	275	26	80	203	204	83	26	26	26	177	3
	10	51.2	51.4	286	30	37	153	194	121	32	30	30	218	2
	11	58.7	29.2	292	33	34	81	161	146	52	33	33	244	1
	12	61.6	0.0	294	33	33	36	108	155	108	36	33	253	12
	Half-day Totals				153	509	969	1003	489	196	146	145	962	
					N	NW	W	SW	S	SE	E	NE	Horizontal	←⏐↑ P.M.

(For convenience, only a portion of Table 2–15 is shown here. The full table is located at the end of the chapter.)

Source: Reprinted by permission from ASHRAE (www.ashrae.org).

Heat from lights and appliances can be calculated using the factor for conversion of electrical to thermal energy:

$$Q = 3.41 \times P \qquad (2\text{–}18)$$

where P = power input to light fixture or appliance (watts)

Often, a precise figure for the heat from building lighting and appliances is unavailable at the time the air-conditioning system is being designed. In that case, an allowance is assumed in watts per square foot of building or room area. Guidelines for estimating electrical loads for common types of buildings are given in Chapter 13.

2.6.5 Loads in Return Air Plenums

If lighting is recessed in a ceiling that is used as a return air plenum, heat from the back side of the fixtures will not enter the occupied space. Heat to the plenum will still need to be removed by the air-conditioning system but will not affect the amount of supply air delivered to cool the occupied space. Similar adjustments should be

TABLE 2–16
Shading coefficient for single glass and insulating glass

Type of Glass	Normal Thickness	Shading Coefficient[a]
Single Glass		
Regular sheet	⅛	1.00
Regular plate/float	¼	0.95
	⅜	0.91
	½	0.88
Grey sheet	⅛	0.78
	¼	0.86
Heat-absorbing plate/float	3⁄16	0.72
	¼	0.70
Insulating Glass		
Regular sheet out, regular sheet in	⅛	0.90
Regular plate/float out, regular plate/float in	¼	0.83
Heat-absorbing plate/float out, regular plate/float in	¼	0.06

[a]Wind velocity 7.5 mph.

Source: Reprinted by permission from ASHRAE (www.ashrae.org).

made for roof loads that occur above return air plenum ceilings. The actual load to the space should be calculated by considering conduction from the plenum through the ceiling. Space loads and plenum loads can be distinguished by comparing the two cases shown in Figure 2–14.

2.6.6 Sample Problems 2–2 and 2–3

The following problems illustrate the calculation of cooling loads for a small office building, shown in Figure 2–15. Criteria and physical properties of the building are as follows:

- Design conditions—inside (78°F, 50% RH); outside (95°F DB and 78°F WB).
- Construction—wall, type F; U-factor, 0.103 roof—2″ insulation over metal deck; U-factor, 0.16 windows—U-factor, 0.56; shading coefficient, 0.65 doors—U-factor, 0.64 ceiling (Problem 2–2)— U-factor, 0.30
- Outside air—infiltration, ½ air exchange per hour; ventilation, 500 CFM.
- Lighting—30 fluorescent fixtures with four 40-W lamps each; ballast factor, 1.2: all heat to occupied space in Problem 2–2 and 50% in Problem 2–3.
- Appliances—allowance of 1.5 W per square foot.
- Occupants—(10) adults, general office work.

TABLE 2–17
Rates of heat gain from occupants of conditioned space[a]

Degree of Activity	Typical Application	Total Heat Adults, Male, Btu/hr	Total Heat Adjusted,[b] Btu/hr	Sensible Heat, Btu/hr	Latent Heat, Btu/hr
Seated at rest	Theater—matinee	390	330	225	105
	Theater—evening	390	350	245	105
Seated, very light work	Offices, hotels, apartments	450	400	245	155
Moderately active office work	Offices, hotels, apartments	475	450	250	200
Standing, light work; or walking slowly	Department store, retail store, dime store	550	450	250	200
Walking; seated	Drugstore, bank				
Standing; walking slowly		550	500	250	250
Sedentary work	Restaurant[c]	490	550	275	275
Light bench work	Factory	800	750	275	475
Moderate dancing	Dance hall	900	850	305	545
Walking 3 mph; moderately heavy work	Factory	1000	1000	375	625
Bowling[d]	Bowling alley				
Heavy work	Factory	1500	1450	580	870

[a]*Note:* Tabulated values are based on 75°F room dry-bulb temperature. For 80°F room dry bulb, the total heat remains the same, but the sensible heat values should be decreased by approximately 20 percent and the latent heat values increased accordingly.
[b]*Adjusted total heat gain* is based on normal percentage of men, women, and children for the application listed, with the postulate that the gain from an adult female is 85 percent of that for an adult male and that the gain from a child is 75 percent of that for an adult male.
[c]Adjusted total heat value for *sedentary work, restaurant,* includes 60 Btu per hour for food per individual (30 Btu sensible and 30 Btu latent).
[d]For *bowling,* figure one person per alley actually bowling and all others sitting (400 Btu per hour) or standing (550 Btu per hour).

Source: Reprinted by permission from ASHRAE (www.ashrae.org).

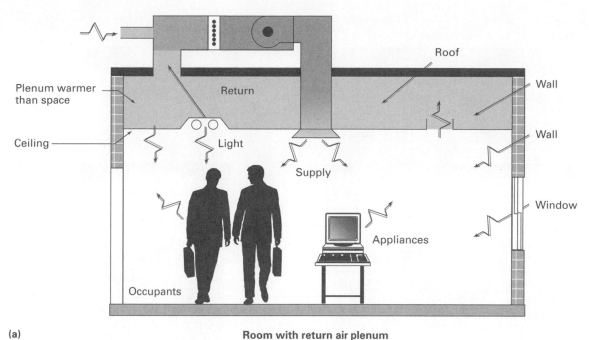

(a)   **Room with return air plenum**

⤳ → Denotes load to space

→ Denotes load to plenum

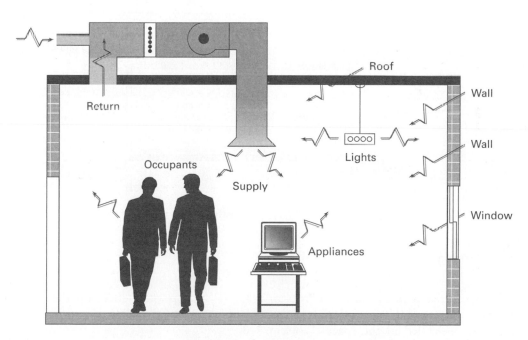

(b)   **Room with direct return**

■ **FIGURE 2–14**

Effect of return air plenum on load to conditioned space. (a) Room with return air plenum. (b) Room with direct return.

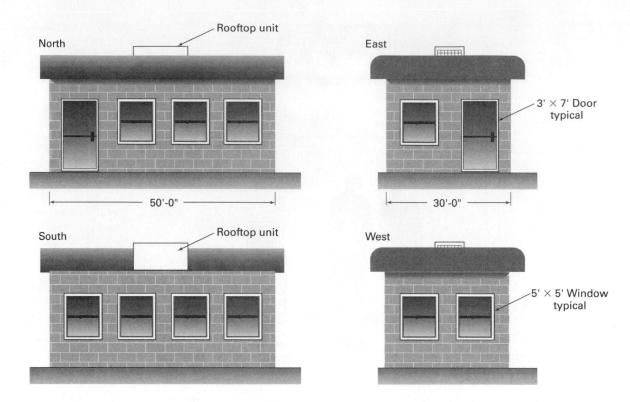

Elevations

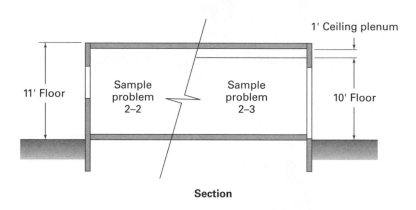

Section

■ **FIGURE 2–15**
A small office building.

Loads are calculated for two cases: In Problem 2–2, the occupied space extends to the underside of the roof, and lighting is suspended in the space. Problem 2–3 includes a dropped ceiling used as the underside of a return air plenum. Lighting fixtures are recessed into the plenum.

Completing the problems shows that the overall air-conditioning load will be about the same, regardless of whether there is a return air plenum; however, the load in the occupied space will be considerably lower if the design includes a return air plenum. This will have

an important effect on the amount of supply air required for cooling.

2.7 REFERENCE TABLES AND FIGURES

Tables and figures that are too lengthy to be included in the body of the text are included at the end of this chapter. The text contains samples sufficient for the discussion at hand. Unless otherwise indicated, all the tables and figures are reproduced with permission from publications of the American Society of Heating, Ventilating and Air Conditioning Engineers, Inc. (ASHRAE).

Additional data on weather, ventilating air, and thermal and solar properties can be found from ASHRAE handbooks, Atlanta, Georgia.

2.8 BUILDING LOAD CONTROL

2.8.1 Energy-Effective Envelope Decisions

The size, initial cost, and energy consumption of HVAC systems are affected by the magnitude of loads for which the systems are designed. Architectural decisions regarding materials, form, and orientation affect heating and cooling loads. Understanding the basics of load estimation presented in Sections 2.4 through 2.7 allows architects to identify ways to save on the initial cost of systems and to produce long-term energy savings. This section (continued on page 51) discusses techniques for load control using the tools from early sections, highlighting parameters for making energy-efficient decisions. Owing to the complexity of interactions among architecture, climate, and occupancy schedules, the easiest and most reliable way of evaluating options is to use computer simulation programs such as DOE2. Understanding the basics is essential to formulating those options that are most likely to be beneficial.

2.8.2 Insulation Levels

Well-designed insulation is one of the least expensive ways to achieve better thermal performance. Insulation requires no maintenance and lasts as long as the structure it serves. Energy analysis can identify an optimal economic level of building insulation for any climate, although these optimized levels need to be rounded off to commonly available construction dimensions.

Applying insulation beyond the optimal level results in diminishing returns. For instance, a poorly insulated wall with thermal resistance of R-8 can be improved greatly by adding insulation to increase thermal resistance to R-12. The additional four units of insulation will decrease thermal loads by 33 percent. Adding another four units of insulation to produce R-16 will further decrease thermal loads by only 25 percent; another four units to R-20 by 20 percent. Each successive addition of insulation becomes less effective in improving performance. The cost of additional insulation eventually becomes uneconomical, and the additional

investment can be better used elsewhere in the building design to save energy more effectively. Making these decisions requires the ability to estimate heating energy and cost, as shown in the following example.

Example 2.1 A 900-sq-ft rectangular building in a cold climate has dimensions of 20 ft by 45 ft, and its net wall area is 1100 sq ft. The location has a heating design temperature of –5°F, and the HVAC system must be designed to maintain 70° inside. The walls are lightweight stud construction insulated with fiberglass batts for a U-factor of 0.07. The average temperature during the 3 months that constitute the location's typical winter is 20°F. At 80% furnace efficiency, how many therms of gas are needed to supply the heat? What will the fuel cost be at $0.50 per therm?

Design heating load due to walls

$$Q = U \times A \times TD$$

$$= 0.07 \times 1100 \times (70) - (-5) = 5800 \text{ Btuh}$$

Average load for the winter

$$Q_{avg} = 0.07 \times 1100 \times (70 - 20) = 3900 \text{ Btuh}$$

Winter heating energy required owing to the wall

$$E = 3900 \text{ Btuh (avg)} \times 3 \text{ months} \times 30 \text{ days/month}$$
$$\times 24 \text{ hours/day} = 8,400,000 \text{ Btu}$$

Gas consumption

$$8,400,000 \text{ Btu} / (100,000 \text{ Btu/therm} \times 80\%)$$
$$= 105 \text{ therms}$$

Fuel cost

$$105 \text{ therms} \times \$0.50/\text{therm} = \$53$$

Example 2.2 What if the wall in the preceding example were better insulated, resulting in a U-factor of 0.04?

The results of using the same equations as in the prior example are shown for both wall options.

	$U = 0.07$	$U = 0.04$	Savings
U-factor	0.07	0.04	
Heating load (Btuh)	5,800	3,300	2,500
Energy loss (Btu)	8,400,000	4,800,000	3,600,000
Therms gas per year	105	60	45
Fuel cost	$53	$30	$23

Annual energy savings and reduction in equipment size can be used to evaluate the economics of adding insulation.

Answer to Problem 2–2
Load for a Simple Office Building

		Heat Gains Sensible	Latent

Heat Gains to Space

Roof

$Q = U \times A \times \text{TETD}$
$Q = 0.16 \times (30 \times 50) \times 81$ — 19,440 | —

Walls in space (below plenum)#

$Q = U \times A \times \text{TETD}$

North	$Q = 0.103 \times (11 \times 50) \times 15$	850	—
South	$Q = 0.103 \times (11 \times 50) \times 26$	1473	—
East	$Q = 0.103 \times (11 \times 30) \times 37$	1258	—
West	$Q = 0.103 \times (11 \times 30) \times 19$	646	—

Doors

$Q = U \times A \times \text{TETD}$

North	$Q = 0.64 \times (3 \times 7) \times 21$	282	—
East	$Q = 0.64 \times (3 \times 7) \times 28$	376	—

Windows

$Q = U \times A \times \text{TETD} + (A \times SC \times \text{SHGF})$

North	$Q = (.56 \times (3 \times 5 \times 5) \times (95 - 78)) + ((3 \times 5 \times 5) \times 0.65 \times 28)$	2079	—
South	$Q = (.56 \times (4 \times 5 \times 5) \times (95 - 78)) + ((4 \times 5 \times 5) \times 0.65 \times 29)$	2837	—
East	$Q = (.56 \times (1 \times 5 \times 5) \times (95 - 78)) + ((1 \times 5 \times 5) \times 0.65 \times 26)$	661	—
West	$Q = (.56 \times (2 \times 5 \times 5) \times (95 - 78)) + ((2 \times 5 \times 5) \times 0.65 \times 216)$	7496	—

Lighting

$Q = \text{Wattage} \times 3.41 \times \text{portion to space}$
Wattage = fixtures × lamps/fixtime × watts/lamp × ballast factor
$Q = 30 \times 4 \times 40 \times 1.2 \times 3.41 \times 1.0$ — 19,642 | —

Appliances

$Q = \text{Wattage} \times 3.41$
Wattage = watts/SF × area
$Q = 1.5 \times (30 \times 50) \times 3.41$ — 7673 | —

Occupants, sensible load

$Q = \text{Btuh, sensible/occupant} \times \text{no. occupants}$
$Q = 250 \times 10$ — 2500 | —

Occupants, latent load

$Q = \text{Btuh, latent/occupant} \times \text{no. occupants}$
$Q = 250 \times 10$ — — | 2500

Infiltration, sensible load

$Q = 1.1 \times \text{CFM} \times \text{TD}$
CFM = (air exchanges per hour × volume)/60 minutes per hour
$Q = 1.1 \times ((0.5 \times (30' \times 50' \times 11'))/60) \times (95 - 78)$ — 2571 | —

Infiltration, latent load

$Q = 4500 \times \text{CFM} \times (W_{\text{room}} - W_{\text{oa}})$
CFM = (air exchanges per hour × volume)/60 minutes per hour
$Q = 4500 \times ((0.5 \times (30' \times 50' \times 11'))/60) \times (0.017 - 0.010)$ — — | 4331

Subtotals, space heat gains* = 69,784 | **6831**

#For simplicity, areas of doors and windows are not deducted from wall area calculations in this example. Adjustment is appropriate if doors and window areas are significant in comparison with the respective wall areas.

*At this point, the space sensible heat gains would be used to calculate the required airflow rate for cooling. The equation is

$$\text{CFM supply} = \text{space sensible load}/(1.1 \times (T_{\text{space}} - T_{\text{supply}})) = 69{,}784/(1.1 \times (78 - 55)) = 2758$$

Answer to Problem 2–2   (*Continued*)

		Heat Gains Sensible	Latent

Heat Gains from Outside Air for Ventilation

Sensible load $Q = 1.1 \times \text{CFM} \times \text{TD}$

$Q = 1.1 \times 500 \times (95 - 78)$ → Sensible: 9350, Latent: —

Latent load $Q = 4500 \times \text{CFM} \times (W_{\text{out}} - W_{\text{in}})$

$Q = 4500 \times 500 \times (.017 - .010)$ → Sensible: —, Latent: 15,750

*Fan Heat** $Q = (\text{horsepower} \times 2545\ \text{Btuh/brake horsepower})/\text{motor efficiency}$

$Q = (2 \times 2545)/.95$ → Sensible: 5358, Latent: —

Load Summary

	Sensible	Latent
Space	69,784	6831
Outside air for ventilation	9350	15,750
Fan heat	5358	—
Total Air-Cooling Load (Btuh)	**84,492**	**22,581**
	(9 tons)[†]	

**The fan horsepower would be obtained from catalogued data after a preliminary selection of the air-conditioning unit.

[†]Refrigeration system and equipment are frequently expressed in *tons,* a term that originated in the ice-making industry. One ton of refrigeration is equal to the capacity to make 1 ton (2000 lb) of ice in 24 hours. The total heat to be removed from 2000 lb of water at 32°F is 288,000 Btu in 24 hours, or 12,000 Btu/hr.

2.8.3 Thermal Mass of Walls and Roof

Sunlight that strikes opaque walls heats them above ambient temperature. During summer, the heat gains contribute to air-conditioning load. During winter, these heat gains reduce heating requirements. The thermal effect of solar heat on opaque walls and roofs depends on orientation and on materials of construction.

Equations presented in this chapter use TETD and *U*-factors to determine total (solar plus conducted) heat flow. The *U*-factor is a constant for a given wall or roof, whereas TETD for a given assembly varies with changes in temperature and sun position. The *U*-factor affects the amount of heat transfer for a given TETD. Variations in TETD during the day affect the rate and timing of the heat transfer.

Understanding TETD can be useful in designing to minimize air-conditioning loads, thus reducing the size, expense, and energy consumption of the system. As the sun strikes a wall or roof the outer surface of the wall or the roof is warmed. Over time, the warmth is transmitted toward the interior of the building. The heat will transfer more slowly and with less intensity through a heavy wall or roof (high thermal capacity) than through a light wall or roof (low thermal capacity).

Thermal capacity is the product of the masses and specific heats of the materials making up the assembly. A heavy assembly such as a wall of concrete or multilayer masonry will impose significantly different air-conditioning loads than will a light assembly such as stud framing construction even if *U*-factors are identical.

The graph of TETD in Figure 2–16 shows how the orientation of walls and selection of wall and roof materials affect air conditioning loads. Note that loads are significantly lower and occur later for heavy assemblies than for light assemblies. Walls and roofs of heavy construction result in lower load.

Generally, peak loads for buildings occur in late afternoon, when air temperatures are highest, and solar effects are experienced to their maximum extent. Using wall materials and planning orientation properly to avoid peak loads during the afternoon will reduce overall load.

Example 2.3   What is the difference in air-conditioning load for a 10,000-sq-ft single-story building with a heavy roof versus a light roof with 2-in. insulation and identical *U*-factors of 0.10? Assume 4 P.M. peak air-conditioning load for the building. What will be the initial cost savings for an air-conditioning system at an incremental cost of $1000 per ton?

Answer to Problem 2–3
Cooling Load for a Simple Office Building with Return Air Plenum

		Heat Gains Sensible	Latent

Heat Gains to Space

Walls in space (below plenum)#

	$Q = U \times A \times \text{TETD}$		
North	$Q = 0.103 \times (10 \times 50) \times 15$	773	
South	$Q = 0.103 \times (10 \times 50) \times 26$	1339	
East	$Q = 0.103 \times (10 \times 30) \times 37$	1143	
West	$Q = 0.103 \times (10 \times 30) \times 19$	587	

Doors

	$Q = U \times A \times \text{TETD}$		
North	$Q = 0.64 \times (3 \times 7) \times 21$	282	—
East	$Q = 0.64 \times (3 \times 7) \times 28$	376	—

Windows

	$Q = U \times A \times \text{TETD} + (A \times SC \times \text{SHGF})$		
North	$Q = (.56 \times (3 \times 5 \times 5) \times (95 - 78)) + ((3 \times 5 \times 5) \times 0.65 \times 28)$	2079	—
South	$Q = (.56 \times (4 \times 5 \times 5) \times (95 - 78)) + ((4 \times 5 \times 5) \times 0.65 \times 29)$	2837	—
East	$Q = (.56 \times (1 \times 5 \times 5) \times (95 - 78)) + ((1 \times 5 \times 5) \times 0.65 \times 26)$	661	—
West	$Q = (.56 \times (2 \times 5 \times 5) \times (95 - 78)) + ((2 \times 5 \times 5) \times 0.65 \times 216)$	7496	—

Lighting to space

$Q = \text{Wattage} \times 3.41 \times \text{portion to space}$
Wattage = fixtures × no. lamps/fixture × watts/lamp × ballast factor
$Q = (30 \times 4 \times 40 \times 1.2) \times 3.41 \times 0.5$ — 9821

Appliances

$Q = \text{Wattage} \times 3.41$
Wattage = Watts/SF × area
$Q = (1.5 \times (30 \times 50)) \times 3.41$ — 7673

Occupants, sensible load

$Q = \text{Btuh, sensible/occupant} \times \text{no. occupants}$
$Q = 250 \times 10$ — 2500 —

Occupants, latent load

$Q = \text{Btuh, latent/occupant} \times \text{no. occupants}$
$Q = 250 \times 10$ — — 2500

Infiltration, sensible load

$Q = 1.1 \times \text{CFM} \times \text{TD}$
CFM = (air exchanges per hour × volume)/60 minutes per hour
$Q = 1.1 \times ((0.5 \times (30' \times 50' \times 10'))/60) \times (95 - 78)$ — 2338

Infiltration, latent load

$Q = 4500 \times \text{CFM} \times (W_{out} - W_{in})$
CFM = (air exchanges per hour × volume)/60 minutes per hour
$Q = 4500 \times ((0.5 \times (30' \times 50' \times 10'))/60 \times (0.017 - 0.010)$ — — 3938

Space gain from plenum (assume plenum temperature 10° higher than space)

$Q = U \times A \times \text{TD}$
$Q = 0.30 \times ((30 \times 50) - (30 \times 2 \times 4)) \times (10)$ — 3780 —

Subtotals, Space Heat Gains = 43,684*   6438

#For simplicity, areas of doors and windows are not deducted from wall area calculations in this example. Adjustment is appropriate if doors and window areas are significant in comparison with the respective wall areas.

*At this point, the space sensible heat gains would be used to calculate the required airflow rate for cooling. The equation is

CFM supply = space sensible load/$(1.1 \times (T_{space} - T_{supply}))$ = 43,684/$(1.1. \times (78 - 55))$ = 1727

Answer to Problem 2–3 *(Continued)*

		Heat Gains Sensible	Latent

Heat Gains to Return Air Plenum **

Roof	$Q = U \times A \times TETD$		
	$Q = 0.16 \times (30 \times 50) \times 71$	17,040	—

Walls#
(above ceiling) $Q = U \times A \times TETD$

North	$Q = 0.103 \times (1 \times 50) \times 15$	77	—
South	$Q = 0.103 \times (1 \times 50) \times 26$	134	—
East	$Q = 0.103 \times (1 \times 30) \times 37$	114	—
West	$Q = 0.103 \times (1 \times 30) \times 19$	59	—

Lighting to plenum $Q = \text{wattage} \times 3.41 \times \text{portion to plenum}$

Wattage = fixtures $\times$ no. lamps/fixture $\times$ watts/lamp $\times$ ballast factor

	$Q = (30 \times 4 \times 40 \times 1.2) \times 3.41 \times 0.5$	9821	—

Subtotal, plenum sensible heat gain = 27,245   =

Fan Heat ***

$Q = (\text{horsepower} \times 2545 \text{ Btuh/brake horsepower})/\text{motor efficiency}$

	$Q = (1.5 \times 2545)/.95$	4018	—

Heat Gain from Outside Air for Ventilation

Sensible	$Q = 1.1 \times CFM \times TD$		
	$Q = 1.1 \times 500 \times (95 - 78)$	9350	—

Latent	$Q = 4500 \times CFM \times (W_{out} - W_{in})$		
	$Q = 4500 \times 500 \times (0.017 \quad 0.010)$	—	15,750

Load Summary

Space		43,684	6438
Plenum		27,245	—
Fan Heat		4018	—
Outside Air Heat Load		9350	15,750

Total Air-Cooling Load (Btuh) = 84,297   22,188

**The roof TETD is adjusted to account for a higher plenum temperature.
***The fan horsepower would be obtained from catalogued data after a preliminary selection of the air-conditioning unit.

Light roof $Q = U \times A \times TETD$
     $= 0.1 \times 10,000 \times 58$
     $= 81,000$ Btuh (about 6.8 tons)

Heavy roof $Q = U \times A \times TETD$
     $= 0.1 \times 10,000 \times 58$
     $= 58,000$ Btuh (about 4.8 tons)

Difference in air conditioning load
     $= 6.8 - 4.8 = 2$ tons

Initial cost savings
     = 2 tons $\times$ \$1000 per ton = \$2000

In addition, there will be energy savings. Based on computer modeling, annual savings will be approximately 3000 kWh, which might be worth \$300 per year in the Midwest.

     Air-conditioning load due to walls can virtually be eliminated in some cases. For instance, an office building may have its air-conditioning systems turned off

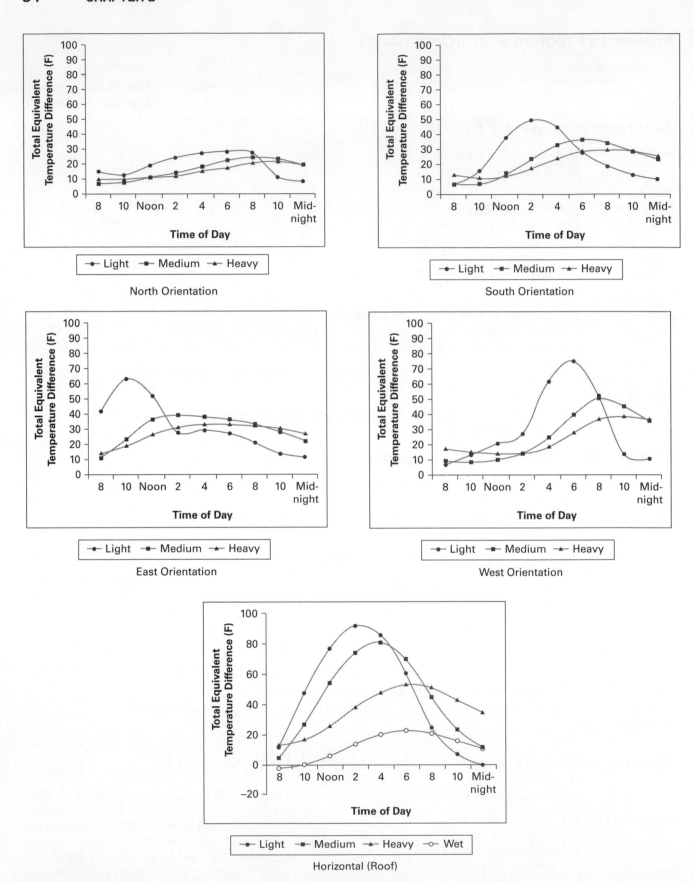

FIGURE 2–16

Heat gains for walls and roof, by orientation and thermal mass.

after the business day. If west-facing walls are designed with high thermal capacity, their peak heat transfers can be deferred until after business hours, when the system is off.

2.8.4 Surface Color (Absorptivity)

TETD is also affected by surface radiant absorptivity of roofs and walls, loosely termed "color." Absorption of incident radiation heats exterior building surfaces. Light colors absorb less radiant energy than dark colors and result in lower air-conditioning loads. An assembly with a given U-factor and thermal capacity will contribute less load if it is light in color (see Table 2–12).

Example 2.1   What is the difference in air-conditioning load for a 10,000-sq-ft, single-story building with a light-colored roof versus a dark, medium-weight roof with 2-in. insulation and identical U-factors of 0.10? What will be the initial cost savings for an air-conditioning system at an incremental cost of $1000 per ton?

Light roof $Q = U \times A \times TETD$
$$= 0.1 \times 10,000 \times 79$$
$$= 79,000 \text{ Btuh (about 6.6 tons)}$$

Dark roof $Q = U \times A \times TETD$
$$= 0.1 \times 10,000 \times 43$$
$$= 43,000 \text{ Btuh (about 3.6 tons)}$$

Difference in air-conditioning load
$$= 6.6 - 3.6 = 3 \text{ tons}$$

Initial cost savings
$$= 3 \text{ tons} \times \$1000 \text{ per ton} = \$3000$$

In addition, there will be energy savings. Based on computer modeling, annual savings will be approximately 5000 kWh, which might be worth $500 per year in the Midwest.

2.8.5 Windows

The insulating quality of windows is key to a building's thermal performance. Typical double-glazed windows conduct heat at 25 to 50 times the rate of a well-insulated wall. Despite their small area in comparison with walls and roof, they represent the majority of envelope thermal load owing to poor insulating quality and infiltration from air leakage.

Conventional glazing materials achieve better winter performance with double or triple layers, relying on the insulating quality of the trapped air spaces for insulation. The best available performance is approximately R-3, for plain, triple-layered glass. Good summer performance is achieved by tints or by reflective films that block the sun's rays. Tints and films also reduce light transmission and impair a clear, natural view. High-performance, "low-E" glazing transmits visible light well but reflects infrared radiation. Windows using the low-E concept are more effective for lighting and view but keep heat in during cold weather and keep heat out during hot weather. New glass and glazing materials have made R-8 readily obtainable along with good blockage of solar heat.

Heat transfer through windows is affected by U-factors and shading coefficients. The U-factor determines the amount of conducted heat leaving or entering owing to the temperature difference between inside and outside. The shading coefficient determines the amount of solar heat transmitted owing to the radiant energy impinging on the window, which depends on latitude, orientation, season, time of day, and sky clearness. Decisions on the location, size, U-factor, and shading coefficient for windows are the most important aspects of building envelope thermal performance.

2.8.6 Conduction through Windows

Window U-factors generally range from 0.3 to 1.0, whereas walls inherently are better insulators, with U-factors in the general range of 0.05 to 0.3. Typically, window U-factors are five times higher than wall U-factors, so windows lose or gain five times as much heat by conduction as equal wall areas. In addition, windows leak air through their construction and through their joints with walls, which further increases their contribution to heat loss and gain owing to differences between inside and outside air temperatures.

2.8.7 Window-to-Wall Ratio

In climates where temperature differences are great, the relative amount of glass versus wall surface has a great effect on loads and on the size of heating and cooling systems. In mild climates the effect is less important. In mild and warm climates the TD is very low, and poor insulating qualities may not result in inordinately high heat transfer. In cold climates TD is high, and large amounts of glass will result in high heat losses and significantly larger heating systems.

Example 2.2   A 1000-sq-ft wall is 70% opaque construction and 30% window. The opaque construction has a U-factor of 0.10. The window construction has a U-factor of 0.50. What is the overall U-factor of the wall? What is the conducted heating load for the wall if the design temperature is 0°F outside and 75°F inside? How much energy will be used annually to offset conduction if the average winter (2000 hours) temperature is 30°F? What will be the equivalent fuel usage for an 80%-efficient gas-fired boiler? How much will the fuel cost at $0.50 per therm?

$$U_{avg} = 30\% \times 0.50 + 70\% \times 0.10 = 0.22$$

$$\begin{aligned} Q_{load} &= U \times A \times TD \\ &= 0.22 \times 1000 \times (75 - 0) = 16{,}500 \text{ Btuh} \end{aligned}$$

$$\begin{aligned} Q_{annual\ avg} &= U \times A \times TD_{avg} \\ &= 0.22 \times 1000 \times (75 - 30) = 9900 \text{ Btuh} \end{aligned}$$

$$\begin{aligned} \text{Energy, annual} &= \text{Load} \times \text{Time} = 9900 \times 2000 \\ &= 19{,}800{,}000 \text{ Btu per year} \end{aligned}$$

$$\begin{aligned} \text{Gas usage} &= \text{Energy}/(100{,}000 \text{ Btu/therm} \times 80\% \text{ eff.}) \\ &= 19{,}800{,}000/(80{,}000) = 248 \text{ therms} \end{aligned}$$

$$\begin{aligned} \text{Annual fuel cost} &= 248 \text{ therms} \times \\ &\quad \$0.50 \text{ per therm} = \$124 \end{aligned}$$

How much more glass could be added for the same fuel budget if the average winter temperature were 40°F?

$$\begin{aligned} Q_{annual\ avg} &= 9900 \\ &= U_{avg} \times 1000 \times (75 - 40) \end{aligned}$$

$$U_{avg} = 0.283$$
$$\begin{aligned} U_{avg} &= \% \text{ glass} \times 0.50 \\ &\quad + (1 - \% \text{ glass}) \times 0.10 = 0.283 \end{aligned}$$

$$\% \text{ glass} = 46\%$$

The difference in climate would allow about 50 percent more glass for the same heating cost, which is not to promote the use of glazing but rather to point out the influence of climate on the energy economics of architectural decisions.

2.8.8 Window Placement and Solar Control

Orientation of walls and windows affects solar heat gains during both the heating and the cooling season. Shading coefficient was defined earlier in the chapter as the ratio between solar heat admitted through a given type of glass to that which would be admitted through single-strength clear glass under identical conditions. The amount of solar heat admitted depends on latitude, orientation, season, time of day, and sky clearness. For a given type of glazing, the solar heat gain will be proportional to the solar heat gain factor (SHGF; see Table 2–15). SHGF varies with latitude, time of day, time of year, and orientation.

Figure 2–17 shows solar heat peaks and totals for the cardinal orientations and for a horizontal exposure. The data and observations are for 40° north latitude, temperate climate (cold winter, hot summer).

Example 2.3 A 100-sq-ft window has a SC of 0.85. What is its peak solar heat gain in Btuh and in tons if the window faces west? Use data from Table 2–15 for July 4 P.M., which will probably be the time of peak air-conditioning load for the building.

$$\begin{aligned} Q &= A \times SC \times SHGF = 100 \times 0.85 \times 216 \\ &= 18{,}300 \text{ Btuh (about 1.5 tons)} \end{aligned}$$

What if the same window were oriented north?

$$\begin{aligned} Q &= A \times SC \times SHGF = 100 \times 0.85 \times 28 \\ &= 2400 \text{ Btuh (about 0.2 ton)} \end{aligned}$$

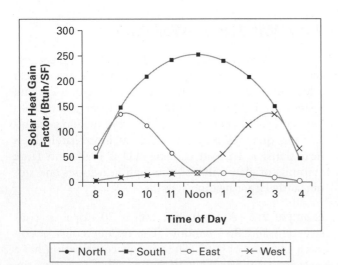

Winter

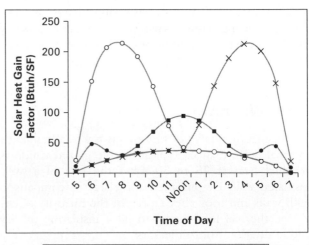

Summer

■ **FIGURE 2–17**
Solar heat gain factors by season and orientation.

What is the difference in system cost if the incremental cost of HVAC is $1000 per ton?

HVAC system cost savings
$$= (1.5 - 0.2) \times \$1000 = \$1300$$

Thus, initial cost savings by avoiding a west orientation are significant. In addition, the cost of electricity to power cooling systems will be saved. From a computerized energy simulation, which is beyond the scope of this book, energy savings for typical Midwest climate are estimated at 1200 kWh per year and are worth approximately $140 annually (about $1.40 per square foot of glass). Over a 20- to 50-year life cycle, the value of the initial and ongoing savings from orientation decisions regarding windows are many times the cost of the window itself.

The foregoing example illustrates that north orientation is much superior to west. North is also to a lesser extent superior to east and south. South orientation does have the advantage of moderate air-conditioning loads and the benefit of solar heat gains during winter, which can reduce heating energy consumption. This solar heat gain is the basis for passive solar heating design, and south orientation is useful especially for some buildings, especially residential, however, for most commercial and institutional buildings, the need for heat in perimeter spaces is low during occupied periods coincident with the short days of winter because of the heating assist from lighting, appliances, and occupants during the normal workweek.

For commercial and institutional buildings the preferred window orientation is north. North glazing is also an excellent source of daylight and view without glare from direct sun. Given the reality that all windows cannot face north, solar shading of direct sun during the air-conditioning season is a way of making windows of any orientation perform as north-facing glass.

Notwithstanding south exposure's potential for beneficial winter heat gains, care is needed to avoid glare from direct sun through windows in late morning through early afternoon from about November through February. During summer, when the sun is high in the sky, south exposure results in moderate heat gains, which is easy to eliminate with simple shading devices such as overhangs.

East and west exposures are problematic, offering little significant potential heating benefit during winter and resulting in high air-conditioning loads during summer. Moreover, east and west are exposed to sun at low angles at various times of day all year. Low sun angles are difficult to shade with horizontal architectural elements. Vertical fins can be used, or operable internal sun control, to avoid unacceptable glare and thermal discomfort. Heavy tints or reflective coatings can also be used to control sun, but they generally affect light transmission, view, and mood in the space.

■ **FIGURE 2–18**
Classic south overhang on this 1960s building shades south glass during summer and admits solar heat during winter. Note the sparing use of fenestration on the adjacent east wall, which eliminates high air-conditioning loads. (Courtesy of WTA.)

■ **FIGURE 2–19**
Skywell at the Williams Technology
Center in Tulsa (a) provides shading
and temperature buffer for the east
side. The west side (b) is shaded from
afternoon sun by their adjacent tower.
(All images courtesy of WTA.)

(a)

(b)

2.8.9 Shading by Deciduous Plants

Deciduous plants have leaves in summer, when shade is desired, and are bare in winter, when solar heat can be beneficial. Trees or vines can be used as a strategy to save energy, reduce glare, and improve views from buildings. Most engineers, however, will not consider the beneficial effects of plants in sizing systems or components. Obviously, plants take time to mature, and they may not be permanent accessories to the site.

■ **FIGURE 2–20**

Shading devices are a key design element for the Danforth Plant Science Center in Saint Louis. The megashade prevents summer sun penetration in the central hall and creates a pleasantly shaded front yard to the building. All windows have solar-control louvers. (Courtesy of WTA.)

■ **FIGURE 2–21**

South-shading device shades upper and lower windows during midday. Deep reveals prevent west and east penetration morning and late afternoon at lower level. (Courtesy of WTA.)

2.8.10 Effects of Internal Heat Gains

During cold weather heat is lost through windows, which generally imposes a load on the heating system. During mildly cold weather that occurs when the building is occupied, heat loss through windows may be offset by internal heat gains, thereby imposing no load on the heating system. Heat loss in this case may actually be beneficial by reducing the need for cooling.

A typical office building may not need heating during normal occupied hours until outdoor temperatures are fairly cold owing to the inherent heating effect of lighting, people, and appliances. The outdoor temperature at which the room needs heating is the *thermal balance temperature* and depends on whether the space is occupied or unoccupied. During unoccupied hours, when lights and appliances are off, internal heat gains are low, and the building will need heating at higher temperatures.

For buildings with very low internal heat gains, such as residences, the breakeven temperature may be as high as 50° to 60°F. For buildings with significant internal heat gains, breakeven temperatures will be much lower. Accordingly, high-gain buildings will not need much heat when they are occupied, and the thermal insulating quality of the building is not as important as for a low-gain building.

The timing of building occupancy also affects the relative importance of building envelope insulation. Most office buildings are occupied 50 to 70 hours per week, which is only 30 to 40 percent of the time. Some buildings, such as hospitals, are occupied around the clock. Thus, on average the hospital will need less heating because internal gains will partially offset conducted heat loss on a continuous basis.

The interactions among thermal envelope design, building occupancy, and weather are difficult to estimate manually, and general guidelines are difficult at best. Computer simulation using programs such as DOE2 is the most reliable method for evaluating alternatives and determining provisions for optimal investment in envelope performance.

■ **FIGURE 2–22**
Deciduous foliage provides extra insulation, solar shading, and evaporative cooling to this wall in an urban area. (Courtesy of WTA.)

TABLE 2–2A
Climatic conditions for the United States

Col. 1	Col. 2		Col. 3		Col. 4	Col. 5		Col. 6			Col. 7	Col. 8		
						Winter,[b] °F			Summer,[c] °F					
	Lat.		Long.		Elev.	Design Dry Bulb		Design Dry Bulb and Mean Coincident Wet Bulb			Mean Daily	Design Wet Bulb		
State and Station[a]	°	'	°	'	Feet	99%	97.5%	1%	2.5%	5%	Range	1%	2.5%	5%
ALABAMA														
Birmingham AP	33	34	86	45	620	17	21	96/75	94/75	92/74	21	78	77	76
Mobile Co	30	40	88	15	211	25	29	95/77	93/77	91/76	16	80	79	78
ALASKA														
Anchorage AP	61	10	150	01	114	−23	−18	71/59	68/58	66/56	15	60	59	57
Fairbanks AP	64	49	147	52	436	−51	−47	82/62	78/60	75/59	24	64	62	60
ARIZONA														
Phoenix AP	33	26	112	01	1112	31	34	109/71	107/71	105/71	27	76	75	75
Tucson AP	32	07	110	56	2558	28	32	104/66	102/66	100/66	26	72	71	71
ARKANSAS														
Little Rock AP	34	44	92	14	257	15	20	99/76	96/44	94/77	22	80	79	78
Pine Bluff AP	34	18	92	05	241	16	22	100/78	97/77	95/78	22	81	80	80
CALIFORNIA														
Los Angeles AP	33	56	118	24	97	41	43	83/68	80/68	77/67	15	70	69	68
San Francisco AP	37	37	122	23	8	35	38	62/64	77/63	73/62	20	65	64	62
COLORADO														
Denver AP	39	45	104	52	5283	−5	1	93/59	91/59	89/59	28	64	63	62
Pueblo AP	38	18	104	29	4641	−7	0	97/61	95/61	92/61	31	67	66	65
CONNECTICUT														
Bridgeport AP	41	11	73	11	25	6	9	86/73	84/71	81/70	18	75	74	73
Norwalk	41	07	63	25	37	6	9	86/73	84/71	81/70	19	75	74	73
DELAWARE														
Dover AFB	39	08	75	28	28	11	15	92/75	90/75	87/74	18	79	77	76
Wilmington AP	39	40	75	36	74	10	14	92/74	89/74	87/73	20	77	76	75
FLORIDA														
Miami AP	25	48	80	16	7	44	47	91/77	90/77	89/77	15	79	79	78
Orlando AP	28	33	81	23	100	35	38	94/76	93/76	91/76	17	79	78	78
GEORGIA														
Atlanta AP	33	39	04	26	1010	17	22	94/74	92/74	90/73	19	77	76	75
Gainesville	34	11	83	41	50	24	27	96/77	93/77	91/77	20	80	79	78
HAWAII														
Hilo AP	19	43	155	05	36	61	62	84/73	83/72	82/72	15	75	74	74
Honolulu AP	21	20	157	55	13	62	63	87/73	86/73	85/72	12	76	75	74
IDAHO														
Boise AP	43	34	116	13	2838	3	10	96/65	94/64	91/64	31	68	66	65
Burley	42	32	113	46	4156	−3	2	99/62	95/61	92/66	35	64	63	61
ILLINOIS														
Champaign/Urbana	40	02	88	17	777	−3	2	95/75	92/74	90/73	21	78	77	75
Chicago, Midway AP	41	47	87	45	607	−5	0	94/74	91/73	88/72	20	77	75	74
INDIANA														
Huntington	40	53	85	30	802	−4	1	92/73	89/72	87/72	23	77	75	74
Indianapolis AP	39	44	86	17	792	−2	2	92/74	90/74	87/73	22	78	76	75
IOWA														
Iowa City	41	38	91	33	661	−11	−6	92/76	89/76	87/74	22	80	78	76
Sioux City AP	42	24	96	23	1095	−11	−7	95/74	92/74	89/73	24	78	77	65
KANSAS														
Dodge City AP	37	46	99	58	2582	0	5	100/69	97/69	95/69	25	74	73	71
El Dorado	37	49	96	50	1282	3	7	101/72	98/73	96/73	24	77	76	75

(Continued)

TABLE 2–2A *(Continued)*

Col. 1	Col. 2		Col. 3		Col. 4	Winter,[b] °F		Col. 6 Summer,[c] °F			Col. 7	Col. 8		
	Lat.		Long.		Elev.	Col. 5 Design Dry Bulb		Design Dry Bulb and Mean Coincident Wet Bulb			Mean Daily	Design Wet Bulb		
State and Station[a]	°	′	°	′	Feet	99%	97.5%	1%	2.5%	5%	Range	1%	2.5%	5%
KENTUCKY														
Lexington AP	38	02	84	36	966	3	8	93/73	91/73	88/72	22	77	76	75
Louisville AP	38	11	85	44	477	5	10	95/74	93/74	90/74	23	79	77	76
LOUISIANA														
Alexandria AP	31	24	92	18	92	23	27	95/77	94/77	92/77	20	80	79	78
Baton Rouge AP	30	32	91	09	64	25	29	95/77	93/77	92/77	19	80	80	79
MAINE														
Augusta AP	44	19	69	48	353	−7	−3	88/73	85/70	82/68	22	74	72	70
Bangor, Dow AFB	44	48	68	50	192	−11	−6	86/70	83/68	80/67	22	73	71	69
MARYLAND														
Baltimore AP	39	11	76	40	148	10	13	94/75	91/75	89/74	21	78	77	76
Salisbury	38	20	75	30	59	12	16	93/75	91/75	88/74	18	79	77	76
MASSACHUSETTS														
Boston AP	42	22	71	02	15	6	9	91/73	88/71	85/70	16	75	74	72
Clinton	42	24	71	41	398	−2	2	90/72	87/71	84/69	17	75	73	72
MICHIGAN														
Battle Creek AP	42	19	85	15	941	1	5	92/74	88/72	85/70	23	76	74	73
Detroit	42	25	83	01	619	3	6	91/74	88/72	86/71	20	76	74	73
MINNESOTA														
Minneapolis/ St. Paul AP	44	53	93	13	834	−16	−12	92/75	89/73	86/71	22	77	75	73
Rochester AP	43	55	92	30	1297	−17	−12	90/74	87/72	84/71	24	77	75	73
MISSISSIPPI														
Biloxi, Keesler AFB	30	25	88	55	26	28	31	94/79	92/79	90/78	16	82	81	80
Jackson AP	32	19	90	05	310	21	25	97/76	95/76	93/76	21	79	78	78
MISSOURI														
Kansas City AP	39	07	94	35	791	2	6	99/75	96/74	93/74	20	78	77	76
St. Louis AP	38	45	90	23	535	2	6	97/75	94/75	91/74	21	78	77	76
MONTANA														
Glasgow AP	48	25	106	32	2760	−22	−18	92/64	89/63	85/62	29	68	66	64
Great Falls AP	47	29	111	22	3662	−21	−15	91/60	88/60	85/59	28	64	62	60
NEBRASKA														
Lincoln Co	40	51	96	45	1180	−5	−2	99/75	95/74	92/74	24	78	77	76
Omaha AP	41	18	95	54	977	−8	−3	94/76	91/75	88/74	22	78	77	75
NEVADA														
Las Vegas AP	36	05	115	10	2178	25	28	108/66	106/65	104/65	30	71	70	69
Reno AP	39	30	119	47	4404	5	10	95/61	92/60	90/59	45	64	62	61
NEW HAMPSHIRE														
Concord AP	43	12	71	30	342	−8	−3	90/72	87/70	84/69	26	74	73	71
Manchester, Grenier AFB	42	56	71	26	233	−8	−3	91/72	88/71	85/70	24	75	74	72
NEW JERSEY														
Newark AP	40	42	74	10	7	10	14	94/74	91/73	88/72	20	77	76	75
New Brunswick	40	29	74	26	125	6	10	92/74	89/73	86/72	19	77	76	75
NEW MEXICO														
Albuquerque AP	35	03	106	37	5311	12	16	96/61	94/61	92/61	27	66	65	64
Farmington AP	36	44	108	14	5503	1	6	95/63	93/62	91/61	30	67	65	64
NEW YORK														
NYC-Kennedy AP	40	39	3	47	13	12	15	90/73	84/72	84/71	16	76	75	74
Rochester AP	43	07	77	40	547	1	5	91/73	88/71	85/70	22	75	73	72

(Continued)

TABLE 2–2A *(Continued)*

Col. 1	Col. 2		Col. 3		Col. 4	Col. 5		Col. 6			Col. 7	Col. 8		
						Winter,[b] °F			Summer,[c] °F					
	Lat.		Long.		Elev.	Design Dry Bulb		Design Dry Bulb and Mean Coincident Wet Bulb			Mean Daily	Design Wet Bulb		
State and Station[a]	°	'	°	'	Feet	99%	97.5%	1%	2.5%	5%	Range	1%	2.5%	5%
NORTH CAROLINA														
Asheville AP	35	26	82	32	2140	10	14	89/73	87/72	85/71	21	75	74	72
Charlotte AP	35	13	80	56	736	18	22	95/74	93/74	91/74	20	77	76	76
NORTH DAKOTA														
Bismarck AP	46	46	100	45	1647	−23	−19	95/68	91/68	88/67	27	73	71	70
Fargo AP	46	54	96	48	896	−22	−18	92/73	89/71	85/69	25	76	74	72
OHIO														
Cleveland AP	41	24	81	51	777	1	5	91/73	88/72	86/71	22	76	74	73
Columbus AP	40	00	82	53	812	0	5	92/73	90/73	87/72	24	77	75	74
OKLAHOMA														
Norman	35	15	97	29	1181	9	13	99/74	96/74	94/74	24	77	76	75
Oklahoma City AP	35	24	97	36	1285	9	13	100/74	97/74	95/73	23	78	77	76
OREGON														
Eugene AP	44	07	123	13	359	17	22	92/67	89/66	86/65	31	69	67	66
Portland AP	45	36	122	36	21	17	23	89/68	85/67	81/65	23	69	67	66
PENNSYLVANIA														
Philadelphia AP	39	53	75	15	5	10	14	93/75	90/74	87/72	21	77	76	75
Pittsburgh AP	40	30	80	13	1137	1	5	89/72	86/71	84/70	22	74	63	72
RHODE ISLAND														
Newport	41	30	71	20	10	5	9	88/73	85/72	82/70	16	76	75	73
Providence AP	41	44	71	26	51	5	9	89/73	86/72	83/70	19	75	74	73
SOUTH CAROLINA														
Greenville AP	34	54	82	13	957	18	22	93/74	91/74	89/74	21	77	76	75
Greenwood	34	10	82	07	620	18	22	95/75	93/74	91/74	21	78	77	76
SOUTH DAKOTA														
Aberdeen AP	45	27	98	26	1296	−19	−15	94/73	91/72	88/70	27	77	75	73
Rapid City AP	44	03	103	04	3162	−11	−7	95/66	92/65	89/65	28	71	69	67
TENNESSEE														
Greeneville	36	04	82	50	1319	11	16	92/73	90/72	88/72	22	76	75	74
Jackson AP	35	36	88	55	423	11	16	98/76	95/75	92/75	21	79	78	77
TEXAS														
Dallas AP	32	51	96	51	481	18	22	102/75	100/75	97/75	20	78	78	77
Houston AP	29	58	95	21	96	27	32	96/77	94/77	92/77	18	80	79	79
UTAH														
Salt Lake City AP	40	46	111	58	4220	3	8	97/62	95/62	92/61	32	66	65	64
Vernal AP	40	27	109	31	5280	−5	0	91/61	89/60	86/59	32	64	63	62
VERMONT														
Barre	44	12	72	31	600	−16	−11	84/71	81/69	78/68	23	73	71	70
Burlington AP	44	28	73	09	332	−12	−7	88/72	85/70	82/69	23	74	72	71
VIRGINIA														
Richmond AP	37	30	77	20	164	14	17	95/76	92/76	690/75	21	79	78	77
Roanoke AP	37	19	79	58	1193	12	16	93/72	91/72	88/71	23	75	74	73
WASHINGTON														
Port Angeles	48	07	123	26	99	24	27	72/62	69/61	67/60	18	64	62	61
Seattle-Boeing Field	47	32	122	18	23	21	26	84/68	81/66	77/65	24	69	67	65
WEST VIRGINIA														
Charleston AP	38	22	81	36	939	7	11	92/74	90/73	87/72	20	76	75	74
Wheeling	40	07	80	42	665	1	5	89/72	86/71	84/70	21	74	73	72

(Continued)

TABLE 2–2A *(Continued)*

Col. 1	Col. 2		Col. 3		Col. 4	Winter,[b] °F Col. 5		Summer,[c] °F Col. 6			Col. 7	Col. 8	
	Lat.		Long.		Elev.	Design Dry Bulb		Design Dry Bulb and Mean Coincident Wet Bulb			Mean Daily	Design Wet Bulb	
State and Station[a]	°	′	°	′	Feet	99%	97.5%	1%	2.5%	5%	Range	1%	2.5%	5%
WISCONSIN														
La Crosse AP	43	52	91	15	651	−13	−19	91/75	88/73	85/72	22	77	76	74
Milwaukee AP	42	57	87	54	672	−8	−4	90/74	87/73	84/71	21	76	74	73
WYOMING														
Casper AP	42	55	106	28	5338	−11	−5	92/58	90/57	87/57	31	63	61	60
Cheyenne	41	09	104	49	6126	−9	−1	89/58	86/58	84/57	30	63	62	60

[a]AP, AFB, following the station name designates airport or military airbase temperature observations. Co designates office locations within an urban area that are affected by the surrounding area. Undesignated stations are semirural and may be compared to airport data.
[b]Winter design data are based on the 3-month period, December through February.
[c]Summer design data are based on the 4-month period, June through September.

Source: Reprinted by permission from ASHRAE (www.ashrae.org).

TABLE 2–2B
Climatic conditions of Canada

Col. 1	Col. 2		Col. 3		Col. 4	Winter,[b] °F Col. 5		Summer,[c] °F Col. 6			Col. 7	Col. 8	
	Lat.		Long.		Elev.	Design Dry Bulb		Design Dry Bulb and Mean Coincident Wet Bulb			Mean Daily	Design Wet Bulb	
State and Station[a]	°	′	°	′	Feet	99%	97.5%	1%	2.5%	5%	Range	1%	2.5%	5%
Calgary AP	51	06	114	01	3540	−27	−23	84/63	81/61	79/60	25	65	63	62
Prince George AP	53	53	122	41	2218	−33	−28	84/64	80/62	77/61	26	66	64	62
Winnipeg AP	49	54	97	14	786	−30	−27	89/73	86/71	84/70	22	75	73	71
Edmundston Co	47	22	68	20	500	−21	−16	87/70	83/68	80/67	21	73	71	69
St. John's AP	47	37	52	45	463	3	7	77/66	75/65	73/64	18	69	67	66
Fort Smith AP	60	01	111	58	665	−49	−45	85/66	81/64	78/63	24	68	66	65
Halifax AP	44	39	63	34	83	1	5	79/66	76/65	74/64	16	69	67	66
Ottawa AP	45	19	75	40	413	−17	−13	90/72	87/71	84/70	21	75	73	72
Charlottetown AP	46	17	63	08	186	−7	−4	80/69	78/68	76/67	16	71	70	68
Montreal AP	45	28	73	45	98	−16	−10	88/73	85/72	83/71	17	75	74	72
Prince Albert AP	53	13	105	41	1414	−42	−35	87/67	84/66	81/65	25	70	68	67

[a]AP, AFB, following the station name designates airport or military airbase temperature observations. Co designates office locations within an urban area that are affected by the surrounding area. Undesignated stations are semirural and may be compared to airport data.
[b]Winter design data are based on the 3-month period, December through February.
[c]Summer design data are based on the 4-month period, June through September.

Source: Reprinted by permission from ASHRAE (www.ashrae.org).

TABLE 2–2C
Climatic conditions of the world (other than USA and Canada)

Country and Station[a]	Winter,[b] °F								Summer,[c] °F						
	Col.1		Col. 2		Col. 3	Col. 4			Col. 5			Col. 6	Col. 7		
	Lat.		Long.		Eleva-tion, ft	Mean of Annual Extremes			Design Dry Bulb			Mean Daily Range	Design Wet Bulb		
	°	′	°	′			99%	97.5%	1%	2.5%	5%		1%	2.5%	5%
ARGENTINA															
Buenos Aires	34	35S	58	29W	89	27	32	34	91	89	86	22	77	76	75
AUSTRALIA															
Melbourne	37	49S	144	58E	114	31	35	38	95	91	86	21	71	69	68
Sydney	3	52S	151	12E	138	38	40	42	89	84	80	13	74	73	72
AUSTRIA															
Vienna	48	15N	16	22E	644	−2	6	11	88	86	83	16	71	69	67
BELGIUM															
Brussels	50	48N	4	21E	328	13	15	19	83	79	77	19	70	68	67
BRAZIL															
Rio de Janeiro	22	55S	43	12W	201	56	58	60	94	92	90	11	80	79	78
BULGARIA															
Sofia	42	42N	23	20E	1805	−2	3	8	89	86	84	26	71	70	69
CHILE															
Santiago	33	24S	70	47W	1555	27	30	32	90	88	86	37	68	67	66
CHINA															
Chungking	29	33N	106	33E	755	34	37	39	99	97	95	18	81	80	79
Shanghai	31	12N	121	26E	23	16	23	26	94	92	90	16	81	81	80
RUSSIAN REPUBLIC (formerly SOVIET UNION)															
Moscow	55	46N	37	40E	505	−19	−11	−6	84	81	78	21	69	67	65
St. Petersburg (Leningrad)	59	56N	30	16E	16	−14	−9	−5	78	75	72	15	65	64	63
CUBA															
Havana	23	08N	82	21W	80	54	59	62	92	91	89	14	81	81	80
CZECH Republic															
Prague	50	05N	14	25E	662	3	4	9	88	85	83	16	66	65	64
DENMARK															
Copenhagen	55	41N	12	33E	43	11	16	19	79	76	74	17	68	66	64
EGYPT															
Cairo	29	52N	31	20E	381	39	45	46	102	100	98	26	76	75	74
FINLAND															
Helsinki	60	10N	24	57E	30	−11	−7	−1	77	74	72	14	66	65	63
FRANCE															
Paris	48	49N	2	29E	164	16	22	25	89	86	83	21	70	68	67
GERMANY															
Berlin	52	27N	13	18E	187	6	7	12	84	81	78	19	68	67	66
HONG KONG															
Hong Kong	22	18N	114	10E	109	43	48	50	92	91	90	10	81	80	80
INDIA															
New Delhi	28	35N	77	12E	703	35	39	41	110	107	105	26	83	82	82
INDONESIA															
Djakarta	6	11S	106	50E	26	69	71	72	90	89	88	14	80	79	78
ISRAEL															
Jerusalem	31	47N	35	13E	2485	31	36	38	95	94	92	24	70	69	69
ITALY															
Milan	45	27N	9	17E	341	12	18	22	89	87	84	20	76	75	74

(Continued)

TABLE 2–2C   *(Continued)*

Country and Station[a]	*Winter,[b] °F*							*Summer,[c] °F*						
	Col.1		*Col. 2*		*Col. 3*	*Col. 4*			*Col. 5*			*Col. 6*	*Col. 7*	
	Lat.		*Long.*		*Eleva-tion, ft*	*Mean of Annual Extremes*			*Design Dry Bulb*			*Mean Daily Range*	*Design Wet Bulb*	
	°	′	°	′			99%	97.5%	1%	2.5%	5%		1%	2.5%	5%

Country and Station[a]	°	′	°	′	Elev. ft	Mean	99%	97.5%	1%	2.5%	5%	Range	1%	2.5%	5%
JAPAN															
Tokyo	35	41N	139	46E	19	21	26	28	91	89	87	14	81	80	79
KOREA															
Pyongyang	39	02N	125	41E	186	−10	−2	3	89	87	85	21	77	76	76
Seoul	37	34N	126	58E	285	−1	7	9	91	89	87	16	81	79	78
MALAYSIA															
Kuala Lumpur	3	07N	101	42E	127	67	70	71	94	93	92	20	82	82	81
MEXICO															
Mexico City	19	24N	99	12W	7575	33	37	39	83	81	79	25	61	60	59
Monterrey	25	40N	10	18W	1732	31	38	41	98	95	93	20	79	78	77
NETHERLANDS															
Amsterdam	52	23N	4	55E	5	17	20	23	79	76	73	10	65	64	63
NEW ZEALAND															
Auckland	36	51S	174	46E	140	37	40	42	78	77	76	14	67	66	65
NORWAY															
Bergen	60	24N	5	19E	141	14	17	20	75	74	73	21	67	66	65
Oslo	59	56N	10	44E	308	−2	0	4	79	77	74	17	67	66	64
PHILIPPINES															
Manila	14	35N	120	59E	47	69	73	74	94	92	91	20	82	81	81
POLAND															
Krakow	50	04N	19	57E	723	−2	2	6	84	81	78	19	68	67	66
Warsaw	52	13N	21	02E	394	−3	3	8	84	81	78	19	71	70	68
SAUDI ARABIA															
Dhahran	26	17N	50	09E	80	39	45	48	111	110	108	32	86	85	84
SINGAPORE															
Singapore	1	18N	103	50E	33	69	71	72	92	91	90	14	82	81	80
SOUTH AFRICA															
Cape Town	33	56S	18	29E	55	36	40	42	93	90	86	20	72	71	70
SPAIN															
Barcelona	41	24N	2	09E	312	31	33	36	88	86	84	13	75	74	73
SWEDEN															
Stockholm	59	21N	18	04E	146	3	5	8	78	74	72	15	64	62	60
SWITZERLAND															
Zurich	47	23N	8	33E	1617	4	9	14	84	81	78	21	68	67	66
TAIWAN															
Tainan	22	57N	120	12E	70	40	46	49	92	91	90	14	84	83	82
Taipei	25	02N	121	31E	30	41	44	47	94	92	90	16	83	82	81
THAILAND															
Bangkok	13	44N	100	30E	39	57	61	63	97	95	93	18	82	82	81
TURKEY															
Istanbul	40	58N	28	50E	59	23	28	30	91	88	86	16	75	74	73
UNITED KINGDOM															
London	51	29N	0	00	149	20	24	26	82	79	76	16	68	66	65

[a]AP, AFB, following the station name designates airport or military airbase temperature observations. Co designates office locations within an urban area that are affected by the surrounding area. Undesignated stations are semirural and may be compared to airport data.
[b]Winter design data are based on the 3-month period, December through February.
[c]Summer design data are based on the 4-month period, June through September.
Source: Reprinted by permission from ASHRAE (www.ashrae.org).

TABLE 2–3A
Thermal properties of typical building and insulating materials

Description	Density, lb/ft³	Conductivity, λ Btu.In/h·ft²°F	Conductance, C Btu/h·ft²°F	Residence, R Thickness per inch,(1/λ) h·ft²°F/Btu	Residence, R For thickness listed 1/C h·ft²°F/Btu
BUILDING BOARD					
Boards, Panels, Subflooring, Sheathing					
Woodboard Panel Products					
Asbestos-cement board	120	4.0	—	0.25	—
Asbestos-cement board ....0.25 in.	120	—	16.50	—	0.06
Gypsum or plaster board ....0.5 in.	50	—	2.22	—	0.45
Plywood (Douglas fir)	34	0.80	—	1.25	—
Plywood (Douglas fir) ....0.25 in.	34	—	3.20	—	0.31
Plywood (Douglas fir) ....0.5 in.	34	—	1.60	—	0.62
Vegetable fiberboard					
Sheathing, regular density ....0.5 in.	18	—	0.76	—	1.32
Nail-base sheathing ....0.5 in.	25	—	0.88	—	1.14
Shingle backer ....0.375 in.	18	—	1.06	—	0.94
Sound-deadening board ....0.5 in.	15	—	0.74	—	1.35
Tile and lay-in panels, plain or acoustic	18	0.40	—	2.50	—
Laminated paperboard	30	0.50	—	2.00	—
Hardboard					
Medium density	50	0.73	—	1.37	—
High density, std. tempered	63	1.00	—	1.00	—
Particle board					
Low density	37	0.54	—	1.06	—
Medium density	50	0.94	—	1.06	—
Wood subfloor ....0.75 in.		—	1.06	—	0.94
BUILDING MEMBRANE					
Vapor-permeable felt	—	—	16.70	—	0.06
Vapor-seal, plastic film	—	—	—	—	Negl.
FINISHED FLOORING MATERIALS					
Carpet and fibrous pad	—	—	0.48	—	2.08
Carpet and rubber pad	—	—	0.81	—	1.23
Terrazzo ....1 in.	—	—	12.50	—	0.08
Tile-asphalt, linoleum, vinyl, rubber			20.0	—	0.05
Wood, hardwood finish ....75 in.	—	—	1.47		0.68
INSULATING MATERIALS					
Blanket and Batt					
Mineral fiber, fibrous form processed					
from rock, slag, or glass approx. 3.5 in.........	0.3–2.0	—	0.077	—	13
Board and Slabs					
Cellular glass	8.5	0.35	—	2.86	—
Glass fiber, organic bonded	4–9	0.25	—	4.00	—
Expanded rubber (rigid)	4.5	0.22	—	4.55	—
Expanded polystyrene, extruded					
cut cell surface	1.8	0.25	—	4.00	—
Cellular polyurethane (R-11 exp.)(unfaced)	1.5	0.16	—	6.25	—
Cellular polyisocyanurate (R-11 exp.) (foll-faced,					
glass fiber–reinforced core)	2.0	0.14	—	7.20	—
Nominal 1.0 in	15.0	0.29	—	3.45	—

(Continued)

TABLE 2–3A *(Continued)*

Description	Density, lb/ft³	Conductivity, λ Btu.In/h·ft²°F	Conductance, C Btu/h·ft²°F	Residence, R	
				Thickness per inch,(1/λ) h·ft²°F/Btu	For thickness listed 1/C h·ft²°F/Btu
Mineral fiber with resin binder	15.0	0.29	—	3.45	—
Mineral fiberboard, wet felted					
Core or roof insulation	16–17	0.34	—	2.94	—
Acoustical tile	18.0	0.35	—	2.86	—
Mineral fiberboard, wet molded					
Acoustical title	23.0	0.42	—	2.38	—
Cement fiber slabs (shredded wood					
with Portland cement binder)	25–27.0	0.50–0.53	—	2.0–1.89	—
LOOSE FILL					
Wood fiber, softwoods	2.0–3.5	0.30	—	3.33	—
Perlite, expanded	2.0–4.1	0.27–0.31	—	3.7–3.3	—
....	4.3–7.4	0.31–0.36	—	3.3–2.8	—
....	7.4–11.0	0.36–0.42	—	2.8–2.4	—
Mineral fiber (rock, stag, or glass)					
approx, 3.75–5 in.	0.6–2.0	—	—		11.0
approx, 6.5–8.75 in.	0.6–2.0	—	—		19.0
Mineral fiber (rock, slag, or glass)					
approx, 3.5 in. (closed sidewall application)..	2.0–3.5	—	—	—	12.0–14.0
Vermiculite, exfollated	7.0–8.2	0.47	—	2.13	—
FIELD APPLIED					
Polyurethane foam	1.5–2.5	0.16–0.18	—	6.25–5.26	—
Spray cellulosic fiber base	2.0–6.0	0.24–0.30	—	3.33–4.17	—
PLASTERING MATERIALS					
Cement plaster, sand aggregate	116	5.0	—	0.20	—
Sand aggregate ....0.375 in.	—	—	13.3	—	0.08
Gypsum plaster:					
Lightweight aggregate ....0.5 in.	45	—	3.12	—	0.32
Lightweight aggregate on metal lath 0.75 in.	—	—	2.13	—	0.47
PLASTERING MATERIALS					
Sand aggregate	105	5.6	—	0.18	—
Sand aggregate ....0.5 in.	105	—	11.10	—	0.09
MASONRY MATERIALS					
Concretes					
Cement mortar	116	5.0	—	0.20	—
Lightweight aggregates including	120	5.2	—	0.19	—
slags; cinders; purmice; vermiculite	80	2.5	—	0.40	—
Perlite, expanded	40	0.93	—	1.08	—
Sand and gravel or stone aggregate					
(oven dried)	140	9.0	—	0.11	—
Sand and gravel or stone aggregate					
(not dried)	140	12.0	—	0.08	—
Stucco	116	5.0	—	0.20	—
MASONRY UNITS					
Brick, common	120	5.0	—	0.20	—
Brick, face	130	9.0	—	0.11	—

(Continued)

TABLE 2–3A *(Continued)*

Description	Density, lb/ft³	Conductivity, λ Btu.In/h·ft²°F	Conductance, C Btu/h·ft²°F	Residence, R Thickness per inch,(1/λ) h·ft²°F/Btu	For thickness listed 1/C h·ft²°F/Btu
Clay tile, hollow:					
1 cell deep ....3 in.	—	—	1.25	—	0.80
2 cells deep ....6 in.	—	—	0.66	—	1.52
3 cells deep ....12 in.	—	—	0.40	—	2.50
Concrete blocks, three oval core:					
Sand and gravel aggregate ....4 in.	—	—	1.40	—	0.71
....8 in.	—	—	0.90	—	1.11
cinder aggregate ....4 in.	—	—	0.90	—	1.11
....8 in.	—	—	0.58	—	1.72
Lightweight aggregate					
(expanded shale, clay, slate ....4 in.	—	—	0.67	—	1.50
or slag; pumice) ....8 in.	—	—	0.50	—	2.00
Stone, lime, or sand	—	12.50	—	0.08	—
ROOFING					
Asbestos-cement shingles	120	—	4.76	—	0.21
Asphalt roll roofing	70	—	6.50	—	0.35
Asphalt shingles	70	—	2.27	—	0.44
Built-up roofing ....0.375 in.	70	—	3.00	—	0.33
Slate ....0.5 in.	—	—	20.00	—	0.05
SIDING MATERIALS (on flat surface)					
Shingles					
Asbestos-cement	120	—	4.75	—	0.21
Wood, 16 in., 7.5-in. exposure	—	—	1.15	—	0.87
Wood, plus insulating backer					
board, 0.3125 in.	—	—	0.71	—	1.40
Siding					
Asbestos-cement 0.25 in., lapped	—	—	4.76	—	0.21
Asphalt roll siding	—	—	6.50	—	0.15
Asphalt insulating siding (0.5 in. Bed.)	—	—	0.69	—	1.46
Wood, bevel, 0.5 + 8 in. lapped	—	—	1.23	—	0.81
Aluminum or steel, over sheathing					
Hollow backed	—	—	1.61	—	0.61
Insulating-board backed nominal 0.375 in.	—	—	0.55	—	1.82
Architectural glass	—	—	10.00	—	0.10
WOODS (12% moisture content)					
Hardwoods					
Oak	41.2–46.8	1.12–1.25	—	0.89–0.80	—
Maple	39.8–44.0	1.09–1.19	—	0.94–0.88	—
Softwoods					
Southern pine	35.6–41.2	1.00–1.12	—	1.00–0.89	—
Douglas fir-larch	33.5–36.3	0.95–1.01	—	1.06–0.99	—
California redwood	24.5–28.0	0.74–0.82	—	1.35–1.22	—

TABLE 2–9A
Outdoor air requirements for ventilation

	Outdoor Air Requirements			
Application	Estimated Maximum Occupants per 1000 sq ft	CFM/ person	CFM/ sq ft	CFM/ room
Dry Cleaners, Laundries[a]				
Commercial laundry	10	25		
Commercial dry cleaner	30	30		
Food and Beverage Service				
Dining rooms	70	20		
Cafeteria, fast food	100	20		
Hotels, Motels, Resorts, Dormitories				
Bedrooms				30
Living rooms				30
Baths[c]				35
Conference rooms	50	20		
Assembly rooms	120	15		
Offices				
Office space[d]	7	20		
Reception areas	60	15		
Conference rooms[b]	50	20		
Public Spaces				
Corridors and utilities			0.05	
Public restrooms, cfm/wc or cfm/urinal[e]		50		
Locker and dressing rooms			0.5	
Retail Stores, Sales Floors, and Showroom Floors				
Basement and street	30		0.30	
Upper floors	20		0.20	
Malls and arcades	20		0.20	
Warehouses	5		0.05	
Specialty Shops				
Beauty	25	25		
Clothiers, furniture			0.30	
Supermarkets	8	15		
Pet shops			1.00	
Sports and Amusement				
Spectator areas	150	15		
Ice arenas (Playing areas)			0.50	
Swimming pools (pool and deck area)[f]			0.50	
Playing floors (gymnasium)	30	20		
Theaters[g]				
Lobbies	150	20		
Auditoriums	150	15		
Stages, studios	70	15		
Transportation[h]				
Waiting rooms	100	15		
Vehicles	150	15		
Workrooms	10	15		
Education				
Classroom	50	15		
Laboratories[i]	30	20		
Libraries	20	15		
Locker rooms			0.50	
Auditoriums	150	15		

(Continued)

TABLE 2–9A   (Continued)

	Outdoor Air Requirements			
Application	Estimated Maximum Occupants per 1000 sq ft	CFM/ person	CFM/ sq ft	CFM/ room
Hospitals, Nursing and Convalescent Homes				
Patient rooms[j]	10	25		
Operating rooms	20	30		
Autopsy rooms[k]			0.50	
Physical therapy	20	15		

[a]Dry-cleaning processes may require more air.
[b]Supplementary smoke-removal equipment may be required.
[c]Installed capacity for Intermittent use.
[d]Some office equipment may require local exhaust.
[e]Normally supplied by transfer air. Local mechanical exhaust with no recirculation recommended.
[f]Higher values may be required for humidity control.
[g]Special ventilation will be needed to eliminate special stage effects (e.g., dry ice vapors, mists, etc.)
[h]Ventilation within vehicles may require special considerations.
[i]Special contaminant control systems may be required for processes or functions, including laboratory animal occupancy.
[j]Special requirements or codes and pressure relationships may determine minimum ventilation rates and filter efficiency. Procedures generating contaminants may require higher rates.
[k]Air shall not be recirculated into other spaces.

Source: Reprinted by permission from ASHRAE (www.ashrae.org).

TABLE 2–15A
Solar position and intensity and solar heat gain factors for 40° north latitude

Date	Solar Time, A.M.	Solar Position		Direct Normal Irradiation, Btuh/sq ft	Solar Heat Gain Factors, Btuh/sq ft									Solar Time, P.M.
		Altitude	Azimuth		N	NE	E	SE	S	SW	W	NW	Horizontal	
Jan 21	8	8.1	55.3	141	5	17	111	133	75	5	5	5	13	4
	9	16.8	44.0	238	11	12	154	224	160	13	11	11	54	3
	10	23.8	30.9	274	16	16	123	241	213	51	16	16	96	2
	11	28.4	16.0	289	18	18	61	222	244	118	18	18	123	1
	12	30.0	0.0	293	19	19	20	179	254	179	20	19	133	12
	Half-day Totals				59	68	449	903	815	271	59	59	353	
Feb 21	7	4.3	72.1	55	1	22	50	47	13	1	1	1	3	5
	8	14.8	61.6	219	10	50	183	199	94	10	10	10	43	4
	9	24.3	49.7	271	16	22	186	245	157	17	16	16	98	3
	10	32.1	35.4	293	20	21	142	247	203	38	20	20	143	2
	11	37.3	18.6	303	23	23	71	219	231	103	23	23	171	1
	12	39.2	0.0	306	24	24	25	170	241	170	25	24	180	12
	Half-day Totals				81	144	634	1035	813	250	81	81	546	
Mar 21	7	11.4	80.2	171	8	93	163	135	21	8	8	8	26	5
	8	22.5	69.6	250	15	91	218	211	73	15	15	15	85	4
	9	32.8	57.3	281	21	46	203	236	128	21	21	21	143	3
	10	41.6	41.9	297	25	26	153	229	171	28	25	25	186	2
	11	47.7	22.6	304	28	28	78	198	197	77	28	28	213	1
	12	50.0	0.0	306	28	28	30	145	206	15	30	28	223	12

(Continued)

TABLE 2–15A   *(Continued)*

Date	Solar Time, A.M.	Solar Position Altitude	Solar Position Azimuth	Direct Normal Irradiation, Btuh/sq ft	N	NE	E	SE	S	SW	W	NW	Horizontal	Solar Time, P.M.
		Half-day Totals			112	310	849	1100	692	218	112	112	764	
Apr 21	6	7.4	98.9	89	11	72	88	52	5	4	4	4	11	6
	7	18.9	89.5	207	16	141	201	143	16	14	14	14	61	5
	8	30.3	79.3	253	22	128	225	189	41	21	21	21	124	4
	9	41.3	67.2	275	26	80	203	204	83	26	26	26	177	3
	10	51.2	51.4	286	30	37	153	194	121	32	30	30	218	2
	11	58.7	29.2	292	33	34	81	161	146	52	33	33	244	1
	12	61.6	0.0	294	33	33	36	108	155	108	36	33	253	12
		Half-day Totals			153	509	969	1003	489	196	146	145	962	
May 21	5	1.9	114.7	1	0	0	0	0	0	0	0	0	0	7
	6	12.7	105.6	143	35	128	141	71	10	10	10	10	30	6
	7	24.0	96.6	216	28	165	209	131	20	18	18	18	87	5
	8	35.4	87.2	249	27	149	220	164	29	25	25	25	146	4
	9	46.8	76.0	267	31	105	197	175	53	30	30	30	196	3
	10	57.5	60.9	277	34	54	148	163	83	35	34	34	234	2
	11	66.2	37.1	282	36	38	81	130	105	42	36	36	258	1
	12	70.0	0.0	284	37	37	40	82	112	82	40	37	265	12
		Half-day Totals			203	643	1002	874	356	194	171	170	1083	
June 21	5	4.2	117.3	21	10	21	20	6	1	1	1	1	2	7
	6	14.8	108.4	154	47	142	151	70	12	12	12	12	39	6
	7	26.0	99.7	215	37	172	206	122	21	20	20	20	97	5
	8	37.4	90.7	246	29	156	215	152	29	26	26	26	153	4
	9	48.8	80.2	262	33	113	192	161	45	31	31	31	201	3
	10	59.8	65.8	272	35	62	145	148	69	36	35	35	237	2
	11	69.2	41.9	276	37	40	80	116	88	41	37	37	260	1
	12	73.5	0.0	278	38	38	41	71	95	71	41	38	267	12
		Half-day Totals			242	714	1019	810	311	197	181	180	1121	
July 21	5	2.3	115.2	2	0	2	1	0	0	0	0	0	0	7
	6	13.1	106.1	137	37	125	137	68	10	10	10	10	31	6
	7	24.3	97.2	208	30	163	204	127	20	19	19	19	88	5
	8	35.8	87.8	241	28	148	216	160	29	26	26	26	145	4
	9	47.2	76.7	259	32	106	194	170	52	31	31	31	194	3
	10	57.9	61.7	269	35	56	146	159	80	36	35	35	231	2
	11	66.7	37.9	274	37	39	81	127	102	42	37	37	255	1
	12	70.6	0.0	276	38	38	41	80	109	80	41	38	282	12
		Half-day Totals			211	645	986	850	347	197	177	176	1074	
Aug 21	6	7.9	99.5	80	12	67	82	48	5	5	5	5	11	6
	7	19.3	90.0	191	17	135	191	135	17	15	15	15	62	5
	8	30.7	79.9	236	23	126	216	180	40	22	22	22	122	4
	9	41.8	67.9	259	28	82	197	196	79	28	28	28	174	3
	10	51.7	52.1	271	32	40	149	187	116	34	32	32	213	2
	11	50.3	29.7	277	34	35	81	156	140	52	34	34	238	1
	12	62.3	0.0	279	35	35	38	105	149	105	38	35	247	12

(Continued)

TABLE 2–15A   *(Continued)*

Date	Solar Time, A.M.	Solar Position Altitude	Solar Position Azimuth	Direct Normal Irradiation, Btuh/sq ft	N	NE	E	SE	S	SW	W	NW	Horizontal	Solar Time, P.M.
								Solar Heat Gain Factors, Btuh/sq ft						
	Half-day Totals				161	503	936	961	471	202	154	153	945	
Sep 21	7	11.4	80.2	149	8	84	146	121	21	8	8	8	25	5
	8	22.5	69.6	230	16	86	205	199	71	16	16	16	82	4
	9	32.8	57.3	263	22	47	195	226	124	23	22	22	138	3
	10	41.6	41.9	279	26	28	148	221	165	30	26	26	180	2
	11	47.7	22.6	287	29	29	77	192	191	77	29	29	206	1
	12	50.0	0.0	290	30	30	32	141	200	141	32	30	215	12
	Half-day Totals				116	300	803	1045	672	221	117	116	738	
Oct 21	7	4.5	72.3	48	1	20	45	41	12	1	1	1	3	5
	8	15.0	61.9	203	10	49	173	187	88	10	10	10	43	4
	9	24.5	49.8	257	17	23	180	235	151	18	17	17	96	3
	10	32.4	35.6	280	21	22	139	238	196	38	21	21	140	2
	11	37.6	18.7	290	23	23	70	212	224	100	23	23	167	1
	12	39.5	0.0	293	24	24	26	165	234	165	26	24	177	12
	Half-day Totals				83	143	610	989	783	245	84	83	535	
Nov 21	8	8.2	55.4	136	5	17	107	128	72	5	5	5	14	4
	9	17.0	44.1	232	12	13	151	219	156	13	12	12	54	3
	10	24.0	31.0	267	16	16	122	237	209	50	16	16	96	2
	11	28.6	16.1	283	19	19	61	218	240	116	19	19	123	1
	12	30.2	0.0	287	19	19	21	176	250	176	21	19	132	12
	Half-day Totals				61	71	442	884	798	267	62	61	353	
Dec 21	8	5.5	53.0	88	2	7	67	83	49	3	2	2	6	4
	9	14.0	41.9	217	9	10	135	205	151	12	9	9	39	3
	10	20.7	29.4	261	14	14	113	232	210	55	14	14	77	2
	11	25.0	15.2	279	16	16	56	217	242	120	16	16	103	1
	12	26.6	0.0	284	17	17	18	177	253	177	18	17	113	12
	Half-day Totals				49	54	380	831	781	273	50	49	282	
					N	NW	W	SW	S	SE	E	NE	Horizontal	↕ P.M.

Data shown in table is for 40° north latitude. Solar intensity is given in the Direct Normal column. Solar heat gain factor is the peak heat gain of the hour calculated with clearness factor = 1.0 and ground reflectance = 0.20. Half-day total is the average heat gain based on Simpson's rule on 10-minute interval. See ASHRAE *Handbook of Fundamentals* for data of other latitudes (16, 24, 32, 48, 56, and 64).

Source: Reprinted by permission from ASHRAE (www.ashrae.org).

Wind Chill Problems:

Problem 2–4	What is the wind chill factor of the weather when the dry-bulb temperature is 20°F, with a wind condition at 30 mph?
Answer:	From conventional units section of Table 2–1, the WCF is −18°F.
Problem 2–5	What is the WCF if the dry-bulb temperature is 35°F, with a wind condition at 15 mph?
Answer:	From conventional units section of Table 2–1, interpolating between 30 and 40°F and 15 mph, the WCF is 9°F for 30°F and 22°F for 40°F. Thus, the WCF is (9 + 22)/2 = 15.5°F.
Problem 2–6	What is the WCF if the dry-bulb temperature is −35°C and the wind velocity is 60 km/h? How severe is this condition?
Answer:	From SI units section of Table 2–1, the WCF is −74°C. The condition is very severe. Exposed flesh may freeze within 30 seconds.

QUESTIONS

2.1 State the factors that affect environmental comfort and their general ranges of values, where applicable.

2.2 Briefly describe the difference between sensible and latent heat.

2.3 For air at 80°F, 50 percent relative humidity, what will happen to relative humidity if the air is cooled to 70°F? To 55°F?

2.4 How much water is present in 1 lb of air at 70°F, 50% relative humidity? How much water is present in 1 lb of air at 100°F DB (dry bulb), 78°F WB (wet bulb)?

2.5 If 50 lb/hr of water is allowed to evaporate into an airstream, what will be the effect on the dry-bulb temperature? What will be the effect on the wet-bulb temperature?

2.6 What is the rate of removal of moisture required to reduce a 1000-CFM airstream from the higher to the lower condition in Question 2.4?

2.7 How much heat is required to warm 1500 gallons of water from 70°F to 140°F?

2.8 How much heat is liberated when 1000 gallons of water cools from 160°F to 120°F?

2.9 What is the heat liberation rate for a 1000-gpm water flow cooling from 150°F to 110°F?

2.10 A heating system load is 300,000 Btuh. How much heating water flow is required to satisfy the load if the system is designed for a 25°F temperature drop? How much flow is required for a 35°F drop?

2.11 How much steam flow would be required for a 350,000 Btuh heating load?

2.12 A cooling system load is 48,000 Btuh sensible. How much chilled air is required to satisfy the load if the system is designed for a 25°F temperature rise? How much flow is required for a 10°F rise?

2.13 Your client desires an interior temperature of 74°F for the design of his HVAC system in St. Louis (close to the airport weather station). Assuming that a 97.5 percent design will be satisfactory, what is the design temperature difference?

2.14 Your client desires an interior temperature of 75°F for the design of her HVAC system in St. Louis (close to the airport weather station). Assuming that a 5 percent design will be satisfactory, what is the design temperature difference?

2.15 What will be the U-factor for a wall constructed as follows?
- Concrete wall 6″ thick.
- The wall is finished on the interior with 1/2″ drywall, adhesively applied to the surface.
- The drywall is applied over 1.5″ furring.
- The furring space is filled with insulation at R-4 per inch.

2.16 What will be the heating load for a wall 8′ high by 600 long that is constructed as per the last description in Question 2.15 under the conditions specified in Question 2.13?

2.17 What will be the cooling load if the wall in Question 2.16 faces south? North? West?

2.18 What will be the July cooling load for a 6′ × 6′ west-facing window constructed of single-pane clear glass according to the criteria in Question 13? Double-pane clear glass? Double-pane tinted glass with a shading coefficient of 0.75?

2.19 What will be the heating load for a 6′-by-6′ window constructed of single-pane clear glass according to the criteria in Question 13? Assume U = 1.05. Double-pane clear glass? Assume U = 0.55.

2.20 What internal heat gain will result from each of the following in an office?
- 15 people, moderate activity
- 5 copy machines, 500 W each
- 15 suspended light fixtures, two 50-W lamps each

2.21 What minimum outside air quantity will be required for the room in Question 20?

2.22 What heating load will result from the minimum outside air in Question 21 under the criteria of Question 13? What cooling load?

2.23 What is the wind chill factor of an outdoor environment if the measured dry-bulb temperature is 35°F with a wind velocity of 45 mph?

2.24 If the wind chill factor of the outdoor temperature is said to be −15°C and the dry-bulb temperature is 0°C, what is the expected wind velocity?

2.25 Absolute humidity can be expressed in grains of water per pounds of air or pounds of water per pound of air. Write the equation relating the following variables using both units for absolute humidity: airflow (cfm), heat flow (Btuh), absolute humidity (grains of water per pound of air), and pounds of water per pound of air).

2.26 For St. Louis, Missouri, what is the percentage difference in heating load between 99 percent design dry-bulb and 97.5 percent design dry-bulb heating criteria? Assume an interior space criteria of 70°F.

2.27 In reference to Question 2.26, is the percentage difference in cooling load the same for 1 percent summer design conditions as for 2.5 percent? Why or why not?

2.28 How much more air-conditioning supply air is needed in a room with a 40-sq-ft single-strength clear window facing north compared with an identical room facing west? Assume supply air temperature of 50°F and a room temperature of 70°F.

2.29 How much more air-conditioning supply air is needed in a room with a 40-sq-ft single-strength clear window facing east compared with an identical room facing west? Ignore conducted load. Assume supply air temperature of 50°F and a room temperature of 75°F.

2.30 How will the orientation (east versus west) of the room in Question 2.29 affect the overall building load?

2.31 How much moisture will a supply air quantity of 500 cfm absorb in warming from 50°F, saturated to 75°F, with 50 percent relative humidity? How many people would liberate this amount of humidity?

2.32 If there is no latent load for the supply air in Question 2.29, what is the room humidity? If there are 10 people in the room, what is the relative humidity? Assume that the sensible load still results in a 75°F room.

2.33 How much heating load (Btu per year) will be saved by changing from R-4 to R-12 insulation in a 10,000-sq-ft wall with an original U-factor of 0.20? Assume the wall is located in a region with 4 months of winter, average outside temperature 20°F, inside temperature 75°F.

2.34 If the net cost of heating energy is $5.00 per million Btu, what will the utility cost saving be for the wall change in Question 2.33?

2.35 How much heating load (Btu per year) will be saved by changing from R-4 to R-20 insulation in a 10,000-sq-ft wall with an original U-factor of 0.20? Assume the wall is located in a region with 4 months of winter, average outside temperature 20°F, inside temperature 75°F.

2.36 If the net cost of heating energy is $5.00 per million Btu, what will the utility cost savings be for the wall change in Question 2.35?

2.37 What would the heating load savings be for Question 2.33 in a different location with an average winter temperature of 35°F? What would the annual heating savings be at $5.00 per million Btu?

2.38 Would savings in Question 2.33 be higher or lower if the facility was operated around the clock rather than only during regular business hours? Why?

2.39 A proposed building design has 15,000 sq ft of glass with shading coefficient 0.6, equally distributed north, south, east, and west. What is the total solar load in tons on July 21 at 4:00 P.M.?

2.40 What would the solar load in tons for the building in Question 2.39 be if the glass were oriented 40% north, 40% south, 10% east, and 10% west?

2.41 What would the solar load in tons for the building in Question 2.39 be if all the glass faced north? How could this be accomplished while maintaining views in all four directions?

2.42 If the initial incremental cost of air-conditioning equipment is $2500 per ton, how much less would the configuration in Question 2.41 cost, compared with the configuration in Question 2.39?

HVAC DELIVERY SYSTEMS

3

COMPLETE HVAC SYSTEMS ARE MADE UP OF SUB-systems that produce heating and cooling, move heat transfer fluids, and control delivery to a space to maintain stable conditions. Heating and cooling are performed by head-end equipment, including refrigeration devices, furnaces, and boilers. Heat transfer fluids are moved by air-handling equipment, ductwork, grilles, and diffusers for air, and by pumps and piping systems for water. Heating and cooling is delivered according to concepts and by equipment described in this chapter.

Delivery methods vary greatly in their ability to maintain space conditions, in the complexity of their operation and maintenance, and in their energy consumption. Selecting appropriate delivery systems is critical to the successful performance of systems.

3.1 CONTROL OF HEATING AND COOLING

Heating and cooling loads vary with time; therefore, the amount of heating or cooling supplied to a space must vary to keep the temperature constant within certain limits. Heating and cooling must be controlled to match the load and vary the flow of energy. Load conditions are measured at a control device, generally a space thermostat.

Spaces with similar load characteristics will require either heating or cooling, but not both at the same time. Thus, HVAC services can be delivered by single packages or systems not capable of simultaneous heating and cooling. Examples include residential combination furnace and air conditioners, and rooftop heating and cooling units for small office buildings.

Large buildings are made up of interior and perimeter spaces. Interior spaces experience heat gains year-round, owing to lights, appliances, and people, and have no exterior exposure from which to lose heat during winter. Accordingly, interior spaces need air conditioning in both summer and winter. By contrast, perimeter spaces

have walls and windows exposed to the outside. Hence, they will need heating during cold weather if the interior gains are not sufficient to compensate for heat losses to the outside.

Large buildings can be served by multiple small systems capable of providing either heating or cooling to individual spaces. They can also be served by larger central systems capable of simultaneously delivering heating and cooling as required by individual spaces.

Central systems most often use air-handling units with coils to raise or lower the temperature of air supplied to spaces. The air is delivered by fans, either directly to the space or through a distribution system with ducts. When the air-handling unit serves multiple spaces with different load conditions, there must be means for providing various degrees of heating and/or cooling to individual areas.

Air systems offer several methods of control to vary the amount of heating or cooling supply:

- Varying the temperature of the air supplied while holding the flow constant
- Varying the flow of warm or cold air supplied while holding the temperature constant
- Varying both the temperature and the flow of air supplied, or turning the system on or off.

These functions may be accommodated within the air-handling unit or by control devices located downstream of the unit.

Air systems are sometimes combined with convection and radiation devices for heating. These are generally controlled by turning them on and off, as in the case of electric heat, or by modulating the temperature or flow of the heat source, as in the case of hot-water or steam heat.

3.2 ZONING

In HVAC terminology, a *zone* is an area for which temperature is controlled by a single thermostat. For instance, a house with one furnace controlled by a single thermostat would be termed a *single-zone system*. If a larger house had two furnaces, each controlled by a separate thermostat, the house would have two zones. Zones are not to be confused with rooms. Several rooms, or even an entire building, might be controlled as a single zone. The term *zone* can also apply to areas of humidity control.

Complex buildings require many zones of control to accommodate load variations among their spaces.

Many systems use air-terminal devices to serve individual spaces or groups of spaces. In general, each terminal will correspond to a zone of control. Terminals can be physically located as shown in Figure 3–1.

3.3 CONTROLS AND AUTOMATION

3.3.1 Definition

Controls and automation provide the intelligence of mechanical and electrical systems. Automation is the function of having equipment react, without any intervention by an operator, to satisfy preset conditions. Control occurs when a signal to the equipment causes the movement or adjustment of a component to produce the desired result. Equipment is selected predominantly to provide adequate capacity for the needs of the building under the design conditions, whereas controls and automation make the equipment operate under all anticipated conditions. Once the equipment is selected, the design of the controls and automation system begins in earnest.

3.3.2 Basic Control Systems and Devices

All HVAC systems require some form of controls, either manual or automatic. Automatic controls enable the equipment or the entire system to operate more precisely and reliably for comfort, safety, and energy efficiency. They accomplish their task by controlling one or more of the following properties of the transporting medium, such as air or water, and the related equipment:

- *Temperature*—with sensors set for an operating temperature, a differential, or temperature limits
- *Pressure*—with sensors set for an operating pressure, a differential, or pressure limits
- *Flow rate*—with sensors set for an operating rate, a differential, or flow rate limits
- *Humidity*—with sensors set for an operating level, a differential, or humidity limits
- *Speed*—with sensors to control the equipment so that it is either on or off or has variable or multiple speeds
- *Time*—with a clock or a program to control the duration of operation of the equipment

Basic Control Systems

Control systems may be electric, electronic, pneumatic, direct digital, or a combination of these.

Electric controls use line voltage (120 volts) or low voltage (12 to 24 volts) to perform the basic functions. A low-voltage control system is more sensitive and is therefore preferred.

■ **FIGURE 3–1**
Typical placement of air terminals. (a) Sill line—applicable to fan-coil units, unit ventilators, induction units, water-source heat pumps, etc. (b) Ceiling-suspended—applicable to all systems. (c) Concealed in soffit—applicable to fan-coil units and water-source heat pumps. (d) Concealed in ceiling—applicable to all systems.

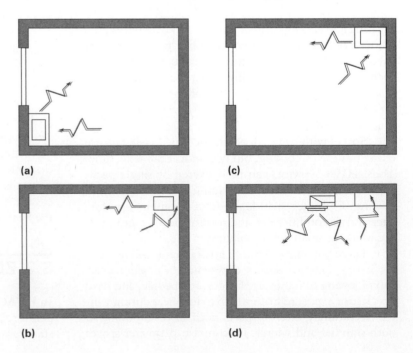

(a)　　　(c)

(b)　　　(d)

Pneumatic controls use 5- to 30-psi compressed air and receiver controllers with force-balance mechanisms. The receiver controllers constantly adjust output air pressure to actuators in response to input pressure from sensors to produce the desired result. Electronic control uses a similar form of controller, except that the signals are electronic rather than pneumatic. These basic systems require manual calibration and adjustment.

With the rapid development of computer chips and network technology during the 1990s, direct digital control (DDC) systems displaced pneumatic controls and to a lesser extent electric controls. With DDC, as with pneumatic or electric controls, an input signal to a controller results in an output to the appropriate device. The major difference is in the controller: instead of physically adjusting the controller components to result in the same reaction again and again, DDC system controllers contain microprocessors that are programmed to interpret the input signal, process the data in resident programs, and intelligently decide on the appropriate response. Figure 3–2 shows the typical hierarchy of a distributed control system.

Basic Control Devices

Control devices include sensors, controllers, and actuators. Sensors measure the monitored or controlled variable. The signal from the sensors is input to a controller for processing and decision making. The controller determines if a signal should be sent to a monitoring station or to an actuator. The controller also determines if the input or output signal is two-position or proportional, or direct- or reverse-acting. Actuators manipulate the equipment to meet the desired set point of the controlled variable.

Two-position input signals are used to indicate the operating status of the equipment, such as on or off, normal or alarm, open or closed. Two-position output signals are used to start, stop, open, or close the controlled equipment. Proportional signals are used to monitor and control variables that change continuously, such as temperature, pressure, or flow. Proportional signals can provide multiple levels of control and alarm through the use of a single sensor or actuator.

Thermostats are devices that sense and respond to temperature, combining the functions of the sensor and controller. Room (space) thermostats may be designed for heating or cooling alone or for heating and cooling in combination. The simplest combination thermostat normally contains a mode selection switch. If this switch is set on the heating mode, the thermostat allows more or less heating energy to be added to the space until a certain level is reached. If the space temperature is already higher than the thermostat setting, owing to an uncontrollable heat source, such as solar heat gain

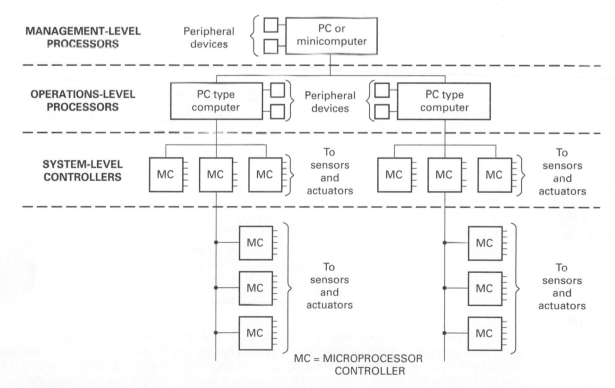

■ **FIGURE 3–2**
Configuration of a hierarchical control system.

through the windows, then lowering the thermostat set point will not bring down the space temperature. The reverse will be true when the mode switch is set in the cooling position. These simple facts are often misunderstood by most users or occupants of buildings.

More sophisticated, programmable electronic thermostats can be seasonally switched if they have 365-day programs. This, of course, requires the programmer to anticipate the beginning of the heating and cooling seasons in order to identify when the switch from heating to cooling should occur. A warm winter day or a cold spring or fall day can contribute to occupants' discomfort with such a system. DDC thermostats are capable of being tied into distributed communication networks and can share information, such as what the outdoor temperature is, to "decide" whether the HVAC system should be in heating or cooling mode, without any intervention by an operator.

Some thermostats can automatically change over from heating to cooling or vice versa, however, such changeovers usually require a more sophisticated HVAC system. A simple heating system using a thermostat is shown in Figure 3–3(a). Adding other control components to a simple heating system can enhance the system control. An example of an expanded heating control arrangement is shown in 3–3(b).

Humidistats are devices that sense and respond to humidity—either relative or absolute. A space humidity sensor sends the appropriate signal to the controller to determine whether humidity should be added to or removed from the supply air. In order to control the space humidity, the airflow can have humidity added by a humidifier or removed by overcooling the air to remove excess humidity and then reheating the air to meet the requirements of the thermostat. All these functions are performed through controls, without any action by the occupants other than adjusting the set point of the humidistat to the appropriate level for space functions.

Other control sensors include pressure switches and transmitters, which respond to pressure; flow switches and transmitters, which respond to rates of flow; speed switches, which respond to flow, pressure, or a program in order to control the speed of the equipment; and position switches, which respond to signals to open, close, or modulate dampers, valves, etc.

Controllers provide the decision-making function of the control system. Controllers are available for all types of control systems: electric, pneumatic, electronic, and DDC. Output signals from these devices are typically two-position, proportional, direct-acting or reverse-acting. DDC controllers are available as preprogrammed or fully programmable. Various levels of control that are identified in Figure 3–2 are provided at the controller. Zone-level controllers are used to operate terminal units or small unitary equipment. Many control system manufacturers provide dedicated controllers for specific types of zone equipment. Preprogrammed

(a)

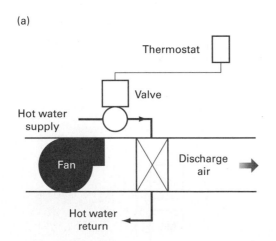

(b)

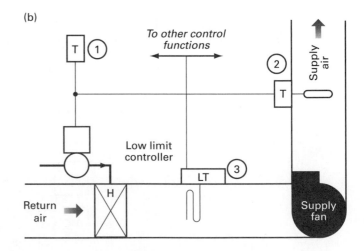

■ **FIGURE 3–3**

(a) Air-handling system with a hot-water coil. The thermostat modulates a two-way valve to control the flow of hot water through the coil for heating the discharge air. (b) Schematic of an expanded heating control system. (1) Space thermostat modulates a two-way coil valve to maintain space temperature. (2) Low limit discharge temperature controller overrides the thermostat to prevent the discharge air temperature from falling below a preset minimum temperature. (3) Separate controller may be installed to initiate a sequence of operations: to close a damper, to turn off the fan, to open the valves, etc. (Courtesy of Honeywell International, Inc.)

controllers for unitary equipment such as heat pumps or VAV terminal units can be installed very inexpensively. System-level controllers are typically provided without any programming because each installation is unique. These controllers are provided with additional input and output capabilities over the zone controllers. Also, the signals are usually "universal"; this results in a completely flexible system for adding or modifying inputs and outputs.

Actuators provide the physical control of the equipment, typically dampers and valves. Movement of the actuator results in a two-position or proportional response of the controlled equipment. Actuators are provided as normally open or normally closed. Normally open actuators return to the open position if the control signal is off; normally closed actuators return to the closed position.

Control valves are used to control the flow of fluids. The valve can be either direct-acting or reverse-acting. A direct-acting valve allows flow with the stem up; a reverse-acting valve shuts off flow with the stem up. The combination of valve body and actuator (called the valve assembly) determines the valve stem position. Figure 3–4 (left) shows a direct-acting two-way valve; Figure 3–4 (right) shows a reverse-acting two-way valve. A two-way valve is a valve with one inlet and one outlet port; a three-way valve may have either one inlet and two outlets or two inlets and one outlet port.

3.3.3 Energy Management

Automation and controls play a major role in managing a building's energy costs. As a rule, the more segregated the equipment (numerous zones, multiple thermostats, etc.), the less control technology is required to make the systems perform at their most efficient energy level. In order to manage energy usage effectively, the controllers of the energy management system must be able to "speak" to one another. Addition of a local area network for dedicated communication among all controllers is the most common arrangement. Communication by telephone lines is a method more commonly used between buildings or sites. A common head-end computer is the interface for an operator to monitor, control, program, and generate reports.

The most basic option for managing energy is to shut down unnecessary equipment. Scheduling equipment on and off is easy with multiple HVAC systems, and it can be done without disturbing areas served by other systems. Hotels, for instance, can schedule room air conditioners on and off based on occupancy. On the other hand, large central HVAC systems provide needed service to all areas covered by the system. With a more sophisticated automation and control system, zone terminal units can be shut down through the automation and control equipment. With proper programming, similar energy savings can result.

Operating equipment at optimum conditions is another way to save energy with controls. Evaluation of loads and conditions leads to a decision about what to operate at that time. An excellent example is operating multiples of a single type of equipment. The energy management system can determine if it is more economical to operate one at high loading, two at low loading, or some other combination that results in minimum energy consumption.

Energy can also be saved through monitoring demand and shutting down equipment to prevent consumption from exceeding a preset limit. This requires an interface with the main electrical service and individual panels or circuits to achieve the desired results.

Another method of energy management is to record equipment trends. When a mechanical or electrical system begins operating outside its normal range, a typical indicator is increased energy usage. By tracking the energy usage of the equipment, it is possible to identify when operations are outside the normal range, and then to perform the required maintenance.

3.3.4 Human Safety

Although safety is not a required part of automation and controls systems, one must be aware of the potential uses

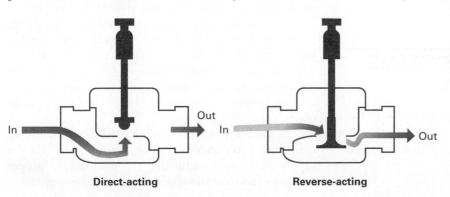

Direct-acting　　　　**Reverse-acting**

■ **FIGURE 3–4**
Two-way valves: the selection of valve body, whether direct-acting or reverse-acting, depends on the combination of valve body and actuator.

of such systems in that regard. The most significant example is providing an interface with the building fire alarm system to activate stair pressurization fans, smoke exhaust fans, and sophisticated air pressure "sandwiches" around fire floors in a high-rise building. Most buildings have fire alarm systems that can be either directly interfaced with the automation and controls or indirectly interfaced through a simple relay cabinet. Most current systems will identify the exact location of the event. On the basis of that information, the automation and controls system can "decide" whether it is necessary to activate any predefined sequences for removing and containing smoke and, if so, which areas will be affected. "Sandwiching" of fire floors requires information on the exact location of the event and floor-to-floor control of the HVAC systems through equipment zoning or zone segregation by dampers. This arrangement creates a negative pressure on the fire floor and a positive pressure on the floors immediately above and below the fire floor. The intent is to allow occupants to evacuate the fire floor and to minimize the opportunity for smoke to migrate to surrounding floors. Other typical interfaces with the fire alarm system include a smoke exhaust fan and a stair pressurization fan. Also, a shutdown of equipment due to smoke or fire can be monitored by both the fire alarm and automation system and the control system. Such monitoring allows HVAC operators to know whether the abnormal off condition is due merely to a mechanical failure or to a potentially life-threatening situation.

3.3.5 Equipment Protection

Equipment protection is typically provided by monitoring specific components for their hours of operation, temperature, and other parameters and then providing the appropriate maintenance response. Additional safety features of the control system include devices in air systems that prevent their operation if damage could occur. Typical examples of these devices are "freezestats" to protect water coils from low temperatures, and smoke detectors or "firestats" to prevent equipment from operating if a fire occurs in the motor.

3.4 COMMONLY USED SYSTEMS FOR ZONE CONTROL

A wide variety of systems are available for building applications. They differ with respect to comfort, cost, energy efficiency, maintainability, and flexibility in terms of altering them.

3.4.1 Constant-Temperature, Variable Volume (On-Off)

A simple residential system will provide hot or cold air. The supply temperature in either the cooling mode or the heating mode is fairly constant. Controls can be set to allow the fan to operate only when heating or cooling is being furnished. As the space temperature falls below or rises above the thermostat set point, the system is activated, and hot or cold air is supplied to the space at a constant rate until the space warms or cools and trips the thermostat. The volume of air supplied to the space will depend on how long the fan is activated. When the proper amount of warm or cold air has been supplied, the system deactivates.

3.4.2 Single-Zone Constant Air Volume

In commercial buildings, fans must generally run continuously to provide ventilation while the building is occupied. This mode of operation is called *constant volume*. Single-zone constant volume is the simplest of systems. Air is supplied to the space at a constant rate. The air-handling unit includes a cooling coil and/or a heating coil that varies the temperature of the air in response to a space thermostat. Heating and cooling may be on-off or proportional. (See Figure 3–5.)

If the space is composed of rooms or areas with different load characteristics, single-zone constant volume will not be a good choice. Good control can be achieved at the location of the thermostat, but the temperature will not be controlled elsewhere. In general, single-zone constant volume is satisfactory only for small simple buildings consisting of spaces with similar load characteristics or for large open spaces; however, single-zone constant-volume systems used in multiples can provide satisfactory control for large or complex buildings.

3.4.3 Single-Zone Reheat

Single-zone reheat is a constant-volume system used for air conditioning when humidity control is especially important. Like the single-zone constant-volume system, the single-zone reheat system is satisfactory only for single spaces with similar load characteristics or for large open spaces. The single-zone reheat system uses a cooling coil to cool and dehumidify air, along with a reheat coil for temperature control. Humidifiers may also be used. (See Figure 3–6.)

A space humidistat may be used to adjust the air temperature to condense enough moisture so that the supply air is sufficiently dry for maintaining proper humidity levels in the space. Since the cooling coil is

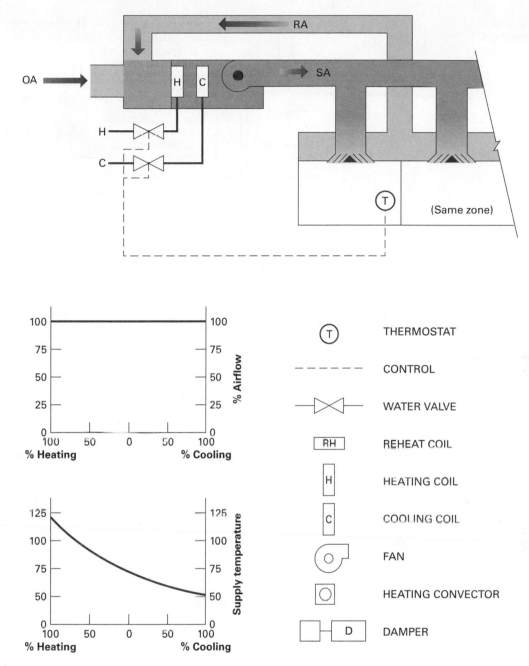

■ **FIGURE 3–5**
Single-zone constant air volume system.

controlled in order to maintain the humidity level, the air temperature will generally be too cold to maintain a proper space temperature. Therefore, a heating coil is placed downstream of the cooling coil to reheat the air. The heating coil is controlled by the space thermostat.

Single-zone reheat systems are characterized by high energy usage. Air may be overcooled to maintain humidity and reheated to compensate for the overcooling. These systems are used only for special applications, such as hospital operating rooms and computer rooms.

3.4.4 Constant-Volume Terminal-Reheat, Multiple Zones

Constant-volume terminal-reheat systems (Figure 3–7) are similar in principle to single-zone reheat systems, except that multiple zones of temperature control can be achieved. The air-handling unit contains a cooling coil that chills all the air supplied to the various zones. One or more main trunk ducts are used to distribute air throughout the area served, and a terminal box

■ **FIGURE 3–6**
Single-zone constant-
volume reheat system.

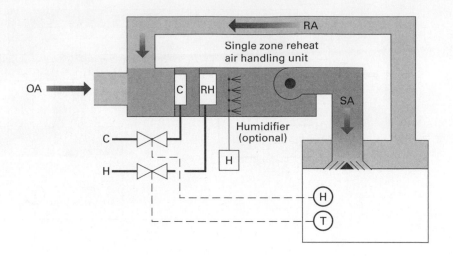

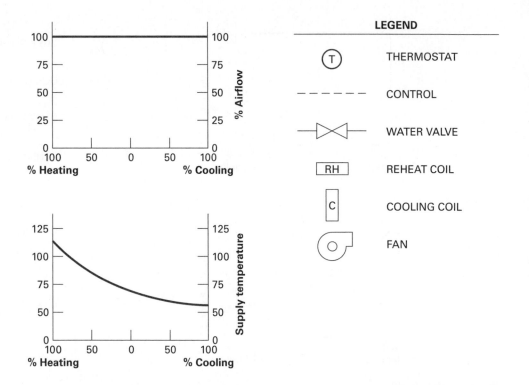

containing a reheat coil is installed for each zone. The cooling coil chills air to the same temperature for all zones. The reheat coils respond to their respective thermostats to maintain temperature control in each zone.

Terminal-reheat systems are capable of serving multiple zones from a single air-handling unit. Reheat terminals, often called *boxes,* are modular devices that are available in sizes from approximately 50 cfm to 3000 cfm. Other types of terminals are available in similar sizes. These sizes are adequate for spaces ranging from individual small offices to large open areas up to approximately 5000 sq ft. A typical reheat box with hot-water and

electric coils is shown in Figure 3–8. Reheat coils are normally installed above the ceiling. Hot-water coils require piping and accessories, which may leak.

Constant-volume terminal-reheat systems are flexible enough that one may add or rearrange reheat boxes on their main trunk as space is reconfigured. Excellent humidity control is another characteristic. These advantages led terminal-reheat systems to be used extensively in commercial and institutional buildings designed from the 1950s through the early 1970s. During the mid-1970s, reheat systems lost favor because of concern for energy conservation and operating costs.

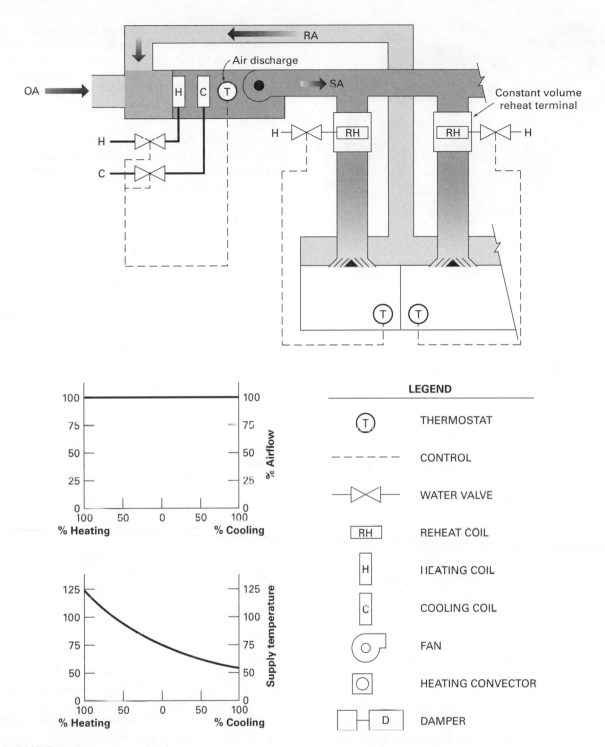

LEGEND

(T)	THERMOSTAT
– – – – –	CONTROL
▷◁	WATER VALVE
RH	REHEAT COIL
H	HEATING COIL
C	COOLING COIL
⊙	FAN
▣	HEATING CONVECTOR
▢─ D	DAMPER

■ **FIGURE 3–7**
Constant-volume terminal-reheat system.

3.4.5 Constant-Volume Dual Duct

Constant-volume dual-duct systems (Figure 3–9) send warm and chilled air through a pair of main trunk ducts to the areas being served. At the air-handling unit, a single fan blows air through a heating coil and a cooling coil into a warm-air trunk duct and a chilled-air trunk duct. Constant-volume dual-duct systems are capable of controlling multiple zones, with each zone served by an individual mixing box. (See Figure 3–9.)

Mixing boxes are controlled with dampers to vary the quantity of warm and chilled air in response to the

■ **FIGURE 3–8**
A typical single-duct terminal with hot-water reheat coil. The coil may also be electric. (Reproduced with permission from Carnes Company.)

zone thermostat. A typical box is shown in Figure 3–10. If a zone needs more cooling, the dampers will be positioned to deliver more chilled air and less warm air. Conversely, the dampers will be positioned to deliver more warm air and less chilled air if a space needs heating or less cooling. The total airflow, consisting of warm and chilled air, stays fairly constant, since each damper closes as the other opens.

During warm weather, the cooling coil operates to provide air conditioning, but the heating coil can be inactive, as there will be no spaces requiring heating. The return air will be sufficiently warm to mix with chilled air for space temperature control. During cold weather, the heating coil must be activated to provide warmer air. At the same time, interior spaces will require cooling. When cooling and heating are operated simultaneously, the mixing of warm and cold air wastes energy, a problem similar to that caused by reheating.

The constant-volume dual-duct system has the flexibility to add or rearrange mixing boxes on the main trunks as space is reconfigured. It also has the advantage of being an all-air system. Unlike the constant-volume terminal-reheat system, it has no coils installed over occupied spaces; thus it reduces maintenance and avoids the possibility of water leaks.

3.4.6 Multizone

Multizone systems are very similar in operation to dual-duct systems. Warm and cold air are mixed to produce the right air temperature for conditioning, according to signals from the zone thermostats; however, in multizone systems, unlike dual-duct systems, the mixing is done by

dampers located at the air-handling unit rather than at terminals located near the spaces served. Figures 3–11 and 3–12a show the operating concept and configuration of a multizone air-handling unit. An individual duct is run from the air-handling unit to each zone of control. A single multizone unit can serve up to approximately 12 zones.

Like dual-duct systems, multizone systems waste energy because they mix warm and cold air. Multizones can have two sets of dampers (warm air and cold air) or three sets (warm, cold, and bypass). The latter, called a *triple-deck multizone,* saves energy by mixing warm and bypass or cold and bypass rather than warm and cold air.

Multizone systems have the disadvantage of being inflexible with regard to change. Rearranging or adding zones will involve considerable ductwork and modification of the unit. Multizone does, however, have the advantage of centralizing controls within the equipment room for easier maintenance and less intrusion into occupied spaces. This is important where security is an issue or where ceiling terminals may be difficult to access.

3.4.7 Single-Zone Variable Air Volume

Single-zone variable air volume (VAV) systems gained popularity during the mid-1970s as an energy-efficient alternative to constant-volume terminal-reheat and constant-volume dual-duct systems. VAV systems used for cooling vary the quantity of air supplied to the space in response to the space thermostat. The air-handling unit supplies chilled air. If the cooling load is high, the chilled airflow will be high. As the load diminishes, the airflow is reduced accordingly. No energy is wasted by reheating or mixing warm and chilled air. In addition, varying the air quantities offers energy savings in terms of operation of the fan, in comparison with constant-volume systems.

Single-zone VAV systems use a cooling coil to produce chilled air at a constant temperature. Airflow is increased or decreased in response to the space thermostat by adjusting dampers or the speed of the fan. Such a unit may also be equipped with an additional coil for heating. Because the unit can serve just one zone, it is suitable only for small, simple buildings or, in multiples, for large buildings. (See Figure 3–13.)

3.4.8 Multiple-Zone Variable Air Volume

Multiple-zone VAV systems use a VAV air-handling unit to supply chilled air into a main trunk duct feeding multiple VAV terminals that in turn serve individual zones of control. VAV terminals, or boxes, contain dampers that vary the airflow to individual zones of control in response to their thermostats. A typical box is shown in

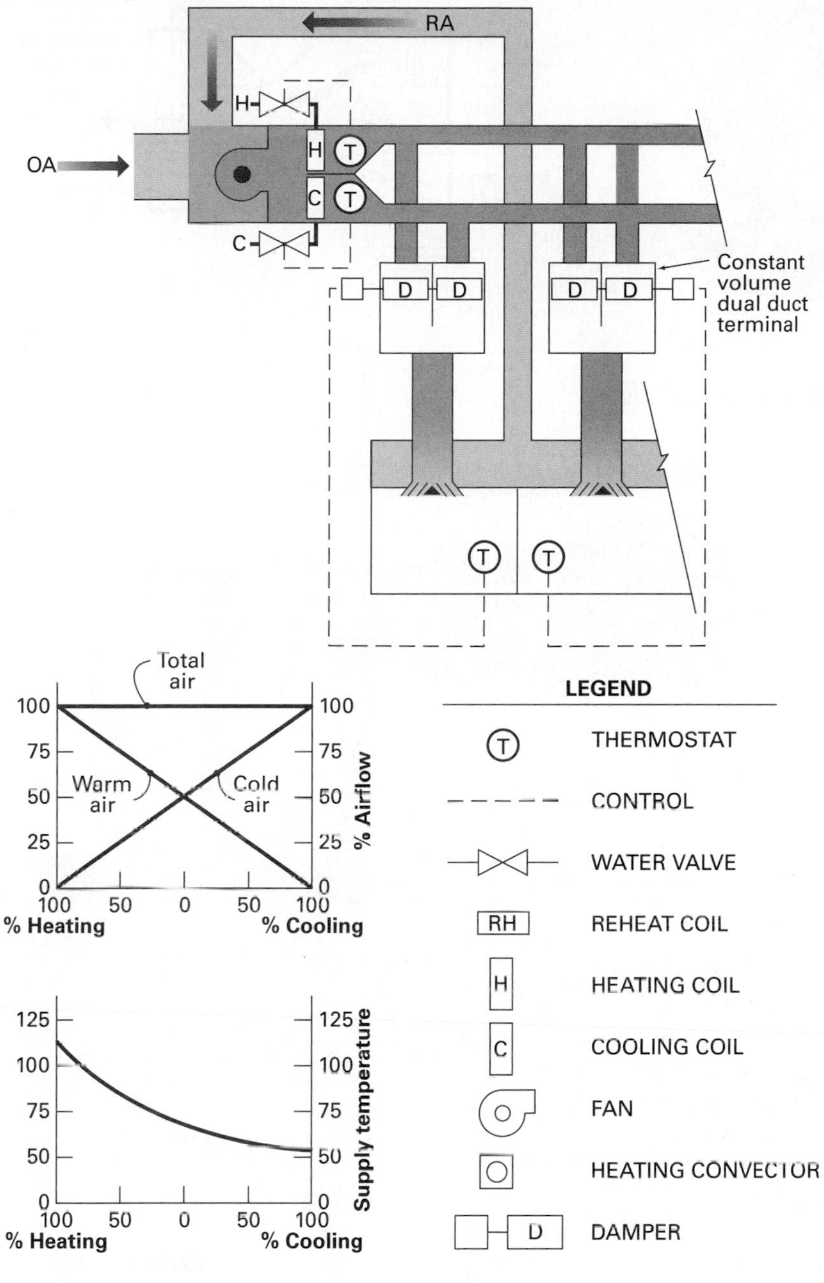

■ **FIGURE 3–9**
Constant-volume dual-duct system.

■ **FIGURE 3–9**
Constant-volume dual-duct system.

LEGEND

(T)	THERMOSTAT
– – – – –	CONTROL
⊲⋈	WATER VALVE
RH	REHEAT COIL
H	HEATING COIL
C	COOLING COIL
⊙	FAN
▢	HEATING CONVECTOR
▢—D	DAMPER

Figure 3–14. This form of VAV system can serve large, complex buildings, offering the flexibility of being able to add or rearrange zones by adding or rearranging terminals on the main trunk ducts.

VAV terminals are inexpensive and energy-efficient. Despite complaints that they produce stuffiness at low airflow, they are used extensively in commercial and institutional buildings. VAV terminals are designed to provide only cooling. For buildings with heating loads, they must be used in combination with other devices to provide heat. Figure 3–15 shows the operation of a typical multizone VAV terminal.

Interior spaces of large buildings need cooling year-round, and VAV systems often meet this need. Perimeter spaces need heating during cold weather and can be served by several means, including convectors, radiant panels, or air terminals with heating capability as well as cooling.

Perimeter convectors are often used as a source of heat in conjunction with VAV terminals for cooling. Perimeter convectors use electric resistance elements or hot-water piping equipped with fins to enhance heat transfer. Often called *finned tube*, this equipment is placed in a low, linear enclosure called a *baseboard* or in

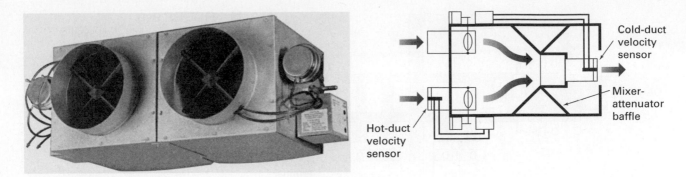

Hot-duct velocity sensor

Cold-duct velocity sensor

Mixer-attenuator baffle

■ FIGURE 3–10

Typical dual-duct terminal and its internal section. (Courtesy: Titus.)

■ FIGURE 3–11

Constant-volume multizone system. (Sketch shown is a triple-zone, double-deck multizone unit.)

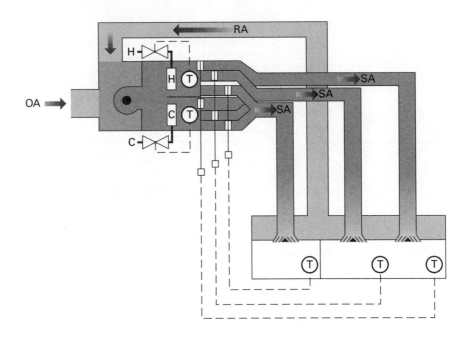

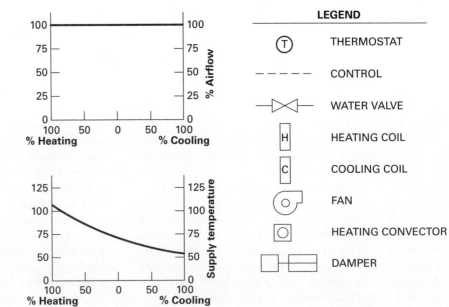

LEGEND

(T)	THERMOSTAT
– – –	CONTROL
⧓	WATER VALVE
H	HEATING COIL
C	COOLING COIL
⌀	FAN
⊡	HEATING CONVECTOR
▭▭	DAMPER

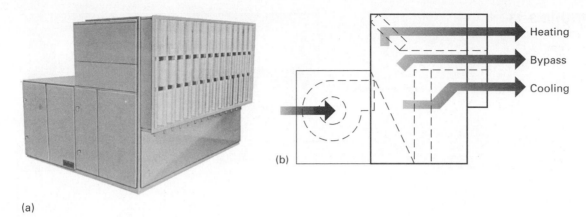

■ FIGURE 3–12

(a) Triple-deck multizone unit. (b) Schematic of the triple-deck unit. The three decks are the standard hot and cold decks plus a bypass deck. This configuration offers significant energy conservation by allowing return or outside air to bypass both coils, and the thermal inefficiency of mixing heated and cooled air is eliminated. (Reproduced with permission from McQuay International.)

a taller configuration called a *sill line.* Baseboard or sill line convectors are excellent at preventing downdrafts under windows. They have the disadvantage, however, of taking up several inches of perimeter floor space and sometimes impairing the flexibility of furniture arrangement. (See Figure 3–16 and Figure 3–17.) A combination system, including perimeter convectors with VAV, is shown in Figure 3–18.

Radiant panels are generally located at the ceiling and use electric resistance elements or hot-water piping to warm their surface and radiate infrared heat downward. Infrared heat offers little protection against downdrafts and is directional. Areas under desks and tables may not be adequately exposed to the heating effect.

The top story of a building poses another perimeter problem: heat lost through the roof. Convectors or unit heaters can be installed to prevent cold air from entering occupied spaces served by cooling-only systems.

3.4.9 Variable Air Volume Reheat Terminals

VAV reheat terminals are equipped with electric or hot-water heating coils (see Figure 3–19) and are similar to constant-volume reheat terminals except for control of the airflow. While the space requires cooling the VAV reheat terminal supplies chilled air. As the cooling load diminishes, the amount of chilled air is reduced accordingly, to a preset minimum quantity. When heating is required, the heating coil warms the minimum quantity of chilled air to a temperature that is suitable for heating the space. The energy waste due to reheat is much less than in constant-volume reheating. Figure 3–20 illustrates the operation of a VAV reheat terminal.

VAV reheat terminals are equipped with electric or hot-water heating coils, and VAV reheat terminals are often used for perimeter heating in combination with simple VAV terminals for the interior (Figure 3–17). VAV reheating uses more energy than VAV with perimeter convectors and offers little resistance to downdrafts. On the other hand, VAV reheating is generally less expensive and does not require locating equipment on the floor, so that space can be used more flexibly.

3.4.10 Variable Air Volume Dual-Duct Terminals

VAV dual-duct terminals are arranged for operation at a variable rather than a constant air volume in order to conserve energy. They are served from a warm-air duct and a chilled-air duct. While the space requires cooling the VAV dual-duct terminal supplies chilled air through a damper from the chilled-air duct. As the cooling load diminishes, the amount of chilled air is reduced accordingly, while the damper from the warm-air duct remains closed. When heating is required, the chilled-air damper will be at or near the closed position, and the warm-air damper will start to open and increase the flow of warm air as the heating load increases. With this mode of operation, there is minimal overuse of energy due to mixing of warm and cold air at the terminal. Figure 3–21 illustrates the operation of a VAV dual-duct terminal.

VAV dual-duct terminals are another common choice for perimeter heating with an interior system using cooling-only VAV terminals. The central air-handling unit is arranged to provide both warm- and chilled-air. VAV terminals serving interior spaces are connected to

■ **FIGURE 3–13**
Single-zone variable air volume
(VAV) system.

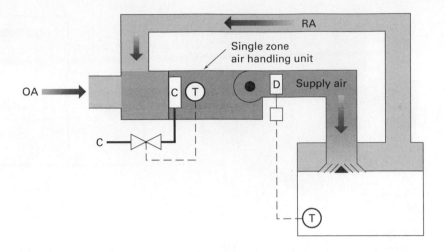

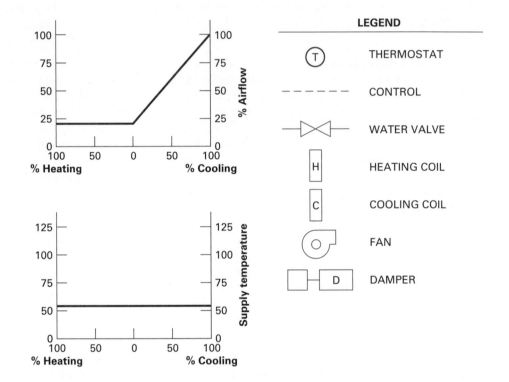

the chilled-air supply. Dual-duct terminals are connected to both the warm- and the chilled-air supply.

VAV dual-duct terminal systems are relatively energy-efficient and have several advantages over VAV with perimeter convectors and VAV reheat. VAV dual duct does not require any electric or hot-water devices to be maintained in occupied spaces at the perimeter; all heating equipment is confined to the mechanical room.

3.4.11 Variable Air Volume Multizone

VAV multizone is similar in operation to VAV dual duct. Separate dampers are used for the warm and cold

portions of the mixing section. (See Figure 3–22.) The cold-air damper is almost closed before the warm-air damper is allowed to open. This prevents energy waste due to mixing.

3.4.12 Fan Terminal Units

Fan terminal units are designed to overcome several of the complaints about the performance of VAV terminals. The nature of VAV is to decrease the flow of cooling air under light load conditions. This can create stuffiness in occupied spaces. The fan terminal unit contains a VAV damper that responds to the space

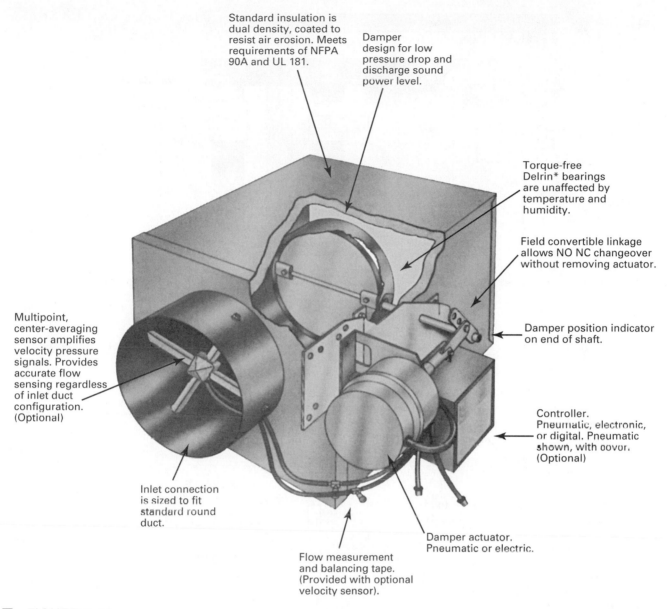

Standard insulation is dual density, coated to resist air erosion. Meets requirements of NFPA 90A and UL 181.

Damper design for low pressure drop and discharge sound power level.

Torque-free Delrin* bearings are unaffected by temperature and humidity.

Field convertible linkage allows NO NC changeover without removing actuator.

Multipoint, center-averaging sensor amplifies velocity pressure signals. Provides accurate flow sensing regardless of inlet duct configuration. (Optional)

Damper position indicator on end of shaft.

Controller. Pneumatic, electronic, or digital. Pneumatic shown, with cover. (Optional)

Inlet connection is sized to fit standard round duct.

Flow measurement and balancing tape. (Provided with optional velocity sensor).

Damper actuator. Pneumatic or electric.

■ **FIGURE 3–14**

Variable air volume terminal or box. Air terminals come in a number of designs, including constant air volume (CAV) or variable air volume (VAV) and single duct or dual duct, and the airflow rate may be dependent on or independent of the system pressure. Control power may be electric, pneumatic, electronic, or direct digital. The flow of air to a room or a zone is in response to a room thermostat or other signaling devices. The illustration shows a pneumatic-controlled single-duct terminal. (Courtesy: Titus.)

thermostat. Chilled air passing through the VAV damper is mixed with return air from above the ceiling and delivered to the space via a small constant-volume fan within the unit. Since the fans within the fan terminal units have only a short length for distribution to the space, their pressure requirements are low, and energy consumption is accordingly modest. The central air-handling unit takes full advantage of reduced chilled-air requirements at partial load, and overall energy savings are achieved at the central fan. Fan terminal units can also be equipped with heating coils and are sometimes used in combination with VAV systems. Figure 3–23 shows a typical fan terminal unit; its operation is diagrammed in Figure 3–24.

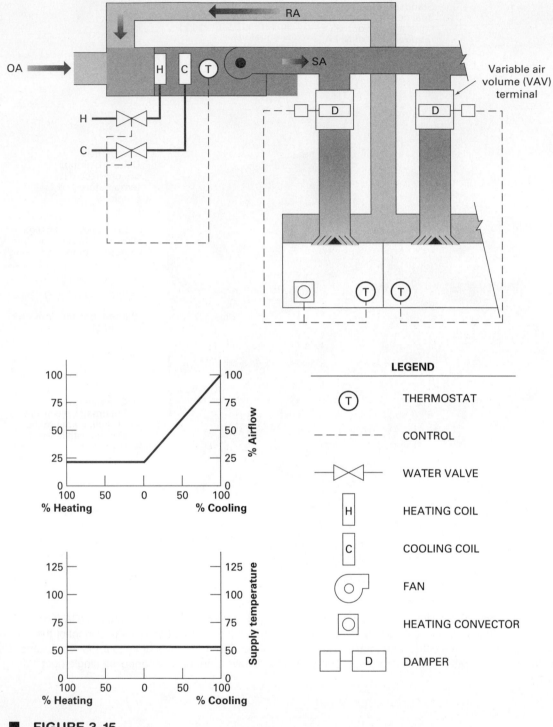

■ **FIGURE 3–15**
Multiple-zone VAV system.

Figure 3–18 and Figure 3–25 illustrate two of many possible choices of heating and cooling systems for a typical building. Figure 3–18, a single-duct VAV interior system with a hot-water convector exterior system, is sometimes described as a "split" system that incorporates both air and water (or steam) as the means of energy distribution. Figure 3–25, a single-duct VAV interior system with dual-duct VAV for the exterior zones, is sometimes described as an "all-air" system which incorporates air only as the means of energy

■ **FIGURE 3–16**
Typical convector installed at the base of a perimeter wall to overcome the downdraft during cold weather.

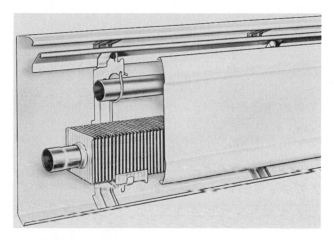

■ **FIGURE 3–17**
Cutaway view of a hot-water convector showing the finned tube and supply pipe. Convectors may use steam or electric power and come in various designs and heights. (Courtesy: Sterling.)

distribution. The application of each system depends on a number of factors, such as the availability of hot water or steam, relative cost of energy, building construction, floor-to-floor height, depth of ceiling cavity, depth of structural members, aesthetics, and the number of exterior zones desired. Although no generalized conclusion can be made, the all-air system is usually lower in cost but requires more space, for both the high-velocity and the low-velocity ducts. In specific building applications, the increased duct sizes and number of terminal units may preclude the use of an all-air system regardless of the economics.

3.4.13 Fan Coils

Fan coils are another system that can be used to provide multiple zones of control. A fan coil consists of a filter,

a cooling and/or heating coil, a fan, and controls. Units can be located above the ceiling, in wall cabinets, or in soffits. Typical fan coil units are shown in Figure 3–26. Fan coils can also be mounted above ceilings or in soffits. Each fan coil can be controlled by an independent thermostat.

Fan coils can be designed with two or four pipes, as shown in Figure 3–27. Two-pipe fan coils are served by a set of supply and return pipes that can carry either hot or chilled water; thus a particular unit is capable of heating or cooling, depending on which water service is available. The piping may be zoned by its exposure to the sun, with some portions of the building capable of cooling while the rest is heating. Obviously, some compromises in comfort can be expected with a two-pipe system.

A four-pipe fan coil system has two sets of piping—one set of supply and return for chilled water and another set for hot water. The four-pipe system is more flexible than the two-pipe system but also more expensive to install.

Fan coil units can be used at building perimeters in combination with VAV cooling-only systems serving interior spaces.

3.4.14 Radiant Panel Heating and Cooling

Radiant heating panels can be used as perimeter heating devices in conjunction with VAV systems, or as heating/cooling devices in conjunction with reduced flow ventilation systems. Used only for heating, radiant panels can be electric or hydronic. Electric panels use heating wire or tape on the back side of drywall or metal.

Hydronic radiant ceiling panels are generally constructed of aluminum with copper tubing bonded to the back surface as shown in Figure 3–28. In the heating mode the tubing carries hot water for perimeter heat. In the cooling mode, the tubing carries chilled water at a temperature controlled to prevent condensation on the surface of the panel. Room cooling results from convection and absorption of heat by radiant heat transfer from the room to the panel. Similar suspended devices are also available.

Systems in cold climates require the same attention to interior/perimeter zoning as conventional air systems. Perimeter piping needs to provide either heating or cooling. Piping serving interior zones need only cool when the building is occupied but may also need to heat to restore temperature after winter night setback.

Radiant cooling panels must be used in combination with ventilation air systems, which provide latent cooling and supplementary sensible cooling. With

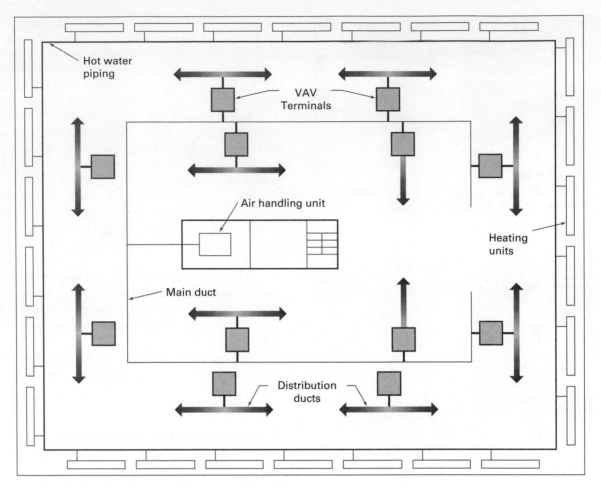

■ FIGURE 3–18

Multiple VAV interior and hot-water convector exterior zone system.

■ FIGURE 3–19

Construction of an electrical reheat coil in the ductwork. The coil may be a hot-water or an electric coil. (Courtesy: Titus.)

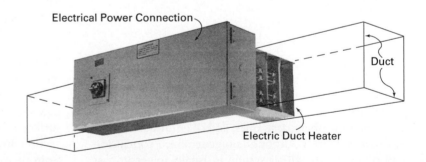

cooling capacities of 30 to 60 Btuh/sq ft of surface, radiant panels covering 15 to 30 percent of the ceiling can provide up to 100 percent of the sensible load for typical office space.

The advantage of radiant cooling panels is that they are energy-efficient, eliminating fan power for the portion of the load they satisfy. The ventilation air system can be designed with sufficient capacity only to accommodate code-required outside air rates based on occupancy plus sufficient air to pressurize the building. Attention to pressurization is critical to prevent infiltration, which might raise interior humidity and cause condensation on the panels. Radiant cooling systems are not appropriate for buildings with high ventilation needs, such as classroom buildings and laboratories.

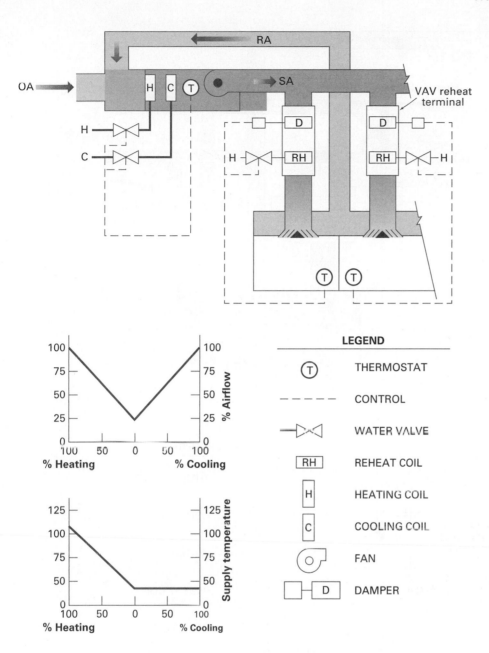

Variable air volume reheat system.

LEGEND

(T)	THERMOSTAT
– – – – –	CONTROL
▷◁	WATER VALVE
RH	REHEAT COIL
H	HEATING COIL
C	COOLING COIL
⊙	FAN
☐–D	DAMPER

Radiant panel systems can be comparable in cost to conventional VAV systems if all factors are considered. The high cost of panels, piping, and controls is offset by smaller air systems, smaller refrigeration equipment, and potentially lower building volume requirements because of the reduced need for air handling and transport. Energy savings, primarily by reducing fan power, are significant, and radiant systems can show a life-cycle cost advantage.

3.4.15 Radiant Floor Systems

Electric resistance cable and hot-water tubing (or glycol antifreeze solutions for outdoor applications) are used

in floor heating system. Outdoor applications are primarily for snow melting while indoor applications are for comfort heating. Most people who have experienced floor heating find it very comfortable due to the evenness of the heat and the absence of drafts that often result from air heating. The only drawbacks are high cost and slow control response. Slow response can be a problem when the heating is in a massive, concrete slab. Heating might be required in the morning; and, as the day warms up, spaces may overheat due to stored heat in the slab.

Chilled water can also be used in floor tubing to provide cooling. Floor cooling systems must be controlled to prevent condensation, which could result in

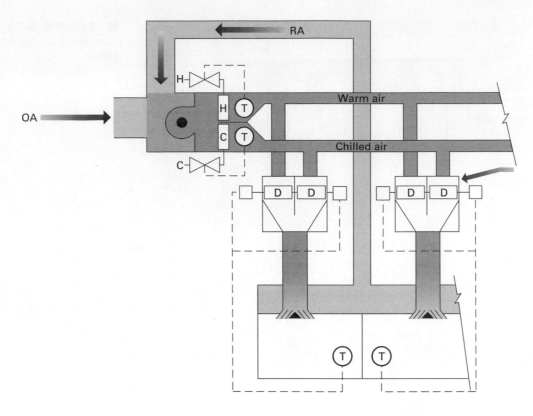

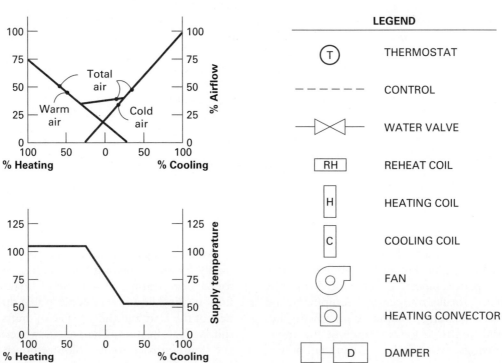

LEGEND

Ⓣ	THERMOSTAT
– – – – – –	CONTROL
⧓	WATER VALVE
[RH]	REHEAT COIL
[H]	HEATING COIL
[C]	COOLING COIL
⊙	FAN
⊙	HEATING CONVECTOR
□—[D]	DAMPER

■ **FIGURE 3–21**
Variable air volume dual-duct system.

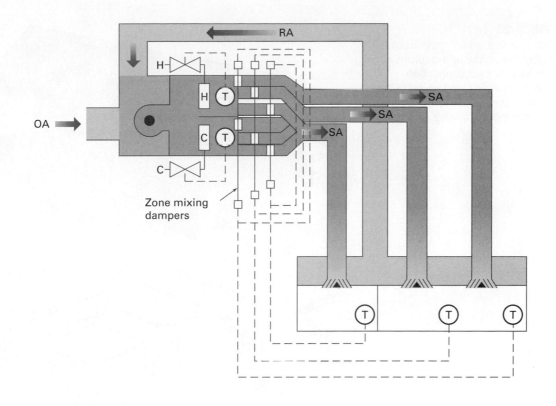

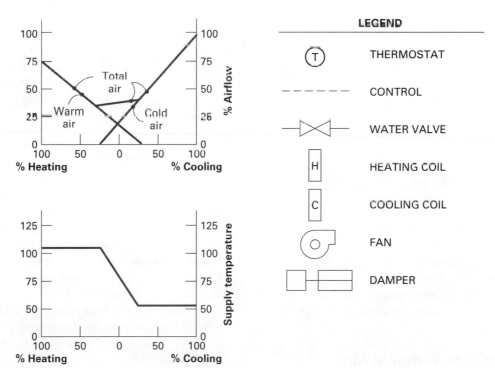

■ FIGURE 3–22
Variable air volume multizone system.

■ **FIGURE 3–23**
(a) Typical fan terminal, which contains a fan drawing air from the space or the return air plenum, a primary air duct connection (behind photo, not visible), and an air outlet. (b) Side and top view of a fan terminal box with arrows indicating the mixing of primary air and induced air. (Courtesy: Titus.)

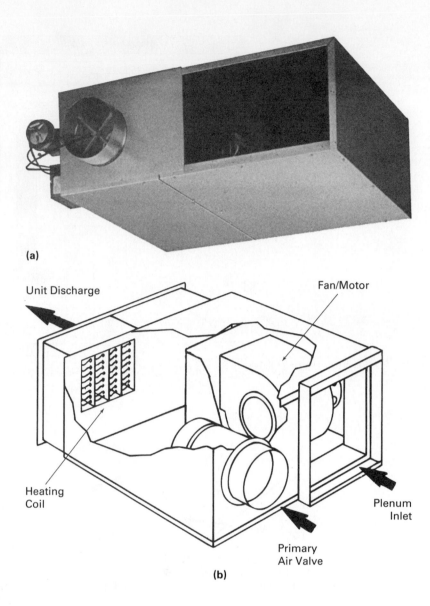

(a)

(b)

mold growth. Properly controlled, floor cooling systems are very energy efficient, since they do not rely on fan power to move cold air. A separate, small air system must, however, be provided for ventilation needs and for dehumidification. Floor cooling systems have capacity limitations and are appropriate only for spaces with relatively low loads, such as residential and light office.

3.4.16 Package Terminal Air Conditioners

Package terminal air conditioners (PTACs) are window or through-wall units containing their own compressors and air-cooled condensers. (See Figure 3–29.) Small package rooftop units with limited ductwork can also be classified as PTACs. PTACs can be installed with or

without electric heat and may be designed as heat pumps for more economical operation. PTACs are limited to rooms with outside exposure or spaces with suitable adjacent places to reject heat from refrigeration. PTACs tend to be noisy and require high maintenance as their compressors age.

Operating as a heat pump, the unit can reverse the refrigeration process to move heat from outdoors to indoors, thus acting as a heater instead of an air conditioner.

3.4.17 Water-Source Heat Pumps

Water-source heat pumps are unitary (contain compressors) terminal devices with reversible-cycle refrigeration, so that they can operate for either heating or cooling. They use refrigerant-to-air coils and water-to-refrigerant

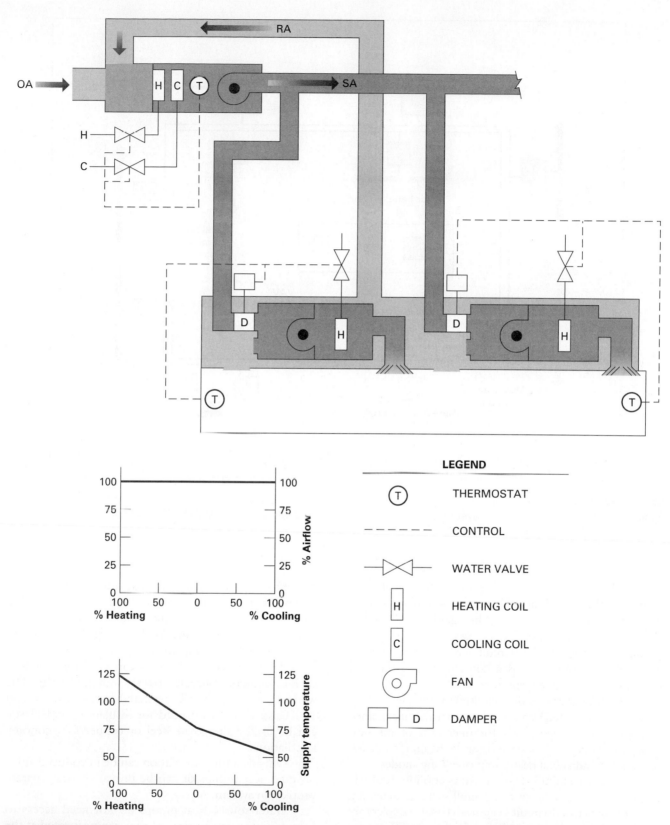

FIGURE 3–24
Fan terminal units.

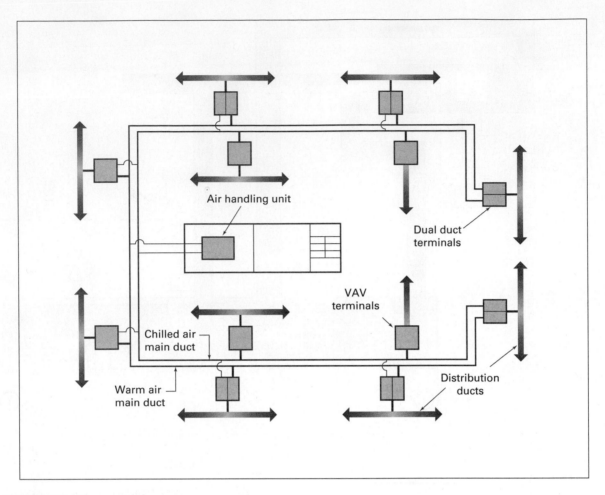

■ FIGURE 3–25

Typical layout of an air distribution system with single-duct VAV for the interior zones and dual-duct constant or variable air volume for the exterior zones.

heat exchangers. In the cooling mode the coil absorbs heat from the air to boil refrigerant. Suction gas is compressed in the compressor, and hot gas is condensed in the refrigerant-to-water heat exchanger. In the heating mode the refrigerant flow circuit is reversed by a reversing valve. The coil acts as a condenser, and the heat exchanger acts as an evaporator.

The heat exchangers of multiple water-source heat pumps are connected by a water loop, which is maintained at appropriate temperatures for operation of the heat pumps in the cooling mode and/or the heating mode depending on individual heat pump operating modes.

If the net HVAC requirement is cooling, heat will need to be rejected from the overall system. Generally, one or more closed-circuit evaporative water coolers are used to reject heat to outdoors. If the net HVAC requirement is heating, heat will be added to the water circuit by one or more boilers. During cold weather some of the heat pumps will require heating and some,

cooling owing to solar loads and internal heat gains. When this is the case, heat from the zones requiring cooling is transferred to the units requiring heating. This heat-reclaim feature is the energy advantage of water-source heat pump systems.

Water-source heat pumps can be floor-mounted cabinet units or concealed ceiling-mounted units. The water circuit is generally operated between 80°F and 110°F, so there is no need for insulation, and plastic can be used rather than steel or copper for economy installations.

Outside air for ventilation can be introduced at individual heat pumps or can be provided by a separate ventilation system.

Water-source heat pumps do not need access to outside air for condensing, so they can be located at the interior of the building. Small units are generally located above ceilings; larger units can be located in equipment rooms.

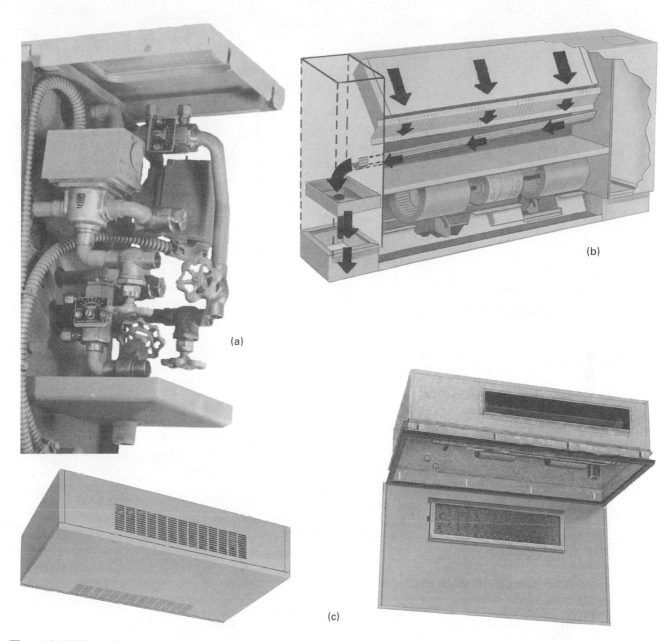

■ **FIGURE 3–26**

(a) Fan coil. A close-up view of the valve assembly showing control and shutoff valves. The small pan under the valves is the auxiliary drain pan, which receives collected condensate from the drain pan under the coil and condensate occurring on the valve assembly. Drain pans require periodic cleaning to avoid clogging. (b) A cutaway view of floor-mounted fan coil units with enclosed coil. The coil is totally enclosed for effective air distribution. (c) Ceiling-surface and ceiling-recess-mounted fan coil units. (Reproduced with permission from Airtherm Manufacturing Co.)

Condensing water for water-source heat pumps is produced by a central evaporative fluid cooler and circulated through the heat pump condensers to absorb heat generated by the air-conditioning process. A boiler is also included to provide heat for units operating in the heating mode by reversal of the refrigeration process. During winter, units serving interior spaces function in the cooling mode, and units serving perimeter spaces operate in the heating mode. Energy rejected from the cooling units is used to supplement heat from the boiler.

Water-cooled condensers are much more effective at heat transfer than air-cooled condensers. In addition, with proper control of the boiler and evaporative cooler refrigerant, condensing temperatures will be more

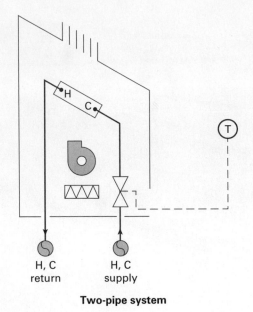

Two-pipe system

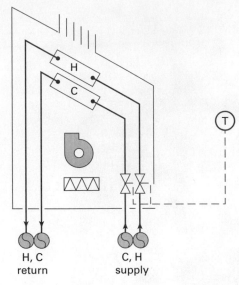

Four-pipe system

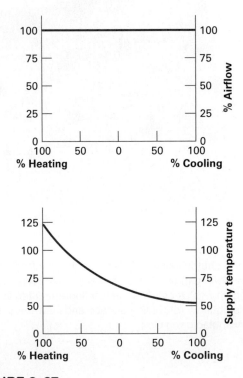

LEGEND

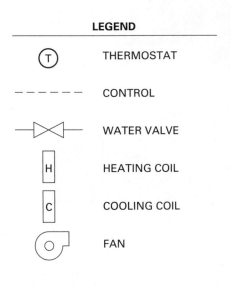

(T)	THERMOSTAT
– – – – –	CONTROL
▷◁	WATER VALVE
H	HEATING COIL
C	COOLING COIL
(fan)	FAN

■ **FIGURE 3–27**
Fan coil systems.

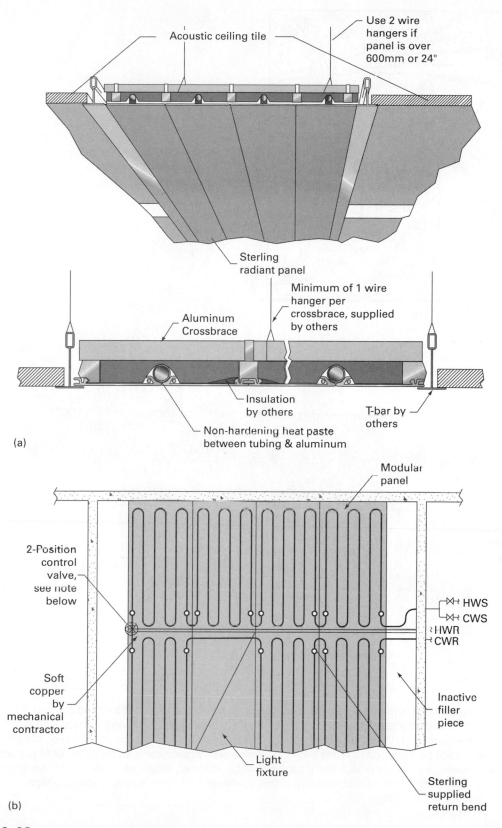

(a)

(b)

■ **FIGURE 3–28**

(a) Section through radiant panel shows water tube bonded to ceiling plate. Top would be insulated in installation.

(b) Radiant panels used to heat and cool at perimeter, cool only at interior.

103

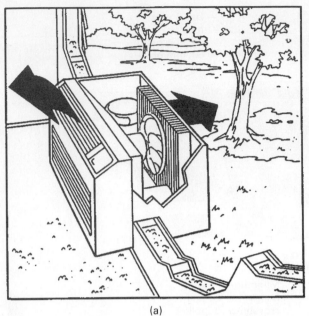

(a)

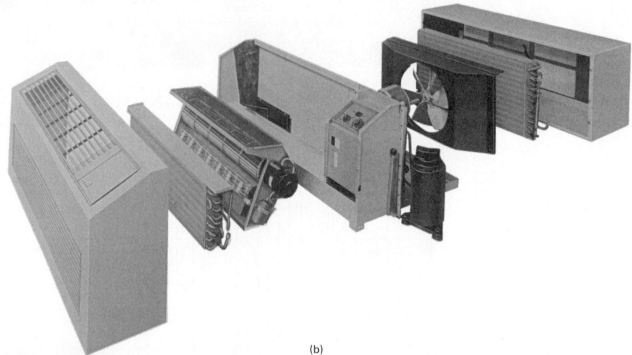

(b)

■ **FIGURE 3–29**

(a) Cutaway view of a package terminal air conditioner (PTAC) shows two sections: indoor and outdoor. The two sections should be completely insulated and separated from each other to increase efficiency and decrease noise. The indoor section contains the evaporator coil, fan, and controls. The outdoor section contains the compressor, condenser coil, and fan. A PTAC unit may also be designed to operate on the reverse refrigeration cycle as an air-to-air heat pump. (b) Exploded view of the PTAC showing all components. From left to right: indoor cabinet, evaporator coil, electric heating coil (optional), separating partition, refrigerant compressor, condenser fan, condenser coil, and outdoor cabinet. (Reproduced with permission from The Trane Company.)

moderate than with outside air. The result is that the compressors need not work as hard, and water-source heat pumps are more energy-efficient and have a longer life than air cooled PTACs.

Water-source heat pump systems are often used in office buildings, schools, and hotels as a low-cost solution for HVAC. Economy is achieved by combining heating and cooling piping into one system, which need not be insulated. Economy is also achieved by avoiding central air-handling equipment and space for fan rooms; however, a heat pump can be a high-maintenance system, since there is a refrigeration system, a heat exchanger, and controls with every unit. Noise from compressors and fans can also be a problem, especially for console units. Ducted concealed units present fewer acoustical problems if supply and return ductwork is designed to attenuate sound.

QUESTIONS

3.1 Describe the basic difference between interior and perimeter space with respect to HVAC loads.

3.2 What principles are used to vary the amount of HVAC service to a space with varying loads?

3.3 What is an HVAC zone, and how might a zone differ from a room?

3.4 Name the HVAC delivery systems that are most flexible for adding or rearranging zones. Why are they effective in this respect?

3.5 Which HVAC delivery systems are best for preventing high humidity in spaces? Why?

3.6 Which HVAC delivery systems are generally avoided because of their high energy usage?

3.7 Which HVAC delivery systems are most energy-efficient? Why?

3.8 Which HVAC delivery systems would be most appropriate in a building in which risk of water damage is a serious concern?

3.9 Which HVAC delivery systems would be most appropriate in a building in which intrusion for maintenance within occupied spaces must be kept at a minimum?

3.10 Of all the HVAC delivery systems discussed in this chapter, which is the most energy-efficient? Why?

3.11 Describe the major difference between electric or pneumatic control and direct digital control.

3.12 When is it appropriate to use two-position control?

3.13 When is it appropriate to use proportional control?

3.14 A hotel developer is deciding whether to use fan coils or PTAC units. How would you counsel him or her on matters of cost and quality?

3.15 You are helping an owner decide on systems for a new office building. For perimeter offices, VAV reheat or VAV with perimeter convectors are used. The contractor will provide either system for the same cost. How would you counsel the owner on matters of cost and quality?

3.16 You are helping an owner decide between VAV terminals and FTU terminals for interior conference rooms in a new office building. How would you counsel the owner on matters of cost and quality?

3.17 Why would you recommend the extra construction cost of a four-pipe fan coil system versus a two-pipe fan coil system?

3.18 What is the primary energy advantage of using radiant cooling?

3.19 How are water-source heat pumps and PTAC units similar? How are they different?

COOLING PRODUCTION EQUIPMENT AND SYSTEMS

4

COOLING SYSTEMS MAY BE DESIGNED USING ANY OF several methods for producing cooling, including vapor compression, absorption refrigeration, and evaporative cooling. Cooling is delivered for air conditioning by evaporative coolers, by direct refrigerant coils, or by chilled-water coils. Large buildings with central chilled-water plants require considerable attention to the placement and design of the major equipment room and heat rejection equipment. Guidelines are given along with examples at the end of this chapter.

4.1 REFRIGERATION CYCLES

Refrigeration is used to produce cooling. In doing so, it moves heat in a manner unlike any natural mode of heat transfer. Conduction, convection, and radiation move heat from high-temperature sources to low-temperature destinations—like a ball rolling downhill. Refrigeration does the opposite, moving heat from low-temperature sources to high-temperature destinations—like rolling a ball uphill. The ball rolls downhill naturally but requires pushing to go uphill. Similarly, refrigeration requires energy to counter the natural flow of heat.

4.1.1 Vapor Compression Cycle

The refrigeration process most often used in HVAC systems is called the *vapor compression cycle*. The process involves boiling and condensing fluids called refrigerants at temperatures that will produce a cooling effect. We are familiar with the process in other contexts. Consider a pot of water boiling on a stove. As the hot gas of the flame passes by the pot, heat is absorbed by the pot and used to boil the water. The water absorbs heat and boils, like a refrigerant; the gas, losing its heat, is cooled, like air across a cooling coil (see Figure 4–1).

Refrigeration involves different substances at different temperatures. Warm air passes across a cooling coil filled with a refrigerant. The heat of the warm air is absorbed to boil the refrigerant. The air is cooled and the refrigerant is boiled, or evaporated. The basic difference between boiling water on a stove and boiling a refrigerant for air conditioning is the temperatures at which they boil. Boiling water at 212°F can absorb heat and cool a gas stream starting at 1000°F. Boiling refrigerant at 45°F can absorb heat and cool an airstream starting at 80°F. The heat absorbed from the air boils the refrigerant, making it a vapor.

The vapor compression cycle returns the refrigerant gas to its liquid state to be boiled again in the coil. Returning gas to a liquid phase is termed *condensing* and is the opposite of boiling. At atmospheric pressure, steam absorbs heat and boils at 212°F and can be condensed back to water by removing heat at 212°F. The boiling or condensing temperature of a fluid depends on pressure: The higher the pressure, the higher will be the boiling or condensing temperature, and vice versa. Refrigerants that boil at low temperature and pressure to produce cooling can be made to condense at high temperature if the pressure is raised.

In the vapor compression refrigeration cycle, the compressor draws refrigerant from the evaporator at low pressure, resulting in a low boiling temperature. Refrigerant gas drawn by the suction of the compressor has its pressure raised at the discharge of the compressor. The refrigerant has a higher boiling point at higher pressure and can be condensed, releasing heat to another fluid at a temperature higher than that of the fluid that was cooled.

Temperatures for the process vary with specific applications. The refrigerant on the suction side of the compressor might be 40°F, which is cold enough to reduce 80°F air to 55°F. The 55°F air is suitable for cooling the space. At the discharge of the compressor, the boiling point of the refrigerant might be 130°F, owing to higher pressure; this is hot enough to release heat to 100°F outside air. Releasing heat from the compressed gas allows it to condense. Once condensed, the liquid refrigerant can be cycled back to the evaporator and used again for cooling.

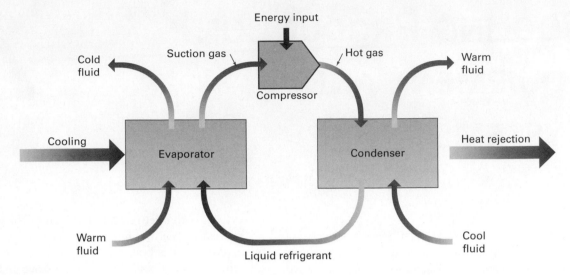

■ FIGURE 4–1
Vapor compression cycle.

Liquid refrigerant from the condenser is still at high pressure. The pressure is reduced at the inlet of the cooling device by an expansion valve or another throttling device, similar to a faucet, that allows a metered amount of fluid to pass from the high-pressure portion of the system to the low-pressure portion.

4.1.2 Absorption Refrigeration Cycle

The absorption refrigeration cycle uses water as a refrigerant. As with all refrigerants, the temperature of boiling and condensation will depend on pressure. At atmospheric pressure, water boils and condenses at 212°F. If the pressure is reduced to a low vacuum (measured in inches of mercury), water will boil and condense at temperatures low enough to produce cooling. At 5–7 mm Hg vacuum, water boils at 40°F and can be used as a refrigerant to produce chilled water or chilled air within the temperature ranges common in air-conditioning systems.

As in the vapor compression cycle, the refrigerant is boiled in an evaporator that absorbs heat from the fluid being cooled. The water vapor produced in the evaporator is conducted to a chamber that contains a strong solution of a hygroscopic chemical such as lithium bromide. The solution absorbs the water vapor, which then becomes liquid. This portion of the process is similar to condensation in the vapor compression cycle.

As the solution becomes more dilute, it must be regenerated by removing water. This is done in a concentrator chamber, which uses heat to boil off excess water and strengthen the solution. The source of the heat can be piped steam, hot water, or direct firing with

gas or oil. The concentrated solution must then be cooled for subsequent return to the absorber chamber, and the excess water that is driven off must be condensed for return to the evaporator. These cooling processes are generally accomplished using water from a cooling tower (see Figure 4–2).

4.1.3 Coefficient of Performance

The energy efficiency of refrigeration processes is measured as the coefficient of performance (COP), which is the ratio of the cooling effect divided by the power used to accomplish the process. The cooling effect is measured in British thermal units per hour (Btuh). Several forms of power might be used to accomplish cooling. Their units must be converted to Btuh for calculating the COP (see Figure 4–3).

Absorption refrigeration is less efficient than the vapor compression cycle. The COP of the absorption cycle ranges from 0.5 to 1.0, in comparison with values of 2.5 to almost 7 for vapor compression. Most of the energy used in the absorption process is heat, with a small amount of electricity used for pumps and accessories. Most vapor compression cycle machines rely solely on electricity as a source of energy for refrigeration. In general, electricity is a more expensive form of energy than heat produced by burning fuels, however, its COP is much higher.

Under some circumstances, absorption refrigeration can be economical despite its poor COP. Absorption is also the logical choice of cooling for systems that rely on waste heat from industrial processes or heat that

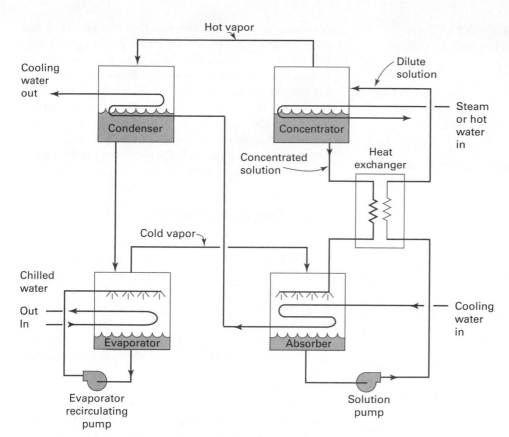

■ **FIGURE 4-2**
Operation of an absorption cycle for generation of chilled water.

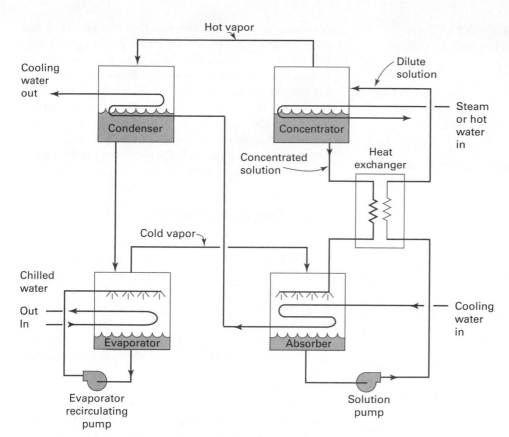

■ **FIGURE 4-3**
Definition of coefficient of performance: COP = cooling effect ÷ power input in constant units. For example, cooling effect = 10,000 Btuh; power input = 1 kW = 3414 Btuh; COP = 10,000/3414 = 2.6.

is a by-product of turbines or engines used to generate electricity.

Comparing the operating cost of gas versus electric cooling requires careful consideration of electric and gas rates. Electric cooling results in peak building electrical demands, and most electric utility rates have demand charges that cause the cost of electricity used for cooling to be much higher than electricity used for lighting, appliances, and ventilation. The actual cost of electricity used for cooling may be two to three times that amount if the effects of demand charges are included.

The opposite demand scenario is true of gas cooling. Gas-supply infrastructure in most climates is sized for winter heating peak, and there is excess capacity during the summer cooling season. Just as electricity suppliers charge more during summer based on capacity, gas suppliers charge less to stimulate sales and better utilize their capital investment. The cost of gas used for cooling may be 10 to 50 percent lower than the winter rate owing to market conditions and incentives.

Example 4.1 What are energy costs for a 500-ton system using electric chillers with total kW per ton of 0.8, assuming 1000 equivalent full load hours (EFLH) per year? What about a gas-fired absorption system using 15,000 Btuh (0.15 therm) per ton of gas and 0.3 kW per ton electric? Assume a cost of $0.15/kWh and $0.30/therm.

Electric chiller system usage = 0.8 kW/ton ×
500 tons × 1000 EFLH = 400,000 kWh

Annual cost = 400,000 kWh/yr × $0.15/kWh
= $60,000/yr

Absorption chiller gas usage = 0.15 therm/ton ×
500 tons = 1000 EFLH = 75,000 therms

Absorption system electricity usage = 0.3 kW/ton ×
500 tons × 1000 EFLH = 150,000 kWh

Annual cost = 75,000 therms × $0.30/therm +
150,000 kWh × $0.15/kWh = $45,000

4.1.4 Evaporative Cooling

Under favorable outdoor conditions, evaporative cooling is an economical alternative or supplement to vapor compression or absorption refrigeration. Evaporative air coolers, common in arid regions, pass dry outside air through a water spray or wetted medium (see Figure 4–4). The air is cooled and humidified for direct use in interior spaces or indirect use to cool another airstream via an air-to-air heat exchanger. The indirect method is preferred if the addition of humid outside air to the space is undesirable.

Evaporative cooling is also a part of many systems that use vapor compression or absorption refrigeration. A typical large system will include a cooling tower to produce condensing water during warm weather.

Evaporative cooling can also be used for making chilled water during cool weather. Many buildings and processes need air conditioning year-round, regardless of weather. During cool weather, the cooling tower will have sufficient capacity to generate water temperatures that are cold enough for air conditioning. The water can be filtered and used directly in the chilled-water system, or it can be used with a heat exchanger to transfer cold to the chilled-water system. The tower water system contains water that is exposed to oxygen and dirt, so

corrosion and fouling can result if the system is not properly tended. The use of a heat exchanger to keep this water out of the entire building system is more expensive but reduces the concern about damage. Cooling can also be generated by the direct migration of the refrigerant from the condenser coil to the evaporator coil when the temperature in the condenser is colder than the evaporator during cold weather. Figure 4–5 shows the various methods for producing chilled water by the evaporative cooling process.

4.1.5 Desiccant Cooling

Desiccant wheels are yet another option for using gas rather than electricity as a fuel for cooling systems. Desiccant wheels absorb humidity. The wheel is configured to rotate between two airstreams. In one of the airstreams, the wheel's desiccant coating absorbs humidity. In the other airstream, usually outside air heated by a burner or heating coil, the desiccant coating is dried out to release the moisture absorbed from the first airstream.

Desiccant wheels are capable of dehumidification, which is only one part of the cooling process. The other part, reduction in temperature, must be performed by another component in the system, namely, with conventional electric vapor compression, absorption, or even evaporative cooling.

4.1.6 Refrigerants

Refrigerants transfer heat (cooling) by changing phase from liquid to vapor. Absorption cooling, evaporative cooling, and desiccant cooling all rely on water as the refrigerant. The vapor compression cycle uses other chemicals selected for their physical and thermodynamic properties, such as latent heat of vaporization, thermal

■ **FIGURE 4–4**
Evaporative cooling.

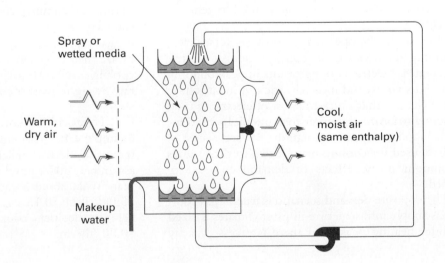

Spray or wetted media

Warm, dry air

Cool, moist air (same enthalpy)

Makeup water

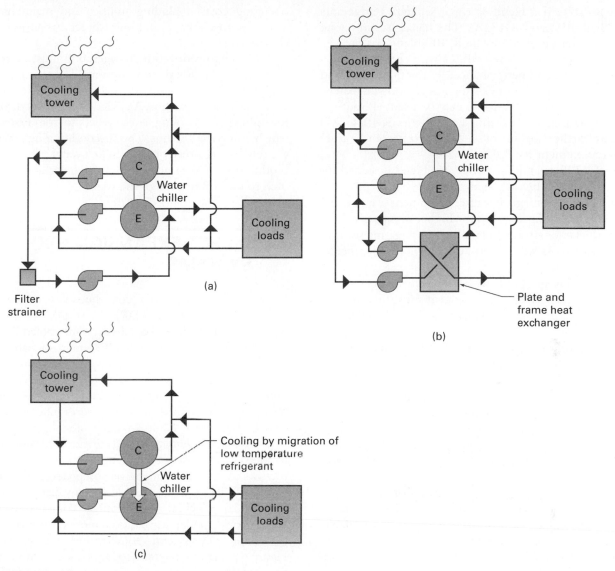

■ FIGURE 4–5

Three chilled-water economizer cycles:
(a) Condenser water as chilled water.
(b) Heat exchange between condensing water and evaporator (chilled) water.
(c) Heat exchange between condensing water and refrigerant in the evaporator.

conductivity, viscosity, and boiling pressures and temperatures. Chemical stability, flammability, and toxicity are also important, along with compatibility with compressor seals and lubricants.

In the late 1960s long-term effects of refrigerants on the ozone layer were discovered, which eventually led to the phaseout of halogenated compounds such as chlorofluorocarbons (CFCs) including R-11 and R-12. The compounds are very efficient refrigerants, and they are also very stable, which translates into longlasting harm. These chemicals, regulated by the Montreal Protocol of 1987, are no longer produced in developed countries.

Hydrochlorofluorocarbons (HCFCs), such as R-22, another very popular refrigerant, are less stable, and their harmful effects are not as longlasting. R-22 is used in direct expansion coils for air-cooling applications. R-22 is a high-pressure refrigerant, which results in compact, economical equipment. Due to its ozone depletion potential, R-22 is scheduled for phaseout in 2020. R-407C and R-410A are two candidates being developed for replacing the market for R-22.

R-407C is a blend of three separate refrigerants (R-32, R-125, and R-134A). The blend has an ozone depletion rating of zero. The R-407 blend has properties very similar to those of R-22 with respect to pressure, capacity, and performance. In addition, the blend is compatible with most existing gasket and seal materials. This means that R-407C may be used without extensive redesign of existing equipment. R-407C will require further applications research to become a practical replacement for R-22.

R-410A is another refrigerant blend designed for replacing R-22, but it operates at even higher pressure. This is an advantage because equipment can be designed more compactly, but the higher pressures will require thicker or stronger components, negating some of the savings. As with R-407C, much developmental work will be required.

R-123 is an HCFC with lower environmental impact than R-22. R-123 is the current substitute for low-pressure water chillers, which used R-11 before its phaseout.

Hydrofluorocarbon (HFC) refrigerants are not regulated by the Montreal Protocol. The most popular is R-134A, used for most medium-pressure refrigeration equipment. R-134A was devised as a replacement for R-12 or R-500, which are obsolete owing to environmental concerns. R-134A was developed for its low ozone depletion potential; however, it has a significant global warming potential. With current concerns about global warming, there is reason to question whether R-134A will be a good long-term solution.

Other refrigerants with good thermodynamic properties and no environmental impact are available, but generally, they are inappropriate for use in buildings owing to flammability or toxicity. Ammonia has historically been the refrigerant of choice for industrial applications, such as in breweries and dairies. Ammonia's thermal performance is excellent, and it is very inexpensive compared with other refrigerants, however, ammonia is extremely toxic and is classified by code as a flammable refrigerant. For these reasons, the best performing and most environmentally friendly refrigerant (besides water!) cannot be used in buildings. Potentially, however, ammonia could be used in large central cooling plants making chilled water for occupied buildings.

4.2 COOLING PRODUCTION EQUIPMENT

The selection of a cooling production and distribution system will depend on the size of the project; the budget; and concerns about costs over the life cycle and operating costs, including utilities and maintenance. There are two basic options: direct expansion and chilled water.

Cooling produced by refrigeration must be transferred to a space for air conditioning. The transfer is accomplished at a cooling coil, which is constructed of tubes surrounded by fins. A chilled fluid passes through the tubes, and cooling is conveyed to the airstream, which flows over the fins. The air-conditioning coil may use refrigerant directly or use chilled water. The latter is produced by refrigerant in a water chiller and distributed to the coil or coils of the system.

4.3 DIRECT EXPANSION (DX) SYSTEMS

If refrigerant is used in the coil, the system is generally termed *direct expansion* (DX). Most DX systems are equipped with air-cooled condensers. Air-cooled DX systems with reciprocating or scroll compressors are commonly used for applications involving small tonnage or multiple small packages. These systems can be very economical at first; however, their power demand is high, owing to high condensing temperatures inherent in air-cooled condensers. Chilled-water systems generally use water-cooled condensers, which are more efficient owing to a lower condensing temperature. Reciprocating compressors are least efficient. Scroll compressors are a more energy-efficient option for small systems. Larger DX systems utilizing helical or centrifugal compressors with water-cooled condensing are even more energy-efficient.

Basic DX system components are shown in Figure 4–6. DX refrigerant coils are used in most small air-conditioning equipment and can be used in large equipment. If the compressor is included with the equipment, the resulting device is termed *unitary*. Unitary devices include small-package terminal units such as window air conditioners and larger units such as rooftop air conditioners.

If the compressor is remote, and refrigerant is piped to the unit, the resulting system is termed a *split system*. Most residences and many light commercial facilities use split systems. The cooling coil or evaporator coil is contained in an air-handling unit or attached to a furnace. The compressor is placed in a package outside at grade or on the roof. The condenser is generally in the same package, and the assembly is called a *condensing unit*.

Some systems include the compressor with the air-handling equipment and use a remote condenser. This relationship between the coil and the compressor is termed unitary, rather than split, despite the fact that the components are separated. The terminology is illustrated in Figure 4–7.

DX systems generally use compressor equipment at each air-conditioning system. If the associated condensers are air cooled, the systems must be in close proximity to the outdoors. Otherwise, long runs of refrigerant piping will create pressure drops, large volumes of expensive refrigerant, and the potential for costly leaks.

For this reason, DX systems with air-cooled condensers are generally used only for small air-conditioning applications, such as in residences. Larger DX equipment is practical only if it can be conveniently located outside, as rooftop units for low-rise commercial buildings can.

DX systems are frequently used for individual tenants in large building complexes, such as shopping malls, apartments, and low-rise buildings. Having a

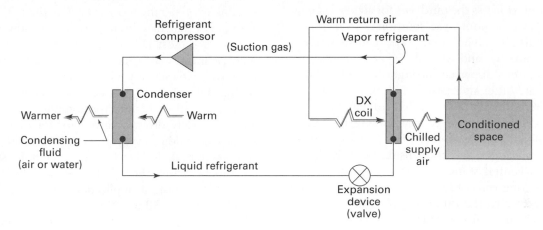

■ FIGURE 4–6
Basic DX system components.

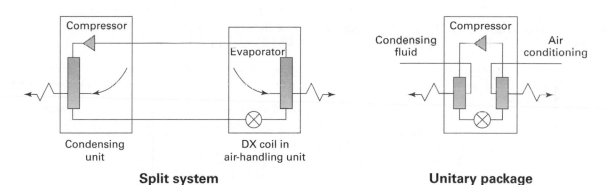

Split system

Unitary package

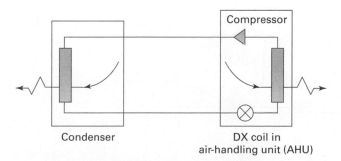

Unitary with remote condenser

■ FIGURE 4–7
Alternative DX system configurations (fans not shown).

separate system for each tenant is generally less expensive than having a central system and allows separate metering and flexible operating schedules, however, maintenance is higher, and smaller, separate systems are not as energy-efficient. Some DX systems (window air conditioners, heat pumps, or through-the-wall units) require openings in the windows or walls; such systems are not aesthetically desirable and can be very noisy.

For massive buildings or high-rise projects, it may be impractical to access the outdoors for air-cooled condensers. Most commonly, the solution is to distribute cooling by using chilled water. The chilled water is produced centrally and distributed to various air-conditioning units dispersed throughout the building.

There are other alternatives. For example, unitary DX equipment can use water-cooled condensers. Cooling water is furnished by a central cooling tower. For this type of system, the condenser is a refrigerant-to-water heat exchanger, and condensing water is distributed from a central source. This concept avoids the necessity for long runs of refrigerant when DX systems have poor access to the outdoors. In addition, water-cooled DX is more efficient than air-cooled DX.

4.4 CHILLED-WATER SYSTEMS

Water is a low-cost thermal transfer medium that can be pumped from a central cooling plant to dispersed

air-conditioning equipment in a large facility. Water chillers may utilize vapor compression or absorption cycle refrigeration, and condensing may be accomplished by air cooling or water cooling. Basic components of chilled-water systems are illustrated in Figure 4–8.

4.4.1 Chilled Water versus DX

On the basis of total refrigeration capacity, chilled-water systems are generally more expensive than DX but have advantages that may be important for individual projects. Unlike most DX systems, water chillers can serve multiple individual air-conditioning units. Since not all units will experience peak load at the same time, the central chilled-water system can be sized for the net demand load rather than for the accumulated peak design loads. Typically, diversity factors for commercial and institutional buildings vary between 70 and 90 percent of combined peak loads. This can result in substantial equipment savings through the use of chilled-water systems. For industrial applications, the diversity factor may be as low as 30 to 50 percent.

Chilled-water systems can be quieter and less visible than DX systems. Water chillers and heat rejection equipment are generally located in a mechanical area remote from the air-handling equipment. This confines the noisiest and most visually objectionable components of the system to a space that can be isolated from occupied areas of the building.

■ **FIGURE 4–8**
Basic chilled-water system.

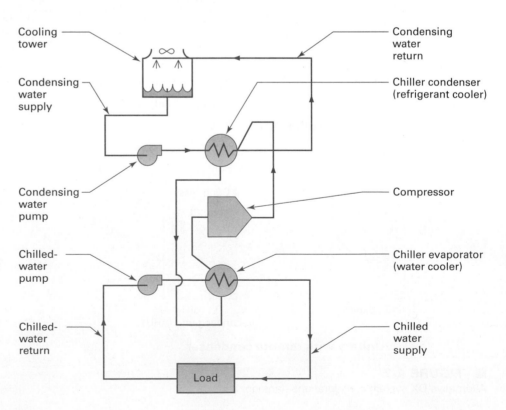

Controlling leaks of refrigerant is an important environmental and safety issue. Chilled-water systems confine the refrigerant to fewer pieces of equipment, so they are less subject to leaks. DX systems, which generally have extensive external refrigerant piping and valves between the compressor, condenser, and evaporator, are more susceptible to leaks. Leak resistance enhances the reliability of chilled-water refrigeration, as well as reducing the likelihood of costly recharging of the refrigerant.

Chilled-water systems are more flexible than DX systems. Chilled-water distribution can be considered an in-building utility, which can be easily tapped for the addition or rearrangement of air-conditioning units. This is extremely important for facilities that undergo frequent remodeling. High-technology buildings in health care, research, and manufacturing will greatly benefit from the flexibility that is inherent in chilled-water systems.

Centralized chilled-water plants are easier to maintain than decentralized or distributed plants, owing to their convenient accessibility for maintenance personnel. This is especially important if security is an issue. With a centralized plant, maintenance personnel and outside contractors can be restricted to central plant areas, which are secured from other areas.

4.4.2 Vapor Compression Chillers

Chillers may operate on either the vapor compression or the absorption cycle. Characteristics of water chillers are summarized in Table 4–1.

When a liquid refrigerant is evaporated or vaporized, it absorbs heat from the surrounding medium. Conversely, when a vapor refrigerant is condensed, it releases heat. Heat from the building (cooling load) is absorbed by the refrigerant in the evaporator, transported by the compressor, and rejected by the condenser.

The evaporator and condenser are usually of the shell-and-tube design, with water circulating through the tubes. In the evaporator, heat to vaporize the liquid refrigerant is absorbed from the circulated chilled water. In the condenser, heat to condense the gaseous refrigerant is absorbed by water from a cooling tower or another source. Air-cooled condensers are sometimes used if water is a scarce commodity or if energy efficiency is not a high priority, or if the ambient air temperature is low.

Chillers operating on the vapor compression cycle are classified according to the type of compressor they use, including reciprocating, rotary (scroll or helical), and centrifugal compressors. Chillers of these basic types are illustrated in Figures 4–9 through 4–11.

Reciprocating Chillers

Reciprocating chillers are among the oldest type of direct expansion-type heat transfer devices and use positive displacement compressors, which may contain two or more cylinders. Compressors operate above atmospheric pressure. Due to inherently high losses associated with reciprocating motion, this type of compressor is generally noisier and uses more power than rotary-type compressors.

Capacity control is achieved by cycling compressors or deactivating cylinders in steps. Cycling and step control will result in varying supply water temperatures, and reciprocating chillers are not appropriate if tight temperature control is required.

Most major manufacturers no longer make reciprocating chillers as they have been replaced by more efficient rotary (helical screw) and cost-effective scroll-type compressors.

Hermetic Scroll Chillers

Hermetic scroll compressor chillers generally range from 10 to 150 tons cooling capacity and can be either water-cooled air-cooled remote, evaporative-cooled remote, or of the air-cooled packaged variety. The compressor is centrifugal in nature and functions similar to a rotary-type Wankel combustion engine in that a rotating eccentric fixture compresses the refrigerant vapor against a stationary elliptical nonmoving assembly. The principal advantages are multiple circuiting, cost-effective and flexible 5- to 25-ton modular compressor sizes, longer component life than reciprocating chillers, lower compressor sound levels, compact footprint, and reasonably efficient performance. Also, scroll compressors have been upgraded to accept both R-134A and R-410A from the new families of refrigerants.

Rotary Helical Chillers

Rotary helical chillers use screw-type compressors, which consist of either one or two intermeshing helical grooved rotors. Operation of this compressor type is relatively vibration-free. Operating pressure is usually above atmospheric pressure. Capacity control is achieved by altering the length of rotors available for service using a slide valve or six-step-type control. Unlike scroll and reciprocating chillers, some model rotary helical chillers can offer continuous modulation rather than only discreet stepped-type control. The refrigerant can operate at high condensing temperatures (up to 220°F). Accordingly, helical chillers are well suited for heat recovery applications.

TABLE 4–1
Characteristics of water chillers, typical refrigerants, operating pressures, and COP

Type	Speed, rpm	Capacity Range, tons	Typical Type	Evaporator Pressure, psia	Condensing Pressure, psia	Condensing Media	Theoretical Performance, COP*	Power Demand, kW/ton	Applications
Reciprocating									
Semihermetic	1750	20–200	HCFC-22	70	211	Water	4.7	0.7–0.8	Commercial/Industrial cooling systems up to 200 tons with water or air-cooled condensers
Open	1750	20–150	HCFC-22	70	211	Water	4.7		
Semihermetic	1750	20–200	HCFC-22	62	278	Air	4.7	1.2–1.4	
Open	1750	20–100	HFC-134a	30	186	Air	4.4		
Rotary (Scroll & Helical)									
Hermetic	3500	70–450	HCFC-22	70	211	Water	4.7	0.6–0.8	Commercial cooling systems
Open	3500	100–450	HFC-134a	10	136	Water	4.4	0.6–0.8	Industrial cooling systems
			HCFC-22	70	211	Water	4.7	0.6–0.8	
			HFC-134a	10		Water	4.4		
Hermetic	3500	20–400	HCFC-22	70	278	Air	4.7	1.2–1.4	
Centrifugal									
Direct-driven hermetic	3500	100–3900	HCFC-123	−4.5	8	Water	4.8	0.5–0.65	Commercial/Industrial cooling systems over 100 tons
Open gear drive	5000–7500	150–8000	HCFC-123	−4.5	8	Water	4.8	0.55–0.7	Commercial/Industrial cooling systems over 150 tons
Hermetic gear drive	6000–20,000	100–2000	HFC-134a	10	112	Water	4.4	0.55–0.7	Commercial/Industrial cooling systems over 100 tons
			HCFC-123	−4.5	8	Water	4.8		
			HFC-134a	10	112	Water	4.4		
Absorption									
Direct-fired	—	100–1500	Water	6 mm Hg absolute	50 mm Hg absolute	Water	1.0	—	Commercial/Industrial cooling: some units can be equipped with heat exchangers for generating or heating hot water
Indirect-fired (Steam, hot water)	—	75–2000	Water	7 mm Hg absolute	7.5 mm Hg absolute	Water	0.8	—	Commercial/Industrial cooling: where waste heat or steam is available
						Two-stage	1.2	—	
						Single-stage	0.6	—	

*COP of indirect-fired absorption units does not include boiler efficiency.

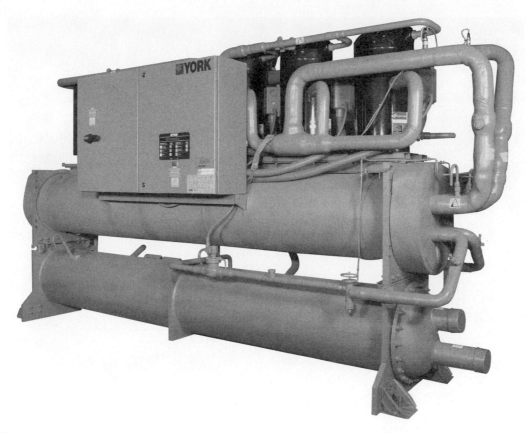

■ **FIGURE 4–9**
Packaged chiller with rotary scroll compressor, evaporator, and condenser assembly. Capacity is 20 to 60 tons. (Courtesy of Johnson Controls, Inc.)

Centrifugal Chillers

Centrifugal chillers use turbocompressors with impellers that operate on a centrifugal principle. Volumetric capacities are higher than those of equivalent-sized reciprocating-type compressors because the flow of the refrigerant is continuous rather than intermittent. There is very little vibration and only minimum wear due to the nonpulsating gas flow and minimum contacting surfaces. Capacity control is most often achieved by modulation of one or more sets of inlet vanes at the inlet of the compressor in concert with a motor speed modulating-type device.

4.4.3 Design Variations

Chillers have several design variations:

- Open versus hermetic
- Direct versus gear drive
- Refrigerant operating pressure

Open-drive chillers differ from hermetic in the placement of the compressor motor. The motor of an open-type chiller is externally coupled to the compressor, while the motor of a hermetic-type unit is housed with the compressor in a common enclosure and cooled by the surrounding refrigerant.

Motors may drive compressors directly or by gears. Direct-drive machines can have one, two, or three stage-type compressors operating at 3600 rpm. These units have fewer moving parts than gear-driven units, which operate through gear trains at speeds ranging from 6000 to 30,000 rpm. High-speed, gear-driven units have been improved over the years, and many of the early objections such as reliability, shaft seal leaks, etc., have been overcome. Gear-driven machines offer the economic advantage of providing a series of available capacities with fewer basic compressor sizes and are more adaptable to the new ozone-free families of refrigerants.

Centrifugal chillers may use refrigerants that operate at pressures above or below atmospheric pressure. Chillers that use refrigerants operating below atmospheric pressure are susceptible to air leaks into the refrigerant circuit. Air in the system can cause serious compressor surging and loss of capacity. For this reason,

■ **FIGURE 4–10**
Packaged chiller assembly with reciprocating compressor. Capacity is 70 to 120 tons. (Courtesy of Johnson Controls, Inc.)

■ **FIGURE 4–11**
Packaged chiller with cutaway view of the helical-rotary compressor. Capacity is from 100 to 450 tons. Large chillers up to 5000 tons are made with multistage centrifugal compressors. (Courtesy of Johnson Controls, Inc.)

there must be a provision for air purging. A small amount of refrigerant is lost along with the air. Due to concern for CFC in the environment, purge units on chillers must be designed for very low refrigerant losses.

With refrigerants that operate above atmospheric pressure, refrigerant leaks can occur. Even so, there is no opportunity for air to leak into a unit. Therefore, an air purge system is not required. It is necessary, however, to have a built-in cycle arranged to store the refrigerant in separate pressurized drums whenever the chiller must be evacuated for maintenance.

4.4.4 Absorption Chillers

Chillers based on absorption refrigeration use water as the refrigerant. In lieu of using a compressor to facilitate alternate evaporating and condensing, as in the vapor compression cycle, the absorption cycle depends on operating the unit at near perfect vacuum conditions and on the affinity lithium bromide has for water. Absorption equipment is factory-evacuated and sealed to preserve the required vacuum for proper operation.

Depending on the source of applied heat to regenerate the lithium bromide solution, absorption chillers may be classified as indirect- or direct-fired. Figure 4–12 shows typical absorption chillers. Figure 4–13 shows the operating features of a single-stage absorption chiller.

Indirect-fired chillers are available up to 2000 tons for use with 5- to 150-psi steam or with 150° to 400°F hot water as the energy source. Direct-fired chillers, available up to 1500 tons, use natural gas combustion or hot-process waste gas as the energy source.

The cooling tower capacity for absorption refrigeration must be adequate to reject the heat absorbed as a result of the refrigeration effect, plus the heat added to the system for the purpose of regeneration. The rejected heat is approximately double that of a vapor compression cycle.

An absorption chiller operates in an extremely high vacuum. Proper operation is dependent on a very tight system to prevent the infiltration of air into the unit. Special care and procedures are implemented during the manufacture and operation of such a chiller.

Lithium bromide is a salt and will crystallize when it is in concentrated form and subcooled. Crystallization will occur whenever the source of heat is interrupted during a normal operating cycle or at very low load conditions. Safety controls are normally incorporated to prevent crystallization. The plant operator should be aware of the abnormal conditions that cause crystallization. Once a unit has crystallized, elaborate procedures are required for decrystallization, and downtime may be as much as two or three days.

Lithium bromide is highly corrosive to some materials, so caution must be exercised in the design and operation of absorption units. Some manufacturers depend entirely on corrosion-resistant materials. Others depend on both corrosion-resistant materials and an inhibitor. When an inhibited solution is required, it is important that the concentration of the inhibitor be periodically checked and maintained.

The COP of an absorption chiller is considerably lower than that of a vapor compression chiller. The COP varies from 0.5 for small, single-stage, indirect-fired absorption chillers to 0.8 for large, two-stage, indirect-fired absorption chillers. A two-stage, direct-fired unit has a COP of approximately 1.0. The nominal COP values for vapor compression chillers range between 2.5 and 7.0.

4.4.5 Engine-Driven Chillers

Using a fossil-fuel-fired or steam-driven prime mover (engine) for vapor compression–cycle chillers is an alternative to using absorption for reducing reliance on electricity for cooling. Engines can be turbine or reciprocating using steam, liquid, or natural gas. The shown in Figure 4–14 uses a material gas fueled V-8. These chillers are often used in heat-recovery applications, owing to the high temperature of their exhausts and the potential for heat recovery from jacket cooling of reciprocating engines.

4.5 HEAT REJECTION FROM COOLING SYSTEMS TO THE ENVIRONMENT

Air-conditioning systems use refrigeration processes to move heat from the indoor to the outdoor environment. The refrigeration cycle absorbs heat by the evaporation of liquid refrigerant in the evaporator (indoor coil), and rejects heat by the condensation of vapor refrigerant in the condenser (outdoor coil). The heat rejected from a refrigeration system is the sum of the actual cooling load and the energy added by the refrigeration equipment.

The condenser may use either water (produced by a cooling tower) or ambient air as the heat rejection medium. The lower the heat rejection temperature, the less power (and energy) is required by the refrigerant compressor. Since cooling tower water is normally cooler than air during the summer, water-cooled condensers are generally more energy-efficient.

Performance characteristics and applications of mechanical heat rejection equipment are given in

(a)

(b)

(c)

■ **FIGURE 4–12**

(a) Direct-fired absorption chiller. (b) Single-stage indirect-fired absorption chiller. (c) Two-stage indirect absorption chiller. Two-stage absorption chillers have a higher COP than the single-stage design. (Courtesy of Johnson Controls, Inc.)

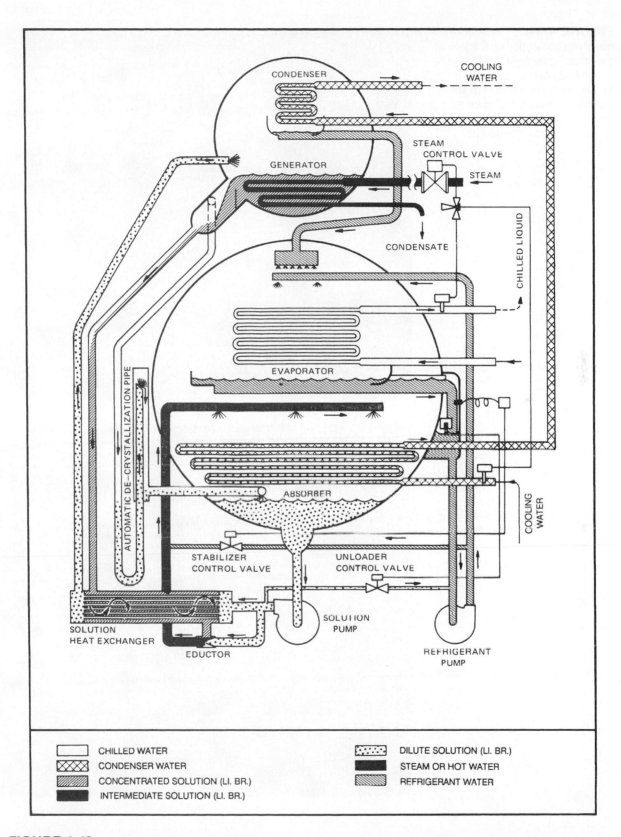

■ **FIGURE 4–13**

Operation of a single-stage indirect-absorption chiller. (Reproduced with permission from York International.)

■ **FIGURE 4–14**
Engine-driven chiller (far unit, figure a) using a natural gas–fired V-8 (figure b) to drive an R-22 compressor. This machine is used in combination with electric chillers (near unit, figure a), sequenced to cover peak loads and to reduce electric demand charges.

(a)

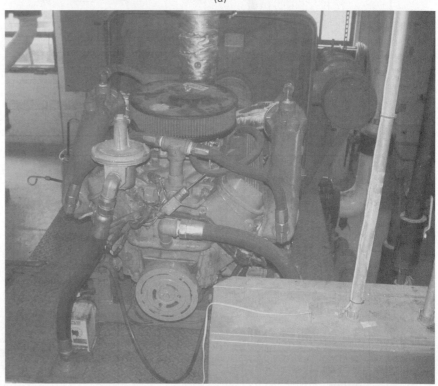

(b)

Table 4–2. The indicated data are useful for preliminary evaluation but should not be used for sizing equipment, owing to substantial variations among manufacturers.

4.5.1 Air-Cooled Systems

Air-cooled condensers use air to cool and condense refrigerant gas to the liquid state. These systems normally require 600 to 900 cfm of air per ton of refrigeration. Condensing temperatures range from 115° to 140°F, depending on climatic conditions. In general, higher condensing temperatures will increase the power requirement and decrease the cooling system's performance. Despite their higher energy cost, air-cooled systems offer several advantages over water-cooled systems:

- No problem of water freezing
- No water treatment required
- No concern about water drift
- No condensing water pump required
- Lower initial cost

Air cooling of condensers may be appropriate for small systems that do not warrant the complexity and expense of water systems or in areas where energy is more available than water. A typical rooftop-mounted air-handling unit with air-cooled refrigeration system is shown in Figure 4–15.

4.5.2 Water-Cooled System

Sources of condensing water include spray ponds, domestic supply systems, wells, surface water, and cooling towers, which are the most common solution. Water from these sources is lower in temperature than air is during hot weather. Lower condensing temperatures

TABLE 4–2
Characteristics of heat rejection equipment

Type	Capacity Range, tons	Airflow, cfm per ton	kW Input per mm Btuh	Applications
Air-cooled				
Air-cooled condensers (propeller fans)	3–150 (1)	500–700	9.0–12.0	Commercial cooling systems where towers are not practical and for year-round systems where freezing of cooling towers is difficult to control, as in extreme climates.
Dry Coolers (propeller fans)	3–65 (2)	1000–1400	10.0–13.0	Can be used to cool condenser water or to directly cool chilled water in northern climates; not economical in southern climates.
Water-cooled				
Cooling towers				
Packaged induced draft (propeller fan)	5–1600 (3)	200–250	2.0–3.0	Ideal for small to large cooling plants. Somewhat less expensive than built-up towers, but usually have a shorter life span.
Packaged forced draft (centrifugal fan)	10–400 (3)	200–250	4.0–6.0	Application to commercial and industrial cooling systems; especially adapted to indoor and restricted outdoor installations; very compact units, relatively quiet.
Field-erected induced draft (propeller fans)	200–1800 (3)	200–250	1.5–2.0	For use with medium to large water-cooled systems. Towers can be built up to 20,000 tons per cell; however, the most commonly used preengineered sizes range from 200 to 1800 tons.
Water-spray type of fluid coolers (centrifugal fans)	5–150 (3)	500–700	14–18	Minimizes water treatment requirements. Eliminates condenser freeze protection when used with water glycol solution. Can be used for direct chilled-water cooling during cold weather.

(1) Larger units available when factory coupled to air-cooled chillers. Data based on 95°F ambient, 115°F condensing temperature.
(2) Data based on cooling condenser containing water glycol solution from 100° to 110°F at 95°F ambient.
(3) Based on one cell; multiple cells available. Data based on cooling from 95° to 85°F at 78°F WB.

■ **FIGURE 4–15**
Typical rooftop model of an air-cooled cooling and heating unit consisting of all indoor and outdoor components. (Reproduced with permission from The Trane Company.)

allow higher refrigeration efficiency, and water is generally preferred to air on this basis.

Using domestic water in condensers is wasteful and expensive. This method is undertaken only for small systems as a convenience when other options are not practical.

Spray ponds are used in lieu of cooling towers when the site conditions are favorable, however, spray ponds are more costly than cooling towers and require large surface areas (approximately 20 to 30 sq ft/ton). Ponds can be aesthetically pleasing when properly designed, but maintenance is high.

Wells and river water are practical sources of cooling water when they are readily available. Shipboard installations offer a good example of this application. Obviously, the cost of cooling towers and similar devices is eliminated. It is necessary, however, to provide more elaborate water treatment and filtration facilities than with conventional heat rejection methods.

4.5.3 Cooling Towers

Used in most applications, cooling towers produce water at an appropriate temperature for condensing by evaporation. A portion of the circulated water is evaporated in the cooling tower to cool the remainder, thus minimizing the use of water and any associated expense for processing. Approximately 7 percent of the circulation rate is evaporated and lost to bleed-off in the tower. New water is introduced to make up for that which is lost. Evaporation is achieved by flowing the water over a fill material in the tower, which is designed to effect good contact between water and air.

The typical, economical cooling tower is capable of producing water within 5°F of the ambient wet-bulb temperature. In typical temperate climates, this results in water temperatures about 10°F cooler than ambient dry-bulb temperature.

Selecting a Tower

Selection of the right cooling tower depends on many factors, including the available space, access for installation and possible replacement of the tower, the initial budget, and expected costs over the life cycle, particularly energy, maintenance, and replacement costs.

Airflow

There are three basic operating concepts for moving air through a cooling tower: gravity draft, induced draft, and forced draft. Some larger towers used primarily in utility power plants create a chimney effect to make the air flow past the water. These towers, generally hyperbolic in configuration, are bulky and tall and thus are not used in air-conditioning applications.

Towers used in building applications generally have fans. The fans can be located to blow air through the fill, an arrangement termed *forced draft*; or they can be located to draw air, an arrangement termed *induced draft*. Air can enter on one side of the tower, termed *single inlet*; or two sides, termed *double inlet*. Air discharge can be vertical or horizontal.

Airflow may be horizontal or vertical through the fill. Horizontal airflow results in a crossflow of air and water; vertical airflow results in a counterflow. (See Figure 4–16.)

Placement of Cooling Tower

Free circulation of air is needed for efficient operation of the tower. Discharge from the tower should be unimpeded, and any obstructions at the inlet should be held at a distance. A general rule for induced-draft towers is to allow a clear air passage equal to the tower intake height; a general rule for forced-draft towers is to allow air passage space equal to 1.5 times the tower width. Violations of these guidelines can cause recirculation of humid discharge into the inlet or impede the airflow. Both effects will degrade the tower's performance, resulting in

the need for a larger tower or more powerful fans. Airflow at the inlet must be considered in the placement of towers and design of visual screening elements.

Discharge air from towers is humid and can contain droplets of water with high concentrations of minerals and water treatment chemicals. Surfaces exposed to tower water carryover may become corroded, soiled, or discolored. Wind can blow humid air and water droplets horizontally from the tower discharge or through the tower body. This phenomenon is most likely during conditions of light load when the tower

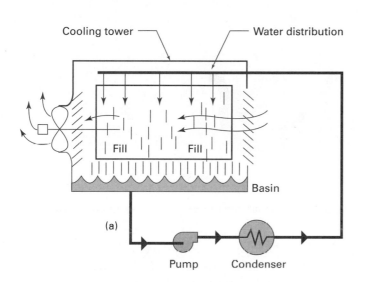

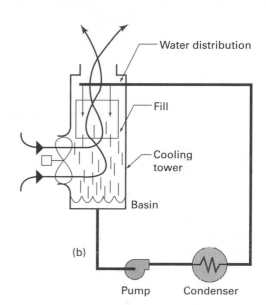

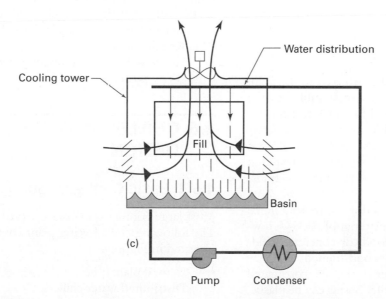

■ **FIGURE 4–16**

Typical cooling towers for building applications. (a) Induced-draft crossflow (draw-through horizontal discharge design). (b) Forced-draft counterflow (single inlet and vertical discharge design). (c) Induced-draft counterflow (double inlet and vertical discharge design).

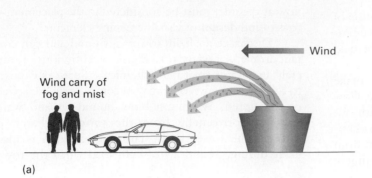

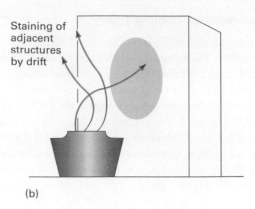

■ FIGURE 4–17

Factors to consider in locating cooling towers. (a) Check for wind directions and drift. (b) Be concerned about staining adjacent structures. (c) Avoid a blocked air intake.

fans are off or running slowly during cool weather. Under these conditions, humid air from the tower may also condense into fog or cause ice on paved adjacent surfaces. All these possibilities must be considered in locating a cooling tower. (See Figure 4–17.)

Configurations and Materials

Towers can be either packaged or erected in the field. Package tower cells are delivered preassembled to the job site. Package towers are used for small to large applications, and cells are available up to approximately 1600 tons. A large tower may be assembled from a number of single-cell packages, or it may be erected in the field if the packaged cells are too large for easy transport to the job site.

Several options are available for the tower's frame and fill material. The framing material of package towers can be fiberglass, metal, wood, or concrete. Field-erected towers are generally wood-framed, built by specialty carpenters. Fill material can be corrugated plastic, wood lath, or ceramic tile. Basins can be plastic, metal, or wood.

Most small cooling towers are steel-framed package units with plastic fill. Despite coatings to protect against corrosion, metal framing will eventually corrode

in the humid environment of a cooling tower. Fiberglass, stainless steel, wood, or concrete framing is a higher-cost option with a longer life. Plastic fill is most economical, but it is fragile and eventually needs replacement. Wood or ceramic tile fill has a longer life at a higher initial cost.

Figure 4–18 shows a variety of cooling towers used with HVAC systems.

4.6 CHILLED-WATER PLANT DESIGN

4.6.1 Basic Configurations

Most large cooling systems are served by chilled water. The following chilled-water plant configurations are illustrated in Figure 4–19:

- Central plant
- Distributed water chillers
- Single- and double-pipe chilled-water loop
- District plant

Each design offers solutions to specific problems.

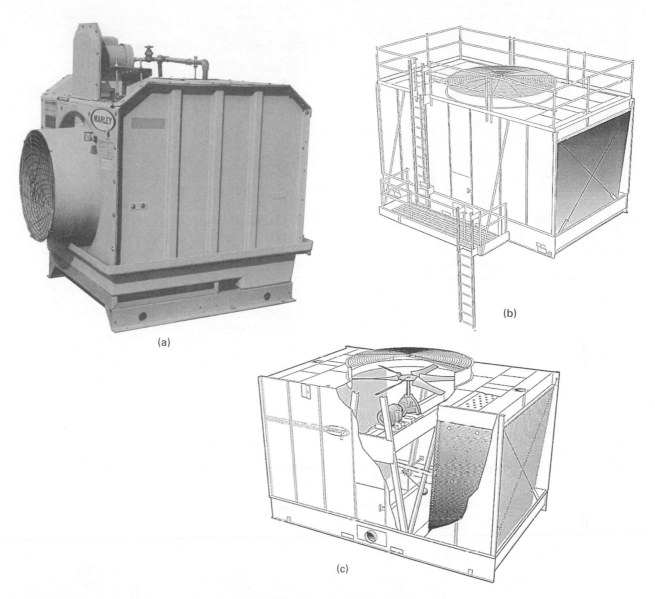

(a)

(b)

(c)

(d)

■ **FIGURE 4–18**

(a) Typical draw-through, horizontal flow tower with fiberglass construction. The flow rate is from a few gallons per minute (GPM) to several thousand GPM.
(b) Large towers require a service platform and safety guardrails. (c) Typical double-flow vertical discharge tower with sizes from 500 to 30,000 GPM. (d) Large custom-designed masonry towers may be constructed of brick, stone, metal, or other materials to match the building material and can also be constructed as a part of the building. (Reproduced with permission from Marley Cooling Tower Co.)

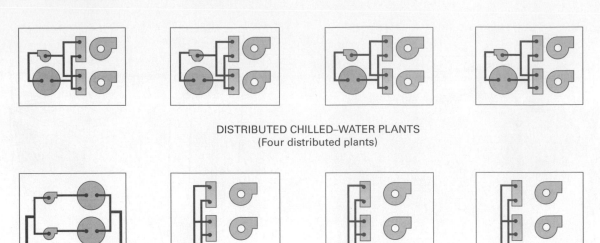

DISTRIBUTED CHILLED–WATER PLANTS
(Four distributed plants)

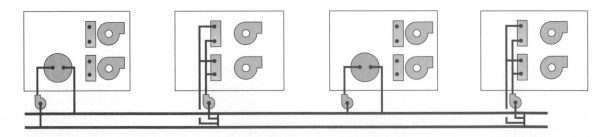

CENTRAL CHILLED–WATER PLANT
(DISTRICT PLANTS SIMILAR)
(One central plant, three distributed)

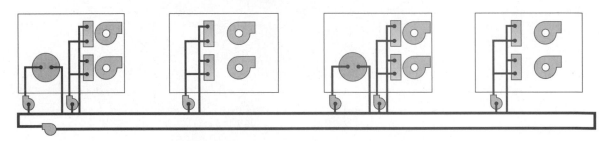

SINGLE-PIPE CHILLED–WATER LOOP
(Two central plants, four distributed)

DOUBLE-PIPE CHILLED–WATER LOOP
(Two central plants, four distributed)

■ **FIGURE 4–19**
Options for chilled-water plants.

Central Plant

Compared with multiple plants, a central plant offers the advantage of requiring less space and fewer auxiliary systems. The greatest advantage is the size of equipment that can take advantage of the diversified load, which may be as low as 50 percent of the loads of individual buildings or areas of buildings.

Central plants also offer operating flexibility, inexpensive redundancy, and centralized maintenance. In addition, the expense of special design provisions is confined to a single location. Among such provisions are sound treatments, ventilation, access, and a consideration of fogging or drift from cooling towers. The central plant is generally the least expensive and best overall solution for a large single building or closely coupled building complex.

Distributed Water Chillers

When growth of the cooling load is unknown or the load is physically separated, distributed plants are appropriate to serve separate areas. Plants can be added as needed, with little commitment of initial investment in space or other provisions for the future.

Distributed chillers will result in smaller piping than central plants will, since less capacity needs to be delivered from a given location, however, the sum of multiple plant capacities is usually substantially greater than the capacity of a central plant, since the diversity of loads will not be as effective as for a central plant.

Another application of distributed chiller plants is in high-rise buildings of over 50 stories, when two or more vertical zones are necessary to limit static pressure in chilled-water piping systems.

Chilled-Water Loop

The loop consists of common distribution piping that connects a number of plants. The scheme is applicable to single buildings or a group of buildings within reasonable distance of one another. The loop will allow equipment to operate as if it were all installed in a single plant.

The loop shares some of the advantages of the central plant. Reliability is enhanced, since a chiller in one location can act as a standby for chillers in other locations. Economy of operation can result from the ability to shift the load from one plant to another to take advantage of the most efficient equipment. The life of the equipment can be extended by concentrating the load in the fewest number of machines possible and minimizing hours of operation for the remaining equipment. Because maintenance is often scheduled on the basis of elapsed run time, less maintenance will be needed.

A loop can be used to supply chilled water to buildings that lack physical space for chilled-water plants. This makes the concept ideal for upgrading older campus buildings.

Loops can be constructed with single- or double-pipe distribution. Single-pipe loops have a distribution pipe, generally underground, that passes by each building served by the system. Some buildings will have water chillers, and some will not. The water chillers serve to reduce the temperature of the water in the loop. Chilled water used by the buildings is returned to the loop at a higher temperature. Ideally, the chillers and loads are distributed evenly so that adequate chilled water is available at all locations. Figure 4–19 shows typical piping arrangements for buildings on a single-pipe loop.

With a single-pipe loop, simply having enough capacity is not sufficient to serve the buildings; the capacity also needs to be located properly. Double-pipe chilled-water loops use separate supply and return piping so that the same water temperature is available to all buildings. Because of the extra piping, a double-pipe loop will generally be more expensive to install, but its performance is superior to that of the single-pipe loop.

District Plants

A large central chilled-water plant may serve a campus, district, or community. Such plants rely on economies of scale to provide low cost per unit of capacity. These savings are used to offset the expense of distribution piping. The district plant will generally occupy a structure and site built expressly for this purpose; therefore, the design can be optimized with respect to spacing of columns, room height, and location.

The district plant centralizes maintenance, keeps noise and cooling tower moisture away from other buildings, and can accommodate improvements without intruding on functions of other buildings. If chilled-water generation is combined with power or heat generation, a district plant is ideal for incorporating cogeneration, which uses the waste heat from electric power generation for heating or cooling.

4.6.2 Sizing Equipment

Several principles must be considered in sizing equipment. Among these are growth, redundancy, partial redundancy, and capacity modulation.

Cooling requirements grow as buildings are more fully utilized and expanded. Allowance for future growth in the size of equipment and the provision of space for future units should be considered.

Redundancy is important for critical applications of cooling service such as computer space, research laboratories, and critical portions of health care facilities. A single unit is obviously less costly than multiple units, however, if a single unit is installed, there will be a total loss of service if the unit fails. Loss of service will also occur during routine maintenance. If continuity of service is critical, some degree of redundancy must be considered. To ensure full load service, a standby chiller must be provided with a capacity equal to the capacity of the largest chiller in the bank of chillers.

The cooling environment in most commercial or office spaces can be compromised for short durations. Partial redundancy can be achieved economically by spreading the design load among two or more units. For example, using two chillers, each sized at half the total load, will ensure 50 percent capacity if one unit fails. Installing three and four units for the total load will ensure 66 percent and 75 percent capacity, respectively, if one unit fails. Each unit may be slightly oversized at only nominal increase in cost. This will result in added spare capacity for the future, as well as additional redundancy in the event of equipment failure.

Turndown capability to meet varying demand loads is another sizing consideration. Modern large chillers can generally operate down to approximately 15 percent of the load without cycling off. If there are periods when the plant must operate at light load, then the chillers must be selected accordingly. If the machines are too large, they will cycle on and off at light load, and the temperature of the chilled water will fluctuate. Also, the machines' lives may be shortened by frequent starting.

Chiller units need not be identical in size. A smaller chiller in the group is often useful for light load conditions during colder weather, on weekends, and at night.

4.6.3 Combination and Fossil-Fuel Alternatives

Electric and absorption or engine-driven chillers can be used in combination to control energy cost in chilled-water plants. Typically, electric chillers are operated to cover the base load, and absorption or engine-driven chillers are used to meet peak loads, reducing demand charges for electricity.

Standby power generators are desirable for many critical building types, such as data centers and health care facilities. Electric chillers constitute a major portion of the critical load to be maintained in the event of a power outage. Using combination plants or using absorption or engine-driven chillers can significantly reduce the size and expense of generators.

4.6.4 Chiller Circuiting Arrangements

Multiple water chillers can be circuited in series, in parallel, or in a combined series/parallel arrangement, as shown in Figure 4–20.

Series Arrangement

A series arrangement reduces water temperature successively through two chillers. The discharge chiller will produce low-temperature leaving water, and the inlet chiller will produce water at a temperature between the return and supply conditions. The advantage of this arrangement is that only the second chiller suffers the

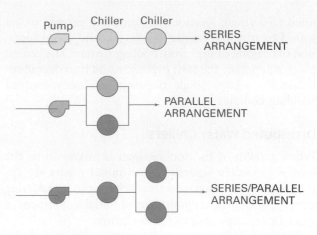

■ **FIGURE 4–20**
Chiller piping arrangements: series, parallel, and series/parallel.

performance penalty of low-temperature discharge. On the other hand, all the chilled water is pumped through both chillers. This results in higher pumping energy than for parallel arrangements. In addition, water flow will be maintained through both chillers regardless of their operating status. This can accelerate the erosion of tubes in the evaporator.

The series arrangement can be very economical in its initial cost, since piping is simple and only one pump is required for each set of chillers in series, however, a standby pump and bypass piping are recommended to ensure continuity of operation if one machine requires service. Once these items are considered, the parallel arrangement can be just as economical to install.

Parallel Arrangement

Chillers in parallel should each have an associated chilled-water pump. The use of a single pump for multiple chillers in parallel will allow mixing of return water with supply water, unless all the chillers are operating. Under this condition, supply water temperature may be too high for good humidity control by the air-conditioning systems. Using one pump per chiller is preferred.

Any number of chillers can be operated in parallel. In this arrangement, unlike the series arrangement, each will be required to produce the same leaving water temperature. The compressor power will be slightly higher; however, the pumping power will be lower.

The flexibility and reliability of the parallel arrangement are excellent. When the load dictates, chillers and their respective pumps can be shut down. In addition, some machines can be withdrawn from service for maintenance without affecting the operation of others.

Combination Series/Parallel Arrangement

A combination series/parallel arrangement may be appropriate for certain applications. For instance, if absorption and centrifugal chillers are used in the same plant, it may be advisable to place the absorption machine in series ahead of the electric centrifugal machines. This design can result in selecting equipment that is more economical overall by choosing electrical centrifugal machines for producing lower leaving water temperatures and absorbers for higher leaving water temperatures.

4.6.5 Distribution Pumping Arrangements

Chillers can serve loads by piping arranged in three basic configurations, as shown in Figure 4–21. The simplest configuration is a unit loop in which chilled water is pumped through the chiller directly to the load.

Unit Loop

To avoid freezing of evaporator tubes, the water flow through most chillers cannot be reduced significantly. Therefore, the unit loop arrangement is generally designed for a flow at constant volume. This will require three-way valves at loads or two-way valves at loads and a single valve to bypass unused water. Chillers can be selected, however, to operate at varying chilled water flows. For such systems two-way valves can be used for a portion of the loads.

If multiple chillers are arranged in a unit loop configuration, the flow from the chiller plant will vary in steps, depending on the number of chillers required to satisfy the load.

The unit loop is lowest in initial cost and can be economical to operate for applications with minimal distribution piping and low requirements for pumping power. If distribution is extensive, pumping costs will be high because a constant flow is required.

Primary Loop

The addition of a primary loop can save pumping energy. The unit loop acts to maintain a constant flow through individual chillers, and the primary pumps draw and return water from the unit loop as the load dictates. To save energy, the primary loop must vary the flow according to the load. Variable-frequency pump motor drives are commonly used for this purpose. Most buildings have considerable load variations owing to occupancy and weather. Therefore, a primary loop distribution has a large potential for saving energy and is commonly used, especially in large applications.

If the distribution is extensive, there may be considerable differences in the available head between the near and far ends of the system. Selecting control valves may then prove difficult. In addition, the head requirement for the primary pumps may be excessive. When this is the case, secondary loops may be added to the system.

Secondary Loops

A secondary loop pump withdraws water from a bridge circuit between primary supply and return piping. Secondary loops are generally controlled for variable flow to save pumping energy. Secondary pumping systems can serve entire buildings from a campus primary distribution system. Space and maintenance requirements for secondary pump stations within the building are minimal.

4.6.6 Energy Features

Heat Recovery

Chillers move heat from one fluid stream to another. If the heat can be used instead of rejected, energy will be saved. This is the principle of heat recovery. There are many potential applications of heat recovery in process and HVAC chilled-water systems.

Most large buildings require cooling year-round, owing to heat gains from lights, people, and appliances in the interior of the building. During cold weather, air conditioning is provided either by cooling with outside air or by refrigeration. If refrigeration is used, the warm condensing water can serve to heat the building, to heat hot water, or to preheat outside air.

Virtually any building with internal heat gains and simultaneous heating requirements will be a candidate for reclaiming heat. Data-processing equipment is a strong heat source owing to its high power usage and continuous operation. Other process cooling requirements should also be considered.

A heat-reclaiming chiller generally has one condenser that uses cooling tower water and an additional closed-circuit condenser for the reclaimed heating water, as shown in Figure 4–22.

The available hot-water temperature is a limiting factor in reclaiming heat. Centrifugal equipment designed for HVAC duty is generally selected to be capable of generating hot water between 105° and 110°F. Reciprocating and rotary helical machines are capable of temperatures up to 140°F, but they are limited in size.

Process cooling, as for computer rooms, often does not require the low evaporator temperatures normally encountered in air-conditioning applications. At higher evaporator temperature, the process cooling system is more energy-efficient and can produce hotter water in a heat-reclaiming system.

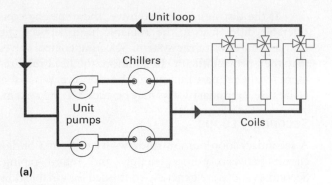

(a)

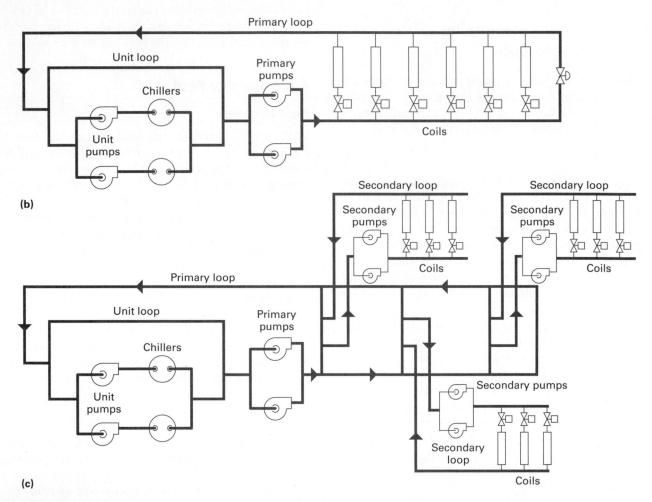

(b)

(c)

■ **FIGURE 4–21**

Alternative configurations of chilled-water distribution systems. (a) Unit loop distribution for small systems. (b) Unit loop with primary distribution for large single buildings. (c) Unit loop with primary and secondary distribution, most often used for groups of buildings such as college and corporate campuses, large retail malls, and industrial complexes.

Thermal Storage

Cooling loads for buildings vary during the course of the day because of weather and occupancy. Conventional chilled-water plants are designed to serve the peak load and have excess capacity during off-peak periods. With a thermal storage system, cooling can be generated during low-load periods, stored, and then retrieved during peak load periods.

Since storage reduces the peak load on chillers, their size can also be reduced. Ideally, chillers can be sized to produce the average output requirement over the course of a day. When the load is less than average, the chillers will contribute to storage. A typical application is analyzed in Figure 4–23; modes of operation are shown in Figure 4–24.

Water can be stored by chilling it or freezing it so that it becomes ice. Chilled-water storage systems have the advantage of using conventional refrigeration equipment, which will chill water to approximately 42°F. The water temperature will generally rise in the system by 15° to 18°F. The required water volume will range between 70 and 100 gal/ton-hr of stored cooling, depending on the rise in temperature and the efficiency of heat retrieval.

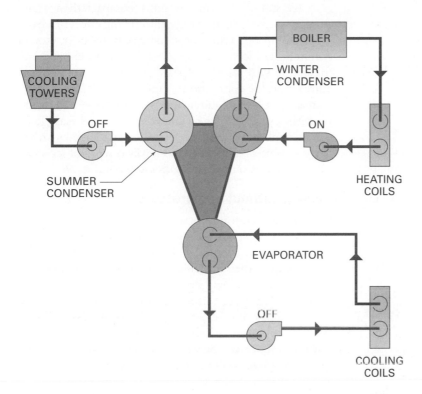

■ **FIGURE 4–22**
Operation of a heat-reclaiming chiller with winter condenser as preheat to the hot-water boiler (heater). The winter condenser may be a separate shell-and-tube heat exchanger or a second tube bundle within the summer condenser shell.

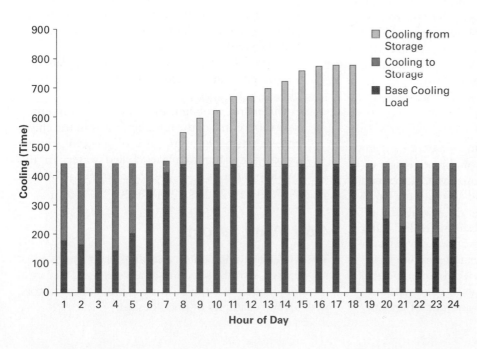

■ **FIGURE 4–23**
Load profile of a 780-ton peak-load cooling system. With thermal storage, it is possible to reduce the capacity of the chiller to 450 tons. During peak hours, the chiller will be supplemented by the stored chilled water that was generated by using the chiller during off-peak and evening hours. From the load profile between 8 A.M. and 6 P.M., it can easily be determined that the chilled-water storage capacity should be a minimum of 2800 ton-hr.

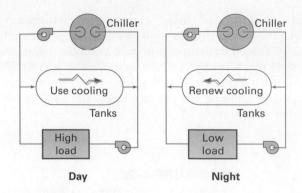

■ FIGURE 4–24

Chilled-water storage operations, day (heavy load) cycle and night (light load) cycle.

If a storage system is installed, chillers, unit pumps, condensing-water pumps, and cooling towers will be smaller and will require less power. This will result in lower utility demand charges, which are a significant portion of overall electrical costs. Both operating cost savings and initial cost savings in equipment can often justify the expense of the storage system.

Other factors should also be considered in the feasibility analysis. If a water storage system is used, it can serve as a secondary water source for fire protection. It can also be used for emergency cooling during a power outage; this is valuable during an orderly shutdown of data-processing equipment.

The total storage volume will depend on the load profile, rate of cooling release, and limit placed on the amount of power demand. For a typical office building, approximately 1000 gal of storage will be required for each ton of reduction in chiller capacity. Storage volume can be substantial, and space needs should be considered early in the design process. Water can be stored in closed tanks or in open vats. Vats are generally less expensive for large systems and can often be integrated with construction of the foundation to reduce costs further.

Ice Storage

Storing ice is an alternative to storing water. Ice is generated by low-temperature brine or direct refrigerant heat exchange and stored in either open vats or enclosed housings. Chilled water is circulated through the ice or closed storage modules, causing the ice to melt and release its latent cooling effect. The resulting chilled water will be colder than is normal in conventional systems. It can either be blended into a higher-temperature circulating system or used to produce colder-than-normal supply air. The advantage of colder air is reduced flow requirements for delivering a given amount of cooling.

This feature offers cost savings from smaller fans and ductwork. Energy is also saved by moving less air.

Ice systems can be used for applications requiring low humidity. Colder water temperatures associated with ice systems allow coil leaving-air dew points as low as 40°F, which corresponds to approximately 30 percent relative humidity at normal room temperatures. Low humidity is necessary in some applications to preserve artifacts and control the growth of mold.

Ice storage systems are more compact than water storage systems. Theoretically, the cooling effect in ice requires about one-tenth the volume of water to be stored (latent heat of ice = 144 Btu/lb). On the other hand, refrigeration systems must operate at much lower temperatures than for conventional building applications. This increases power requirements by 30 to 80 percent. Nonetheless, ice storage systems have proved feasible in certain applications and are available as packaged systems from a number of manufacturers. Two types of ice storage systems are shown in Figures 4–25 and 4–26.

Cooling without Compressors

Outside air economizers are most commonly used to produce cooling for air conditioning during mild and cold weather, however, air economizers may not be desirable where the air quality is poor, where loss of humidity is a concern, or where intake of snow is a problem. There are various means of producing chilled water without the use of a refrigerant compressor. These systems find applications at outdoor temperatures below 40°F. They can be designed using methods involving evaporative cooling. (See Figure 4–5.) Among such systems are the following:

■ Cooling towers can be controlled to produce water as low as 45°F. With appropriate filtration, the water can be introduced directly into the chilled-water system.

■ In lieu of the direct use of tower water, plate heat exchangers can be used to transfer cold into the chilled-water system.

■ A chiller-free cooling feature uses the natural pressure differential of the refrigerant within the chiller that is attainable by the use of low-temperature condenser water to make chilled water. In this mode of operating, chilled water can be produced without operating a compressor.

4.6.7 Plant Layout

The layout of equipment in a cooling plant shares equal importance with the selection of a system and the specification of equipment. Following are the salient factors, with several examples of good design.

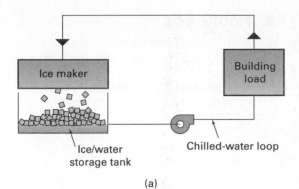

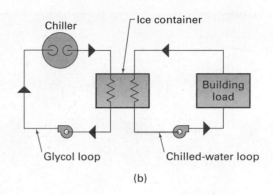

(a)

(b)

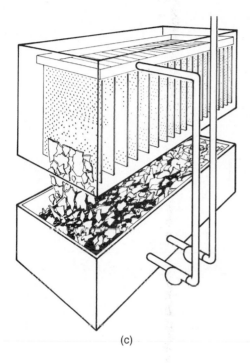

(c)

■ **FIGURE 4–25**

(a) Schematic of an open ice storage system. (b) Schematic of a closed ice storage system. (c) Actual product design showing ice maker on top and ice/water tank below. (Reproduced with permission from Paul Mueller Co.)

Arrangement of Equipment

The orientation and relative position of equipment are important to achieve a workable and efficient plant. The major criteria are as follows:

- Adequate aisle and circulation space for service and maintenance of equipment must be provided.
- Code clearance from electrical gear must be maintained (i.e., 3 ft minimum in front of wall-mounted panels and 3 to 5 ft at the front and rear of freestanding switchgear). The minimum clearance also depends on the voltage class.
- Means of removing and replacing equipment must be provided.
- Vertical clearance must be maintained. Some equipment may be 10 to 12 ft in height; associated pipe and valving may be at higher elevations. Access platforms, ladders, etc., should be provided.

- Equipment should be arranged in an orderly fashion to allow efficient piping, which is essential to low cost and good maintenance.
- Efficient pipe routing is largely dependent on the arrangement of equipment. Associated pieces of equipment should be grouped to allow short, direct runs of piping. Also, piping should be grouped as much as possible in runs of parallel piping at one or more elevations. This will allow the use of trapeze-type hangers and will result in a neat, orderly arrangement of piping. Vertical piping should be arranged to clear tube pulling and maintenance spaces.
- Heat rejection equipment, including cooling towers and air-cooled condensers, requires ready access to a source of air and is therefore usually located on the roof or on a site adjacent to the plant. Nuisance drift and fog should be considered

(a)

■ **FIGURE 4–26**

(a) Multiple-tank closed ice system. (Reproduced with permission from The Trane Company.)
(b) Primary/secondary ice storage system. (Reproduced with permission from Baltimore Aircoil Company (B.A.C.).)

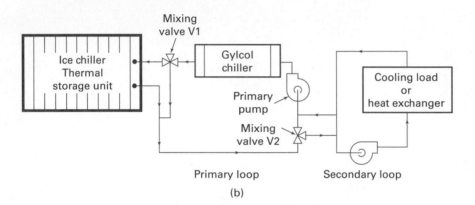

(b)

in locating equipment. If visual screening is necessary, adequate clearances must be maintained to ensure that the equipment performs as desired. Note that even though heat rejection equipment can be concealed within the building, such installations usually result in higher usage of power due to air movement against added static pressure.

Maintenance and Service

Maintenance and service must be considered early in space planning. Essential issues include the following:

- Equipment should be arranged so as to provide for the removal and replacement of those elements that may require major maintenance. In most cases, boilers, chillers, engines, generators, and other equipment in this category will not be replaced in their entirety. It will be necessary, however, to remove major components of these items, such as tube bundles, compressors, impellers, and motors. Adequate aisle space and paths of egress should be provided, as well as provisions for hoisting. Consideration must be given not only to the mechanical spaces themselves, but also to adjoining corridors, vestibules, elevators, and areaways through which the equipment must be removed.

- Storage space for tools, materials, and spare parts should be considered. In accordance with the preference of the owner, shelves and cabinets can be included to allow for organized storage.

- Shop space should be provided for minor repairs to plant equipment. Minimum provisions include a workbench, tool cabinets, machine tools, etc.

- Space should be provided for maintaining plant records and for housing such items as building automation equipment, monitoring, and annunciator panels. The space should be constructed to limit the penetration of noise and to provide security.

- Toilet and washroom facilities with locker space are necessary in plants that require extensive monitoring of their equipment.

4.6.8 Case Study 1

A large building complex requires three 1500-ton chillers. Figures 4–27 and 4–28 illustrate the layout and actual installation of these chillers. The room is 24 ft high, twice the height of a normal building floor. The room appears spacious with adequate service access as a result of good planning.

4.6.9 Case Study 2

A large computer center demands high cooling capacity because the computers generate high amounts of heat. Load analysis indicates that the high heat gain can be economically utilized to heat other buildings without using additional heating energy in the winter. Thus, the plant includes a heat-reclaiming chiller (Figure 4–29, chiller No. 4). The floor plan also illustrates the large amount of space needed for the chiller's transformers and starters, which require nearly the same amount of floor space as the chillers. Figure 4–30 shows the arrangement of pumps and conduits.

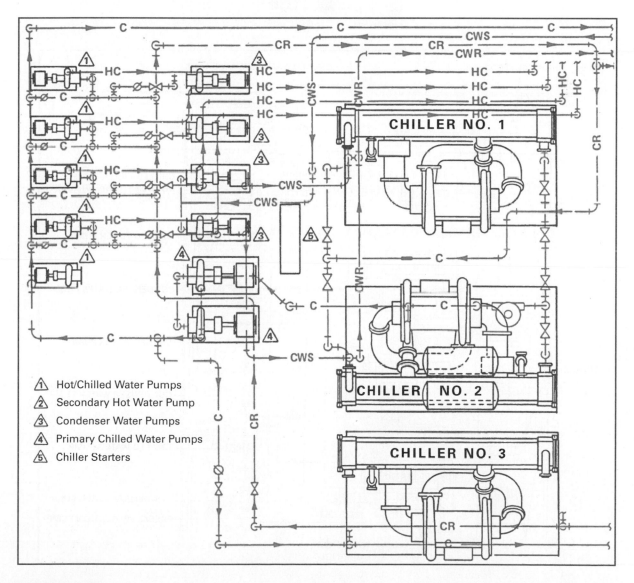

1. Hot/Chilled Water Pumps
2. Secondary Hot Water Pump
3. Condenser Water Pumps
4. Primary Chilled Water Pumps
5. Chiller Starters

■ **FIGURE 4–27**
Floor plan of a large chilled-water plant with three chillers and their associated pumps.

■ FIGURE 4–28
Large chilled-water plant.

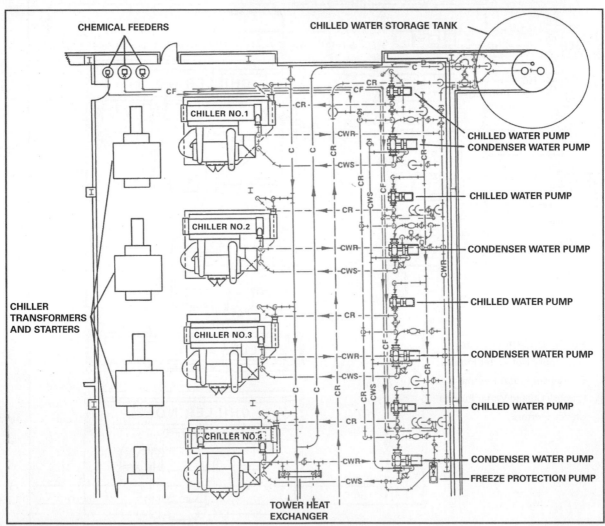

■ FIGURE 4–29
Floor plan of a large chilled-water plant with heat-reclaiming feature.

■ FIGURE 4-30
Lineup of pumps and major piping headers and conduits over the equipment access aisle. This arrangement enhances the visibility of pipe and conduit routing, which is important for better plant maintenance.

QUESTIONS

4.1 How do the vapor compression and the absorption cycle differ in their methods of condensing refrigerant?

4.2 A vapor compression refrigeration machine uses 15 kW of electric power to produce 25 tons of cooling. What is its COP?

4.3 An absorption refrigeration machine uses 15 kW and 600 lb per hour of steam to produce 50 tons of cooling. What is its COP?

4.4 What is the difference between direct evaporative air cooling and indirect evaporative air cooling?

4.5 Describe the basic difference between unitary and split DX systems.

4.6 What are the limitations of DX equipment that prevent its application to large systems?

4.7 What types of compressors are typically installed on water chillers?

4.8 Which compressors are appropriate for smaller machines? Which are appropriate for large machines?

4.9 Rank compressors in terms of their typical energy efficiency.

4.10 Under what circumstances would an absorption water chiller be an economical choice with respect to the energy cost of its operation?

4.11 What is the most widely used type of cooling tower for small- and medium-capacity applications, and why?

4.12 What advantages are offered by air-cooled condensers in comparison with water-cooled systems using cooling towers?

4.13 Electricity demand charges are a large portion of the cooling bill for large buildings. (True/False)

4.14 What design options are available to reduce cooling demand charges?

4.15 Why are refrigerants R-11 and R-12 no longer used in new installations?

4.16 What are the most commonly used replacement options for R-11 (low-pressure applications) and R-12 (high-pressure applications)?

4.17 What is the status of replacement refrigerants for R-22?

4.18 If ammonia is a high-performance, environmentally friendly refrigerant, why is it not used in most applications?

4.19 What are the pros and cons of low-pressure versus high-pressure refrigerants?

4.20 What would be the advantage of using electric chillers and gas-fired chillers in combination? How would you sequence their operation?

4.21 What options are available for using natural gas rather than electricity as an energy source for cooling?

4.22 What is the most widely used type of cooling tower for large-capacity applications, and why?

4.23 How do primary and secondary chilled-water loop arrangements save energy in comparison with unit loop arrangements?

4.24 What are the major advantages of ice as a thermal storage medium in comparison with water?

4.25 How are vapor compression, absorption, and evaporative cooling processes similar?

4.26 What advantage is offered by plate heat exchangers when used in a system to generate chilled water by the operation of cooling towers?

4.27 Approximately how much chilled-water storage would be required to reduce chiller load by 100 tons for a period of 5 hours? How much ice storage would be required?

4.28 Why would chillers of unequal size be installed in a chilled-water plant?

4.29 What factors need to be considered in locating a cooling tower?

4.30 What factors need to be considered in mechanical plant layout for future removal and replacement of components and equipment?

4.31 What nonmechanical support spaces should be provided near a major mechanical plant?

HEATING PRODUCTION EQUIPMENT AND SYSTEMS

5

An UNRELIABLE HEATING SYSTEM CAN RENDER A building uninhabitable and result in a loss of productivity far greater than any savings that might be realized by accepting a marginally designed plant. More than any other portion of the HVAC system, heating plant designs need to consider location, space, reliability, and operating efficiencies. This chapter describes available heating equipment and systems, with an emphasis on the most commonly used systems—air, water, and steam. Other systems, such as infrared and heat pump systems, are only briefly mentioned.

5.1 TYPES OF HEATING SYSTEMS

Heating systems may be classified according to the following criteria:

1. *Energy source.* Coal, electric, gas, oil, biomass, district heat energy from air or water via a heat pump cycle, or a combination of these sources.
2. *Energy transport medium.*
   - *Steam* Low-pressure (below 15 pounds per square inch gauges, psig), medium-pressure (15 to 100 psig), and high-pressure (100 to 350 psig).
   - *Water* Low-temperature (below 250°F), medium-temperature (250° to 350°F), and high-temperature (over 350°F).
   - *Air* Gas, oil, or electric furnaces; in-space heaters, etc.
   - *Direct radiation* Gas or electric infrared heaters, etc.

The basic components of a heating system are as follows:

1. *Steam systems.* Steam boilers, heat transfer equipment (exchangers, coils), combustion air supply and preheating, makeup air supply and preheating, flue gas venting (breaching, flue stack, or chimney), condensate (return, deaeration), water (makeup, deaeration, treatment, and pumping), fuel (supply, pumping, heating, burners, and storage), control and safety devices, etc.
2. *Water systems.* Hot-water boilers (generators) and circulating pumps. Other components are the same as in the steam system, except that equipment for condensate return and deaeration is not required.
3. *Air systems.* Furnaces, in-space air heaters, ductwork, fuel, combustion air, and flue gas components are similar to those in steam or hot water systems. Electric furnaces do not require flue gas removal or fuel systems.
4. *Infrared systems.* Heaters (electric, gas), flue, gas venting.
5. *Heat pump systems.* Air-to-air, air-to-water, water-to-water and air-to-refrigerant systems; pumps; and compressors.

5.2 HEATING ENERGY SOURCES

Heat for HVAC systems is commonly produced by combustion of fuels or electric resistance. Heat pumps (reverse refrigeration) and heat reclaiming from processes are alternative methods of producing heat that can conserve fuels and electricity.

Fuels vary in cost and convenience. Coal was once a common fuel even for small residential heating systems. Now, because of the inconvenience of transporting and handling coal, and because of the impact of its products of combustion on the environment, coal is appropriate only for large-volume users who can afford the high equipment cost to take advantage of the low energy cost. Similarly, heavy oils are economical only for large-volume users. Most residential, commercial, and institutional HVAC systems use light oil or gas as fuel for combustion.

Gas and light oil are burned to produce heat in several types of HVAC equipment. Furnaces and space heaters heat air, which is distributed by either natural convection or forced air; the latter uses a fan. Fuels can also be burned in boilers to heat water or produce steam used in convectors or forced-air devices.

Electric resistance elements can similarly be used to produce heat directly in convectors, in coils to heat air, or in boilers to distribute heat by hot water or steam.

Several factors must be considered before a fuel can be selected:

1. *Availability.* Adequate supplies must be maintained for normal operating conditions and during emergencies in critical facilities such as hospitals.
2. *Fuel storage.* Natural gas is normally supplied by a utility company through piping to the site; therefore, no storage is required. All other fuels, including liquid petroleum gas, oil, and coal, must be stored on the site.
3. *Code restrictions.* Building and boiler codes limit the use of certain types of fuels for reasons of safety and their impact on the environment. Heavy oils, coal, and other solid fuels must be used in a manner that will meet the requirements of the Environmental Protection Agency (EPA) and local ordinances.
4. *Cost.* Cost is a major factor in selecting a fuel, in terms of both capital investment and annual energy costs. Cost is also affected by the degree of maintenance required. In general, electric heating will result in a low initial cost and low maintenance; however, the cost of energy is high. Gas heating and light oil heating are higher in initial cost but will be less expensive overall. Heavy-oil heating and coal heating are most expensive to install and prone to high maintenance but are significantly less expensive than light oil, gas, or electric heating.

The fact that coal is lower in cost than either gas or oil compensates somewhat for the higher cost of the plant. On the other hand, electric heating may be economical in some cases because of its simplicity and favorable regional, seasonal, and time-of-day rates. There is no need for combustion air, chimneys, or fuel storage, which can be expensive, especially in high-rise buildings.

Some properties of energy sources, including gas, oil, coal, and electricity, are shown in Table 5–1.

5.3 FURNACES AND AIR HEATERS

Furnaces are packaged air heaters that are generally used in small buildings, in residences, or in small decentralized systems for large buildings. There are, however, large furnaces that produce up to millions of British thermal units per hour (Btu/hr, or Btuh) for industrial plants or warehouses, where the hot air can be distributed into open spaces. They are often installed in combination with an air conditioner. Natural gas, oil, and electricity are the most common energy sources.

It is not practical to transfer heat by air for long distances, owing to the physical space required for ductwork and the cost of electricity to run fans. In applications requiring extensive distribution of heat from a central plant, boilers are used to generate steam or hot water, which can then be distributed in a compact piping system.

5.3.1 Type of Furnaces

Furnaces are classified and described according to the following categories:

- *Type of fuel or energy* Gas, oil, gas–oil, or electricity.
- *Process of combustion* Open combustion chamber (atmospheric or power blower) or a sealed chamber (impulse).

TABLE 5–1
Energy value of common sources

Type of Energy	Unit of Measure	Conventional Unit, Btu	IS Unit, kcal
Electricity	Kilowatt-hour (kWh)	3413	860
Gasoline	Gallon (gal)	128,000	32,000
Fuel oil (No. 2)	Gallon (gal)	140,000	35,000
Residual oil (No. 6)	Gallon (gal)	147,000	37,000
Natural gas	Cubic feet (cf)	1000	250
	Therms (tm)	100,000	25,000
Liquid petroleum gas (butane)	Gallon (gal)	102,600	26,600
Liquid petroleum gas (propane)	Gallon (gal)	93,000	23,300
	Cubic feet (cf)	2500	625
Steam (at 14.7 psia)	Pound (lb)	1150	290
Steam (at 200 psia)	Pound (lb)	1200	300
Coal (average)	Pound (lb)	10,000	2500
Wood (12% moisture)	Pound (lb)	8000	2000

Btu = British thermal unit; kcal = kilocalorie. Energy values shown in the table may vary with the source.

- *Design and construction* Cabinet configuration (vertical or horizontal), airflow (up or down flow), air delivery (ducted or free delivery), construction (indoor or outdoor, pad-mounted or rooftop-mounted).
- *Services* Heating only or heating and cooling combined.

5.3.2 Combustion Efficiencies

The open-chamber furnace usually has an annual fuel utilization efficiency (AFUE) between 75 and 80 percent. It uses either gas or oil. The impulse type of furnace is designed for gas only; it has a high AFUE of from 92 to 95 percent. The principle of operation of the impulse furnace is shown in Figure 5–1.

5.3.3 Typical Furnace Arrangements

Figure 5–2 illustrates a typical open-chamber furnace and a number of its design arrangements.

5.4 BOILERS

5.4.1 Types of Boilers

In terms of their design and construction, boilers can be classified and described as follows:

- *Source of energy* Gas, oil, gas–oil, electric, coal, biomass, etc.
- *Heat transfer surface* Combustion type, including water tube or fire tube boilers; and electric type, including resistance or electrode boilers.
- *Design of combustion chamber* High-firebox or low-firebox (Scotch marine) type.
- *Construction* Cast-iron sectional or packaged, steel or copper tubes, etc.

5.4.2 Fire Tube Boilers

Fire tube boilers are so named because the combustion gases travel through the firebox and through the insides of steel tubes, usually in two or more passes, before being

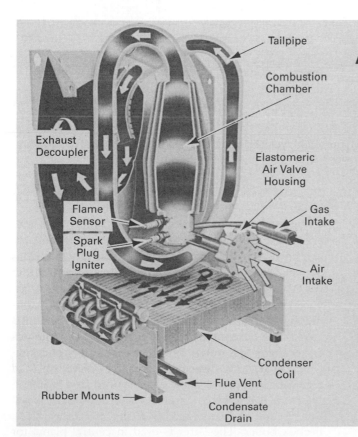

PROCESS OF COMBUSTION

The process of combustion begins as gas and air are introduced into the sealed combustion chamber with the spark plug igniter. Spark from the plug ignites the gas/air mixture, which in turn causes a positive pressure build-up that closes the gas and air inlets. This pressure relieves itself by forcing the products of combustion out of the combustion chamber through the tailpipe into the heat exchanger exhaust decoupler and on into the heat exchanger coil. As the combustion chamber empties, its pressure becomes negative, drawing in air and gas for the next pulse of combustion. At the same instant, part of the pressure pulse is reflected back from the tailpipe at the top of the combustion chamber. The flame remnants of the previous pulse of combustion ignite the new gas/air mixture in the chamber, continuing the cycle. Once combustion is started, it feeds upon itself, allowing the purge blower and spark plug igniter to be turned off. Each pulse of gas/air mixture is ignited at a rate of 60 to 70 times per second, producing from one-fourth to one-half of a Btu per pulse of combustion. Almost complete combustion occurs with each pulse. The force of these series of ignitions creates great turbulence which forces the products of combustion through the entire heat exchanger assembly, resulting in maximum heat transfer.

■ FIGURE 5–1
Cutaway section of the sealed combustion chamber of an impulse type of furnace. (Reproduced with permission from Lennox Industries, Inc.)

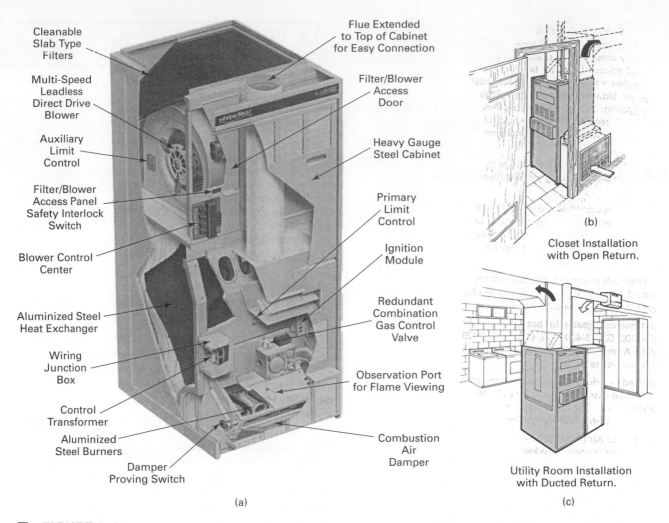

Cleanable Slab Type Filters

Multi-Speed Leadless Direct Drive Blower

Auxiliary Limit Control

Filter/Blower Access Panel Safety Interlock Switch

Blower Control Center

Aluminized Steel Heat Exchanger

Wiring Junction Box

Control Transformer

Aluminized Steel Burners

Damper Proving Switch

Flue Extended to Top of Cabinet for Easy Connection

Filter/Blower Access Door

Heavy Gauge Steel Cabinet

Primary Limit Control

Ignition Module

Redundant Combination Gas Control Valve

Observation Port for Flame Viewing

Combustion Air Damper

(a)

(b)

Closet Installation with Open Return.

Utility Room Installation with Ducted Return.

(c)

■ **FIGURE 5–2**

(a) Cutaway view of a typical open-chamber type gas-fired furnace; (b) typical closet installation; and (c) typical utility room installation. (Reproduced with permission from Lennox Industries, Inc.)

exhausted to the stack. In such boilers, the water is contained in the boiler shell, which completely surrounds and submerges the tubes. In hot-water boilers, the shell is completely filled with water. In steam boilers, the water level is lowered to allow a space for evaporation.

Scotch marine-type boilers are fire tube units that have low contours. They were initially designed for use on board ships, where headroom is limited. Low height is an advantage in building spaces that are restricted in height. These boilers are suitable for stable loads similar to the HVAC loads of commercial buildings. Figure 5–3 illustrates a typical Scotch marine boiler.

5.4.3 Water Tube Boilers

Water tube boilers are so named because water is on the inside of the tubes and the combustion gases pass around the outside of the tubes. Water tube boilers generally have collection headers and upper drums. There are, however, other arrangements proprietary to various manufacturers. Convective water flow takes place within the tubes connecting the drums and headers.

Water tube boilers are available in all sizes, ranging from small residential units to very large units used for generating power. This type of boiler generally contains less water than a fire tube boiler of equivalent capacity. It therefore responds quickly to load fluctuations, such as those experienced in a hospital. Figure 5–4 illustrates a typical small, modular type of low-firebox boiler; Figure 5–5 shows a large high-firebox boiler. The latter is used in central plants for very large structures, in industrial plants, and in campus applications.

(a)

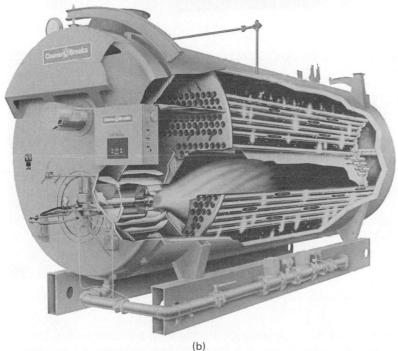

(b)

■ **FIGURE 5–3**
(a) Typical configuration of a low-firebox
Scotch marine-type water heater or boiler;
(b) cutaway view of the Scotch marine-type
unit showing combustion chamber and the
fire tubes. (Reproduced with permission from
Cleaver Books.)

5.4.4 Cast-Iron Sectional Boilers

Cast-iron sectional boilers are units consisting of a series of vertical cast-iron sections filled with water. Each section is shaped like an inverted U with a bottom connecting link. When the sections are connected, the void inside the U forms the furnace or firing volume. These boilers are used primarily in residential and small- to medium-size commercial buildings. They can be assembled in the field, section by section and thus are adaptable to many existing installations where it is not possible to get a completely assembled boiler through the entry to the building. Also, they can be expanded for additional capacity by adding sections. Cast-iron boilers are not applicable to high-pressure systems, owing to the limited strength of cast iron. They are

(a)

(b)

(c)

■ **FIGURE 5–4**

(a) Typical removable tube-type water tube boiler. A portion of the boiler panel is cut out to show the interior tubes. The model shown is a hot-water unit. (b) Bend tube, water, and flue gas passages in the boiler. (c) Manifolding of three identical boilers for modular design. (d) Dual low-temperature, high-efficiency condensing boilers. (Illustrations (a) through (c) reproduced with permission from Bryan Steam Corp.; (d) Courtesy of WTA.)

(d)

■ **FIGURE 5-4**   *(Continued)*

generally rated for 15 psi for steam and a maximum of 100 psi for water. See Figure 5–6 for typical construction and piping arrangements.

5.4.5 Condensing Boilers

Condensation of water vapor is achieved in condensing boilers by operating at lower temperatures, thus allowing the flue gas to condense in the heat exchanger, similar to pulse furnaces described earlier. In most boilers, condensation of flue gases is avoided due to the corrosive nature of the liquids that deposit on the heat exchange surface. Condensing boilers and furnaces have heat exchangers constructed of corrosion-resistant metals, such as stainless steel. In order for condensing boilers to be effective, return water temperatures must be 140°F or cooler. Higher water temperatures will not induce condensation of flue gas moisture. Using lower water temperatures generally means that air-heating coils or convectors, which deliver heat to the space, must be more liberally sized. The extra expense for condensing boilers and associated upsizing of heating delivery components generally has good economics in cold climates due to fuel savings.

In temperate and cold climates, heating accounts for 20 to 40 percent of building energy use, and system efficiency is important to control operating expenses, save energy, and reduce environmental impact. Boiler efficiency in conventional systems is generally 65 to 85 percent. Higher efficiencies of up to 95 percent can be achieved by using condensing boilers, which extract the latent heat of condensation of water vapor in the products of combustion.

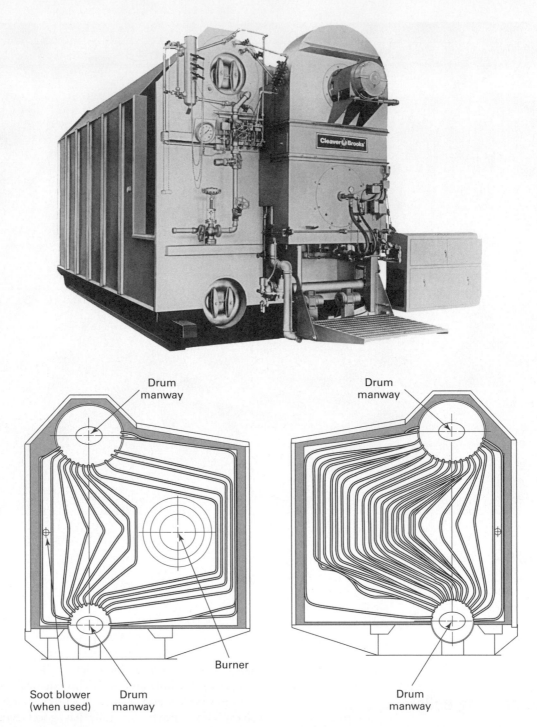

Drum manway

Drum manway

Burner

Soot blower (when used)

Drum manway

Drum manway

■ **FIGURE 5–5**
Typical large high-firebox water tube boiler. (Reproduced with permission from Cleaver Books.)

Example 5.1 A building owner is considering a condensing boiler for his 100,000-sq-ft facility. The local gas company estimates that his heating load will be 30,000 Btu per sq ft per year, based on their benchmarking of facilities in the area. Gas is forecast to cost an average $1.50 per therm over the next 5 years. Assume a conventional boiler system will be 80 percent efficient on an annual basis, and a condensing boiler system will be 90 percent efficient. How much extra can the owner afford to spend if his management needs a 5-year simple payback for justifying capital?

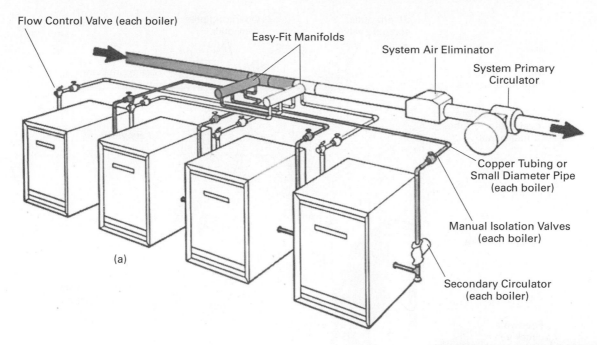

Flow Control Valve (each boiler)

Easy-Fit Manifolds

System Air Eliminator

System Primary Circulator

Copper Tubing or Small Diameter Pipe (each boiler)

Manual Isolation Valves (each boiler)

Secondary Circulator (each boiler)

(a)

(b)

■ FIGURE 5–6

(a) Typical lineup of modular cast-iron boilers with primary and secondary pumping. Each boiler has a pump to circulate water between the supply and return headers. The system's primary circulator provides circulation to all loads. With the primary and secondary pumping system, operation of the boiler can be sequenced to conserve energy. (b) The simple modular design of a cast-iron boiler. (Reproduced with permission from Weil-McLain/A United Dominion Company.)

How much extra could an owner justify in a warmer climate with an estimated heating load of 20,000 Btu per sq ft per year?

Answer A 100,000-sq-ft facility using 30,000 Btu/sq ft/year will need 3000 mmBtu/year of net heating energy. At 80 percent efficiency, this will require a gross input of 3000 mmBtu/0.8, or 3750 mmBtu. At 90 percent efficiency, the input will be 3000 mmBtu/ 0.9, or 3333 mmBtu. The difference is 416 mmBtu. At 100,000 Btu per therm, 416 mmBtu is 4160 therms. At $1.50/therm, the annual savings will be 4160 × 1.50 = $6000 per year (rounded to avoid false accuracy). This would justify $30,000 extra investment at 5-year simple payback.

Similar arithmetic renders an allowable investment of $20,000 in the warmer climate (lower usage means lower savings and less allowable investment).

5.4.6 Electric Boilers

Electric boilers are made for either hot water or steam. The most popular type is the immersion resistance boiler, with capacities up to 1500 kW of input. Larger boilers may be of the electrode design, which utilizes water as the current-conducting medium to generate steam by electrolysis. The immersion resistance elements are usually made of a high-temperature alloy and sheathed

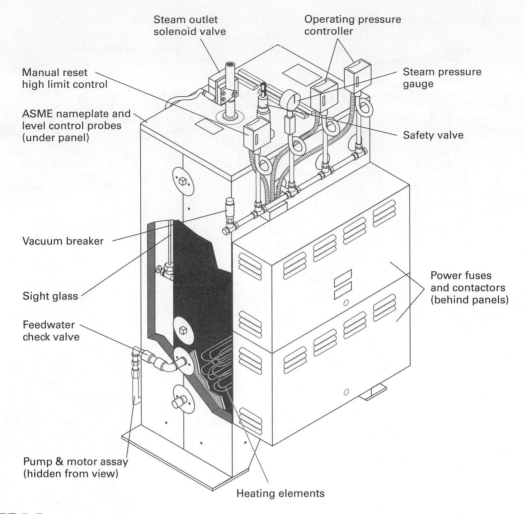

■ FIGURE 5–7

Typical low-resistance-type electric boiler. (Reproduced with permission from Chromalox Wiegland Industrial Div., Emerson Electric Co.)

within Monel or Incoloy alloy tubes to separate the heating element from the water. The tube bundles are usually constructed in U configurations to facilitate wiring and replacement. Figure 5–7 illustrates the construction of a typical low-resistance type of electrical boiler.

With the electric boiler, electrical energy is converted directly into heat energy with nearly 100 percent efficiency. The only losses are from radiation and convection. There is no need for breeching or a flue stack. This is a very important factor in designing high-rise buildings. From an energy-cost point of view, electrical energy is always more expensive than gas or oil. Electric boilers can, however, be economical to operate in conjunction with fuel-fired boilers for off-peak, off-season, and standby operations. The selection of an electric boiler is based on an evaluation of life-cycle costs, distributed between capital and operating costs.

5.5 DISTRICT HEATING

Many metropolitan areas offer district heating as an option to proprietary boiler systems. District heating plants are large boiler installations producing steam for distribution, typically through pipes under the streets of closely packed central business districts. The plants use large boilers fired by coal, oil, or gas and distribute steam at 80 psig to 150 psig, which can be reduced to lower pressure at the building by a pressure-reducing station to provide low-pressure steam for space heating, water heating, or absorption air conditioning. Steam heat can be used directly or converted to hot water in a steam-to-hot water heat exchanger. Typically, condensate is wasted to sewer to avoid the expense of pumps and piping for its return to the central plant.

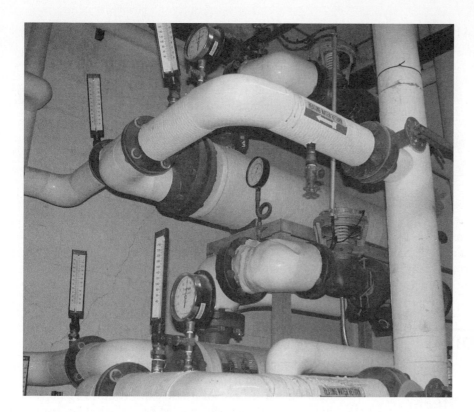

Steam heat exchangers using district heat are a compact, low-maintenance solution compared with boilers.

The building's energy usage is determined by metering incoming steam or wasted condensate. Billing is based on pounds of steam used factored by its Btu content. Cost of energy for district heat includes not only fuel used in the plant but also maintenance, depreciation, investment capital, and profit. Compared purely on the basis of energy cost, district heat is often two to three times the cost of gas; however, district heat can offer an economic advantage to individual building owners. (See Figure 5–8.)

Eliminating boilers results in less investment for mechanical systems including equipment and space for equipment. Boilers require not only space in the traditional basement boiler room but also space on all floors for flue stacks. Eliminating boilers also reduces ongoing maintenance and eventual replacement costs. Reliability may also be improved by using district heat.

Selecting district heat over proprietary boilers is an economic decision that weighs initial cost over operating cost. The decision will depend on the investment criteria of the owner.

Example 5.2   What are the initial costs of a boiler system versus a district steam system to serve a 100,000-sq-ft office building. Include system installation, the cost of space, and engineering fees. The boiler plant will require 1500 sq ft of building space, and the district steam system will require 150 sq ft. Engineering cost will be 8 percent of system cost.

	Boilers	*District Steam*
System construction	100,000	25,000
Cost of equipment space	75,000	7,500
Engineering	8,000	2,000
Total	183,000	34,500

Boilers cost about $150,000 more to install.

What will the annual operating costs be if the building uses 30,000 Btu/sq ft/yr? Gas cost is $0.50 per therm, and district heat rate is $15.00/mmBtu (million Btu's). Assume that boilers are 80 percent efficient. Maintenance contractors have quoted $4000 per year to service the boiler plant, and $800 per year for the district heating plant.

Operating cost, boilers:

Energy load = 100,000 sq ft × 30,000 Btu/sq ft/yr
= 3000 mmBtu/yr

Gas usage = 3000 mmBtu/yr/(100,000 Btu/therm × 80% eff.) = 37,500 therms

Gas cost = 37,500 therms × $0.50/therm
= $18,750

Steam cost = 3000 mmBtu/yr × \$15.00/mmBtu
= \$45,000

Operating Cost Summary:

	Boilers	District Steam
Gas	18,750	45,000
Maintenance	4,000	800
Total	22,750	53,000

Gas-fired boilers would be about \$30,000 less expensive to operate than district steam. This would pay back the additional \$150,000 investment in approximately 5 years at current rates. If a more detailed investment analysis were required, the discounted cash flow method should be used as presented in Chapter 1.

5.6 SELECTION OF MEDIUM AND EQUIPMENT

5.6.1 Selection of Medium: Steam versus Hot-Water Generation

Several factors must be evaluated to decide whether a central heating plant should have hot-water or steam boilers. Hot water is preferred, where practical, owing to its simplicity of design and lower maintenance when compared with steam. If the system supported by the boiler plant is strictly a heating system, the most logical choice is hot water. If the system serves steam loads as well as hot-water loads, it may be desirable to generate steam for the specific requirements and use steam-to-water heat exchangers for heating hot water. Another alternative is to use hot-water generators for the building heating load and a separate steam boiler to serve the steam load.

When the basic system has been selected, operating temperature and pressure must be addressed. Uses that may require steam include kitchens, laundries, sterilizers, industrial processes, and absorption cooling.

Hot-water systems can operate at different temperatures:

- A low-temperature hot-water system (below 250°F) consists of one or more hot-water generators (boilers) normally operating between 180° and 240°F. Such a system is simple to design and economical to operate. It is most commonly used for single-building application, regardless of the size of the building.
- A medium- or a high-temperature hot-water system operates between 250° and 350°F and is commonly referred to as a medium-temperature hot-water (MTW) system. When it operates over 350°F, the system is referred to as a high-temperature hot-water (HTW) system.

At the higher temperature, the heating capacity of a given distribution system is greatly increased. For example, the heat transfer that can be provided with 200°F supply water and 120°F return water is proportional to the difference in temperature at the two conditions (i.e., 80°F). With 360°F supply water and 120°F return water, the temperature difference is 240°F, and the heating capacity transmitted by the distribution system is tripled (i.e., the size of pipe required to carry the same load can be substantially reduced).

MTW and HTW systems are applicable to campus-type facilities such as colleges, research centers, and industrial complexes, where large underground distribution pipes are too costly, however, equipment for MTW and HTW systems must be rated for higher temperature and pressure. In addition, piping must be designed to allow for more thermal expansion; all pipes must be heavily insulated, and a nitrogen pressurization system must be installed to prevent the hot water from flashing into steam. All these extra features result in a higher capital cost, which must be evaluated against the cost savings from the reduction in the size of the pipes.

Steam systems can operate at different pressures:

- A low-pressure steam system (below 15 psig) uses steam as the heating medium with an operating temperature ranging from 220° to 250°F. Steam can easily be transported through the piping system by its own pressure without the need of pumping, however, steam condensate must be pumped, except where it can be returned by gravity. Steam systems require larger sizes of pipe than comparable hot-water systems do. Pipes for both steam and condensate must be properly pitched to facilitate the return of the medium and to avoid water hammer and noise. Steam may not be suitable for buildings in which the floor height is limited or the ceiling space is congested with other mechanical or electrical installations.
- A medium- or high-pressure steam system (between 60 and 350 psig) is preferred when steam loads exist. Medium-pressure systems up to 100 psig are commonly used for hospitals and research buildings. The heating load can be served by hot water through the use of steam-to-water heat exchangers.

Heating capacities of fluids in pipes are governed by the pressure drop per unit length of the pipe (usually psi or feet of water column per 100 ft of length) and by the velocities of flow. Table 5–2 lists the heating capacities, in 1000 Btuh (MBH), for water and steam with different temperature drops. For long-distance

TABLE 5–2
Heating capacity of heat media in pipes

| Heat Medium | Heating Capacity (MBH) @ Pipe Size | | | |
	1″	2″	4″	8″
Low-temperature water (from 160° to 200°F)	120 @ 2.5 fps	700 @ 4.5 fps	5000 @ 6 fps	30,000 @ 10 fps
Medium-temperature water (from 200° to 300°F)	300 @ 2.5 fps	1750 @ 4.5 fps	12,500 @ 6 fps	75,000 @ 10 fps
High-temperature water (from 200° to 400°F)	600 @ 2.5 fps	3500 @ 4.5 fps	25,000 @ 6 fps	150,000 @ 10 fps
Low-pressure steam (from 5 psig to 180°F water)	50 @ 50 fps	300 @ 100 fps	2000 @ 150 fps	11,500 @ 200 fps
Medium-pressure steam (from 50 psig to 180°F water)	200 @ 60 fps	1400 @ 100 fps	6000 @ 150 fps	25,000 @ 130 fps
High-pressure steam (from 150 psig to 180°F water)	800 @ 100 fps	4000 @ 130 fps	15,000 @ 130 fps	60,000 @ 130 fps

MBH = 1000 Btu/hr; MTW and HTW require extra-heavy pipe material; fps = velocity, in feet per second; fpm = 60 fps.

transmission, the pressure drop per unit length of pipe becomes the factor governing the size of the pipe, particularly with low-pressure steam. The reader should notice that the design velocity for steam is about 20 to 40 times faster than that for water. This is because steam, as a vapor, does not have the mass to erode the piping material. The limitation on velocity is governed by the noise level that can be tolerated.

5.6.2 Selection of Equipment

Once the system and the type of fuel have been determined, the next step in design is the selection of boilers and auxiliary equipment, including burners, feedwater heating, water treatment, combustion air heating, safety devices, and other equipment. Good practice dictates that some redundancy be provided for critical equipment that is subject to routine maintenance or mechanical failure. The following represent general criteria for boiler plant design:

1. *Boilers*. Boilers are selected on the basis of their load demand and load profile. Depending on these characteristics, the plant can utilize two or more boilers to carry the major portion of the load when one boiler is out of service. In general, with a two-boiler configuration, each boiler can be sized for 65 percent to 75 percent of the load; with a three-boiler configuration, each boiler can be sized for 50 percent of the load.

   Efficiency varies according to the type of boiler, the type of burner, and the method of control. In general, most boilers with power burners can maintain a fuel-to-heating medium efficiency between 80 and 85 percent under varying load condi-

tions. Atmospheric burners use more excess air and therefore have lower overall efficiencies. With electric boilers, virtually 100 percent of the electrical energy is converted to the heating medium. The only losses are a result of radiation and convection associated with the boiler shell and appurtenances.

2. *Burners*. There are a number of gas, oil, and combination burners that will work with a variety of boilers. Safety provisions for the control of gas and oil burners are subject to building codes and insurance regulations.

3. *Feedwater systems*. Feedwater systems provide treated water to steam boilers. The water quality, the capacity of the system, and adequate pressure are essential.

4. *Fuel supply systems*. When liquid fuels are used, equipment for the transfer of the fuels from storage tanks to the point of usage at the boilers is necessary.

5. *Combustion air supply*. A large quantity of air is required to support proper fuel combustion. The air should be heated to improve the efficiency of the boiler.

6. *Flue gas discharge*. Flue gas must be vented through one or more stacks, which can influence the selection of the fuel and the location of the plant. The stack must extend vertically above the highest part of the building or any adjacent taller buildings. In high-rise construction, the cost of the stack and the net space occupied by the stack passing through each floor can become prohibitively expensive when the boiler is located at the base of the building.

7. *Water treatment systems*. Water treatment systems consist of water softeners and chemical feeders to control the corrosive and scale-forming characteristics of water.

5.7 AUXILIARY SYSTEMS

5.7.1 Burners

Burners are designed to mix fuel and air and ignite the mixture for combustion within the boiler. Burners may be classified as atmospheric (natural draft) and power types.

Atmospheric burners are used for small- and medium-size gas boilers. Such burners depend on the natural draft provided by the stack for the introduction of combustion air. Where adequate stack height cannot be provided, and where the breeching must be abnormally long, induced-draft fans can be used to provide the necessary draft.

Atmospheric burners are generally of the ribbon type, with one or more pilots strategically located to provide a smooth shutoff. These burners are the simplest available and operate on low gas pressure: 3 in. to 5 in. water gauge (WG).

Power burners incorporate a blower that supplies combustion air at a pressure adequate to compensate for the resistance to airflow offered by the boiler. The blower also provides a means of including prepurge and postpurge cycles. This adds an element of safety in that the boiler is scavenged of any possible combustibles before lighting off and after the burning cycle.

Burners used for firing oil are equipped with a means of atomizing the oil for proper combustion. With small burners, pressure atomization is used. With larger, modulating types of burners, atomization is accomplished with the use of compressed air or steam. Steam atomizers are commonly used in large institutional and industrial types of installations firing heavy oil. Air atomizers are used in firing light oil.

Power burners allow the use of smaller-diameter breechings and stacks, which must be gastight because they are under positive pressure. Power burners also provide a means for controlling the percentage of excess air for efficient combustion.

Combination gas and oil burners are a hedge against the uncertainty of the availability of fuel, cost fluctuations, and restrictions imposed by utilities or gas consumption during certain high-demand periods. Such burners incorporate the features of both gas and oil burners and can be easily switched from one fuel to the other.

5.7.2 Boiler Blowdown

Blowdown is the process of draining water from the boiler to reduce the concentration of dissolved solids, remove sludge from the bottom of the boiler shell, and skim off suspended materials at the surface of the boiler water. Sludge is formed as a precipitate composed of waterborne solids combined with water treatment chemicals.

Before any effluent is discharged to the floor drain of a building, the effluent must be cooled to a temperature below 140°F. This is a code requirement and is generally accomplished by using a blowdown separator and cooler designed to separate flashing steam from the effluent and to cool the effluent to an acceptable temperature. The flashing steam is vented, and cooling is accomplished with a heat exchanger using once-through city water or boiler feedwater.

Intermittent blowdown is implemented with a quick-opening valve at the boiler blowdown outlet, piped in series with a slow-opening valve. The quick-opening valve is opened first, followed by the slow-opening valve. When the slow-opening valve is fully opened, it is immediately closed. The quick-opening valve is then closed as well. The frequency of blowdown is a function of the concentration of solids and degree of water treatment required for the boiler water.

Continuous blowdown is common in large systems and serves to remove suspended materials from the surface of the boiler water. The quantity of blowdown is controlled manually or automatically.

5.7.3 Feedwater System

Steam is condensed into water after releasing its latent heat. The water is called *condensate,* and it should be returned to the boiler for reuse. A feedwater system is used for this purpose. In addition, fresh water will have to be added to the boiler to make up for losses due to boiler blowdown and any other drains and leakages. A feedwater system may consist simply of a feedwater pump for small systems, or an assembly of complex equipment for large systems. A complete feedwater system may serve any or all of the following functions:

- To return the condensate
- To make up the shortage due to leaks and blowdown
- To preheat the water to the boiler operating temperature, preventing thermal shock
- To remove the undissolved air and uncondensed gases in the system
- To accumulate the condensate in order to compensate for the time delay between the output of steam and the return of condensate

Figure 5–9 illustrates a typical feedwater unit consisting of a storage tank, steam heater feedwater pumps, and vents.

5.7.4 Fuel Handling

Gaseous fuels, such as natural gas and liquid petroleum gas, are usually supplied with pressure and thus can flow without pumping. Liquid fuels such as oil are usually

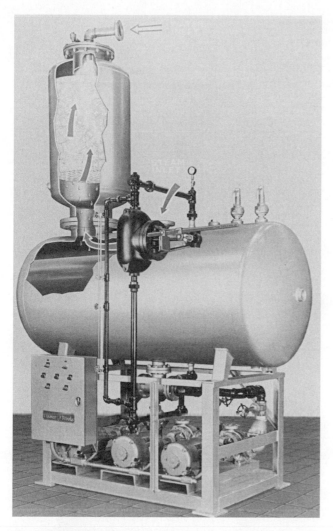

■ FIGURE 5–9
Typical feedwater heater, deaerator, and pumping
assembly. (Courtesy: Containment Solutions, Conroe, TX.)

stored in tanks prior to usage and must be pumped, unless the oil tank is elevated, permitting a gravity flow. Each heating plant must be individually analyzed to determine its pumping requirements. Factors to evaluate include the height of equipment above the storage level and the friction loss of piping and fittings.

There are two basic oil pump systems:

- *Simple suction system* This type of system utilizes the pump that is normally furnished with the boiler to pump oil from the storage tank to the boiler/burner. Such a system works reasonably well as long as the suction lift to the pump is not exceeded and the tanks are reasonably close to the boiler room (within approximately 50 ft). The maximum practical lift for priming a gear-driven pump is generally 12 in. of mercury vacuum, or about 15 ft.

    When priming is lost (usually owing to a leaky check valve or foot valve or an air leak in the piping), it becomes necessary for the pump to purge the line of air and to draw oil from the tank into the line. This can require several minutes and create a start-up problem.

- *Pressurized loop system* A pressurized loop system is used whenever the pumping head and suction lift of the burner pump are exceeded. This system will ensure that oil is always available at adequate pressure at the point of usage. It also eliminates loss of priming at the burner pump, as is often experienced with systems utilizing the burner pump only.

    A pressurized loop system circulates oil from the tank to the point of usage and returns unused oil to the tank. A pressure-relief valve in the loop maintains positive pressure in the segment of the loop supplying the boiler burners. The burner pumps become secondary pumps, deriving oil from the pressurized portion of the loop and returning unused oil to the return side of the loop.

The amount of oil storage depends on the desired frequency of delivery of oil, the availability of oil, and the rate of fuel consumption. A storage capacity equal to three or four weeks of peak consumption is common. The most acceptable method of storage is the use of buried fuel oil tanks located outside the building in an accessible location for servicing from a delivery truck.

- *Direct-burial tanks* Double-walled fiberglass tanks are used in direct-burial applications. These tanks are corrosion resistant, and the double-wall feature traps leaking oil to prevent the surrounding soil from becoming contaminated. The cavity must be monitored for leaks. Steel tanks are more fire-resistant and durable than fiberglass is and can generally withstand more internal pressure. Steel tanks, however, are more susceptible to corrosion, and resulting leaks can cause fire and groundwater contamination. The construction and installation of buried tanks, piping, and anchors must meet the regulations issued by authorities governing hazardous waste disposal. The assembly must be weighted down to avoid buoyancy under all ground water conditions. Figure 5–10 shows the illustration of a typical buried tank.

- *Aboveground tanks* Steel is preferred for aboveground applications.

- *Inside tanks* These tanks are subject to stringent requirements dictated by the National Fire Protection Association and local codes. In general, inside tanks must be housed in three-hour fire-rated enclosures, properly vented and constructed to retain oil within the enclosure in the event of a rupture or leak.

An oil heating system is required for boilers utilizing low-cost heavy oil (generally No. 5 or 6). Heating

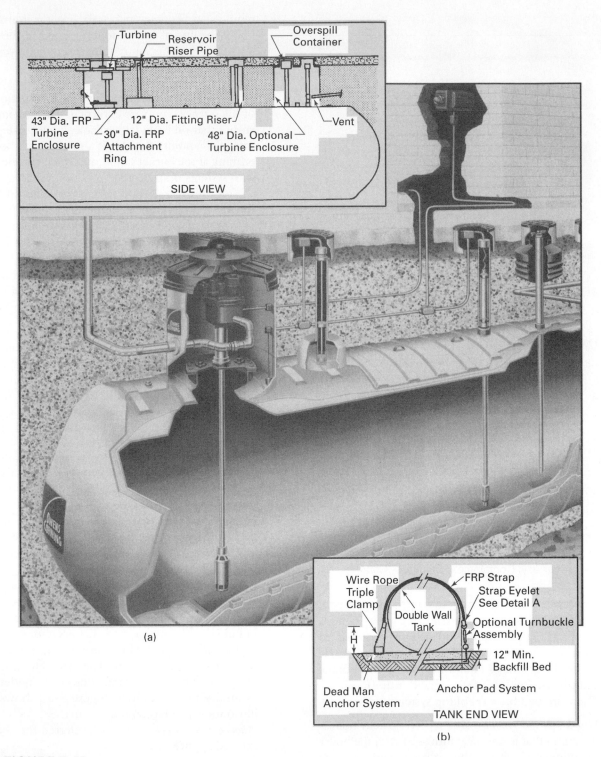

■ FIGURE 5–10

(a) Double-walled buried fiberglass fuel oil tank with manholes with fuel level measurement risers. (b) Typical mounting detail showing a concrete anchor pad with dead-man anchors to prevent the tank from floating during high groundwater conditions. (Courtesy: Containment Solutions, Conroe, TX.)

must be provided at several locations to ensure satisfactory circulation:

- *In the tank*   Electric or steam heaters are needed in the oil tank to reduce the viscosity of oil for easy pumping. The tank temperature should be maintained at approximately 110°F.
- *At the pipes*   Underground or outside fuel oil piping should be electrically or steam traced and insulated. If these lines are not traced, and no circulation occurs for a period during severe weather, oil flow may not be possible.
- *Burner preheater*   This device should be provided at the boiler to raise the oil temperature to between 150° and 180°F for immediate firing. Heaters can be of the combination electric or steam type. The electric heater is deactivated once steam pressure has developed to operate a steam-to-oil heat exchanger.

5.7.5 Combustion Air

Air is a vital part of the combustion process when gas, oil, or coal is the fuel. Basically, all fuels contain varying amounts of combustibles, primarily carbon (C), hydrogen (H), and sulfur (S). Combustion is the result of the following chemical reactions between oxygen (O) and the combustibles:

$$C + O_2 = CO_2 \qquad (5-1)$$

$$2H_2 + O_2 = 2H_2O \qquad (5-2)$$

$$S + O_2 = SO_2 \qquad (5-3)$$

Air contains approximately 21 percent oxygen by volume. Thus, for each cubic foot of oxygen required, 4.79 cu ft of air are required. The density of air at sea level and 70°F is about 13.5 cu ft/lb of dry air. Accordingly, the amount of air required for the stoichiometric combustion of various fuels can be calculated as shown in Table 5–3. In stoichiometric combustion of a hydrocarbon fuel, fuel is reacted with the exact amount of oxygen required to oxidize all combustibles.

Air must be supplied to the combustion chamber in excess of the theoretical quantity required to ensure proper combustion. In fact, there are varying degrees of incomplete combustion, even with excess air. If there is too much air, the combustion efficiency is also lowered, because heat is absorbed by the excess air. The amount of air required for combustion is proportional to the firing rate and thus should be modulated according to this rate or the heating load. Practically speaking, between 10 and 15 percent of excess air is necessary to ensure proper mixing of fuel with air. Some newer burners can operate with as little as 5 percent of excess air with turndown ratios of 5 to 1 and still achieve a high quality of combustion.

Example 5.3   A gas-fired boiler is rated for 3 million Btu/hr with an efficiency of 80 percent. The gas is natural gas having a heating value of 1000 Btu/cf. Calculate the amount of natural gas the boiler requires and the amount of combustion air with 25 percent excess air the boiler would require.

1. Amount of natural gas required:

    Gas = 3,000,000 / (1000 × 80%) = 3750 cu ft/hr

2. Amount of combustion air with 25% excess air:

    Combustion air = 3750 × 12 = 45,000 cu ft/hr

    = 750 cu ft/min (cfm)

When large amounts of combustion air are introduced into the boiler room, the space may become extremely cold during the winter, even to the point where it causes localized freezing of water piping and other components. Thus, combustion air must be heated to a reasonable temperature to prevent portions of the system from freezing. Many designers mistakenly avoid

TABLE 5–3

Approximate amounts of air required for stoichiometric combustion of fuels

Type of Fuel	Basic Content	Theoretical Air for Combustion	Air Required with 25% Excess
Coal (anthracite)	C, S	9.6 lb/lb	160 cf/lb
Coal (bituminous)	C, S	10.3 lb/lb	175 cf/lb
Oil (No. 2)	C, H, S	106 lb/gal	1800 cf/gal
Oil (No. 5)	C, H, S	112 lb/gal	1900 cf/gal
Oil (No. 6)	C, H, S	114 lb/gal	1900 cf/gal
Natural gas	CH_4, C_2H_6	9.6 cf/cf	12 cf/cf
Propane (liquid petroleum)	C_3H_8	24 cf/cf	30 cf/cf

heating combustion air, believing that they will save energy. This is not true; air must be heated to several thousand degrees in the combustion chamber (firebox), regardless of whether it has been preheated in the boiler room. If it is not preheated, then additional energy will be needed, and thus more energy will be expended anyway.

Note that combustion air must be provided in a manner that will not result in the boiler rooms being under negative pressure, which can affect the performance of the burner and be hazardous.

Ventilation is required in boiler rooms to prevent the room from overheating during hot weather. The boiler room should be maintained at positive pressure relative to atmospheric pressure. A well-designed boiler plant should utilize the combustion air as part of the ventilation air, although the requirements of the two do not always coincide. Properly coordinated, combustion and ventilation air can be consolidated into a common system with the redundancy and safety features necessary to satisfy both requirements.

In some boiler rooms, combustion air can be adequately provided by an intake louver communicating directly with the outside. In larger plants, a more active method may be desirable. This can be accomplished by installing a makeup air unit arranged to heat and deliver tempered outside air to the plant. Preheating the air with a tempered air unit will eliminate problems with freezing. Return and outside air dampers can be installed at the makeup air unit, allowing control of both ventilation and combustion air. With this type of system, a stationary louver should serve as a means of relief for any quantity of air delivered by the fan unit that is not used for combustion. The louver will also serve as a secondary combustion air inlet if the fan unit is out of

service. Such a system maintains the boiler room at a neutral or slightly positive pressure.

5.7.6 Fuel Train

Fuel trains consist of regulators, gauges, valves, test cocks, and pressure switches to convey fuel to burners safely. Standardized fuel trains for both oil and gas are designed for optimum safety.

Fuel trains are available from the factory for all types of commercial and industrial boilers. Two separate motor-operated valves are installed, in addition to modulating valves. In the case of gas trains, a motor-operated or solenoid vent valve is provided between the two gas valves. A typical insurer-approved gas train is illustrated in Figure 5–11. This type of train provides a high degree of safety. The two motor-operated, two-position gas valves in the main gas piping are slow-opening, fast-closing valves that close in less than 1 second. These valves respond in unison to boiler start-up and shutdown commands and always close upon a signal that the boiler is malfunctioning, as when there is high pressure, high temperature, low water, or high or low fuel pressure, or when the flame fails.

A main gas pressure regulator should be provided upstream of the gas train. Also, a gas strainer should be considered and may or may not be required, depending on the gas distribution system.

5.7.7 Flue Gas Handling

The products of combustion from boilers and furnaces are called *flue gas*. Flue gas is toxic and must be ducted and vented to the outdoors through flue stacks or

■ **FIGURE 5–11**
Typical gas train of a gas-fired boiler.

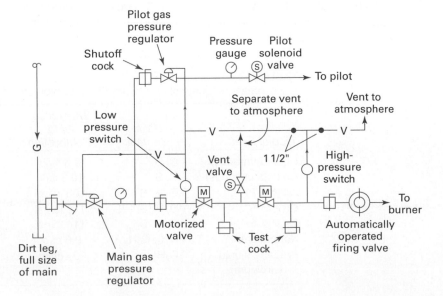

chimneys. The horizontal duct from the flue gas outlet on the boiler or furnace to the vertical uptake is called the *breeching*. The vertical uptake is called the *(flue) stack* or chimney. The terms *flue stack* and *chimney* are used interchangeably, although *chimney* is more correct when this component is constructed as part of a building or on a freestanding structure.

Flue Gas Temperature

For most combustion processes, the temperature of the flue gas is over several thousand degrees—e.g., 1500° to 3000°F for incinerators. At the outlet of heating appliances the actual flue gas temperature, which has been cooled by the heat transfer medium (air, water, or steam), is much lower—e.g., 350°–400°F for gas appliances and 400°–550°F for oil-fired equipment. A mean temperature below 300°F may cause condensation of moisture (containing sulfuric acid) inside the flue stack. Too high a discharge temperature is an indication of energy loss. A proper design should avoid these extreme conditions.

Flue Gas Velocities

In breeching and stacks, flue gas velocities vary from 300 to 3000 fpm. High velocity causes a pressure drop and noise. A natural-draft system is usually designed for 300 to 500 fpm, whereas a forced-draft system is used between 1000 and 2000 fpm. Dispersal of the effluent to improve the ambient air quality may occasionally require a minimum upward chimney outlet velocity of 3000 fpm. A tapered exit cone best meets these requirements.

Draft

Draft is negative static pressure, measured relative to atmospheric pressure. It is normally calculated in inches of water. The draft needed to overcome chimney flow resistance is

$$\Delta p = D_t - D_a \qquad (5\text{–}4)$$

where Δp = draft, inches of water column (in. w.c.)

D_t = theoretical draft—i.e., natural draft produced by the buoyancy of the hot gas in the chimney relative to the cooler gases in the atmosphere, in. w.c.

D_a = available draft—i.e., draft needed at the equipment outlet, in. w.c.

The approximate theoretical draft at sea level (14.7 psig, or 30 in. Hg) may be calculated by using Equation 5–5. The theoretical draft at a given elevation is proportional to the atmospheric pressure at that elevation:

$$D_t = 7.7\text{H}(T_m - T_0)/(T_m \times T_0) \qquad (5\text{–}5)$$

where H = height of chimney above equipment outlet, ft

T_m = mean chimney temperature, °R (Rankine)

T_o = ambient temperature, °R

The available draft D_a needed at the appliance outlet should be provided by the appliance manufacturer. If the theoretical draft D_t created by the chimney is insufficient, then an induced- or forced-draft fan will be required. If the draft is too high, then a pressure balancing damper should be installed. The complete procedure for calculating the chimney size and height of a combustion system is given in Chapter 30 of the systems and equipment volume of the *ASHRAE Handbook* (1996).

Mechanical Draft

When natural draft is insufficient to produce the required draft, mechanical draft-producing equipment is used. There are two basic types of such equipment:

- *Forced-draft type* This is normally incorporated into the fuel burner installed at the front of the boiler or furnace to provide a positive pressure (or negative static pressure) in the combustion chamber.

- *Induced-draft type* This is a fan assembly installed with the breeching to make up for the shortage of draft.

Flue Stack Height

The height of the flue stack is determined by the draft required, but in no case is it lower than the building. Building codes also provide guidelines regarding stack heights in relation to the surrounding building, rooflines, or other obstructions.

Construction

Breeching (see Figure 5–12) is commonly fabricated of 10-gauge steel and covered with high-temperature insulation. Manufactured breechings are also available.

Breechings are subject to variations in operating temperature, ranging from ambient temperature to the temperature of the flue gases. Because of the wide temperature range, provisions to compensate for expansion must be considered. Manufactured expansion joints are available for this purpose and should be used whenever the breeching design cannot otherwise accommodate the calculated changes in length.

Steel, masonry, and manufactured flue stacks are commonly used. Manufactured stacks are generally fabricated of double-wall steel with an air gap separating the two walls, or of a steel casing with a refractory lining. For boilers with power burners, the stack and breeching are under positive pressure and must be gastight.

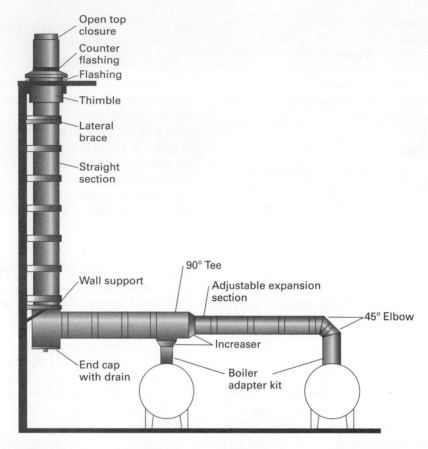

Open top closure

Counter flashing

Flashing

Thimble

Lateral brace

Straight section

Wall support

90° Tee

Adjustable expansion section

45° Elbow

Increaser

End cap with drain

Boiler adapter kit

(a)

(b)

(c)

■ **FIGURE 5–12**

(a) Double-wall sheet metal flue components and installation of breeching and stack. (Note: Horizontal portion of flue is called the *breeching* and the vertical portion of the flue is called the *stack.*) (b), (c). Step-by-step construction of an indoor stack. The view in (c) shows the acid-resistant refractory lining, which varies from 2 to 6 in. (Courtesy: Van Packer Company.)

In the case of atmospheric burners, the draft developed by the stack must satisfy the requirements of the boiler-burner unit or be supplemented by an induced-draft fan. The height of the stack and the temperature of the flue gas determine the theoretical available draft.

Figure 5–12 illustrates the construction and installation of a typical double-wall metal stack used for furnaces or low-temperature hot-water heaters, as well as a heavy-duty factory-insulated metal stack for boiler and high-temperature equipment, such as an incinerator.

5.7.8 Water Treatment

All heating (and cooling) systems involve a temperature change that affects the solubility of solids (minerals) and oxygen in water. The precipitation of dissolved solids and the oxygen released from water can cause three symptoms: (1) a reduction in the heat transfer rate, (2) reduced water or steam flow, and (3) corrosion or destruction of the equipment. Good water treatment is important for maintaining the capacity of the equipment, avoiding breakdown of the equipment, and prolonging the life of the boiler. Water treatment chemicals are used to control corrosion, scaling, embrittlement, and foaming.

Principles of Treatment

- *Removal of solids in water* Scale is a hard substance deposited on the interior surface of pipes and heat-exchanging equipment. It is caused primarily by the precipitation of calcium (Ca) salts in water. One method of preventing scale is to replace calcium ions in makeup water with sodium (Na) ions through the zeolite ion exchange process. Blowdown is also performed to reduce the concentration of total dissolved solids, which build up when water is lost from the system through evaporation.
- *Removal of oxygen* Oxygen is the primary cause of corrosion, particularly in a steam condensate system. Deaeration by heating boiler feedwater is an effective way to remove oxygen.
- *Reduction of acidity in water* Acidity is another cause of corrosion. Adjustment of the water to a pH of 7.0 (neutral) is not sufficient to stop corrosion. A pH of 10.5 (alkalinity) is desired.
- *Inhibition* Chromate has been a very effective agent for inhibiting corrosion, however, it is believed to be a carcinogen (cancer-causing substance) and hence has been banned in most countries. Sodium nitrite has been used as a substitute.

Planning

Water treatment is a must for steam systems and is desirable for closed-circuit hot-water systems. Automatic makeup provisions should be metered and monitored.

Often, leaks occur through pump packings, gaskets, vents, etc. The lost water is automatically made up with fresh, untreated water that can cause scaling and oxygen corrosion.

Because of the varying characteristics and quality of water throughout the country, a specialist should be consulted to ensure that the proper types and quantities of chemicals are used. Overtreatment can be just as damaging to a system as undertreatment.

5.8 OPERATING AND SAFETY CONTROLS

Whether a furnace or a boiler, any heating equipment involves the combustion of fuels. A malfunction of any equipment involved in combustion fuel can cause fire, destruction of property, or loss of life. Building codes and insurance regulations must be strictly followed. Boilers are required to have pressure, temperature, water level, fuel, and flame controllers to ensure their safe operation. The following controls are applicable for commercial and industrial installations:

1. Operating temperature and pressure controls are two-position or modulating types that are arranged so as to start and stop or modulate the burner when necessary.
2. Operating limit controls serve to shut down the burner if the temperature or pressure of the system continues to increase with the boiler in the low-fire or minimum-fire mode.
3. High-limit temperature and pressure controls have manual reset devices that are arranged so as to shut down the burner when abnormal conditions occur.
4. Low-water control is used for steam boilers to cycle a boiler feedwater pump at a given water level and to shut down the burner when the water level drops below that.
5. High-/low-gas/oil pressure controls are arranged to shut down the burner when abnormal pressures occur. These are manual reset devices. A time delay may be required on the low-oil-pressure switch to allow time for pressure to build up.
6. Flame failure controls are coupled with a programmer and arranged to shut down the burner when either the pilot or the main flame fails. Insurers and underwriters' laboratories require a manual reset for this control function. When applied to a forced-draft burner, the programmer also sets the firing cycle.
7. Pressure or pressure/temperature relief valves are mechanical devices without any electrical connection and are designed to open at the rated boiler design pressure or temperature. These valves should be ASME-stamped and sized to relieve the full boiler capacity.

8. With power burners, the combustion air damper can be controlled by the position of the fuel valves through a linkage and cam arrangement. As the fuel valves modulate in response to the firing demand, the air damper modulates proportionally. This type of control provides reasonably good results. To improve the efficiency of the burner and to control the quality of the flue gas emissions even more, independent control of the combustion air damper is desirable.

9. Flue gas control is used for continuous measurement of the products of combustion in terms of O_2, CO_2, CO, H_2, etc., on a continuous basis. The controller can be designed to adjust the combustion air damper and fuel valve for maximum efficiency of combustion.

10. When two or more boilers are installed in parallel, it may be desirable to sequence the boilers on line automatically as required by the load demand. This can be accomplished by utilizing a steam flowmeter or a water temperature sensor to energize an additional boiler when the load cannot be met by the boilers that are already on line. This portion of the control system would serve only to activate the controls of an additional boiler.

11. Additional pressure or temperature sensors located in the common steam or hot-water heater are used as a controller to modulate all boilers on line in parallel. With this type of control, all the boilers will fire at approximately the same rate and will feed the common header in a uniform manner.

5.9 BOILER PLANT DESIGN

The rules for good heating plant design are similar to those set forth in Chapter 4 for cooling production equipment and systems.

5.9.1 Location of the Plant

The heating plant may be located at the lower levels (ground level or basement) or upper levels (top floor or penthouse) of the building, or in a separate or attached building. In general, heating and cooling plants are integrated and thus should be in the same location but not necessarily the same room. (See Chapter 19 for a discussion locations of mechanical rooms.)

5.9.2 Factors to Consider in Selecting Heating Plant Locations

Chimneys for combustion-type boilers must be extended above the roof of the building and must be clear of surrounding buildings. Locating the boiler plant at the base of a high-rise building involves considerable cost and loss of floor space, compared with locating it near the top of the building or in an ancillary building attached to or separate from the main building or complex.

While the boiler plant enclosure can be attractively designed in harmony with the main building itself, many projections from the building, such as chimneys, exhaust fans, vents, etc., are difficult to conceal and may limit the desirability of having the plant close to the main building or complex. For a campus-type design, the boiler plant itself can be located in a totally separate building remote from the load and utilizing underground piping that is directly buried or that runs through tunnels to serve all building loads.

The proximity of the heating plant to the fuel supply and to the combustion air must also be considered in selecting a location for the heating plant, as must the effect of the location on the equipment rating. For example, a boiler located at the base of a high-rise building is subject to the high-static pressure of the piping system and therefore increases the required pressure rating of the equipment.

Finally, the boiler plant generates noise and products of combustion, and both of these effects should be considered in locating the plant. The plant should not be located where noise will be transmitted to other spaces. Also, the stack discharge should be coordinated to avoid the reentry of combustion gases through the outside air intakes.

5.9.3 Heating Piping Circuits

Piping Distribution

Hot-water systems normally consist of a closed loop, and the arrangement is relatively simple; however, care must be exercised regarding the proper selection of expansion tanks, air elimination devices, and pumping arrangements. Steam return systems are more complicated and thus require more attention to detail.

Heating-Only Steam Plants

In heating-only steam plants, essentially all the steam generated is converted to condensate and returned to the boiler. The only losses are those that occur from leaks in the system, such as from pump seals, evaporation through vents, blowdown, etc. Since very little water is lost from the system, only a nominal quantity of makeup water is required, and preheating of the condensate is not necessary.

Steam Condensate

The condensate is collected in receivers or surge tanks and pumped back to the boiler. Calculations must be made to determine the necessary capacity of the receiver tanks to accommodate the amount of condensate required to fill the distribution system. There will be a period between the start-up of the system and the return of condensate to the receiver task. During this time, there should be enough reserve capacity in the receiver to satisfy the makeup water requirement of the boiler. This approach will minimize the required amount of city makeup water and prevent any waste of water due to the tanks overflowing.

Deaerating Feedwater Heaters

These heaters are used for boiler plants operating at 75 psi over and plants using 25 percent or more makeup water. The devices are designed to operate at slightly above atmospheric pressure by injecting steam to the deaerating tank. A deaerating heater, as the name implies, both heats and eliminates oxygen from the feedwater prior to its introduction into the boiler.

QUESTIONS

5.1 If gas is available at $0.50 per therm and No. 2 oil is available at $1 per gallon, which will be the more economical fuel source on the basis of cost per Btu? (Assume equal efficiencies.)

5.2 If gas is available at $0.50 per therm and the efficiency of utilization is 80 percent, what is the net cost of heating energy in dollars per million Btu?

5.3 If electricity is available at $0.1/kWh, what is the cost of electric for heating in dollars per million Btu?

5.4 Name several factors that should be considered in locating a central boiler plant for a large building.

5.5 What is the chief limitation preventing air furnaces from being used in heating systems for large buildings?

5.6 When might steam boilers be preferable to hot-water boilers for a large building system?

5.7 What characteristics of a Scotch marine boiler should be considered in selection for a building application?

5.8 What characteristics of a water tube boiler should be considered in selection for a building application?

5.9 What characteristics of a cast-iron sectional boiler should be considered in selection for a building application?

5.10 Describe some advantages and disadvantages of power burners in comparison with atmospheric burners.

5.11 What special provisions must be made for a heavy-oil storage and distribution system?

5.12 What special provisions must be made for underground oil tanks to comply with environmental regulations?

5.13 When will a pressurized-loop oil supply system be preferable to a simple suction system?

5.14 What is the purpose of boiler blowdown?

5.15 Name the chief safety features applicable to commercial and industrial boilers.

5.16 What is the nominal AFUE of impulse-type furnaces?

5.17 Which of the fire tube and water tube boilers are more responsive to load changes, and why?

5.18 Why is nitrogen pressurization required for MTW and HTW systems? Why not air?

5.19 A heating plant requires 4000 MBH (1000 Btuh) of heating capacity. Select the approximate size of pipe for the main if the plant is a
a. 200°F hot-water system
b. 5-psig low-pressure steam system
c. 150-psig high-pressure steam system

5.20 For the heating plant in Question 5.19, but using natural gas with a heating content of 1000 Btu/cf, what should be the gas flow rate (cf/hr)? How much combustion air is required with 25 percent excess air?

5.21 What are the major chemical components of natural gas?

5.22 What is the natural draft (in. w.c.) created by a chimney 120 ft tall with a mean temperature of 400°F and an ambient temperature of 65°F? *Note:* Equation (5–5) is in degrees Rankine, which is (°F + 460).

5.23 Name the factors considered in selecting the location of a heating plant.

5.24 What are the four basic principles of water treatment?

5.25 Hydronic heating systems are normally closed loop. What does this imply regarding changes in temperature of the water?

5.26 How are multiple boilers controlled to assure that they fire at approximately equal output?

5.27 Why does water treatment by replacement of calcium salts with sodium salts prevent scaling in boilers?

5.28 What are general criteria for sizing boilers for a two-boiler plant? For a three-boiler plant?

5.29 What are the functions of a complete feedwater system as shown in Figure 5–9?

5.30 What factors are so important that heating plants require special care in design?

5.31 What factors are considered in deciding the height of a flue stack?

AIR-HANDLING EQUIPMENT AND SYSTEMS

6

$\mathbf{A}$S MENTIONED IN CHAPTERS 3 AND 4, AIR IS THE MOST common medium used to deliver a heating or cooling effect to a space. In such a case, the delivery system is called an *air-handling system*. The air-handling system consists of air-handling units, ductwork, and air devices.

6.1 AIR-HANDLING EQUIPMENT

6.1.1 General Construction

The air-handling system consists of one or more fan sections, heat-exchange sections for heating and/or cooling, an air filtration section, a section for mixing return air with outside air, and a discharge air plenum. A small air-handling unit may simply contain a supply air fan, coils, and an air filter. Such small units are generally referred to as *fan coil units*. A large air-handling unit up to 100,000 to 300,000 cfm may be custom-built to contain an array of components. The following components can be found in air-handling units:

- Fan sections, for supply air and return air/relief air fans
- Cooling section, for chilled-water or refrigerant cooling coils
- Heating section, for hot-water or steam coils, a gas heat exchanger, or an electrical coil
- Humidification section for extra humidity, if required
- Filter sections, for prefiltering, filtering, and post-filtering
- Air-mixing section, for outdoor air to mix with recirculated air
- Discharge air plenum
- Other components, for electrical power, controls, operating a motor, drainage, etc.

Depending on their size, air-handling units can be delivered as a single package or assembled from modular components. Custom applications may even be built on the job site from panels or sheet metal applied to a metal frame.

Construction materials must be suitable for the environment. Outdoor units generally are galvanized and painted to resist corrosion or are constructed of aluminum. Indoor units may be simply painted or constructed of galvanized steel. Casings are insulated to prevent thermal losses and possible condensation on the exterior surface of the cooling section. If the air-handling units are large enough for walk-in maintenance, their insulation may be protected by woven wire fabric or a sheet metal liner.

Casings are subject to positive and negative pressures with respect to the surrounding air. High-pressure units (up to 10 in. w.c.) require adequate bracing to resist pressure and need to be well sealed to prevent excessive leakage.

6.1.2 Air-Handling Unit Configurations

The components of air-handling units can be arranged in many configurations. Major distinctions regard whether the fan blows through or draws through the coils and whether the housing is in a horizontal or vertical arrangement. Figure 6–1 illustrates the various arrangements.

6.2 HEAT TRANSFER

Heat transfer occurs at the heat-exchange section of the air hauling unit. Several heating and cooling media are common, including water, steam refrigerant, and electric. Depending on the medium, heat is transferred through one or a combination of components.

6.2.1 Water Coils

Water coils are the most common component for transferring heat with circulating air. The coils are normally constructed of copper tubes with aluminum fins. A special coating and special materials such as stainless steel may be appropriate for corrosive environments,

■ FIGURE 6–1

(a) Basic fan coil unit. The unit may be in a vertical or horizontal configuration for an exposed or concealed mounting. (b) Packaged horizontal air-handling unit with a single supply air fan. The unit shown has one return air inlet and one air supply outlet. (c) Built-up horizontal air-handling unit showing one return air fan with mixing dampers, combination filters, and cooling and heating coils; and one supply air fan with a sound-attenuating section and a discharge plenum. (Reproduced with permission from York International.)

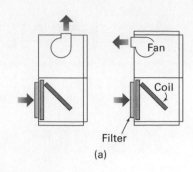

(a)

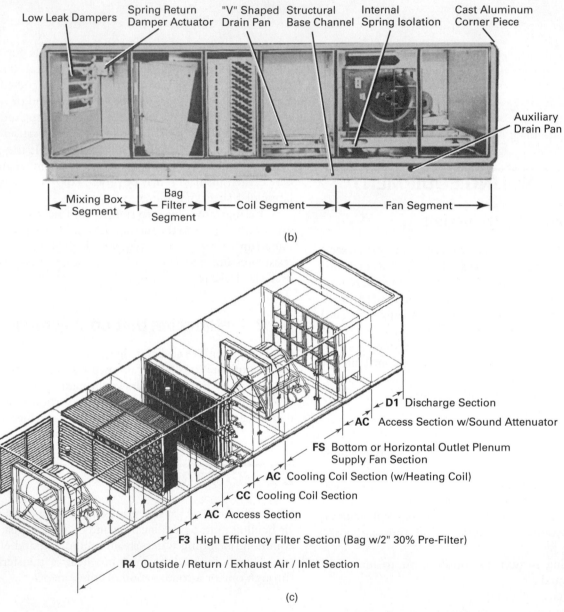

(b)

D1 Discharge Section

AC Access Section w/Sound Attenuator

FS Bottom or Horizontal Outlet Plenum Supply Fan Section

AC Cooling Coil Section (w/Heating Coil)

CC Cooling Coil Section

AC Access Section

F3 High Efficiency Filter Section (Bag w/2" 30% Pre-Filter)

R4 Outside / Return / Exhaust Air / Inlet Section

(c)

including salt spray and industrial pollution. Drain pans, required under cooling coils to collect condensate, are coated with a corrosion-resistant material or made of stainless steel. Figure 6–2 illustrates the construction of a typical water coil.

The performance of a heating or cooling coil depends on the design of its tubes and fins, as well as the size of the coil, including the depth and face area. The depth of the coils is expressed in rows, which represent layers of tubes that conduct the heating or cooling fluid. In general, deeper coils have more capacity because air is in contact with them for a longer time and the temperature of the air can more closely approach the temperature of the heating or cooling medium.

The number of rows not only is significant for the performance of the coil but also affects the location of inlet and outlet piping. A one-row coil would consist of a single layer of tubes conducting the heating or cooling from one side of the coil to the other. A two-row coil would have two layers of tubes with supply and return on the same side. Odd numbers of rows result in opposite-side piping, even rows in same-side piping. Coils constructed with even numbers of rows are preferred for simplicity of piping.

Heating coils can be designed with a high temperature difference between the heating fluid and the heated air. Heating coils are generally one to four rows deep. The temperature difference between the cooling medium and the cooled air is generally small, so cooling coils are deeper, requiring four to eight rows of depth.

The height and width of a coil establish its face area. The area must be sufficient to handle the required airflow with acceptable velocities across the coil. If the velocity is too high, the pressure drop on the air side will be excessive, and condensate moisture will carry over into the airstream from the cooling coils.

Heating coil face velocities of up to 1200 fpm can be accommodated with a reasonable pressure drop, since heating coils are generally shallow, rarely exceeding four rows. For cooling coils, however, face velocities

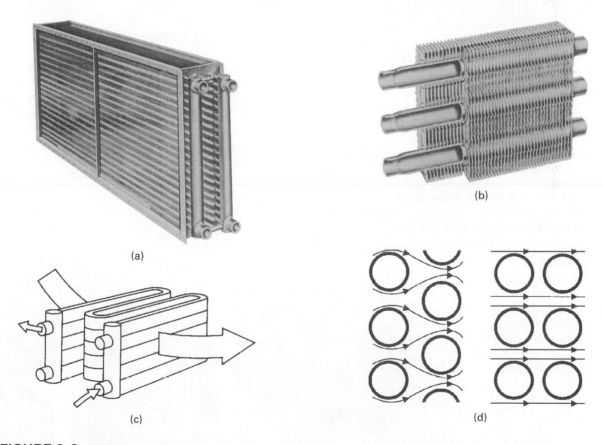

(a)

(b)

(c)

(d)

■ **FIGURE 6–2**

(a) Typical serpentine, finned tube, multirow water coil assembly. (b) Cutaway view of the finned tube construction. The fins are rippled and corrugated to improve the rate of heat transfer. (c) Coils are normally circuited for counterflow—i.e., the inlet air and inlet water flow in opposite directions to have maximum overall heat transfer. (d) Tubes of adjacent rows are staggered to gain more contact between air and coil (left); when tubes are lined up (right), the surface between tubes and air is reduced. (Reproduced with permission from McQuay International.)

should be kept well below 600 fpm to avoid carryover of condensate. Often, heating coils in series with cooling coils have similar face areas for convenience of construction.

6.2.2 Steam Coils

In steam coils, the tubes are designed for easy drainage of the condensate. Figure 6–3 illustrates two types of steam coil: the general-purpose or the conventional type, with supply (steam) and return (condensate) at different ends of the coil; and the steam-distributing type, in which steam is distributed evenly from an inner orifice tube within the outer heat-transfer tube. The latter design ensures an even temperature across the face of the coil and is most effective in preventing freezing of the coil in below-freezing temperatures.

6.2.3 Electrical Coils

Electrical coils may be designed as a part of the air-handling unit or may be installed on the ductwork exterior to the air-handling unit. The heating elements are usually made of a nickel–chromium alloy. Open electrical heating elements must be provided with safety features, such as a flow switch and thermal cutouts. They may be circuited for single-step on-off heating, heating with two or more steps, or modulated heating. Figure 6–4 illustrates a typical electrical duct heater.

Electric coils have very low resistance to airflow, so higher velocities can be used than with water or steam coils.

6.2.4 Direct Expansion (DX) Coils

When the cooling medium is a refrigerant (e.g., Freon), the cooling coil is designed to allow the refrigerant to vaporize in the coil, thus absorbing heat from the air. A typical DX coil consists of a refrigerant header and many distribution tubes. The tubes may be single circuited for small coils or horizontally and vertically split for even distribution of the refrigerant to the coil depending on the performance required. Figure 6–5 illustrates a typical DX coil and circuiting.

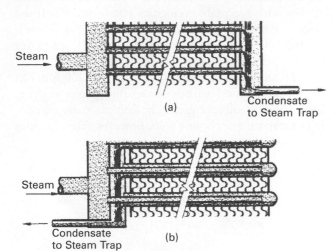

■ **FIGURE 6–3**
(a) Conventional steam coil with a single-tube design. Coil temperature is likely to be lower at the far end of the coil, as steam is condensed inside the finned tubes. (b) Steam distribution-type coil with tube-in-tube design. Steam is evenly distributed. This type of steam coil is ideal for low-temperature preheating and special process applications. (Reproduced with permission from The Trane Company.)

■ **FIGURE 6–4**
(a) Typical electric duct heater, showing electrical fuses and three-step contactors.
(b) Heater with slide-in construction. (Reproduced with permission from Brasch Manufacturing Co., Inc.)

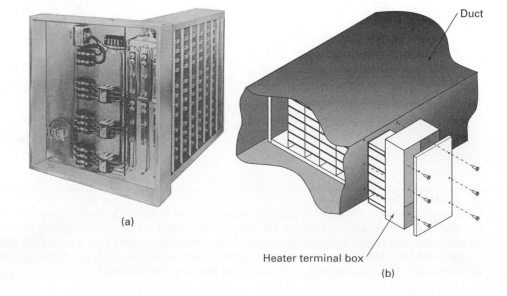

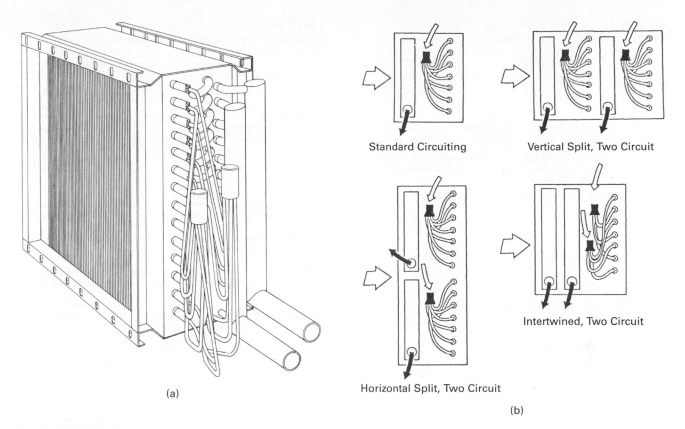

Standard Circuiting

Vertical Split, Two Circuit

Horizontal Split, Two Circuit

Intertwined, Two Circuit

(a)

(b)

■ **FIGURE 6–5**

(a) Typical DX coil construction and (b) circuiting arrangements for effective coil temperature controls. (Reproduced with permission from The Trane Company.)

6.3 AIR CLEANING

Mountain air, although apparently clean and pure, actually contains impurities. Air in urban environments contains even more impurities, in the form of gas, liquid, and solid particulates, and many of these particulates are classified as pollutants, such as smog, smoke, and pollen. In addition, air may contain bacteria and viruses. All these pollutants are detrimental to health. It is imperative that air be cleaned to maintain an acceptable indoor air quality. Table 6–1 lists some common atmospheric particulates that should be evaluated in selecting among filtration media.

6.3.1 Means of Cleaning Air

Air can be cleaned by passing it through a liquid curtain or spray or through a dry filter medium. A liquid curtain or spray may use water or chemical solutions to remove the air particulates, but these solutions usually serve other functions, such as cooling or humidification. The dry type of filtration is by far the most commonly used method for cleaning air, and accordingly, we shall limit our discussion to this variety of filter.

6.3.2 Rating and Testing Air Filters

There are three major operating characteristics of air filters: efficiency, resistance to airflow, and dust-holding capacity. No single test can adequately specify all three characteristics, so, in general, four types of tests are used for rating the efficiency of air filters:

1. *Dust weight arrestance test.* Particles of synthetic dust of various sizes are fed into the air cleaner (filter), and the fraction by weight of the dust removed is determined.
2. *Dust spot efficiency test.* Atmospheric dust is passed into the air cleaner, and the discoloration is observed.
3. *Fractional efficiency or penetration test.* Uniform-sized particles are fed into the air cleaner, and the percentage removed by the cleaner is determined.

TABLE 6–1
Common suspended particulates in urban air

Particulate	Normal Size (μm)*	Particulate	Normal Size (μm)
Fumes	0.001–1	Tobacco smoke	0.01–1
Smog	0.001–2	Bacteria	0.3–30
Dust	0.001–20	Pollen	10–100
Viruses	0.003–0.05	Human hair	40–200

*μm = micron or micrometer (10^{-6} meter).

Source: Reprinted by permission from ASHRAE (www.ashrae.org).

4. *Particle size efficiency test.* Atmospheric dust is fed into the air cleaner, and air samples taken upstream and downstream are counted to determine the efficiency of removal of each size of particle.

6.3.3 Types of Air Filters

Filters may be classified according to the following criteria:

1. *Filtration principle.* Filtration by the medium or by electrostatic precipitation.
2. *Impingement.* Dry medium or viscous impingement.
3. *Configuration.* Flat or extended surface (pockets, V-shaped or radial pleats).
4. *Service life.* One-time disposable or renewable.
5. *Performance.* Low and medium efficiency, high-efficiency particulate air (HEPA), or ultrahigh efficiency (UEPA).
6. *Special features.* Odor absorption, disposal of radioactive material, etc.

Table 6–2 compares the performance of the various types of filter, based on the ASHRAE standard 52.1 test methods. For convenience, the filters are placed in four groups.

6.3.4 Typical Air Filters

Filters are available in a wide variety of designs to suit particular applications. Figure 6–6 shows a range of designs for general applications. Figure 6–7 shows special filters for applications requiring odor control or high degrees of air cleanliness and electrostatic filters.

6.3.5 Application of Air Filters

Filters are selected on the basis of their efficiency, but no less important are the resistance to airflow and the life

TABLE 6–2
Comparative performances of viscous impingement and dry filters

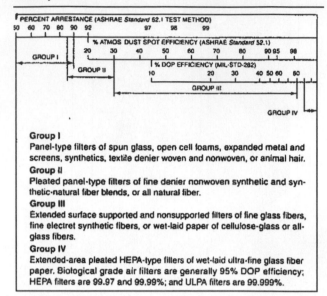

Group I
Panel-type filters of spun glass, open cell foams, expanded metal and screens, synthetics, textile denier woven and nonwoven, or animal hair.
Group II
Pleated panel-type filters of fine denier nonwoven synthetic and synthetic-natural fiber blends, or all natural fiber.
Group III
Extended surface supported and nonsupported filters of fine glass fibers, fine electret synthetic fibers, or wet-laid paper of cellulose-glass or all-glass fibers.
Group IV
Extended-area pleated HEPA-type filters of wet-laid ultra-fine glass fiber paper. Biological grade air filters are generally 95% DOP efficiency; HEPA filters are 99.97 and 99.99%; and ULPA filters are 99.999%.

1. Group numbers have no significance other than their use in this table.
2. Correlations among the test methods shown are approximations for general guidance only.
3. DOP = di-octylphthalate, an oily substance used to test the penetration property of a filter medium.
Source: Reprinted by permission from ASHRAE (www.ashrae.org).

cycle cost of the system. High-resistance filters require more power and thus more energy. The resistance can be reduced when the design airflow rate is lowered. In general, the filter manufacturer provides a range of face velocities and initial and final resistances for each type of filter. Table 6–3 gives an overview of filter applications for various kinds of buildings. In general

1. *For residential and commercial buildings,* low-efficiency and medium-efficiency filters are adequate. Typical low-efficiency filters are made of fiberglass installed in a rigid frame of cardboard. Bag-type or pleated filters are used for higher efficiency. These designs provide more surface area for airflow, resulting in a lower velocity through the filter medium and, accordingly, less resistance to airflow and less pressure drop across the coil. Excessive pressure drops should be avoided to conserve energy.
2. *For health care facilities and laboratories,* the filtration requirements are often dictated by codes and regulations. Other special applications include clean rooms and special laboratories. HEPA (high-efficiency particulate air) filters are used in these applications for very clean environments or removal of hazardous particles. HEPA filters are expensive

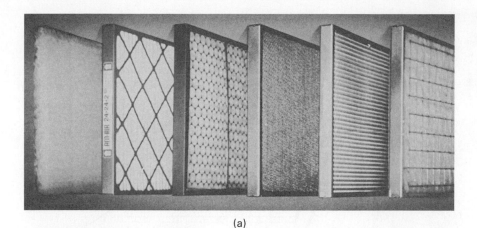

(a)

(b)

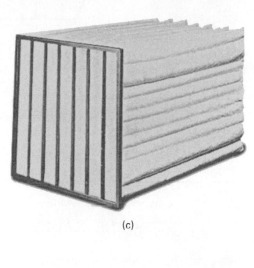

(c)

■ **FIGURE 6–6**
(a) Low-efficiency filter with replaceable medium; low-efficiency washable aluminum filter; low-efficiency disposable fiberglass filter; low- to medium-efficiency pleat-type panel filter. (b) High-efficiency HEPA filters in a rack. (c) Medium-efficiency extended-surface bag filter.

and exhibit a high resistance to airflow, so their use is limited to these special applications.

3. *For hazardous materials,* the filter housing may be constructed so that the filter can be replaced without exposing it to the environment. The housing is designed so that the contaminated filter can be pulled directly into a plastic bag; hence the term "bag out" for this feature.

4. *For odor removal,* adsorption-type filters are used to remove gaseous contaminants from the airstream. Often used for odor control, these filters rely on extremely porous activated charcoal to collect the contaminant. They may also contain oxidant chemicals, such as potassium permanganate. Adsorption filters can be used as an alternative to maintaining higher quantities of fresh air to save energy for conditioning outside air. They are also effective at eliminating hazardous contaminants from exhaust or ventilation air. When combined with HEPA filters, adsorption filter

assemblies can be rated for chemical, biological, and radioactive (CBR) applications.

6.4 AIR MIXING

Outside air required for a building is usually ducted to the inlet of an air-handling unit by mixing with the return air. The two airstreams must be balanced with dampers to introduce sufficient outside air for ventilation, but not so much as to require excessive conditioning during extremes of weather. The mixing box portion of the air-handling unit must be carefully designed to prevent stratification of cold outside air in winter, which could freeze the tubes of coils. Dampers are arranged to force the two airflows to collide, with turbulence for good mixing. A typical mixing box is shown in Figure 6–8. Accessories or rotating vanes can also be used to create turbulence and promote mixing.

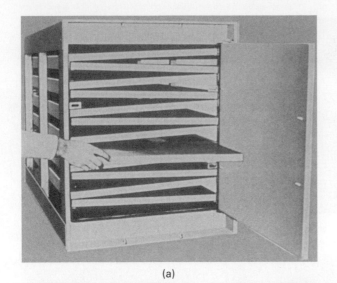

(a)

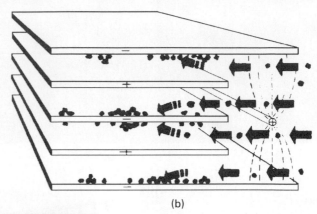

(b)

(c)

■ **FIGURE 6–7**

(a) Odor-removal carbon filters in a rack. (b) Principle of operation of an electrostatic filter. Dust and fumes first are positively charged at 14,000 V and then enter a second electric field, where they are attracted to the collector plates. (c) Assembly of an electrostatic filter. (Courtesy: AAF International, Louisville, KY.)

Even during cold weather, some buildings require cooling. To avoid operating refrigeration equipment, air-handling units can be designed to allow increased amounts of outside air for free cooling. This feature, called an *economizer cycle*, involves controlling the position of the outside and return air dampers so as to result in a mixed air temperature that will supply cooling for the building—generally, 55°F to 60°F.

Care must be taken in the design of mixing sections for air-handling units used in variable air volume applications. Low airflow during cold weather results in low velocity through the mixing sections, stratification of cold air, and the risk of freezing coils.

Introducing large quantities of outside air for ventilation or free cooling will tend to overpressurize a building, unless provisions are made for relieving air. Gravity- or motor-operated dampers can be used to

TABLE 6–3
Application of air filters in air-handing systems[a]

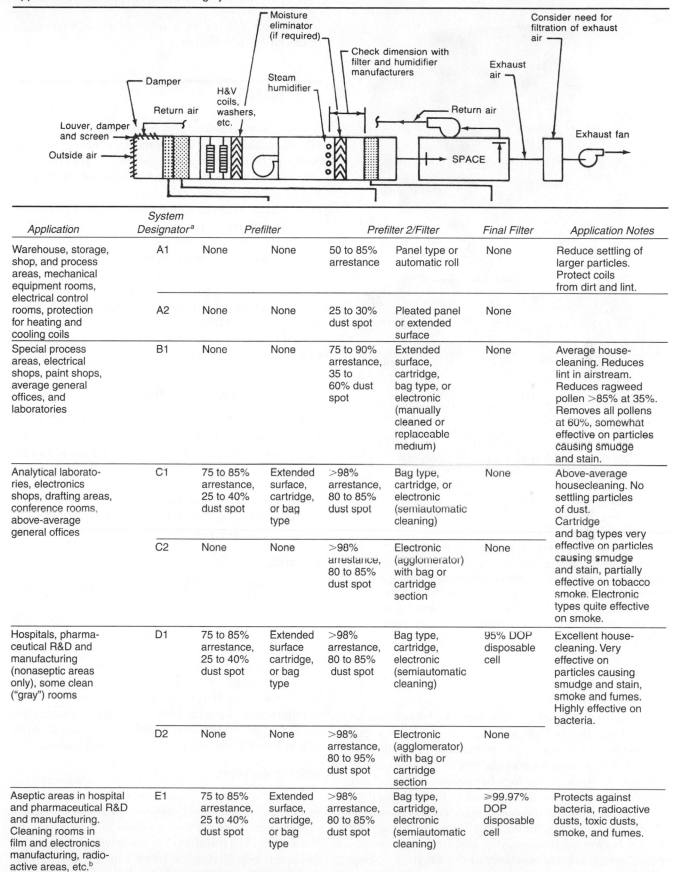

Application	System Designator[a]	Prefilter		Prefilter 2/Filter		Final Filter	Application Notes
Warehouse, storage, shop, and process areas, mechanical equipment rooms, electrical control rooms, protection for heating and cooling coils	A1	None	None	50 to 85% arrestance	Panel type or automatic roll	None	Reduce settling of larger particles. Protect coils from dirt and lint.
	A2	None	None	25 to 30% dust spot	Pleated panel or extended surface	None	
Special process areas, electrical shops, paint shops, average general offices, and laboratories	B1	None	None	75 to 90% arrestance, 35 to 60% dust spot	Extended surface, cartridge, bag type, or electronic (manually cleaned or replaceable medium)	None	Average house-cleaning. Reduces lint in airstream. Reduces ragweed pollen >85% at 35%. Removes all pollens at 60%, somewhat effective on particles causing smudge and stain.
Analytical laboratories, electronics shops, drafting areas, conference rooms, above-average general offices	C1	75 to 85% arrestance, 25 to 40% dust spot	Extended surface, cartridge, or bag type	>98% arrestance, 80 to 85% dust spot	Bag type, cartridge, or electronic (semiautomatic cleaning)	None	Above-average housecleaning. No settling particles of dust. Cartridge and bag types very effective on particles causing smudge and stain, partially effective on tobacco smoke. Electronic types quite effective on smoke.
	C2	None	None	>98% arrestance, 80 to 85% dust spot	Electronic (agglomerator) with bag or cartridge section	None	
Hospitals, pharmaceutical R&D and manufacturing (nonaseptic areas only), some clean ("gray") rooms	D1	75 to 85% arrestance, 25 to 40% dust spot	Extended surface cartridge, or bag type	>98% arrestance, 80 to 85% dust spot	Bag type, cartridge, electronic (semiautomatic cleaning)	95% DOP disposable cell	Excellent house-cleaning. Very effective on particles causing smudge and stain, smoke and fumes. Highly effective on bacteria.
	D2	None	None	>98% arrestance, 80 to 95% dust spot	Electronic (agglomerator) with bag or cartridge section	None	
Aseptic areas in hospital and pharmaceutical R&D and manufacturing. Cleaning rooms in film and electronics manufacturing, radio-active areas, etc.[b]	E1	75 to 85% arrestance, 25 to 40% dust spot	Extended surface, cartridge, or bag type	>98% arrestance, 80 to 85% dust spot	Bag type, cartridge, electronic (semiautomatic cleaning)	≥99.97% DOP disposable cell	Protects against bacteria, radioactive dusts, toxic dusts, smoke, and fumes.

[a]System designators have no significance other than their use in this table.
[b]Electronic agglomerators and air cleaners are not usually recommended for clean room applications.
Source: Reprinted by permission from ASHRAE (www.ashrae.org).

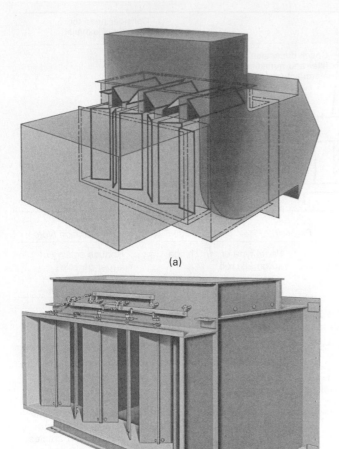

(a)

(b)

■ **FIGURE 6–8**

(a) Typical air-mixing box. In this design, the opposed blade dampers between the two airstreams are staggered to provide proper mixing. Other blade configurations, such as parallel blades, are also applied, depending on the duct inlet conditions. (b) Same mixing box as in (a), with the dampers shown for both airstreams. (Reproduced with permission from The Trane Company.)

relieve air from small buildings. Large buildings require more positive control by relief fans. Relief fans can be located on the roof or at perimeter walls. They can also be designed as part of the air-handling unit.

If the pressure drop in the return ductwork is significant, a return air fan may be required. This fan can also be used to relieve the excess air in the building during the operation of the economizer cycle. Figure 6–9 shows three schemes for air relief from a building. The choice between a return air fan and a relief air fan depends on the relative resistance of the air paths. A long relief air path that causes a high resistance to airflow

would favor the use of a relief air fan. Otherwise, a return air fan is sufficient.

6.5 FANS

A fan is a mechanical device that moves air used in HVAC systems to ventilate or transport heat or cooling. Although heating energy can also be delivered by means of radiation, conduction, or convection without a forced airflow, there is little choice but to use air to deliver cooling. Without air movement created by the fan, cold air tends to be stagnant and may result in condensation of moisture on room surfaces.

All fans have a rotating impeller with blades; this increases the kinetic energy of air by changing its velocity. The increased velocity is then converted to pressure. There are two basic fan designs—centrifugal and axial—each with many variations but the same operating principle. Figure 6–10 illustrates the general configuration of these two types of fans.

6.5.1 General Classification of Fans

Both centrifugal and axial fans are used in HVAC systems. Centrifugal fans draw air into the center of a rotating wheel. Vanes on the wheel project the air radially and tangentially toward the outside of the fan housing, which directs the air to the fan discharge. Axial fans move air through their housings in a direction parallel to the wheel axle. The following commonly used fans are illustrated in Figure 6–11.

Centrifugal Fans

General-purpose centrifugal fans can be freestanding or housed within cabinets or air-handling units. Double-inlet fans are most common, but the geometry of the system may require a single-inlet fan. In-line centrifugal fans and plug fans are used when saving space and simplicity of installation are important considerations.

The design of the blade affects its performance. Small centrifugal fans most often use a forward-curved configuration. Backward-inclined blades are common on larger fans, and blades can be constructed with an efficient airfoil shape rather than uniform thickness.

Roof Ventilators

These are centrifugal exhaust fans suitable for outdoor installation. For general exhaust applications, the discharge opening of the fans normally face downward toward the roof so that the fan housing will not accumulate rainwater. For kitchen or other exhausts that contain grease or odor, upblast configurations are normally used.

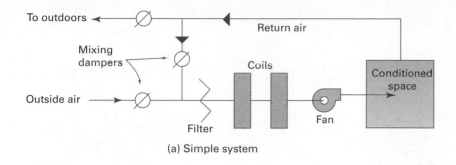

(a) Simple system

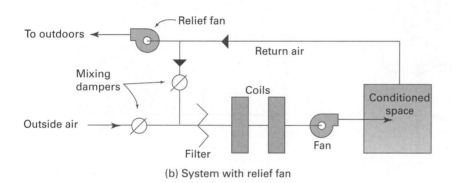

(b) System with relief fan

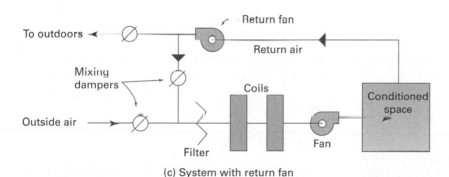

(c) System with return fan

■ **FIGURE 6–9**
Three relief air schemes are commonly used to utilize outside air to cool a space when the outside air temperature is below the set-point temperature of the space.

(a) Simple damper control scheme without using fans. This scheme is applicable to small air-handling systems within a large building where return air can easily be expelled from the building without overly pressurizing the building. This scheme is simple and low in cost.

(b) Relief air fan scheme. A relief air fan is used when the relief air path to the outdoors after the mixing dampers is long and restrictive.

(c) Return air fan scheme. A return air fan is used when the return air path to the mixing dampers is long and restrictive. In large systems, both relief air and return air fans are often used.

The upblast configuration prevents grease from depositing locally near the fan and helps to disperse the exhaust air into a higher elevation, above the roof level.

Utility Sets

Utility sets are weatherproof, single-inlet, generally roof-mounted centrifugal fans used in exhaust applications. Horizontal discharge is preferred for general exhaust, upward discharge for kitchen or fume exhaust. Special materials or coatings will be required for corrosive exhaust, such as the exhaust that might be discharged from laboratories. If the exhaust is potentially toxic, a duct should be installed at the discharge with its opening at least 7 ft above the roof level.

Axial and Propeller Fans

These types of fans can be used for general-purpose applications and have the advantage of being compact when installed in line with ductwork. Propeller fans are generally installed at building walls and are suitable for large-volume, nonducted applications, as in greenhouses and for ventilation of manufacturing space. Propeller fans can be installed at the roof if proper attention is paid to covering the discharge and preventing rain from entering the space.

6.5.2 Application of Fans

Air-Handling Unit Fans

Air-handling fans are generally centrifugal, although axial fans are occasionally used. The supply fan section of the air-handling unit can be placed in the draw-through or blow-through position with respect to the coils. Many factors should be considered in choosing between draw-through and blow-through configurations.

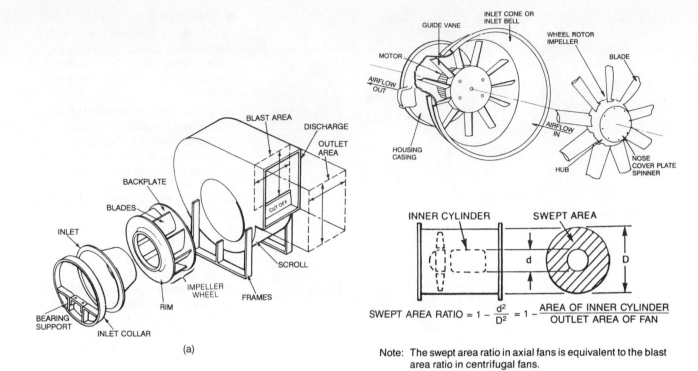

(a)

$$\text{SWEPT AREA RATIO} = 1 - \frac{d^2}{D^2} = 1 - \frac{\text{AREA OF INNER CYLINDER}}{\text{OUTLET AREA OF FAN}}$$

Note: The swept area ratio in axial fans is equivalent to the blast area ratio in centrifugal fans.

(b)

■ **FIGURE 6–10**

Typical construction and components of fans. (a) Centrifugal fan components. (b) Axial fan components. (Reprinted by permission from ASHRAE (www.ashrae.org).)

Draw-through units are generally more compact and less expensive to construct than blow-through units. However, since the fans are closer to the supply duct system, the ducts tend to be noisy. In addition, the energy introduced by the fan produces heat downstream of the cooling coil, requiring lower discharge temperatures for a desired supply temperature. This can result in slightly higher power requirements for refrigeration equipment.

Blow-through units have the advantage of promoting mixing of air upstream of the coils, minimizing stratification and the risk of freezing during cold weather; however, special provisions must be made to promote even airflow distribution across the entire face of the coil. Diffuser plates or screens are often installed at the fan discharge for this purpose.

Return Air Fans

These fans may be part of the air-handling unit or may be installed separately in ductwork upstream of the unit. Return fans can be centrifugal or axial and need not be the same type as the supply fans.

6.5.3 Volume (Flow Rate) Controls

Fans may be required to deliver a constant or a variable volume of air. Variable-frequency motor speed controllers are energy-efficient and have become standard accessories in recent years. Figure 6–12 shown a drive along with performance curves.

6.5.4 Fan Drives

Motors can be placed within the fan cabinet or air-handling unit or can be mounted externally. Fan wheels can be coupled directly to the motor or can be driven by belts and pulleys (sheaves).

Direct-drive fans operate at the same speed as the motor and are not adjustable. Sheaves and belts allow fans to be designed for slower, quieter operation, and the performance of the fans can be tailored by proper selection of the sheave diameter. Some sheaves are adjustable, and sheaves can be replaced to slow down or speed up the operation of the fan when higher or lower performance is needed. Belt guards are needed if the belt drive is exposed.

6.5.5 Fan Performance and Fan Laws

The performance of a fan is measured by the following characteristics:

- Volume of air delivered per unit time (airflow rate), normally in cfm

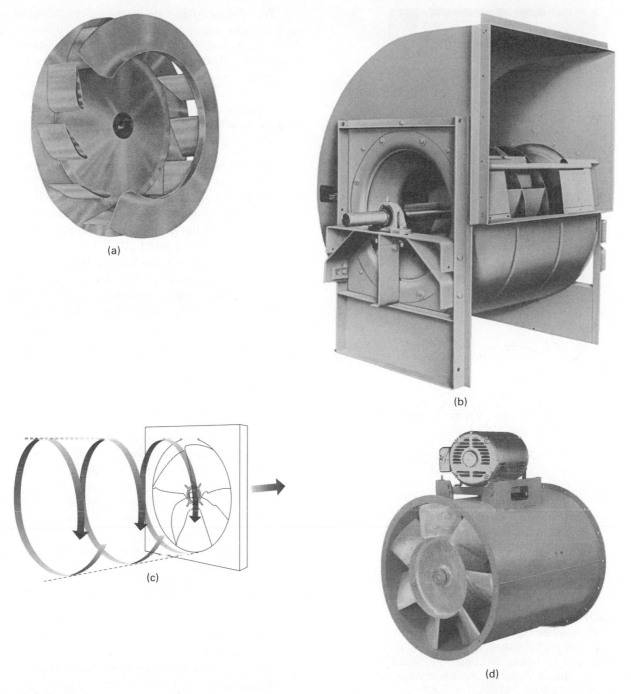

(a)

(b)

(c)

(d)

■ **FIGURE 6–11**

Types of fans. (a) Typical backward-inclined centrifugal fan wheel. The blades may be either straight radial, backward, or forward inclined. (b) Assembly of a centrifugal fan with horizontal discharge, which may also be vertical or down discharge. The scroll housing may have either single or double inlet. (Reproduced with permission from The Trane Company.) (c) Movement of discharge air of a typical propeller fan. (d) Vane axial fan used in line for compact installation. Note: motor, belt, and grease fittings, which need access for maintenance. (Courtesy: Loren Cook Company.)

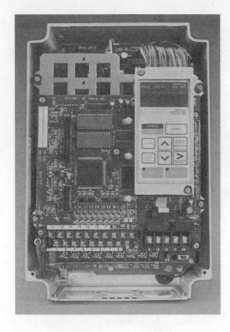

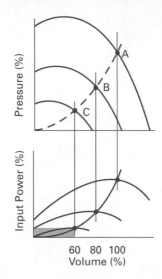

■ FIGURE 6–12

IGBT (insulated-gate bipolar transistor) drives control volume by lowering fan speed. Because input power varies with the cube of fan speed, as speed is reduced, input power is substantially reduced. *At 60% volume, input power is 28%.* (Courtesy: MagneTek, Inc.)

- Pressure created (static, velocity, and total pressure), normally in inches of water column (in. w.c.), inches of mercury column (in. Hg), or pounds per square inch (lb/sq in., or psi). (See Appendix C for conversion to SI units.)
- Power input, horsepower (hp).
- Mechanical and static efficiency, percentage.
- Other factors, such as sound level in noise criteria (NC) or decibels (dB), etc.

Performance data are provided by the fan or equipment manufacturer either in tabulated form or in charts. A typical fan performance chart is shown in Figure 6–13. Plots represent air-handling system curves with resistance or static pressure drop displayed as a function of the flow rate. In general, for the same system (equipment and ductwork), the resistance or the pressure drop varies with the square of the flow rate.

The fan laws most frequently used are as follows:

- The system resistance increases with the square of the airflow rate.
- For the same fan size and system, the airflow rate varies directly with the speed, the pressure varies as the square of the speed, and the horsepower varies as the cube of the speed.
- For constant speed and varying fan size, the flow rate varies as the cube of the fan size, the pressure varies as the square of the fan size, and the horsepower varies as the fifth power of the fan size.

For purposes of estimation, the following formulas can be used to determine airflow rate, pressure required, and fan power:

- *Relation between pressures:*

$$TP = SP + VP \qquad (6\text{--}1)$$

where TP = total pressure developed or required by the fan, in inches of water
VP = velocity pressure, that is, the pressure created by the air velocity at a point in the air system selected for calculations, in inches of water
SP = static pressure, that is, the pressure exerted on the sides of the duct at a point in the air system selected for calculations, in inches of water

- *Velocity of flow to velocity pressure:*

$$V = 4005 \times \sqrt{VP} \qquad (6\text{--}2)$$

where V = velocity f the airflow at a point in the air system

- *Power required at the rated flow and pressure (at sea level):*

$$AHP = \frac{cfm \times TP}{6356} \qquad (6\text{--}3)$$

$$BHP = \frac{cfm \times TP}{(6356 \times FE)} \qquad (6\text{--}4)$$

$$EHP = \frac{cfm \times TP \times 0.746}{6356 \times FE \times ME} \qquad (6\text{--}5)$$

$$= \frac{0.000117 \, cfm \times TP}{FE \times ME}$$

where AHP = air horsepower at 100% fan efficiency, in hp

BHP = brake horsepower at the selected fan efficiency, in hp

EHP = electrical power at the selected motor efficiency, in kW

FE = mechanical efficiency of the fan, per unit (Pu)

ME = motor efficiency of the selected motor, per unit (Pu)

6.5.6 Examples of Fan Performance

An air-handling system is designed to circulate 15,000 cfm at a system pressure of 2.7 in. w.c. A fan is selected on the basis of the graph in Figure 6–14. Plotted on this graph is also the air-handling system curve, which is constructed according to the fan laws.

a. With this particular choice of fan, what should be the fan speed?
b. What is the BHP operating at this condition?
c. If the same fan is operating at 825 rpm, what will be the fan delivery, the air-handling system static pressure, and the BHP required?

(*Note:* Instead of using the graph, you may use the fan laws to obtain the same results.)

Answers

a. From the graph, the intersection of 15,000 cfm with the system curve demands that the fan run at 1125 rpm.
b. The fan should be driven by a motor having a minimum 8.55 BHP.
c. At 825 rpm, the fan can deliver 11,000 cfm, operating at 1.46 in. w.c. and drawing 3.39 BHP.

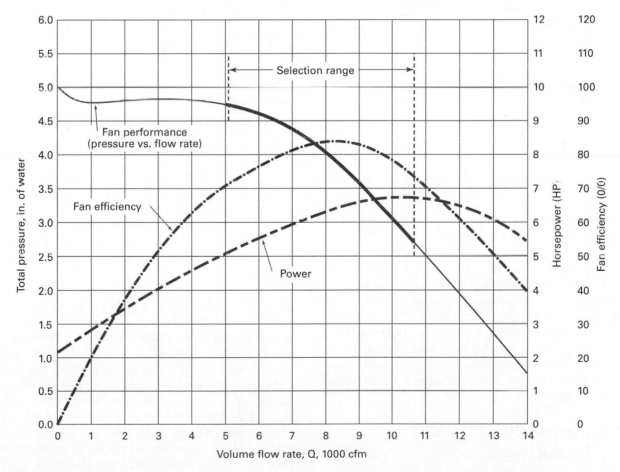

■ **FIGURE 6–13**
Typical fan performance curves.

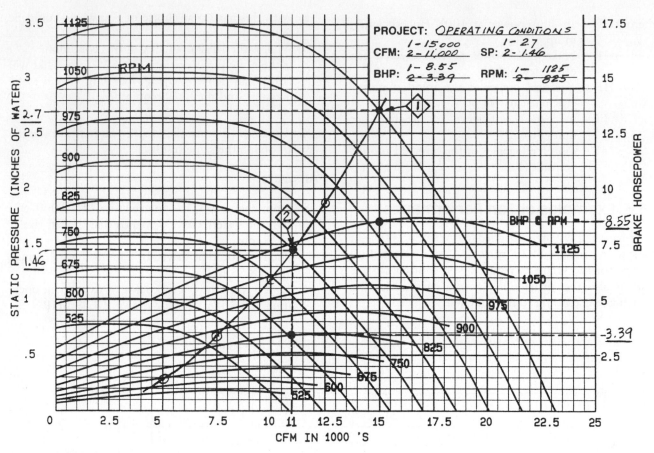

■ **FIGURE 6–14**
Fan performance curves for the fan in Section 6.5.6. (Courtesy of PennBarry.)

6.6 DUCT SYSTEMS

Ductwork is part of the air-handling system and includes the supply, return, outside air, relief air, and exhaust air ducts. Whereas the supply and return air ducts must be connected to the air-handling units, the other ducts may be run independently. Ducts are usually fabricated from sheet metal, such as galvanized steel, aluminum, or stainless steel; thus ductwork is also called *sheet metal work,* although some ducts are made with nonmetals, such as plastics.

6.6.1 General Classifications for Ductwork

Duct systems for supplying air may be classified as low pressure/velocity, medium pressure/velocity, and high pressure/velocity. For a given airflow, lower velocities reduce friction in the duct, and power for distribution. In addition, lower velocities reduce air noise.

Low-velocity ductwork is used for small airflow requirements, generally at final branches of the system. It can also be used for large air quantities where space is available for larger ductwork and the initial cost for the extra duct material is warranted. High airflow requirements generally dictate higher velocities to conserve space and duct materials. Table 6–4 shows typical ranges of acceptable velocities for various airflows through ductwork. According to the Sheet Metal and Air Conditioning Contractors' National Association, Inc. (SMACNA), there are seven velocity/pressure classifications for duct systems. There are also three types of sealing requirements that govern the type of joints and sealant.

TABLE 6–4
Pressure/velocity classification of ductwork

Class Type	Class Pressure	Operating Pressure, in. W.C.	Maximum Velocity, ft per min (fpm)	Design Velocity, ft per min (fpm)	
				Main	Branch
1	½″ w.c.	−½″ to +½″	2000	Up to 2000	Up to 800
2	1″ w.c.	−1″ to +1″	2500	Up to 2500	Up to 1200
3	2″ w.c.	−2″ to +2″	2500	Up to 2500	Up to 1500
4	3″ w.c.	−3″ to +3″	4000	Up to 4000	Up to 2000
5	4″ w.c.	−3″ to +4″	4000	Up to 4000	Up to 3000
6	5″ w.c.	5″ to +6″	5000	Up to 5000	Up to 4000
7	6″ w.c.	0″ to +10″	5000	Up to 5000	Up to 4000

6.6.2 Symbols for Sheet Metal Work

There are various standard symbols for sheet metal work to be used in drawings. One of the commonly accepted standards is the SMACNA Symbols for Ventilation and Air Conditioning, which includes sheet metal work and component devices related to ventilation and air conditioning systems. These standard symbols are found in most engineering handbooks.

6.6.3 Duct Shapes and Insulation Methods

Low-velocity ductwork is most often rectangular in cross section. Insulation is often applied to the interior of the duct and provides acoustical absorption as well as thermal benefits. When applied internally, insulation is called *duct liner*. Liner, which is applied in the fabrication shop, reduces heating or cooling loss from the duct. In cooling applications, liner also prevents condensation on the outer surface of the duct.

Round or flat oval ducts can be used for low-velocity applications, but these shapes are generally reserved for medium- and high-velocity ductwork. If insulation is required, it is usually applied externally in the form of a fiberglass blanket wrap with an external vapor barrier for cooling applications. The vapor barrier is intended to prevent the migration of humid air through the insulation, which could result in condensation on the surface of a cold duct. External insulation is more expensive than liner because of the field labor required for installing it.

Internal insulation is also available for round or flat oval ductwork. Internal insulation is preferred if ductwork is to be exposed to view. The material is scored and formed to the interior contour of round and flat oval ducts. Fitting this type of insulation is difficult, especially at elbows and fittings.

6.6.4 Materials of Construction

Galvanized steel is the material most widely used for general-purpose ductwork. Aluminum, stainless steel, or plastic may be used for ducts installed in humid environments or ducts that carry moist air, such as a dishwasher exhaust. Heavyweight, fire-resistant steel ductwork is used for ducts exhausting kitchen hoods over ranges and fryers. Laboratory fume hoods are often called upon to carry corrosive materials. Stainless steel, plastic, or regular steel ductwork with corrosion-resistant coatings are used in fume exhaust applications. Figure 6–15 illustrates typical low- and high-velocity duct construction and fittings.

All but the smallest rectangular ductwork requires bracing for rigidity. Bracing consists of angles or channels attached at intervals transverse to the duct. Bracing prevents the duct walls from deflecting excessively under pressure. Requirements for bracing depend on the cross-sectional dimension and internal pressure of the duct.

6.6.5 Duct Assembly

Figure 6–15 shows duct assemblies in an industrial plant, the method of connecting round ducts, and various duct configurations. Round ducts have the most cross-sectional area per unit area of sheet metal and thus are most economical. When ceiling space is limited, oval ducts fit the need. Round or oval ducts are available in double-wall construction, which incorporates perforated interior duct and insulation material to reduce the air-transmitted and duct-radiated noises.

There are many standard fittings designed to create smooth airflow, change direction, or accommodate changes in duct dimensions. All fittings cause air turbulence, resulting in pressure drop and power loss. These problems can be minimized by using low-loss fittings such as long-radius or segmented elbows, or special accessories such as turning vanes or extractors.

■ **FIGURE 6–15**
(a) Assembly of high-pressure/velocity round ducts with low-pressure rectangular ducts.
(b) Connecting round ducts. (Reproduced with permission from United McGill Corporation.)
(c) Typical construction and configuration of double-wall flat and round ducts. The perforated inside duct is separated from the outside duct by acoustic insulation material. (Reproduced with permission from United McGill Corporation.)

(a)

(b)

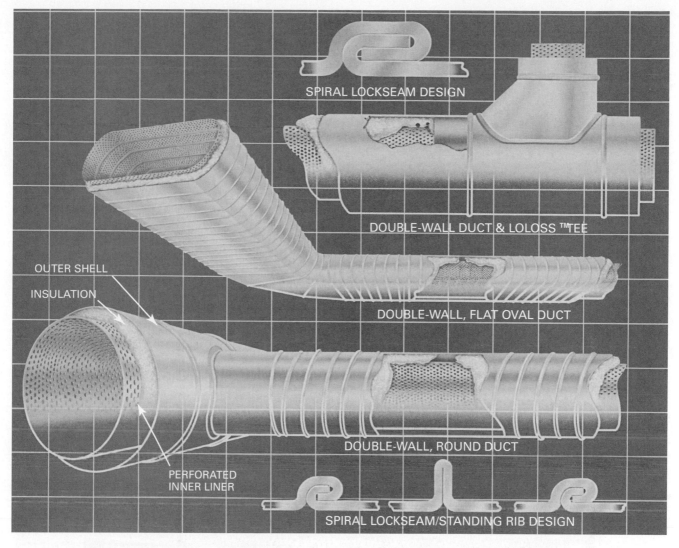

(c)

■ **FIGURE 6–15** *(Continued)*

Plastic Film and Fabric Ductwork

Plastic film and fabric ductwork is a good alternative to sheet metal ductwork in high-humidity spaces such as indoor swimming pools and greenhouses. The rigidity and tubular shape of these ducts is achieved by internal air pressure, and air is diffused through holes (plastic duct) or through pores (fabric). Both materials can achieve good diffusion at low velocity even with high air volumes. This feature is beneficial in preventing uncomfortable drafts on swimmers and turbulence on plants. Transparent-film plastic ductwork is also beneficial for maximizing light for plants (see Figure 6–16).

Fabric ductwork can be an inexpensive alternative to sheet metal in virtually any application where exposed round ductwork is desired for appearance. Color can be achieved without field painting, further reducing cost (see Figure 6–17).

6.6.6 Coordination of Ductwork with Other Building Elements

In most commercial buildings, ductwork shares space above the ceiling with other elements, including structural supports, fireproofing, electrical conduits, sprinkler piping, and light fixtures. All these elements must be coordinated to fit together and allow flexibility for future changes in their layout and function.

Clearances must be provided between ductwork and the lighting fixtures below them to allow the fixtures to be removed. Generally, 3 in. is adequate to raise and remove a fixture from a standard tee-bar grid. The clearance envelope around ductwork must allow for external insulation and bracing. Bracing may extend 1 to 3 in. from the surface of the duct, depending on the duct's size and pressure classification.

■ **FIGURE 6–16**
Plastic-film ductwork, used primarily in greenhouses for its transparency and water resistance, provides high-volume, low-draft ventilation. (Courtesy of WTA.)

■ **FIGURE 6–17**
Colored-fabric ductwork, especially appropriate for high-humidity spaces such as this pool area, can be used in any application where colorful, exposed ductwork is desired to complement the interior design concept. (Courtesy of Ductsox.)

Detailed planning should recognize that buildings are not constructed to tight tolerances. Also, a structural element will deflect under a load. A minimum of 2 in. of extra clearance should be planned. Figure 6–18 illustrates clearances and coordination with other elements in the ceiling.

6.6.7 Materials and Fittings for Sound Control

Sound attenuators, also called *sound traps*, are special duct fittings containing sound-absorbing material faced with perforated metal. They are designed to provide a

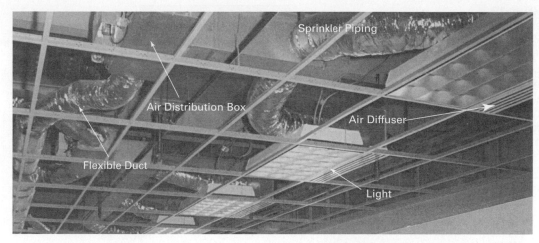

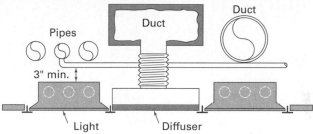

■ **FIGURE 6–18**
Congested ceiling space and the importance of coordination between systems.

large surface area to be in contact with the airstream. Sound traps are installed at the discharge of air-handling units to prevent fan noise from being transmitted into the distribution system. Sound traps are also used at return air openings or ductwork between ceiling plenums and noisy equipment rooms.

Double-walled ductwork (Figure 6–15c) is also used for sound attenuation. The inner wall is constructed of perforated metal, the outer wall of regular sheet metal. The cavity between the walls is packed with sound-absorbing material. Double-wall duct installed for 20 to 50 ft downstream from the air-handling unit will help absorb fan noise.

6.7 AIR DEVICES

6.7.1 General Classifications

Air devices, used for supplying air and removing air from spaces, are among the few parts of the HVAC system that are visible to the occupants of a building. The physical appearance of the devices and their organization with other building elements are important considerations, as are their capacity, performance, and maintainability. Air devices include diffusers, grilles and registers, flow control devices, and other accessories. Figures 6–19 and 6–20 illustrate a few of these devices.

6.7.2 Grilles and Registers

Grilles and registers are air devices equipped with vanes for directing airflow. They can be located in ceilings, walls, or floors. Vanes may be fixed or adjustable. Grilles simply contain vanes; registers contain a control damper behind the vanes.

Grille faces may be constructed of blades, bars, or egg crates of various finishes and materials. Heavy-duty grilles are available for floor locations and other applications that are subject to excessive wear and tear.

6.7.3 Ceiling Diffusers for Air Conditioning

Diffusers are ceiling-mounted air supply devices with louvers, slots, or vanes. Their function is to mix or diffuse supply air with room air without undue draft or localized hot or cold areas. Diffuser accessories and options include volume dampers (air quantity), pattern control mechanisms (direction of discharge), and means for removal of face elements to facilitate cleaning.

■ FIGURE 6–19

(a) Perforated ceiling diffuser with its duct connection through a flexible duct. (b) Wall-mounted diffuser (register) and its adjustable air distribution patterns. (c) Air distribution patterns of the perforated ceiling diffuser. (d) Coanda effect, which allows the supply air (during cooling operation) to mix gradually with the room air, thereby avoiding cold spots below the diffuser. (Courtesy of Titus.)

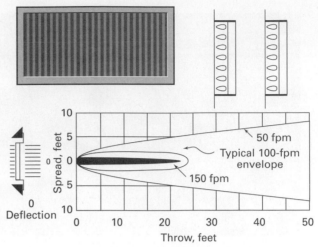

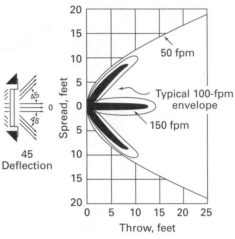

(b)

Spread versus throw at 0° horizontal deflection angle. (Plan view of air jet leaving wall-mounted grille or register.)

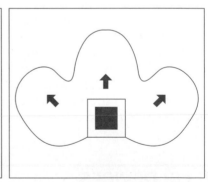

Spread versus throw at 45° horizontal deflection angle. (Plan view of air jet leaving wall-mounted grille or register.)

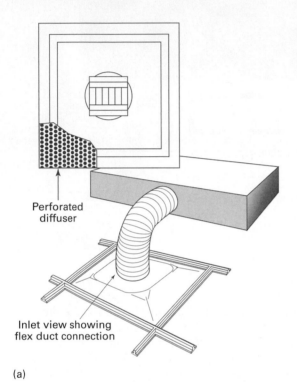

Perforated diffuser

Inlet view showing flex duct connection

(a)

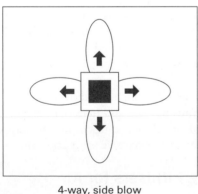

4-way, side blow plan view

4-way, corner blow plan view

3-way, adjustable blow plan view

(c)

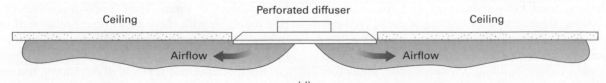

Ceiling   Perforated diffuser   Ceiling

Airflow ←   → Airflow

(d)

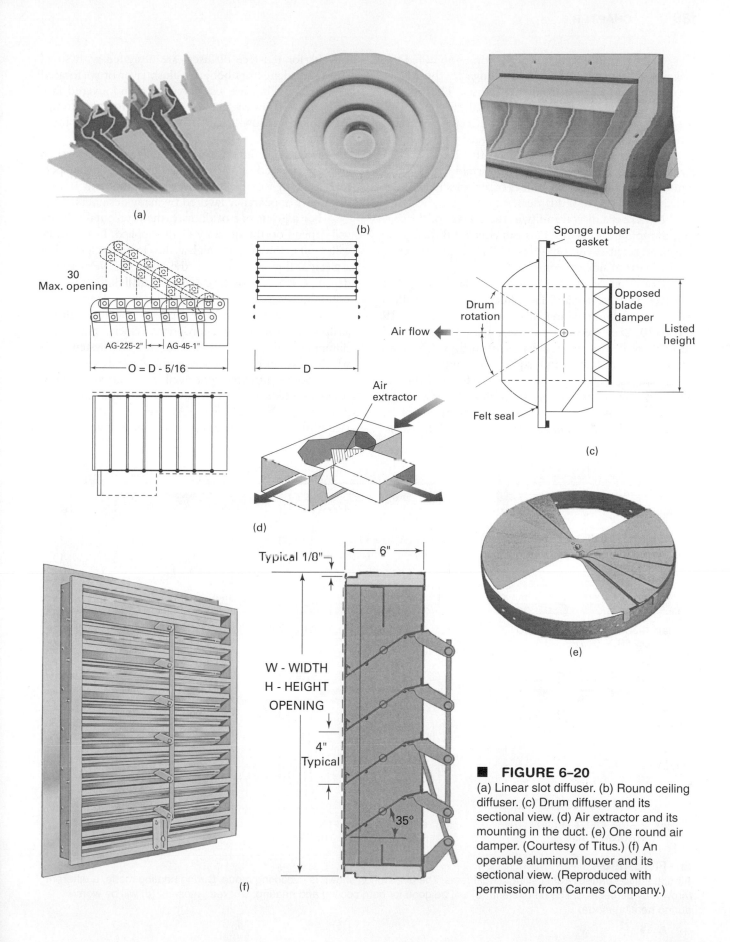

■ **FIGURE 6–20**
(a) Linear slot diffuser. (b) Round ceiling diffuser. (c) Drum diffuser and its sectional view. (d) Air extractor and its mounting in the duct. (e) One round air damper. (Courtesy of Titus.) (f) An operable aluminum louver and its sectional view. (Reproduced with permission from Carnes Company.)

Ceiling diffusers rely on a phenomenon called the *Coanda effect,* in which cold air clings to the ceiling upon being discharged from the diffuser. The Coanda effect allows the airstream to fall gradually and mix with the room air over a large area. If the effect is not established, the chilled air will fall downward from the diffuser and create a cold spot beneath. This is called *dumping* and results in an uncomfortable cold draft. Figure 6–21 illustrates the effect of air velocity and the distribution pattern of diffusers.

If there is a discontinuity in the surface of the ceiling, dumping can result. This can occur if diffusers are placed adjacent to light fixtures with parabolic lenses or on the bottom side of soffits.

Common types of diffusers include those with a louvered face, perforated face, linear slot, and troffers placed over lighting fixtures, all illustrated in Figures 6–19 and 6–20. Louvered-face diffusers are economical and perform well over a wide range of airflow. They are available round, half-round, square, rectangular, and in other shapes, with mounting frames for virtually all types of ceiling construction. In modular lay-in ceilings, they may be installed in the face of a ceiling tile or be provided with a flat extension to fill the module.

Perforated-face diffusers are designed with small louvers or deflectors behind a flush layer of perforated sheet metal. They are less obtrusive than louvered-face diffusers. Perforated-face diffusers are prone to dumping at low airflows, which occur in variable-volume applications.

Linear-slot diffusers are generally more costly than either perforated or louvered-faced diffusers. They perform well in variable-volume applications and have a distinctive appearance favored by many designers.

For a given size of diffuser, the horizontal velocity will depend on the quantity of air supplied. Low airflow will result in low velocity, high airflow in high velocity. For a given airflow requirement, the horizontal velocity will depend on the size of the diffuser. A small diffuser will result in high velocity, a large diffuser in low velocity.

The size of the diffuser must be matched with the airflow to develop an appropriate velocity and avoid dumping. This can be a problem with VAV systems, for which a diffuser is required to perform over a range of airflows. Some diffusers cope well with varying airflow, whereas others do not. Louvered-face and linear-slot diffusers work well for VAV systems. Perforated-face diffusers tend to dump at low airflow and should

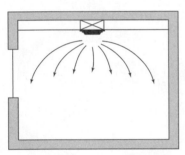

(a) GOOD COVERAGE BASED ON PROPER THROW

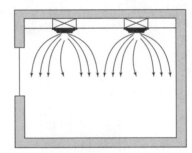

(b) GOOD COVERAGE WITH PROPERLY SELECTED MULTIPLE DIFFUSERS

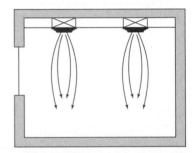

(c) POOR COVERAGE. DUMPING OF COLD AIR. THROW IS NOT ESTABLISHED

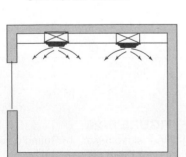

(d) INSUFFICIENT THROW. STAGNANT AIR

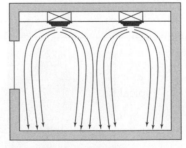

(e) COLD DOWNDRAFTS FROM EXCESSIVE THROW

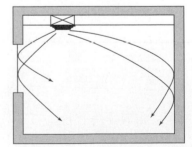

(f) EXCESSIVE THROW. AIR TURBULENCE

■ **FIGURE 6–21**

Air distribution pattern from ceiling air diffusers. The distribution shown is in cooling mode. During heating mode, a different throw may be desired. The diffusers in (b) will be good for both cooling and heating, whereas those in (d) will be worse during heating mode.

be avoided in spaces where large load variations are expected.

Different types of diffusers can be used in the same space. For example, a large office might effectively use perforated-face diffusers at the interior, which exhibits a relatively constant load, and linear slot diffusers at the perimeter wall, where load variations are significant.

6.7.4 Special Concerns for Warm-Air Supply

Areas close to cold surfaces, such as large windows, are subject to uncomfortable downdrafts. If air is supplied at the floor or windowsill under the cold surface, the downdraft can be offset. If ceiling devices are used to deliver warm air, the velocity must be sufficient to create turbulence and mixing with cold air at the offending surface.

Using the same ceiling diffuser to deliver heating and air conditioning can be tricky. Warm air has a tendency to stay at the ceiling, owing to its low density. If the warm air simply floats to the nearest return opening, the space will be deprived of any heating effect. Again, adequate velocity is essential because it will promote turbulence and mixing with room air to distribute heat.

Generally, the amount of warm air required to heat a perimeter space is less than the amount of chilled air required to cool the same space. If the diffuser is selected solely on the basis of good chilled-air distribution, it may be too large to impart adequate velocity to the smaller quantity of warm air. In this case, the warm air may simply float at the ceiling and be ineffective. The problem can be avoided by compromising on a diffuser that is slightly smaller than ideal for cooling airflow and slightly larger than ideal for heating airflow.

6.7.5 Spacing, Distribution, and Area of Coverage

The area that can be effectively served by a diffuser is affected by a parameter called *throw*, which is a measure of the distance from the diffuser at which the room air velocity is within a specified range for a given airflow. In most applications, a room velocity of 25 to 75 feet per minute is desirable. Minimum velocity is required to avoid complaints of stuffiness. If the velocity is too high, drafts will result. (See Figure 6–21.)

Diffusers must be selected and spaced to result in good coverage based on throw. If they are too far apart, areas of dead airflow will result; if they are too close together, airflows may impinge at the ceiling and create downdrafts. Uncomfortable downdrafts may also occur if air is distributed too close to a wall.

Diffusers are placed and selected to deliver airflow in the right directions for good distribution. The direction is adjustable, as in the case of linear-slot diffusers, which contain blades to vary the amount of air delivered from either side. Directionality can also be accomplished by the selection and adjustment of louvers or pattern control devices, as in the case of louvered-face or perforated-face diffusers.

The direction of blow should be considered in selecting the right design based on the placement of the diffuser in the room or its position with respect to other devices. A four-way blow device may be placed in the center of a room or distributed uniformly in a large space to provide even air distribution. Other options for the geometry of placement and distribution are shown in Figure 6–21.

Linear-slot diffusers perform well in variable-air-volume applications. Owing to their shape, they are strong visual elements, requiring attention to their organization with other ceiling devices, including light fixtures and sprinkler heads.

Troffers placed over light fixtures deliver air from slots designed integrally with the light fixtures. This method of air distribution is least obtrusive visually. Distribution characteristics are satisfactory for variable-air-volume applications, but good workmanship is essential to avoid leaks at the joint between the troffer and the light fixture.

6.8 GENERAL GUIDELINES FOR DUCT SYSTEM DESIGN

A well-designed duct system will result in the lowest installation cost and most energy-efficient operation of the air-handling system. Duct sizes are selected that will produce the highest possible velocity consistent with a reasonable friction loss. Extensive duct systems in multistory buildings tend to have higher duct velocities in order to minimize the size and associated cost of the ductwork. A penalty is paid in the form of a higher operating cost, because the fans operate against a higher pressure. Smaller duct systems typical of one-story buildings tend to have lower duct velocities, lower system pressures, lower installation costs, and lowest operating costs.

Duct installations are most economical when maximum use is made of straight duct runs, and the number of fittings is minimized. Design of fittings is important to reduce pressure requirements. Elbow fittings should have turning vanes. Butt-head tee fittings should be avoided. Branch ducts in higher-velocity duct systems should be connected to the main duct using factory-made aerodynamically designed fittings.

Duct systems that are not properly designed will create objectionable noise. Insulation placed inside the ductwork will attenuate noise radiating from the system. Lined ductwork is particularly cost-effective for cold-air ducts, in that the lined duct is an insulator as well as a sound attenuator. Return-air and exhaust-air ducts are at a neutral temperature and are lined only when acoustical considerations dictate doing so.

Doors should be provided in the duct walls to allow access for cleaning inside the duct and for maintaining components that are sometimes installed in the ductwork. Figure 6–22 (p. 191) is a simple supply air duct drawing with design criteria and notes to identify the ductwork, airflow rate, air velocity in ducts, friction loss, and fittings. Figure 6–23 is the same simple ductwork coordinated with the ceiling grid and lighting fixtures. Coordination of the ductwork with the lighting layout is an important criterion in the architectural design process.

6.9 UNDERFLOOR AIR SYSTEMS

6.9.1 Access-Floor Distribution

Delivering air from underfloor is an alternative to conventional overhead air distribution. Underfloor distribution for commercial buildings generally uses 2- × 2-ft access-floor panels supported by pedestals on the structural slab to form an air plenum, as shown in Figure 6–24. When the floor is used for air distribution, conventional ceiling plenums can be eliminated or reduced in height, so that the overall height of the building can remain the same or even be shorter.

Air is introduced to the space through floor-mounted diffusers such as those shown in Figures 6–25 and 6–26, which are specially designed with cleanable receptors for dust and debris. Because air is delivered at the floor, velocities must be low and temperatures high enough to prevent uncomfortable drafts. Supply temperatures for air conditioning are generally 60°F to 65°F.

6.9.2 Effect on Room Loads

Underfloor distribution can be used to create a displacement effect whereby air is stratified in the space with cool, comfortable temperatures in the occupied zone and warmer air higher in the space. Convective heat from light fixtures does not enter the occupied zone, and heat from people and appliances rises to the warm upper zone locally in plumes around the heat source.

The advantage of the displacement phenomenon is that much of the heat gain from space sources is conducted to the return without entering the occupied space. The result is that less supply air is needed to offset these loads. Conversely, conventional ceiling distribution systems rely on diffusion (mixing) of supply air with room air, and these loads are introduced into the space rather than being conducted upward and out of the space.

6.9.3 Humidity Control

In climates with low humidity, high supply air temperatures can be used without humidity problems. In climates with humid weather, primary air must be cooled to 55°F or colder to achieve dehumidification. To attain higher air temperatures suitable for floor distribution, the air must be reheated or blended with return air. Reheat wastes energy not only by introducing heat but also by making it necessary to remove the heat from the return air. Blending of return air is the preferred energy solution. Air can be blended at the central air-handling unit or at fan terminals, which are generally located in the floor plenum. Desiccant cooling is another option for achieving dehumidification, though it is more expensive to install.

6.9.4 Air Quality Advantage

The upward movement of air also has the advantage of carrying indoor air pollutants such as body odors and small particulate matter directly to the return rather than allowing it to mix in the overall space. The occupied breathing zone is thus cleaner, and fresh, filtered supply air is used more effectively to maintain good indoor air quality.

6.9.5 Applications

The plenum must be tall enough to accommodate airflow and can be as short as 6 in., though, generally, it is higher to allow room for ductwork to conduct air from shafts to remote parts of the space and terminal boxes if required for zoning or terminal temperature blending.

Underfloor systems can have a major life-cycle advantage in office buildings where the floor plenum is also used for distribution of electrical and data wiring, which must be run in conduit or be otherwise code rated for air-plenum application. Power and data outlets, as well as supply air devices, are affixed to modular floor panels and can be rearranged.

Low-velocity floor-mounted air supply devices are also an effective solution for fixed-seating applications such as auditoriums and tiered classrooms. Low velocity results in low noise, and the upward movement of air results in good air quality in the occupied space. These

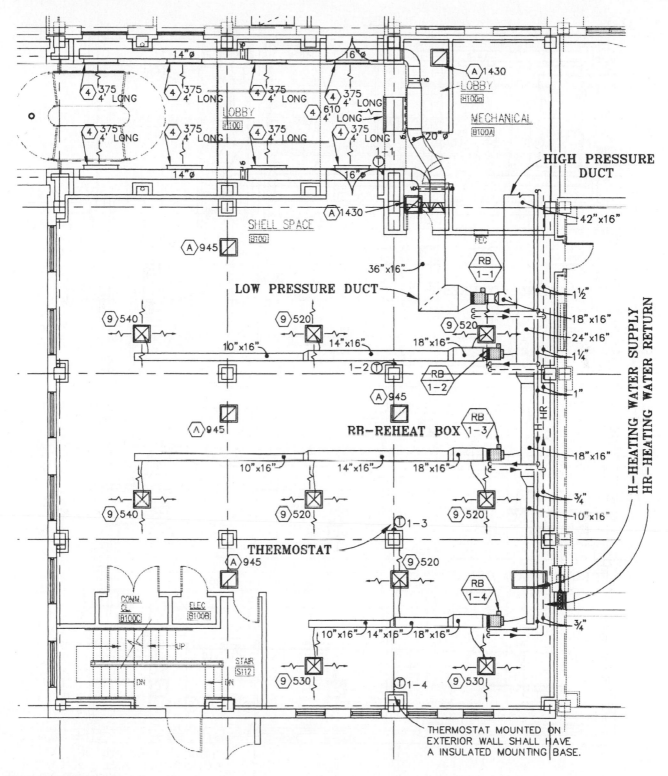

■ FIGURE 6–22

Ductwork plan of a simple HVAC system showing supply air ducts and heating water piping. Design criteria notes are added for reference only.

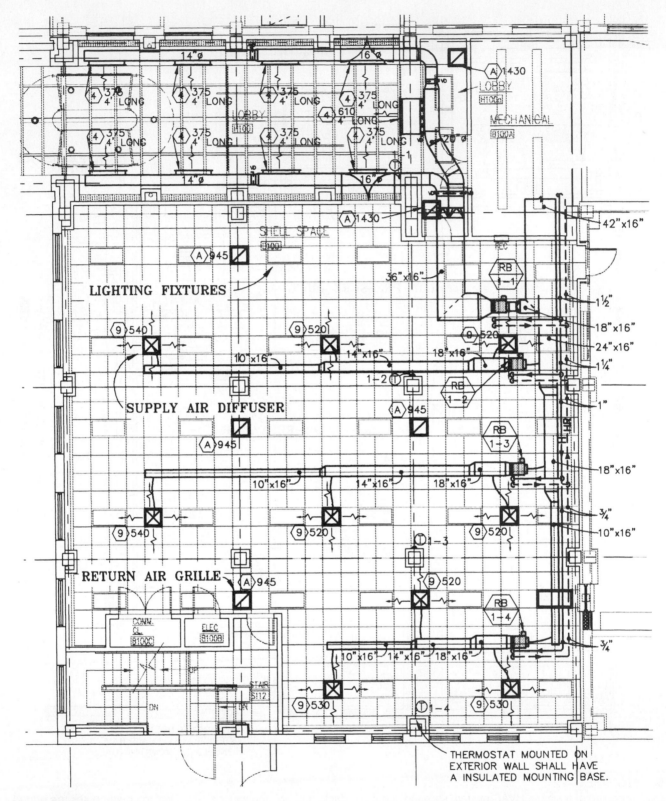

■ FIGURE 6–23

Ductwork plan of the HVAC system shown in Figure 6–22. Ceiling grid and lighting fixtures are added for coordination between different space elements.

Conventional System

Access Floor System

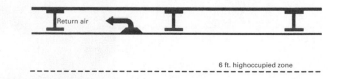

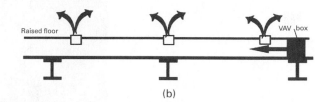

(a)   (b)

■ **FIGURE 6–24**

Conventional system (a) using overhead distribution, diffusing pollutants and heat into the occupied zone. Access-floor system (b) introduces clean, filtered air into the occupied zone and assists with convection to push convected heat upward to returns. (Courtesy of Titus.)

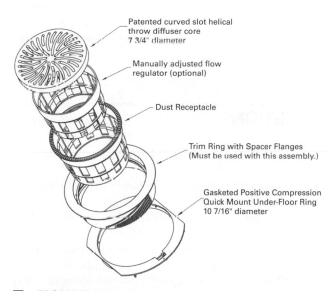

■ **FIGURE 6–25**

Floor diffuser fully mixes air within a short radius, is adjustable for personal comfort, and is equipped with dust receptacle for housekeeping. (Courtesy of Titus.)

spaces generally have high ceilings, so the energy advantage of temperature statisfaction is also beneficial.

Monumental high spaces such as lobbies and atriums are likewise good applications for floor-mounted supply devices owing to the potential air stratification by temperature. A comfortable environment can be maintained in the occupied zone while allowing warmer air to stratify in the high spaces.

6.10 NATURAL VENTILATION

6.10.1 Historical Perspective

Prior to widespread use of air conditioning in the early 1950s, natural ventilation was not only commonplace but necessary as a means of limiting interior temperatures during warm weather. The use of natural ventilation was facilitated by high ceilings, large windows, transoms, and relatively narrow floor plans, which allowed cross ventilation of buildings. The use of ducted cooling systems allowed spaces to be kept more comfortable and, fueled by cheap energy, became the norm for building design in subsequent years. The widespread use of air conditioning has led to architecture with lower ceilings and large, deep floor plates, and nonoperable windows.

6.10.2 Common Arguments against Operable Windows

Owners have become so accustomed to air conditioning in commercial and institutional buildings that the notion of operable windows is often met with opposition. There are several arguments against operable windows. First, open widows interfere with controls of air-conditioning systems. For instance, several perimeter offices may be served by a single VAV terminal controlled by a single thermostat in one of the offices. If the window is open in

■ **FIGURE 6–26**
Floor diffuser on access floor with carpet tiles. (Courtesy of WTA.)

the office with the thermostat, that office's temperature will not be typical of the others, and good temperature control will not be possible. Another common argument against operable windows is the possibility that pipes will freeze in the building if windows are left open in cold weather. Reliance on automatic controls may have withered occupants' willingness to operate windows.

A more valid concern is that open windows allow dust and allergens unfiltered into occupied space, which is certainly a housekeeping problem and can have health consequences. Moreover, operable windows can work against the control of building airflow and pressure relationships, which is functionally necessary in some applications. For instance, a research laboratory might need to control building pressure relationships to prevent the spread of odors or contaminants in the building. With operable windows, airflow control is difficult if not impossible. Tall buildings are subject to the *stack effect*, which draws cold air low into the building during winter and warm air into the building during summer.

6.10.3 Energy Savings

Nonetheless, natural ventilation can be an effective strategy for energy conservation. Energy conservation accrues from natural cooling and ventilation. In addition, occupants are tolerant of wider temperature and humidity ranges when the windows are open. The sense of control and connection with the outdoors make natural ventilation a desirable feature for most building users.

In proper applications energy savings can range from 5 to 25 percent depending on climate and control

strategy. In climates with short mild seasons in spring and fall, savings will be small. In mild climates properly controlled natural ventilation can be used most of the year, saving not only energy for ventilation but also energy for cooling.

QUESTIONS

6.1 What is the difference between an air-handling unit and a fan coil?

6.2 Why are the casings of air-handling equipment insulated?

6.3 Why can heating coils be designed for higher face velocity than cooling coils can?

6.4 What construction materials are appropriate for indoor air-handling unit casings? Outdoor?

6.5 What is the advantage of bag-type filters over flat filters?

6.6 What type of air filters would be appropriate for hazardous particulate materials?

6.7 What are the advantages of blow-through air-handling units in comparison with draw-through units?

6.8 What is the chief consideration in designing air-handling unit sections for mixing outside air and return air?

6.9 Why is plastic-film ductwork especially appropriate for greenhouses?

6.10 What is the advantage of belt-drive fans compared with direct-drive fans?

6.11 What is the limitation of low-velocity ductwork in a large distribution system?

6.12 When might aluminum or stainless steel be preferred over galvanized steel duct construction?

6.13 When might a plug fan be considered instead of a centrifugal fan?

6.14 When is fabric ductwork an appropriate alternative to sheet metal ductwork?

6.15 What is the function of a diffuser in an HVAC distribution system?

6.16 Name several conditions that could cause uncomfortable dumping of cold air from a diffuser.

6.17 Briefly compare the benefits of commonly used diffusers.

6.18 What potential problems do variable-air-volume systems pose in the selection of diffusers?

6.19 What special provisions should be made to prevent heating supply air from stratifying at the ceiling?

6.20 What are the benefits of using light fixture slots for air return from the room to the ceiling?

6.21 What is the significance of NC levels listed in catalogs? Will NC levels be higher or lower as the airflow increases for a given diffuser? Why?

6.22 A fan with the curve shown in Figure 6–14 is required to deliver 10,000 cfm at a static pressure of 2.2 in. w.c. What will be the required fan speed and brake horsepower?

6.23 If the motor driving the fan in question 6.22 is 95% efficient, what is the electric power consumption for the fan?

6.24 How might floor air distribution affect thermal loads compared with overhead air distribution?

6.25 Why is the use of a return fan or a relief fan good practice in the use of the design of large air-handling systems?

6.26 Compare the air quality of coventional overhead air distribution with floor distribution.

6.27 Name the three commonly used methods for varying fan output. Which is the most efficient?

6.28 What applications might be appropriate for considering floor distribution? Why?

6.29 Briefly discuss the pros and cons of natural ventilation for a low-rise faculty office wing.

6.30 Briefly discuss the pros and cons of natural ventilation for your current residence.

6.31 Briefly discuss the pros and cons of natural ventilation for a high-rise office building.

PIPING EQUIPMENT AND SYSTEMS

7

In addition to air-handling systems, piping systems are another means of conveying (transporting) heating or cooling energy. In fact, for most applications, air-handling and piping systems must complement each other. For example, without piping to carry chilled water or refrigerants to coils, air could not be cooled by the coils and, consequently, could not cool a building.

Hot and chilled water, refrigerants, steam and condensate, and gas and oil are fluids transported through piping in HVAC systems. This chapter addresses the basic components, accessories, and materials of construction for piping used in the systems described in other chapters.

The flow of fluid in a piping system depends on a number of factors, such as properties relating to fluidity (viscosity and specific gravity), phase (liquid, vapor, or gas), and physical status (temperature, pressure, and velocity).

Like ductwork, piping should be large enough to prevent a high pressure drop (friction) and high noise generation. There are special accessories designed for piping systems, such as valves, couplings and air controls in water piping, and oil separators in refrigerant piping systems. Other special considerations in piping design include corrosion control, pressure control, flow control, and control of pipe expansion and contraction.

7.1 PIPING SYSTEMS AND COMPONENTS

Piping systems are classified as follows:

1. *Service provided.* Heating, cooling, oil supply, condensate return, etc.
2. *Medium conveyed.* Steam, hot water, chilled water, refrigerant, oil, gas, condensate, chemicals, etc.
3. *Pressure class.* Low-, medium-, or high-pressure classes. The pressure ranges vary with the types of media to be conveyed, such as steam, water, etc.

4. *Temperature class.* Low-, medium-, or high-temperature classes. The temperature ranges vary with types of media to be conveyed.
5. *Piping arrangement.* One-pipe (monoflow) or two-pipe system, direct or reverse return, series or parallel flow, etc.
6. *Hydraulics.* Gravity or forced flow, open or closed.
7. *Piping materials.* Steel, copper, plastic, nonmetallic, etc.

7.1.1 Forced or Gravity Flow

Most piping systems use pumps or compressors to maintain the flow of the fluid. This is called a *forced system*, however, pumps are not always required. Steam flows in a piping system because of the pressure differential between the steam source and the devices that use the steam. Steam condensate flows by gravity or, if adequately cooled, can be pumped. Refrigerant is generally forced through a piping system by the action of a compressor if the refrigerant is in the vapor phase or a pump if it is in the liquid phase.

7.1.2 Open or Closed System

In an open system, the fluid surface is exposed to the atmosphere. See Figure 7–1(a) to (c). The flow may be either by gravity or by pumping. In a closed system, fluid is not in contact with the atmosphere. The flow must be by pumping. Most HVAC piping systems are closed systems. Figure 7–1(d) to (h) illustrates various closed-system piping arrangements.

In an open system, the net pressure for the pump to overcome is frictional pressure plus the pressure equivalent of the height of the fluid being lifted. In a closed system, the net pressure for the pump to overcome is only the frictional pressure. It makes no difference whether various pieces of equipment are at different elevations, since the elevation is balanced at both sides of the pump. Closed systems are always preferred in designing a pumping system.

■ FIGURE 7–1

Hydraulic circuit of water systems. (a) Open system with gravity flow and no return.
(b) Open system with forced flow. (c) Open system with forced flow and no return.
(d) Simple closed system. (e) Closed system with reverse return. (f) Closed system with direct
return. (g) Closed system with series pumping. (h) Closed system with parallel pumping.

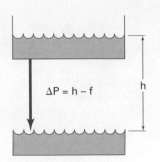

(a)

*h – height, ft.
f – friction, ft.
P – pump head, ft.
ΔP – Net pressure, ft.

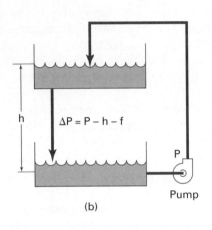

$\Delta P = P - h - f$

(b)

Pump

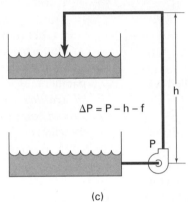

$\Delta P = P - h - f$

(c)

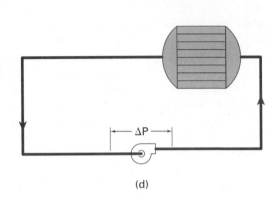

ΔP

(d)

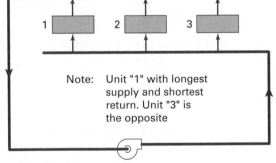

Note: Unit "1" with longest supply and shortest return. Unit "3" is the opposite

(e) Reverse return

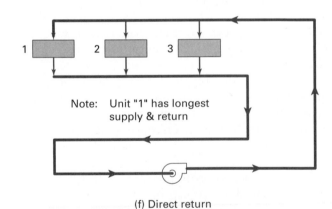

Note: Unit "1" has longest supply & return

(f) Direct return

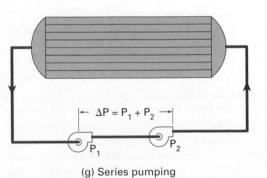

$\Delta P = P_1 + P_2$

(g) Series pumping

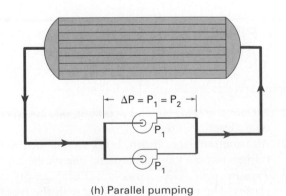

$\Delta P = P_1 = P_2$

(h) Parallel pumping

In the direct-return system in Figure 7–1(f), equipment closest to the pump has the least pressure loss, and thus may have more or imbalanced water flow. The reverse-return piping arrangement in Figure 7–1(e) is preferred, since the water flows through multiple units of HVAC equipment and is inherently balanced. Reverse-return piping systems usually cost more because more pipe is needed.

For equal-size pumps, a series arrangement doubles the head, as in Figure 7–1(g), and a parallel arrangement doubles the flow. Either arrangement can be used to provide continuity of service and save energy if one of the two pumps is shut off. (See Section 7.2.)

In an open, pumped system, the pressure that causes the fluid to flow is the pump pressure minus the static height and the frictional loss. This relation is expressed as

$$\Delta P = h \div f \qquad (7-1)$$

where ΔP = pressure rise through pump, ft
$\quad h$ = static head or difference in elevation, ft
$\quad f$ = frictional loss at the rated flow, ft

Pressure expressed in feet (of water column) can be converted to other units by the following relations:

- 1 psi = 34 ft/14.7 psi = 2.31 ft w.c.
- 1 psi = 29.9 in./14.7 psi = 2.03 in. Hg
- 1 psi = 2.31 ft × 12 in./ft = 27.7 in. w.c.
- 1 in. w.c. = 0.036 psi
- 1 atm = 14.7 psi

In a closed system, static height is not a factor, because it is balanced on both the suction side and the discharge side of the pump. This pressure relation is expressed as

$$\Delta P = f \qquad (7-2)$$

By comparing Eqs. (7–1) and (7–2), it should be evident that a closed piping system is preferred when fluid must be circulated at different elevations.

7.1.3 Water as the Basic Medium

Water is the most common fluid used in HVAC systems, because it is chemically stable, nontoxic, and economical. Water has a density of 62.4 lb per cubic foot and is considered to have a specific heat of 1.0. This means that 1 Btu will raise the temperature of 1 lb of water 1°F.

7.1.4 Basic HVAC Piping Systems

Common piping systems for HVAC applications are water, steam (water in the vapor phase), refrigerant (Freon, ammonia, water), fuel (oil, gas), and other fluids. Pipes are used to transport the fluids to and from system components, which may be pumps to provide power for transport or heat-exchanging equipment to transfer the fluid's thermal energy. Valves that govern the flow of fluid in the pipes are important components in a piping system and will be discussed later in the chapter. Another important component is the instrumentation. This includes temperature, pressure, and flow gauges. (See Figure 7–2.)

(a)

(b)

(c)

■ **FIGURE 7–2**
Some basic instruments. (a) Pressure gauge.
(b) Thermometer. Without proper instrumentation, the performance of a fluid system cannot be confirmed, tested, balanced, or adjusted. (c) Temperature instruments are usually installed in thermal wells to separate the instruments from direct contact with the fluid. (Reproduced with permission from Palmer Instruments Inc.)

7.2 PUMPS

7.2.1 Classification

Pumps are basic equipment in a hydronic system. Depending on their operating principles, construction, and functions, they may be classified and used as follows:

1. *Operating principles.* Positive displacement or centrifugal type.
   - For positive-displacement type—reciprocating piston, reciprocating plunger, rotary blades, rotary roots, screw types, etc.
   - For centrifugal type—radial flow, axial flow, or mixed flow, etc.
   - For turbine type—axial flow, mixed flow, etc.
2. *Casing design.* Horizontal split case, vertical split case, submerged, etc.
3. *Mounting.* Base mounted, in-line mounted, etc.
4. *Connection with driver (motor or engine).* Flexible coupled, close coupled, belt driven, etc.
5. *Construction material.* Cast iron, stainless steel, bronze, concrete, plastic, fiberglass, etc.
6. *Service functions.* Hot-water supply, condensate return, vacuum, oil supply, boiler feedwater, etc.

Centrifugal pumps are used in most HVAC applications. Commonly used pumps are shown in Figure 7–3. Small pumps can be mounted in-line with the piping system; larger pumps are mounted on the floor. Virtually all pumps are flexibly coupled with the motor that drives them.

Turbine pumps can be used effectively for applications requiring water to be pumped from a sump into a piping system. The most common application of a turbine pump is pumping water from a cooling tower sump to water-chiller condensers.

Positive-displacement pumps use pistons or rotary gears to pump oil for boilers and engine generators. Because they are self-priming, positive displacement pumps can be used to draw oil up from underground tanks.

(*Note:* The term *self-priming* means that the pump can start pumping the fluid without the need to prefill its intake.)

Pumps are selected to meet flow criteria and to develop appropriate pressure to move fluid through the system. Water flow is measured in gpm; pressure, often called "head," is measured in psig or, more commonly, feet of water. Pumps are usually driven by motors rated at standard induction speeds. For 60-Hz power, the common induction motor speeds are 1150, 1750, and

3500 rpm. These speeds correspond to the synchronous speed at 1200, 1800, and 3600 rpm. The difference between the induction speed and the synchronous speed, called *slip*, is due to torque under load, which slows the motor slightly.

7.2.2 Pump Performance

Performance Curves

A specific size of a pump operating at a given speed can deliver a range of flows and pressures, depending on the diameter of the impeller installed in the casing. Impellers can be trimmed to match the precise requirements for a specific application. When the impeller becomes too small or too large, a new pump casing size is selected. A set of pump performance curves for different impeller trims is shown in Figure 7–4(a).

System Curve

The system curve is the pressure requirement of the piping system at various flow rates. For a given flow rate, the pressure is the sum of the pressure drop of the system components, including pipes, fittings, equipment, control devices, and static head. If the flow rate is changed, the pressure will change in proportion to the square of the flow. For a closed system:

System pressure = Pressure drop of
(piping + fittings + equipment + controls) (7–3)

$$(q_2/q_1)^2 = p_2/p_1$$
$$\text{or } p_2 = p_1 \times (q_2/q_1)^2$$
$$\text{or } (q_2)^2 = (q_1)^2 \times (p_2/p_1) \quad (7\text{–}4)$$

where p_1 = system pressure required at flow rate q_1
p_2 = system pressure required at flow rate q_2
q_1 = system flow rate at condition 1
q_2 = system flow rate at condition 2

A system curve can be plotted as long as one of the operation conditions (e.g., q_1 and p_1) is known. All other points on the system curve can be plotted using Eq. (7–4) and the known q_1 and p_1.

An example of system flow characteristics is plotted on the pump curves shown in Figure 7–4(b). The pump operating point is the point where the pump curve intersects the system curve. At this point, the head and flow characteristics of the pump match those of the system.

Power Requirement

Another set of curves on the pump performance curve in Figure 7–4(a) indicates horsepower requirements. This information is used to select the proper size of

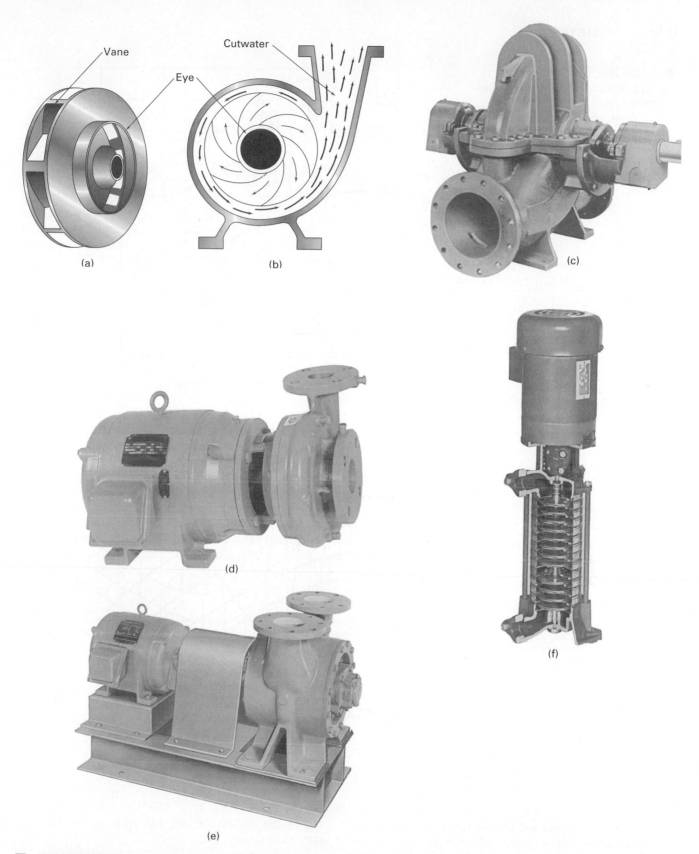

■ **FIGURE 7–3**

Centrifugal and turbine pumps. (a) Basic impeller. (b) Directional arrows indicate the flow pattern and illustrate why the impeller direction must be toward the discharge opening for proper operation. Reversing the direction of rotation causes turbulence and reduces pump capacity, at the same time increasing motor loading. (c) Horizontal split-case centrifugal pump. (d) End-suction, direct-coupled centrifugal pump. (e) Pump-motor assembly with mechanical coupling. (The coupling is covered by the protection guard and is not visible.) (f) Multistage turbine pump. The impeller of each stage may be removed individually. (Reproduced with permission from ITT Fluid Handling Sales.)

■ **FIGURE 7–4**

(a) Typical pump performance curves indicating head, flow rate, brake horsepower, and efficiency for different size impellers. (*Note:* NPSH = net positive suction head.) (b) Same pump performance curves with system curve plotted.

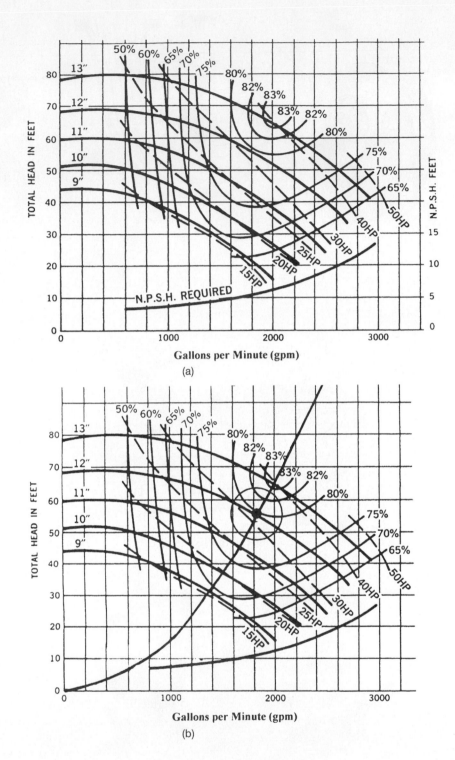

motor. Often, a motor will be selected that is capable of the maximum requirement indicated by the pump curve, regardless of the normal system operating conditions. If the motor is grossly oversized (or underloaded), it will operate at a low power factor. This situation should be avoided by selecting a smaller motor. Standard integral HP motor sizes are 1, 2, 3, 5, 10, 15, 20, 25, 30, 40, 50, and up to several hundred HPs. (See Section 7.3.)

Example 7.1 Given the pump performance curves shown in Figure 7–4(a) with a 12-in. impeller, the pump should deliver about 1800 gpm at about 55 ft of head. Operating at this condition, the pump will require a 32-BHP motor and is operating at about 80 percent efficiency. The pump performance curve is shown in Figure 7–4(b). A standard 40-BHP rated motor will be used.

(*Note:* BHP = brake horsepower, the mechanical equivalent of electrical power.)

Net Positive Suction Head

When fluid enters the impeller "eye" of a centrifugal pump, fluid pressure is reduced, owing to the greatly increased velocity at the impeller. The reduced pressure may cause the fluid to "flash" or "vaporize," inside the pump. This phenomenon causes cavitation and damages the pump. *Net positive suction head* (NPSH) is the characteristic of a centrifugal pump that must be considered during the design of the piping system. Each centrifugal pump has a specific NPSH pressure rating that must be ensured by the piping system design. For a given pump, the required NPSH increases with the flow rate, varying from a few feet up to 30 or 40 ft. NPSH is not normally critical in closed systems, however, in open systems, such as those for cooling tower water, the placement of the pumps in the piping system is critical.

Capacity Control

Capacity control for variable-flow water systems can be accomplished by throttling (restricting) valves, staging of multiple pumps, variable-speed motors, or combinations of these methods. In any case, the pump or pumps are selected for the peak design flow. When throttling occurs, the system pressure changes, and a new system curve is generated, as shown in Figure 7–5(a) The new system curve will intersect the pump curve at a lower flow and higher pressure. Another strategy for variable-flow systems is to use variable-speed pumping. In this case, the system curve remains constant, and the pump curve expands or contracts according to the change in speed, as shown in Figure 7–5(b).

Arranging two or more pumps in parallel or in series is another method of varying the performance of the pumps. Individual pump heads are additive, as shown in Figure 7–5(c) for series operation, and the pump flow rate is additive for parallel operation, as shown in Figure 7–5(d). These figures show the performance of the pump for single- and dual-pump operation against a given system curve.

7.2.3 Basic Hydronic Formulas

The flow of fluid in a piping system is similar to that of airflow in a duct system. In both cases fluid power is proportional to flow and pressure. For piping systems the fluid power is provided by a pump. Power requirements will vary, depending on the fluid viscosity and specific gravity.

The basic formulas for determining power are as follows:

$$\text{WHP} = (q \times h \times \text{SG})/3960 \qquad (7\text{–}5)$$

$$\text{BHP} = \text{WHP}/\text{Ep} \qquad (7\text{–}6)$$

$$\text{Ep} = \text{WHP}/\text{BHP} \qquad (7\text{–}7)$$

$$Q = 500 \times \text{TD} \times q \qquad (7\text{–}8)$$

where
- q = flow rate, gpm
- h = system head, ft w.c.
- SG = specific gravity. For water = 1.0; for other fluids, SG is determined by the ratio of the specific weight of water to the specific weight of the other fluids
- Ep = pump efficiency, per unit
- WHP = water horsepower (theoretical power required)
- TD = temperature difference, °F
- Q = heat, Btu/hr (Btuh)
- BHP = brake horsepower (the actual power required, including the pump's efficiency)

Affinity laws provide methods for approximating the effects of changed impeller diameters and rotational speed on the pump curve. These laws are as follows:

- Pump capacity (gpm) varies directly as the speed (rpm) or the ratio of impeller diameters.
- Pump head varies directly as the square of the speed (rpm) or the ratio of impeller diameters.
- BHP varies directly as the cube of the speed (rpm) or the ratio of impeller diameters.

Some useful formulas are given in Table 7–1.

7.2.4 Examples

The examples cover flow rate calculations (Section 7.4.4), pipe sizing (7.4.4), and pump selection (7.2.2 and 7.2.3).

A hot-water system consists of a boiler (hot-water generator) and several air-handling units (AHUs), shown in Figure 7–6(a). Copper pipes and fittings are used. The total heating load of all the AHUs is 1,500,000 Btuh.

The maximum pressure required to pump water through the AHU which has the highest pressure drop is 40 ft in total head. The inlet and outlet temperatures at the AHUs are designed to be 180°F and 150°F, respectively.

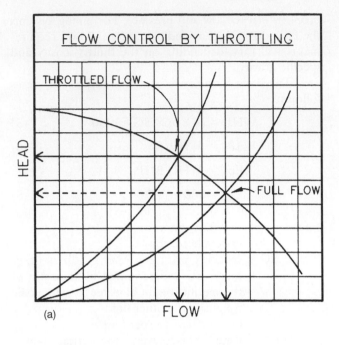

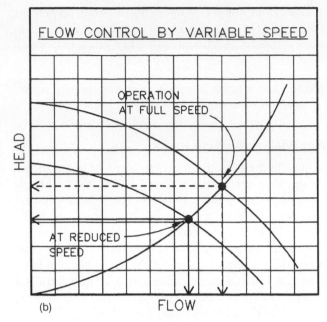

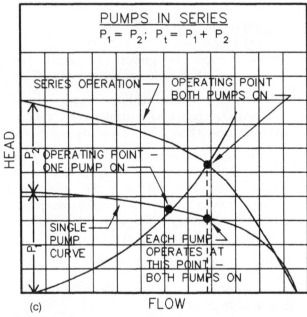

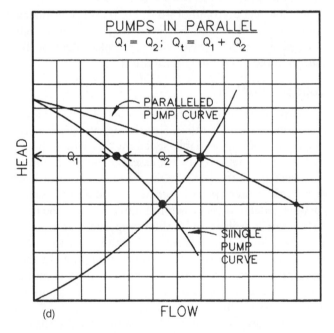

■ **FIGURE 7–5**
Method of capacity control for pumping operations.

TABLE 7–1
Effect of change of impeller diameter and speed

Change of	Capacity	Head	BHP
Impeller diameter	$q_2 = \dfrac{D_2}{D_1} q_1$	$H_2 = \left(\dfrac{D_2}{D_1}\right)^2 H_1$	$P_2 = \left(\dfrac{D_2}{D_1}\right)^3 P_1$
Speed	$q_2 = \dfrac{\text{rpm}_2}{\text{rpm}_1} q_1$	$H_2 = \left(\dfrac{\text{rpm}_2}{\text{rpm}_1}\right)^2 H_1$	$P_2 = \left(\dfrac{\text{rpm}_2}{\text{rpm}_1}\right)^3 P_1$

where q = flow rate, gpm; D = diameter, in.

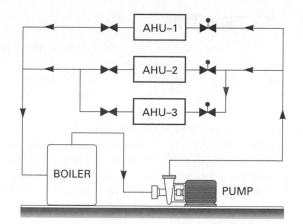

(a)

Examples. (a) Piping diagram for the examples. (b) Pump performance and system curves. (c) Pump performance and system curves under revised condition.

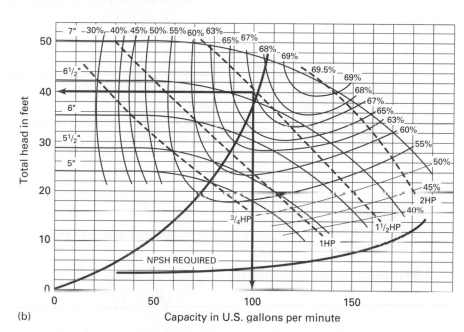

(b)

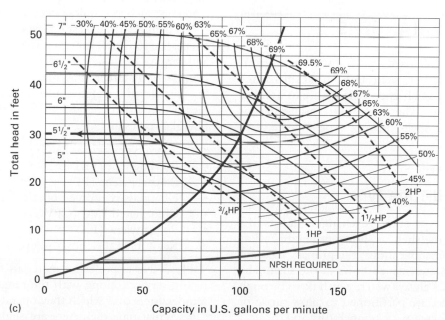

(c)

205

7.2 What is the required total flow rate (gpm)?

7.3 Allowing 30% extra capacity for load pickup, what should be the boiler capacity?

7.4 What should be the main pipe size, based on a flow velocity of 4.5 fps? (*Note:* Use Figure 7–11(b))

7.5 Select a pump from the manufacturer's pump performance curves that will fit the need. (*Note:* The pump selected is running at 1750 rpm.) Plot the system curve on the pump curves. What size impeller is required for this pump operating at this condition?

7.6 If the piping system is redesigned to lower the total system pressure to 30 ft, what pump impeller should be used?

7.7 What is the brake horsepower required at both operating conditions?

Answers

7.2 From Equation (7–8), 100 gpm is required.

7.3 The boiler should be rated for 1,950,000 Btuh.

7.4 From Figure 7–11(b) for copper tubing, a 3-in. pipe may be used.

7.5 From Figure 7–6(b), the impeller size should be slightly larger than 6½ in. in diameter.

7.6 The flow still should be 100 gpm. At 30-ft total head, a new system curve is established in Figure 7–6(c).

7.7 From Figure 7–6(b), the horsepower required is 1½ BHP, when the pump operates at 40-ft head and is reduced to about 1¼ BHP when it operates at 30-ft head.

7.3 HEAT EXCHANGERS

Heat exchangers are used to transfer heat from one medium to another, such as from steam to hot water, or from water at a higher temperature to water at a lower temperature. There are two basic types of heat exchangers: the shell-and-tube type and the plate type.

7.3.1 Shell-and-Tube Type

The shell-and-tube type of heat exchanger consists of a bundle of tubes in the shell, as shown in Figure 7–7(a) and (b). For HVAC applications, the primary medium is either steam or water, which flows in the shell. The secondary medium is always water, which flows through the tubes. The tubes are partitioned to allow single or multiple passes to increase the temperature and the heat transfer.

7.3.2 Plate Type

Plate-type heat exchangers consist of thin metal plates forming alternate compartments to allow the primary water to flow in one compartment and the secondary water in the other. Figure 7–8(a) illustrates the typical construction of this type of heat exchanger. Such a heat exchanger is ideal for heat transfer when the temperature difference between two water media is very close. A typical application of a plate-type heat exchanger is the economizer cycle of a chilled-water system during winter, as shown in Figure 7–8(b). In this application, the chilled water is cooled through a heat exchanger by cold water from the cooling tower, thereby conserving electrical energy without using the chiller compressor.

7.4 PIPING

Piping for fluid systems is analogous to ducts for air systems. Both serve to transport energy from equipment to equipment. Piping includes pipes, fittings, and accessories, which differ with the type of fluid, services, and application.

7.4.1 Materials for Pipes

General-purpose piping for HVAC systems can be constructed of plain (black) or galvanized steel, of copper, or of cast iron. Special applications requiring corrosion resistance may call for nonmetallic materials, such as fiberglass, plastic, or glass, or more expensive metallic materials, such as stainless steel or aluminum.

Fiberglass and plastic pipe can be an economical alternative to metal; however, these materials are fragile and require extra design attention to prevent damage from water hammer. Thermal expansion also needs to be considered carefully; plastics expand at a much higher rate than metals. Commonly used plastic or fiberglass pipes include:

- Polyvinyl chloride (PVC)
- Polybutylene (PB)
- Polypropylene (PP)
- Fiberglass-reinforced plastic (FRP)
- Chlorinated PVC (CPVC)
- Polyethylene (PE)
- Reinforced thermosetting resin (RTR)

Table 7–2 lists preferred materials for various HVAC piping services, along with the range of pressures and temperatures over which they operate.

Actual pipe diameters are slightly different from their nominal dimensions and will vary depending on

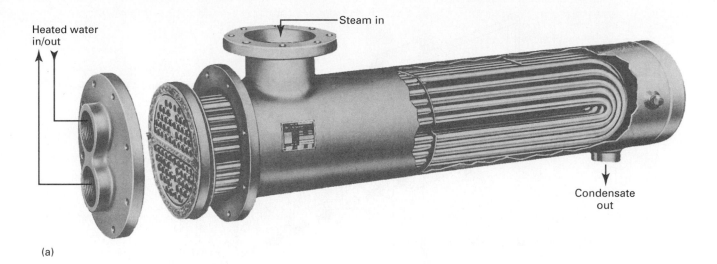

(a)

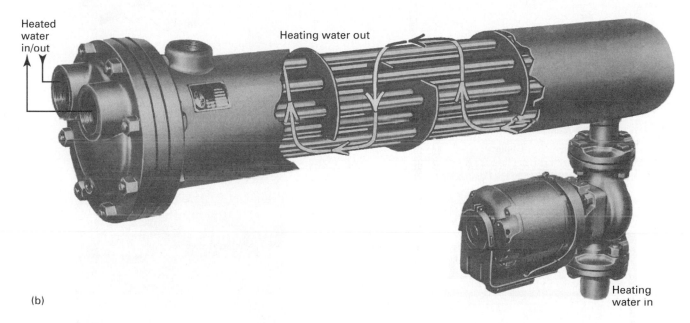

(b)

■ FIGURE 7–7

(a) Typical steam-to-water U-tube heat exchanger. (b) Typical water-to-water U-tube heat exchanger with water circulating pump shown. (Reproduced with permission from ITT Fluid Handling Sales.)

the material of construction. Plastic piping is manufactured in the same dimensions as steel.

7.4.2 Methods for Joining Pipes

Joints can be soldered, brazed, welded, screwed, flanged, solvent-welded, or mechanically coupled. Table 7–3 shows joining methods applicable to piping of various sizes and materials. Figure 7–9 shows several pipe joints, including a popular mechanical coupling system that is less expensive than welding and has the added advantage of inherent flexibility to relieve stress and thermal expansion.

7.4.3 Valves and Accessories

Valves are used in piping systems to adjust and control the flow of the medium. They may be on-off valves or adjustable either manually or automatically. Figure 7–10 shows a variety of valves commonly used in HVAC piping systems.

Functions of Valves

- *Shutoff valves* are installed to isolate components and portions of systems. These valves avoid the inconvenience of draining and refilling entire

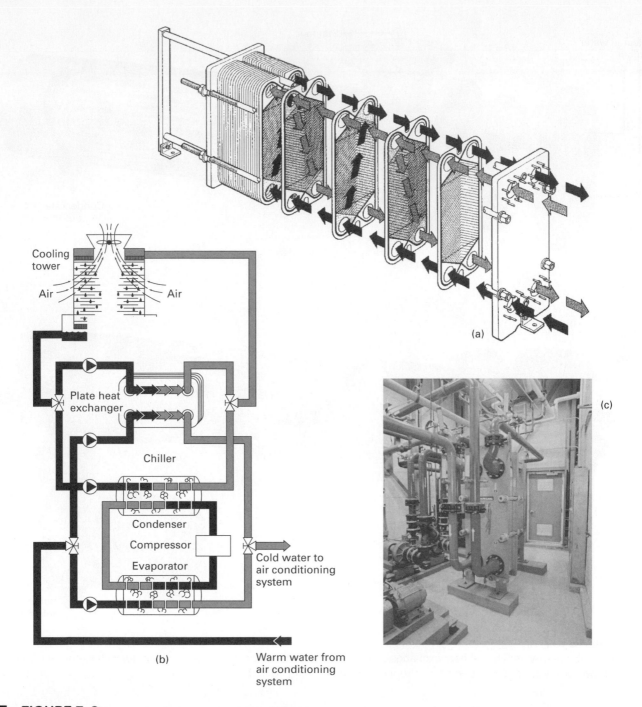

(a)

(c)

Cooling tower

Air   Air

Plate heat exchanger

Chiller

Condenser

Compressor

Evaporator

Cold water to air conditioning system

(b)

Warm water from air conditioning system

■ **FIGURE 7–8**

(a) Operating principle of a plate-type heat exchanger. (b) Piping arrangement of an economizer chiller cycle (free cooling cycle) during cold weather (winter in Northern Hemisphere and summer in Southern Hemisphere). (c) Typical installation of a plate-type heat exchanger. (Reproduced with permission from Alfa Laval Thermal Inc.)

systems or portions of systems during maintenance or repairs. If properly planned, they also allow continued service to portions of the system not affected by the work.

▪ *Drain valves* are installed at equipment and at the low points of piping systems. Hose end fittings are normally used, and fluid is conducted to the

nearest floor drain. Floor drains are required in mechanical rooms and should be placed close to condensate discharge from cooling coils to avoid piping over the floor. This is generally a convenient location for draining hot- and chilled-water coils.

▪ *Balance valves* are installed to adjust the flow through the system in proper amounts to serve

TABLE 7–2
Preferred materials for HVAC piping and their operating ranges

Services	Material	Temperature, °F	Pressure, psi
Water	Black steel, copper, plastic	40° to 250°	To 400 for steel
Steam	Black steel, copper		
Steam condensate	Steel, copper, black steel		
Condensing water	Galvanized, black steel	40° to 250°	To 150 for copper
Refrigerant	Copper		
Fuel oil	Black steel	40° to 200°	To 125 for FRP, PVC
Gas	Black steel, copper		

TABLE 7–3
Piping material and normal methods of joining

Piping material	Joint type							
	Welded	Threaded	Flanged	Soldering	Bracing	Solvent	Flared/Compression	Grooved Pipe
Black steel, up to 2″		•	•					•
Black steel, 2½″ and up	•		•					•
Galvanized steel		•	•					•
Type M or L copper			•	•	•			•
Type K copper			•		•			•
Stainless steel	•		•					•
Rigid aluminum	•	•	•					
Rigid plastic			•			•		
Fiberglass				•			•	

loads. These valves need to be adjusted when a building is commissioned and readjusted periodically as needs change. With "memory stops" to restore a set position, they can also be used as shutoff valves, reducing the number of components required and the expense of construction.

- *Control valves* are installed to modulate the flow of fluids through heat transfer devices in response to fluctuating loads. These valves are positioned automatically by actuators that receive input from control systems. Unlike shutoff and balance valves, control valves are in continuous operation. They are designed to wear well in movement and be accurate over a wide range of operating conditions. There are two types of control valves:
  - A *two-way* control valve is installed in series with the load device and simply throttles the

flow in response to the load. This control method results in a variation of flow.
- A *three-way* control valve is arranged to force fluid through the load device or to bypass fluid around the load device, resulting in a constant system flow.

Two-way valve arrangements are simpler and less expensive to install; but under light load conditions, choking effects on the overall system can result in high pressures and poor control. (See the discussion of pump and system curves on p. 200.) Often, a combination of two-way and three-way valves is designed for economical installation with acceptable control. The three-way valves maintain enough flow during light load conditions to prevent an undue rise in system pressure by the choking effect of the two-way valves.

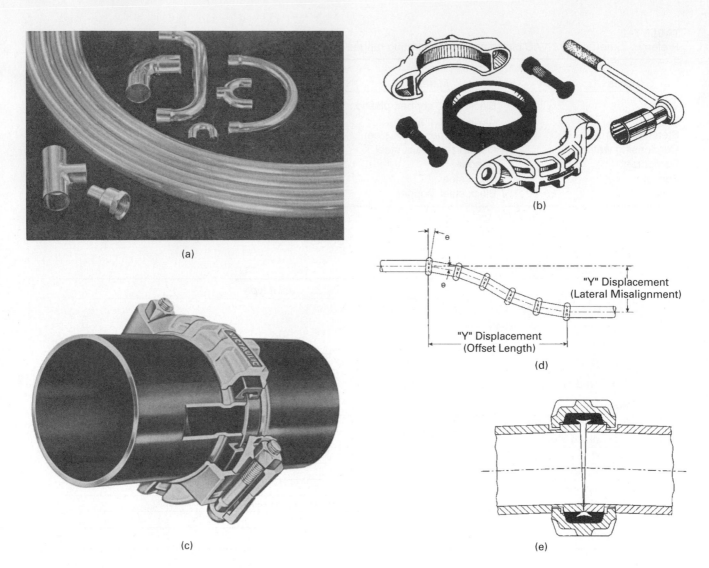

■ FIGURE 7–9

Two of many methods of joining pipes. (a) Soldering fittings for copper, stainless steel, and aluminum pipes. (Courtesy: Mueller Brass Co., Port Huron, MI.) (b) Mechanical joints (couplings) on grooved steel, iron, or other heavy wall pipes. (c) A mechanical coupling is typically constructed of a split-ring malleable iron housing, a synthetic rubber gasket, and nuts, bolts, locking toggle, or lugs to secure the unit together. (d) A unique characteristic of mechanical joints is their flexibility, which allows the pipes to deflect without causing leakage. (e) Cutaway view of the pipe and coupling showing a deflected or misaligned pipe joint. (Courtesy, (b) through (e): Victaulic Company of America, Easton, PA.)

- *Back-pressure regulators* can also be used to avoid unacceptably high system pressure at light load conditions. These regulators are valves designed to modulate toward the open position when there is a rise in system pressure.
- *Check valves* allow flow to occur in only one direction. They are installed in multiple pump applications to prevent backflow through pumps that are not running. The most common construction is a flapper arrangement. The flapper is held closed by pressure opposite the desired direction of flow.

- *Pressure-reducing valves* (PRVs) are installed in steam systems to reduce high-pressure distribution steam to a lower, safer pressure in building systems. These valves are installed in pressure-reducing stations along with other accessories.

Types of Valves

Valves are made in various designs and of various materials and types of valve construction. Each has its specific

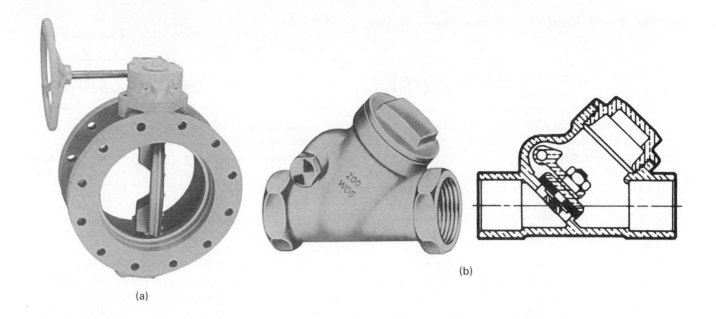

(a)

(b)

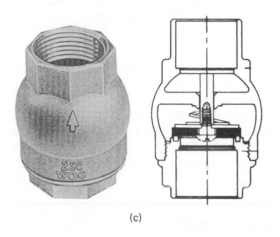

(c)

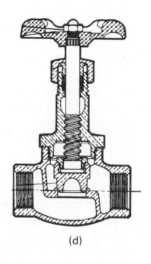

(d)

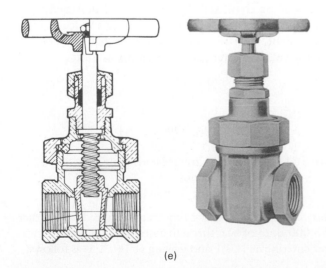

(e)

■ **FIGURE 7–10**
Commonly used valves.
(a) Flanged butterfly valve. (b) Swing check valve.
(c) Spring-loaded center-post check valve. (d) Globe
valve. (e) Gate valve. (Reproduced with permission from
Grinnell Corp.)

working characteristics. The following common types are shown in Figure 7–10:

- *Gate valves* are used to isolate flow. These valves have a low resistance to flow when they are wide open. Gate valves are used for shutoff duty and are not intended to regulate flow.
- *Globe valves* are used to isolate as well as regulate flow. They have a high resistance to flow even when wide open.
- *Ball and butterfly valves* are used to regulate flow. They can also be used for isolating flows, although gate or globe valves are better suited for this purpose.

Fittings and Accessories

Many fittings and accessories are required to make a complete operational piping system. Some are used in all systems, while others are applicable only to a particular system. Following is a list of some common fittings and accessories:

- *Pipe fittings* include elbows, tees, flanges, couplings, unions, etc. A union is used on threaded pipe joints to facilitate disconnecting the pipes for maintenance or repair.
- *Strainers* are installed to prevent foreign materials from damaging sensitive components that have moving parts or close clearances, mainly pumps and control valves. Strainers contain a perforated basket to entrain objects large enough to cause damage. They are designed with provisions for removal and inspection of the baskets.
- *Pressure and temperature taps* are installed to allow the insertion of gauges to check the performance of systems and components. The flow through a piece of equipment can be determined if the pressure drop across the equipment can be measured. Measurements of inlet and outlet pressures are also useful for determining whether pumps are operating effectively and for diagnosing possible system blockages. Knowing inlet and outlet temperatures of heat-transfer equipment is essential for verifying the performance of the equipment and diagnosing possible problems. Taps can be equipped with permanent thermometers and gauges for the convenience of maintenance personnel.

7.4.4 Water-Piping Systems

Pipe Sizing Criteria

Water piping is sized on the basis of flow, velocity, and friction losses through the pipes, fittings, and equipment. Friction in pipes causes a pressure drop, which is

TABLE 7–4
Typical and maximum recommended pressure drops and velocities in water systems

Applications	Velocity, fps		Pressure drop, ft/100 ft	
	Typical	Maximum	Typical	Maximum
Hot/chilled mains	1.5–4	6	6	10
Hot/chilled branches	1–3	4	4	6
Heat exchangers				10′ total
Coils				10′ total

normally expressed in feet of head per 100 ft of straight pipe. The flow velocity in the pipe should be fast enough to move water with its entrained air through the system, yet slow enough to prevent undue noise and erosion. Table 7–4 shows typical and maximum recommended pressure drops and velocities based on the type of service. These guidelines can be used to select reasonable pipe sizes for a given application.

Pipe Sizing Data

The flow of water in pipes is affected by the material of the pipe, the smoothness of the interior surface, the types of fittings in the pipe, and the velocity of the water. Figure 7–11 gives fundamental data relating to pipes. The pressure drop of a fitting or a valve may be as much as many linear feet of straight pipe. Table 7–5 gives data on pressure drop for valves and fittings. Additional data can be found in the *ASHRAE Handbook of Fundamentals*.

Flow Rate Calculations

Water is heated or cooled to transport the energy for heating and cooling systems. The amount of water that needs to be circulated depends on the designed temperature difference of the flow entering and leaving the equipment, such as a coil or a heat exchanger. The basic formula for water flow calculations is

$$Q = 500 \times FL \times (t_2 - t_1) \qquad (7–9)$$

or

$$FL = Q/500(t_2 - t_1)$$

where Q = heat being transferred, Btu/hr
FL = flow rate, gpm
$t_2 - t_1$ = temperature change (rise or drop), °F

For example, the water flow rate required to transfer 25,000 Btuh of heat through a heating unit with hot water entering at 180°F and leaving at 140°F is as follows:

$$FL = 25,000/500 \times (180 - 140) = 1.25 \, gpm$$

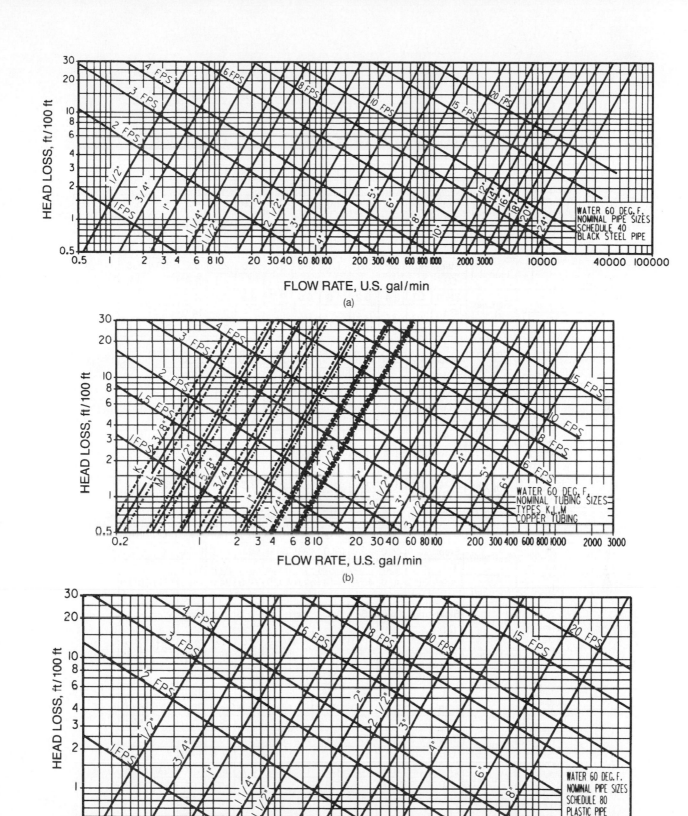

FIGURE 7–11

Friction loss for water in straight pipes. (a) For steel pipes (Schedule 40). (b) For copper tubing (Types K, L, M). (c) For plastic pipes (Schedule 80). (Reprinted by permission from ASHRAE (www.ashrae.org).)

Fittings			1/2	3/4	1	1 1/4	1 1/2	2	2 1/2	3	4	5	6
										Pipe size			
Regular 90° elbow	Screwed	Steel	3.6	4.4	5.2	6.6	7.4	8.5	9.3	11.0	13.0		
		C.I.								9.0	11.0		
	Flanged	Steel	.92	1.2	1.6	2.1	2.4	3.1	3.6	4.4	5.9	7.3	8.9
		C.I.								3.6	4.8		7.2
Long Radius elbow	Screwed	Steel	2.2	2.3	2.7	3.2	3.4	3.6	3.6	4.0	4.6		
		C.I.								3.8	8.7		
	Flanged	Steel	1.1	1.3	1.6	2.0	2.3	2.7	2.9	3.4	4.2	5.0	5.7
		C.I.								2.8	3.4		4.7
45° elbow	Screwed	Steel	.71	.92	1.3	1.7	2.1	2.7	3.2	4.0	5.5		
		C.I.								3.3	4.5		
	Flanged	Steel	.45	.59	.81	1.1	1.3	1.7	2.0	2.6	3.5	4.5	5.6
		C.I.								2.1	2.9		4.5
Tee inline flow	Screwed	Steel	1.7	2.4	3.2	4.6	5.6	7.7	9.3	12.0	17.0		
		C.I.								9.9	14.0		
	Flanged	Steel	.69	.82	1.0	1.3	1.5	1.8	1.9	2.2	2.8	3.3	3.8
		C.I.								1.9	2.2		3.1
Tee branch flow	Screwed	Steel	4.2	5.3	6.6	8.7	9.9	12.0	13.0	17.0	21.0		
		C.I.								14.0	17.0		
	Flanged	Steel	2.0	2.6	3.3	4.4	5.2	6.6	7.5	9.4	12.0	15.0	18.0
		C.I.								7.7	10.0		15.0
180° Return bend	Screwed	Steel	3.6	4.4	5.2	6.6	7.4	8.5	9.3	11.0	13.0		
		C.I.								9.0	11.0		
	Reg. Flanged	Steel	.92	1.2	1.6	2.1	2.4	3.1	3.6	4.4	5.9	7.3	8.9
		C.I.								3.6	4.8		7.2
	Long Rad. Flanged	Steel	1.1	1.3	1.6	2.0	2.3	2.7	2.9	3.4	4.2	5.0	5.7
		C.I.								2.8	3.4		4.7

Fittings			1/2	3/4	1	1 1/4	1 1/2	2	2 1/2	3	4	5	6
										Pipe size			
Globe valve	Screwed	Steel	22.0	24.0	29.0	37.0	42.0	54.0	62.0	79.0	110.0		
		C.I.								65.0	86.0		
	Flanged	Steel	38.0	40.0	45.0	54.0	59.0	70.0	77.0	94.0	120.0	150.0	190.0
		C.I.								77.	99.0		150.0
Gate valve	Screwed	Steel	.56	.67	.84	1.1	1.2	1.5	1.7	1.9	2.5		
		C.I.								1.6	2.0		
	Flanged	Steel						2.6	2.7	2.8	2.9	3.1	3.2
		C.I.								2.3	2.4		2.2
Angle valve	Screwed	Steel	15.0	15.0	17.0	18.0	18.0	18.0	18.0	18.0	18.0		
		C.I.								15.0	15.0		
	Flanged	Steel	15.0	15.0	17.0	18.0	18.0	21.0	22.0	28.0	38.0	50.0	63.0
		C.I.								23.0	31.0		52.0
Swing check	Screwed	Steel	6.5	6.3	7.0	8.4	9.3	12.0	13.0	16.0	21.0		
		C.I.								18.0	17.0		
	Flanged	Steel	.38	.53	.72	1.0	1.2	1.7	2.1	2.7	2.8	5.0	6.3
		C.I.								2.2	3.1		5.2
Coupling or union		Steel	.21	.24	.29	.36	.39	.45	.47	.53	.65		
		C.I.								.44	.52		
Bell mouth inlet		Steel	.10	.13	.18	.26	.31	.43	.52	.67	.95	1.3	1.6
		C.I.								.55	.77		1.8
Square mouth inlet		Steel	.96	1.3	1.8	2.6	3.1	4.3	5.2	6.7	9.5	13.0	16.0
		C.I.								5.5	7.7		13.0
Sudden enlarge-ment			$h = \dfrac{(V_1 - V_2)^2}{2g}$ feet of fluid; if $V_2 = 0$ $\quad h = \dfrac{V_1^2}{2g}$ feet of fluid										

(Courtesy of Hydraulic Institute, www.Pumps.org.)

Air Management

A primary concern in hydronic system design is dissolved air in water. The solubility of air in water decreases as the water temperature rises. Air migrates to high points of the piping system and may curtail the water flow to portions of the system. Air in heat-transfer equipment can also reduce the effective heat transmission. Therefore, undissolved air must be eliminated from the system.

There are two approaches to air management—air control and air elimination. In the air-control approach, all air separated from water (undissolved, or free, air) must be returned to a tank where air is accumulated at the top of the tank with water at the bottom of the tank. The tank also acts as a cushion to accept thermal expansion of the water. The tank is known as an *air/water interface compression tank*. Figure 7–12(a) shows a

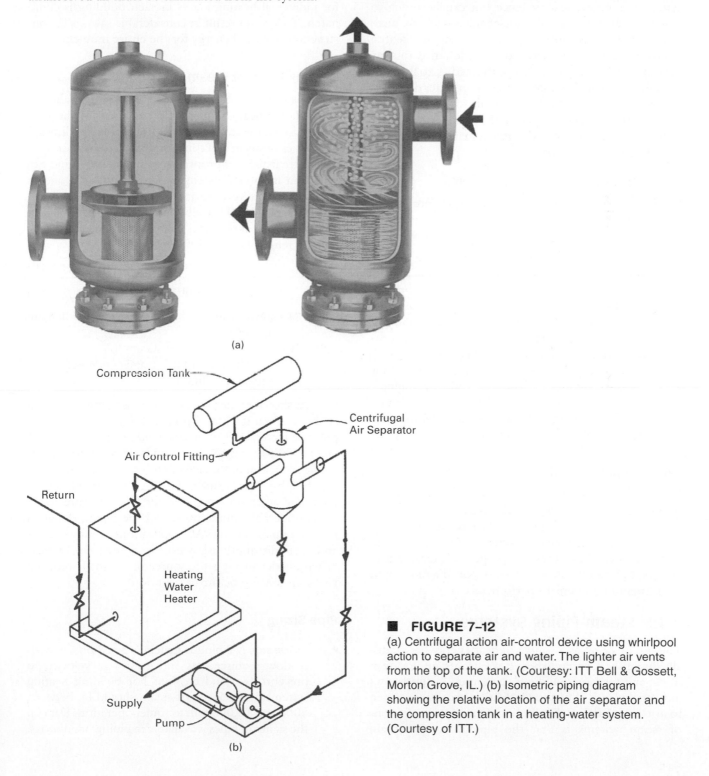

(a)

(b)

■ **FIGURE 7–12**

(a) Centrifugal action air-control device using whirlpool action to separate air and water. The lighter air vents from the top of the tank. (Courtesy: ITT Bell & Gossett, Morton Grove, IL.) (b) Isometric piping diagram showing the relative location of the air separator and the compression tank in a heating-water system. (Courtesy of ITT.)

centrifugal action–type air separator, and (b) shows a typical piping diagram.

In the air elimination approach, a diaphragm-type tank is used to maintain the pressure head (see Figures 7–13(a) and (b). Excessive free air is vented to the atmosphere through separate air vents.

Air vents may be manual or automatic. Manual vents require periodic maintenance to release air. Automatic vents reduce maintenance but can be prone to nuisance water leakage. An air separator with an automatic air vent is normally installed at the central mechanical room. Air separators are designed to slow the water flow locally and allow entrained air to float upward for removal. The solubility of air in water decreases as the water temperature rises, so air separators are installed in the warmest portion of the piping system. This is the return side of chilled-water systems and the supply side of hot-water systems.

Specialty Devices for Water System

In addition to valves and pipe fittings, there are a number of specialty devices unique to water systems. Some of these, illustrated in Figure 7–13, are as follows:

- Because water expands when heated, an *expansion tank* must be installed as a reservoir for the expanded water. The expansion is usually accommodated by compression of an air volume. In an open tank the air is on the top half of the tank, and in a diaphragm type of tank the air and water are separated from each other by a rubber bladder.
- *Suction diffusers* are devices that allow even flow and low resistance to flow at the inlet of a pump.
- *Airtrol fittings* are devices that allow the air in a tank or equipment to escape, without allowing the water to leak out of the system.

Typical Piping Details

Figure 7–14 illustrates several piping details of water system equipment. (Note the flexible hose connections at the pump inlet and outlet, as well as the air vents at top of the coils.) Manual air vents (MAVs) are shown. Note the location of pressure gauges (*P*), temperature gauges (*T*), and balance devices for diagnosis and adjustment of equipment performance.

7.4.5 Steam-Piping Systems

The use of steam as a source of heat energy has gradually diminished, except where steam plants are already in place, and for large buildings, campuses, hospitals, and industrial plants. When used, steam is usually converted to hot water for final distribution. One of the limitations of steam systems is that the piping must maintain a correct slope so that the condensate can drain properly. Therefore, such systems require higher floor-to-floor space. One popular source of steam is a central or district supply, where steam is piped underground from remote steam-generating plants. The steam may be a by-product of an industrial process, an electric power plant, or a waste treatment plant. By providing district steam in the central distribution of a major city, the need for individual heating plants in each building is eliminated. This could result in considerable savings in construction costs and energy for the entire municipality.

Heating Value of Steam and Condensate

Steam is water in the vapor phase. Its heating value includes latent heat of vaporization and sensible heat associated with the steam temperature. Table 7–6 lists the properties of steam and condensate at various pressures. Complete listings of properties of steam and liquid can be found in numerous engineering handbooks and in the Keenan and Keyes steam tables. For preliminary design purposes, 1000 Btu/lb can be used as the total heat of 1 lb of steam at pressures from 0 to 15 psig and a condensate temperature of 200°F. Thus,

$$H = 1000 \times W \qquad (7\text{–}10)$$

$$h = 1000 \times w \qquad (7\text{–}11)$$

where H = total heating value of steam and condensate at 200°F Btu
     W = mass of condensed steam, lb
     h = total heat rate of steam and condensate, Btuh
     w = steam rate, lb/hr

As shown in the table, the specific volume of steam is about 20 cu ft/lb at 5 psig and decreases rapidly as pressure increases. The specific volume at 30 psig is 9.4 and at 100 psig is 3.88. It should be clear that a considerable reduction in the size of piping can be realized by using high-pressure steam. On the other hand, equipment and accessories required for high-pressure steam are considerably more costly, and the piping design is more complex. For HVAC applications, steam systems and their distribution are usually limited to 30 psig. When district steam is the source, it is usually reduced to lower pressure before distribution.

Pipe Sizing

- *Steam supply* Piping is generally sized on the basis of flow requirements and maximum velocity, to prevent noise and erosion. For building heating systems, the velocity of steam should be between 8000 and 12,000 fpm for quiet operation. Data on the sizing of pipes is available in piping handbooks.

■ **FIGURE 7–13**

Several air-control fittings and their operating principles. (a) Tank fitting with built-in vent tube. (b) Tank fitting with separate vent tube. (c) Boiler fitting. (d) Cutaway section of a tank fitting as in (a). (e) The diaphragm tank assembly consists of an air purger to separate air from water and a float vent to bleed off air. (f) The diaphragm that separates air from water flexes to balance the air and water pressure. (Source: (a)-(d), reproduced with permission from ITT Fluid Handling Sales; (e) and (f), courtesy of AMTROL, Inc.)

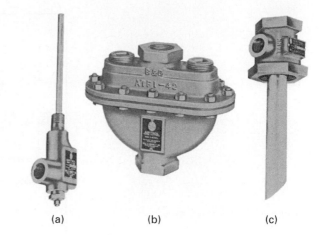

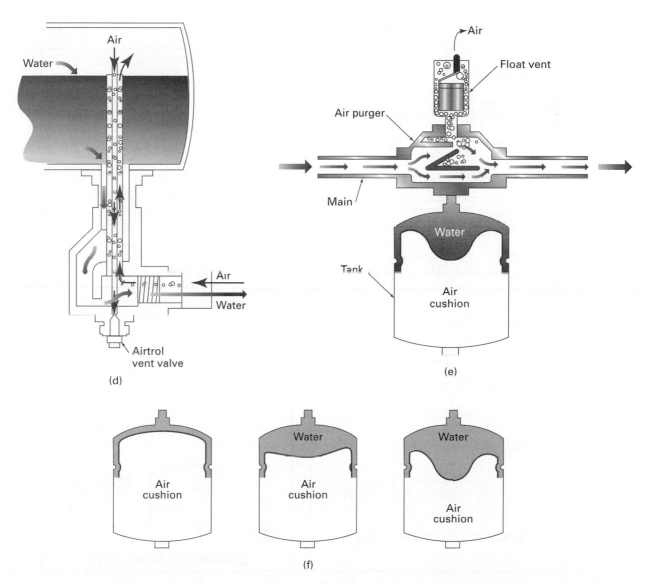

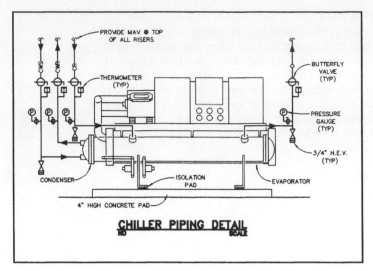

CHILLER PIPING DETAIL
NO SCALE

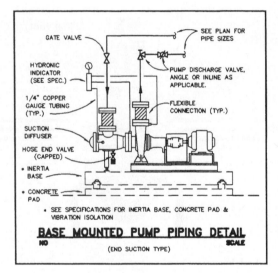

BASE MOUNTED PUMP PIPING DETAIL
NO SCALE
(END SUCTION TYPE)

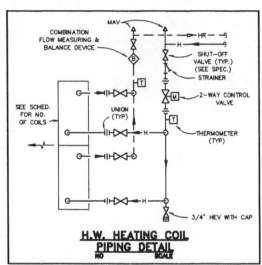

H.W. HEATING COIL
PIPING DETAIL
NO SCALE

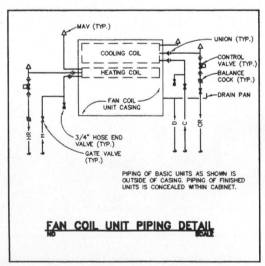

FAN COIL UNIT PIPING DETAIL
NO SCALE

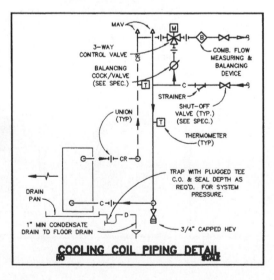

COOLING COIL PIPING DETAIL
NO SCALE

■ **FIGURE 7–14**

Typical piping details of equipment and coils.

TABLE 7–6
Properties of saturated steam

Gauge Pressure, psig	Absolute Pressure, psia	Steam Temperature, °F	Heat of Saturated Liquid, Btu/lb	Latent Heat Btu/lb	Total Heat of Steam, Btu/lb	Specific volume, cu ft/lb
0.0	14.696	212.00	180.07	970.3	1150.4	26.80
1.3	16.0	216.32	184.42	967.6	1152.0	24.75
2.3	17.0	219.44	187.56	965.5	1153.1	23.39
5.3	20.0	227.96	196.16	960.1	1156.3	20.09
10.3	25.0	240.07	208.42	952.1	1160.6	16.30
15.3	30.0	250.33	218.82	945.3	1164.1	13.75
20.3	35.0	259.28	227.91	939.2	1167.1	11.90
25.3	40.0	267.25	236.03	933.7	1169.7	10.50
30.3	45.0	274.44	243.36	928.6	1172.0	9.40
40.3	55.0	287.07	256.30	919.6	1175.9	7.79
50.3	65.0	297.97	267.50	911.6	1179.1	6.66
60.3	75.0	307.60	277.43	904.5	1181.9	5.82
70.3	85.0	316.25	286.39	897.8	1184.2	5.17
80.3	95.0	324.12	294.56	891.7	1186.2	4.65
90.3	105.0	331.36	302.10	886.0	1188.8	4.23
100.0	114.7	337.90	308.80	880.0	1188.8	3.88

Source: Condensed from Keenan and Keyes, *Thermodynamic Properties of Steam,* by permission of John Wiley and Sons, Inc.

If the allowable pressure drop is limited, larger pipe sizes may be required. Pipes should slope downward 1 in. in 40 ft in the direction of flow to allow free drainage of liquid condensate which forms in the pipe.

- *Condensate return* Piping from heating devices may flow by gravity in a system that is open to atmospheric pressure. A downward slope is required for piping toward the condensate tank. Most designers attempt to maintain a pipe slope of 1 in. per 40 ft. Closed pumped condensate piping need not be sloped. Generally, for low-pressure steam systems, condensate piping will be half the diameter of the steam service upstream.

Steam Traps

Steam traps are used to separate and remove condensate from steam piping and heat-transfer devices. There are many designs, such as the inverted bucket, float, and thermostatic thermodynamic. Steam traps are selected on the basis of the condensate temperature, steam pressure, flow rate, and other properties. It is advisable to get recommendations from manufacturers for different applications. Figure 7–15 illustrates several steam traps.

Accessories and Details

Steam-piping systems require considerably more accessories than water systems do. Piping must be properly pitched to avoid noise, vibration (steam hammer), and reduction in capacity. Illustrated in Figure 7–16 are

typical connection details at coils, steam traps, pressure-reducing stations, and heaters. Note in particular the steam heating coil piping. The coil is divided into two sections with multiple traps to ensure a uniform steam distribution. Pressure-reducing valves are used to reduce the steam pressure from up to 150 psig down to 50 psig for medium pressure–rated equipment and 5 to 10 psig for HVAC heating equipment.

7.4.6 Refrigerant Piping Systems

Refrigerant flows in three forms through refrigeration systems: cold suction gas, compressed hot gas, and liquid. Most refrigeration piping is copper or brass, with the notable exception of piping carrying ammonia, which is not often used in HVAC systems. Ammonia is low in cost and very efficient; however, it is toxic and has a strong smell. It is prohibited for use in building HVAC systems by many municipalities. Recently, concerns about depleting the ozone layer with CFC-type refrigerants have revived ammonia applications in central plant designs. Ammonia pipe is normally black steel.

Refrigerants

With the concern about ozone depletion and global warming, several of the popular refrigerants, such as Freon nos. 11 and 12, are being phased out. Refrigerant R-22 is still in use but will be phased out completely in 2020. Newer types of refrigerants, such as HCFC-123

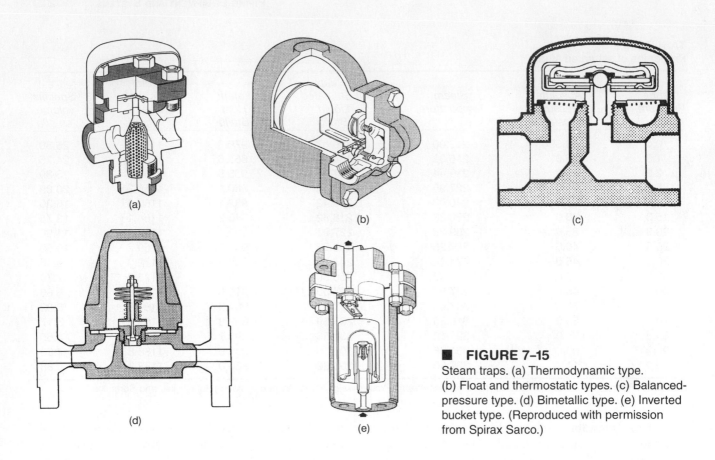

(a)

(b)

(c)

(d)

(e)

■ **FIGURE 7–15**
Steam traps. (a) Thermodynamic type.
(b) Float and thermostatic types. (c) Balanced-pressure type. (d) Bimetallic type. (e) Inverted bucket type. (Reproduced with permission from Spirax Sarco.)

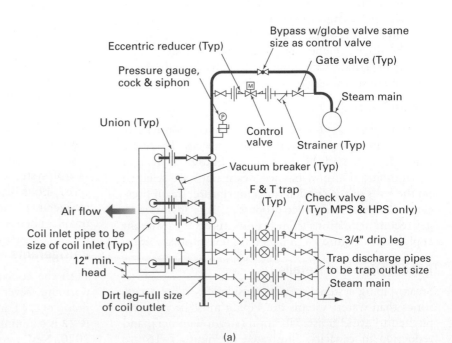

(a)

Bypass w/globe valve same size as control valve

Eccentric reducer (Typ)

Gate valve (Typ)

Pressure gauge, cock & siphon

Steam main

Control valve

Strainer (Typ)

Union (Typ)

Vacuum breaker (Typ)

Air flow

F & T trap (Typ)

Check valve (Typ MPS & HPS only)

Coil inlet pipe to be size of coil inlet (Typ)

3/4" drip leg

12" min. head

Trap discharge pipes to be trap outlet size
Steam main

Dirt leg–full size of coil outlet

■ **FIGURE 7–16**
Typical piping details for steam and condensate systems. (a) Steam heating coil piping detail.

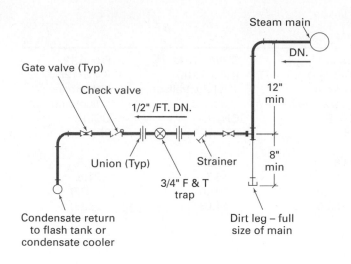

Gate valve (Typ)

Check valve

1/2" /FT. DN.

Union (Typ)

3/4" F & T trap

Strainer

Condensate return to flash tank or condensate cooler

Steam main

DN.

12" min

8" min

Dirt leg – full size of main

Note:
Drip leg piping to be trap size

(b)

■ **FIGURE 7–16** (*Continued*)
(b) Medium- and high-pressure steam drip leg detail.
(c) Steam pressure-reducing valve (SPRV) and relief valve
(RV) schedule. (d) Duplex condensate pump piping detail.
(e) Steam unit heater piping detail.

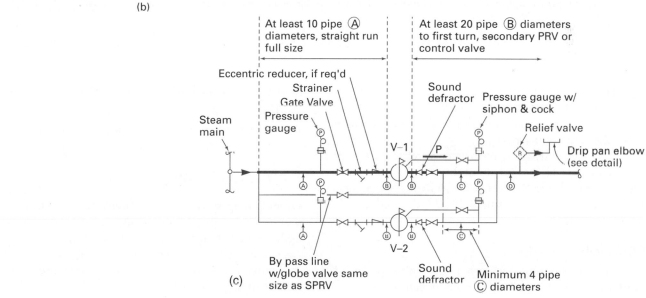

At least 10 pipe Ⓐ diameters, straight run full size

At least 20 pipe Ⓑ diameters to first turn, secondary PRV or control valve

Eccentric reducer, if req'd
Strainer
Gate Valve
Pressure gauge

Sound defractor
Pressure gauge w/ siphon & cock

Relief valve

Steam main

V–1

P

Drip pan elbow (see detail)

By pass line w/globe valve same size as SPRV

V–2

Sound defractor

Minimum 4 pipe Ⓒ diameters

(c)

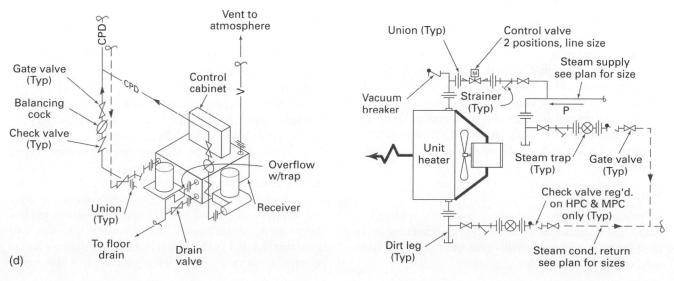

CPD

CPD

Vent to atmosphere

Gate valve (Typ)

Balancing cock

Check valve (Typ)

Control cabinet

Overflow w/trap

Union (Typ)

To floor drain

Drain valve

Receiver

(d)

Union (Typ)

Control valve 2 positions, line size

Steam supply see plan for size

Vacuum breaker

Strainer (Typ)

P

Unit heater

Steam trap (Typ)

Gate valve (Typ)

Check valve req'd. on HPC & MPC only (Typ)

Dirt leg (Typ)

Steam cond. return see plan for sizes

221

TABLE 7–7
Physical properties of Du Pont SUVA™ and FREON® refrigerants

Refrigerant	SUVA® 123 (HCFC-123)	Freon 11 (CFC-11)	SUVA® 134a (HFC-134a)	Freon 12 (CFC-12)
Chemical name	2,2-dichloro, 1,1,1, trifluoroethane	trichloro-fluoromethane	1,1,1,2-tetrafluoro-ethane	dichlorodi-fluoromethane
Chemical formula	$CHCl_2CF_3$	CCl_3F	CH_2FCF_3	CCl_2F_2
Vapor pressure (psia) @ 25°C (77°F)	14	16	96	95
Boiling point °C	27.9	23.8	−26.5	−29.8
@ 1 atm °F	82.2	74.9	−15.7	−21.6
Freezing point °C	−107.0	−111.0	−101.0	−158.0
°F	−161.0	−168.0	−149.8	−252.0
Flammability Limits in air vol%	None	None	None	None
Ozone depletion potential	0.02	1.0	0	1.0
Global-warming potential	120	4600	1300	10,600
Toxicity	50 (AEL)*	1000 (TLV)**	1000 (AEL)*	1000 (TLV)**

*AEL (allowable exposure limit) is a preliminary toxicity assessment established by Du Pont; additional studies may be required.
**TLV (threshold limit value) is a registered trademark of the American Conference of Government Industrial Hygienists.
Source: Courtesy of Du Pont Company.

and HFC-134A, have gained acceptance as replacements for R-11 and R-12, respectively. Table 7–7 is a comparison of their physical properties compared with their obsolete counterparts.

The thermodynamic properties of refrigerants vary widely. In general, it takes about 2.9 and 3.3 lb/min. of refrigerant to produce 1 ton (12,000 Btuh) of refrigeration effect for the refrigerants CFC-22 and SUVA-123, respectively.

Pipe Sizing Criteria

Most refrigerant piping is contained within packaged equipment. Field-installed piping between system components is generally limited to the suction and liquid lines from the condensing unit to the coil or the hot gas and liquid lines from a unitary device to an outdoor condenser. Sizing is best left to manufacturers' recommendations. Losses due to friction are a critical criterion for preventing undue loss of compressor capacity. The velocity must be high enough for the flowing refrigerant gas to draw oil back to the compressor. When the flow is upward, a specially designed oil trap detail is generally used. Figure 7–17 is a DX coil piping riser connection.

Joints

In refrigerant piping, joints are generally soldered or brazed, although special screwed connectors are often used for connection to small residential and light commercial air-conditioning equipment.

Refrigeration Piping Diagram and Accessories

The basic DX refrigeration system illustrated in Figure 7–17 shows the arrangement of components and accessories.

7.4.7 Other Piping Systems

Other piping systems in buildings include gas and oil piping for fuel-fired equipment such as water heaters, boilers, and furnaces. Gas and oil are combustible fuels. Leaks could be very hazardous. Building codes and utility regulations have strict rules governing the installation of these piping systems.

Gas Piping

Gas is usually measured by volume. The normal heating value for natural gas is approximately 1000 Btu/cu ft. Heating values are measured in therms; 1 therm equals 100,000 Btu. The heating value may vary with the gas composition. The exact heating value should be obtained from the supplier. For preliminary design purposes, 1000 Btu/cu ft can be assumed.

Black steel or type-K copper pipes are used. Pipes should be exposed for ready inspection and repair. When installed above ceilings, pipes should be enclosed in a duct that is vented to the atmosphere, or they should have all welded joints. Liquid petroleum gas has been used for residences when natural gas was not available. Liquid petroleum gas is heavier than air and is generally odorless. Leaks will vaporize, but the vapor

■ **FIGURE 7-17**

(a) Typical DX system diagram showing major components. (b) One method of capacity control of DX system is to unload the compressor cylinders and to install the corresponding number of solenoid valves (SV) and thermal expansion valves (TEV) to control the evaporator coils. This diagram shows a single evaporator coil with two SVs and two TEVs, which can be controlled in sequence, reducing capacity. (Reproduced with permission from Sporlan Valve Company.)

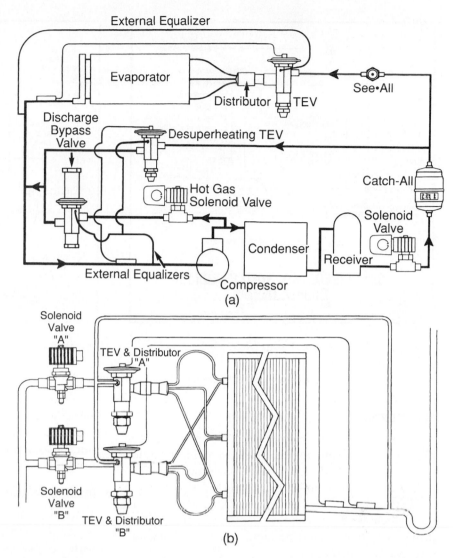

Fuel Oil Piping

Like gas piping, fuel oil piping is made of black steel or copper. The heating value of oil varies with the classification of the oil. No. 1 and No. 2 (diesel) are the lightest. No. 5 and No. 6 (bucket fuel) are the heaviest and require heating before they can be pumped in piping systems.

7.4.8 Thermal Expansion

Once filled, an HVAC piping system will be subject to changing temperatures and accompanying expansion and contraction of the water volume. Expansion

tends to stay near the floor or migrates to the lower part of a building. It is extremely hazardous. The space where liquid petroleum gas pipes are located must be properly inspected and ventilated. Figure 7–18 illustrates some equipment connection details, including many safety provisions.

compensation must be provided to prevent damaging pressure; similarly, contraction must be provided to compensate for negative pressures, which might draw air into the system.

Expansion Tanks

Expansion tanks are installed to provide high and low limits on the pressure. Tanks can be of the open gravity variety or closed and pressurized. Gravity tanks need to be placed at the very top of the system and are open to the atmosphere. As water in the system expands and contracts, the level in the expansion tank rises and falls. The surface exposed to the atmosphere is a source of dissolved air, which can cause corrosion and air binding.

Closed tanks need not be placed at the high point of the system. They can be constructed with the water surface exposed to a charge of air, or they can be separated from water by a rubber bellows. Bellows-type expansion tanks are preferred, to avoid problems with dissolved air.

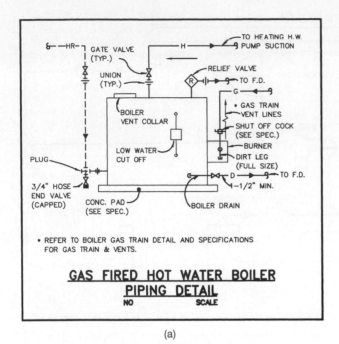

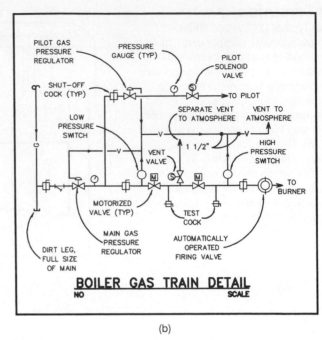

(a)   (b)

■ **FIGURE 7–18**

(a) Gas piping connection of a gas-fired boiler. (b) Gas train details.

Piping Expansion

Heating system piping must operate over a wide variety of temperatures, and thermal expansion must be accommodated. Good design uses anchorage to firm structures at regular intervals and provides for expansion between anchors. Expansion compensation devices include bellows fittings and concentric slip fittings. Slip fittings require seals and packing, which need maintenance to prevent leakage.

Elbows and Z-shaped and U-shaped expansion loops can also be designed to use the inherent flexibility of the piping materials to absorb expansion. Elbows and expansion loops are generally the least expensive method for accommodating expansion, and they require the least maintenance, but they also require supports designed for lateral movement of piping and clearance for this movement. Care is also required to size expansion legs adequately to prevent undue flexure stress.

7.4.9 Insulation

Purpose of Insulation

Insulation is desirable to prevent thermal losses, to reduce the hazard of touching hot piping surfaces, and to prevent condensation on cold piping. Materials commonly used for piping insulation include calcium silicate, foam glass, fiberglass, and foam rubber. Calcium silicate and foam glass are durable, rigid materials that are suitable and much less expensive for hot or cold piping applications. Fiberglass is less durable but equally suitable and much less expensive for the range of temperatures encountered in HVAC systems.

Insulating Materials

Most insulating materials require a protective jacket. Materials for the jacket include fabric wrap applied with wet plaster, thin plastic casing with taped joints, and sheet aluminum. Plastic and aluminum function as barriers to vapor, to prevent condensation on the surface of cold piping. For small-diameter piping, foam rubber insulation can be used without a jacket, but it should be avoided in high temperatures and in locations subject to excessive mechanical wear and tear. Refer to the manufacturer's data on the properties of insulation material.

QUESTIONS

7.1 What factors are considered in selecting the size of pipe for fluid flow applications?

7.2 What feature most differentiates turbine pumps from centrifugal pumps? How are turbine pumps applied in HVAC systems?

7.3 What feature most differentiates positive-displacement pumps from centrifugal pumps? How are positive-displacement pumps applied in HVAC systems?

7.4 For a given water piping system, a pressure of 50 ft of head is required to produce 30 gpm water flow. How much pressure will be required to flow 60 gpm?

7.5 Compare the differences in flow developed for two identical pumps operating in parallel versus operating in series.

7.6 What are the main advantages and disadvantages of plastic piping in comparison with steel?

7.7 What factors limit maximum velocities in piping systems? Minimum velocities?

7.8 What are the main differences in the characteristics of calcium silicate, fiberglass, and foam rubber insulation that would affect their application in piping systems?

7.9 What is the functional difference between a control valve and a balance valve?

7.10 Under what circumstances would three-way control valves be used rather than two-way control valves, despite the higher expense of three-way valves?

7.11 What is the purpose of a check valve?

7.12 What are some limitations and disadvantages of open expansion tanks for hot-water systems?

7.13 What are the advantages and disadvantages of expansion loops in comparison with slip joints for compensating for thermal expansion?

7.14 What provisions are recommended for air control in hydronic piping?

7.15 What provisions are recommended for condensate control in steam piping?

7.16 Why is a reverse-return piping system more desirable than a direct-return system?

7.17 What advantage does a closed-loop system have over an open-loop system?

7.18 Is a water-cooling-tower piping system an open- or closed-loop system?

7.19 One psi is equal to 2.31 ft of water. (True) (False)

7.20 Calculate the WHP (water horsepower) of a water-piping system with a flow rate of 100 gpm and 30 ft of head.

7.21 What BHP (brake horsepower) is required if the pump selected has an efficiency of 75% operating as specified in Question 7.20?

7.22 What are the two types of fluid heat exchangers?

7.23 Which type of heat exchanger is ideal for a low temperature differential?

7.24 Grooved-pipe mechanical joints can be used for steel pipe only. (True) (False)

7.25 The advantage of grooved-pipe mechanical joints is their extremely strong coupling. (True) (False)

7.26 Soldered joints are used for type-K copper. (True) (False)

7.27 What is the equivalent straight pipe length of a 2-in. screwed globe valve? A 2-in. gate valve?

7.28 Why is it necessary to have air venting in a hot-water system?

7.29 What is the latent heat of steam at 10.3 psig?

7.30 The total heat of steam at 50 psig (including the sensible heat of condensate at 200°F and the latent heat of vapor) is _____.

7.31 What is the absolute pressure (psia) at sea level with a gauge reading of 60.3 psig?

7.32 What is the temperature of saturated steam at 65 psia?

7.33 Name one or more refrigerants that have low potential for depleting ozone.

7.34 What refrigerant has the least potential for contributing to global warming?

7.35 How many pounds of refrigerant per minute are needed to produce 1 ton of refrigeration?

7.36 What are the advantages of using district heating plants instead of individual building heating systems?

7.37 What are the main advantages and disadvantages of distributing steam at higher pressures?

7.38 The contractor offers a fair cost credit for deleting isolation valves from the new heating system being constructed for you. Why should you insist that valves be installed?

7.39 Why is it important to repair leaky steam traps as soon as possible?

PLUMBING EQUIPMENT AND SYSTEMS

8

Broadly defined, plumbing systems include water and drainage systems exterior to and within buildings:

- Water supply, distribution, treatment, quality, and temperature conditioning
- Selection and installation of plumbing fixtures and drainage devices
- Waste and soil collection
- Treatment and disposal of drainage
- Collection, retention, and disposal of storm water
- Building equipment provisions, including water and waste for HVAC, food service, pools, fountains, processing, etc.

The design directions for each system or subsystem can be found in appropriate codes and handbooks published by technical and engineering organizations. This chapter will not duplicate the available data but rather will present sample data sufficient for illustrating the design methodology. The following are some leading organizations that maintain a plumbing database:

- American Society of Plumbing Engineers (ASPE)
- American National Standards Institute (ANSI)
- American Society of Heating, Refrigerating and Air Conditioning Engineers (ASHRAE)
- American Society of Sanitary Engineers (ASSE)
- American Society for Testing Materials (ASTM)
- American Water Works Association (AWWA)
- Building Officials and Code Administration (BOCA)
- Basic National Plumbing Code (BOCA/NPE)
- Cast Iron Soil Pipe Institute (CISPI)
- National Standard Plumbing Code (NSPC)
- Southeast Building Code Conference (SBCC)
- Uniform Plumbing Code (UPC)

Since plumbing systems and installations directly affect personal hygiene and public health, plumbing codes are quite specific regarding the design and installation of sanitary facilities and their piping systems. As a rule, all applicable codes specify the minimum number of plumbing fixtures that must be provided in a building and the specifics of piping connections. These specific code requirements have greatly simplified the design of a sanitary piping system, to the extent that detailed engineering calculations are seldom necessary; however, the simplified process cannot be applied to the design of water systems, which depend strictly on the design engineer's decisions. The method of dealing with water-piping systems will be addressed in this chapter.

Graphic symbols are used extensively in illustrations of plumbing systems to explain their operational principles and construction details. Commonly used graphic symbols are shown in Figure 8–1.

8.1 WATER SUPPLY AND TREATMENT

Providing water is among the most critical services in a modern building. Without water, a building is not suitable for human occupation or most industrial operations. Water is also needed for protection against fire.

Generally, potable water is supplied from a public water system. For buildings remote from the public water system, an alternative source of water must be found, such as a well or a lake. Potable water supplied to a building must meet the quality standards prescribed by the governing public health agencies. If the water does not meet these standards, physical and chemical treatment of the source is needed.

Units of measure frequently used to describe properties relating to water and its quality are as follows:

- 1 liter (L) = 1000 cubic centimeters (cc)
- 1 pound (lb) = 454 grams (g)
- 1 kilogram (kg) = 1000 g = 1,000,000 milligrams (mg)
- 1 lb = 7000 grains (gr)
- 1 gr = 0.0648 g
- 1 gallon (gal) of water weighs 8.33 lb
- 1 cc of water weighs 1 g

Commonly used plumbing and piping symbols

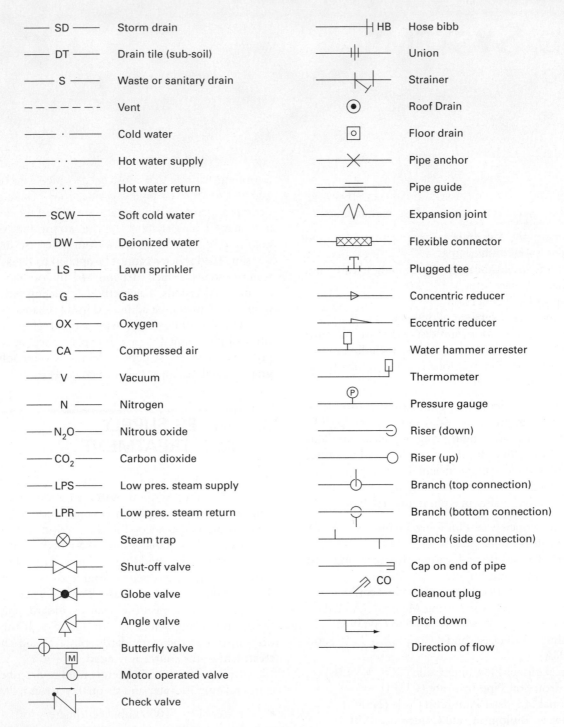

Symbol	Name	Symbol	Name
—— SD ——	Storm drain	——⊣⊢ HB	Hose bibb
—— DT ——	Drain tile (sub-soil)	——‖‖——	Union
—— S ——	Waste or sanitary drain	——⋈——	Strainer
– – – – –	Vent	⊙	Roof Drain
—— · ——	Cold water	▢	Floor drain
—— · · ——	Hot water supply	——✕——	Pipe anchor
—— · · · ——	Hot water return	——≡——	Pipe guide
—— SCW ——	Soft cold water	——⋀——	Expansion joint
—— DW ——	Deionized water	——▨▨——	Flexible connector
—— LS ——	Lawn sprinkler	——⊤——	Plugged tee
—— G ——	Gas	——▷——	Concentric reducer
—— OX ——	Oxygen	——◸——	Eccentric reducer
—— CA ——	Compressed air	——▢——	Water hammer arrester
—— V ——	Vacuum	——▢——	Thermometer
—— N ——	Nitrogen	——Ⓟ——	Pressure gauge
—— N₂O ——	Nitrous oxide	——↺	Riser (down)
—— CO₂ ——	Carbon dioxide	——○	Riser (up)
—— LPS ——	Low pres. steam supply	——⊕——	Branch (top connection)
—— LPR ——	Low pres. steam return	——⊖——	Branch (bottom connection)
——⊗——	Steam trap	—L——	Branch (side connection)
——⋈——	Shut-off valve	——⊐	Cap on end of pipe
——◀●▶——	Globe valve	——⟋ CO	Cleanout plug
▲	Angle valve	—┐→	Pitch down
——⊘——	Butterfly valve	———→	Direction of flow
——M ○——	Motor operated valve		
——⊼——	Check valve		

■ **FIGURE 8–1**
Graphic symbols commonly used in illustrating plumbing and piping installation.

- 1 cubic foot (cu ft) of water = 28.3 L
- 1 cu ft of water ≈ 7.5 gal
- 1 cu ft of water at room temperature weighs 62.4 lb
- 1 L of water at room temperature weighs 1000 g (1 kg)
- 1 part per million (ppm) of chemical contents in water = 1 mg/L = 1 mg/kg
- 1 g per gallon (gpg) = 17.1 ppm

8.1.1 Water Quality

For use in a building, the water supply must meet a minimum quality based on several major characteristics:

1. *Physical characteristics.* The water supply may contain only a limited amount of suspended material, as measured in terms of cloudiness, clarity, color, acceptable taste, odor, and temperature. To qualify for drinking, the water supply should be lower than 5 turbidity units (TU), which is a measure of cloudiness (turbidity), or clarity.
2. *Chemical characteristics.* The water supply may contain no more than the maximum content prescribed by health standards pertaining to hardness and dissolved matter, such as minerals and metals. The preferred hardness of a water supply is lower than 200 ppm of calcium carbonates.
3. *Biological and radiological characteristics.* The water supply should be practically free of bacteria, viruses, and radioactive materials.

Upon request, the public water utility will provide a typical analysis of a building's water supply. The utility is responsible for treating water to meet the quality standards of the local health department. Typical characteristics of a public water supply in the United States are shown in Table 8–1. Plainly, the chemical and physical properties of water vary considerably among geographical regions. To qualify as suitable for drinking from the tap without further treatment or filtration, water must be better in quality than the minimum standards set forth by the Environmental Protection Agency and local health departments. The minimum standards shown in Table 8–1 are consolidated values in the United States. Standards may differ in other countries, owing to the unique properties of their water sources and local lifestyles.

TABLE 8–1
Analysis of typical water supplies and minimum quality standards

Substance	Chemical Symbol	Water Samples* (1)	(2)	(3)	(4)	(5)	(6)	Typical Quality Standard (7)
Silica	SiO_2	2	12	37	10	22	—	—
Iron	Fe_2	0	0	1	0	0		0.3
Calcium	Ca	6	36	62	92	3	400	—
Magnesium	Mg	1	8	18	34	2	1300	—
Sodium	Na	2	7	44	8	215	11,000	—
Potassium	K	1	1		1	10	400	—
Bicarbonate	HCO_3	14	119	202	339	549	150	—
Sulfate	SO_4	10	22	135	84	11	2700	250
Chloride	Cl	2	13	13	10	22	19,000	250
Nitrate	NO_3	1	0	2	13	1	—	10
Dissolved solids	—	31	165	426	434	564	35,000	500
Carbonate hardness	$CaCO_3$	12	98	165	287	8	125	—
Noncarbonate hardness	$CaSO_4$	5	18	40	58	0	5900	—

Numbers indicate location or area, as follows:
(1) Catskill supply, New York City
(2) Niagara River (filtered), Niagara Falls, New York
(3) Missouri River (untreated), average
(4) Well waters, public supply, Dayton, Ohio, 30 to 60 ft
(5) Well water, Smithfield, Virginia, 330 ft
(6) Ocean water, average
(7) Minimum quality standards for drinking water based on the composite values of the Environmental Protection Agency (EPA) and health departments in the United States.
*Condensed from Table 3, Chapter 43, Application Volume, *ASHRAE Handbook*, 1991. Reprinted by permission from ASHRAE (www.ashrae.org).
**All units are in ppm or mg/L, rounded to the nearest whole number.

8.1.2 Processes Commonly Used to Improve Water Quality

Depending on the initial quality of water and the purpose for which the water will be used in a building, one or more of the following processes may be used to improve the quality of the water:

1. *Sedimentation*. Allows suspended matter to settle out of water by precipitation. Reduces the turbidity of the water.
2. *Coagulation*. Removes suspended matter from water by means of chemicals, such as alum (hydrated aluminum sulfate), to reduce turbidity and improve the color and taste of the water.
3. *Aeration*. Introduces air into water to oxidize impurities and improve its taste and color.
4. *Filtration*. Removes suspended material and bacteria from water to improve its overall quality, including its turbidity, potability, color, and taste.
5. *Disinfection*. Uses chemicals such as chlorine gas, hypochlorite solids, or ultraviolet light to disinfect whatever bacteria remain in water after previous treatment.
6. *Fluoridation*. Adds a fluoride chemical in water, primarily to prevent tooth decay.
7. *Softening*. Replaces calcium and magnesium ions with sodium ions to reduce scale formation in the hot-water piping system and to improve sudsing for doing laundry.

8.1.3 Water Conditioning

When water quality does not meet the minimum standards for which it is to be used, it must be conditioned by the various treatment processes on the site, either within or outside the building. The treatment may be applied to the entire water source or to only a portion of the water supply. For example, even if the water supply to a hospital is good enough for general use, a portion of the water may need to be softened for laundry use, filtered for food preparation, deionized for laboratory use, and distilled for use in research. Figure 8–2 is a schematic diagram of water treatment systems for a research hospital.

Water Softening System

Water from rain or snow usually contains no minerals; however, flowing through mountains or rivers dissolves minerals such as calcium or magnesium salts in the form of carbonates or sulfates. When the water is heated, carbonates precipitate from the solution, forming hard scale. This scale can reduce the efficiency of heat exchangers and the net cross-sectional area of the pipe, as well as corrode the equipment. Carbonates in water are termed *temporary hardness*. When they are in excess of 200 ppm, they should be removed from the water.

The most popular water-softening process is the zeolite system. In this system, water is softened by an ion exchange process in which calcium or magnesium ions are replaced by sodium ions. Sodium salts are much more soluble than magnesium or calcium salts and do not precipitate as easily. The softener consists of vertical tanks filled with small beads of exchange resin saturated with sodium cations (positive ions). The resin preferentially removes calcium and magnesium cations from the water while releasing sodium cations to the water. Thus, the water becomes softened. When the resin beads are saturated with calcium or magnesium cations, they are replenished with sodium cations from salt through the regenerative cycle. Figure 8–3 illustrates the construction of typical water softener tanks. While the zeolite system effectively reduces the hardness of water by replacing calcium or magnesium with sodium, it does not reduce the total dissolved minerals in the water.

Water Purification Systems

A *deionizing system* is similar to the zeolite softener system, except that the tank or tanks contain a mixed bed of both cation- and anion-absorbing resins made of porous polymer minerals. The deionizing system can produce water that approaches the theoretical limits of purity. The system reduces both the mineral content and the hardness of water. Figure 8–4 illustrates the operating principle of deionizers.

A *reverse osmosis* (RO) system is another effective way to purify water and reduce its hardness. The system operates on the principle of diffusion rather than ion exchange. Osmosis is a natural phenomenon that occurs when water solutions of different concentrations are separated by a semipermeable membrane. Water tends to flow from lower concentrations to higher concentrations of impurities. RO works on the principle of reversing water flow by applying high external pressure (200 to 400 psi) to the side with the higher concentration (the more impure side) to force pure water to flow into the side with the lower concentration (the more pure side). As a result, water downstream of the membrane is highly purified. The RO system is used to obtain highly purified water that is normally associated with research or manufacturing and is generally considered too costly for general building use. Figure 8–5 illustrates the principles of the system.

Distillation is a traditional method of obtaining highly purified water. The water is heated to water vapor and condensed into highly purified water that is practically free of impurities. Distillation systems are

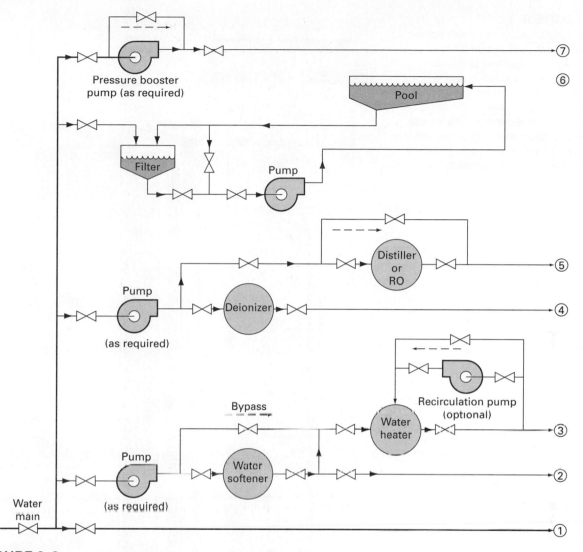

FIGURE 8–2

Schematic diagram of a water distribution system for a large hospital.

1. Cold-water supply to lower floors for general-purpose use
2. Soft cold-water supply for building heating system
3. Soft domestic hot water for laundry and plumbing fixtures
4. Deionized cold water to laboratory
5. Distilled or reverse osmosis (RO) water for laboratory testing services
6. Filtered water for swimming pool, therapeutic pool, etc.
7. Cold-water supply to upper floors, with pressure booster pump as required to overcome static height and frictional losses

used in research, hospitals, and manufacturing; however, they are less energy efficient than the other two types of purification systems.

Water Filtration Systems

Water *filtration* systems differ from water *purification* systems in that the former remove (filter out) only the undissolved matter, such as dirt, suspended matter, and debris, and do not alter the dissolved matter in water,

such as chlorine, calcium carbonates, and salt. Filtration systems are used primarily for large bodies of water, such as swimming pools, reflection pools, and fountains. They are also used for drinking water systems if the water turbidity is unacceptable.

There are two basic filtration systems for large bodies of water:

1. *Sand filters* can be of either the pressure or the gravity variety. The main equipment consists of one

■ **FIGURE 8–3**
Typical construction of a zeolite water softener showing (1) a plastic-lined steel tank, (2) immersed sensor to monitor the zeolite solution and to signal the controller when recharging is needed, (3) zeolite resin in the tank, (4) brine tank to regenerate the zeolite, and (5) refill system with float shutoff. (Reproduced with permission from Culligan International Company.)

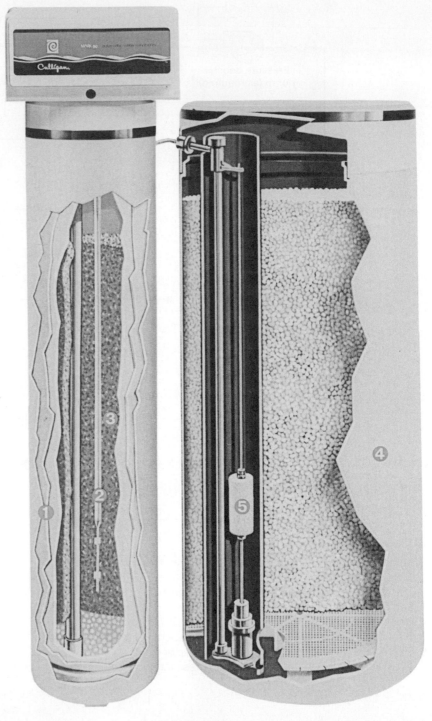

or more large tanks containing graduated layers of sand and pumps to circulate the water through the layers. The water passes through the layers, but the solids are trapped in them. When the layers are clogged with trapped solids, the pump reverses its direction of circulation and washes out the solids. The reversed flow process is called the *backwash cycle*. Figure 8–6 shows the construction of a typical pressurized sand filtering system. Figure 8–7 illustrates the operation of the filtration and back-washing cycles.

2. *Diatomaceous earth filters* use a fine mineral powder (silica, or SiO_2) found in nature. In the filtering process, water is pumped through the filter bed to trap the solids, and the cycle is reversed when backwash is needed. A diatomaceous earth filtering system is more compact than sand filters and requires less physical space, but it is more costly to operate.

Mineral-laden Water

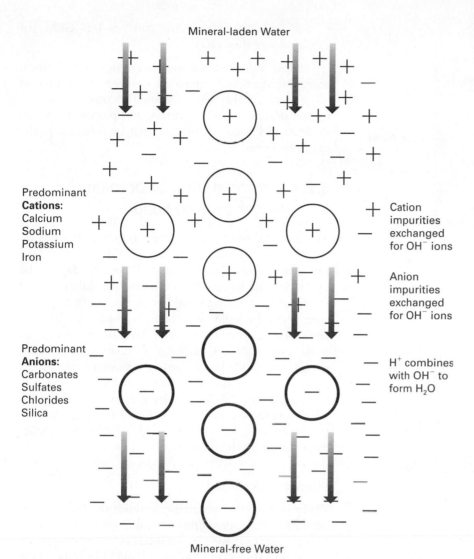

Predominant
Cations:
Calcium
Sodium
Potassium
Iron

Predominant
Anions:
Carbonates
Sulfates
Chlorides
Silica

\+ Cation
impurities
exchanged
for OH⁻ ions

Anion
impurities
exchanged
for OH⁻ ions

H⁺ combines
with OH⁻ to
form H_2O

Mineral-free Water

■ **FIGURE 8–4**
Operating principle of a mixed-bed
deionizer.

8.2 DOMESTIC WATER DISTRIBUTION SYSTEMS

In addition to its traditional use for plumbing facilities, water is also used for other building equipment and systems. A water distribution system is often called the "domestic water system," to differentiate it from fire protection and industrial-processing water systems. Domestic water system loads may be grouped into the following categories:

1. Plumbing facilities
2. Food service—preparation, refrigeration, washing, dining, etc.
3. Laundry
4. Heating and cooling systems
5. Exterior—lawn and plant irrigation, reflecting pools, fountains, hose bibbs, etc.
6. Pools—swimming pools, whirlpools, therapeutic pools

7. Research and process—laboratory equipment, commercial or industrial processes, computer equipment
8. Fire protection (if combined with the domestic system)
9. Others

Water required for all of these categories must be accounted for in the design of a domestic water system. In general, water demand for plumbing is the highest in most building occupancies except some industrial plants where water may be used in the production process. Accordingly, this chapter will concentrate on the design of plumbing facilities.

Units of measure commonly associated with water distribution systems are:

- 1 gallon per minute (GPM) = 0.063 liter per second (L/s)
- 1 liter per second (L/s) = 15.87 GPM

(a) Osmosis

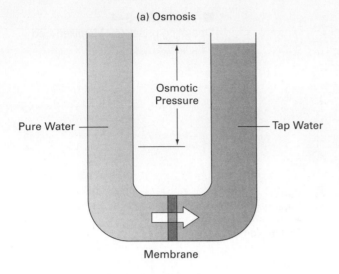

Osmotic
Pressure

Pure Water — | — Tap Water

Membrane

(b) Reverse Osmosis

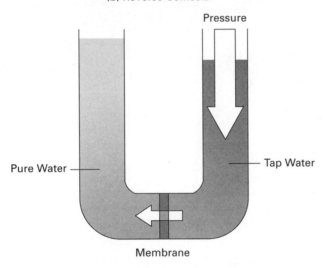

Pressure

Pure Water — | — Tap Water

Membrane

■ **FIGURE 8–5**
Principle of (a) osmosis and (b) reverse osmosis.

- 1 foot per minute (fpm) = 5.08 millimeters per second (mm/s)
- 1 foot per second (fps) = 0.305 meter per second (m/s)
- 1 cubic foot per minute (CFM) = 0.472 (L/s)
- 1 liter per second (L/s) = 2.12 CFM
- 1 pound per square inch (psi) = 6.9 kilopascals (kPa)
- 1 psi = 2.04 inches of mercury column (in. Hg)
- 1 psi = 2.31 feet of water column (w.c.)
- 1 atmosphere (atm) = 14.7 psi = 101.4 (kPa)
- 1 atmosphere (atm) = 30 in. Hg
- 1 water supply fixture unit (wsfu) ≈ a numerical weighing factor to account for the water demand of various plumbing fixtures, using the lavatory as 1 wsfu (1 to 1.5 GPM water flow rate)

- 1 drainage fixture unit (dfu) ≈ 0.5 GPM (of drainage flow rate)

Most water distribution system materials, fittings, control devices, and pumps are similar to those used in other mechanical systems. Basic information on the selection of pipes, valves, controls, and pumps is covered in detail in Chapter 7 and will not be repeated in the present chapter.

8.2.1 Determination of Domestic Water System Load

The required water capacity of a building depends on the coincidental peak load demand (CPLD) of all load categories, based on an assumed time of day in the heavy-demand season. For example, the highest CPLD for an office building would be noontime in the summer, when the building is fully occupied, plumbing facilities are in heavy use, and air conditioning is near its peak. The highest CPLD for an apartment building would be around dinnertime in the summer, when most people are home taking showers, washing, and preparing meals. Table 8–2 provides a convenient form for tabulating the estimated loads of various categories and the method used to determine the gross system capacity required for the building.

Plumbing Facilities

Water demand for plumbing facilities depends on the number and type of fixtures actually installed. Section 8.3 provides data on the various loads of different types of plumbing fixtures, and Section 8.4 provides data on the minimum number of fixtures required by codes. In practice, the actual number of plumbing fixtures in modern buildings usually exceeds the minimum required by codes, particularly in public assembly facilities and high-rise office buildings.

Each plumbing fixture is assigned a wsfu rating, representing the relative water demand for its intended operating functions (see Section 8.3.4 for more details). For example, a lavatory that does not demand a heavy flow of water is given a wsfu of 1, and a flush-valve-operated water closet that demands a heavy flow of water (even for only a few seconds) is given a wsfu of 10. The wsfu values of all other fixtures are shown in Table 8–6 (page 257). For special plumbing equipment that does not carry a wsfu rating, a wsfu of 1 may be used for each GPM of flow rate.

It is recognized that in most installations, only a fraction of the total number of connected plumbing fixtures are expected to be in use concurrently. The portion of these fixtures that could be concurrently active

■ **FIGURE 8-6**

Construction of typical water softener tanks. The cutaway sections illustrate the following:

1. Layers of minerals with lightweight chips at top to retain large, flat debris. Second layer collects coarse particles, third layer removes finer particles, and bottom layer polishes water down to micron-size insolubles.
2. Timer controller assembly.
3. Automatic valve system to accomplish service, backwash, and rinse cycles.
4. Pressure-rated steel tanks.
5. Conical distributor system for small systems.
6. Hub-radial distributor for large systems. (Reproduced with permission from Culligan International Company.)

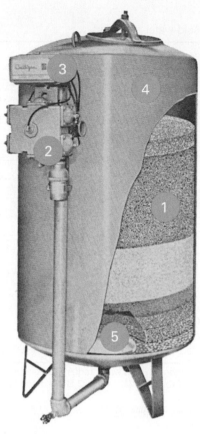

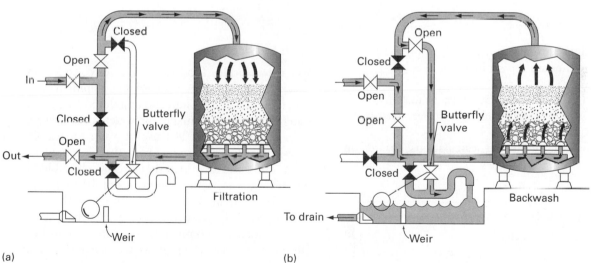

■ **FIGURE 8-7**

(a) Water flow of filtration cycle. (b) Water flow of backwash cycle with backwash water drains to building drainage system.

is a function of the occupancy of the building, the pattern of use of the fixtures, and other parameters.

An early study by Roy B. Hunter, of the National Institute of Standards and Technology (formerly the National Bureau of Standards), established a statistical base for estimating the peak demand flow rate of plumbing systems having mostly intermittent (noncontinuous) flows. These peak demand values are shown in Table 8–3(a). The values are conservative, but they are useful for sizing risers or branches of a water distribution system. Recent studies have resulted in a considerably lower set of *maximum probable flow values* for

TABLE 8–2
Determination of demand load for domestic cold-water systems[a]

	Connected Load		Net Load Demand of Each Category[b]		Coincidental Peak Demand[c]	
No.	Load Category	Description	wsfu	GPM	GPM	Remarks
1.	Plumbing facilities					
2.	Food service					
3.	Laundry					
4.	HVAC systems					
5.	Exteriors					
6.	Pools					
7.	Data processing					
8.	Research/process					
9.	Fire protection					
10.	Other ( )					

Coincidental peak demand . . . . ____ GPM
Spare capacity anticipated . . . . ____ GPM
Gross system capacity required . . . . ____ GPM
Net system capacity installed[d] . . . . ____ GPM

[a]Domestic cold-water demand of each category includes the cold water required to generate hot water.
[b]Net load demand for plumbing facilities is based on Tables 8–3(A) and 8–3(B); demand for other categories is determined from code and design references or the inputs of user/consultants.
[c]The coincidental peak demand is based on the selected season and time of day when peak water demand is likely to occur.
[d]Explain if different from gross system capacity.

residential, light commercial, motel, and similar services. These values are shown in Table 8–3(b).

Example 8.1 If an office building is installed with 10 water closets, 4 urinals, and 8 lavatories, what is the total wsfu installed?

We have the following calculations, based on Table 8–6:

		wsfu
Water closets (flushometer)	10×10	100
Urinals, blowout (flushometer)	4×5	20
Lavatories	8×2	16
Total plumbing fixtures installed		136 wsfu

Note: According to most building codes, flushometer valves must be used in public buildings, such as offices.

Example 8.2 For the office building in Example 8.1, what is the estimated demand for cold water, in GPM?

On the basis of Table 8–3(a) (office building), the estimated demand for cold water for these installed plumbing facilities is between 68 GPM for 100 wsfu and 83 GPM for 160 wsfu. By interpolation, the demand for cold water for 136 wsfu is about 77 GPM.

Food Services

Water demand for food services varies considerably between residential and commercial equipment. In general, food preparation and cooking do not require much water. The major demand for water is for washing in sinks or dishwashers. The use of water for washing in sinks has been accounted for in the plumbing fixture units. Thus, the only load that needs to be added is for the dishwashers, which require from 10 to 15 gal per wash for residential units. In estimating the demand load, 2 to 3 GPM per machine may be used. The demand load for commercial machines could be 5 to 10 times greater. Data on water demand should be obtained from the product manufacturer.

Laundry Services

Water demand for laundry also varies between residential and commercial equipment. Residential clothes washers require 20 to 40 gal of water per wash, depending on the size of the machine and the design of the wash cycle. In estimating water demand, 4 to 6 GPM per machine may be used. The demand load for a commercial laundry should be provided by the laundry manufacturer or a consultant.

TABLE 8–3
Estimation of probable demand for water

(a) For building occupancies other than residential, light commercial, motel, and similar services, and for sizing individual risers of multiriser water distribution system

For Predominantly Flush Tanks		For Predominantly Flush Valves		For Either Tank or Valve Systems	
wsfu	Demand, GPM	wsfu	Demand, GPM	wsfu	Demand, GPM
1 to 6	5			1000	208
10	8	10	27	1250	240
15	11	15	31	1500	270
20	14	20	35	1750	300
25	17	25	38	2000	320
30	20	30	41	2250	350
40	25	40	47	2500	375
50	29	50	52	2750	400
80	38	80	62	3000	430
100	44	100	68	4000	525
160	57	160	83	5000	600
200	65	200	92	6000	645
250	75	250	101	7000	685
300	85	300	110	8000	720
400	105	400	126	9000	745
500	125	500	142	10,000	770
750	170	750	178		

(b) For residential, light commercial, motel, and similar services

For Predominantly Flush Tanks		For Predominantly Flushometer		For Either Tank or Flushometer Systems[a]	
wsfu	Demand, GPM	wsfu	Demand, GPM	wsfu	Demand, GPM
up to 6	8	up to 6	30	1000	126
10	12	10	35	1250	152
15	13	15	40	1500	175
20	13	20	44	1750	200
25	14	25	48	2000	220
30	15	30	50	2250	240
40	16	40	55	2500	260
50	17	50	59	2750	282
80	21	80	68	3000	300
100	24	100	72	4000	352
150	31	150	79	5000	395
200	35	200	85	6000	425
250	42	250	91	7000	445
300	48	300	96	8000	456
400	60	400	102	9000	461
500	71	500	107	10,000	462
750	100	750	118		

[a]For large public assemblies, such as a concert hall, sports arena, or stadium, the demand for water would be greater, owing to the pattern of use of the facilities. In such applications, Table 8–3(a) should be used.

Heating and Cooling Systems

Normally, heating and cooling systems are closed-circuit systems that do not require constant water replacement. One exception to this rule is water-cooling towers for condenser water. Water is used to make up for the evaporation and drift losses in cooling towers. In general, 3 to 4 GPM of condensing water is required for every refrigeration ton of cooling load. A 1 percent makeup rate will require about 0.03 GPM per ton, or about 2 gal per ton-hour. This flow rate may not seem significant, but the annual consumption basis is astonishing.

Example 8.3   If the office building in Example 8.1 has a gross floor area of 25,000 sq ft and a 100-ton chiller is installed, what are the water demands for the cooling system and the annual water consumption if the system operates 12 hours a day for 200 days?

Assuming that the circulating rate of the condensing water is 3 GPM/ton with a 4 percent water makeup, the demand (continuous) and annual water consumption will be as follows:

- Water demand: 100 tons × (3 GPM/ton × 4%) = 12 GPM
- Annual consumption: 12 GPM × 60 min/hr × 12 hr/day × 200 days/year = 1,728,000 gallons

Note that this estimate is conservative, since the chiller will not operate at full capacity all the time. Evaporation losses will therefore be lower most of the time.

Exteriors

Water usage for building exteriors depends on the size of the lot and the portion that is landscaped. In some luxury residences, campus-type institutions, or country clubs, water demand for irrigation could be far greater than the usage within buildings. No generalization can be made for a load of this nature. The demand load must be determined on a project-by-project basis:

- For manual watering of plants and lawns with ½-in. to ⅝-in. hoses, water demand is between 5 and 15 GPM per hose outlet and may be neglected in the overall calculations if watering occurs at off-peak hours.
- For landscape sprinkler systems, the demand water flow rate of sprinkler heads ranges from 1 to 10 GPM for ½-in.-diameter pipe inlet to 30 GPM for 1-in.-diameter pipe inlet utilizing 20 to 60 psi of water pressure. Usually, a landscape sprinkler system is timer controlled to operate in off-peak hours.
- For fountains, each nozzle may have a flow rate from several GPM for a fine spray pattern to several thousand GPM for 3-in. or larger pipe geysers. Fountains are usually designed for recirculation. A 10 percent makeup capacity should be provided.

Swimming Pools

Swimming pools vary widely in size, from residential pools containing a few thousand gallons of water to Olympic-size pools containing several hundred thousand gallons. Normally, the flow rate of the circulating pump is designed to turn over (circulate) the entire volume of water in the pool in 6 to 8 hours, or 3 to 4 times in 24 hours. About 1 or 2 percent of the pumped circulation rate should be provided as continuous makeup water demand to overcome losses from evaporation, bleed-off, and spillage. To fill the pool initially, a separate quick-fill line should be provided to do the job in 8 to 16 hours; however, filling is usually done at off-peak hours. Thus, the demand flow rate need not be considered in the system demand calculations, unless it outweighs the demand of all other demands even during the off-peak hours.

Example 8.4 If a swimming pool contains 100,000 gallons of water, and the circulating pump is designed to change the water in 6 hours, what is the capacity of the pump, and what is the water demand load for makeup?

If the turnover rate of the filtering system pump is 6 hours, then:

- Pump circulation rate = 100,000 gal/6 hr = 16,600 gph; 16,600/60 = 277 GPM
- Demand load for makeup water = 277 × 2% = 5.5 (use 6 GPM)

Research and Processing

The use of water for research and processing in special buildings could be very high. The demand water load should be obtained from building staff members who are familiar with the operations or by metering actual usage.

Fire Protection

Normally, the water supply for fire protection is not included in the domestic water system; however, two components of a fire protection system may be combined with the domestic water system.

When a *standpipe system* is connected to a domestic water system, the domestic system must be capable of supplying a minimum of 100 GPM of additional demand for small buildings, using 1½-in. hose, to 500 GPM or more for large buildings. Design criteria will depend on the building code requirements.

If a limited area within a building is to be protected with automatic sprinklers (fewer than 20 heads) even though the rest of the building lacks sprinklers, these *limited-area sprinklers* may be connected to the domestic water supply system.

The demand water flow for a sprinkler head depends on the pressure available (residual pressure) at the head and the size of the orifices in the head, expressed by a *K* factor (varying between 2.7 and 8). For estimating demand loads, the flow rate of each sprinkler can be based on 30 GPM (at 30 psi with $K = 5.5$). These calculations are explained in detail in Chapter 9.

If the building is installed with a standpipe (and hose) system, the limited-area sprinklers are connected to the standpipe system risers. Such a riser is normally rated for 500 GPM for the first standpipe and 250 GPM for each additional standpipe. For determining the water demand load, only the greater of the two loads (standpipe and sprinklers) needs to be considered.

When a single water service (main) is used to serve both the fire protection system and the domestic system, the combined water main must be installed with a special bypass meter that consists of two components—one small meter to measure the normal domestic water flow, and an oversized detector check valve to allow a large volume of water flow during a fire. The detector check valve assembly usually consists of a (normally) closed check valve that opens automatically when the pressure drop across the small domestic meter reaches a preset value, say 1.5 to 4.5 psi.

Example 8.5 If the office building in Example 8.1 is required to have one standpipe (500 GPM) and two rooms with 15 sprinklers each, what demand load should be included in the domestic water system?

The demand water load for limited-area fire protection as part of the domestic water system is as follows:

- For the standpipe = 500 GPM
- For water demand in one of the two areas with sprinklers = 15 sprinklers at 30 GPM = 450 GPM

Since the sprinkler demand is lower than that of the standpipe system, only the standpipe system load is added to the overall demand load (500 gpm) of the domestic water system.

Example 8.6 For the office building in Example 8.5, estimate the gross and net system demand flow rate by considering the following:

- Plumbing facilities: 77 GPM (Example 8.1)
- Food service: There will be a future small dining and kitchen area with two kitchen sinks (3 wsfu) and one dishwasher (10 wsfu)
- Exterior: Two hose bibbs at 4 wsfu (5 GPM) each, to be used during office hours
- Air conditioning (HVAC): 12 GPM (Example 8.3)
- Fire protection: 500 GPM (Example 8.5)

The demand load is determined by using Table 8–2 as a basis (see page 236). It is assumed that the coincidental peak demand for this office building occurs during a summer weekday and includes water for air conditioning.

8.2.2 Determination of Water Pressure

A water system must be maintained with positive pressure to establish a flow in the distribution system and through the plumbing fixtures or equipment. Furthermore, positive water pressure prevents water from being contaminated by external sources, since at a positive pressure, water tends to leak out of the pipe, thus preventing it from being infiltrated by foul water external to the pipe, such as groundwater in a buried piping system.

Water pressure should be sufficient to overcome any pressure loss due to friction, differences in elevation, and flow pressure at outlets or equipment.

Fixture or Equipment

Every plumbing fixture or connection that uses water must have the proper pressure to maintain the required flow. The minimum flow pressures required at standard plumbing fixtures are given in Table 8–4. Flow pressure is defined as the pressure at the plumbing fixture or equipment while water is flowing at the required flow rate. It is different from static pressure, which is pressure at no-flow conditions only. The flow pressure required for equipment varies widely, and the requirements must be obtained from manufacturers.

When the pressure required for a specific equipment exceeds the system pressure that can be economically provided, a booster pump can be installed for that equipment. This is preferable than raising the overall system pressure to meet the special need. Pressure higher than 80 psig should be reduced by pressure-reducing devices connected to standard plumbing fixtures. This is the current upper limit specified in most plumbing codes.

TABLE 8–4
Minimum flow pressure required for typical plumbing fixtures and equipment

Fixture or Equipment	Minimum Flow Pressure, psi
Lavatory, sink, bathtub, shower, bidet, drinking fountain, water closet (tank)	8–10
Water closet and urinal (flushometer)	20–25
Garden hose, lawn sprinkler, dishwasher, clothes washer	15–20
Commercial dishwasher (self-contained pump)	30–50
Fire protection sprinkler	25–30
Fire hose (1½ in.)	65
Fire hose (2½ in.)	65

Water Meter and Backflow Preventer

Depending on the design and size, a pressure drop between 5 and 10 psi can be expected for a water meter, and similarly for the backflow preventer.

Piping

Pressure drop for water flow in the pipes is given in terms of psi per unit length, usually psi/100 ft. The data are readily available in most code and engineering handbooks. Pressure drop depends on the piping material (e.g., copper, steel, PVC), the nominal internal diameter, and the flow rate selected for design. The pressure-drop chart for fairly smooth (Type L copper) pipes is shown in Figure 8–8. The chart shows the pressure loss due to friction, in psi/100 ft, corresponding to the flow rate, in GPM, and the resulting flow velocity in the pipe. Good practice should limit the flow velocity to below 10 fps to avoid noise.

Example 8.7   If the coincidental peak demand of a water system in a commercial building is 120 GPM, what is the main water service size, based on copper pipe and a flow velocity not to exceed 8 fps?

Working Table for Example 8.6 (using Table 8–2)

	Connected Load		Net Load Demand of Each Category		Coincidental Peak Demand	
No.	Load Category	Description	wsfu	GPM	(GPM)	Remarks
1.	Plumbing facilities	Plumbing fixtures	136	77	77	TABLE 8–3(a)
2.	Food service	Kitchen equipment	—	40	0	Future
3.	Laundry	—	—	—	—	
4.	HVAC systems	100 tons cooling	—	12	12	
5.	Exteriors	Two hose bibbs	—	10	5	one in use
6.	Pools	—	—	—	—	
7.	Data processing	—	—	—	—	
8.	Research/process	—	—	—	—	
9.	Fire protection	Standpipe + sprinkler	—	500	500	Standpipe only
10.	Other ( )					

Coincidental peak demand . . . .   <u>594</u> GPM
Spare capacity anticipated . . . .   <u>40</u> GPM   Note (2)
Gross system capacity required . . . .   <u>694</u> GPM   (<u>500</u> used)
Net system capacity installed . . . .   <u>600</u> GPM

Notes: (1) The standpipe water supply is connected ahead of the domestic water meter through a detector check valve.
(2) The water main need be sized only for 500 GPM, since the domestic water load will not be active during a fire. The domestic water supply from the water meter to the building is sized for 132 GPM.
Remarks: 600 GPM is adequate for standpipe (500 GPM) and other loads (totaling 100 GPM) during a fire.

From Figure 8–8, interpolating between 100 and 200 GPM and the 8-fps velocity line, the service should be a 2½-in. line for a pressure loss of 5 psi/100 ft. If the velocity is limited to 6 fps, then a 3-in. line should be selected for a pressure loss of 2.5 psi/100 ft.

Pipe Fittings and Control Devices

Pressure loss in pipe fittings, such as elbows, tees, valves, and controls, is significantly higher than in straight pipe. In fact, an improperly selected control valve will impose a higher pressure loss than all the straight pipes in the entire system. The pressure loss for fittings and devices is usually given in the unit of "equivalent length of straight pipe," "equivalent length" (EL) for short. For example, a 1½-in. gate valve has a pressure drop of 1 EL, whereas the same size globe valve has a pressure drop of 40 EL. A 45° elbow has about 2 EL, and a 90° elbow 4 EL. Values for standard pipe fittings can be found in most codes and handbooks. In engineering design, the number and character of all fittings and control devices are identified, and their pressure loss in EL is accounted for. For preliminary estimating purposes, generally 50 percent to 100 percent is added to the linear pipe

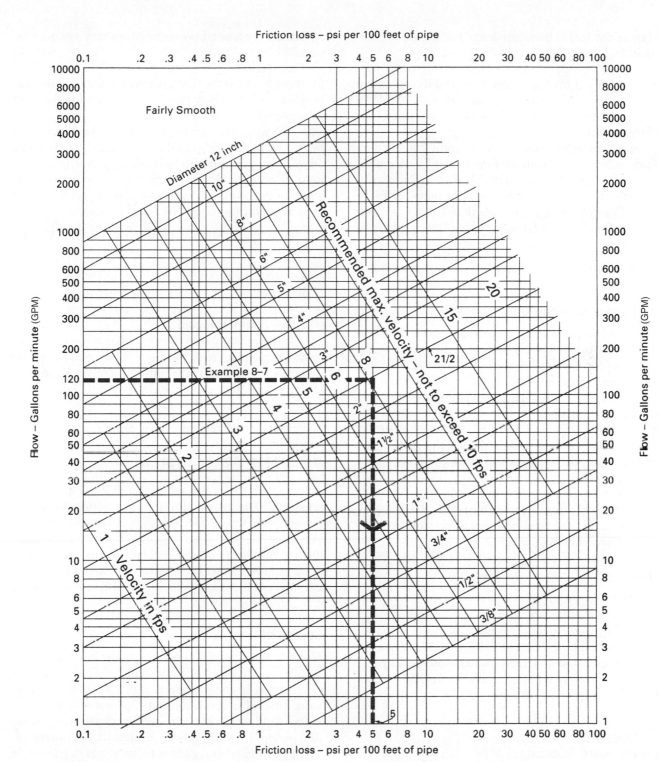

Friction loss – psi per 100 feet of pipe

■ FIGURE 8–8

Pressure loss (psi/100 ft) versus flow rate (GPM) for fairly smooth pipe (e.g., copper, PVC).

length as the total equivalent length of the piping system, including fittings.

Example 8.8   If the most remote part of the plumbing system for the commercial building is about 200 ft from the service entrance, what is the pressure loss due to the piping system?

Since the piping system for a commercial building is fairly simple, in that no unusual control device is required, it can be safely assumed that the EL for pipe fittings will be 50 percent of the straight pipes, or EL = 100 ft. Thus, the total piping length is 300 EL, and the pressure loss to the piping system at a nominal 5 psi/100 ft is

$$300\,\text{ft EL} \times 5\,\text{psi}/100\,\text{ft} = 15\,\text{psi}$$

Differences in Elevation

Water required at floors higher than the water service entrance to the building must overcome the force of gravity, which provides *static pressure* owing to the difference in elevation. From simple units of conversion, 1 psi is equivalent to 2.31 ft w.c. Thus, in a building 231 ft high, the plumbing fixtures on the top floor will require 100 psi of static water pressure to reach that elevation. This static pressure must be added to the pressure required to cover friction losses and flow pressure.

Example 8.9   If a commercial building is 70 ft in height above the water service entry to the building, what should be the minimum water pressure to serve the top-floor plumbing fixtures?

The water pressure required at the entry is as follows:

- Water meter: 15 psi
- Backflow preventer: 15 psi
- Piping pressure loss (see Example 8.8): 15 psi
- Static height (elevation) = 70 ft/2.31: 30 psi
- Flow pressure at fixtures (water closets): 25 psi
- Total pressure required at the service: 15 + 15 + 15 + 30 + 25 = 100 psi

Available Water Pressure and Pressure Boosting

Normally, the flow pressure of underground water mains from the utility company is between 50 and 100 psi; however, in municipalities where the infrastructure lags the development, water pressure may be lower than that required, particularly during peak hours. Other cases of low pressure are often found downtown in older cities, where the water mains are clogged with scale or are too fragile to stand higher internal pressure. In such a case,

an interior pressure-boosting system may need to be installed.

Example 8.10   If the water pressure available at the water main is 60 psi, how can the commercial building in Example 8.7 be served?

The water pressure required for this building is 100 psi (see Example 8.9). Since the pressure at the main is only 60 psi, water pressure must be boosted, or the requirement must be reduced. To boost the pressure, pumps are inserted in the system at the base or at the upper floor level to make up the shortage. The added pressure is

$$\text{Booster pressure required} = 100 - 60 = 40\,\text{psi}$$

Life-cycle cost analysis will determine whether the proper solution is to reduce the pressure required by installing oversized components to reduce pressure drop, or to insert a booster pump system, or to have a combination of both techniques.

Water pressure may be boosted by installing one or more pumps at strategic locations. These pumps may be designed to run constantly, intermittently with storage tanks, or constantly with variable speed. Figure 8–9 illustrates a water distribution riser diagram and the pumping schemes of a high-rise building.

8.2.3 Hot-Water System

A hot-water system is a subsystem of the domestic water system. The demand for hot water is included in that for domestic water. The use of hot water in buildings varies considerably, from very little in office-type buildings to high in residences, restaurants, and hotels. The design of a hot-water distribution system closely follows that of a cold-water system, but with several additional considerations.

Source of Energy

Hot water is normally generated in the building by the installation of water heaters using oil, gas, steam, or electricity as an energy source. Figure 8–10 shows a cutaway view of a residential type of water heater, and Figure 8–11 shows three heaters in parallel using manifold piping. In general, water heaters have a storage capacity of from several gallons to hundreds of gallons. It is more economical to preheat water in storage than to generate hot water on demand. Furthermore, storing hot water makes it easier to maintain an even water temperature.

Hot-Water Demand

The demand for hot water depends on the user. For example, a person may take a 3-minute shower or

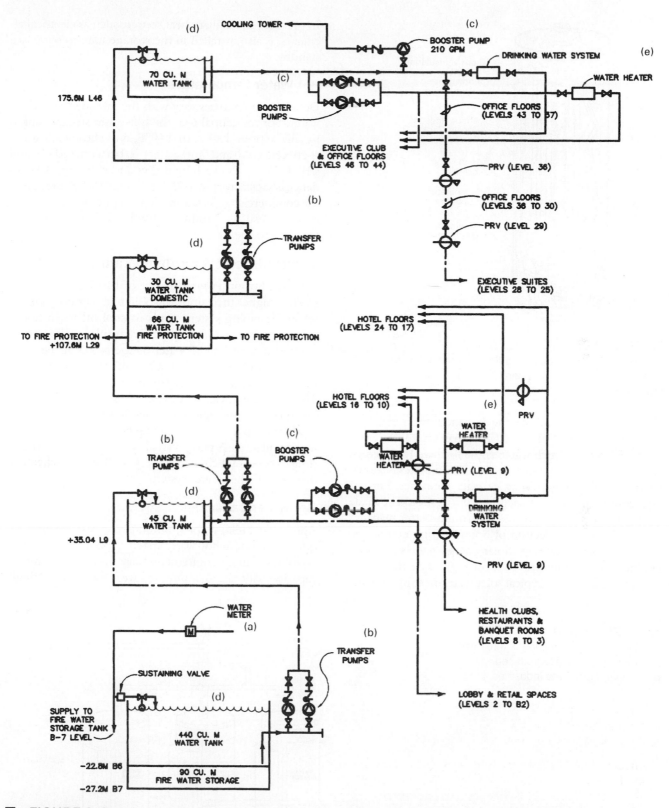

FIGURE 8–9

Schematic and riser diagram of cold- and hot-water system of a high-rise building. (a) Water main, (b) three sets of transfer pumps, (c) three sets of water-pressure booster pumps, (d) house water tanks at three upper floor levels. Lower portion of the tanks at levels B6 and L29 is reserved for fire protection. (e) Water heaters at three upper levels.

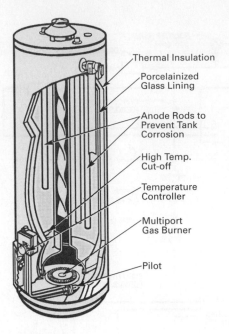

Thermal Insulation

Porcelainized Glass Lining

Anode Rods to Prevent Tank Corrosion

High Temp. Cut-off

Temperature Controller

Multiport Gas Burner

Pilot

■ **FIGURE 8–10**

Cutaway view of a typical gas-fired residential water heater. (Reproduced with permission from State Industries, Inc.)

a 15-minute shower. While the demand flow rate (GPM) of hot water remains the same in both cases, the latter requires 5 times as much water as the former. The average amount of hot water usage in the United States is shown in Table 8–5. Usage differs in other countries. For energy conservation, a conscientious effort should be made to control the use of hot water, through better management and proper selection of water-heating equipment and flow control devices. Solar energy panels are both practical and economical in most regions of the world. Figure 8–12 shows a typical solar water-heating scheme.

In practice, an auxiliary heat source, such as electrical elements, is also installed in the storage tank to serve as a standby.

Hot-Water Temperature

Desired temperatures vary with the intended use of the water. For residential use, the hot-water storage tank is usually kept at 130°F to 140°F, and the mixed water (between cold and hot water) is between 100°F and 110°F. Water for washing dishes and clothing in a residence is satisfactory at 140°F, whereas 180°F is required for commercial applications. This higher water temperature is usually produced by local booster heaters attached to the machines.

Maintenance of Water Temperature

Since the hot-water temperature of 140°F is about 60°F to 70°F higher than ambient room temperature, hot water in the piping system tends to cool off when not in constant use, even if the pipes are insulated. For better maintenance of desired water temperatures at the point of use (plumbing fixtures), a second pipe known as the *hot-water circulation line* (or *re-circ. line*) is installed. The *re-circ.* line can be designed for gravity circulation based on the principle that when water is cooled off, it tends to drop down the second pipe, thus creating a natural circulation. A pump can also be provided to force a slow but constant circulation. Figure 8–13 is a schematic diagram of a hot-water system.

Hot-Water Capacities

Applicable building or plumbing codes frequently require a minimum hot-water system capacity in residential buildings. These requirements vary from 20-gal storage capacity with a minimum recovery rate for a single

■ **FIGURE 8–11**

Manifold piping connection of three water heaters in parallel. (Reproduced with permission from State Industries, Inc.)

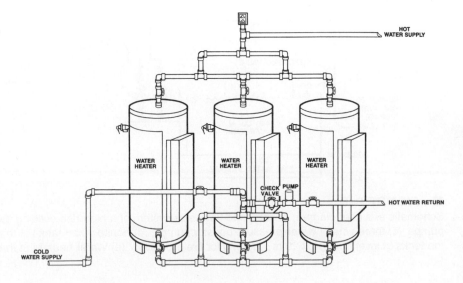

TABLE 8–5
Hot-water consumption

Usage	Demand, GPM	Total Consumption, gal
Shower, 5 min	2–6	10–30
Bath, in tub	5–15	25–30
Dishwashing, residential	1–2	10–15
Dishwashing, commercial	10–30	50–200
Clothes washer, residential	2–4	10–50

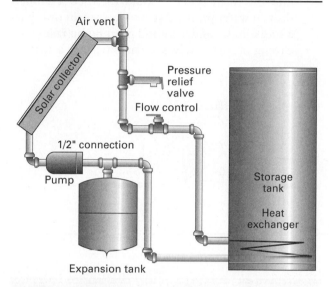

■ FIGURE 8–12
Piping diagram of a solar-energy-heated water-heating system. (Reproduced with permission from State Industries, Inc.)

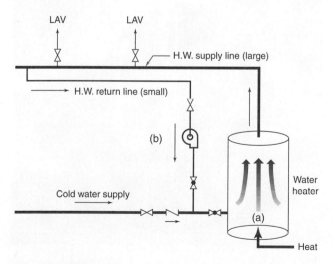

■ FIGURE 8–13
Schematic diagram of a hot-water system, showing the circulation pattern from the water heater to typical plumbing fixtures. (a) Hot water rises within the water heater, because it is lighter than cold water. (b) An electrical pump may be used (optional).

bedroom to 80-gal storage capacity with a correspondingly higher recovery rate for three or more bedrooms. The appropriate hot-water capacity for residences may be higher than that required by code and will have to be analyzed on an individual basis.

8.2.4 Design Considerations for Water Distribution Systems

Piping Material

Copper, plastic, galvanized steel, and stainless steel are approved materials for potable water services.

Copper is the most commonly used water-piping material because of its physical strength, durability, and resistance to corrosion. It is made in three classifications: type K (heavy wall gauge), type L (medium gauge), and type M (light gauge). Type L is the normal choice for interior distribution systems. It is joined with lead-free soldering compounds.

Stainless steel is sometimes used in lieu of copper when the sulfur content in the water or air is high, as in the area of hot springs. The joints are either welded or threaded. Stainless steel pipes and fittings are more expensive than copper pipes and fittings.

Hot-dipped galvanized steel is standard low-carbon steel dipped into molten zinc. Joints must be threaded or of the mechanical coupling type. Galvanized steel pipe is an alternative to copper and may be more economical, depending on its price, which fluctuates. It is generally more economical to use for larger pipes. If steel and copper are mixed in a system, dielectric couplings must be used to avoid galvanic corrosion.

Plastic is used increasingly for water distribution because of its lower cost, corrosion resistance, and low potential for scaling. Plastic pipes must be specially rated for potable water service. Polyethylene (PE), acrylonitrile butadiene styrene (ABS), and polyvinyl chloride (PVC) are rated only for cold-water use; polyvinyl dichloride (PVDC) is approved for either cold- or hot-water services.

Thermal Insulation

Pipes are insulated with thermal material, such as fiberglass, mineral wool, or foam plastic, to maintain the temperature of water for either chilled or hot water. Insulation material varies from ½ in. to 2 in. in thickness, depending on the desired heat transfer resistance. Cold-water pipes should be insulated to avoid condensation in spaces with high relative humidity, such as an industrial plant with no air conditioning, or where usage is very frequent, as in a public toilet.

Acoustic Isolation

When the noise of water flowing in a pipe is annoying or disturbing in a quiet space, such as a conference room or residence, both cold- and hot-water piping should be insulated for both thermal and acoustical purposes.

Expansion or Contraction

When the ambient temperature to which the piping system is exposed is changed, and the relative coefficient of expansion between the building and the piping material is different, a differential movement will be created between the piping and the building. Differential movement will also occur when the water temperature in the pipe changes. Flexibility must be built into the piping system to allow for such movement. The methods commonly used involve installing expansion loops or joints to compensate for the physical expansion (or contraction) of the pipes. Figure 8–14 illustrates a typical bellows type of expansion compensator.

Preventing Backflow

The water distribution system must be safeguarded against contamination. This provision is governed strictly by code. Some of the techniques used are as follows:

- A check valve allows water to flow in one direction only. Figure 8–15 illustrates several check valves.
- A vacuum breaker is installed at the branch connection to an equipment item or plumbing fixture, such as a sink, dishwasher, boiler, water closet, or urinal. A vacuum breaker will automatically open

the piping system to atmospheric pressure (14.0 to 14.7 psig) when pressure in the piping drops below the atmospheric pressure level, to prevent foreign material or foul water in the equipment or fixture from being siphoned into the piping system. Figure 8–16 illustrates the construction and operating principle of a vacuum breaker.

A backflow preventer (BFP) is used at the entrance to the main water supply and at critical branches to stop water from flowing backward into the main system should there be a sudden drop in water pressure at the water main owing to a break in the underground main pump failure, or closing of supply valves for maintenance activities.

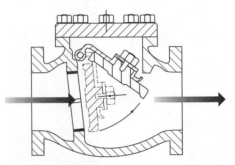

(a)

(b)

■ FIGURE 8–15

(a) Wing-type check valve showing the check in closed and open positions. (b) Center-pivoted design with reduced pressure drop. (Reproduced with permission from Watts Regulator Co., N. Andover, Mass.)

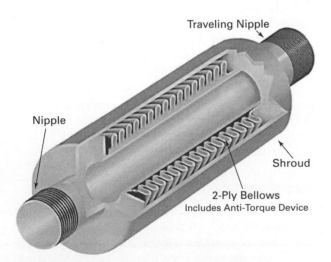

Traveling Nipple

Nipple

Shroud

2-Ply Bellows
Includes Anti-Torque Device

■ FIGURE 8–14

Cutaway view of a metal bellows type of expansion compensator showing the externally pressurized accordion-type bellows and threaded pipe ends. (Reproduced with permission from Flexonics, Inc.)

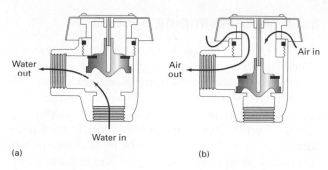

(a)   (b)

■ **FIGURE 8–16**

(a) Water pressure pushes the spring-loaded plunger up. The air port is closed and the water port is open. (b) When water pressure is below atmospheric pressure, the plunger is down, the water port is closed, and the air port is open. (Reproduced with permission from Sloan Valve Co.)

A backflow preventer is essentially a double check valve with pressure-test fittings to ensure that the valves are tight. The reduced-pressure-zone-relief BFP incorporates a relief valve between the check valves for a more positive action against failure of the check valve. Figure 8–17 illustrates the construction of a backflow preventer.

■ Perhaps the most positive way to prevent backflow of foreign material into the water system is by means of the air gap required on plumbing fixture installations at the end of a water outlet, such as a valve or faucet. An air gap must be installed at least two pipe diameters higher than the receptor (bowl or basin), so that in case water overflows above the

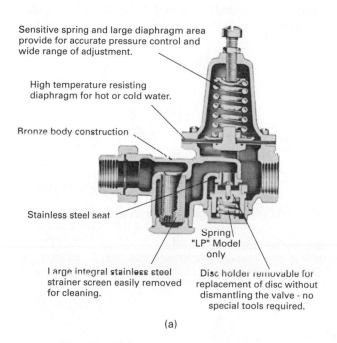

(a)

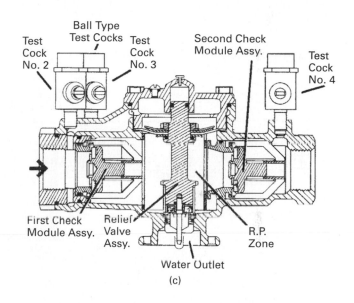

(c)

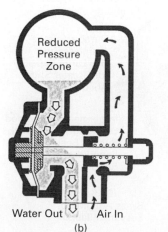

(b)

■ **FIGURE 8–17**

(a) Cutaway section of a typical spring-loaded water pressure reducing valve.
(b) Schematic of the operating principle of the reduced-pressure-zone relief valve.
(c) Cutaway section showing the internal construction of a reduced-pressure-zone backflow preventer. When the supply pressure is reduced, the first check valve will close and the relief valve will open to drain off any water in the assembly. Should both check valves fail, the relief valve will allow air to enter the valve to relieve the negative pressure in the water supply side and to allow backflow (potentially foul water) to circulate to the drain.
(Reproduced with permission from Watts Regulator Co., N. Andover, Mass.)

receptor rim, foreign material will not get into the piping system. Figure 8–18 (a) and (b) illustrate air gap requirements.

Shock Absorption

When the flow of water in a pipe is abruptly stopped, as by the closing of a faucet or a flushometer, the dynamic (kinetic) energy in the water must be absorbed. If it is not, the energy will be converted into a loud noise and vibration known as *water hammer*. To avoid the disturbing water hammer, pipes at the end of a branch or next to a quick-closing fixture must be installed with an air chamber in which the air acts as a cushion. If space is limited, mechanical shock absorbers can be used instead. They have a large neoprene chamber that operates on the same principle. Figure 8–18(c) illustrates the construction of an air and shock absorber, also known as a *water hammer arrester*.

8.2.5 Water-Pumping System

Water pressure required for water distribution is determined from a number of design parameters: water demand (flow rate), elevation, pipe sizes, material, routing, type of fittings, accessories, etc. All these factors contribute to pressure loss. If the water supply does not have sufficient pressure to overcome the total pressure loss, then the pressure must be boosted by a pump; however, a pump consumes energy and requires maintenance; thus, it should be avoided whenever possible. (See Chapter 7 for a discussion of the various types of pumps and a guide to their selection.)

A special need for pumping is manifested in the hot-water recirculation system, where a low-flow in-line pump is installed to ensure the immediate delivery of hot water at the far end of a hot-water piping system. Figure 8–19 illustrates a typical in-line centrifugal pump and its pressure-flow characteristics.

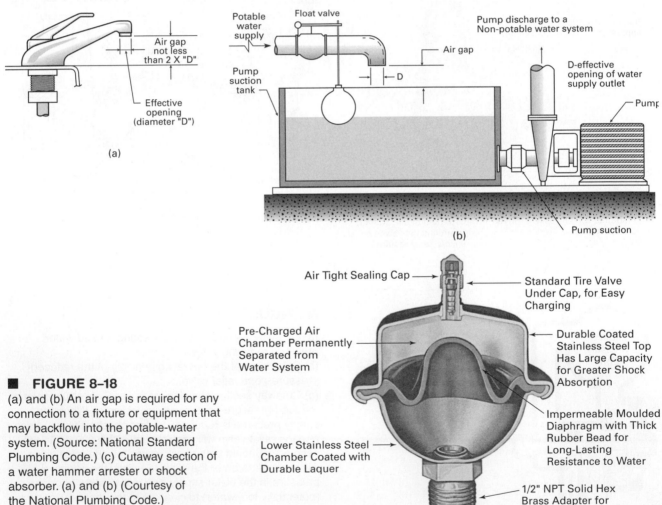

■ FIGURE 8–18

(a) and (b) An air gap is required for any connection to a fixture or equipment that may backflow into the potable-water system. (Source: National Standard Plumbing Code.) (c) Cutaway section of a water hammer arrester or shock absorber. (a) and (b) (Courtesy of the National Plumbing Code.)
(c) (Reproduced with permission from Watts Regulator Co., N. Andover, Mass.)

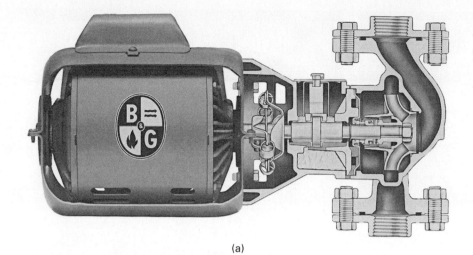

(a)

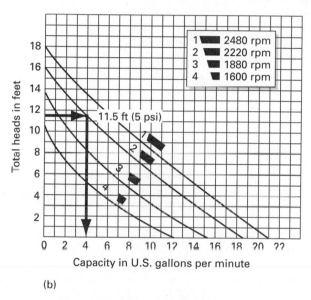

1	2480 rpm
2	2220 rpm
3	1880 rpm
4	1600 rpm

11.5 ft (5 psi)

Total heads in feet

Capacity in U.S. gallons per minute

(b)

■ FIGURE 8–19

(a) Cutaway view of an in-line centrifugal pump.
(b) Pressure-flow chart of a typical multiple-speed small in-line pump operating at 1600, 1880, 2220, and 2480 rpm. *Example:* Select the pump speed that will provide a 4-GPM flow rate at 5 psi. *Answer:* From the chart, at 11.5 feet (5 psi) and 4 GPM, the two lines meet at speed "2." The pump should accordingly be selected to operate at 2220 rpm. (Reproduced with permission from ITT Fluid Handling Sales.)

- Showers (SH)
- Drinking fountains (DF)

8.3.1 Plumbing Fixtures

As illustrated in Figures 8–20 and 8–21, plumbing fixtures are normally made of dense, impervious materials, such as vitreous china, enameled cast iron, stainless steel, or some other acid-resistant material. Vitreous china is most suitable for water closets, urinals, bidets, and lavatories, whereas sinks are more often made of stainless steel for durability and abrasion resistance.

Water Closets

Water closets are normally made of vitreous china with hollow interior walls to direct the passage of water and an integral water-seal trap to separate the fixture from the drainage system. Water closets are usually the most prevalent plumbing fixture in a building, both in number

8.3 PLUMBING FIXTURES AND COMPONENTS

Plumbing fixtures are receptacles, devices, or appliances that are supplied with water or that receive liquid-borne wastes and then discharge wastes into the drainage system. The following are some plumbing fixtures commonly used for building services, with their common abbreviations in diagrams:

- Water closets (WC)
- Sinks (SK)
- Bidets (BD)
- Urinals (UR)
- Bathtubs (BT)
- Service sinks (SS)
- Kitchen sinks (KS)
- Lavatories (LAV or LV)

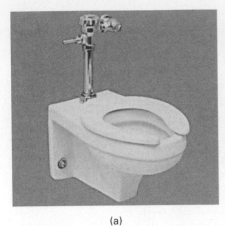

(a)

(b)

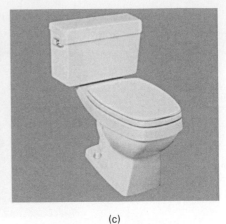

(c)

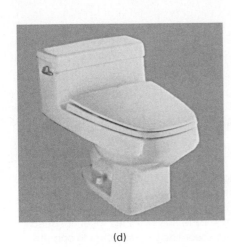

(d)

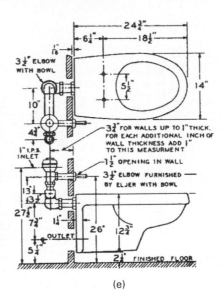

(e)

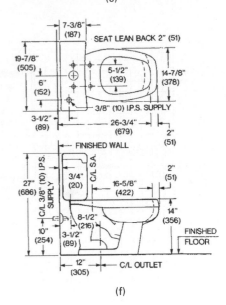

(f)

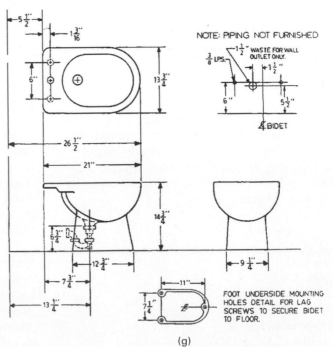

(g)

■ **FIGURE 8–20**

Typical plumbing fixtures.

(a) Water closet with elongated bowl and flushometer valve.
(b) Bidet with hot-cold water-mixing valve.
(c) Water closet with standard flush tank.
(d) Water closet with low-profile flush tank.
(e) Top and side views of water closet in (a).
(f) Top and side views of a floor-mounted water closet with flush tank.
(g) Top, side, and front views of a bidet.
(Reproduced with permission from Eljer Plumbing Products Co.)

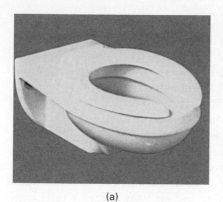
(a)

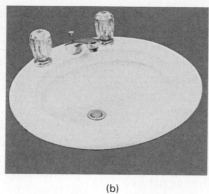
(b)

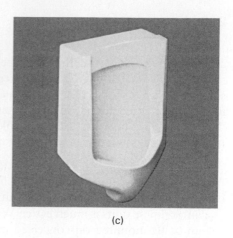
(c)

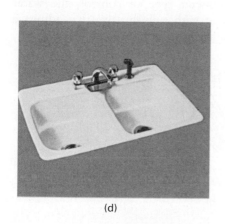
(d)

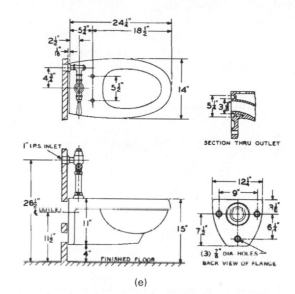
(e)

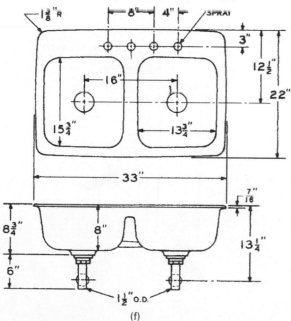
(f)

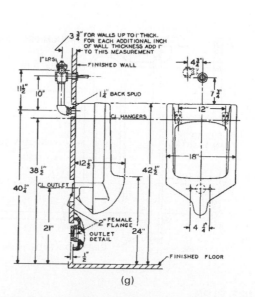
(g)

■ **FIGURE 8–21**
Typical plumbing fixtures.
(a) Water closet with elongated bowl.
(b) Countertop-mounted lavatory.
(c) Wall-hung urinal.

(d) Double-compartment service sink.
(e) Top and side views of water closet in (a).
(f) Top and side views of a service sink.
(g) Side and front views of a wall-hung urinal. (Reproduced with permission from Eljer Plumbing Products Co.)

251

and in water demand; thus, they have the most impact on the capacity of water and drainage systems. For this reason, the designer should thoroughly understand the various types of water closets available, their similarities and differences, and their operating principles. Water closets may be classified in terms of the following features:

1. *Method of mounting.* There are floor-mounted and wall-mounted (-hung) models. The wall-hung variety is more costly to install, but cleaning floors under these units is much easier. Wall-hung fixtures must be mounted on concealed carriers, supports in the wall.

2. *Cleansing action of the bowl.* Siphon-jet and washdown varieties are the quietest in operation and are universally used in private and residential applications. The blowout type used in public facil-

ities is noisier and more positive in cleansing action. Figure 8–22 illustrates the flushing action of siphon-jet and blowout-type bowls. There are also air-water and air-vacuum bowls.

3. *Method of water control.* There are four methods: gravity tank, flushometer, pressure tank, and vacuum type. In the tank type of control, the water closets generally have 1.0- to 4-gal water storage tanks depending on age and design of the unit. The water supply is connected to the tank through a vacuum breaker and air gaps. Water is discharged into the bowl by gravity. Figure 8–23 illustrates the operating principle of the tank type of water closet.

In the flushometer type of control, the water closets are equipped with a flush valve that admits a time-measured (adjustable between 5 and 10 seconds)

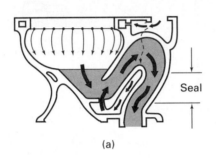
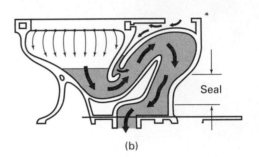

(a) (b)

■ **FIGURE 8–22**

Flushing action of water closets. (a) Blowout design: Water enters through the rim for cleansing, and a strong jet at the base blows out the contents. (b) Siphon-jet design: Water enters through the rim for cleansing and jets in up the leg of the trapway to create a siphon.

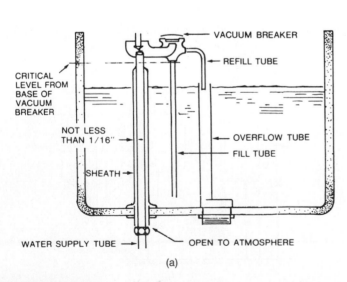
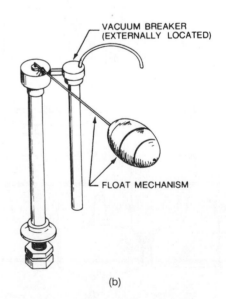

(a) (b)

■ **FIGURE 8–23**

Operating principle of a flush tank. (a) Water feeds through the supply tube and down to the fill tube through a vacuum breaker. A portion of the water bleeds into the overflow tube (or compartment) to refill the bowl. (b) Float valve rises to stop the water supply until the next flushing. (Courtesy of National Standard Plumbing Code.)

amount of water into the bowl under the water pressure. The amount of water admitted is comparable to that of the tank type of control; however, the instantaneous water demand rate is considerably higher (around 20 to 30 GPM). The flushometer type of water closet can be ready for use immediately after flushing and therefore is required in all public occupancies. Figure 8–24 illustrates the operating principle of a diaphragm type of flushometer. It has become increasingly popular to use automatic sensing devices to operate the flushometer in public facilities, such as airports, restaurants, hotels, theaters, arenas, and sport stadiums. Automatic flushing design and construction with solid-state components have become very reliable and cost-effective and are a more positive way to ensure sanitary conditions. Figure 8–24 (a)– (d) illustrates several flushometer designs and their operations.

In the pressure-tank type of control, the water closets are equipped with a pressurized tank within a conventional gravity tank. The pressurized tank is charged with air and water under 25 psi water pressure. When the plunger at the base of the tank is released, the air–water mixture is forced into the bowl to blow out its contents. Because the blowout action is under pressure, 1.5 gal of water is sufficient; thus, a considerable amount of water is conserved. With water shortages in many parts of the world, water conservation should be of primary concern to all building owners and designers. The pressure-tank type of water closet is becoming increasingly popular; however, pressure tanks are noisy, which must be taken into consideration, especially in multitenant application. Figure 8–25 illustrates the construction and operating principle of the air–water type of water closet.

The vacuum type of water closet operates on a central vacuum piping system. When the valve below the bowl is opened, the contents of the bowl are sucked into the drainage piping system under a vacuum using only 0.3 gal of water per flushing. This type of water closet is commonly used on airplanes and oceangoing ships. Figure 8–26 illustrates its construction and operating principle.

Example 8.11 In the office building in Example 8.1, if the water closets are of the pressure-tank type with a wsfu rating of 2, what size should the domestic water riser be to serve the plumbing fixtures?

- If the water closet is rated for 2 wsfu, and the other fixtures remain the same, then the total wsfu installed will be 48 in lieu of 128, as calculated in Example 8.1.
- From Table 8–3(b), the demand flow will be about 17 GPM (this value taken from the "predominantly flush tank" column), instead of 59 GPM (from the "predominantly flushometer column").

- From Figure 8–8, the pipe size for 17 GPM (at 8 fps velocity) is between ¾-in. and 1-in.-diameter pipe, whereas that for 59 GPM (at 8 fpm) is a 1½-in.-diameter pipe.

Urinals

Similar to water closets in construction and operating principle, urinals come in siphon-jet, blowout, and wash-down varieties. They are either wall-mounted or floor-mounted. Wall-mounted units are most commonly used, because they require less space and are easier to clean. Urinals should have a visible water seal with no strainer above the seal. Water-flushing action should thoroughly clean the entire interior fixture surface. Construction deviating from these principles is prohibited by many governing codes.

Lavatories

Lavatories are designed in a variety of sizes and shapes. Fittings, such as faucets, drains, and other accessories, are nearly unlimited in design and material.

Sinks

General-purpose sinks and kitchen sinks are available in single-, double-, and triple-compartment models. Stainless steel sinks are preferred because they are durable and easy to clean. As a special application, service sinks are installed in public buildings for cleaning crews. They may be mounted on a wall, on the floor, or in a recess on the floor; and they may be made of stainless steel, enameled cast iron, precast terrazzo, or reinforced glass fiber.

Drinking Fountains

Drinking fountains can be wall- or floor-mounted and are usually nonrefrigerated; however, they may be piped from a central chilled-water supply with a separate filtering system.

Electric Water Coolers

Electric water coolers are individual drinking fountains with self-contained water-chilling units. The individual electric water cooler is preferred over the central system because of lower installation and operating cost.

Bathtubs

The most popular bathtub is 5 ft long with a net water basin dimension of 4'–6'. Recent trends show an increased use of models 6 ft or longer, as well as wider, with built-in whirlpool nozzles and pump. Bathtubs are made of enameled cast iron, porcelain enamel on pressed steel, or fiberglass-reinforced plastics. A showerhead with combined water-control fitting is usually incorporated with a bathtub. The base of a tub should be designed with a nonslip surface.

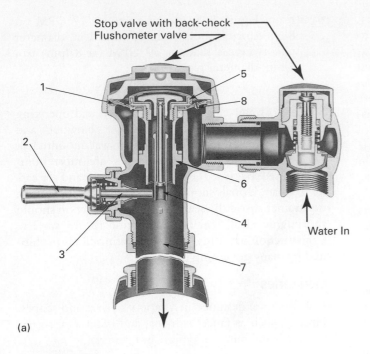

Stop valve with back-check
Flushometer valve

1
2
3
5
8
6
4
7

Water In

(a)

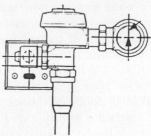

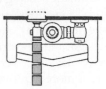

1.
A continuous, invisible light beam is emitted from the OPTIMA Sensor.

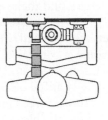

2.
As the user enters the beam's effective range (15" to 30") the beam is reflected into the OPTIMA's Scanner Window and transformed into a low voltage electrical circuit. Once activated, the Output Circuit continues in a "hold" mode for as long as the user remains within the effective range of the Sensor.

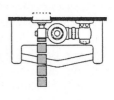

3.
When the user steps away from the OPTIMA Sensor, the loss of reflected light initiates an electrical "one-time" signal that operates the Solenoid (24V AC) and initiates the flushing cycle to flush the fixture. The Circuit then automatically resets and is ready for the next user.

(b)

(c)

■ FIGURE 8–24

(a) Principle of operation of a diaphragm type of flushometer. (Courtesy of Sloan Valve Co., Franklin Park, IL.) When the flush valve is in the closed position, the segment diaphragm (1) divides the valve into an upper and lower chamber with equal water pressure on both sides of the diaphragm. The greater pressure on top of the diaphragm holds it closed on its seat. Movement of the handle (2) in any direction pushes the plunger (3), which tilts the relief valve (4) and allows water to escape from the upper chamber. The water pressure in the lower chamber below the segment diaphragm (1), now being greater, raises the working parts (1), (4), (5), and (6) as a unit, allowing water to flow down through the valve outlet (7) to flush the fixture. While the valve is operating a small amount of water flows through the bypass (8) of the diaphragm, gradually refilling the upper chamber and equalizing the pressure once more. As the upper chamber fills, the diaphragm (1) returns to its seat to close the valve. (b) Electronic sensor (Optima model) and its principle of operation. (c) Battery-operated model operates on 4 AA 1.5-V batteries. (d) Hydraulic-operated model with remotely mounted push button. (Reproduced with permission from Sloan Valve Co.)

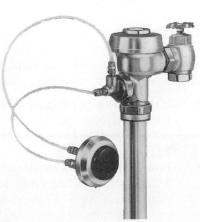

(d)

254

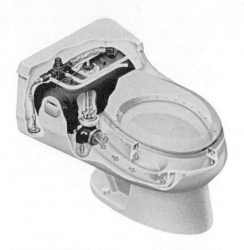

FIGURE 8-25

Pressure tank system operating on a water supply system 25 psi or higher. The tank contains a 1.5-gal pressurized air–water reservoir. When the flush lever is activated, it allows the mixture of air and water to rush into the bowl and thus expels (blows out) the contents of the bowl under pressure. After flushing, a charge cycle takes place to refill the reservoir until the 25 psi (or higher) pressure is reached. (Reproduced with permission from Kohler Company.)

Showers

Showers can be integrated with a bathtub or be independently constructed into shower stalls or a battery of showers. One major concern is the control of the water temperature by a mixing valve between cold and hot water. Thermostatic valves mix the water for the desired outlet water temperature. Better-designed control valves also have a pressure-balance feature so that the water temperature will not fluctuate when there is a sudden change in the relative water pressure between the cold- and hot-water systems.

Bidets

Bidets are small baths, the size and shape of a water closet, used primarily for personal hygiene. Hot or cold water enters around the flushing rim, with a spray rinse as an optional feature. The rinse is also referred to as a *douche*. Bidets are popular in Europe.

8.3.2 Low-Water-Use Fixtures

Water usage for many commonly used plumbing fixtures is mandated by the Energy Policy Act (EPAct), which is periodically updated. Maximum flow rates for typical fixtures are as follows:

Water closets	1.6 gal per flush
Urinals	1.0 gal per flush
Lavatories	2.0 gal per minute
Showers	2.2 gal per minute

Fixtures are avaiable that use significantly less water, and overall savings of 30 percent or more can easily be achieved in most buildings by using low-water-use urinals, water closets, faucets, and showers.

Urinals manufactured according to EPAct standards are rated at 1.0 gallons per flush. The amount of water used is regulated by a diaphragm in the flush valve. Optional diaphragms are available from several manufacturers, which can reduce water usage to 0.5 gallon per flush, a savings of 50 percent.

Waterless urinals are also available to reduce consumption. Their design contains an oil layer through which urine, which is heavier than oil, sinks and flows into the waste line. The urinal body is similar in appearance to a conventional unit, but there is no flush valve or water piping, which results in cost savings as well as water conservation. They are prohibited by code in

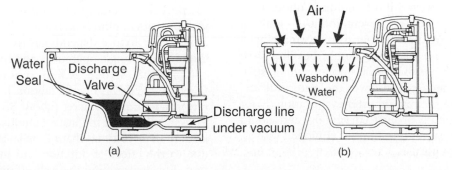

FIGURE 8-26

Cross-sectional views of a vacuum type of water closet. (a) Water seal in the bowl when the water closet is not in use. (b) When an electrical or pneumatic button is pushed, it opens the discharge pipe, and air at atmospheric pressure forces the contents of the bowl through the discharge valve (open for 4 seconds) and into the discharge line under the vacuum. Concurrently, a small amount of water (0.4 gal) opens (for 7 seconds) to provide wash-down and water seal in the bowl. (Reproduced with permission from Envirovac Inc.)

some jurisdictions, so the designer should check with local authorities before specifying these units.

Water closets manufactured according to EPAct standards are rated at 1.6 gallons per flush. Models are available that use as little as 1.0 gallons per flush using pressure-assist devices. Pressure-tank water closets (see above) are also available below EPAct minimum standards. The dual-flush flushometer is yet another method for saving water used by water closets. The handle of the flushometer is labeled "up for liquid waste" and "down for solid waste." The same concept can be applied to tank toilets with a similar two-button feature, which offers a choice of two flush volumes depending on need.

Low-water-use fixtures are available for faucets and showers. Conventional fixtures can also be converted to use less water by inserting flow restrictors.

8.3.3 Components of Fixtures

Traps

To prevent the backup of sewer gas into a building through the drainage connection of plumbing fixtures, every plumbing fixture must be connected by means of a trap seal. As a rule, all water closets, bidets, and urinals are manufactured to have an integral trap, whereas other plumbing fixtures use external traps.

The trap is a portion of piping in a U shape and filled with a water seal. It operates on the principle that two columns of water balanced within the two legs of the trap prevent sewer gas, noxious fumes, or vermin from passing from the sewer side to the building. The water column in a trap is normally between 2 and 4 in. deep; it thus requires a minimum of 2-in. w.c. pressure differential to break the water seal. There are various trap designs. P-traps are most common, and are normally used for standard plumbing fixtures.

A water seal is always filled with water if the fixture is regularly used; however, if a building is unoccupied for long periods of time—say, several months—water in the trap of any fixture may dry out from normal evaporation. In such a case, the trap should be filled with some lighter-than-water but nonevaporative fluid, such as glycerine or mineral oil, to cover the water-seal surface, thus retarding the rate of water evaporation. Floor drain traps are especially prone to drying out and should be filled with nonevaporating fluid or equipped with an automatic trap-filling valve.

Air Gap

To avoid the possibility of soil or waste contamination of the water supply system, connections between water supply components and the plumbing fixtures must be separated vertically through an air gap. The gap is normally designed to have a minimum separation between the bottom opening of the water supply and the fixture rim, which is the highest point of the fixture. With an air gap, any backflow of the waste will overflow at the fixture rim, and the potential for contaminating water is eliminated (see Figure 8–18(a)).

Prevention of Backflow

Water flowing in the pipe is normally under positive pressure. When the water pressure within the piping system drops suddenly, it may be possible for soil or waste, to which the piping is connected, to be sucked or siphoned into the water system. Figure 8–16 illustrates the operation of pipe-mounted vacuum breakers. Figure 8–23 shows the vacuum breaker required on a tank-type water closet. Figure 8–24 shows the combination stop valve and backflow preventer ahead of a flushometer. Prevention of backflow is required on any equipment that is connected to water piping without an air break (gap). Components of fixtures, such as faucets with "hose-end" connections and dishwashers, are required by code to be installed through vacuum breakers.

8.3.4 Fixture Units

The demand for water and the load imposed on the drainage system of plumbing fixtures differ considerably. In fact, a fixture with low water demand may impose a high drainage load, depending on its mode of operation. For example, a bathtub usually does not demand a high water flow, but it could impose a high drainage load. It normally takes over 10 minutes to fill a tub but only 2 to 3 minutes to drain it. For this reason, a fixture may have a water supply rating different from its drainage rating.

The *water supply fixture unit* (wsfu) is a measure of the probable hydraulic demand on the water supply by various types of plumbing fixtures. The value of wsfu for a particular fixture depends on the rate of supply, the duration of a single operation, and the frequency of operation of the fixture. A ½-in. residential-type lavatory faucet is rated for 1 wsfu (about 1 to 1.5 GPM flow rate). A 1-in. flushometer (about 35 GPM) is assigned 10 wsfu, although the instantaneous demand for water is more than 10 times that of the lavatory faucet. This is due to the effect of duration and frequency of operation. For determining the wsfu of nonstandard plumbing fixtures, one wsfu rating may be used for each 1.5 GPM of flow rate. The wsfu ratings of standard plumbing fixtures are given in Table 8–6.

The *drainage fixture unit* (dfu) is a measure of the probable discharge into the drainage system by various

TABLE 8–6
Drainage and water fixture unit values[1] and minimum pressure and pipe size requirements

Fixture	Drainage		Water				
	Load Value, dfu[2]	Minimum[3] Pipe Diameter	Load Values, wsfu[2]			Minimum at Fixture[4]	
			Cold	Hot	Total	psig	Pipe Diameter
Public or General							
Water closet, flushometer	6	4	10	—	10	20–25	1
Water closet, gravity tank	4	3[7]	5	—	5	8–10	½
Water closet, pressure tank[5]	2	3[7]	2	—	2	25–30	½
Urinal, flushometer	4	2–3	5	—	5	15–20	¾
Urinal, tank	2	2	3	—	3	8–10	½
Lavatory	1	2	1.5	1.5	2	8–10	⅜
Bathtub	2	2	3	3	4	8–10	½
Shower	2	2	3	3	4	8–10	¾
Service sink	3	3	2.25	2.25	3	8–10	¾
Kitchen sink	3	2	3	3	4	8–10	¾
Electric water cooler	0.5	2	0.25	—	0.25	8–10	⅜
Private							
Water closet, flushometer	6	4	6	—	6	20–25	1
Water closet, gravity tank	4	3[7]	3	—	3	8–10	½
Water closet, pressure tank[5]	2	3[7]	2	—	2	25–30	½
Lavatory	1	2	0.75	0.75	1	8–10	⅜
Bathtub	2	2	1.5	1.5	2	8–10	½
Shower	3	2	1.5	1.5	2	8–10	½
Bathroom group, flushometer[6]	8	4	6	6	8	20–25	1
Bathroom group, tank[6]	6	4	4	4	6	15–20	1
Kitchen sink	2	2	1.5	1.5	2	8–10	½
Combination fixtures	3	4	2.25	2.25	3	10–15	½
Dishwasher	2	2	—	1	1	10–15	½
Laundry, clothes washer (8 lb)	3	2	1.5	1.5	2	10–15	½

		Drainage				Load values, wsfu	
Fixtures not listed above, but with trap size given	1¼ in.	1	2	Fixtures not listed above, but with pipe size given	Pipe Size	Private	Public
	1½ in.	2	2				
	2 in.	3	2		⅜ in.	1	2
	2½ in.	4	2½		½ in.	2	4
	3 in.	5	3		¾ in.	3	6
	4 in.	6	4		1 in.	6	10

[1]For fixtures with continuous water flow, the load is added without a demand factor.
[2]dfu = drainage fixture unit; wsfu = water supply fixture unit.
[3]Minimum drainage-pipe diameter, in inches, for fixture branches. Fixture outlet tailpieces may be smaller in diameter, as per manufacturer's specifications.
[4]Minimum pressure and pipe size should be verified with fixture manufacturer.
[5]dfu and wsfu are estimated for the pressure-tank type.
[6]A bathroom group is defined as consisting of not more than (1) WC, (1) LV, (1) BT, and (1) SH, or not more than (1) WC, (2) LV, and (1) BT or (1) separate SH.
[7]It has been common practice to install 4-in. soil pipe for all water closets to minimize the chance of blockage.

types of plumbing fixtures. The value of dfu for a particular fixture depends on the rate of drainage discharge, the duration of a single operation, and the frequency of operation of the fixture. Laboratory tests show that the rate of discharge of an ordinary lavatory with a nominal 1.25-in. outlet, trap, and waste is about 7.5 GPM (0.5 L/s). It is assigned a dfu = 1. The value is low, owing to frequency of operation. The dfu equivalent for continuous-flow (or semicontinuous-flow) equipment is higher, at 1 dfu for each 0.5 GPM of continuous drain. The dfu ratings of typical plumbing fixtures are given in Table 8–6.

8.4 PLANNING PLUMBING FACILITIES

Plumbing facilities for buildings are primarily toilets, bathrooms, washrooms, and lockers for private or public use. The requirements for such facilities are normally analyzed and planned by the architect to fulfill the needs of the building's occupants; however, planned facilities must equal or exceed the minimum code requirements. In practice, the plumbing facilities in high-quality buildings are more than are required by the code, for the following reasons:

- *To improve convenience* Facilities should be dispersed and thus frequently duplicated, so that occupants need not travel too far on one floor or to other floors to use them.
- *To accommodate the fluctuation in a building's occupants* The proportion of a building's occupants of each sex may change over time; thus, extra facilities must be built in to accommodate this

fluctuation. This is particularly true for assembly and sports facilities, where different events may draw a different mix of participants.

- *To avoid congestion* The use of a building's plumbing facilities is usually concentrated at certain times of the day (e.g., at lunchtime, at the end of an event, and during intermissions of events of one kind or another). To avoid congestion, extra plumbing fixtures beyond the number specified as a minimum by the code should be provided. It is quite common for modern sports arenas and stadiums to have more than twice the minimum number of facilities required by code to provide satisfactory services.

8.4.1 Minimum Requirements by Code

Nationally recognized plumbing codes normally include the minimum numbers of plumbing fixtures that must be installed in a building, based on the building's population (occupants) and use. Table 8–7 lists such

TABLE 8–7
Minimum plumbing fixture requirements in a building (Values shown are consolidated from different codes. Follow the applicable code requirements in actual design.)

Building Use Group		Recommended Maximum Number of Occupants per Fixture					
		Water Closets (WC) (Urinal see note a)		Lavatories (LV)	Bathtubs (BT) Showers (SH)	Drinking Fountains (DF)	Service Sinks (SS)
		Male	Female				
Theaters, churches	First fixture	50	50	½ of WC	—	1000	1/floor
	Additional	150	150				
Stadiums/arenas, ballparks, terminals	First 3 fixtures	50	50	½ of WC	—	200	1/floor
	Additional	300	150				
Restaurants Convention halls	First 3 fixtures	50	50	½ of WC	—	200	1/floor
	Additional	200	100				
Recreational Nightclubs	First fixture	40	40	½ of WC	—	75	1/floor
	Additional	40	20				
Businesses/stores, banks Offices, shopping centers	First fixture		15	½ of WC	—	100	1/floor
	Additional		25				
Educational	Preschools		15	½ of WC	—	30	1/floor
	Elementary		25			40	
	Secondary		30			50	
Hospitals	Ward	1:8 patients		½ of WC	1:20 patients	100	1/floor
	Private rooms	1/room		1/room	—		
Nursing homes		15		15	1/units	100	1/floor
Institutional, prisons		1/cell		1/cell	1:6 inmates	—	1/floor
Hotels/motels		1/unit		1/unit	1/units	—	1/floor
Dwelling units		1/unit		1/unit	1/units	—	(c)
Dormitories		20		½ of WC	8	100	1/floor
Industrial		30		½ of WC	(b)	75	1/floor

(a) One urinal may be installed in lieu of one water closet in male toilets for up to 50 percent of the required water closets.
(b) See applicable code for detail requirements.
(c) 1 kitchen sink and 1 clothes washer connection per dwelling unit.

requirements. For determining the number of occupants within the building, the architect or engineer should make the estimate based on the following factors (subject to local codes):

1. The maximum number of occupants intended should be calculated; but this figure should not be less than that specified in the governing building code. Table 8–8 is one method commonly used to determine the number of occupants of a building. For special or unlisted occupancies, the occupant load shall be established by the architect or engineer, subject to the approval of the building official.

2. The plumbing fixtures shall be distributed equally between the sexes, based on the percentage of each sex anticipated in the occupant load, and shall be composed of 60 percent of occupants for each sex if the estimated occupants are evenly divided.

3. In commercial buildings and institutions, facilities for employees or workers may need to be calculated separately from those for customers or inmates.

TABLE 8–8
Method for calculating number of occupants of a building

Occupancy	Occupant/ Seat or Floor Area/Occupant
Assembly, with fixed chairs	1 occupant/seat
with fixed seating benches	1 occupant/18 in.
with fixed seating booths	1 occupant/24 in.
with movable chairs	7 NFA[a]
with movable chairs and tables	15 NFA
with standing space only	3 NFA
Business area in a building	100 GFA[b]
Educational, classroom area	20 NFA
Shops and other vocational areas	50 NFA
Library reading rooms	50 NFA
Library stack areas	100 GFA
Institutional, inpatient areas	240 GFA
Outpatient area	100 GFA
Sleeping area	120 GFA
Mercantile, basement, and grade floor area	30 GFA
Area on other floors	60 GFA
Storage, stock, shipping areas	300 GFA
Parking garages	200 GFA
Residential	200 GFA
Miscellaneous (storage, mechanical equipment, etc.)	300 GFA

[a]NFA (net floor area) is defined as the actual occupied area of the space and shall not include unoccupied accessory areas or the thickness of walls.
[b]GFA (gross floor area) is defined as the floor area within the perimeter of the outside walls of the building under consideration. (Courtesy of International Code Council.)

8.4.2 Substitution for Water Closets

Urinals may be substituted for water closets in male toilets on a one-for-one basis, but the substitution should not be more than 50 percent of the required water closets.

Example 8.12 If a 100,000-GFA office building has 80 percent of the GFA for office occupancy and 20 percent for miscellaneous spaces (storage and mechanical rooms), determine the number of building occupants and the minimum number of plumbing fixtures required.

- According to Table 8–8, the number of building occupants shall be:

Business (office) Area:

80,000 sq ft/100	800 occupants
Miscellaneous spaces:	
20,000 sq ft/300	67 occupants
Total occupants	**867**
Assumed male occupants, 50% (use 60%)	520
Assumed female occupants, 50% (use 60%)	520

- From Table 8–7, the minimum required plumbing fixtures are as follows:

	WC	UR	LV	BT	DF	SK
Men	11	11	11	0	—	—
Women	22		11	0	—	—
For building	—		—	0	10	1/floor

8.4.3 Planning Guidelines

Plumbing fixtures shall be installed so as to afford easy access for cleaning and maintenance. More space is required for the physically disabled. For example, a 42-in. clearance on one side of a WC stall is necessary to make room for a wheelchair.

Figure 8–27 shows typical layouts of residential bathroom groups with piping in one or more walls. Toilets and washrooms should be located where they are easily accessible or likely to be expected: near building entries, waiting areas, elevator lobbies, stairways, telephone stations, etc.

Men's and women's toilets should be located in the same general area. They may be placed back to back or with some separation between them. When they are placed back to back, the cost of piping is substantially reduced for small and low-rise buildings. The cost of piping for large and high-rise buildings may not be significantly affected, since the large number of fixtures to be installed will probably require multiple stacks and risers in any case.

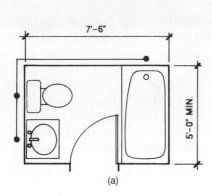

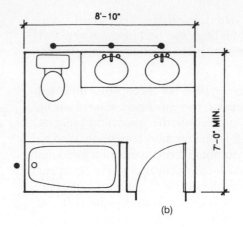

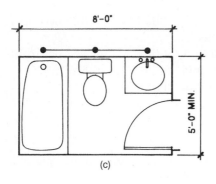

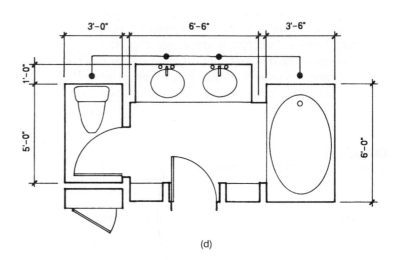

■ **FIGURE 8–27**

Typical residential bathroom plans.

(a) Basic 3-fixture plan with piping on two walls.

(b) More spacious plan with piping on two walls.

(c) Basic 3-fixture plan with piping on one wall (most economical).

(d) Deluxe plan including whirlpool and private water closet compartment. (Courtesy: Tao and Lee Associates, Inc., Architects, St. Louis, MO.)

The designer must be careful to provide a visual barrier between the public passageway and the interior of the toilets when the door to the room is opened and there is a possibility that a reflected image may be seen through a mirror. Allow ample distance between fixtures and room surfaces. Refer to codes and standards for the minimum space requirements.

Figure 8–28 shows several examples of public toilets and many of the design insuls facing the architect.

8.5 SANITARY DRAINAGE SYSTEMS

Drainage in buildings consists of three major components: sanitary waste, storm water, and specialty waste,

such as toxic, radioactive, chemical, or other processing wastes. Sanitary waste and storm water may be piped separately or combined, depending on the public sewer system to which the drainage is connected. Combined storm and sanitary sewer systems still exist in some major cities, carried over from the early practice of discharging untreated sewage into rivers. This practice is no longer permitted in the United States and most other developed, as well as developing, countries. Because of their pollutants, specialty wastes must be piped and treated separately.

Following are definitions of several terms commonly used in drainage systems:

- *Waste (liquid)* Liquid discharged from water-consuming equipment.
- *Sanitary waste* Liquid discharged from plumbing fixtures.

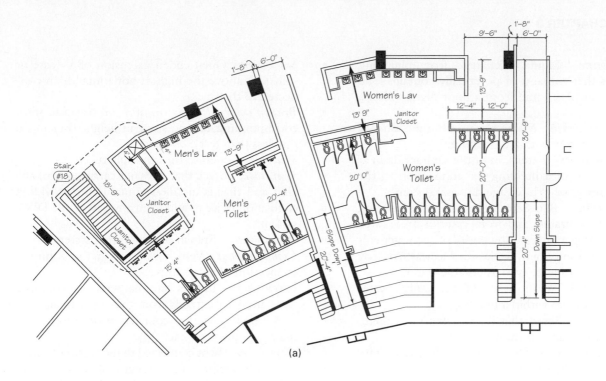

(a)

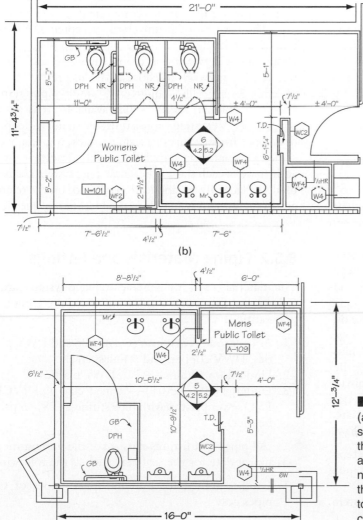

(b)

■ **FIGURE 8–28**

(a) Public toilets in a large sports arena. Note the separate entry and exit for each toilet, the toilets for the handicapped, and the janitors' closets. The men's and women's toilets are separated by a vomitory for noise control. (b) Typical public toilet layouts. Note the drinking fountain at the entry to the women's toilet, the screen walls for privacy, and the water closets for the handicapped.

- *Soil (waste)* Liquid discharged from plumbing fixtures that contains or potentially contains fecal matter, such as liquid from a floor drain within a toilet.
- *Sanitary drain* Main drain of the sanitary drainage system within a building.
- *Sanitary sewer* Extension of the sanitary drain at the exterior of a building for connection to the public sewer or to a sewage disposal system.
- *Storm water* Rainwater collected from building roofs (roof drains) and from exterior areas (area drains). Depending on the purity of the water collected, rainwater may be classified as waste or may be stored for reuse, even as potable water.
- *Storm drain* Main drain of the storm water drainage system within a building.
- *Storm sewer* Sewer that is exterior to a building and that contains storm water only.
- *Combination sewer* Sewer that contains sanitary waste and storm water.

A sanitary drainage system is thus a drainage system designed to carry away sanitary wastes (including soil wastes) from within a building to a public sewer or to a sewage disposal plant. The system may be designed to flow by gravity without mechanically or electrically powered equipment, or it may be designed to flow under pressure by pumping. The gravity system is, of course, more reliable and more economical to operate and thus should be used whenever feasible. The discussions in this section will deal with gravity flow only.

8.5.1 Drainage-Waste Venting

When the sanitary drainage system is connected to a public sewer, the entry of sewer gas, insects, or rodents through the system must be avoided. To overcome this problem, all drainage equipment (including all plumbing fixtures) connected to the sanitary drainage system must be separated by a liquid seal trap that acts to separate the building from the sewer. Each trap must be adequately vented to the atmosphere to prevent the liquid seal from being siphoned or sucked dry if a pressure differential is created between the building and the sewer owing to the flow of the drainage. Figure 8–29 illustrates the operating principle of a liquid seal trap. Because of the importance of venting, a sanitary drainage system is often referred to as a *drainage-waste-venting* (DWV) system. Some of the important terms associated with a DWV system and their definitions are as follows:

- *Stack* Vertical portion of a DWV-piping system.
- *Waste stack* Vertical portion of a waste-piping system.
- *Soil stack* Vertical portion of a soil-piping system.

- *Stack vent* Open-ended extension of a waste or soil stack above the highest horizontal drain connected to the stack.
- *Branch interval* Section of a soil or waste stack corresponding to one story in height (but in no case less than 8 ft).
- *Vent* Pipe open to the atmosphere.
- *Vent stack.* Stack that does not carry waste of any kind and that is installed primarily for providing circulation of air to and from any part of the DWV system.
- *Branch vent* Branch of the venting system.
- *Common vent* Vent connected at the common connection of two fixtures.
- *Circuit vent* Branch vent that serves two or more traps and that extends from the downstream side of the highest fixture connection of a horizontal branch to the vent stack.
- *Crown vent* Vent connected to the crown of a trap.
- *Developed length* Total length of a pipe, measured along the centerline of the pipe.

Figure 8–30 is an isometric diagram of a DWV system identifying the various components of the system.

The conventional universally accepted DWV system is essentially a two-pipe system consisting of a waste-piping network that drains the waste and a separate venting network to equalize the air pressure in the system. A specially engineered single-pipe system known as a *sovent system* is also available. It is an all-copper DWV system consisting of standard copper drainpipes from fixtures to a single stack, a specially designed aerator fitting at each floor, and a deaerator fitting at the base of each stack. The single-pipe system has been used satisfactorily. Its application depends on the relative costs of labor and materials in the localities in question.

8.5.2 Piping Materials and Fittings

The materials used for sanitary drainage systems are as follows, in the general order of their preference or popularity:

1. Aboveground—cast iron (C.I.), ABS and DWV plastics, DWV copper, and stainless steel
2. Underfloor (drains)—C.I., DWV plastics
3. Underground (sewer)—vitrified clay pipe (VCP), PVC sewer plastics with pipe stiffness (PS) 35 psi or heavier.

All pipes and fittings must have the joining methods appropriate for the material, such as the following:

1. Caulked joints, for hub-and-spigot types of C.I. pipes only

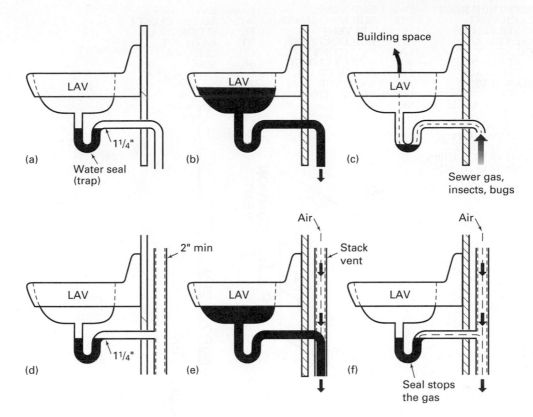

■ FIGURE 8–29

Liquid seal traps.

(a) With no vent, a trap may be started with a seal at no flow.

(b) During heavy drainage flow, the momentum of the flow can siphon out the liquid from the seal.

(c) As a result, the trap will be only partially filled and will allow sewer gas and other substances to enter the building.

(d) With proper venting, a trap is filled with liquid from the previous drainage flow.

(e) Even under heavy flow, air enters the vent pipe to break up the siphoning effect.

(f) As a result, the trap will continue to maintain a full depth of liquid.

2. Threaded joints, for steel or heavy-wall plastic pipes
3. Soldered joints, for copper DWV pipes
4. Brazed joints, for heavy copper pipes only
5. Compression joints, for no-hub C.I., VCP, or plastic pipes
6. Gasket joints, for hub-and-spigot C.I. pipes
7. Flexible compression joints, for VCP pipes
8. Solvent joints, for plastic pipes
9. Transition joints, specially made for joining different materials

See Figure 8–31 for illustrations of the various types of joints.

8.5.3 Sizing of Drain Lines

The capacity of (waste or soil) drainpipes of the DWV system depends on two major factors: the slope of the pipes and the dfu they serve. All horizontal drain lines should have a uniform downward slope in the direction of flow. If the slope is not steep enough, solid contents in the liquid may drop out. If the slope is too steep, turbulence flow and erosion of the pipes may occur. In general, the slope shall not be less than ¼ in. per foot (approximately a 2 percent slope) for 3-in.-diameter and smaller horizontal lines and ⅛ in. per foot (approximately a 1 percent slope) for 4-in.-diameter and larger horizontal lines. To determine the drainage capacity and velocity of horizontal main and branches:

1. If the flow GPM is known, use Table 8–9 to find the flow velocity in fps for various slope and pipe sizes.
2. If the diameter is known, use Table 8–10 to determine the maximum number of dfu that can be connected for horizontal branches and stacks of various sizes.
3. If the diameter of the building drain or sewer is known, use Table 8–11 to determine the maximum number of dfu that may be connected.

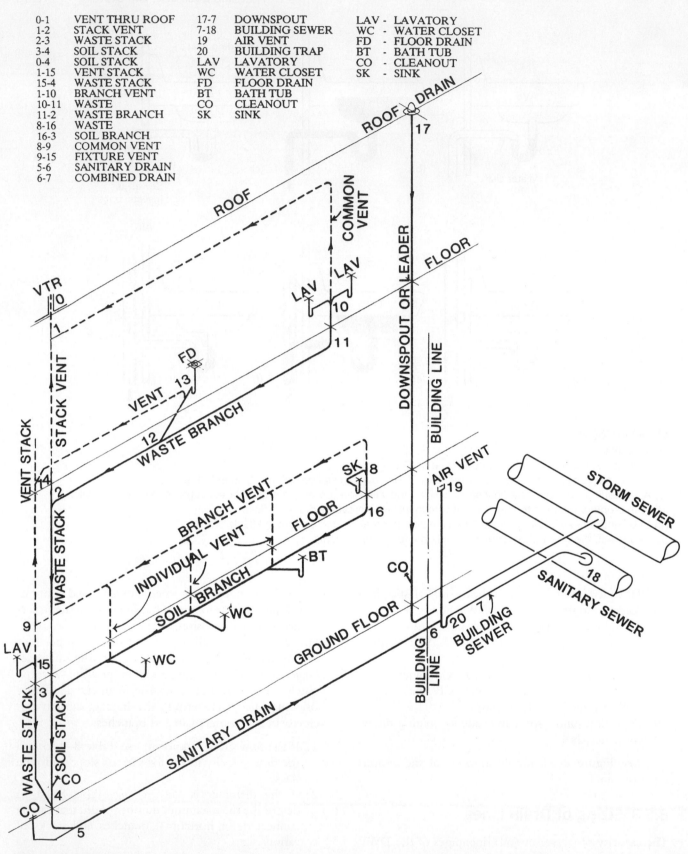

0-1	VENT THRU ROOF	17-7	DOWNSPOUT	LAV - LAVATORY
1-2	STACK VENT	7-18	BUILDING SEWER	WC - WATER CLOSET
2-3	WASTE STACK	19	AIR VENT	FD - FLOOR DRAIN
3-4	SOIL STACK	20	BUILDING TRAP	BT - BATH TUB
0-4	SOIL STACK	LAV	LAVATORY	CO - CLEANOUT
1-15	VENT STACK	WC	WATER CLOSET	SK - SINK
15-4	WASTE STACK	FD	FLOOR DRAIN	
1-10	BRANCH VENT	BT	BATH TUB	
10-11	WASTE	CO	CLEANOUT	
11-2	WASTE BRANCH	SK	SINK	
8-16	WASTE			
16-3	SOIL BRANCH			
8-9	COMMON VENT			
9-15	FIXTURE VENT			
5-6	SANITARY DRAIN			
6-7	COMBINED DRAIN			

■ **FIGURE 8–30**

Isometric diagram of a DWV drainage system identifying the various components of the system. (Note: When the public sewer is a combined sanitary and storm sewer, building sewers may also be combined at the site line.)

264

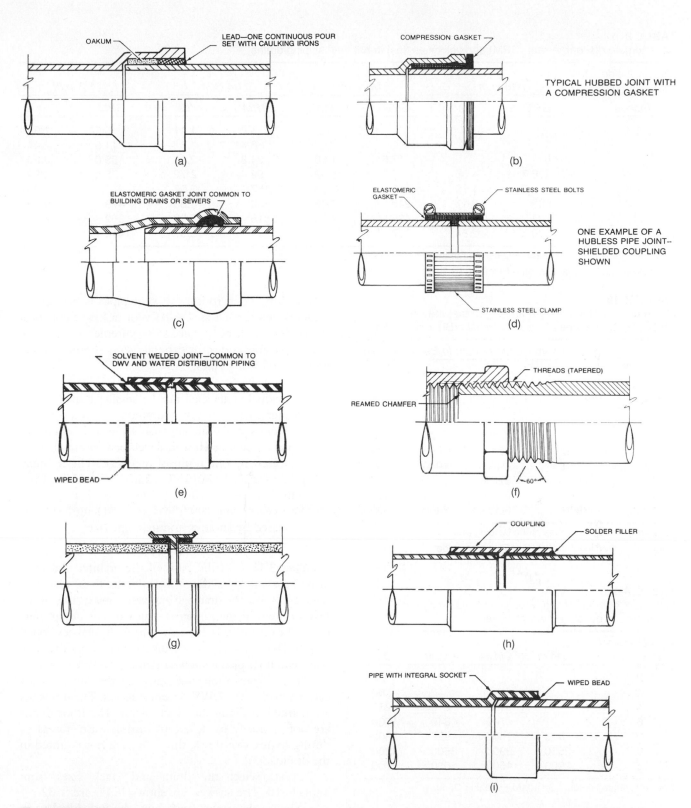

■ **FIGURE 8–31**

Typical methods of joining pipes. (a) Cast iron (C.I.), caulked; (b) C.I., compression gasket; (c) DWV, plastic elastomeric gasket; (d) C.I./plastic, joint-shielded coupling; (e) plastic, solvent; (f) steel/plastic, threaded; (g) clay (VCP), plastic bell; (h) copper, soldered; (i) plastic, solvent with socket. (Courtesy of National Standard Plumbing Code.)

TABLE 8–9
Approximate discharge rate (GPM) and velocities (fps) in half-full[1] sloping drains

Actual Inside Pipe Diameter	1/16 in./ft		1/8 in./ft		1/4 in./ft		1/2 in./ft	
Inches	GPM	fps	GPM	fps	GPM	fps	GPM	fps
1½					3.9	1.41	5.52	1.99
2					8.41	1.73	11.9	2.44
3			17.5	1.60	24.8	2.26	35.0	3.20
4	26.7	1.36	37.7	1.92	53.4	2.72	75.5	3.84
6	78.7	1.79	111	2.53	157	3.57	223	5.05
8	169	2.17	240	3.06	339	4.33	479	6.13
10	307	2.51	435	3.55	615	5.02	869	7.10
12	500	2.84	707	4.01	999	5.67	1413	8.02

[1]*Half full* means filled to a depth equal to half of the inside diameter.
Courtesy of the National Standard Plumbing Code.

TABLE 8–10
Maximum number of fixture units (dfu) that may be connected to branches, branch intervals (BI), and stacks

Diameter of Pipe, Inches	Maximum Number of dfu to Be Connected			
	Horizontal Branches[1]	Stacks with < or = 3BI	Stacks with > 3BI	
			Total	Each BI
1½	3	4	8	2
2	6	10	24	6
4	160	240	500	90
6	620	960	1900	350
8	1400	2200	3600	600
10	2500	3800	5600	1000
12	3900	6000	8400	1500

[1]Does not include branches of the building drain.
Courtesy of the National Standard Plumbing Code.

TABLE 8–11
Maximum number of fixture units (dfu) that may be connected to building drain or sewer

Pipe Diameter, Inches	$\frac{1}{16}$ in./ft	$\frac{1}{8}$ in./ft	$\frac{1}{4}$ in./ft	$\frac{1}{2}$ in./ft
2			21	26
3			42	50
4		180	216	250
6		700	840	1000
8	1400	1600	1920	2300
10	2500	2900	3500	4200
12	2900	4600	5600	6700

Courtesy of the National Standard Plumbing Code.

8.5.4 Sizing of Vent Pipes

The capacity of vent pipes depends on three major factors: the size of the stack, the number of dfu connected on the stack, and the developed length of the vent pipe.

In addition, when routing of vent pipes is not direct, auxiliary vents, such as a relief vent, a loop vent, or a crown vent, will be required. Requirements may vary with applicable codes. In general, vents should comply with the following rules:

1. Individual vents shall not be smaller than half the diameter of the required drainpipe served.
2. No vent shall be smaller than 1¼ in. in diameter.
3. Vents exceeding 40 ft in developed length shall be increased by one nominal pipe size. The nominal pipe sizes are 1¼ in., 1½ in., 2 in., 2½ in., 3 in., 4 in., 5 in., 6 in., 8 in., 10 in., 12 in., and 15 in.
4. Vents exceeding 100 ft in developed length shall be increased by another nominal pipe size.

Example 8.13 On the basis of the architectural floor plan of two office building toilets, Figure 8–32(a), determine and show the drainage and vent piping sizes of the DWV system in the isometric diagram. For design purposes, the vent stack is assumed to be 30 ft total developed length (DL), and the floor drains are 2 in. outlet size. The plumbing floor plan is shown in Figure 8–32(b).

First, from Table 8–6, identify the dfu load for each section of the DWV system it serves. The dfu loads are marked in Figure 8–32(c). (*Note:* The floor drains are not normally used, except during wash-downs or during fixture overflows; thus, they are not counted in the dfu loads.)

Next, select the drain and stack sizes from Table 8–10. The answers are shown in Figure 8–32(c).

Finally, determine and show the individual vent, circuit vent, and vent stack sizes in accordance with the rules given in this section. These answers are shown in Figure 8–32(d). (*Note:* 2-in. minimum vent size is used throughout. A 2-in. C.I. pipe is actually cheaper than smaller pipes.)

8.5.5 Design Guidelines for a DWV System

1. The minimum size of DWV pipes is governed strictly by the plumbing code. When C.I. pipe is used, it is good practice to use minimum 2-in. pipe for branch drain and vent pipes, even though the code permits the use of 1¼-in. and 1½-in. pipes.
2. DWV piping design should be optimized through the proper coordination of pipe chase sizes and locations.
3. The layout of the toilet room is usually initiated by the architectural plan; however, it is the responsibility of the plumbing engineer to determine the need for floor drains and to coordinate with the architect on the location of cleanouts, etc.
4. Indirect waste shall be provided for all equipment that contains toxic or harmful chemicals. The drainage shall be piped to a separate receptor for sedimentation, neutralization, or filtration before being discharged into the public sewer system.
5. Other than intermittent discharges into the drainage system from dishwashing and laundry equipment with water at a temperature of 140°F or above, no high-temperature waste or steam pipe shall discharge into the drainage system without subcooling the effluent prior to connecting to the sanitary sewer.
6. All plumbing fixtures or drainage equipment without a built-in trap must be connected through an external trap.
7. All traps must be vented.
8. All horizontal drainage piping shall be installed in alignment at a uniform slope not less than ¼ in. per foot for a diameter of 3 in. and less, and not less than ⅛ in. per foot for a diameter of 4 in. or more. In rare cases, ¹⁄₁₆ in. per foot is permitted by agreement with the code-enforcing authority. A slope greater than 1 in. per foot should be avoided when the waste contains solid matter.
9. Cleanouts shall be installed at the base of drainage stacks and at the beginning of main horizontal branches so that the entire DWV system can be cleaned and cleared to prevent clogging.
10. Grease-laden waste from kitchens should be piped directly to the building drain or stack whenever practical. A grease trap shall be installed for commercial kitchens, prior to connection to the waste pipe.
11. Waste containing high volumes of insoluble matter, such as sand, plaster, etc., shall be intercepted by sediment basins or catch basins prior to discharging into the sewer.
12. Waste containing oil, such as drains from a commercial garage, shall be connected through an oil interceptor
13. A 4-in. waste (soil) horizontal branch shall be used for a WC outlet even though plumbing codes do allow the use of a 3-in. branch for a tank type of WC in public buildings and up to two bathroom groups in private residences. Experience indicated that the increased horizontal branch will substantially reduce the chance of blockage.

8.6 SEWAGE TREATMENT AND DISPOSAL

To protect water resources and the greater environment, all waste from buildings and industrial processes must be treated to meet certain standards of quality. Domestic sewage from dwellings and DWV systems in buildings are permitted to be discharged into the public sewer system, which provides the necessary treatment prior to its discharge into nature. When public sewers are not accessible, or when there is no public sewer system in the vicinity of a building or buildings, a private sewage treatment system will have to be constructed.

Following are the definitions of some commonly used terms related to the subject of sewage treatment methods and disposal processes:

- *Digestion* That portion of the sewage treatment process in which biochemical decomposition of organic matter takes place, resulting in the formation of simple organic and mineral substances. Also known as *aerobic (bacterial) digestion*.
- *Influent* Untreated sewage flowing into a treatment system.
- *Effluent* Treated or partially treated sewage flowing out of a treatment system.
- *Septic tank* Watertight tank that receives influent from a DWV system designed to separate solids from the liquid, to digest organic matter through a period of detention, and to discharge the effluent to an approved method of disposal.
- *Sedimentation* Formation of layers of heavy particulates in the influent.
- *Aerobic (bacterial) digestion* Digestion of the waste through the natural bacterial digestive action in a septic tank or digestion chamber.
- *Active sludge* Sewage sediment, rich in destructive bacteria, that can be used to break down fresh sewage more quickly.
- *Dosing tank or chamber* Component of the septic tank system that periodically discharges effluent to an approved method of disposal.

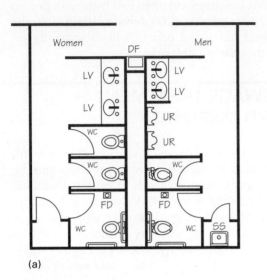

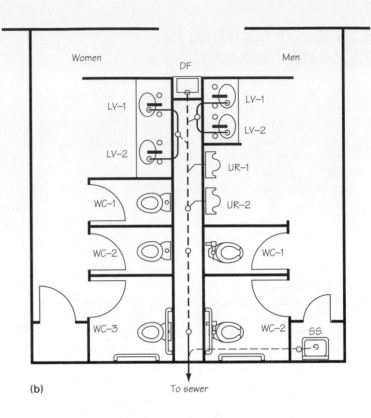

(a)   (b)   To sewer

■ **FIGURE 8–32**
Architectural floor plan of two toilets and plumbing drawings showing a DWV piping system. (a) Architectual floor plan. (b) Plumbing floor plan.

- *Filtration*   Means of filtering out any solid matter from the effluent.
- *Disinfection*   Process to disinfect the effluent with chemicals.
- *Percolation*   Flow or trickling of a liquid downward through a filtering medium or soil.
- *Drain field or leaching field*   Set of trenches containing open-ended, or perforated, pipes designed to allow the treated effluents to percolate into the ground.

8.6.1 The Sewage Treatment Process

The sewage treatment process may be divided into three major steps:

1. *Primary treatment,* which is subdivided into:
   - *Sedimentation and retention*   Raw sewage is retained for the preliminary separation of indigestible solids and the start of aerobic action
   - *Aeration*   Introduction of air through natural convection or mechanical blowers to accelerate the decomposition of organic matter
   - *Skimming*   Removal of scum that floats on top of the partially treated sewage

- *Sludge removal*   Disposal of heavy sludge at the bottom of treated sewage
2. *Secondary treatment,* The removal of fine suspended matter from the effluent through a filtration process, such as the use of sand filters, drain fields, or seepage pits.
3. *Tertiary treatment,* The disinfection of effluent by the addition of chemicals, such as chlorine.

8.6.2 Sewage Treatment Plants

The design of sewage treatment plants for large buildings, building complexes, and municipalities follows precisely the same processes described in Section 8.6.1. Modern treatment plants require considerable mechanized equipment and controls in order to be efficient and reliable. These treatment plants are designed by engineers who specialize in the subject.

8.6.3 Septic Tank System

Rather than a sewage treatment plant, the septic tank system is most commonly used in rural areas for small-capacity applications. It consists of a septic tank serving as primary treatment and a method of filtering the

■ **FIGURE 8–32** *(continued)*
(c) DWV isometric indicating dfu load on
waste branches. (d) DWV isometric
indicating DWV pipe sizes. (*Note:* 2-in.
minimum vent pipe is used for C.I. pipes.)

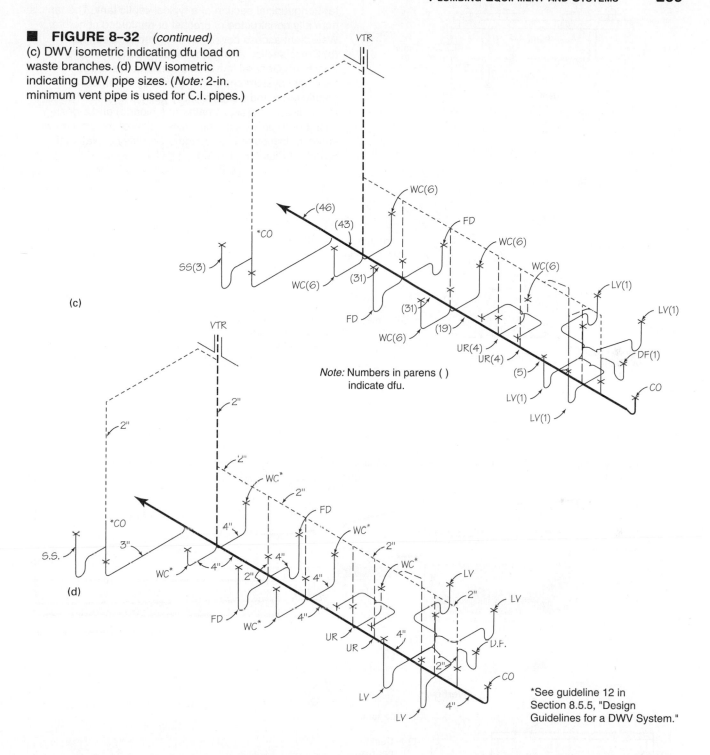

Note: Numbers in parens ( )
indicate dfu.

*See guideline 12 in
Section 8.5.5, "Design
Guidelines for a DWV System."

effluent as secondary treatment. The septic tank is usually constructed of concrete and has a retention volume
from 1000 to 10,000 gal. A typical design is shown in
Figure 8–33(a). A septic tank system is normally designed to flow by gravity, although ejector pumps are
used when the topography does not permit gravity flow.

Filtering of effluent may be by means of a filter pit,
sand filters, or drain fields. A drain field, which consists

of multiple runs of underground trenches, is popular
because it requires less maintenance. Typical construction of a drainage trench is shown in Figure 8–33(b)
and that for a drain field is shown in Figure 8–33(c).
Figure 8–33(d) illustrates the addition of a tertiary
treatment for a septic tank system.

To design a septic tank system, one begins with an
analysis of the sewage load based on the number of

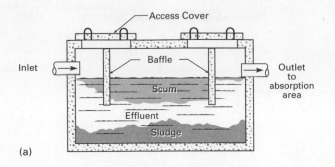

(a)

(a) Longitudinal section of a typical septic tank. The tank is normally constructed of precast or reinforced concrete. Watertight access covers are required for cleanup.
(b) Cross section of a typical underground drainage trench. (c) General layout of a septic tank sewage treatment system showing the primary treatment (septic tank) and the secondary treatment (drain field).
(d) Addition of tertiary treatment (chlorine) and a dosing chamber to provide an intermittent flow of the effluent will improve distribution. (a, b, and d Courtesy of National Standard Plumbing Code.)

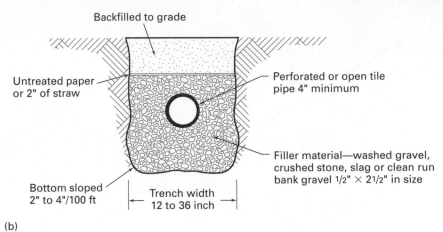

(b)

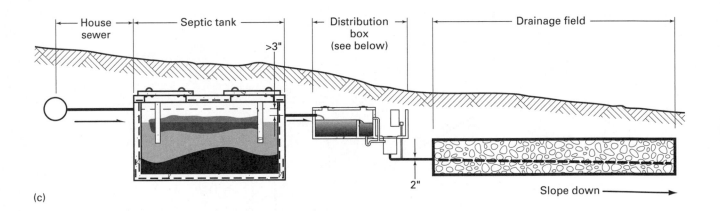

(c)

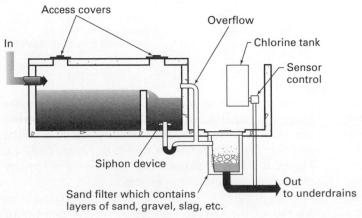

(d)

occupants in and type of occupancy of the building on a 24-hour basis. Table 8–12 provides the estimated load for various occupancies. The data are applicable to most regions in the United States and should be modified for other cultures and countries.

Next, a suitable location is selected. Normally, septic tanks and their disposal fields are located away from heavy traffic and maintain at least a minimum distance from the building, wells, water services, etc., as prescribed in the local health or building codes. In general, septic tanks and disposal fields are a minimum of 10 feet from building water lines and property lines, 50 feet from deep wells, and 100 feet from shallow wells.

The required size of a septic tank for use in residences is determined from either the number of fixture units (dfu) served or the number of bedrooms in the residence. Table 8–13 provides a sizing guide.

The size of the septic tank for a nonresidential building is given by the following general rule: Use a multiplier of 1.5 applied to the calculated daily sewage rate (gpd), with gradual reductions of the multiplier for systems larger than 1500 gpd.

The drain field includes one or more rows of drainage trenches complying with the following guidelines:

- *Trench construction* Perforated or open-jointed tile surrounded with washed gravel or crushed stone and covered with backfill. See Figure 8–33(b).
- *Drain tile size* Minimum 4-in.-diameter VCP or plastic.
- *Trench slope* Bottom of trench shall be sloped downward between 2 and 4 in. per 100 ft.
- *Trench depth* Below the frost line of the area, but in no case less than 18 in. below grade.

TABLE 8–12
Average daily sewage load for occupancies in the Untied States (in gallons/day/person unless otherwise indicated)

Occupancy	Load	Occupancy	Load
Schools (without cafeteria and showers)	15	Hospital (general)	150
Schools (with cafeteria and showers)	35	Public institutions	100
Schools (boarding)	100	Restaurants (per serving)	25
Day camps	25	Motels	60
Trailer parks, etc.	50	Stores (per toilet)	400
Swimming pools and beaches	10	Airport (per passenger)	5
Luxury residences/estates	150	Assembly hall (per seat)	2
Hotels (2 persons/room)	100	Churches (small)	3–5
Factories	25	Churches (large)	5–7
Nursing homes	75	Subdivision or homes	75

TABLE 8–13
Required capacity of septic tanks for residential units

Single-Family Dwellings (Number of Bedrooms)	Multiple Dwelling Units or Apartments (One Bedroom Each)	Other Uses: Maximum Fixture Units Served	Minimum Septic Tank Capacity, Gallons*
1–3		20	1000
4	2	25	1200
5 or 6	3	33	1500
7 or 8	4	45	2000
	5	55	2250
	6	60	2500
	7	70	2750
	8	80	3000
	9	90	3250
	10	100	3500
Extra bedroom			150 gal ea.
Extra dwelling units over 10			250 gal ea.
Extra fixture units over 100			25 gal/dfu

*Septic tank sizes in this table include sludge storage capacity and the connection of domestic food waste disposal units without further volume increase.

- *Trench width* Between 12 and 36 in. Wider trenches have a higher percolation capacity. (See Table 8–14 for capacity ratings.)
- *Trench spacing* Minimum 6 ft between adjacent trenches for 12- to 24-in.-wide trenches and up to 9 ft for a 36-in.-wide trench.
- *Length of trench* Not to exceed 100 ft.

The minimum size of the drain field depends on the width of the drain trench, makeup of the soil, absorption capacity of the soil, and total length of the drain tiles. The absorption capacity of the soil is determined by a percolation test to determine the time taken for water in the drain field trenches to drop 1 in. The total length of drain pipe required can be found from Table 8–14. When the water table of the field is too close to the bottom of the drain field trenches, alternative filtering methods, such as mounded drain fields and sand filters at higher elevations, should be used.

Example 8.14 Design a septic tank system for an elementary school based on the following criteria:

1. The school has a cafeteria and showers.
2. The total student and faculty population is 200.
3. A drain field system with 2-ft-wide trenches shall be used as secondary treatment.
4. A percolation test of the soil indicates that the absorption rate is 2 min/in. of drop.

   Answers:

- From Table 8–12, the estimated daily sewage flow is 35 gal per person, and the total daily flow for 200 persons is 35 × 200, or 7000 GPD.
- Based on 1.5 times the daily flow rate, the septic tank volume should be 1.5 × 7000, or 10,500 gal.

TABLE 8–14
Drain tile required (in linear feet for each 100 gal of sewage/day)

Time for 1-in. Drop in Percolation, Minutes	Tile Length for Trench Widths of		
	1 ft	2 ft	3 ft
1	25	13	9
2	30	15	10
3	35	18	12
5	42	21	14
10	59	30	20
15	74	37	25
20	91	46	31
25	105	53	35
30	125	63	42

Courtesy of the National Standard Plumbing Code.

- Based on Table 8–14, the number of linear feet of drain field should be 15 ft per 100 GPD; thus

$$\text{Total length of trench} = \frac{10{,}500 \times 15}{100}$$
$$= 1575\,\text{ft (use}\,1600\,\text{ft)}$$

- The plan would favor the use of two 5000-gal tanks, each with 800 ft of drain tiles divided among eight 100-ft-long trenches on 6-ft centers. Figure 8–34 is a preliminary layout of the system prior to a consideration of other concerns.

8.6.4 Grey Water Systems

Potable water is treated and fit for human consumption. Due to the availability and low cost of potable water, it is used not only for consumption, cooking, washing, and bathing, but also for functions that do not need so high a quality and purity, such as flushing and irrigation. Using relatively clean wastewater, termed "gray water," for these functions can reduce overall water consumption and sewer load, which is good for the environment and saves utility cost for the owner. Due to the low cost of water and sewer in most areas, gray water systems are slow to catch on.

Water from sanitary fixtures, kitchen sinks, and dishwashers is unsafe to reuse; however, water from showers, lavatory sinks, and laundries contains only small amounts of potentially harmful materials, especially soaps and detergents, and can be used as gray water. Effluent can be used immediately upon discharge or stored for future use (see Figure 8–35). If the gray water is left to stand, it should be filtered to reduce the risk of bacteria.

Several precautiond should be noted. Most authorities advise against using gray water for irrigating food crops or for any application that might create a mist or spray, such as lawn sprinklers. Codes and regulations on gray water systems vary widely, and local authorities should be consulted prior to considering for a project.

8.7 STORM DRAINAGE SYSTEM

A storm drainage system conveys rainwater or melting snow from a building or site to the points of disposal. Among the locations to be drained are roofs, patios, and areaways of buildings and parking lots, and roadways, lawn, and gardens on the site. In general, except for small or incidental areas, all exterior storm drainage should be connected externally to the building storm drainage system.

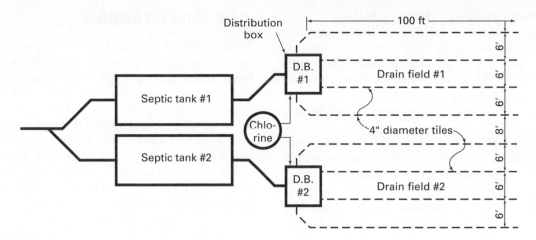

■ FIGURE 8–34
Flow diagram of a typical septic tank sewerage treatment and disposal system.

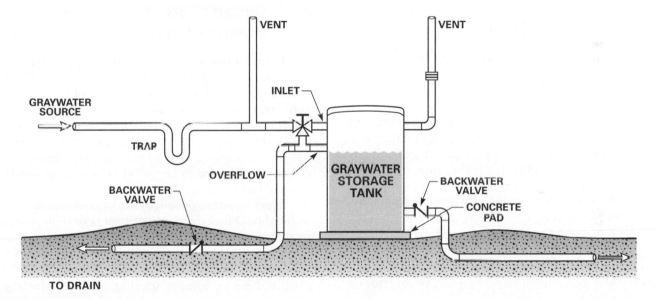

■ FIGURE 8–35
Gray water collection detail.

Following are some of the commonly used terms related to storm drainage systems:

- *Roof drain* Device installed on a roof to receive water collected on the surface of the roof and to discharge it to the point of disposal through a conductor (or downspout).
- *Area drain* Similar to a roof drain, except that it is installed in an exterior area, not over a building space.
- *Conductor or downspout* Vertical portion of a storm drainage system installed in the interior of a building.

- *Gutter* Exterior trough, installed below a roof, that collects rainwater from the roof and discharges it to the point of disposal through a leader.
- *Leader* Vertical portion of a storm drainage system installed at the exterior of a building.
- *Subsoil drain* Drainpipe that collects subsurface water and conveys it to a place of disposal.
- *Controlled storm drainage system* Storm drainage system that collects storm water on a roof and releases the flow slowly to the drainage system to allow the load to drain in a longer time.
- *Primary drainage system* Basic storm water drainage system designed for normal use.

- *Secondary drainage system* Additional storm drainage system that will handle any storm water overflow that may occur when heavier storms occur.
- *Sump* Receiver or pit that receives liquid waste, storm water, or groundwater located below the elevation of a gravity system.
- *Sump pump* Pump that removes liquid from a sump.
- *Projected roof area or horizontal projected roof area* Horizontal component of a sloping or pitched roof. The projected roof area is in effect equal to a flat roof at the base of the pitched roof. When rain falls at an angle, one side of the pitched roof will intercept more rain than the other side; however, the average rainfall on both sides of the roof is still no more than that of a flat roof. (See Section 8.7.3 for a discussion of the sizing of storm drainage pipes.)

8.7.1 Design Principle

The fundamental principle behind the design of storm water systems is to install a piping or conductor system to lead the storm water away from the building and site in a reasonable time. The size of the system depends on the rate of rainfall, rather than the total rainfall in a day or in a year. The rate of rainfall varies with intensity and frequency of occurrence—for example, inches of rainfall per hour during a 15-minute period as it may happen once per 100 years (15 min per 100-year return). It should be understood that "per 100-year return" actually means "1 percent probability," and "per 10-year return" means "10 percent probability," that the stated intensity of rainfall may occur. As with any statistical data, the same probability may occur in consecutive years or may not occur at all for many years.

Although the design of an exterior storm drainage system is similar in principle to that of an interior system, the circumstances surrounding each are considerably different. Exterior drainage must be closely coordinated with the capability of the public sewer or public drainage system. It will not be cost-effective to drain a large area, such as a parking lot or a park, on the same basis as a building area. In many instances, the instantaneous drainage of storm water on a large exterior area may overtax the public sewer system. If this occurs, a method of delaying the runoffs may be necessary or may be directed by the public sewer district. Delaying methods include undersizing the drainpipes, letting the ground gradually absorb water, and creating a storm water holding area, a technique known as "ponding." Exterior storm drainage systems are normally designed by civil engineers who specialize in the topic.

8.7.2 Data on Rainfall

The selection of a design value for the intensity of rainfall is usually governed by code and the economic and safety factors of the project. Overdesigning for intensity will ensure faster drainage of storm water from the building or the site, whereas underdesigning will cause inconvenience, safety hazards, or possible structural failures. Careful coordination between the architect and structural and plumbing engineers will result in an optimum design. Most codes specify the design rainfall rate to be based on 1-hour and 15-minute duration rates in a 100-year return period for the primary and secondary drainage systems, respectively.

Recognized rainfall rates of selected cities in each of the 50 states are shown in Table 8–15.

8.7.3 Design of Storm Drainage Systems

To design a storm drainage system, first determine where drainage is required; other than roofs, areaways, driveways, and walkways toward spaces of lower elevation, entrances into and exits from buildings are areas that should be considered. Whenever possible, all exterior storm drains should be connected to a storm sewer outside the building. Drains lower than the elevation of the sewer may have to be collected in a sump and be pumped out automatically.

Next, determine the location of roof drains. Roof drains should be located at lower spots or in depressed areas of the roof. The locations should be determined by the architect and structural engineer. A 0.5 percent minimum slope toward each roof drain is necessary to ensure positive drainage without standing water or ponding. Where controlled discharge is desired, in order to use the roof as a reservoir, the roof structure must be designed to carry the extra weight of water (at 62.4 lb/cu ft). In general, a minimum of two roof drains should be installed for a building, and a minimum of four for buildings over 10,000 sq ft.

Next, determine the roof drain design criteria, based on or exceeding the code requirements for the rate of rainfall for primary and secondary roof drain systems. Secondary drainage is a safety provision during an extremely heavy rainfall or in the event that drains become clogged. A secondary system may be piped separately or may be designed to spill over the roof, as is permitted by the code.

Then, select the appropriate drainage fittings. Numerous designs of roof and area drains are available to suit the construction methods. Typical roof and area drains are shown in Figure 8–36. Drainage receptors and trenches are shown in Figure 8–37.

TABLE 8–15
Rate of rainfall for U.S. cities (rainfall rates, in inches per hour (in./hr), based on a storm of 1-hr duration and a 100-year return period)

State	City	in./hr	State	City	in./hr
Alabama	Mobile	4.6	Montana	Ekalaka	2.5
Alaska	Fairbanks	1.0	Nebraska	Omaha	3.8
Arizona	Phoenix	2.5	Nevada	Las Vegas	1.4
Arkansas	Little Rock	3.7	New Hampshire	Concord	2.5
California	Los Angeles	2.1	New Jersey	Newark	3.1
	San Fernando	2.3	New Mexico	Albuquerque	2.0
Colorado	Denver	2.4	New York	New York	3.0
Connecticut	Hartford	2.7	North Carolina	Charlotte	3.7
Delaware	Wilmington	3.1	North Dakota	Bismarck	2.8
District of Columbia	Washington	3.2	Ohio	Cincinnati	2.9
Florida	Miami	4.7	Oklahoma	Oklahoma City	3.8
Georgia	Atlanta	3.7	Oregon	Eugene	1.3
Hawaii	Hilo	6.2	Pennsylvania	Philadelphia	3.1
	Honolulu	3.0	Rhode Island	Providence	2.6
Idaho	Boise	0.9	South Carolina	Charleston	4.3
Illinois	Chicago	3.0	South Dakota	Rapid City	2.9
Indiana	Indianapolis	3.1	Tennessee	Memphis	3.7
Iowa	Des Moines	3.4	Texas	Dallas	4.0
Kansas	Wichita	3.7		Houston	4.6
Kentucky	Lexington	3.1	Utah	Salt Lake City	1.3
Louisiana	New Orleans	4.8	Vermont	Burlington	2.1
Maine	Portland	2.4	Virginia	Norfolk	3.4
Maryland	Baltimore	3.2	Washington	Seattle	1.4
Massachusetts	Boston	2.5	West Virginia	Charleston	2.8
Michigan	Detroit	2.7	Wisconsin	Milwaukee	3.0
Minnesota	Minneapolis	3.1	Wyoming	Cheyenne	2.2
Mississippi	Biloxi	4.7			
Missouri	Kansas City	3.6			
	Saint Louis	3.2			

Source: U.S. National Weather Service.

Finally, select the piping materials and method of installation. Different piping materials are used for above- and underground locations. The following are commonly used piping materials:

- *Aboveground (interior)* C.I. with caulked or mechanical joints, galvanized steel with mechanical joints (welded joints are not permissible), copper pipe (types K or L), copper tube (DWV), aluminum, plastic (ABS or PVC).
- *Aboveground (exterior)* C.I., galvanized steel, copper leader (round or rectangular), aluminum leader (round or rectangular).
- *Below ground (within building)* C.I., copper (type K), reinforced concrete, plastic (ABS or PVC).
- *Below ground (exterior to building)* Extra-strength VCP, reinforced concrete, plastic (ABS or PVC).
- *Subsoil drainage* Open-jointed or perforated VCP, plastic, bituminized fiber, concrete pipes

embedded in crushed stone, rock, or gravel. See details in Figure 8–37(b).

When plastic pipes are used, consideration should be given to contraction and expansion of the material, which is more severe than in metal. Thermal insulation may be needed for metal pipes to avoid external condensation of moisture on the surface of the pipes, which may cause staining on ceilings or walls.

In all the preceding situations, the size of the drainage pipe is determined by three factors: the equivalent horizontal projected area (EHPA), the rate of rainfall, and the slope of the drainpipe. The EHPA is determined as follows:

- For a horizontal roof, use the actual roof area.
- For a two-sided pitched roof, use the horizontal projected roof area.
- For a single-sided pitched roof, use 1.4 times the horizontal projected roof area.

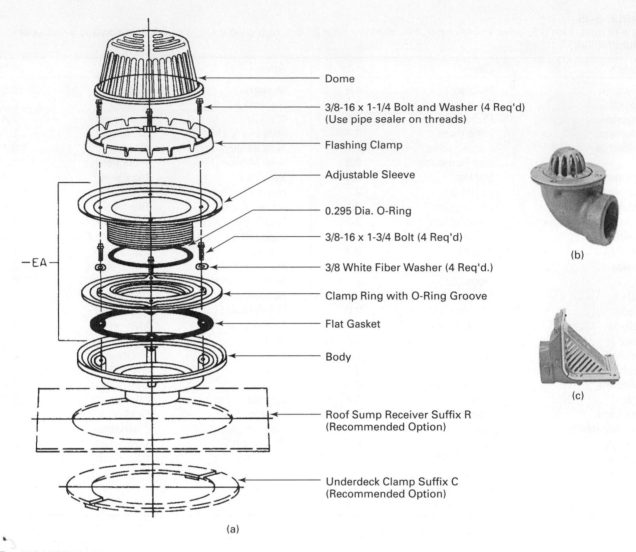

Dome

3/8-16 x 1-1/4 Bolt and Washer (4 Req'd)
(Use pipe sealer on threads)

Flashing Clamp

Adjustable Sleeve

0.295 Dia. O-Ring

3/8-16 x 1-3/4 Bolt (4 Req'd)

3/8 White Fiber Washer (4 Req'd.)

Clamp Ring with O-Ring Groove

Flat Gasket

Body

Roof Sump Receiver Suffix R
(Recommended Option)

Underdeck Clamp Suffix C
(Recommended Option)

(a)

(b)

(c)

■ **FIGURE 8–36**

(a) Exploded view of a roof drain showing the drain body (sump), clamp ring, adjustable roof deck sleeve, flashing clamp/gravel guard, and dome strainer. (b) Cornice drain. (c) Oblique scupper drain. (Reproduced with permission from Zurn Industries, Inc.)

- For a flat roof adjacent to a vertical wall, use the flat roof area plus 50 percent of the vertical wall area above the roof.
- For a flat roof adjacent to two or more vertical walls, use the flat roof area plus 25 percent of the average of the two largest walls.

Size of storm drainpipes is determined from the following:

- For horizontal drainpipes (interior), use Table 8–16.
- For gutters (exterior), use Table 8–17.
- For conductors (interior) or leaders (exterior), use Table 8–18.

Example 8.15   Determine the size of horizontal drainpipes and the main vertical conductor (downspout) for a building having two roof drains. Each roof drain serves a flat roof area of 2000 sq ft. The building is located in Saint Louis, Missouri.

- From Table 8–15, the design rainfall rate for Saint Louis, Missouri, is 3.2 in./hr, based on a 1-hr duration and a 100-year return period.
- Based on a ¼-in.-per-foot slope, each roof drain shall be connected with a 4-in. horizontal drainpipe. (*Note:* A 4-in. drain line at a ¼-in. slope can serve a 5300- and 2650-sq-ft EPRA for 2-in. and 4-in. rainfall rates, respectively. Interpolating for the

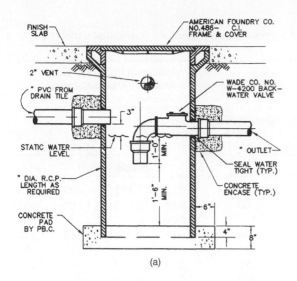

(a)

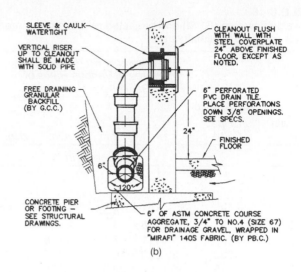

(b)

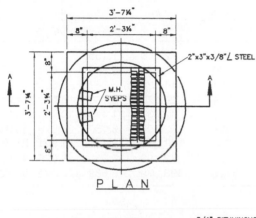

PLAN

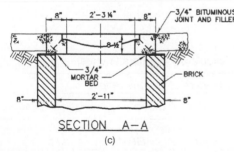

SECTION A—A

(c)

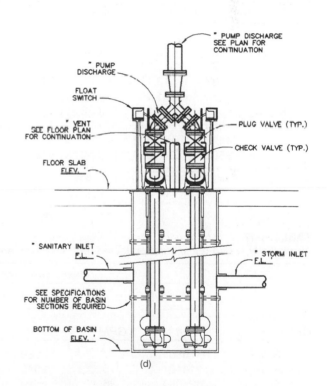

(d)

■ **FIGURE 8-37**

Typical storm drainage.

(a) Sand interceptor where sand entrained in the storm water is settled at the bottom of the basin with only clear effluent to flow out.

(b) Subsoil drainpipe installation detail with drainpipe embedded in 6 in. of granular fill. Wall cleanouts are provided at frequent intervals.

(c) Storm water drainage manhole or catch basin with C.I. grating.

(d) Combination storm and sanitary basin with duplex submersible sump pumps.

TABLE 8–16

Allowable flow rate (GPM) and projected roof or surface area (sq ft) for horizontal storm drainpipes (inches) at various slopes and design rainfall rates

Diameter of Drain, in.	Design Flow in Pipe, GPM	Allowable Projected Roof Area at Various Rates of Rainfall per Hour (sq ft)				Diameter of Drain, in.	Design Flow in Pipe, GPM	Allowable Projected Roof Area at Various Rates of Rainfall per Hour (sq ft)			
		1 in.	2 in.	4 in.	6 in.			1 in.	2 in.	4 in.	6 in.
		Slope 1/16 in./ft						Slope 1/4 in./ft			
2						2	17	1,636	818	409	273
3						3	50	4,813	2,406	1,203	802
4	53	5,101	2,551	1,275	850	4	107	10,299	5,149	2,575	1,716
5	97	9,336	4,668	2,334	1,556	5	194	18,673	9,336	4,668	3,112
6	157	15,111	7,556	3,778	2,519	6	315	30,319	15,159	7,580	5,053
8	339	32,629	16,314	8,157	5,438	8	678	65,258	32,629	16,314	10,876
10	615	59,194	29,597	14,798	9,866	10	1,229	118,292	59,146	29,573	19,715
12	999	96,154	48,077	24,039	16,026	12	1,999	192,404	96,202	48,101	32,067
						15	3,625	348,907	174,454	87,227	58,151
Size	GPM	1 in.	2 in.	4 in.	6 in.	Size	GPM	1 in.	2 in.	4 in.	6 in.
		Slope 1/8 in./ft						Slope 1/2 in./ft			
2						2	24	2,310	1,155	578	385
3	35	3,369	1,684	842	516	3	70	6,738	3,369	1,684	1,123
4	75	7,219	3,609	1,805	1,203	4	151	14,534	7,267	3,633	2,422
5	137	13,186	6,593	3,297	2,198	5	274	26,373	13,186	6,593	4,395
6	223	21,464	10,732	5,366	3,577	6	445	42,831	21,416	10,708	7,139
8	479	46,104	23,052	11,526	7,684	8	959	92,304	46,152	23,076	15,384
10	869	83,641	41,821	20,910	13,940	10	1738	167,283	83,641	41,821	27,880
12	1413	136,002	68,001	34,000	22,667	12	2827	272,099	136,050	68,025	45,350
15	2563	246,689	123,345	61,672	41,115	15	5126	493,379	246,689	123,345	82,230

Courtesy of the National Standard Plumbing Code.

TABLE 8–17

Allowable flow rate (GPM) and projected roof or surface area (sq ft) for semicircular roof gutters (inches)

Diameter of Gutter, in.	Maximum Projected Roof Area for Gutters of Various Slopes							
	1/16 -in./ft Slope		1/8 -in./ft Slope		1/4 -in./ft Slope		1/2 -in./ft Slope	
	sq ft	GPM	sq ft	GPM	sq ft	GPM	sq ft	GPM
3	680	7	960	10	1360	14	1920	20
4	1440	15	2040	21	2880	30	1080	42
5	2500	26	3520	37	5000	52	7080	74
6	3840	40	5440	57	7680	80	11,080	115
7	5520	57	7800	81	11,040	115	15,600	162
8	7960	83	11,200	116	14,400	165	22,400	233
10	14,400	150	20,400	212	28,800	299	40,000	416

Courtesy of the National Standard Plumbing Code. Data are based on design rainfall rate of 1 in. per hour and shall be multiplied by the appropriate rainfall rate of the locality.

TABLE 8–18
Allowable flow rate (GPM) and projected roof or surface area (sq ft) for various conductors or leaders (inches) at various design rainfall rates

Diameter of Pipe or Leader, in.	Design Flow in Pipe, GPM	Allowable Projected Roof Area at Various Rate of Rainfall per Hour (sq ft)			
		1 in.	2 in.	4 in.	6 in.
2	23	2176	1088	544	363
2½	41	3948	1974	987	658
3	67	6440	3220	1610	1073
4	144	13,840	6920	3460	2307
5	261	25,120	12,560	6280	4187
6	424	40,800	20,400	10,200	6800
8	913	88,000	44,000	22,000	14,667

3.2-in. rainfall rate, a 4-in. pipe should be selected for the horizontal drainpipes.)

- The main horizontal drainpipe should be a 5-in. pipe.
- From Table 8–18, interpolating between 2 in. and 4 in. for the 3.2-in. rainfall rate, we find that a 4-in. conductor can serve up to 4844 sq ft, however, a 5-in. downspout still should be used, since reducing the size of pipe downstream of the flow is prohibited.

8.7.4 Subsoil Drainage System

When the lower level of a building (e.g., the basement) is belowground, there are two potential problems: surface water may flow into the building through cracks or seepage of the basement walls, and groundwater may push into the building owing to hydrostatic pressure if the normal groundwater level (known as the *water table*) is higher than the lowest floor level.

To avoid water seepage, all belowground walls must first be waterproofed, and the grade next to the building is often sloped away from the building. If the anticipated water table is close to or higher than the building floor level, a subsoil drainage system consisting of 4- to 6- in.-diameter drain tiles should be installed.

Note also the following:

- Minimum 4-in.-diameter VCP or plastic pipe laid end to end with about a ½-in. gap between sections or perforated pipes should be used.
- The drainpipes should be located on the exterior of foundation wall if the water table is higher than the footing. Figure 8–37(b) shows the construction of a subsoil drain outside of the foundation wall.
- Pipes should be surrounded by graded, crushed rocks, or gravel and be screened on the outer area

by a mesh material to prevent dirt, sand, or back-fill material from washing into the tiles.

- Tiles should be sloped to drain, if that is achievable, or to a sump from which water can be pumped. A sand interceptor (trap) may be necessary at the end of drain tile to allow solid material to drop out prior to disposal of the liquid. Figure 8–37(a) is a typical layout of a sand interceptor.

8.8 PLUMBING SERVICES FOR OTHER BUILDING EQUIPMENT

Plumbing services for other building equipment are part of the plumbing work that must be designed and/or coordinated by the plumbing engineer. These services include water supply, treatment, drainage, and waste disposal for the following:

- *HVAC equipment* Boilers, heaters, chillers, cooling towers, etc.
- *Fire protection equipment (for buildings without an independent fire service)* Fire hose cabinet, limited-area automatic sprinklers, etc.
- *Food service equipment* Kitchen sinks, dishwashers, disposals, grease interceptors, etc.
- *Laundry equipment* Clothes washers, oil and fluid separators, etc.
- *Landscape* Irrigation sprinklers, reflecting pools, fountains, etc.
- *Athletic and health equipment* Health club whirlpools, baths, steam baths, swimming pools, etc.
- *Research processing equipment* High-quality water, specialty gases such as oxygen and nitrogen, compressed air, and a vacuum-piping system in hospitals and research laboratories.

Design criteria for the foregoing equipment are usually provided by the owner/architect, manufacturers, or specialty consultants and evaluated for implementation by the plumbing engineer.

QUESTIONS

8.1 Do National Plumbing Codes provide sufficient data and guidelines for designing a sanitary drainage system? If so, explain.

8.2 Do National Plumbing Codes provide sufficient data and guidelines for designing a domestic water distribution system? Explain.

8.3 What are the equivalents between pounds and grains, gallons and cubic feet, ppm and mg/L?

8.4 Name the five substances (impurities) that are limited in minimum quality standards for drinking (potable) water systems in the United States.

8.5 Name the major influences on the quality of water.

8.6 Name seven processes used to improve the quality of water.

8.7 What mineral content is used to determine the hardness in water?

8.8 Describe the zeolite water-softening process.

8.9 What is osmosis? What is reverse osmosis?

8.10 What are the two commonly used water filtration systems?

8.11 A light commercial building has four flush-tank-type water closets, two urinals with flushometers, and four lavatories. What is the total water supply fixture units (wsfu)? What is the probable water demand?

8.12 What are the equivalents of gpm versus L/s, cfm versus L/s, psi versus kPa?

8.13 A building has 300 tons of air conditioning. The air-conditioning unit is water-cooled through a cooling tower. How much makeup water should be provided?

8.14 What is the pressure drop due to friction loss for 400 ft of 1-in. PVC pipe with a flow of 10 gpm?

8.15 What is usually assumed to be the equivalent length (EL) of pipe fittings in a plumbing piping system? (*Note:* This procedure applies to preliminary design only.)

8.16 What is the pressure at the base of a water pipe feeding a plumbing facility 500 ft above the base due to static height alone?

8.17 What is the hot-water demand for a residential dishwasher in gpm and total consumption?

8.18 What type of copper pipe is normally used for a domestic water system?

8.19 What is the function of a vacuum breaker?

8.20 What is the function of a shock absorber? Where should it be installed?

8.21 What is the most effective method used to prevent backflow?

8.22 Given the pump shown in Figure 8–18, what pump speed should be selected for a flow of 10 gpm requiring a total head of 8 ft of water column (w.c.)? If the same pump is used requiring a flow of 15 gpm, what total head can the pump develop?

8.23 Why is a flushometer, rather than flush tanks, required for water closets in public facilities?

8.24 What is the water demand for a pressure tank water closet per flushing? For standard gravity tanks?

8.25 What is the purpose of a drainage trap? Do water closets require traps?

8.26 What is the nominal rating in drainage fixture units (dfu) of a water closet with flushometer? With gravity tank?

8.27 What is the difference between a waste pipe and a soil pipe?

8.28 What is a bathroom group? What is the wfu of a bathroom group with flush tank? What should be the minimum soil pipe diameter?

8.29 A school is planned for 2000 students, teachers, and staff. Assuming that the population is 50 percent male and 50 percent female, determine the minimum number of plumbing fixtures required. In your judgment, is this adequate?

8.30 Calculate the minimum number of plumbing fixtures for a 500-seat movie theater. In your judgment, is this adequate?

8.31 What is a stack? A vent? A stack vent? A vent stack?

8.32 What is the purpose of a stack vent?

8.33 What piping materials are commonly used in drainage piping systems?

8.34 What are the joining methods for cast-iron pipes? Copper? Plastic?

8.35 Normally, municipalities in the United States have separate sanitary and storm sewer systems; however, several major cities still have combination storm and sanitary sewer systems. In such cases, can the sanitary and storm piping systems within the building be combined?

8.36 Give the major steps in a sewage primary treatment system, and the secondary treatment process.

8.37 What is the purpose of a dosing chamber in a septic tank sewage treatment system?

8.38 What is the maximum number of dfu that can be connected to a 4-in. horizontal branch?

8.39 What is the maximum number of dfu that can be connected to a 4-in. building drain with ⅛-in.-per-foot fall? ½-in.-per-foot fall?

8.40 A day camp has 150 occupants. Determine the size of the septic tank. Determine the linear feet of drain tile required based on a 10 min/in. drop from percolation test and 2-ft-wide trenches.

8.41 What is project roof area?

8.42 What is the rainfall rate of Saint Louis, Missouri, based on a storm of 1-hour duration and a 100-year return period?

8.43 A roof area is 1500 sq ft, and the roof is adjacent to three vertical walls of 100, 200, and 500 sq ft each. What shall the roof drain or drains be designed for?

8.44 What is the minimum number of roof drains for a building of any size? For a building over 10,000 sq ft of roof area?

8.45 What size horizontal storm drainpipe should be installed for a building with 16,000 sq ft of projected roof area, in a region having a 2-in. rainfall rate, and the drainpipe laid with 1/16 in./ft slope? With ¼ in./ft slope?

8.46 What size roof drain leader should be installed for a 1000-sq-ft roof area in a region with 5 in./hour rainfall rate?

8.47 In addition to the sanitary facilities, what other water drainage should be considered in plumbing system design for a large building complex?

8.48 What is the recommended minimum slope of a 4-in. drainline? What is it for a 5-in. drainline?

8.49 What is the equivalent slope in percentage of ¼ in. per foot?

8.50 If the flow rate of an industrial waste is 180 GPM, what size drain pipe shall be used with 2 percent slope?

8.51 What is the maximum number of fixture units (dfu) that may be connected to a 4-in. drainline with ¼-in.-per-foot slope?

REFERENCES

National Standard Plumbing Code/1984. Falls Church, Virginia: National Association of Plumbing/Heating/Cooling Contractors, 1992.

Basic Plumbing Code/1993. BOCA International, Inc., Chicago: Building Officials and Code Administrators International, Inc.

National Bureau of Standards, Report *BMS 79*, 1970 (now the National Institute of Building Sciences).

ASHRAE Handbook and Product Directory, Systems, 1992. Atlanta: American Society of Heating, Refrigerating and Air-Conditioning Engineers, Inc.

Building Officials and Code Administrators International, Inc., *National Building Code*, 1993.

Fundamentals of Plumbing Design, ASPE Data Book, 1992, vol. 1. Sherman Oaks, California: American Society of Plumbing Engineers, 1992.

Tao, William, and Karl Norman, "Plumbing System Design." *Heating, Piping and Air Conditioning*, March 1989.

FIRE PROTECTION EQUIPMENT AND SYSTEMS

9

A FIRE PROTECTION SYSTEM IS A SYSTEM THAT IN-CLUDES devices, wiring, piping, equipment, and controls to detect a fire or smoke, to actuate a signal, and to suppress the fire or smoke. The two primary objectives of fire protection are to save lives and protect property. A secondary objective is to minimize interruptions of service due to a fire. Imagine that a fire is detected on the ground floor of a high-rise hotel. If the fire can be quickly isolated and extinguished, then there will be no need to disturb the guests on the upper floors. Otherwise, the entire building must be evacuated. Statistics indicate that to evacuate a 40-story building with 100 persons per floor through a single stairway will require one full hour. With two stairways, it will still take more than half an hour. The chaos and financial consequences can hardly be overstated.

The cost of a fire protection system depends on a number of factors, such as the fire resistivity of the building, type of occupancy, number of floors below the ground level, height of the building, adequacy of egress, and degree of protection desired. The initial cost of a fire protection system may be as low as 1 percent, and is seldom more than 5 percent, of the total building construction cost. In many applications, the additional investment for a better fire protection system can have a positive payback in that the building's fire insurance premium will be reduced in addition to achieving better protection of lives and property.

Current trends in building design and modern lifestyles make for serious fire hazards. Some of the contributing factors are the following:

1. *High-rise buildings.* Most building codes define a high-rise building as a building more than 75 feet in height to the floor of the highest occupied level. The figure is more or less determined by the reach of fire department ladders in most cities. With rapid population growth in congested urban areas, buildings become taller and more densely situated. Super-high-rise buildings and skyscrapers from 50 to over 100 stories in height are a common sight in all regions. Thus, the design of fire protection systems must be closely coordinated with the local fire department's capabilities.

2. *Architectural design.* For better communication and productivity, modern buildings are also designed with larger areas and open spaces. Quite often, a single floor exceeds 50,000 sq ft without a separation. In such a case, building codes may require fire walls or partitions with separate zoning of the fire protection system.

3. *Controlled interior environment.* The exterior fenestration of modern buildings is usually constructed of fixed glass panels in lieu of operable windows in order to provide security and an interior environment in which temperature, humidity, and air quality are controlled. Furthermore, opening a window on the upper floor of a high-rise building may create a drastic updraft within the building in the winter, with air rushing out of the upper floors owing to a stack effect. Concurrently, outside air will rush in on the lower floors. The building fire code usually specifies the installation of operable, but locked, windows for access to fight fires and to remove smoke. These windows should not be operated when there is no fire.

4. *Increased use of combustible materials.* While a building's construction materials are well controlled, its furnishings, equipment, and decorative finishes are not easy to manage. Materials such as plastic and synthetics are a source of toxic gas and smoke during a fire. Serious consideration should be given to selecting these materials.

Most building codes specify the type of construction methods and materials that can be used for building occupancies. In addition, building codes usually refer to the National Fire Codes as the basis for criteria and technical specifications. The National Fire Codes are published by the National Fire Protection Association (NFPA) and include more than 100 separate codes, standards, and recommended practices dealing with diversified subjects related to fire protection. Some of

283

the codes and standards related to building design and construction are the following:

- *NFPA 10*   Standards for portable fire extinguishers
- *NFPA 11*   Standards for foam extinguishing systems
- *NFPA 12*   Standards for carbon dioxide extinguishing systems
- *NFPA 13*   Standards for the installation of sprinkler systems
- *NFPA 14*   Standards for the installation of standpipe and hose systems
- *NFPA 70*   National Electrical Code
- *NFPA 78*   Lightning Protection Code
- *NFPA 101*   Life Safety Code
- *BOCA*   National Fire Prevention Code

This chapter covers only the principles of fire protection. The reader should refer to the above codes and standards for more specific requirements.

Following are some commonly used terms pertaining to fire protection, and their definitions:

- *Automatic fire suppression system*   Engineered system using an automatic sprinkler system (wet or dry), gas (carbon dioxide or inert gases), or chemicals (wet or dry) to detect and suppress a fire through fixed piping and nozzles. (*Note:* A standpipe-and-hose system is operated manually; thus it is not an automatic fire suppression system.)
- *Fire area*   Aggregate floor area enclosed and bounded by fire walls.
- *Fire partition*   Vertical assembly of wall material designed to restrict the spread of fire. The overall assembly shall have a fire resistance rating (in hours) equal to or greater than that specified in the governing code. Depending on the use of the building and whether it has a sprinkler system, a fire partition shall have a rating up to 2 hours.
- *Fire door*   Door assembly that provides protection against the passage of fire.
- *Fire wall*   Fire-resistant wall that extends continuously from the foundation of a building up to or through the roof. Fire walls are rated in hours. Depending on the use of the building and the fire separation distance, required fire resistance ratings may vary from 1 hour to 4 hours.
- *Fire separation distance*   Distance in feet or meters measured from the face of a building to the adjacent building or the nearest public way.
- *Fire protection rating*   Time in hours or fractions thereof that an opening protective assembly, such as a wall or a partition, will resist exposure to fire.
- *Fire separation assembly*   Horizontal or vertical fire-resistance-rated assembly of materials designed to restrict the spread of fire.

- *Flame spread*   Propagation of flame over a surface of construction materials for building construction and interior finishes.
- *Flame spread rating (FSR)*   Rating describing how quickly fire will spread in certain materials. The values (Class I, 0–25; Class II, 26–75; Class III, 76–200) are relative and are based on the testing procedure developed by the American Society for Testing Materials (ASTM). Materials with a rating of over 200 shall not be used in any building with human occupancy.
- *Means of egress*   Continuous and unobstructed path of horizontal and vertical travel from any point in a building to a public way, such as a street, an alley, or open land.
- *Use group*   Classification of building occupancy in accordance with the intended use of the building.
- *Fire loading*   Based used by building codes to estimate the nominal amount of combustibles for determining the required fire resistivity of construction for various building use groups, expressed in pounds per square foot. Heat liberated from the total fire loading can, therefore, be calculated on the basis of the caloric value of the materials and the weight of the combustible materials.
- *Smoke-developed rating (SDR)*   Numerical rating developed by ASTM to express the smoke developed by interior finishes and materials. With this method, the densities of smoke developed by various materials under identical testing (burning) conditions are compared. The method assigns SDR = 0 for asbestos cement board and SDR = 100 for red oak. An SDR rating of 50 is considered safe for most building occupancies, and an SDR rating over 450 is prohibited for use in any buildings.

9.1 CLASSIFICATION OF FIRE AND CONSTRUCTION HAZARDS

Fire is the rapid combustion of materials in the presence of oxygen. Without oxygen, a fire cannot be sustained. The products of fire have two components: the thermal elements, which produce flame and heat; and the nonthermal elements, which produce smoke and toxic gases. Smoke is always associated with a fire. Indeed, statistics indicate that smoke is the primary cause of fire-related deaths. Thus, a fire protection system is used not just to extinguish a fire but also to remove smoke. To simplify this discussion, the term "fire" shall mean "fire and smoke" to the extent applicable in the text.

9.1.1 Classification of Fires

According to the NFPA, there are four classes of fires:

1. *Class A.* Fires of ordinary combustible materials, such as wood, cloth, paper, rubber, and many plastics.
2. *Class B.* Fires in flammable liquids, oils, greases, tars, oil-base paints, lacquers, and flammable gases.
3. *Class C.* Fires that involve energized electrical equipment. In such fires, it is important that the extinguishing medium not be a conductor of electricity.
4. *Class D.* Fires of combustible metals, such as magnesium, titanium, zirconium, sodium, lithium, and potassium.

9.1.2 Classification of Hazards

According to the NFPA, fire hazards may be grouped into three classes:

1. *Light (low) hazard.* Locations (buildings or rooms) where the total amount of Class A combustible materials, including furnishings, decorations, and other contents, is minor. Among these locations are offices, classrooms, churches, assembly halls, etc.
2. *Ordinary (moderate) hazard.* Locations where Class A combustibles and Class B flammables, in total, are present in greater amounts than expected under light hazard occupancies. Ordinary hazards are divided into two groups: Group 1, stockpiles lower than 8 ft; Group 2, stockpiles lower than 12 ft. These locations generally consist of mercantile shops and allied storage, light manufacturing, research operations, auto showrooms, garages, etc.
3. *Extra (high) hazard.* Locations with large quantities of highly combustible materials and conditions are such that fires could develop quickly with high heat release. Extra-hazard Group 1 occupancies have few or no flammable liquids. Group 2 has significant amounts of flammable liquids or are locations where combustibles are shielded from suppression.

Fire protection systems must be designed specifically to be most effective for each class of fire and hazard.

9.1.3 Type of Construction

According to BOCA, building construction is divided into types 1 through 5, with type 1 being the most fire-resistant. Walls, partitions, ceilings, floors, roofs, structural system, and exit envelopes shall all be constructed of noncombustible material having at least the minimum fire resistance rating specified in the code.

9.1.4 Use or Occupancy

The requirements for fire protection in a building also are governed by how the building is being used or occupied. Most building codes, including BOCA, UBC, and SBC (see Appendix A), classify building occupancy by use groups and subgroups. When a building houses more than one occupancy, each portion of the building shall conform to the requirements for the occupancy housed therein. For example, an office building that is primarily used for administrative work, typing, and data processing is classified as Use Group B. If the building also contains an auditorium, the auditorium portion shall be classified as a place of assembly to comply with Use Group A. Following is the general classification of buildings by use group or occupancy:

- *Group A: Assembly* Occupied by more than 1000 people (A–1), less than 1000 people (A–2), and other situations (A–3, A–4, and A–5).
- *Group B: Business* Used for offices, professions, or service-type transactions.
- *Group E: Educational* Elementary schools (E–1, E–2), daycare (E–3).
- *Group F: Factory* Moderate hazard (F–1), low hazard (F–2).
- *Group H: Hazard* Groups H–1 through H–7, depending on the hazardous material being handled or stored.
- *Group I: Institutional* Nurseries, hospitals, nursing homes (I–1), others (I–2, I–3).
- *Group M: Mercantile* Used for display, storage, and sale of merchandise.
- *Group R: Residential* Hotels, motels, or boarding houses (R–1); multifamily dwellings (R–2); one- or two-family dwellings (R–3); child care (R–4).
- *Group S: Storage* Moderate hazard (S–1), low hazard (S–2), repair garage (S–3), open parking garage (S–4), aircraft (S–5).
- *Group U: Utility* Buildings not covered by the above groups.

In addition, there are requirements for special types of buildings such as malls, high-rise buildings, atriums, and underground structures. Refer to the applicable codes for specific requirements.

9.2 PLANNING FOR FIRE PROTECTION

Planning for fire protection starts with architectural and engineering design in all disciplines. Figure 9–1 illustrates the decision steps. Once the plan is formalized

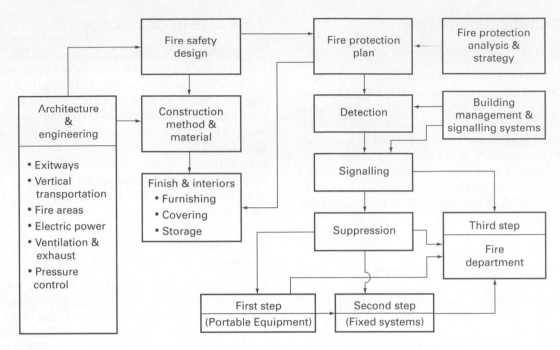

■ **FIGURE 9–1**
Fire protection decision tree.

and implemented, a well-planned fire protection system usually operates sequentially as follows:

- *Step 1, detection* The presence of a fire is detected manually or automatically.
- *Step 2, signaling* The building's management, its occupants, and the fire department are notified of the presence of the fire. The occupants are advised of the actions to take.
- *Step 3, suppression* Manual or automatic fire suppression equipment and systems are used to extinguish the fire and remove the smoke.
  - *3A, initial effort* Portable and manual firefighting equipment, such as fire extinguishers, fans, and a first-aid fire hose, are used to extinguish the fire and to remove smoke by dilution or exhaustion.
  - *3B, main effort* Fire suppression systems, such as automatic sprinklers, fire hoses, and other systems, are used to extinguish the fire, and smoke control systems are activated to remove or contain the spread of smoke.
  - *3C, last effort* The fire department takes over the firefighting effort when all previous efforts are ineffective.

Note that a fire may spread rapidly within minutes, and all three steps of fire protection may have to be initiated or activated concurrently to prevent it from spreading.

9.3 FIRE SAFETY DESIGN

Fire safety starts with the design and construction of a fire-safe building. Building codes contain specific design requirements for fire safety. Some of the fundamental criteria are as follows:

1. *Fire-resistant construction.* The construction of walls, partitions, ceilings, and floors shall meet or exceed the fire resistance ratings specified in the governing codes. The required ratings vary with building occupancy, size, and height. A 3- to 4-hour-rated building assembly is considered a highly resistive design. For example, a 4-in.-thick, reinforced concrete slab without additional ceiling material below it has only a 1-hour fire resistance rating.
2. *Smoke controls.* In addition to fire-resistant construction, a building of any size must have proper smoke control by removal, dilution, and/or confinement. Such control could be as simple as opening windows or as complicated as exhausting smoke with automated mechanical systems.
3. *Length of travel.* All exits shall be located so that the maximum length of travel to access the exit, measured from the most remote point to an approved exit along the natural and unobstructed line of travel, shall not exceed the distances given in Table 9–1.

TABLE 9–1
Typical building fire protection requirements according to occupancy

Typical Occupancies	Required Protection		Maximum Distance to Exit, ft	
	Fire Alarm	Fire Suppression	Not Sprinkled	Sprinkled
Theaters, TV studios	—	Yes	200	250
Amusement, entertainment	—	Yes	200	250
Churches and religious services	Yes	Yes	200	250
Business (2 or more stories)	Yes[a]	—	200	250
Education	Yes	Yes[b]	200	250
Factories (low hazard)	—	Yes[c]	200	250
Factories (moderate hazard)	—	—	300	400
Penal or correction institutes	Yes	Yes	150	200
Hospitals, child care	Yes	Yes[d]	150	200
Prisons, detention centers	Yes	Yes	—	—
Mercantile	—	Yes[e]	200	250
Hotels, motels	Yes	Yes[c]	200	250
Apartments	Yes	—	200	250
One- or two-family dwellings	—	—	200	250
Storage (low hazard)	—	—	300	400
Miscellaneous	—	—	—	—

(*Note:* The requirements are based on the interpretation of several prevailing building codes, which vary slightly. In actual building design, the requirements of the prevailing code shall govern.)
[a]Manual boxes are not required for buildings below 75 ft.
[b]Except if the fire area is less than 20,000 sq ft.
[c]Except less than three stories.
[d]Except child care facilities with 100 or fewer children.
[e]Except less than 12,000 sq ft for each fire area or less than 3 stories.

4. *Means of egress.* There shall be two separate means of egress from any space, except where a space is so small and arranged in such a way that a second exit would not provide an appreciable increase in safety.

5. *Exit enclosures.* Exit enclosures, such as stairways used for exit purposes, shall be separated from other portions of the building by appropriate fire-resistive construction. Penetration by ducts, conduits, boxes, and pipes shall be limited and protected.

6. *Adequate lighting.* Egress passages shall be illuminated to a minimum of 1 foot-candle (fc), and preferably 3 fc, with clearly identified and illuminated signs. (*Note:* Signs are normally mounted at a level of 7 ft or higher. Additional signs at a 2-ft level above the floor are effective ways to guide occupants to safety in a smoke-filled corridor.)

7. *Vertical openings (other than elevator shafts).* Vertical openings shall be sealed to limit fires to a single floor.

8. *Vertical transportation.* Elevators are not recognized as exits. Elevator shafts shall be vented or pressurized, depending on the HVAC system. Escalator floor openings shall be protected with fire shutters or protected by water curtains as part of the sprinkler system.

9. *Coordination with mechanical and electrical systems.* Mechanical and electrical systems shall be designed to meet the applicable codes, such as the National Electrical Code (NFPA 70), BOCA National Mechanical Code (BNMC), and BOCA National Fire Protection Code (BNFPC).

10. *Compliance with code requirements for specific use groups.* The classification is generally consistent with that of other building codes. Table 9–1 lists building occupancies and requirements of fire protection systems.

11. *Coordination with fire department.* The fire marshal must be consulted about the required access to the building and the locations of fire hoses, fire hydrants, electrical power disconnects, fire suppression, and alarm systems.

9.4 FIRE DETECTION AND SIGNALING DEVICES

When fire or smoke is detected either by the occupants of a building or by automatic devices, the building's management needs to evaluate the severity of the fire immediately and take appropriate action, such as activating the building alarm system, announcing a total or partial evacuation of the building, and notifying the fire department. For larger buildings, the fire code often requires the installation of automatic sensing (detecting) devices, as well as automatic alarm and signaling systems. Some basic devices and systems are briefly discussed in this section; Figures 9–2 and 9–3 illustrate several detection and signaling devices.

Detection and signaling devices may be addressable or nonaddressable. The addressable type can be individually addressed and identified so that the system can immediately identify the type and location of the initiating devices associated with a given address.

9.4.1 Manual Alarm Station

A manual alarm station is an electrical switch, specially designed for fire protection, that activates an alarm system, such as bells, gongs, and flashing lights. To avoid accidental operation of the switch, the station is usually designed so that a person must break a glass panel or glass rod or must perform other preliminary actions before the alarm can be operated.

9.4.2 Thermal Detectors

Thermal detectors are temperature-activated sensors that initiate an alarm when the temperature in their immediate vicinity reaches a predetermined setting. These detectors are designed to meet various conditions. Thermal detectors are used to provide property (the building) protection only. They are not intended to protect life. Some commonly used sensing devices are the following:

1. *Fixed-temperature type.* These detectors are of either self-restoring or nonrestoring design. The self-restoring type consists of normally open contact held by bimetallic elements that will close the contacts when the ambient temperature reaches a fixed setting. The setting is generally designed for operation at 135°F (57°C), 190°F (88°C), or 200°F (94°C). The contact will return to the normally open position when the ambient temperature drops back to normal. The nonrestoring type usually contains a fusible element which holds open the electrical contacts and will melt to close the electrical contact, initiating a signal. The entire detector must be replaced after activation.
2. *Rate-of-rise type.* This sensor reacts to the rate at which the temperature rises. A sealed but slightly vented air chamber within the device expands quickly when the temperature in the vicinity of the device rises quickly. When the air chamber expands faster than it can be vented, electrical contacts

attached to the chamber begin to close and thus initiate an alarm. A rate-of-rise device reacts to fire much earlier than a fixed-temperature device; however, it may cause false alarms if installed in spaces that are subject to rapid temperature fluctuations.
3. *Combination type.* This device reacts to both a fixed temperature and a rate of rise. It may be designed on the principle of differential expansion between two different metals or on the rate of expansion of the air chamber. The combination type of sensing device is desirable for most applications.

9.4.3 Smoke Detectors

Smoke detectors are quicker to respond than thermal detectors as long as the smoke generated by the fire is within their limits of detectability. Two of the more commonly used smoke detectors, shown in Figure 9–2, are the following:

1. *Photoelectric type.* This variety operates on the principle of the scattering of light. Under normal conditions, the beam generated from a light-emitting diode (LED) is reflected by a mirror within the detector chamber away from a photosensitive diode. When smoke is present in the detector chamber, the light beam is scattered, causing the diode to sense the scattered light and activate the electric circuit to sound an alarm.
2. *Ionization type.* This variety operates on the principle of changing conductivity of air within the detector chamber. The detector contains a minute source of radioactive material that emits alpha particles, which ionize the nitrogen and oxygen molecules in the air. Voltage applied across the ionization chamber causes a very small electrical current to flow to an electrode of opposite polarity. When invisible or visible combustion particles enter the ionization chamber, the conductivity of the air is reduced and the voltage on the electrodes increases. This in turn initiates an alarm when the voltage reaches a preset level. Ionization-type detectors are the most sensitive, although the presence of radioactive material in them has been a concern in terms of health risks as well as reliability, however, improved technology has eliminated both problems. See Figure 9–2 for illustrations.

9.4.4 Flame Detectors

Flame detectors are used to detect the direct radiation of a flame in the visible, infrared, and ultraviolet ranges of the spectrum. There are four basic types: infrared, ultraviolet, photoelectric, and flame flicker detectors.

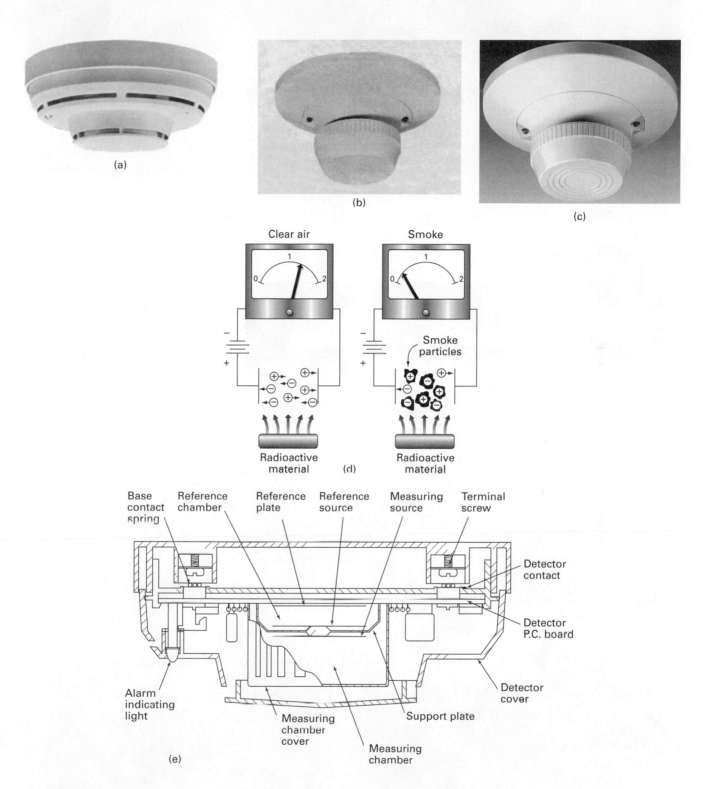

(a)

(b)

(c)

Clear air

Smoke

Smoke particles

Radioactive material

Radioactive material

(d)

Base contact spring

Reference chamber

Reference plate

Reference source

Measuring source

Terminal screw

Detector contact

Detector P.C. board

Detector cover

Alarm indicating light

Measuring chamber cover

Measuring chamber

Support plate

(e)

■ **FIGURE 9–2**

Fire and smoke detection devices. (a) Ionization smoke detector. (Courtesy of Siemens.) (b) Ionization smoke detector. (Courtesy of BRK Brands, Inc.) (c) Ionization detector. (Courtesy Gamewell-FCI.) (d) Principle of operation for ionization smoke detector. (e) Cross-sectional view of an ionization smoke detector. (Reprinted with permission from the Fire Protection Handbook, 19th edition © 2003, National Fire Protection Association, Quincy, MA 02169.)

(a)

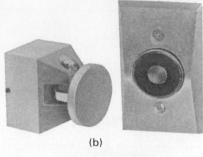

(b)

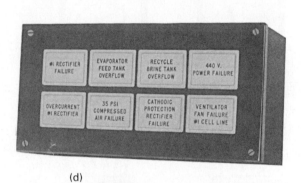

(c)

(d)

(e)

■ **FIGURE 9–3**

(a) Typical signal gongs and bells. (b) Typical magnetic door holder. (c) Typical fire alarm with manual pull station. (d) Typical visual annunciation panel. (e) Typical signal horns, general-distribution and narrow projecting types (from 60 dB to 120 dB at a 10-ft range). (Reproduced with permission from Edwards Co.)

Flame detectors are used chiefly in industrial processes, in mining, and for the protection of combustion equipment. Thermal or smoke detectors would be unreliable and generate false alarms in these environments.

9.4.5 Magnetic Door Release

A magnetic door release is an electromagnetic device that holds a fire door open when the device is energized and releases the door to a closed position in case of fire.

9.4.6 Signal Devices

Signal devices are audio and video devices used to alert and inform of an emergency situation. Among the devices commonly used are the following:

1. *Single-stroke bell.* Device containing a bell that is struck once each time electrical power is applied.
2. *Vibrating bell.* Device containing a bell that rings continuously as long as electrical power is applied.
3. *Buzzer.* Device consisting of an electromagnetic coil that, when electrical power is applied to it, will cause thin metal piece (reed) to vibrate.
4. *Chime.* Device that produces a pleasing or musical tone each time electrical power is applied to it.
5. *Horn.* Device consisting of an electromagnetic coil that causes a metal diaphragm to vibrate and produce a sound that is amplified by a horn.
6. *Siren.* Device consisting of an electric motor that produces a continuous high-pitched sound (up to 100 dB). Sirens are used in places with normally high ambient noise, such as a steel mill or an auto assembly plant.
7. *Light signal.* Flashing strobe or steady illuminated light with wording such as "Fire, leave the building," or "Go to 20th floor," for hearing-impaired persons. Strobe lights are required by most applicable fire codes.

9.4.7 Flow Detectors

Flow detectors are devices that indicate or initiate an alarm when water is flowing in the fire suppression systems. They are either sail-type switches that deflect under a flow of water or pressure switches that sense the pressure differential caused by a flow of water. Their operation depends on prior actuation of a sprinkler head or the operation of a fire hose. Flow detectors monitor the fire suppression system even when a building is unoccupied.

9.4.8 Visual Annunciation Devices

Visual annunciation devices are displays that may consist of single or multiple lights with marked messages such as "Fire," "Fire escape," or "Go to Area B." The lights may be of different colors and may flash or be steadily lit. The use of long-life lamps or fiber-optic light tubes (usually lasting over 30,000 hours) embedded in floors or walls to lead the occupants to safety is an effective means of visual signaling. This type of design is increasingly used in places of public assembly, such as theaters, auditoriums, sports arenas, and airplanes.

The selection and spacing of detectors and signal devices depend on the performance characteristics of the products and are governed by the specifications of manufacturers. Figure 9–3 shows several signaling and activating devices.

9.5 FIRE ALARM SYSTEMS

Fire alarm systems are an integral part of a fire protection plan. They are basically electrical systems that are specially designed to announce the presence of fire or smoke. They are not intended to suppress or extinguish a fire. Fire alarm systems are described in Chapter 12.

9.6 FIRE SUPPRESSION SYSTEMS

Fire suppression is achieved by cooling the combustible material to below its ignition temperature or by preventing oxygen from reacting with the combustible material. Depending on the class of fire and the type of building occupancy, some fire suppression systems are more effective than others. For example, while water is an effective cooling agent, it is detrimental in an electrical fire because water is an excellent electrical conductor. Fire suppression systems may be classified in several ways:

- According to the fire suppression medium—water, foam, chemical, gas, etc.
- According to the action of the device—a portable extinguisher, standpipe and hose, automatic sprinkler system, etc.
- According to the method of operation of the device—manual or automatic.

Depending on the building use group, both manual and automatic suppression systems may be required. Requirements vary with the applicable building codes; the designer should check with the codes for specifics and exceptions. In general, a fire suppression system is required in the building use groups given in Table 9–1.

9.6.1 Water Supply

Water is the universal firefighting medium. It is readily available in large quantities and, in general, is more economical than any other firefighting medium. Public water is preferred over on-site water supply whenever it is available.

For fire protection purposes, the water supply should be separated from a building's domestic water system, even though the two are connected to the same public water main. Usually, the fire main is separated from the domestic main ahead of the water meter. A check valve is often required by the water department to detect the flow of water, but not to measure the quantity.

The pressure at various parts of the system should be higher than the following values when the system is at its rated flow:

- At the water main (residual pressure at maximum flow), 10–20 psig
- At the end of a hose (nozzle), 65 psig
- At each sprinkler head, 15–30 psig
- At the top of the roof, 15–30 psig

When the public water supply is inadequate to serve firefighting needs, alternative water sources such as lakes, ponds, wells, and storage tanks must be provided. The minimum amount of storage required is dictated by the applicable code. A 30-minute storage capacity at the calculated flow rate is the norm. Figure 9–4 illustrates several alternative water supply systems.

9.6.2 Portable Fire Extinguishers

Portable fire extinguishers are used as the first line of fire protection. Often, they avert a major disaster. They are normally precharged with water or chemicals and are hand-operated. Fire extinguisher requirements are governed by applicable building codes and NFPA Standard No. 10. In general, a portable fire extinguisher shall be installed in the following locations:

- In all occupancies in use groups A–1, 2, 3, E, I–2, R–1, and H; and in staff locations of I–3.
- In all special areas containing commercial kitchens, wherever fuels are dispensed or combustible liquids handled, and in laboratories and shops.

Extinguishers shall be selected for a Class A fire in all areas and for Classes B, C, and D fires in special areas.

Extinguishers containing various media are labeled as follows for different classes of fires:

- Water stored-pressure or cartridge-operated   A
- Aqueous film-forming foam (AFFF)   A, B
- Film-forming fluoroprotein (FFFP)   A, B
- Carbon dioxide (CO_2)   B, C
- Halogenated (non-ozone-depleting variety)   B, C
- Dry chemical (potassium-based)   B, C, D
- Dry chemical (ammonium phosphate-based)   A, B, C, D

Typical sizes and effective ranges for various classes of fire extinguishers are summarized in Table 9–2.

The advantage of a water fire extinguisher is its lower cost. Both the AFFF type and the FFFP type work on the principle of film coating the surface material and thus isolating the material from oxygen. The carbon dioxide type displaces oxygen from the electrical fire, but it is toxic and has a short range. The halogenated type is similar to the carbon dioxide type and may be inhaled for a short duration (several minutes). The major advantage of the gaseous type (carbon dioxide and other non-ozone-depleting gases) is that the gas does not deposit any film on the protected surface. This feature is very important for electrical and electronic equipment. Dry chemicals are effective for chemical or burning metals, but they are usually corrosive and difficult to clean.

The size and placement of extinguishers shall comply with the criteria given in Table 9–3.

9.6.3 Standpipe-and-Hose Systems

Standpipe-and-hose systems consist of piping, valves, hose connections, and allied equipment to provide streams or sprays of water for fire suppression. For simplicity, standpipe-and-hose systems are abbreviated as standpipe systems, although a hose is always used in a standpipe system. Typical components and system diagrams are illustrated in Figure 9–5. Most building codes adapt their requirements from NFPA Pamphlet No. 14. In general, standpipes are required in building use groups or areas:

- Where the top floor is higher than 30 ft from access to fire department vehicles.
- Where the lowest floor is lower than 30 ft from access to fire department vehicles.
- Where any portion of the building is more than 400 ft from access to fire department vehicles, except if the building is equipped with an automatic sprinkler system or if the building is less than 10,000 sq ft or if the building is in any of the use groups A–4, A–5, F–2, R–3, S–2, and U, according to BOCA.
- That are malls.
- That are stages of a theater, auditorium, etc., where props are prepared and used.

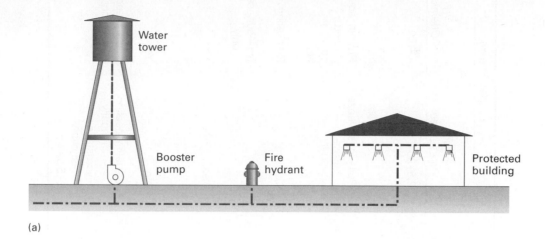

(a)

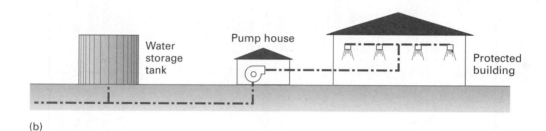

(b)

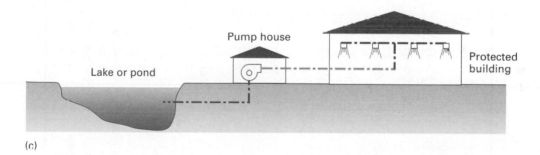

(c)

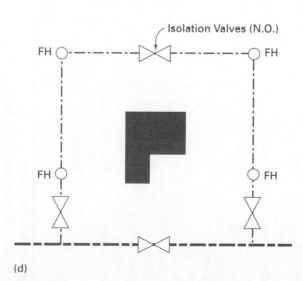

(d)

■ **FIGURE 9–4**

Schematic diagrams of alternative water supply systems. (a) Elevated water tower: At night, the public water supply pressure is adequate to fill the elevated water tower. If the water pressure is insufficient to feed the tower, a booster pump will be required. (b) Groundwater tank: A pump is used to transfer water from the tank. (c) Lake or pond: A pump is used to transfer water from the lake or pond. (d) A fire loop surrounds a building with fire hydrants (FH) and isolation valves. Typically, spacing is 300 to 600 ft depending on building use group and construction type. Water may be shut off from either direction if necessary because of breaks or leaks.

TABLE 9-2
Summary of fire extinguisher applications

Types	Water		Soda Acid	Foam	Gas	Loaded Stream	
	Air Pressurized	Cartridge-Operated			Halogenated/CO_2	Air Pressurized	Dry Chemical
Typical sizes[a] (gal)	2.5–5	2.5–60	2.5–33	2.5–33	2.5–100	2.5–33	1–100
Subject to freezing	Water—yes Antifreeze—no	Water—yes Antifreeze—no	Yes	Yes	No	No	No
Operating method	Pull pin, squeeze handle	Invert and bump	Invert	Invert	Pull pin, squeeze handle	Pull pin, squeeze handle	Pull pin, squeeze handle
Effective range	40–50 ft	40–50 ft	40–50 ft	30–40 ft	4–10 ft	40–50 ft	10–25 ft
Conductor of electricity	Yes	Yes	Yes	Yes	No	Yes	No
Extinguishing effect	Cooling	Cooling	Cooling	Blanketing and cooling	Blanketing, some cooling	Cooling	Blanketing
Class A fires, wood, paper, textiles, etc.	Yes	Yes	Yes	Yes	No—surface only	Yes	No—surface only
Class B fires, flammable liquids, and gases	No	No	No	Yes	Yes	Yes	Yes
Class C fires, electrical	No	No	No	No	Yes	No	Yes
Maintenance	Visual inspection semiannually Hydrostatic test every 5 years	Weigh cartridge yearly Hydrostatic test every 5 years	Recharge yearly Hydrostatic test every 5 years	Recharge yearly Hydrostatic test every 5 years	Weigh semiannually Hydrostatic test every 5 years	Visual inspection semiannually Hydrostatic test every 5 years	Visual inspection semiannually

[a]Wall mounted and on carts.

(Reproduced with permission from Elkhart Brass Mfg. Co., Inc.)

TABLE 9–3
Fire extinguisher size and placement for Class A hazard

Criteria	Light (Low) Hazard Occupancy	Ordinary (Moderate) Hazard Occupancy	Extra (High) Hazard Occupancy
Minimum rated single extinguisher	2-A	2-A	4-A
Maximum floor area per unit of A	3000 ft²	1500 ft²	1000 ft²
Maximum floor area for extinguisher	11,250 ft	11,250 ft	11,250 ft
Maximum travel distance to extinguisher	75 ft	75 ft	75 ft

For SI units, 1 ft = 0.305 m; 1 ft² = 0.0929 m².

Note: For maximum floor area explanation, see E.3.3.

(*Source:* Portions of this table reprinted with permission from NFPA 14-2007, *Installation of Standpipe and Hose Systems*, copyright © 2007, National Fire Protection Association, Quincy, MA 02269. This reprinted material is not the complete and official position of the National Fire Protection Association on the referenced subject, which is represented only by the standard in its entirety.)

NFPA defines three classes of standpipe systems:

- *Class I*   2½-in. hose stations for use by the fire department
- *Class II*   1½-in. hose stations for use by the occupants of the building or the fire department
- *Class III*   Both 1½-in. and 2½-in. hoses

Standpipe systems are further classified as follows:

- *Wet systems*   Systems in which the standpipes are filled with water under pressure. Whenever the system is activated, water will charge into the connected hose immediately. As a rule, wet systems are used for most building occupancies.
- *Dry systems*   Systems in which the standpipes are not filled with water. Each standpipe is normally filled with compressed air. On being opened, the dry pipe valve will automatically admit water into the system. Dry systems may be of the automatic or semiautomatic (deluge) type. The semiautomatic type usually includes a deluge valve that is remotely controlled to allow the water to enter the standpipe. A dry system is used when the building space is subject to freezing or when the use of water may cause other hazards or extreme property damage—for example, in a computer center or a power-generating station.
- *Manual systems*   These systems are used by fire departments exclusively. They may be of the dry or wet variety, depending on the application. A manual dry-type standpipe system is directly connected to a fire department connection (also called a Siamese connection) at the exterior of the building, where water is pumped into the standpipe through the use of a fire pump or via a connection to fire hydrants.

According to BOCA, standpipe systems installed in all buildings shall be of the wet variety, except when the highest floor of the building is:

- Lower than 75 feet (dry allowed)
- *Or* lower than 150 feet and the building is protected by an automatic sprinkler system (dry allowed)
- *Or* lower than 150 feet and the building is an open parking structure (dry or manual allowed)

The number of standpipe risers required depends on the horizontal distance covered by the hoses. In general, a hose connection shall be provided at each intermediate landing between floor levels in every required stairway. The stairway is normally enclosed by fire walls. Thus, the standpipe riser is ideally located within the stairway enclosure. Standpipe risers at some stairways may be omitted if:

- All parts of the building can be reached by a 30-ft "hose stream" (water stream) at the end of a 100-ft hose from other risers. (*Note:* The distance shall be the actual developed length measured along a path of travel originating from the hose connection.)
- *Or* if all parts of the building are within 400 ft from access to fire department vehicles and not more than 200 ft from another hose connection.

For design purposes, the water supply, pressure, and flow rate shall generally be based on the following criteria:

- Water from a public water system or from an alternative water source shall be adequate for a minimum duration of 30 minutes.
- The minimum pressure shall not be less than 65 psig residual pressure at the topmost hose outlet. (*Residual pressure* is the pressure remaining at the outlet when water is discharging at the rated flow. *Static pressure* is the pressure available at the outlet when there is no flow. The difference between the two is the pressure drop.)
- If the water pressure is too high, the extremely high reaction force of the hose owing to a high water flow will be too difficult to handle. If the calculated residual pressure is more than 100 psig, a pressure-regulating device (valve) shall be provided. If no fire hose is provided at the standpipe outlet, as in the manual standpipe system, the maximum pressure may be raised to 175 psig.

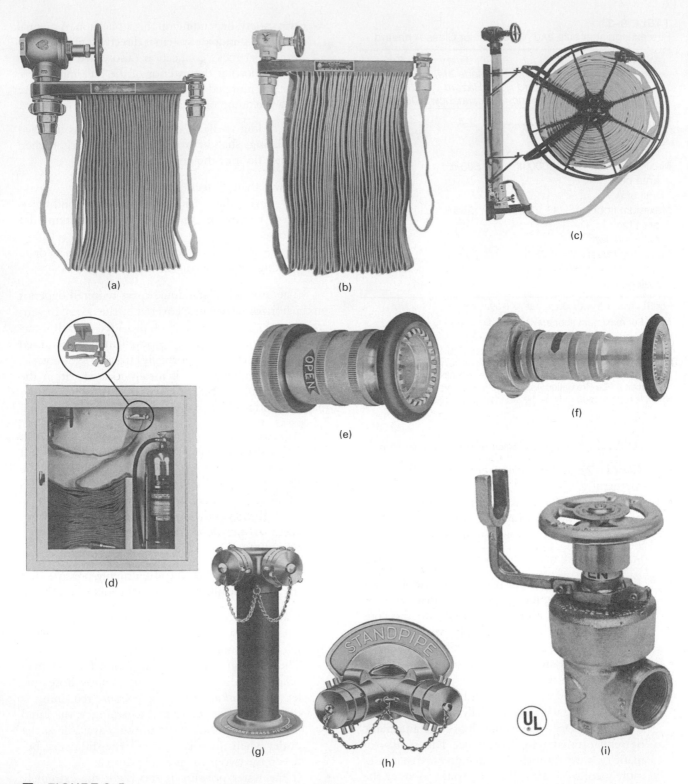

■ FIGURE 9–5

Standpipe-and-hose system components. (a) Hose rack with 2½-in. valve and 1½-in. hose (up to 125 ft). (b) Hose rack with 1½-in. valve and 1½-in. hose (up to 100 ft). (c) Hose reel with 1½-in. valve and 1½-in. hose. (d) Hose cabinet with 1½-in. valve, 1½-in. hose, and fire extinguisher; enlarged detail of the hose clamp. (e) Fog-type nozzle for use on an electrical fire (30° fog pattern). (f) High-capacity (up to 250 GPM) direct-connection nozzle. (g) Sidewalk-type, freestanding two-way Siamese connection. (h) Wall-mounted surface two-way Siamese connection. (i) Automatic pressure-reducing valve to limit the water pressure at the nozzle under both flow and no-flow conditions. (Reproduced with permission from Elkhart Brass Mfg. Co., Inc.)

- In Class I and III systems—i.e., 2½-in. systems—for calculation purposes, piping and risers shall be sized for 500 GPM for the first standpipe and 250 GPM for each additional standpipe, but normally not more than 1250 GPM for the total system, including the sprinkler water demand for buildings that have a limited-area sprinkler system.
- In Class II systems—i.e., 1½-in. systems—pipe sizes shall be based on 100 GPM for each riser and not more than 500 GPM for the total system.
- For buildings in use groups B, I, R–1, or R–2, and with an automatic sprinkler system, the flow rate may be sized for 250 GPM for each standpipe and not more than 750 GPM for all standpipes. The demand flow rate for the automatic sprinkler system shall be calculated independently.

The sizes of standpipes shall not be less than those shown in Table 9–4.

Piping material in standpipe-and-hose systems is usually black steel with welded or mechanical fittings, galvanized steel with mechanical fittings, or copper with solder fittings. Mechanical fittings, which allow for limited pipe movement due to thermal expansion, contraction, or vibration, are desirable for most high-rise buildings.

A fire department connection, also known as a Siamese connection, shall be provided outside the building. A minimum of two connections shall be installed for high-rise buildings. Hoses may be installed on racks, on reels, or in cabinets containing portable fire extinguishers and tools. The cabinet door will be latched.

A water flow alarm shall be provided for automatic and semiautomatic standpipe systems. Also, valves shall be provided for isolation of piping, prevention of back pressure, and shutoff. All valves shall indicate at a distance whether they are open or closed. A standing stem valve is one of the approved varieties and shall be capable of being closed within 5 seconds. Isolation valves shall be locked in the open position.

Each standpipe shall be provided with a means of draining. The drainpipe shall be ¾ in. for a 2-in. standpipe and up to 2 in. for a standpipe 4-in. or larger. The drain line shall be 3 in. for standpipes equipped with pressure-regulating devices.

The standpipe system shall be limited to a vertical height of 275 ft. For high-rise buildings, separate standpipe systems shall be provided for each 275 ft of vertical height. (See Figure 9–6.)

9.6.4 Other Fire Suppression Systems

In addition to automatic sprinkler systems, to be discussed in Section 9.7, there are a number of specialty systems that do not use water.

TABLE 9–4
Pipe schedule, standpipes, and supply piping, in.

Total Accumulated Flow, GPM	Total Distance of Piping from Farthest Outlet		
	<50 ft	50–100 ft	>100 ft
100	2	2½	3
101–500	4	4	6
501–750	5	5	6
751–1250	6	6	6
1251 and over	8	8	8

Source: Portions of this table reprinted with permission from NFPA 14-2007, *Installation of Standpipe and Hose Systems*, copyright © 2007, National Fire Protection Association, Quincy, MA 02269. This reprinted material is not the complete and official position of the National Fire Protection Association on the referenced subject, which is represented only by the standard in its entirety.

Foam Systems

Foam systems are most effective for Class B fires (fires in which flammable liquid, oil, grease, paint, etc., is a factor). The foam is made by generators, which mix water with detergent or other chemicals to produce as much as 1000 gallons of foam for each gallon of water. Sprayed from large nozzles, the foam covers the fire, insulating it from oxygen, and cools down the temperature when water is evaporated into steam. Figure 9–7(a) illustrates the operation of a foam system. Typical applications of foam systems are in printing plants and aircraft manufacturing plants, among others.

Gaseous Fire Suppression Systems

Gaseous systems are most effective for Class C fires (fires caused by electrical equipment). All these gases are stored in liquid state under high pressure. There are three varieties:

The *carbon dioxide system* stores CO_2 in a liquid state in pressurized tanks. When discharged through nozzles, the liquid vaporizes and smothers the fire by displacing oxygen. The evaporation of the liquid to a gas also cools the fire, absorbing about 120 Btu of heat per pound of CO_2. To be effective, carbon dioxide systems should be used in confined and normally nonventilated spaces. The CO_2 is toxic. A few minutes of exposure in an atmosphere containing over 10 percent CO_2 could render a person unconscious. A carbon dioxide fire suppression system is shown in Figure 9–7(b).

Halogenated gas is a generic name for gases containing the elements fluorine, chlorine, bromine, or iodine. The gases most commonly used for fire extinguishing are Halon 1301 and 1211; however, because of their ozone-depleting property, these gases have been banned and are being replaced with a number of

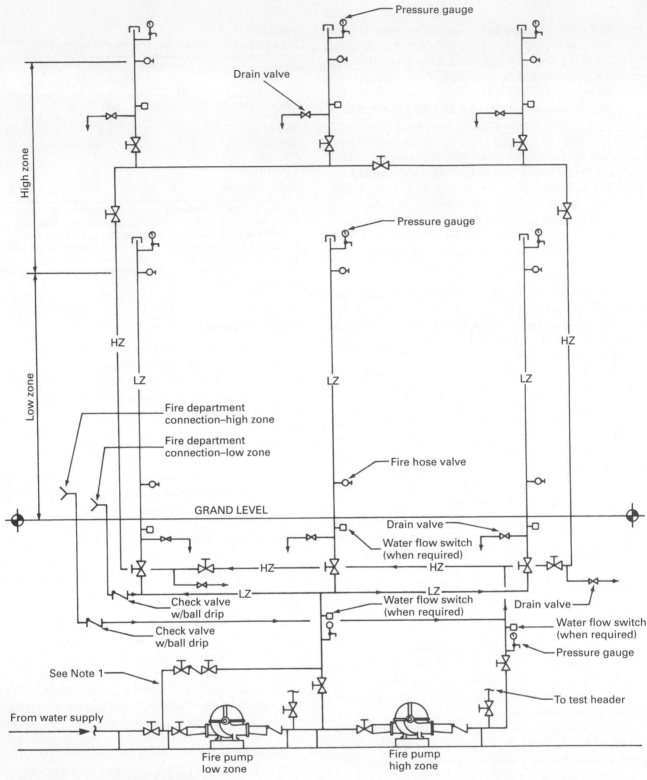

NOTE 1: Bypass subject to NFPA 20.
NOTE 2: High zone pump may be arranged to take suction directly from source of supply.

■ **FIGURE 9–6**

Riser diagram of a three-zone standpipe-and-hose system showing a fire pump for each zone and a storage tank for the high zone. (Reprinted with permission from NFPA 14-2007, *Installation of Standpipe and Hose Systems*, Copyright © 2007, National Fire Protection Association, Quincy, MA 02269. This reprinted material is not the complete and official position of the National Fire Protection Association on the referenced subject, which is represented only by the standard in its entirety.)

(a)

■ **FIGURE 9–7**
(a) Foam fire suppression system. (Courtesy: Kidde-Fenwal, Inc., Ashland, MA.) (b) Carbon dioxide fire suppression system in an industrial plant. (Reproduced with permission from Ansul Incorporated.)

(b)

alternatives. These include FE-25 (pentafluoroeth-ane), FE-232 (dichlorotrifluoroethane), FE-13 (trifluo-roethane), FM-100 (bromodifluoromethane), and FM-200 (heptafluoropropane), developed by a number of leading producers. These agents have zero or negligi-ble ozone-depletion potential (ODP). The effectiveness of several of these chemicals as fire extinguishing agents is still being evaluated.

Atmospheric gas is a mixture of argon, carbon diox-ide, and nitrogen. A typical composition is 40 percent argon, 8 percent carbon dioxide, and 52 percent nitro-gen. The gas mixture is nontoxic, with zero ODP and zero global-warming potential (GWP). This type of gas system is effective but requires more gas per volume of space than the halogenated gases. Figure 9–8 shows a typical application in a data processing space.

Dry Chemicals

Dry chemicals are used especially for Class D fires (fires in which combustible metals or flammable liquids are a factor). Most of the dry chemicals contain bicarbonates, chlorides, phosphates, and other proprietary com-pounds. The use of water should be avoided on burn-ing metals, since hot metal extracts oxygen from water, promotes combustion, and at the same time liberates hydrogen, which ignites readily. When water must be used to fight a Class A fire simultaneously, water must be of sufficient quantity to help cool off the burning metal and thus extinguish the fire.

Mist Systems

Water-mist fire protection systems deliver very small-sized water droplets to a fire to achieve fire suppression principally by thermal cooling and oxygen displace-ment. The tiny droplets generated by mist systems have a total surface area-to-volume ratio such that they rapidly absorb heat from the fire to the point where the water changes phase from liquid to vapor. The mist is often delivered to the fire in pulses to accentuate the heat absorption. The water phase change from liquid to vapor reduces the oxygen content in the fire atmosphere to levels below that which can support combustion.

The amount of water delivered is a small fraction of the amount delivered by conventional sprinkler systems and it is delivered for a maximum duration of 20 minutes. Mist systems are classified by their droplet size and pressure and are listed for specific applications. One manufacturer calls the smallest droplet size a fog. That system is listed for light-hazard non-machine spaces up to a limited height.

Mist systems are sold as engineered fire suppres-sion systems with proprietary components, which in-clude the special nozzles, a specialized high-pressure pumping apparatus, and a delivery controller. Mist pip-ing is very small and must be stainless steel to avoid clogging by corrosion. The water stored in tanks is perfectly clean.

Mist systems pose very little water hazard to room content; therefore, mist systems might be an appropri-ate choice for spaces with high-value contents, such as museums and computer rooms. They are also a good choice for machine spaces, where fires are confined and intense enough to effect the droplet phase change. Mist systems have the advantage of penetrating into and around equipment. Their disadvantages are that they deliver the mist only for a limited time, and they cannot penetrate down through a rising fire plume. Accord-ingly, their fire protection listings have height and volume limits. Building code officials do not always recognize mist system protection as an alternative to automatic sprinkler protection.

9.7 AUTOMATIC SPRINKLER SYSTEMS

Automatic sprinkler systems are integrated fire suppres-sion systems consisting of a water supply and a network of pipes, sprinkler heads, and other components to pro-vide automatic fire suppression in areas of a building where the temperature or smoke has reached a predeter-mined level. Automatic sprinklers are the most effective systems for suppressing a Class A fire in buildings con-taining ordinary combustible materials, such as wood, paper, and plastics. The design and installation of the sys-tem are strictly regulated by insurance companies and in accordance with NFPA Standard 13. Specific require-ments for various system applications are too numerous to be included in this section; the principles and practices described provide an overview of requirements, the scope of such systems, their features, and design approaches. Construction drawings showing all dimensions of piping, sprinklers, etc., as well as detailed hydraulic calculations, are normally prepared by the installation contractors.

9.7.1 Required Locations

According to BOCA and similar building codes, the fol-lowing building use groups or areas within the building are required to be protected by an automatic sprinkler system. When the use of water is detrimental or ineffec-tive, other fire suppression systems, such as dry chemi-cal, foam, or halogenated gas systems, should be used.

1. Assemblies (use groups A–1, A–3, and A–4): All fire areas exceeding 12,000 sq ft, except for audito-riums (A–1 or A–3), naves or chancels (A–4), or

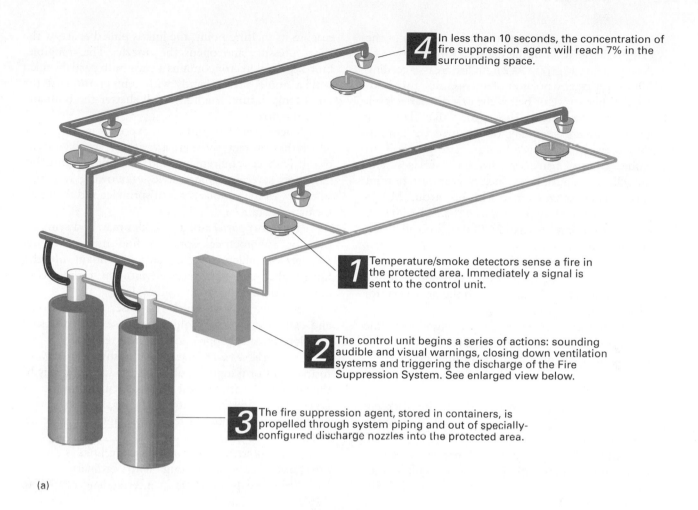

4 In less than 10 seconds, the concentration of fire suppression agent will reach 7% in the surrounding space.

1 Temperature/smoke detectors sense a fire in the protected area. Immediately a signal is sent to the control unit.

2 The control unit begins a series of actions: sounding audible and visual warnings, closing down ventilation systems and triggering the discharge of the Fire Suppression System. See enlarged view below.

3 The fire suppression agent, stored in containers, is propelled through system piping and out of specially-configured discharge nozzles into the protected area.

(a)

MONITORS:

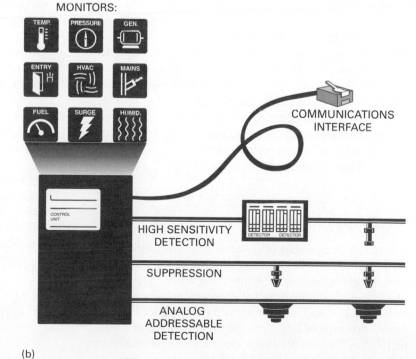

COMMUNICATIONS
INTERFACE

HIGH SENSITIVITY
DETECTION

SUPPRESSION

ANALOG
ADDRESSABLE
DETECTION

(b)

■ **FIGURE 9–8**

(a) Gaseous fire suppression system for a building space. System may be divided into a number of zones. (b) Enlarged view of the control unit. It can be designed to monitor a diversity of functions or equipment, such as temperature, pressure, generator, entry, HVAC system, fuel, voltage surge, or humidity. The system may be interfaced with the building automation system. (Reproduced with permission from Kidde-Fenwal, Inc.)

sport arenas (A–3) where the main floor is at the exit discharge level of the main entrance.

2. Assemblies (use group A–2): All fire areas exceeding 5000 sq ft or any portion of an assembly that is located either above or below the exit discharge level.

3. Business occupancies (use group B): This use group consists generally of occupancies not included in the other use groups. Use group B includes offices, banks, professional services, and miscellaneous businesses, such as a car wash or print shop. (There is an overlap with use group M.) In general, when a building is more than 12,000 sq ft on one floor or more than 24,000 sq ft on all floors, an automatic sprinkler system is required. Furthermore, sprinkler requirements in individual spaces or areas, such as an auditorium, a computer room, or a laboratory, shall comply with the provisions for those other use groups.

4. High-rise buildings (over 75 ft in height above the fire truck access): All buildings and structures shall be installed with an automatic sprinkler system.

5. Educational (use group E): All fire areas exceeding 20,000 sq ft.

6. High hazard (use group H): All fire areas, except magazines (explosive storage areas) if specially constructed.

7. Institutions (use group I): All fire areas, except child care facilities with 100 or fewer children located at the level of the exit discharge and with exit doors in all child care rooms.

8. Mercantile, storage, and factory (use groups M, S–1, and F–1): All fire areas exceeding 12,000 sq ft on one floor, or 24,000 sq ft on all floors, or more than three stories above grade.

9. Residential (use group R–1, hotel, motel, etc.): All fire areas, except when all guest rooms are no more than three stories above grade and each guest room is directly open to the exterior exit.

10. Residential Residential (use group R–2, multifamily dwellings): All fire areas, except when the building is only two stories high with no more than 12 dwelling units per fire area.

9.7.2 Sprinklers

The major component of an automatic sprinkler system is the sprinkler, which discharges water in a specific pattern for extinguishing or controlling a fire. A sprinkler head consists of three major components: a nozzle, a heat detector, and a water spray pattern deflector. The fusible link type of heat detector is constructed of a "eutectic alloy," which melts at a specific temperature rather than gradually softening. When the link temperature reaches its melting point, the link is pulled apart by the water pressure and opens the nozzle. The frangible-bulb type of detector contains a glass bulb partially filled with a liquid that expands with temperature. At the rated temperature, the liquid will shatter the bulb and open the nozzle.

Since sprinklers must be exposed in the space which they protect, there are a variety of sprinkler styles to satisfy space requirements and aesthetics. Figure 9–9 illustrates several of these styles. Following is a brief summary of the various types of sprinkler and their performance data.

The *spray pattern* of a sprinkler may be a symmetrical or asymmetrical spray, a fine mist, or water droplets. Sprinklers may be *mounted* pendant, upright, flush with the ceiling, recessed into the ceiling, concealed in the ceiling, or in a sidewall.

Response may be quick response, quick response and extended coverage, quick response and early suppression, or early suppression and fast response.

The *heat-sensing elements* of a sprinkler may be fusible links or frangible bulbs. Or the sprinkler may be the open type (with no heat-sensing element) or the dry type (with a built-in seal at the end of the inlet to prevent water from entering the nipple until the sprinkler operates).

The *temperature rating* of fusible links is divided into seven groups, starting with ordinary (135°F to 170°F) and proceeding to intermediate (175°F to 225°F), high (250°F to 300°F), extra high (325°F to 445°F), and on up. Sprinklers are color-coded for ease of identification. Sprinklers with ordinary ratings are used in general occupied spaces, those with intermediate ratings are installed in boiler or mechanical rooms, and those with higher ratings operate in spaces where the normal ambient temperature is close to the range of the fusible links, such as industrial spaces.

Ordinary temperature-rated sprinklers shall be used throughout a space, except in locations near a heater, directly exposed to the sun's rays under a skylight, inside unventilated show windows, or near commercial cooking equipment. In these locations, a higher-temperature sprinkler shall be used.

The *flow rate* of a sprinkler depends on the size of its orifice and the residual pressure of the water supply. The flow rate is calculated from the fundamental fluid flow formula,

$$Q = K \times \sqrt{p} \qquad (9\text{--}1)$$

where Q = flow rate, GPM
  K = flow constant, per unit; varies from 1.3–1.5 for a ¼-in. orifice, 5.3–5.8 for a ½-in. orifice, to 13.5–14.5 for a ¾-in. orifice
  p = pressure, psi (residual pressure at rated flow)

(a)

Flow Rate (gpm)	Pressure (psi)	Enclosure Area Width x Length	
		(ft)	(ft)
25	9.3	14 x	18
30	13.4	14 x	20
30	13.4	16 x	18
35	18.2	16 x	20

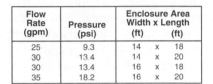

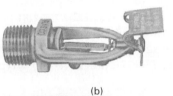

(b)

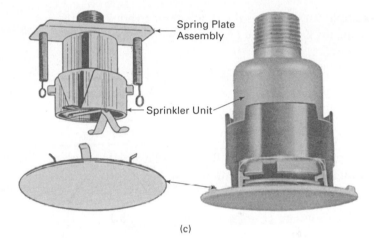

Spring Plate Assembly

Sprinkler Unit

(c)

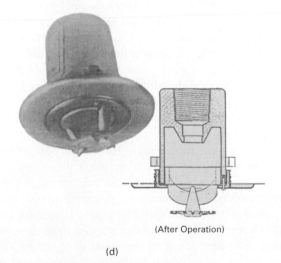

(After Operation)

(d)

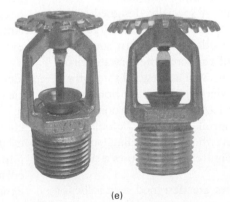

(e)

(f)

■ **FIGURE 9–9**

(a) Standard pendant sprinkler showing water deflected by a deflector plate. (Reproduced with permission from Figgie Fire Protection Systems, Charlottesville, VA.) (b) Sidewall type of sprinkler with flow rate and enclosure area. (Reproduced with permission from Reliable Automatic Sprinkler Co., Inc.) (c) Typical ceiling-concealed sprinkler. (Courtesy of Tyco Fire & Building Products, Lansdale, PA.) (d) Typical ceiling-recessed sprinkler. (Courtesy of Tyco Fire & Building Products, Lansdale, PA.) (e) Frangible-bulb type of sprinklers, upright and pendant. (Courtesy of Tyco Fire & Building Products, Lansdale, PA.) (f) Standard fusible link type of upright sprinkler. (Reproduced with permission from Reliable Automatic Sprinkler Co., Inc.)

The sprinklers most commonly used are for ½-in. pipe thread with a ½-in. orifice (K factor of 5.5). The flow rate at 15 psi residual pressure is about 20 GPM (21.3 GPM, as calculated by Eq. (9–1). With 15 psi at the base, the flow rate will be 28 GPM when the pressure is doubled to 30 psi, and 40 GPM when the pressure is quadrupled. Sprinklers with other orifices or K factors have different flow rates, varying from a ¼-in. orifice ($K = 1.3$) to a ¾-in. orifice ($K = 14.5$). Sprinklers with smaller orifices are ideally suited for residential occupancies, since most residential spaces are relatively small.

Relative to a ½-in. standard orifice as 100 percent, the flow rate of a sprinkler with a ¼-in. orifice is as low as 25 percent, and that of a sprinkler with a ¾-in. orifice is as high as 250 percent.

Example 9.1 What is the flow rate of a sprinkler with a standard ½-in. orifice ($K = 5.5$) with 20 psi residual pressure?

Answer:

$$\text{Flow rate } Q = 5.5 \times \sqrt{20} = 5.5 \times 4.47$$
$$= 24.6 \text{ GPM}$$

Example 9.2 What is the flow rate of a sprinkler with a ¾-in. orifice ($K = 14$) with 65 psi?

Answer:

$$\text{Flow rate } Q = 14 \times \sqrt{65} = 14 \times 8.1$$
$$= 112.9 \text{ GPM}$$

Normally, the flow rates of individual sprinklers need not be calculated in the design of sprinkler systems if the layout and pipe sizes comply with Section 9.7.7.

There are a number of specially designed sprinklers. Sidewall sprinklers are designed to discharge most of the water toward one side, in a pattern somewhat resembling one-quarter of a sphere. The forward throw may be as far as 20 to 25 ft, with a sideward throw from 7 to 8 ft. Sidewall sprinklers are ideal for hotel guest rooms, lobbies, executive offices, dining rooms, etc., where ceiling-mounted sprinklers would be objectionable aesthetically or where there is no ceiling space to run the pipes through. Figure 9–9(b) shows a sidewall sprinkler.

Residential sprinklers are designed especially for residential spaces, which are usually small and narrow and have lower ceilings. The orifice factor K is as low as 1.5. The combination use of sidewall and residential sprinklers can accommodate most unusual architectural features in decorative spaces.

9.7.3 Types of Automatic Sprinkler Systems

There are numerous kinds of automatic sprinkler systems, each ideally suited for certain spaces. Two major varieties are:

1. *Wet systems.* Wet pipe, antifreeze, and circulating closed loop.
2. *Dry systems.* Dry pipe, preaction, deluge, and combined dry-preaction.

A *wet-pipe system* is a piping system containing sprinklers under water pressure so that water discharges immediately from the sprinklers when they are opened by heat from a fire. Wet pipe is the basic automatic sprinkler system for all building use groups. (See Figure 9–10 for a typical piping perspective and Figure 9–11 for the operation of some typical automatic sprinkler systems.)

A *dry-pipe system* is a piping system filled with compressed air (or nitrogen). The air pressure prevents water from entering the pipes beyond a control valve known as a dry-pipe valve. When the air pressure is released at the opening of a sprinkler, the water flows into the piping system and out of the opened sprinkler. The water supply and the size of the dry-pipe valve shall be designed so that water can reach the far end of the system within 60 seconds. Dry-pipe systems are installed in areas that are subject to freezing, such as in a nonheated warehouse. (See Figure 9–10 for a typical piping perspective.)

An *antifreeze system* is a wet-pipe system containing antifreeze solutions used in areas subject to freezing. If the system is connected to the potable water system, then only pure glycerin or propylene glycol solutions are permitted. Suitable glycerin–water solutions and their freezing temperatures are given in NFPA Standard 10. Antifreeze systems are similar to dry-pipe systems, except that they have a faster response.

A *circulating closed-loop system* is a wet-pipe sprinkler system having non-fire-protection connections in a closed-loop piping arrangement for the purpose of utilizing sprinkler piping to circulate water for heating or cooling. Water is not removed from the system. A typical utilization of this system is the unitary hydronic heat pump system, where the sprinkler piping is used for circulating water through either the evaporator or the condensing coils, depending on the mode of operation. To qualify for fire suppression purposes, the operation of the HVAC system should not interfere with the water flow from the sprinkler when the automatic sprinkler function is activated. (See Chapter 7 for a discussion of hydronic heat pump systems.)

Wet-pipe sprinkler systems employ automatic sprinklers attached to a piping system containing water and connected to a water supply so that water discharges immediately from sprinklers opened by a fire. Wet-pipe systems are the most reliable and simple of all sprinkler systems since no equipment other than the sprinklers themselves need to operate. Only those sprinklers which have been operated by heat over the fire will discharge water.

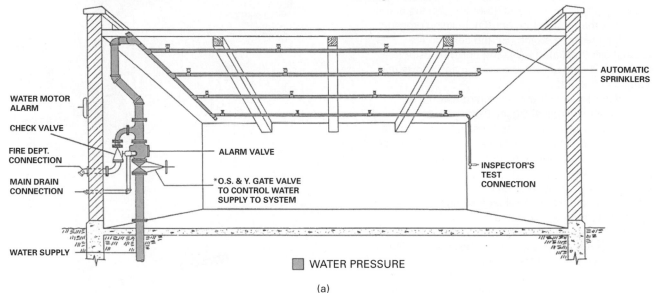

WATER MOTOR ALARM

CHECK VALVE

FIRE DEPT. CONNECTION

MAIN DRAIN CONNECTION

WATER SUPPLY

ALARM VALVE

*O.S. & Y. GATE VALVE TO CONTROL WATER SUPPLY TO SYSTEM

AUTOMATIC SPRINKLERS

INSPECTOR'S TEST CONNECTION

■ WATER PRESSURE

(a)

Dry-pipe sprinkler systems employ automatic sprinklers attached to a piping system containing air or nitrogen under pressure, the release of which, as from the opening of a sprinkler, permits water pressure to open the dry-pipe valve. The water then flows into the piping system and discharges only from those sprinklers which have been operated by heat over the fire. Dry-pipe systems are installed in lieu of wet-pipe systems where piping is subject to freezing.

*O. S. & Y. – Outside screw and yoke gate valve.

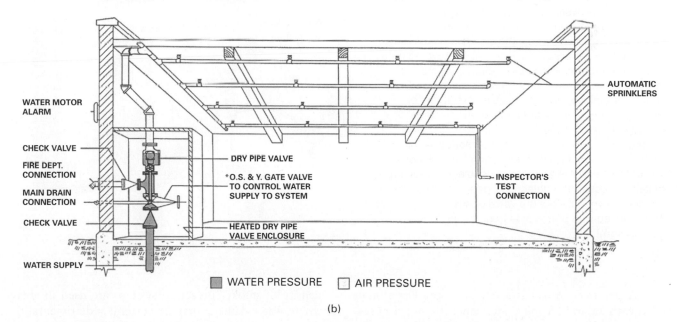

WATER MOTOR ALARM

CHECK VALVE

FIRE DEPT. CONNECTION

MAIN DRAIN CONNECTION

CHECK VALVE

WATER SUPPLY

DRY PIPE VALVE

*O.S. & Y. GATE VALVE TO CONTROL WATER SUPPLY TO SYSTEM

HEATED DRY PIPE VALVE ENCLOSURE

AUTOMATIC SPRINKLERS

INSPECTOR'S TEST CONNECTION

■ WATER PRESSURE   □ AIR PRESSURE

(b)

■ **FIGURE 9–10**

Typical piping perspective of automatic sprinkler systems. (a) Basic wet-pipe system. (b) Basic dry-pipe system. (Courtesy of Tyco Fire & Building Products, Lansdale, PA.)

■ **FIGURE 9–11**

Operation of automatic sprinkler systems.
(a) Wet-pipe system.

The isometric piping illustrates typical components of a wet-pipe system. When sprinkler (A) opens, water pressure lifts the alarm valve clapper (B) from its seat and flows through the alarm port (C) to the retard chamber (D), building pressure under the pressure switch (E) and sending the optional electrical alarm (F). The optional water-flow indicator (G) also activates an alarm and shows the location of the fire.

Water flows to the water motor alarm (H), sounding a mechanical signal. During surges or pressure fluctuations, excess pressure is trapped in the system, allowing only small amounts of water into the alarm port and retarding chamber in order to prevent a false alarm.

(b) Deluge and preaction system.

The isometric piping illustrates typical deluge or preaction system components.

In the deluge system, the sprinklers are always open; there are no fusible links at the sprinkler heads. The upper chamber of the deluge valve (A) is pressurized in the ratio of 2:1 to keep the clapper seal on the inlet (B). When the release (C) detects a fire, the pneumatic actuator (D) opens by an air pressure drop, venting the upper valve chamber and permitting the valve to open. The pressure-operated relief valve (E) continues venting. Alarm (F) sounds, and water flows to the sprinklers. Since the sprinklers are open, water is applied to the fire immediately. Hydraulic and electric release systems may also be used.

In the preaction system, the sprinklers contain fusible links. In other words, the sprinkler heads are closed. When the preaction valve (A) is activated by one or more thermal or smoke detectors, water simply fills the pipes, and no water is discharged from the sprinklers until the fusible links in one or more sprinklers are melted by heat. Air-pressure integrity in the piping is supervised by a monitoring system. Preaction systems may be interlocked singly, so that broken piping or broken sprinklers will not trip the system. It is also possible to double-interlock the system to require the detection system to trip and a sprinkler to open before the valve opens. (Reproduced with permission from Viking Corporation.)

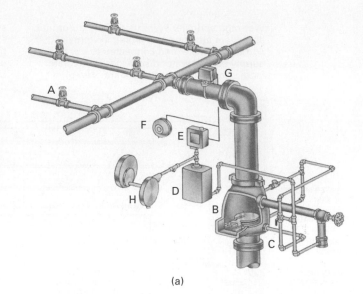

(a)

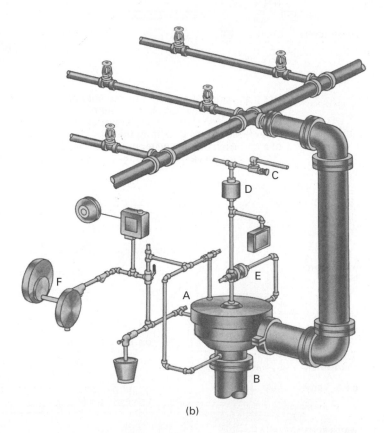

(b)

A *preaction system* is a dry-pipe sprinkler system filled with air and having a supplemental detection system installed in the same area. Actuation of the detection system opens a valve that permits water to enter into the sprinkler piping system and to be discharged from any sprinklers that may be open. The detection system may be sensitive to either the temperature or the density of smoke. Preaction systems are used in areas where water damage may be extremely detrimental to the property or may interfere with business operations. Yet, if a fire is out of contol, water is readily available when any sprinkler fusible link is opened. A typical application of a preaction system is in a computer center of a large business or a central bank, where even a short interruption of services may cause huge financial losses. (See Figure 9–11 for the operation of a typical preaction system.)

A *deluge system* is a dry sprinkler system equipped with open-type sprinklers (no fusible links). The control valve is operated by a supplemental detection system in the same area. Actuation of the detection system opens the control valve, which allows water to flow through all the sprinklers at once. Sprinklers are closely spaced so as to create a "water wall" to separate different fire areas. This system is frequently used in a multilevel atrium, where the construction of fire walls to separate areas is undesirable for operational or aesthetic reasons. (See Figure 9–11 for the typical operation of a deluge system.)

9.7.4 Piping Material and Components

Piping material is usually black steel with welded or mechanical fittings. Galvanized steel is also used, but it should not be welded, as welding will destroy the galvanized (zinc) coating. Copper may be used but is generally too costly. Mechanical fittings of the groove-joining method are ideal for high-rise buildings when the differential expansion between the structure of the building and the piping system is substantial.

The fire department connection, water alarm, valves, and drain lines in an automatic sprinkler system are the same as in the standpipe-and-hose system.

The alarm check valve is the main control valve of every automatic sprinkler system. The valve assembly consists of single or double check valves, pressure gauges, water flow alarm test and drain valves, and one or more check and control valves.

The water motor alarm is a water power–driven gong that produces a loud sound at the exterior of the building when water is flowing in the water-operated motor immediately inside the building. The water motor alarm does not require any electrical power, because such power may be cut off during a fire.

Figure 9–12 shows some components of an automatic sprinkler system.

9.7.5 Fire Pumps

Fire pumps are used when the residual water pressure at the most remote sprinkler heads of an automatic sprinkler system or at the hose outlets of a standpipe system cannot be met by the water supply system. A fire pump serves to boost the pressure to the necessary level and is selected on the basis of its flow rate (GPM), the pressure differential required, control features, and the driving equipment—an electrical motor or engine. The fire pump is usually of the split-case centrifugal variety for

either horizontal or vertical mounting. A vertical pump assembly requires less floor space but more floor-to-ceiling height. The control system is designed to start the pump whenever the system pressure drops below the required operating pressure. To avoid frequent start-up of the fire pump, a small pressure maintenance pump known as a *jockey pump* is installed in parallel with the fire pump. When the system pressure drops below the preset level because of leakage or drainage, the pressure switch will first start the jockey pump to maintain the system pressure. If the system pressure drops rapidly, as in the case of the opening of sprinkler heads or the discharge of fire hoses, the jockey pump will not be able to maintain the required operating pressure, and the fire pump will start automatically. Shown in Figure 9–12(f) and (g), respectively, are a diesel engine–driven fire pump and a side view showing components of a fire pump piping assembly.

9.7.6 Planning Guidelines

Although the final layout of the automatic sprinkler system is prepared by the fire protection contractor, the architect/engineer should be knowledgeable regarding basic requirements of the system, so the layout is coordinated with other systems and is acceptable aesthetically while maintaining a reasonable construction cost. The following is a summary of planning guidelines for sprinkler systems:

1. The maximum floor area of a building that can be protected by a single system shall not exceed 52,000 sq ft for hazards classified as light or ordinary. The maximum area is reduced for extra-hazard-classified buildings.
2. The maximum floor area that can be covered by a sprinkler shall not exceed that given in Table 9–5.
3. In addition to the maximum floor area limitation, sprinklers shall meet the following dimensional limitations:
   - The maximum distance between sprinklers shall be no more than 15 ft for light and ordinary hazards and 12 ft for extra hazards and above, except that the spacing of sidewall-type sprinklers may be according to the approved area of coverage.
   - When sprinklers are spaced less than 6 ft on centers, baffles shall be located between the sprinklers to prevent nonactivated sprinklers from being cooled off by the water discharge from adjacent sprinklers. The distance of a sprinkler from a wall shall be no more than half of the allowed distance between sprinklers or less than 4 in. from a wall.
   - The distance between vertical obstructions and the sprinkler shall not be less than that given in Table 9–6.

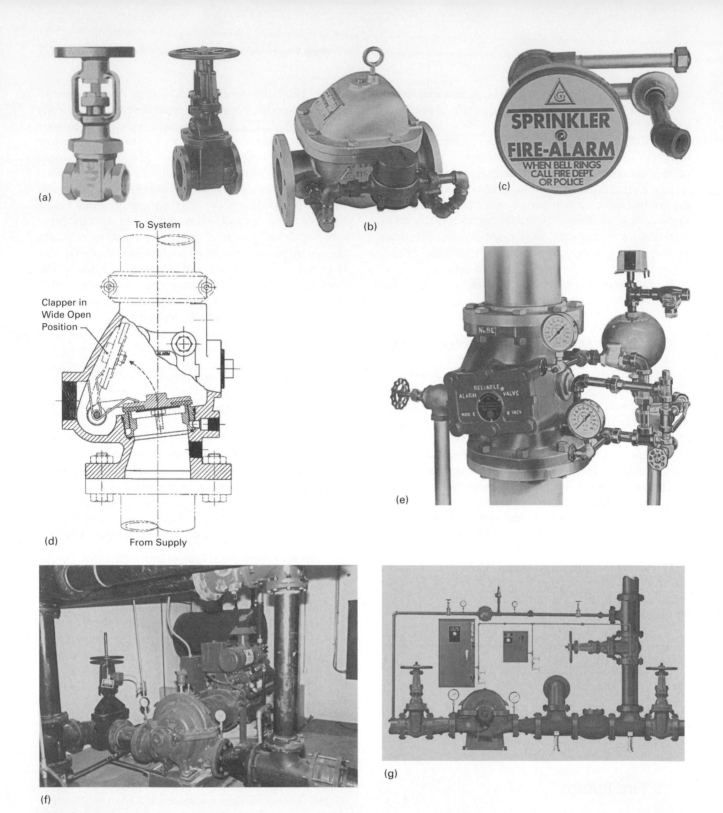

■ FIGURE 9–12

Components of an automatic sprinkler system. (a) Gate valves. (b) Detector check valve. (c) Water motor alarm. (Courtesy of Tyco Fire & Building Products, Lansdale, PA.) (d) Cutaway section of an alarm check valve showing clapper in the closed position. When water is flowing, the clapper will be in the wide-open position. (Reproduced with permission from Reliable Automatic Sprinkler Co., Inc.) (e) Wet-pipe sprinkler alarm check valve assembly showing auxiliary piping, flow control devices, and pressure gauges. (Reproduced with permission from Reliable Automatic Sprinkler Co., Inc.) (f) Diesel engine–driven fire pump, which has the advantage of being available during an electrical power failure and downsizing the emergency generator capacity. Engine-driven fire pump systems have a higher installation cost than those that are electrically driven. (g) Side view of a fire pump piping assembly.

TABLE 9–5

Maximum sprinkler protection area (sq ft) for standard spray sprinklers[d]

Construction	Classification of Hazard		
	Light	Ordinary	Extra
Unobstructed construction[a]	200–225[b]	130	90–130[c]
Obstructed construction:			
Noncombustible	200–225[b]	130	90–130[c]
Combustible	130–225[a]	130	90–130[c]

[a]For light combustible framing members spaced less than 3 ft on centers, the maximum area is 130 sq ft, and for heavy combustible framing members spaced 3 ft or more on centers, the maximum area is 225 sq ft.
[b]If pipe sizing is based on pipe schedule, the maximum protected area shall not exceed 200 sq ft.
[c]If pipe sizing is based on pipe schedule, the maximum area shall not exceed 90 sq ft. If it is based on the hydraulic method with density less than 0.25 GPM/sq ft, the maximum area shall not exceed 130 sq ft.
[d]See Table 5–8.21 of NFPA 13 for extended coverage sprinklers.

TABLE 9–6

Minimum horizontal distance of sprinklers from any vertical obstructions, ft

Maximum Dimension of Obstructions	Minimum Distance
Less than ½ in,	No limit
Between ½ and 1 in.	6 in.
Between 1 and 4 in.	12 in.
More than 4 in.	24 in.

9.7.7 System Design Approaches

Piping for automatic sprinkler systems can be designed by either the pipe schedule or the hydraulic method.

Pipe Schedule Method

The pipe schedule is a traditional method permitted for new light- and ordinary-hazard occupancies of 5000 sq ft or less, and for the modernization of existing systems designed by the pipe schedule. The sizes of branches and risers can be determined from a pipe schedule, assuming that other requirements, such as residual pressure and flow rates, are all in compliance with code. In general, the pipe schedule yields more conservative pipe sizes; it is very useful for preliminary planning and cost estimation. Table 9–7 gives the pipe schedule.

Example 9.3 A two-story office building totaling 4000 sq ft is interpreted as a light hazard (200 sq ft/sprinkler from Table 9–5). After adjustment for the room configurations, it was determined that there shall be 14 sprinklers installed at each floor. If black steel pipe is used, determine the riser size.

TABLE 9–7

Pipe schedule for number of sprinklers allowed in a sprinkler system

Pipe Size (inch)	Hazard Classification			
	Light		Ordinary	
	Steel	Copper	Steel	Copper
1	2	2	2	2
1¼	3	3	3	3
1½	5	5	5	5
2	10	12	10	12
2½	30	40	20	25
3	60	65	40	45
3½	100	115	65	75
4	a	a	100	115
5	—	—	160	180
6	—	—	275	300
8	—	—	b	b

[a]One 4-in. system may serve up to 52,000 sq ft of floor area.
[b]One 8-in. system may serve up to 52,000 sq ft of floor area.

Answer The main riser is for 28 sprinklers. From Table 9–7, a 2½-in. riser that is allowed to serve up to 30 sprinklers will be used. If, instead, the building is interpreted as an ordinary hazard by the building code official, then a 3-in. riser will be used (good for 40 sprinklers).

Hydraulic Method

The hydraulic method calculates the pipe size of the entire piping system, based on the distribution of sprinklers, developed length, fitting losses, size and location of areas within the building, and water density and pressure required.

Calculations

The engineering for hydraulic calculations is based on the following formulas:

- Friction loss, $P_f = \dfrac{4.52\,Q^{1.85}}{C^{1.85} d^{4.87}} = F Q^{1.85}$ (9–2)

   where P_f = frictional resistance, psi/ft of pipe equivalent
      Q = water flow rate, GPM
      d = actual internal pipe diameter, in.
      C = friction loss coefficient of piping material
      F = combined value of $(4.52/C^{1.85} D^{4.87})$; See Table 9–9 for P_f values for pipes with $C = 100, 120,$ and $150.$

- Velocity pressure, $P_v = 0.001123 \dfrac{Q^2}{Q^4}$ (9–3)

  where P_v = velocity pressure, psi
  Q = water flow rate, GPM
  d = inside diameter of pipe, in.

- Normal pressure (acting perpendicular to pipe wall)

  $$P_n = P_t - P_v \qquad (9\text{–}4)$$

  where p_n = normal pressure, psi
  p_t = total pressure, psi
  P_v = velocity pressure, psi

- Elevation pressure, $p_e = 0.433h$ (9–5)

  where p_e = elevation pressure, psi
  h = height of elevation, ft

- Velocity, $V = \dfrac{0.485\, Q}{d^2}$ (9–6)

  where V = velocity, fps
  Q = flow rate, GPM
  d = inside diameter of pipe, in.

For convenience, the equivalent lengths of pipe fittings and controls are given in Table 9–8, and the values of friction loss (P_f), velocity (V), and velocity pressure (P_p) for steel and copper pipes are given in Table 9–9.

Water demand   Statistics indicate that during the early stage of a fire, only a small portion of the sprinklers need to be operated. Thus, fire protection codes normally require that only a small area be calculated for simultaneous flow demand. This area is known as the "area of sprinkler operations," or, more simply, the "design area." The area selected for calculations should be the most hydraulically remote from the water supply source. It varies from a minimum of 1500 sq ft to a maximum of about 5000 sq ft, depending on the hazard classification of the building, as interpreted by the authoritative code or the insurance company. The code also calls for a minimum water flow density per unit area (GPM/sq ft). The minimum water supply for the sprinkler system is determined independent of the actual size of the building. For example, the water demand for a 50,000-sq-ft and a 750,000-sq-ft office building may be the same as far as sprinkler systems are concerned. This is because a fire can be expected to start in only one area, unless it is deliberately set at multiple locations. If the sprinkler system also supplies inside and outside hoses, the water demand of the hoses should be included as well. The water demand for a combined sprinkler and standpipe system can be expressed by the formula

$$TWD = (AOP \times DD \times OVF) + HSD \qquad (9\text{–}7)$$

where TWD = Total water demand, GPM
AOP = Area of operations (see Figure 9–13)
DD = Density demand, GPM/sq ft (see Figure 9–13)
HSD = House stream demand, GPM (see Table 9–10)
OVF = Overage factor (usually 1.1)

Example 9.4   A school has 50,000 sq ft of floor area over which an automatic sprinkler system is to be installed. The building is four stories in height and is equipped with a standpipe system consisting of two fire hose cabinets per floor. The official code classifies the building as a light hazard. Determine the minimum water supply that must be provided or made available to the building.

Answer

Spinkler water demand = AOP × DD × OVF

From Figure 9–13, select AOP = 1500 sq ft and DD = 1.10, so that

Sprinkler water demand = 1500 × 0.10 × 1.17

= 165 GPM

From Table 9–10, for a light hazard, HSD = 100 GPM. Thus,

Total water demand (TWD) = 165 + 100

= 265 GPM

If the building were classified as an ordinary hazard (group 1), sprinkler demand would be 1500 × 0.15 × 1.1 = 247 GPM, HSD would be 250 GPM, and TWD would be 497, or about 500, GPM.

If the water supply is an in-house storage tank, then the water storage should be of at least a 30-minute duration, or

TWD = 265 GPM × 30 min

= 7950 (about 8000) gal

9.7.8 Hydraulic Calculation Procedure

Hydraulic calculations require the following steps:

- Determine the building hazard classification and the allowed maximum sprinkler protection area, as given in Table 9–5.
- Determine the water demand of the sprinkler system and the hose stream system, based on Eq. (9–7).
- Determine the pressure available from the water supply source by a flow test at the street main or storage system.

TABLE 9-8
Equivalent schedule 40 steel pipe length chart

Fittings and Valves	Fittings and Valves Expressed in Equivalent Feet (Meters) for Pipe														
	½ in. (15 mm)	¾ in. (20 mm)	1 in. (25 mm)	1¼ in. (32 mm)	1½ in. (40 mm)	2 in. (50 mm)	2½ in. (65 mm)	3 in. (80 mm)	3½ in. (90 mm)	4 in (100 mm)	5 in. (125 mm)	6 in. (150 mm)	8 in. (200 mm)	10 in. (250 mm)	12 in. (300 mm)
45° elbow	—	1 (0.3)	1 (0.3)	1 (0.3)	2 (0.6)	2 (0.6)	3 (0.9)	3 (0.9)	3 (0.9)	4 (1.2)	5 (1.5)	7 (2.1)	9 (2.7)	11 (3.4)	13 (4)
90° standard elbow	1 (0.3)	2 (0.6)	2 (0.6)	3 (0.9)	4 (1.2)	5 (1.5)	6 (1.8)	7 (2.1)	8 (2.4)	10 (3)	12 (3.7)	14 (4.3)	18 (5.5)	22 (6.7)	27 (8.2)
90° long-turn elbow	0.5 (0.2)	1 (0.3)	2 (0.6)	2 (0.6)	2 (0.6)	3 (0.9)	4 (1.2)	5 (1.5)	5 (1.5)	6 (1.8)	8 (2.4)	9 (2.7)	13 (4)	16 (4.9)	18 (5.5)
Tee or cross (flow turned 90°)	3 (0.9)	4 (1.2)	5 (1.5)	6 (1.8)	8 (2.4)	10 (3)	12 (3.7)	15 (4.6)	17 (5.2)	20 (6.1)	25 (7.6)	30 (9.1)	35 (10.7)	50 (15.2)	60 (18.3)
Butterfly valve	—	—	—	—	—	6 (1.8)	7 (2.1)	10 (3)	—	12 (3.7)	9 (2.7)	10 (3)	12 (3.7)	19 (5.8)	21 (6.4)
Gate valve	—	—	—	—	—	1 (0.3)	1 (0.3)	1 (0.3)	1 (0.3)	2 (0.6)	2 (0.6)	3 (0.9)	4 (1.2)	5 (1.5)	6 (1.8)
Swing Check*	—	—	5 (1.5)	7 (2.1)	9 (2.7)	11 (3.4)	14 (4.3)	16 (4.9)	19 (5.8)	22 (6.7)	27 (8.2)	32 (9.3)	45 (13.7)	55 (16.8)	65 (20)

For SI units, 1 in. = 25.4 mm; 1 ft = 0.3048 m.

Note: Information on 1/2 in. pipe is included in this table only because it is allowed under 8.15.19.4 and 8.15.19.5

*Due to the variation in design of swing check valves, the pipe equivalents indicated in this table are considered average.

Source: Reprinted with permission from NFPA 13-2007, *Installation of Sprinkler Systems,* Copyright © 2007, National Fire Protection Association, Quincy, MA 02169. This reprinted material is not the complete and official position of the NFPA or the referenced subject, which is represented only by the standard in its entirety.

TABLE 9–9
Friction loss, flow velocity, and velocity pressure for steel and copper pipes at the calculated flow rate

Size in.	O.D., in.	Wall Thickness, in.	I.D., in.	P_f, psi/lin. ft Multiply $Q^{1.85}$ by		V, fps Multiply Q by	P_v, psi Multiply Q^2 by
Steel Pipe: ½–6 Sch 40 8–12 Sch 30				For C = 120	For C = 100		
½	0.840	0.109	0.622	6.50×10^{-3}	9.11×10^{-3}	1.056	7.50×10^{-3}
¾	1.050	0.113	0.824	1.65×10^{-3}	2.32×10^{-3}	0.602	2.44×10^{-3}
1	1.315	0.133	1.049	5.10×10^{-4}	7.14×10^{-4}	0.371	9.27×10^{-4}
1¼	1.660	0.140	1.380	1.34×10^{-4}	1.88×10^{-4}	0.215	3.10×10^{-4}
1½	1.90	0.145	1.610	6.33×10^{-5}	8.87×10^{-5}	0.158	1.67×10^{-4}
2	2.375	0.154	2.067	1.87×10^{-5}	2.63×10^{-5}	0.096	6.15×10^{-5}
2½	2.875	0.202	2.469	7.89×10^{-6}	1.11×10^{-5}	0.067	3.02×10^{-5}
3	3.500	0.216	3.068	2.74×10^{-6}	3.84×10^{-6}	0.0434	1.27×10^{-5}
3½	4.000	0.226	3.548	1.35×10^{-6}	1.89×10^{-6}	0.0325	7.09×10^{-6}
4	4.500	0.237	4.026	7.29×10^{-7}	1.02×10^{-6}	0.0252	4.27×10^{-6}
5	5.563	0.258	5.047	2.43×10^{-7}	3.40×10^{-7}	0.0160	1.73×10^{-6}
6	6.625	0.280	6.065	9.91×10^{-8}	1.39×10^{-7}	0.0111	8.30×10^{-7}
8	8.625	0.277	8.071	2.47×10^{-8}	3.45×10^{-8}	0.0063	2.65×10^{-7}
Steel Pipe: 1–3½ Sch. 10 S 4–0.188 Wt.				For C = 120	For C = 100		
1	1.315	0.109	1.097	4.10×10^{-4}	5.75×10^{-4}	0.339	7.75×10^{-4}
1¼	1.660	0.109	1.442	1.08×10^{-4}	1.52×10^{-4}	0.196	2.60×10^{-4}
1½	1.900	0.109	1.682	5.12×10^{-5}	7.17×10^{-5}	0.144	1.40×10^{-4}
2	2.375	0.109	2.157	1.52×10^{-5}	2.13×10^{-5}	0.088	5.19×10^{-5}
2½	2.875	0.120	2.635	5.75×10^{-6}	8.05×10^{-6}	0.059	2.33×10^{-5}
3	3.500	0.120	3.260	2.04×10^{-6}	2.86×10^{-6}	0.0384	9.94×10^{-6}
3½	4.000	0.120	3.760	1.02×10^{-6}	1.43×10^{-6}	0.0289	5.62×10^{-6}
4	4.500	0.188	4.124	6.49×10^{-7}	9.09×10^{-7}	0.0240	3.88×10^{-6}
5	5.563	0.188	5.187	2.12×10^{-7}	2.98×10^{-7}	0.0152	1.55×10^{-6}
6	6.625	0.188	6.249	8.57×10^{-8}	1.20×10^{-7}	0.0105	7.36×10^{-7}
8	8.625	0.188	8.249	2.22×10^{-8}	3.11×10^{-8}	0.0060	2.43×10^{-7}
Copper Tube: Type "K"				For C = 150			
¾	0.875	0.065	0.745	1.79×10^{-3}		0.736	3.65×10^{-3}
1	1.125	0.065	0.995	4.36×10^{-4}		0.413	1.15×10^{-3}
1¼	1.375	0.065	1.245	1.47×10^{-4}		0.264	4.67×10^{-4}
1½	1.625	0.072	1.481	6.29×10^{-5}		0.186	2.33×10^{-4}
2	2.125	0.083	1.959	1.61×10^{-5}		0.106	7.63×10^{-5}
2½	2.625	0.095	2.435	5.59×10^{-6}		0.069	3.19×10^{-5}
3	3.125	0.109	2.907	2.36×10^{-6}		0.0483	1.57×10^{-5}
3½	3.625	0.120	3.385	1.12×10^{-6}		0.0357	8.55×10^{-6}
4	4.125	0.134	3.857	5.95×10^{-7}		0.0275	5.07×10^{-6}
Copper Tube: Type "L"				For C = 150			
¾	0.875	0.045	0.785	1.38×10^{-3}		0.663	2.96×10^{-3}
1	1.125	0.050	1.025	3.78×10^{-4}		0.389	1.02×10^{-3}
1¼	1.375	0.055	1.265	1.36×10^{-4}		0.255	4.39×10^{-4}
1½	1.625	0.060	1.505	5.82×10^{-5}		0.180	2.19×10^{-4}
2	2.125	0.070	1.985	1.51×10^{-5}		0.1037	7.23×10^{-5}
2½	2.625	0.080	2.465	5.26×10^{-6}		0.0672	3.04×10^{-5}
3	3.125	0.090	2.945	2.21×10^{-6}		0.0471	1.49×10^{-5}
3½	3.625	0.100	3.425	1.06×10^{-6}		0.0348	8.16×10^{-6}
4	4.125	0.110	3.905	5.60×10^{-7}		0.0268	4.83×10^{-6}

(Reproduced with permission from Viking Corporation.)

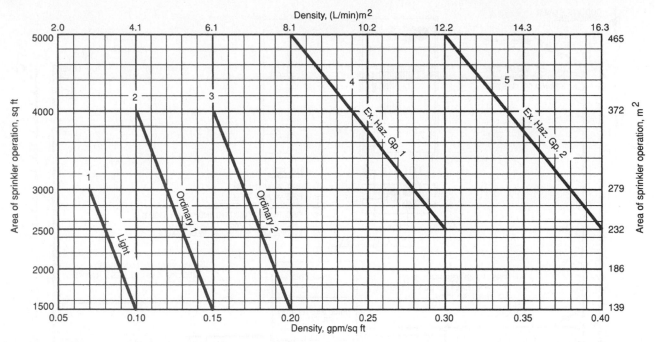

For SI Units: 1 sq ft = 0.0929 m²; 1 gpm/sq ft = 40.746 (L/min)/m².

■ FIGURE 9–13

Area of operations (AOP) versus density demand. (Source: Reprinted with permission from NFPA 13-2007, *Installation of Sprinkler Systems*, Copyright © 2007, National Fire Protection Association, Quincy, MA 02169. This reprinted material is not the complete and official position of the NFPA on the referenced subject, which is represented only by the standard in its entirety.)

TABLE 9–10

Hose stream demand and water supply duration requirements

Hazard Classification	Inside Hose, GPM	Combined Inside and Outside Hose, GPM	Duration in Minutes
Light	0, 50, or 100	100	30
Ordinary	0, 50, or 100	250	60–90
Extra Hazard	0, 50, or 100	500	90–120

- Determine the required flow rate and pressure of the end head (last sprinkler).
- Calculate the residual pressure available from the end head to the water supply main, the required total pressure, and the average pressure loss (psi/ft), including the loss through pipe fittings.
- The sizes of branches, cross-mains, ceiling mains, and underground mains can be determined from the standard chart of friction loss versus flow rate for the selected piping material (steel, copper, etc.), the calculated flow rate (total demand in GPM), the maximum water velocity (fps), and/or the selected friction loss (psi/100 ft). A typical chart for fairly smooth pipes (copper or plastic) is

shown in Figure 8–8 in Chapter 8. Similar charts for other piping materials can be found in most handbooks on water or hydraulics.

In contrast with the design of domestic or HVAC piping systems, which should limit the flow velocity to less than 10 fps, the flow velocity of water in a fire protection piping system is not constrained, since noise and erosion caused by water flow are not of concern during a fire.

For large buildings in which the developed length of pipe main is extensive, the friction loss should be kept low to minimize the total friction loss. Note that the manual pipe-sizing method is useful only for preliminary design calculations. The final design, by computer program, will yield optimum pipe sizes.

Hydraulic calculations for a complete automatic sprinkler system are somewhat tedious, although not complex. Computer software is available through various professional and trade organizations.

9.7.9 Combined Systems

In design applications, a standpipe system and an automatic sprinkler system are often combined into a single system to achieve economy. Figure 9–14 illustrates a combined system for a large building.

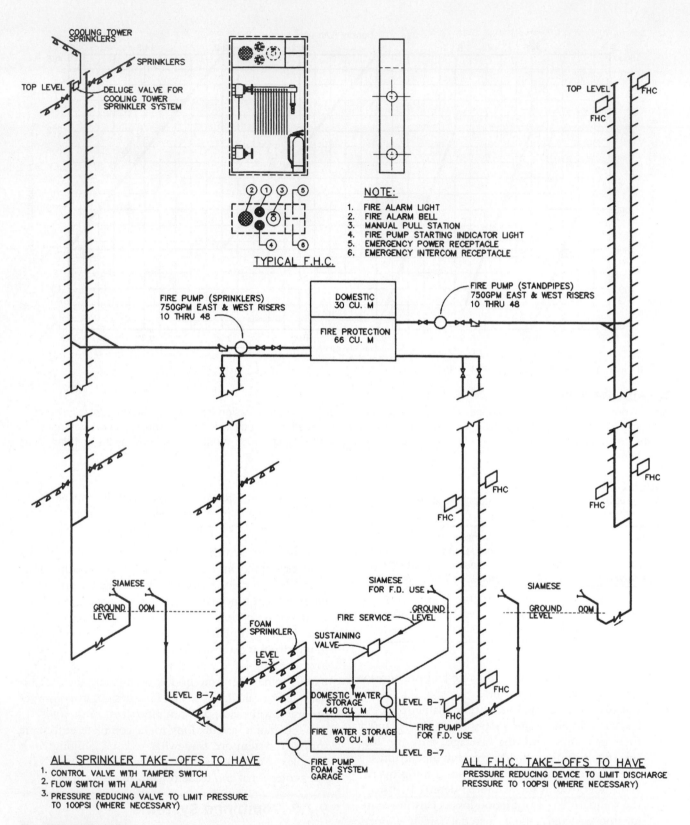

COOLING TOWER SPRINKLERS

SPRINKLERS

TOP LEVEL

DELUGE VALVE FOR COOLING TOWER SPRINKLER SYSTEM

TOP LEVEL

FHC

FHC

NOTE:

1. FIRE ALARM LIGHT
2. FIRE ALARM BELL
3. MANUAL PULL STATION
4. FIRE PUMP STARTING INDICATOR LIGHT
5. EMERGENCY POWER RECEPTACLE
6. EMERGENCY INTERCOM RECEPTACLE

TYPICAL F.H.C.

FIRE PUMP (SPRINKLERS) 750GPM EAST & WEST RISERS 10 THRU 48

FIRE PUMP (STANDPIPES) 750GPM EAST & WEST RISERS 10 THRU 48

DOMESTIC 30 CU. M

FIRE PROTECTION 66 CU. M

FHC

FHC

FHC

SIAMESE

GROUND LEVEL   00M

FOAM SPRINKLER

SIAMESE FOR F.D. USE

FIRE SERVICE

SUSTAINING VALVE

GROUND LEVEL

SIAMESE

GROUND LEVEL   00M

LEVEL B-3

LEVEL B-7

DOMESTIC WATER STORAGE 440 CU. M

LEVEL B-7

FHC

FHC

FHC

FIRE WATER STORAGE 90 CU. M

FIRE PUMP FOR F.D. USE

LEVEL B-7

FIRE PUMP FOAM SYSTEM GARAGE

ALL SPRINKLER TAKE-OFFS TO HAVE

1. CONTROL VALVE WITH TAMPER SWITCH
2. FLOW SWITCH WITH ALARM
3. PRESSURE REDUCING VALVE TO LIMIT PRESSURE TO 100PSI (WHERE NECESSARY)

ALL F.H.C. TAKE-OFFS TO HAVE
PRESSURE REDUCING DEVICE TO LIMIT DISCHARGE PRESSURE TO 100PSI (WHERE NECESSARY)

■ **FIGURE 9–14**

Riser diagram of a combination sprinkler and standpipe system. The system is for a 48-story high-rise building. The system is divided into two vertical zones. The domestic water distribution system shares the same storage tanks, except that the lower part of each tank is reserved exclusively for fire protection purposes and cannot be drawn down by the domestic water pumps.

314

9.8 SMOKE CONTROLS

Smoke is always present when there is a building fire. The degree of smoke generated depends on the combustible material of the fire. Fire from wood and paper generates relatively light smoke, whereas fire from plastic or synthetic materials generates heavy, toxic smoke. It has been proved that loss of life due to smoke is considerably higher than from fire alone. Often, smoke spreads out to great distances from the origin of a fire. Smoke control, therefore, has emerged as an important topic in building design. It requires close coordination between architectural, structural, HVAC, and fire protection systems. The principles of utilizing HVAC systems for smoke control are documented in detail in the *ASHRAE Handbook*.

9.8.1 Stack Effect

Smoke spreads in a building primarily because of two factors: hot air, which makes the smoke rise, owing to its lower density; and pressure differences in the building, which cause air to migrate throughout the building. The following is a direct quotation from the *ASHRAE Handbook*:

> When it is cold outside, air often moves upward within building shafts, such as stairwells, elevator shafts, dumbwaiter shafts, mechanical shafts, or mail chutes. Referred to as normal stack effect, this phenomenon occurs because the air in the building is warmer and less dense than the outside air. Normal stack effect is great when outside temperatures are low, especially in tall buildings. However, normal stack effect can exist even in a one-story building.
>
> When the outside air is warmer than the building air, there is a natural tendency for downward air flow, or reverse stack effect in the building. The pressure difference due to either normal or reverse stack effect frequently exists in shafts. At standard atmospheric pressure, the pressure difference due to either normal or reverse stack effect is expressed as:

$$\Delta p = 7.64 \, (1/T_o - 1/T_i) h \qquad (9\text{--}8)$$

where Δp = pressure difference, inches of water
T_o = absolute temperature of outside air, °R
T_i = absolute temperature of air inside shaft, °R
h = distance above neutral plane, ft

For a building 200 ft tall with a neutral plane at the midheight, an outside temperature of 0°F, and an inside temperature of 70°F, the maximum pressure difference due to stack effect would be 0.22 inch of water. This means that at the top of the building, a shaft would have a pressure of 0.22 inch of water greater than the outside

pressure. At the bottom of the shaft, the shaft would have a pressure of 0.22 inch of water less than the outside pressure.

Stack effect is more serious in cold weather than in hot weather, since the temperature differential between the outdoors and the interior of the building is greater. For example, with maintained at 75°F, the differential is 75°F when the outdoor temperature is 0°F in the winter, and only 20°F when the outdoor temperature is 95°F in the summer. Figure 9–15 illustrates pressure variations in a high-rise building during cold weather. Figure 9–16(a) shows air leaking into a shaft, such as an elevator shaft, when the shaft is vented at the top. Figure 9–16(b) shows an unvented shaft in which air infiltrates the lower floors with diminishing intensity (pressure) and leaks out (exfiltration) with increasing intensity (pressure) above the neutral zone (floor). The total amount of air leaking in must be equal to the total amount of air leaking out. The neutral zone is practically at the midlevel of the building but is normally lower than that because lower floors are usually larger in area, and there are more doors and windows at the ground floor. If a building is perfectly airtight and without exhausts, the air will not travel between floors even with stack effect, however, smoke still may rise to the upper floors because of a buoyancy force or wind pressure. For these reasons, current practice in smoke control depends heavily on the HVAC systems to counteract the natural air movement in the building.

9.8.2 Pressure Control

Air flows only when there is a pressure difference between two areas. The flow is from the area of higher pressure to the area of lower pressure. If the fire area is maintained at a relatively low pressure by exhausting, then air containing smoke will not flow easily to the other areas in the building. The principle can be applied to all types of buildings, low rise or high rise. Following are some common control practices.

Local Exhaust

Exhaust by fans or relief by venting at the floor where fire is started will create low pressure in the fire zone, causing air in the other zones to rush in and thus confine the smoke.

Pressure Sandwich

By the proper control of air supply, return, and exhaust, smoke at the fire zone will have less chance to migrate to the other zones. In general, if the pressure at the fire zone is between 0.1 and 0.15 in. w.c. lower than that at

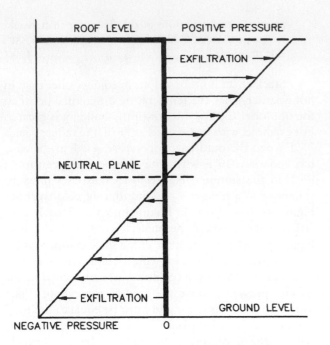

■ **FIGURE 9–15**

Stack effect. During cold weather, outdoor air is heavier than indoor air. Air begins to rise in the building, exfiltrating through the upper floor windows, cracks, or openings. The cold outdoor air is drawn into the building through the lower floors. This is known as the stack effect. If a fire occurs on the lower floor, smoke can easily migrate into the upper floor.

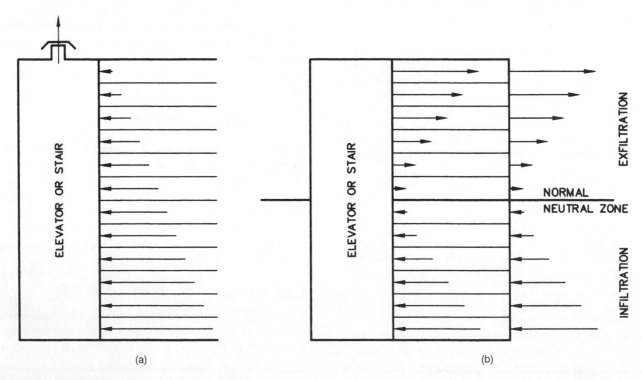

■ **FIGURE 9–16**

Typical building pressure profile. If a building shaft, such as an elevator hoistway, is vented at the top, (a) air will enter the shaft at all floor levels with diminishing amounts at the upper levels, and smoke will be drawn into the shaft during a fire. For this reason, elevator lobbies should have positive air pressure in relation to surrounding spaces, and elevators are deactivated after they are brought down to the ground level. (b) If a shaft is not vented, but not airtight, then air infiltration and exfiltration will behave as in Figure 9–15.

the other zones, smoke can be effectively contained. Figure 9–17 shows the pressure-sandwich principle. This concept is often used in high-rise buildings.

Compartmentation

The building is divided into two or three vertical compartments as if they were separate buildings stacked on top of each other. This practice is used only in buildings taller than about 50 stories. Figure 9–18 illustrates this principle. The drawback is the added cost and inconvenience caused by the required transfer between shafts and stairs of individual zones.

Stair Pressurization

The same concept of pressure differential used between zones applies to stairways, which are the major means of egress from a building. If positive pressure is maintained in stairways by a stair pressurization fan, smoke will not likely migrate into the stairways. Figure 9–19 illustrates a stair pressurization system. The design requires careful analysis and includes the following guidelines:

■ *Performance of the stair pressurization fan* Overpressurization will make it difficult to open the stair doors. A 30-lb force at the door handle is considered the maximum that an average person can

exert. In general, a minimum of 0.05 in. w.c., and not more than 0.25 in. w.c., positive pressure should be maintained.

■ *Capacity of the stair pressurization fan* This depends on the particular analysis and on the assumption as to how many doors may be opened at the same time. Obviously, the more doors are opened at the same time, the more air capacity is required. As a rule, a minimum of three doors, or at least 20 percent of the doors, shall be considered fully open. The amount of air to be delivered for stairs is normally between 15,000 and 25,000 CFM, depending on the above assumptions and the design of the stairway.

■ *Points of air supply* To ensure even pressure at upper levels of the building, a multiple-injection system with air injecting into the stairway at every 5 to 10 floors is preferred. The division of stairs into compartments will further improve the pressure variation at different levels.

Sealing of All Penetrations

All openings for piping, ducts, or structural members in or out of the fire partitions, walls, floors, and shafts are paths of smoke. These openings should be sealed and caulked. Ducts should be equipped with smoke and fire dampers interlocked with the fire protection (signaling and suppression) systems.

Pressure Control in the Elevator Shaft

Normally, exhaust is required at the top of an elevator shaft to relieve any smoke that may leak into the shaft.

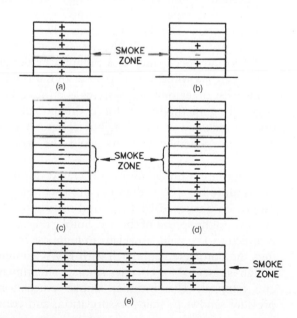

■ FIGURE 9–17

Pressure-sandwich design. The smoke zone is indicated by a minus sign, and pressurized spaces are indicated by a plus sign. Each floor can be a smoke control zone as in (a) and (b), or a smoke zone can consist of more than one floor as in (c) and (d). A smoke zone can also be limited to a part of a floor, as in (e).

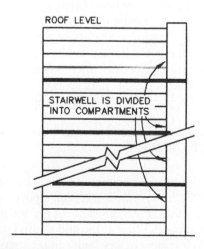

■ FIGURE 9–18

Compartmentation. This can be an effective means of providing stairwell pressurization for very tall buildings when a staged evacuation plan is used. The drawback is the additional landing space required at the transfer floors.

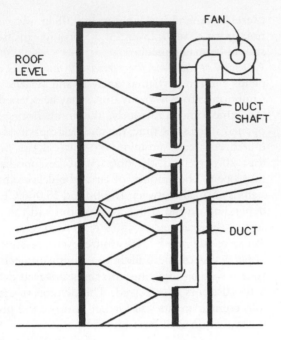

■ FIGURE 9–19

Stair pressurization. Rooftop-mounted fan injects air at every floor. A single point of injection is acceptable for buildings not over 10 stories. The fan may also be mounted at the base of the stairwell.

This concept is, however, in conflict with the pressurization concept for the other areas of the building. Studies indicate that maintaining a positive pressure in the elevator shaft is preferred. This is particularly important for elevators designated for use by firefighters and the physically disabled. Further studies are being conducted by ASHRAE and NFPA for new design guidelines. Present practice is to avoid the use of elevators during a fire. On the other hand, many occupants of a building, such as those in a high-rise hotel, may not be able to walk down many flights of steps, because their physical ability is limited. In such a case, safe-refuge areas or floors may have to be provided until the fire is under control.

9.9 SUMMARY

Although this chapter covers only fundamental principles and guidelines regarding several aspects of fire protection, the information can easily be expanded by using the reference codes listed at the end of this chapter. The following is a review of key issues in planning fire protection systems for buildings.

- Most building and fire protection codes are similar but not identical. While the information presented in this chapter is valid in principle, the designer must be aware of the specific variations in the codes governing the building under consideration.

- Fire protection requirements differ widely for different building use groups. A list of the classifications is given in Table 9–1.

- An effective fire protection system starts with a carefully analyzed operational plan that involves multiple design disciplines, the building management, and the fire department. The operational plan requires the input and implementation of all those mentioned.

- A safe building starts with good architectural layout as to the location and construction of the means of egress.

- Water is the most effective medium for fighting a Class A fire, which is most common among all fires. Adequate quantity and pressure are essential. When the reliability of the water supply is questionable, on-site storage should be considered.

- Halogenated gas is most effective for extinguishing an electrical fire. The halogenated gas must be of a type that will not cause ozone depletion at the upper atmosphere.

- Portable fire extinguishers are economical and effective in extinguishing a fire before it spreads beyond control. Thus they should be provided and installed at more locations than codes require.

- An automatic sprinkler system can be designed using either the pipe schedule method or the hydraulic method. The former is applicable for small light-hazard occupancies only, however, it is very useful for any occupancy as a preliminary design tool when coordination of piping and sprinkler heads with other building systems is critical. The engineering methodology for hydraulic calculations is included in the text, but computer programs are generally used to avoid the tedious calculations.

- Statistics indicate that smoke generated during a fire is more hazardous than the fire itself. Smoke cannot be quenched by water and usually travels far beyond the spread of the fire. Smoke migration is difficult to contain, especially in high-rise buildings. A coordinated approach with HVAC system design is essential for effective smoke containment and removal. This includes smoke evacuation, pressure-sandwich, stair-pressurization, and compartment provisions.

QUESTIONS

9.1 What are the two primary objectives of building fire protection systems?

9.2 Name one of the most important secondary objectives for building fire protection.

9.3 Name the major factors contributing to serious fire hazards in modern buildings.

9.4 What is the normal range of fire protection system costs as a percentage of total building construction cost?

9.5 What is the smoke developed rating (SDR)?

9.6 What are the classes of flame spread rating (FSR)?

9.7 What are the codes governing fire protection in the United States?

9.8 What are the normal classifications of building occupancies?

9.9 What are the four classes of fire, according to NFPA?

9.10 What are the three classes of fire hazard? What is the fire hazard class for office buildings, laboratories, and production facilities?

9.11 Name the operating sequence of a well-planned fire protection system.

9.12 Name 10 fundamental criteria for planning fire safety.

9.13 Does a single-family dwelling require a fire alarm system?

9.14 What is the maximum distance to exit for a hospital protected with sprinklers? Not protected with sprinklers?

9.15 Name as many detection devices in a fire protection system as you can.

9.16 What is the operating principle of an ionization-type smoke detector?

9.17 What is the pressure required for a sprinkler head during its rated flow?

9.18 Name the four types of media used in fire suppression systems.

9.19 What is the maximum floor area that can be served by a portable fire extinguisher with a 2–A rating for a Class A fire hazard?

9.20 What is a Class III standpipe-and-hose system?

9.21 What is the required minimum water flow for standpipe systems?

9.22 Describe the basic automatic sprinkler system.

9.23 What is a Siamese connection?

9.24 What is the building use group classification for an office building? If the building contains a 15,000-sq-ft health club on the second floor, is an automatic sprinkler system required?

9.25 Determine the amount of water that can be delivered by a sprinkler head having a ½-in. orifice with a 5.5 K-factor, and installed in an automatic sprinkler system having 36 psi residual pressure.

9.26 What is the maximum sprinkler protection area for a building classified as light hazard?

9.27 If a room has a soffit 25 in. below the ceiling, what is the minimum distance that a sprinkler can be installed? A beam 4 in. below the ceiling?

9.28 What is the difference between the pipe schedule method and the hydraulic method in sprinkler system design?

9.29 What is the maximum number of sprinklers that can be served by a 3-in. steel pipe in a light-hazard building? (Use the pipe schedule method.)

9.30 Select an area of sprinkler operation (ASO) and the corresponding density demand (DD) for an automatic sprinkler system classified as "ordinary hazard-2." What is the water demand for the sprinkler system (with OVF = 1.1)? If the hose stream demand (HSD) for the standpipe system is 250 gpm, what is the total water demand (TWD) for the combined standpipe and sprinkler system?

9.31 Can a domestic water system and fire protection system be combined in an on-site water storage system? Which system has the priority, and how is this ensured?

9.32 Should smoke control be considered in planning a fire protection system?

9.33 What is stack effect in a building?

9.34 How does one overcome the stack effect of a high-rise building?

9.35 Under what circumstances would a fire pump be included in a fire protection system?

9.36 What types of fire suppression systems would be appropriate for a high-value electrical equipment room?

9.37 What types of fire suppression systems would be appropriate for an open-air loading dock?

9.38 What types of fire suppression systems would be appropriate for an underground parking garage?

REFERENCES

National Fire Codes. Quincy, MA: National Fire Protection Association, 2007.

- NFPA 10, Standards for portable fire extinguishers, 2007
- NFPA 11, Standards for foam extinguishing systems, 2007
- NFPA 12, Standards for carbon dioxide extinguishing systems, 2007
- NFPA 13, Standards for the installation of sprinkler systems, 2007
- NFPA 14, Standards for the installation of standpipe-and-hose systems, 2007
- NFPA 70, National Electrical Code, 2007
- NFPA 78, Lightning Protection Code, 2007
- NFPA 90A, Installation of air-conditioning and ventilating systems, 2007

- NFPA 101, Life Safety Code, 2007

Fire Protection Handbook, 19th ed. Quincy, MA: National Fire Protection Association, 2003.

Building Codes. Includes the current editions of:
- BOCA, Building Officials and Code Administrator's International, Inc.
- SBC, Standard Building Code
- SBCCI, Southern Building Code Congress International, Inc.
- UBC, Uniform Building Code

Fire Resistance Directory. Underwriters Laboratories (UL), Inc.

ASHRAE, *Handbook HVAC applications*, 2007, Chapter 52, "Fire and Smoke Management."

INTRODUCTION TO ELECTRICITY

10

ELECTRICITY IS STEADILY BECOMING THE PREFERRED source of power for lighting, heating, air conditioning, transportation, production equipment, and numerous appliances in all building occupancies. The nominal use of electrical power in offices increased. Statistics indicate that the nominal use of electrical power in offices in terms of Unit Power Density (UPD) has increased from 1 to 3 W/ft² in the 1940s, to 3 to 5 W/ft² in the 1980s, and to 5 to 10 W/ft² in the 1990s. With improved technology and concerted effort in energy conservation designs, the unit power density in modern office building has actually dropped to below the 1990s level while the use of electrical equipment continues to expand.

All matter is made of *atoms*. The atoms contain *electrons* that drift at random through the atomic *structure of matter*. If two dissimilar materials, such as nylon and fur, are rubbed together, the surface of one material will accumulate a surplus of electrons, resulting in a negative overall charge known as *static electricity*. When such a charge, or *electromotive force* (potential), is applied across the two ends of a wire, the electrons within the wire are induced to flow in a certain direction, establishing an electric current. This flow of electrons is known as *current electricity*.

One unique property of electricity is that it can be transmitted through a relatively small space by means of wires and cables. The amount of power that can be transmitted by a particular wire depends on the voltage level at which the transmission occurs. A ½-in. -diameter wire at high voltage—say, 1 million volts—can deliver enough power to serve the needs of a large municipality. The disadvantage of high-voltage power transmission is the inherent danger to human life and property. For efficient utilization with a reasonable degree of safety, utility voltages in the United States are standardized at 120 V for small appliances and up to 480 V for larger equipment. In large buildings, voltages up to 34,000 V have been used directly with spaces designed to accommodate them.

The flow of electrical current (and thus power and energy) is proportional to the available electrical voltage. Electrical systems with voltages below some 30 V are not easy to pass through the human body and thus are relatively safe to touch. On the other hand, household electrical systems at 120 V (used in the United States) and 240 V (used in Europe and some other countries) may cause shocks or even death. For safety reasons, electrical wiring and equipment are strictly regulated. The primary code governing electrical systems and installations in the USA is the National Electrical Code (NEC).

Building electrical systems are normally designed by electrical engineers with specialized knowledge of such systems; however, every member of the building design profession must have a fundamental knowledge of electricity and its operating principles, characteristics, and limitations. The electrical terms, equipment, and units listed in Table 10–1 should be considered as the minimum knowledge for architects, engineers, contractors, and building managers.

The most effective means of illustrating an electrical design or electrical circuit is to use symbols and schematic diagrams. Table 10–2 shows some symbols commonly used for basic electrical circuitry. Additional symbols for electrical equipment and wiring are introduced in Chapters 11, 12, and 13.

10.1 DIRECT CURRENT

Direct current (DC) is an electric current that flows through a circuit in only one direction, although the rate of flow may vary.

10.1.1 Basic Properties of Electricity

The *electric charge* or *quantity of electricity* (Q) is the basis of electricity. It is expressed in coulombs (C). One coulomb is equivalent to 6×10^{18} electrons. This term is rarely used in practical applications. Instead, watt-hour/(Wh) and kilowatt-hour/(kWh) are used. (See Section 10.1.5.)

TABLE 10–1
Electrical terms, equipment, and units

Electrical Fundamentals

1. Alternating current (AC)	17. Contacts, momentary	54. Power, apparent
2. Capacitance	18. Controllers	55. Power, reactive
3. Current (*I*)	19. Demand (power)	56. Power, working
4. Direct current (DC)	20. Demand charge	57. Power, peak
5. Electricity	21. Demand load	58. Raceways
6. Electromagnetic	22. Diagrams, connection	59. Resistors
7. Energy	23. Diagrams, one line	60. Root mean square (RMS)
8. Frequency	24. Diagrams, riser	61. Short-circuit capacity
9. Ground	25. Diagrams, schematic	62. Single phase
10. Impedance	26. Disconnects	63. Switchboard
11. Kirchhoff's laws	27. Diversity	64. Switches, 3-way, 4-way
12. Magnetic Induction	28. Electric metallic tubing (EMT)	65. Switches, double pole
13. Ohm's law	29. Energy, watt-hour (Wh)	66. Switches, double throw
14. Parallel circuits	30. Exist signs	67. Three-phase
15. Power	31. Fault current	68. Transformers
16. Reactance	32. Feeders	69. Twisted pairs
17. Resistance (*R*)	33. Fiber optics	70. Underfloor ducts
18. Series circuits	34. Fuses	71. Voltage drop
19. Voltage (*V* or *E*)	35. Generator	72. Voltage spread
	36. Ground fault interruptors (GFIs)	73. Voltage, equipment
Electrical Systems	37. Ground	74. Voltage, system
1. Alternators	38. Grounding (system, equipment)	
2. American wire gauge (AWG)	39. Interrupting capacity	**Units of Measure/Quantities**
3. Ampacity	40. Lighting fixtures (luminaries)	1. Ampere (A)
4. Auxiliary systems	41. Load factor	2. British thermal unit (Btu)
5. Boxes (junction, outlet)	42. Loads, connected	3. Horsepower (hp)
6. Branch circuit	43. Loads, demand	4. Kilovolt-ampere (KVA)
7. Bus bars	44. Loads, inductive	5. Kilowatt (KW)
8. Bus duct	45. Loads, resistive	6. Volt-ampere (VA)
9. Capacitors	46. Meters, ammeter	7. Volt (V)
10. Cellular floor	47. Meters, volt	8. Watt (W)
11. Circuit breakers	48. Meters, watt-hour	9. Ohm
12. Circular mill (CM)	49. Motor	10. Megawatt (MW)
13. Coax cables	50. Overcurrent	11. Watt-hour (Wh)
14. Conductors	51. Overcurrent protection	12. Kilowatt-hour (KWh)
15. Contactors	52. Panelboards	
16. Contacts, maintained	53. Power factor	

Current (I) is the flow of electricity in an electrical circuit, or the rate of electron flow of 1 coulomb per second. Conventionally, the current flow is from positive to negative voltage, even though the electrons that make up the current flow are from negative to positive voltage. This convention is still the standard in applications such as marking dry-cell or automotive batteries. The positive terminal is considered the supply side, the negative terminal the return side.

The unit of current is the *ampere (A)*. It is defined as flow of one coulomb per second. Smaller units are the *milliampere* (mA) or *microampere* (μA).

Voltage (*V* or *E*) is the electromotive force (EMF) or potential difference that causes an electric current to

flow. The symbols *E* or *V* are used interchangeably, depending on the application. The unit of voltage is the *volt* (V).

Resistance (*R*) is an internal property of matter that resists the flow of electric current. A material with low resistance to electrical flow is called a *conductor;* a material with high resistance to electrical flow is called an *insulator.* Typical insulators are paper, rubber, neoprene, and other synthetic materials. Their selection depends on application, voltage class, physical requirements, etc. There is no rigid demarcation between conductors and insulators. Typical electrical conductors are pure metals. In the order of increasing conductivity per unit volume of material, they are aluminum, copper, silver, gold, and

TABLE 10–2
Basic symbols for electrical circuits*

———————	CONDUCTOR, WIRE	—⋀⋀⋀—	RESISTANCE
	WIRE CROSSING (NOT CONNECTED)	—∿∿∿—	REACTANCE
	WIRE CROSSING (CONNECTED)	—⊣ ⊂—	CAPACITANCE
—⊦⊦⊦⊦—	BATTERY, (DC SOURCE)	—Ⓒ—	COIL (MAGNETIC)
—⊦⊦⊦⊦	GROUNDING	—Ⓖ—	GENERATOR
	SWITCH (OPEN, CLOSE)	—Ⓜ—	MOTOR
—⊣ ⊦^{NO}—	CONTACT (NORMALLY OPEN)	—[LOAD]—	LOAD
—⊬^{NC}—	CONTACT (NORMALLY CLOSE)	—⋛⋚—	TRANSFORMER

*Additional symbols for electrical equipment and wiring are included in later chapters.

platinum. Aluminium is cheap, but it is too soft to be used as fine wire. Silver and gold are too expensive to use for general wiring purposes. This leaves copper as the preferred metal for most applications. Materials with limited conductivity are called *semiconductors*. Typical semiconductors are silicon and germanium. The unit of resistance is the ohm (Ω), named for the scientist George Simon Ohm, who discovered the law also named for him.

Capacitance (C) is the property by virtue of which an electrical circuit or a component stores energy. Capacitance provides an opposition force in a circuit to delay the change of voltage of the circuit. Capacitance is present in every component of an electric circuit when there is dielectric material in the component, such as plastic, paper, rubber, liquids, and even air. Devices specially designed to store electrical charge are called *capacitors*.

Units of capacitance are the farad (f), microfarad (μf), and picofarad (pf). One microfarad is one-millionth of a farad, and one picofarad is one-millionth of a microfarad.

10.1.2 Ohm's Law

In 1827, George Simon Ohm discovered a simple, but very important, relationship among the current, voltage, and resistance of a DC circuit. This relationship may best be visualized by comparing the electric circuit with a hydraulic circuit, as illustrated in Figure 10–1. In a hydraulic circuit, the rate of water flow is directly proportional to the pressure differential produced by the pump and inversely proportional to the resistance of the pipes and valves. Analogously, in an electric circuit, the electric current flowing through the circuit is

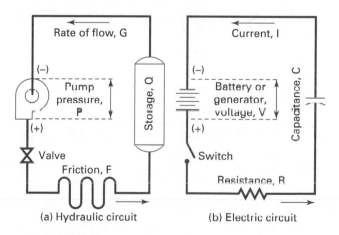

■ FIGURE 10–1
Comparison between (a) hydraulic circuit and (b) electric circuit.

directly proportional to the voltage delivered by the battery and inversely proportional to the electrical resistance inherent in the wires and other components. This relationship, known as *Ohm's law*, may be expressed mathematically.

$$I = E/R \qquad (10\text{–}1)$$

Example 10.1 In a 24-V DC circuit, the flow of current is measured as 8 As. What is the resistance of this circuit?
  Answer From Eq. (10–1), $R = E/I = 24/8 = 3$ ohms.

Example 10.2 What is the current flow in a 12-V DC circuit containing a total resistance of 2 ohms?
  Answer From Eq. (10–1), $I = E/R = 12/2 = 6$ As.

10.1.3 Kirchhoff's Laws

Two other relationships important in the study of electrical circuits are Kirchhoff's laws. According to Kirchhoff's *first law*, the algebraic sum of all electric currents into and out of any junction of an electric circuit is zero. In other words, the total current flowing into a junction must be equal to the total current flowing out of the junction. Using the hydraulic circuit as an analogy, if water flowing into a pipe flows out of two branch pipes, the total outflow from both pipes must be equal to the water flowing into the pipe. (See Figure 10–2.) Mathematically, Kirchhoff's first law may be expressed as

$$\sum I = 0 \qquad (10\text{–}2)$$

According to Kirchhoff's *second law*, the algebraic sum of all votages measured around a closed path in an electric circuit is zero. Again using the hydraulic circuit as an analogy, in a closed pump-and-load circuit the total rising pressure due to the pump must be equal to the total dropping pressure due to the loads. (See Figure 10–3.) Algebraically,

$$\sum E = 0 \qquad (10\text{–}3)$$

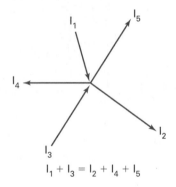

$$I_1 + I_3 = I_2 + I_4 + I_5$$

■ **FIGURE 10–2**
Kirchhoff's first law.

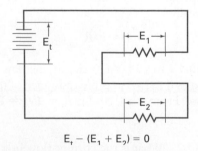

$$E_t - (E_1 + E_2) = 0$$

■ **FIGURE 10–3**
Kirchhoff's second law.

10.1.4 Resistance in Electric Circuits

Resistance used in electric circuits may be arranged in series, in parallel, or in a combination of both. The mathematical relationship of the total circuit resistance to the individual resistances may be derived from Ohm's law and Kirchhoff's laws for either the series or the parallel arrangement.

Series Circuit

Here, the resistances are arranged one after another, or end to end. (See Figure 10–4.)

$$R_t = R_1 + R_2 + R_3 \ldots + R_n \qquad (10\text{–}4)$$

Example 10.3 An electrical circuit consists of one 5-ohm resistance and one 10-ohm resistance in service. What is the total resistance of this circuit?
Answer From Eq. (10–4),

$$R_t = R_1 + R_2 = 5 + 10 = 15 \text{ ohms.}$$

Parallel Circuit

Resistances may be connected side by side and joined at each end as shown in Figures 10–5(a) and 10–5(b).

- For two resistances in parallel, the total resistance is

$$1/R_t = 1/R_1 + 1/R_2 \qquad (10\text{–}5a)$$
$$R_t = (R_1 \times R_2)/(R_1 + R_2) \qquad (10\text{–}5b)$$

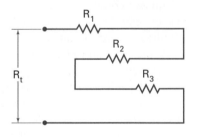

$$R_t = R_1 + R_2 + R_3$$

■ **FIGURE 10–4**
Resistances in series.

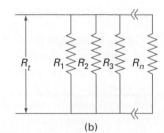

(a)   (b)

■ **FIGURE 10–5**
(a) Two resistances in parallel; (b) More than two resistances in parallel.

- For more than two resistances in parallel, the total resistance is

$$1/R_t = 1/R_1 + 1/R_2 + 1/R_3 \ldots + 1/R_n (10\text{-}5c)$$

When there are more than two resistances connected in parallel, it is more convenient to solve the circuit by reducing the two adjacent resistances into a single equivalent resistance and continuing to reduce the remaining parallel resistances into additional equivalent resistances unit the circuit is reduced into one total resistance. The diagram that keeps track of each step of reduction is known as a '*ladder diagram*,' as illustrated in Figure 10–6 of Example 10.5.

Example 10.4 What is the total resistance of this circuit if the two resistances, 5 and 10 ohms, are connected in parallel?

Answer From Eq. 10–5(a),

$$1/R_t = 1/5 + 1/10 = 0.2 + 0.1 = 0.3,$$

Thus,

$$R_t = 1/03 = 3.34 \text{ ohms}$$

Example 10.5 Three resistances are connected in parallel. The values are 10, 20, and 30 ohms for R_1, R_2, and R_3, respectively. Determine the total resistance of the circuit.

Answer The ladder diagrams of the circuit are shown in Figures 10–6(a) and 10–6(b) illustrating the step-by-step simplification process.

$$R_4 = (R_1 \times R_2)/(R_1 + R_2) = 200/30 = 6.67 \text{ ohms}$$

$$R_t = (R_4 \times R_3)/(R_4 + R_3) = 200.1/36.67$$

$$= 5.46 \text{ ohms}$$

Series-Parallel Circuit

When resistances are in a combination series and parallel arrangement, the circuit can be resolved by the simplification process illustrated in Example 10.6.

Example 10.6 The three resistances, 30, 20, and 10 ohms, are so arranged that R_1 and R_2 are in parallel, and the combination is in series with R_3, See Figure 10–7. What is the total resistance?

Answer

$$R_4 = (R_1 \times R_2)/(R_1 + R_2) = (30 \times 20)/(30 + 20)$$

$$= 600/50 = 12 \text{ ohms}$$

$$R_t = (R_4 + R_3) = 12 + 10 = 22 \text{ ohms}$$

The ladder diagram is also called an *elementary diagram*, which is the simplest way to illustrate the operating principle of an electrical circuit. For example, the circuit for an electric motor should include a power supply and return, a disconnecting switch, and an overload protection device, such as a switch with a fuse or an automatic circuit breaker, to protect the power supply wires (feeder) and the interior wires within the motor (magnetic windings) from being overheated. The elementary diagram shown in Figure 10–8 clearly illustrates the basic components and their operating principles. For any electrical equipment to work, there must be an electrical potential or voltage at an appropriate level between the power supply and return lines. Chapter 13, "Electrical Design and Wiring," discusses these issues.

10.1.5 Power and Energy

Energy exists in many forms, such as mechanical, sound, light, electrical, nuclear, and chemical energy. All energy can be converted from one form to another, and all energy eventually degrades into heat energy. An overview of energy basics is given in Section 1.1 of Chapter 1.

Of all forms of energy, electrical energy is perhaps the most convenient to use. It is readily convertible to other forms: for example, to mechanical energy through a motor, to lighting energy through a lamp, and to heating energy through a resistance heater.

Power is the *rate* of consuming energy. Thus, time is a crucial element in the consumption of energy: A large

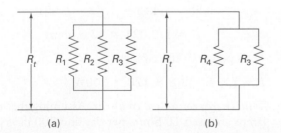

(a)   (b)

■ FIGURE 10–6

Circuit and ladder diagrams of the parallel resistance circuit for determining the total resistance of Example 10.5.

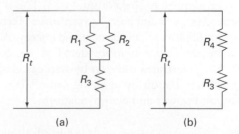

(a)   (b)

■ FIGURE 10–7

Circuit and ladder diagrams for Example 10.6.

■ **FIGURE 10–8**
■ **FIGURE 10–8**

An elementary diagram illustrating the essential components and operating principle of an electrical motor, including the necessary operating and protection devices.

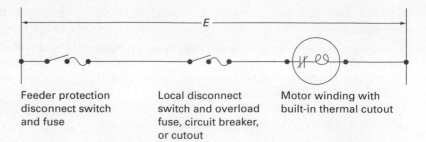

Feeder protection disconnect switch and fuse

Local disconnect switch and overload fuse, circuit breaker, or cutout

Motor winding with built-in thermal cutout

power load may consume less energy than a smaller power load if it is used over a shorter time. For example, a 1000-W lamp operating for 1 hour consumes 1000-Wh of energy. However, a 40-W lamp operating for 30 hours consumes 1200 Wh of energy. In this instance, the small 40-W lamp actually consumes more energy than the large 1000-W lamp.

Power is the product of voltage and current; units of power are the watt (W) or kilowatt, 1000 W (kW).

$$P = \text{power} = \text{voltage} \times \text{current}$$

$$P = E \times I \qquad (10\text{–}6)$$

or

$$P = (I \times R) \times I = I^2 R \qquad (10\text{–}7\text{a})$$

or

$$P = E \times \left(\frac{E}{R}\right) = E^2/R \qquad (10\text{–}7\text{b})$$

Energy is the product of power and time; units of energy are the watt-hour (Wh) or the kilowatt-hour, 1000 watt-hours (kWh)

$$W = \text{Energy} = \text{Power} \times \text{time}$$

$$W = p \times t = I^2 \times R \times t \qquad (10\text{–}8)$$

The terms *power* and *energy* are often misunderstood, even by professionals. The operation of an automobile is a good example. The power of a car is rated in horsepower (hp). (1 hp is equivalent to 746 watts.) It burns gasoline as the source of energy, but the amount of energy it consumes depends on how fast and how far the car is being driven, and the state of maintenance of the engine. The driver can manage the power of the engine to economize its fuel consumption. This is known as *power management* and *energy conservation*. Techniques and practices of power management and energy conservation will be discussed in more detail in Chapter 13. Example 10.7 illustrates the step-by-step calculations of an electrical installation by applying Ohm's law, Kirchhoff's law, and power and energy relations.

Example 10.7 As a practical example of an electrical power application, suppose that one 100-W lamp and one 200-W lamp are plugged into a 120-V circuit. For

the purpose of this example, the power system may be either DC or AC. The two lamps are connected in parallel. Calculate the current flow through each lamp, the total current in the circuit, the resistance of each lamp, the total resistance of the circuit, the total energy consumed in a year, and the cost of electrical energy for the year (based on $0.10/kWh).

- Current flow through each lamp
    From Eq. (10–6), $P = I \times E$, or $I = P/E$. So

    $$I_1 = \text{current through lamp 1}$$
    $$= 100\ \text{W}/120\ \text{V} = 0.83\ \text{A}$$

    $$I_2 = \text{current through lamp 2}$$
    $$= 200\ \text{W}/120\ \text{V} = 1.66\text{A}$$

- Total current through the circuit
    From Kirchhoff's first law,

    $$I = I_1 + I_2 = 0.83 + 1.66 = 2.49\ \text{A}\ (\approx 2.5\ \text{A})$$

- Resistance of lamps
    From Ohm's law, $I = E/R$, or $R = E/I$
    Thus,

    $$R_1(\text{lamp 1}) = 120/0.83$$
    $$= 145\ \text{ohms}$$

    $$R_2(\text{lamp 2}) = 120/1.66$$
    $$= 72\ \text{ohms}$$

- Total resistance of the circuit
    From Eq. (10–5a), $1/R = 1/R_1 + 1/R_2$

    $$1/R = 1/145 + 1/72 = 0.021$$
    $$R = 1/0.021 = 47.6\ \text{ohms}$$

- Total power of the circuit

    $$P = I \times E = 2.5 \times 120 = 300\text{W}$$

    Total energy consumed in a year. Assuming that the lamps are used 10 hours per day and 200 days per year.

    $$W = p \times t = (100 + 200) \times 10 \times 200$$
    $$= 600,000\ \text{Wh} = 600\ \text{kWh}$$

- *Cost of electrical energy for the year.*

$$\text{Cost} = \text{energy used} \times \text{energy rate}$$
$$= kWh \times \$/kWh$$
$$= 600 \times 0.10 = \$60 \text{ per year}$$

10.2 DIRECT CURRENT GENERATION

One way to generate direct current is with batteries that convert chemical energy into electrical energy. Another method is with a generator that converts mechanical energy into electrical energy. This process involves the principle of *electromagnetism*. When electrical current flows in a wire, it induces a magnetic field around the wire, as shown in Figure 10–9(a). Conversely, if a wire moves across a magnetic field, then an electrical voltage (and thus a current) will be induced in the wire. Figure 10–9(b)

illustrates the operating principle of a DC generator that contains a stator and a rotor. The stator consists of a permanent magnet or an electromagnet; the rotor consists of multiple coils of wire that rotate between the magnetic poles. (For simplicity, only a single coil is shown in the illustration.) When the coil rotates between the magnetic poles, it cuts the magnetic field at various angles and thus generates a variable electrical voltage. The generated voltage follows the form of a sinusoidal curve. During the first 180° of the rotation, the voltage generated in the coil starts from zero, goes to a peak value, and then drops to zero. During the second 180° turn (180° to 360°), another sinusoidal curve is generated, but with reverse polarity. When the ends of the rotating coil are connected to an external wiring system through segmented slip rings known as the *commutator*, the voltage becomes rectified; i.e., the polarity remains unchanged. The resulting voltage (and current) characteristics are shown in Figure 10–9(c).

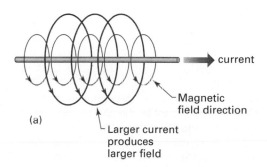

(a)

current

Magnetic field direction

Larger current produces larger field

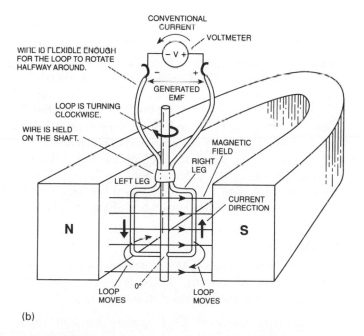

(b)

CONVENTIONAL CURRENT

VOLTMETER

WIRE IS FLEXIBLE ENOUGH FOR THE LOOP TO ROTATE HALFWAY AROUND.

– V +

GENERATED EMF

LOOP IS TURNING CLOCKWISE.

WIRE IS HELD ON THE SHAFT.

MAGNETIC FIELD

RIGHT LEG

LEFT LEG

N

CURRENT DIRECTION

S

LOOP MOVES

0°

LOOP MOVES

(c)

+ Voltage or Current

Peak @ 1.0

Mean @ 0.707

0° 90° 180° 270° 360° 450° 540°

Time (in Degrees Rotation) →

■ **FIGURE 10–9**

(a) Electromagnetic induction: a magnetic field is induced surrounding the flow of an electrical current. (b) Conversely, when an electrical conductor (coil) moves (rotates) across a magnetic field, EMF (voltage) is induced and the flow of electricity (current) is rectified through a commutator. (c) Voltage and current are cyclic in nature, complying with the sinusoidal relations. This cycle repeats itself every 180° rotation of the coil. (Reproduced with permission from *Building Power Supplies.* © 1991 by Master Publishing, Inc.)

10.3 ALTERNATING CURRENT (AC)

An alternating current (AC) system is an electrical system in which voltage and current are reversed periodically or cyclically in the circuit; i.e., the voltage alternates in polarity between the conductors, and so does the current. Nearly all power provided by electrical utilities in the United States is through AC systems, in lieu of the DC systems that were the standard a long time ago. The advantages of AC over DC will be discussed later.

The operating principle in an AC generator is the same as in a DC generator except that the connection between the generator and the external electrical system is through smooth slip rings, without the use of a segmented commutator. Figure 10–10 illustrates the operation of an AC generator (also known as an *alternator*).

10.3.1 Basic Properties of Alternating Current

Because current reverses its direction of flow rapidly in an alternating circuit, there are some properties unique to AC.

AC Resistance

In an AC circuit, the reversing flow of electrons (current) tends to concentrate near the outer diameter of the conductor, resulting in increased resistance of the conducting material compared with a DC circuit. This phenomenon is known as the *skin effect*, and the effective resistance is called *AC resistance*. It is also expressed in ohms (Ω).

Figure 10–11 illustrates the voltage and current relationship of an electrical system containing pure resistance. In a pure resistive circuit, the current and voltage waves increase in unison or in phase.

Reactance

Reactance, which causes AC resistance, opposes the reversing flow of current. There are two types of reactance; each is also expressed in ohms (Ω).

- *Inductive reactance (X_L)* appears through the presence of an electromagnetic field of either an electromagnet or a permanent magnet in the equipment, as in a motor, a relay, or a ballast for lights. Inductive reactance has a delaying effect on current in relation to the voltage. Thus the current lags the voltage. Since most buildings contain lots of motors and electromagnetic devices, building electrical systems are usually more inductive than other electrical systems. Inductive reactance is expressed mathematically as

$$X_L = 2\pi f L \qquad (10\text{--}9)$$

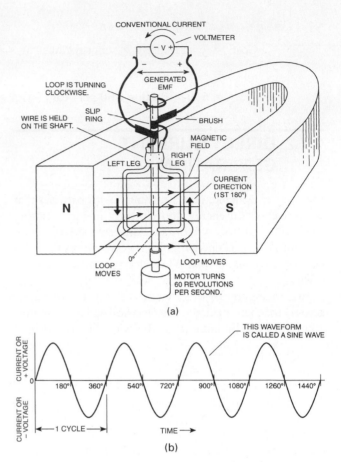

(a)

(b)

■ FIGURE 10–10

(a) The operating principle of an AC generator is identical to that of a DC generator except for the final connection between the generator and the electrical system. The alternator has smooth slip rings whereas a DC generator has segmented slip rings, or a commutator. (b) The generated voltage is cyclic and reverses in polarity every 180° rotation of the coil. The flow of current is similar. For each revolution of the coil, it completes an electrical cycle. The frequency in number of cycles per second is called hertz (Hz). The standard frequency in the United States is 60 Hz; 50 Hz is the standard in most European and Asian countries. (Reproduced with permission from *Building Power Supplies.* © 1991 by Master Publishing, Inc.)

where f = frequency, in hertz (Hz)
and L = inductance, in henrys (H)

Figure 10–12(a) illustrates the voltage and current relationship of an electrical system containing pure inductive reactance with current lagging the voltage wave by 90°.

- *Capacitive reactance (X_C)* is created by the presence of capacitance in the equipment, such as insulation on cables and devices. Capacitive reactance has a storage effect on current in relation to voltage. Thus the voltage lags the current, or the current leads the voltage. In general, the capacitive

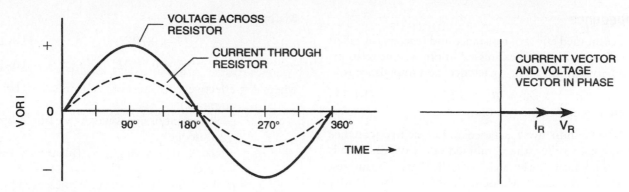

■ **FIGURE 10–11**

Current and voltage relationship in a resistive circuit. The vector angle between current and voltage quantities is zero.

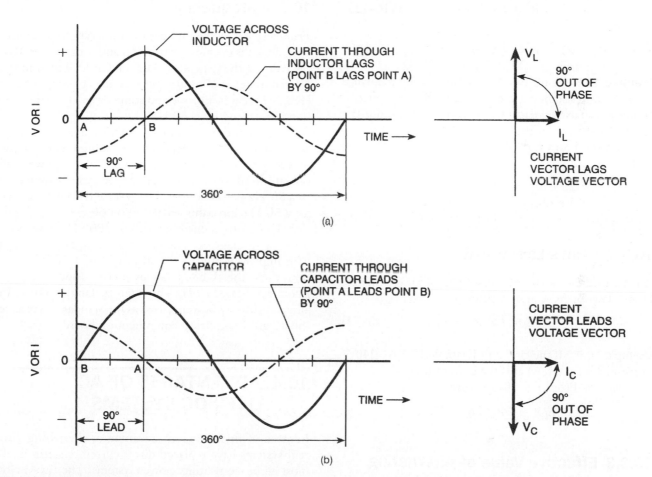

■ **FIGURE 10–12**

Voltage and current relationship in a reactive circuit. (a) Inductive circuit where voltage leads current by 90°; (b) capacitive circuit where current leads voltage by 90°. (Reproduced with permission from *Building Power Supplies.* © 1991 by Master Publishing, Inc.)

reactance of building components is quite small and can be neglected unless the system is equipped with capacitors to overcome the effect of inductive reactances in the system. Algebraically, we have

$$X_C = \frac{1}{2\pi f C} \qquad (10\text{–}10)$$

where f = frequency, in hertz (Hz)
and C = capacitance, in farads (F)

- Figure 10–12(b) illustrates the voltage and current relationship of an electrical system containing pure capacitive reactance, with current leading the voltage wave by 90°.

Impedance

The combined effect of resistance and reactance is called *impedance* and is also expressed in ohms. The relationship between resistance, reactance, and impedance is

$$Z^2 = R^2 + X^2 \qquad (10\text{-}11)$$

where Z = impedance, in ohms

Since inductive reactance and capacitive reactance oppose each other, the combined effect of reactance is the numerical difference of the two quantities, whichever is greater. Mathematically, Eq. (10–11) may be written as (10–12a) or (10–12b) for a circuit containing both inductive reactance and capacitive reactance.

$$Z^2 = R^2 + (X_C - X_L)^2 \qquad (10\text{-}12a)$$

or

$$Z^2 = R^2 + (X_L - X_C)^2 \qquad (10\text{-}12b)$$

Example 10.8   A 60-H_Z AC circuit has the following properties. Resistance = 5 ohms, inductive reactance = 10 ohms, capacitive reactance = 3 ohms. Determine the overall impedance of the circuit.
   Answer From Eq. (10–12a),

$$Z^2 = R^2 + (X_C - X_L)^2 = 25 + (-7)^2$$
$$= 25 + 49 = 74$$
$$Z = 8.6 \text{ ohms}$$

10.3.2 Ohm's Law for AC

Since impedance is the AC equivalent of DC resistance, Ohm's law for AC can be written as

$$E = I \times Z \qquad (10\text{-}13)$$

Example 10.9   If the circuit in Example 10.8 is a 240-V single-phase circuit, what is the current flow?
   Answer From Eq. 10–13, current

$$I = E/Z = 240/8.6 = 27.9 \text{ A}$$

10.3.3 Effective Value of Alternating Current and Voltage

The values of current and voltage in a typical alternating current system varies as the sine wave. Questions arise as to the representative value of the current or voltage in such a system. To compare alternating current with direct current, the *effective* values of current and voltage are used. These correspond to 0.707 (reciprocal $\sqrt{2}$) times their maximum values. The effective values used are frequently referred to as *root-mean-square* (rms) values of current and voltage.

Mathematically,

$$I = 0.707 I_p \qquad (10\text{-}14)$$
$$E = 0.707 E_p \qquad (10\text{-}15)$$

where I = effective rms current, A (measured by an ammeter)
   I_p = peak current of sine wave, A (measured by an oscilloscope)
   E = effective rms voltage, V (measured by a voltmeter)
   E_p = peak voltage of sine wave, V (measured by an oscilloscope)

10.3.4 Frequency

Frequency (f), or the number of cycles repeated per second, depends on the construction and rotation speed of the rotor of the generator. The standard adopted in the United States is 60 cycles per second (cps) or 60 hertz (Hz), whereas in Europe and some other countries, the standard frequency is 50 Hz.

The speed of rotation of electrical apparatus, such as motors or clocks, depends on the frequency of the power system. A motor designed for 60-Hz power will operate at 50/60 of its designed speed when connected to a 50-Hz power source. The reverse ratio will be true for a 50-Hz apparatus operating on 60 Hz.

Higher frequencies of 400 to 3000 Hz have been used for specially designed fluorescent lighting systems with improved efficiency and economy. In fact, utility power in the United States may adopt a higher frequency—say, 120 Hz—as a standard in the future. In aircraft, 400-Hz power is used as the standard to reduce the weight of electrical equipment on board.

10.4 ADVANTAGES OF AC OVER DC SYSTEMS

Because of their inherent advantages, alternating current systems have replaced direct current systems in almost all modern utility power systems. The two major advantages of AC over DC systems are as follows:

1. *Lower generating cost.* An AC alternator is simple to construct, with no need for a complicated split ring commutator. Thus, the AC alternator permits a higher speed of rotation with lower maintenance cost and a lower cost of power generation.
2. *Easier voltage transformations.* The voltage of an AC system can be changed by the use of a simple piece of electromagnetic equipment known as a *transformer*. The transformer permits the transmission of large

amounts of electrical power at higher voltages for long distances with relatively small transmission lines and thus a lower cost of transmission.

For example, utility power is normally generated at between 4000 to 25,000 V at the generating plant and transmitted at 10,000 to 70,000 V for distribution, but it may be higher than 300,000 V for long-distance transmission. These transmission voltages are then stepped down to around 3000 to 4000 V at local substations for distribution to users or buildings. The distribution voltage is then stepped down again at the users' premises to utilization levels, such as 120, 208, 240, 277, and 480 volts, as appropriate for the connected loads.

Figures 10–13(a) and 10–13(b) illustrate the transmission of electrical power from the utility company's generating station through various transmission substations and terminating at the users' properties. Figure 10–13(a) depicts the major transmission components at various stages of transmission. One should visualize that a large utility power system may cover hundreds of miles. Figure 10–13(b) is a simplified line diagram with the nonessential components deleted.

10.5 AC-TO-DC CONVERSION

While AC is the dominant power system used for buildings, DC is needed for electronic equipment—for example, television receivers, computers, batteries, precision controls, and special power equipment, such as elevators and industrial processes. DC power for this type of equipment needs to be connected to an independent DC source or an AC-to-DC converter. AC converted to DC by a rectifier, and DC converted to AC is by an inverter.

10.6 SINGLE-PHASE VERSUS THREE-PHASE ALTERNATOR

A single-phase generator is an alternator with a single set armature coil producing a single voltage waveform. A three-phase alternator has three sets of coils spaced 120° apart and generates three sets of voltage waveforms. The three-phase alternator is the alternator most commonly used in the United States. Rarely used are two-phase and six-phase alternators. When more than one coil is wound in the armature of the alternator, each coil will generate its own independent voltage. Figure 10–14 illustrates the phase relations of a three-phase power system generated by a three-phase alternator.

10.7 POWER AND POWER FACTOR

In a single-phase AC circuit that contains only resistance, the current and voltage will be *in phase*, and the power consumed will simply be the product of voltage and current, as in a DC circuit. In a three-phase circuit that contains only resistance, power is 1.73 times the line voltage and current because the separate powers of the three phases are staggered.

In an AC circuit that contains reactance in addition to resistance, the current and voltage will be out of phase by an angle known as the *power angle* and designated by the Greek letter theta (θ).

For single- and three-phase AC systems, the relationships between power (P_a, P, P_r), line current (I), and line voltage (E) are explained in the next few sections.

10.7.1 Apparent Power

The apparent power P_a in volt-amperes (VA) is the product of voltage and current, which may or may not be in phase with each other. For single-phase power

$$P_a = E \times I \qquad (10\text{--}16)$$

For three-phase power

$$P_a = 1.73 \times E \times I \qquad (10\text{--}17)$$

10.7.2 Reactive Power

The reactive power P_r, in volt-amperes reactive (VAR), is the component of AC power that does not perform useful work because power is alternately stored and released in the reactive components. Thus the net useful or true power is zero. For single-phase power

$$\begin{aligned} P_r &= E \times I \times \sin\theta \\ &= P_a \times \sin\theta \end{aligned} \qquad (10\text{--}18)$$

For three-phase power

$$\begin{aligned} P_r &= 1.73 \times E \times I \times \sin\theta \\ &= 1.73 \times P_a \times \sin\theta \end{aligned} \qquad (10\text{--}19)$$

10.7.3 Working Power

The working power P, in watts (W), is the component of AC power that performs useful work. In practical applications, *power P*, always refers to working power. For single-phase power

$$\begin{aligned} P &= E \times I \times \text{PF} \\ &= E \times I \times \cos\theta \end{aligned} \qquad (10\text{--}20)$$

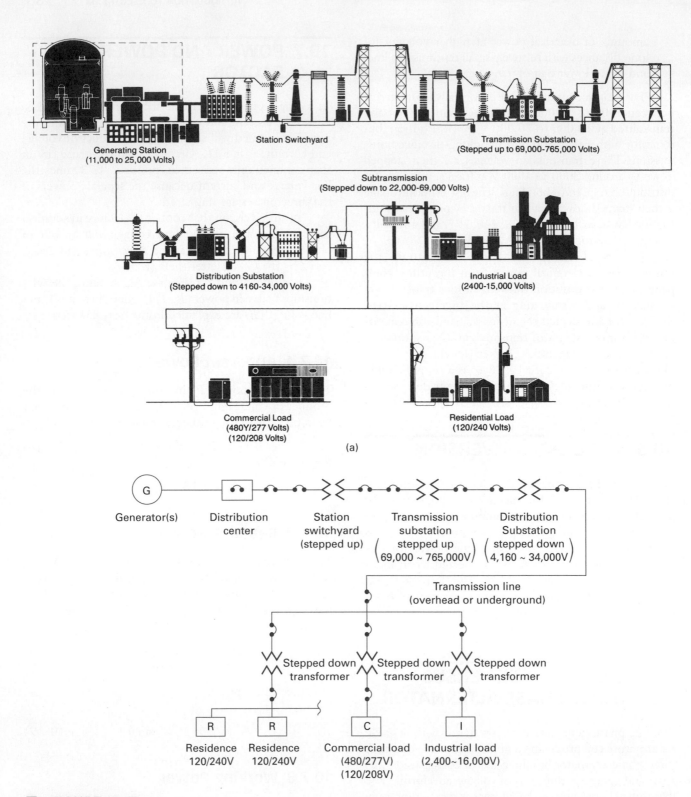

■ FIGURE 10–13

(a) Power transmission of a utility system. (Reproduced with permission from Cutler-Hammer/Westinghouse.)
(b) A simplified one-line diagram of the power transmission of a utility system shown in part (a) with several nonessential components deleted.

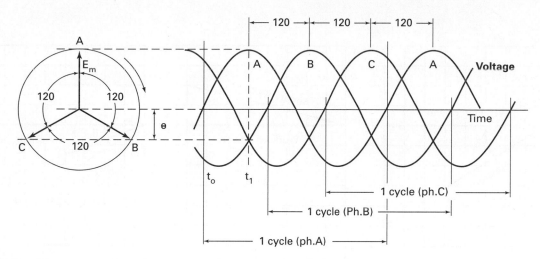

■ **FIGURE 10–14**
Phase relations of a three-phase power system.

For three-phase power

$$P = 1.73 \times E \times I \times PF$$
$$= 1.73 \times E \times I \times \cos\theta$$
$$= 1.73 \times P_a \times \cos\theta \qquad (10–21)$$

where the voltage is the *line-to-line* voltage (E), not the *line-to-neutral* voltage (E_n). The relationship between E and E_n is explained in Chapter 11.

10.7.4 The Power Triangle

The three types of power in an AC system—apparent power (P_a), reactive power (P_r), and working power (P)—can be represented by a power triangle, as in Figure 10–15. Working power (P) is represented horizonatally; inductive reactive power is represented vertically downward; and capacitive reactive power is represented vertically upward. The two reactive components are subtractive. The hypotenuse of the resulting right triangle represents apparent power (P_a). The angle between the hypotenuse and the base of the triangle is the power angle θ. The relationship between the three types of power is expressed in Eq. 10–22:

$$P_a^2 = P^2 + P_r^2 \qquad (10–22)$$

where P_a = apparent power = $I_a^2 \times Z$
    P = working power for one phase = $I^2 \times R$
    P_r = reactive power for one phase = $I_r^2 \times X = I_r^2 \times |X_L - X_C|$

From the power triangle shown in Figure 10–15 it is clear that the angle extended between the hypotenuse and its base can be determined by trigonometry, or

$$\cos\theta = P/P_a \qquad (10–23a)$$
$$\sin\theta = P_r/P_a \qquad (10–23b)$$
$$\tan\theta = P_r/P \qquad (10–23c)$$

or

$$\cos\theta = R/Z \qquad (10–23d)$$
$$\sin\theta = X/Z \qquad (10–23e)$$
$$\tan\theta = X/R \qquad (10–23f)$$

where θ = power angle

Example 10.10   A single-phase 240-V AC power system is measured as having 16 ohms, 10 ohms, and 4 ohms for resistance, inductive reactance, and capacitive reactance respectively. Determine the current flow through the system, and the apparent, working, and reactive power.

Answer The impedance of the system is calculated from Eq. (10–11):

$$Z = \sqrt{R^2 + (X_L - X_C)^2} = \sqrt{16 + [10 - 4]^2}$$
$$= \sqrt{256 + 36} = \sqrt{292} = 17.09 \text{ ohms}$$

From Eq. (10-13), $I = E/Z = 240/17.09$

$$= 14.04 \text{ A}$$

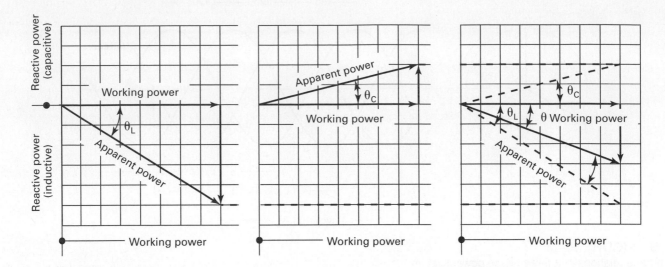

■ **FIGURE 10–15**

The power triangle illustrates the power relationship in vector quantities where a power system contains (a) resistance and inductive reactance, (b) resistance and capacitive reactance, and (c) resistance, inductive, and capacitive reactances.

Apparent power $P_a = E \times I = 240 \times 14.04$
$$= 3369.6 \text{ VA}$$

Working power $P = E \times I \times \cos \theta$
$$= P_a \times \cos \theta$$
$$= 3369.6 \times (16/17.09)$$
$$= 3154.9 \text{W} = 3.15 \text{kW}$$

The relationship between apparent, working, and reactive power can best be illustrated by using the power triangle in Figure 10–15.

10.7.5 Power Factor

The mathematical term $\cos \theta$ is also called the *power factor* (PF). If the power angle of a load circuit is 30°, then $\cos \theta$ is 0.866, and the power factor is said to be 0.866 or 86.6 percent of the unit power.

As can be seen from the preceding equations, for the same values of voltage and current, a higher-power factor will produce more useful power. Conversely, a low-power-factor load will require more current compared with a high-power-factor load. Obviously, a higher-power factor is preferred by both the utility company and the user. For residential and commercial loads, the overall system power factors generally range from 0.8 to 0.9; for industrial plants, the overall power factor is generally lower owing to the extensive use of motors and inductive equipment. In such case, some means to improve the power factor may be necessary.

Example 10.11

1. In a 208-V three-phase motor load, the current drawn by the motor is 25 A, and the PF is 0.6. What are the apparent power and working power of the circuit?

   From Eq. (10-17), $P_a = 1.73 \times E \times I$
   $$= 1.73 \times 208 \times 25 = 8996 \text{VA}$$

   From Eq. (10-21), $P = 1.73 \times E \times I \times \cos \theta$
   $$= 8996 \times 0.6 = 5398 \text{W}$$

2. What will the current be through the circuit if the PF is increased to 0.8 for the same motor load?

   The working power will remain the same, and the current through the circuit (wires and devices) will be reduced proportionately, or

   $$I = 25 \times 0.6/0.8 = 18.75 \text{A}$$

The preceding example serves to identify the potential savings in construction and operating costs of an electrical power system between low- and high-power- factor circuits. A 5-hp motor was used in this example. If this motor were a 50-hp motor, the current would be 250 A at 0.6 PF, and 187.5 A at 0.8 PF, or a 25 percent current reduction through the motor branch circuit, distribution feeders, and their operating components. With lower working current, the cost of electrical energy will be proportionately reduced. The methods for calculating wire and conduit sizes are discussed in Chapter 11.

10.8 VOLTAGE AND VOLTAGE DROP

From Ohm's law, for the same power to be transmitted, a higher voltage will require a lower current, which in turn will require smaller conductors (wires); however, a high-voltage system is more expensive and more hazardous than a low-voltage system. The selection of system voltages is the most important aspect of electrical system design in buildings. Chapters 11 and 13 discuss this topic in detail.

Every electrical component, whether a wire, a switch, or a piece of electrical equipment, contains more or less electrical resistance and reactance. The presence of these properties in an electrical circuit causes a voltage drop (loss) between the supply and the receiving ends of the circuit. The drop is unavoidable but must be minimized.

In general, equipment and appliances are rated at about 5 percent lower voltage than the rated system voltage. For example, a lamp is normally rated at 115 V for use on 120-V systems. If the system voltage becomes lower than the rated equipment voltage, it may lead to a lower capacity, reduced speed, lower efficiency, a shutdown due to overheating, or some other malfunction. For critical equipment, such as TV broadcasting equipment, electron microscopes, research instruments, and computers, separate voltage regulators may be required to maintain the voltage within 1 percent of the rated equipment voltage.

To avoid excessive voltage drop from wiring within a building, the National Electrical Code (NEC) limits the voltage drop for lighting and power loads within the building. Table 10–3 gives the recommended maximum voltage drop for different types of load.

10.9 SUMMARY OF PROPERTIES

Table 10–4 summarizes the various properties of AC and DC circuits relative to their basic quantities, such as current, voltage, resistance, reactance, impedance, power, and energy.

TABLE 10–3
Maximum allowable voltage drops (percent)

Portion of Distribution System	For Lighting and Power Loads	For Electric Heating	For Power Only
Service entrance to switchboard	1	1	2
Feeder to distribution centers	1	1	3
Branch circuits to connected load	3	1	3
Overall maximum voltage drop	5	3	8

TABLE 10–4
Summary of DC and AC properties (for single-phase circuits)

Quantity	Unit	Symbol	Formula
Voltage	Volts (V)	E (or V)	DC: $E = I \times R$
			AC: $E = I \times Z$
Current	Amperes (A)	I	DC: $I = E/R$
			AC: $I = E/Z$
Resistance	Ohms (Ω)	R	DC/AC: $R = E/I$
Inductive reactance[a]	Ohms (Ω)	X_L	AC: $X_L = 2\pi \times f \times L$
Capacitive reactance[b]	Ohms (Ω)	X_C	AC: $X_C = 1/(2\pi \times f \times C)$
Impedance	Ohms (Ω)	Z	AC: $Z^2 = R^2 + (X_C - X_L)^2$
Apparent power	Volt-amperes (VA)	P_a	DC/AC: $P_a = E \times I$
Working power	Watt (W)	P	AC: $P = E \times I \times \cos\theta$
		P	AC: $P = E \times I \times PF$
		P	DC/AC: $P = I^2R$
Reactive power	VA reactive (VAR)	P_r	AC: $P_r = E \times I \times \sin\theta$
Energy	kilowatt-hours (kWh)	W	DC/AC: $W = P \times t$
Power factor	Per-unit or percent	PF	AC: $PF = P/(E \times I) = \cos\theta$

[a]L–inductance, henrys
[b]C–capacitance, farads

QUESTIONS

10.1 The flow of electrical current through a circuit is directly proportional to the EMF (voltage) for both AC and DC systems. (True) (False)

10.2 What is the relationship between current, voltage, and impedance in an AC circuit?

10.3 What is the relationship among energy, power, current, and voltage in an AC circuit?

10.4 What is the standard voltage rating for convenience power (plug-in receptacles) in the United States? In Europe?

10.5 What is Kirchhoff's first law of electrical circuits? Is it possible to have all currents flow into a junction?

10.6 What is Kirchhoff's second law of electrical circuits? What will happen when a 120-V appliance is connected to a 240-V system?

10.7 What is the relation between peak current and effective (rms) current of an AC power system generated by commercial power companies in the United States?

10.8 What are the line-to-line and line-to-neutral voltages in a 120/240-V single-phase AC system? In a 120/208-V three-phase system? (Use System diagrams shown in Figure 11–4(b) and (d) in Chapter 11.)

10.9 What is power factor? What is the power factor of a DC system?

10.10 Is it desirable to maintain a high power factor in an electrical system? Can power factor be improved? If so, how?

10.11 What is the standard AC system frequency in the United States and in some other countries?

10.12 Name the two major advantages of an AC system over a DC system in building applications.

10.13 Can AC and DC systems be interconverted? If so, how?

10.14 What is the advantage of a three-phase system over a single-phase system?

10.15 Owing to the resistance and/or reactance of wiring, voltage is always lower at the end of a distribution system than at its beginning. This is called *voltage drop*. A well-designed system should limit voltage drop to a minimum. Answer the following:

   a. What is the recommended maximum overall voltage drop for a combination lighting and power system?

   b. For an electrical heating system?

   c. What is the nominal voltage rating of single-phase appliances in the United States?

10.16 An electrical load is rated for 1150 W at 115-V, single-phase. What is the load current if the power factor is 1.0, or 100 percent?

10.17 If the current of the load in Question 10.16 is 12 A, what is the power factor?

10.18 If the same load as in Question 10.16 (1150 W) is 202-V, three-phase, what is the current with 0.8 PF?

10.19 What is the apparent power in VA of the load in question 10.18?

10.20 What is the peak current of the load in question 10.18?

10.21 What is the peak voltage of a 120-V, single-phase system? Of a 208-V, three-phase system?

10.22 If the resistance and reactance of a circuit are 2 ohms and 4 ohms, respectively, what is the impedance?

10.23 If three loads rated for 2, 3, and 4 kW are connected in parallel, what is the total load?

10.24 If the system voltage for the loads in Question 10.23 is 480-V, three-phase, what is the total current of the loads? (*Note:* Since PF is not given in the question, it must be assumed.)

10.25 Describe the difference between power and energy.

10.26 If a single-phase electrical resistance load operates at 100 V, how much current is required to consume 800 W of power?

10.27 If the load operates for 20 hours, how much energy has been consumed?

10.28 If the power is a three-phase system with 100 V between each phase, what energy will be consumed after 20 hours of operation of the load?

10.29 What is the current flow in the line supplying the single-phase load and the three-phase load? Assume the load is 100 percent resistance.

POWER EQUIPMENT AND SYSTEMS

11

11.1 POWER DISTRIBUTION SYSTEMS

There are numerous power distribution systems used for buildings. The systems that are most appropriate for a specific building depend on the size of the building and the characteristics of the predominant loads, such as power ratings of the equipment (hp or kW), voltages, and phases. Frequency is another important characteristic of a power system; however, a country or region normally uses a standard frequency, such as 60 hertz (Hz) in the United States and 50 Hz in Europe. The standard frequency should be used except in unusual applications. Most large buildings contain loads with diversified characteristics, such as single-phase lighting and appliances and three-phase motors. Thus it is quite common to have more than one power distribution system in the same building, although multiple systems are costly and difficult to maintain. Cost–benefit studies should be made to compare options. As a rule, if more than one system is necessary, the main system should be selected to satisfy the predominant loads, with one or more subsystems converted from the main system for the minor loads.

Example 11.1 If a building contains 300 kVA of single-phase load and 20 kVA of three-phase load, then the electrical power distribution system selected should be the most appropriate for the single-phase load with a second system or a subsystem to handle the three-phase load. A detailed evaluation of this example will appear at the end of this section after the various system applications and limitations are introduced.

Electrical power systems used in buildings may be divided into three voltage classes. The system voltage is defined as the voltage between lines, or line-to-line voltage.

- *Extra-low-voltage systems (50 V and below)* This class is normally used for control, signal, and communication systems.
- *Low-voltage systems (nominally below 600 V)* The practical range in the United States is between 120 and 600 V. In commercial use, 120 to 250 V have been known as low-voltage systems, and 480 to 600 V as medium-voltage system; however, strictly speaking, the National Electrical Code considers all systems up to 600 V low-voltage systems. This class is normally used for supplying lighting and power equipment, except extremely large equipment such as a single 1000-hp motor.
- *Medium-voltage systems (600 to 100,000 V)* The practical range in the United States is 2300 to 69,000 V. This class is normally used to supply distribution transformers with step-down voltages to building distribution systems and equipment loads. Voltages of 13,800 and below are also used to supply large motors. *Note:* High-voltage systems (greater than 100,000 V) are not normally used in buildings.

There are variations within each voltage class. For power distribution within buildings, low-voltage systems up to 480 V are most common. Medium- and high-voltage systems are used only in large buildings with several million square feet in floor area or in industrial plants. For the purpose of this book, we shall consider only the low- and medium-voltage systems.

The following sections give guidelines for selecting a system and knowing its limitations.

11.1.1 Loads Less Than 100 kVA

Residential and small buildings typically have a demand load less than 100 kVA. Loads are normally rated for 120 or 240 V, single phase. Thus, a 120 to 240-V single-phase three-wire system is most appropriate. (See Figure 11–1(a).)

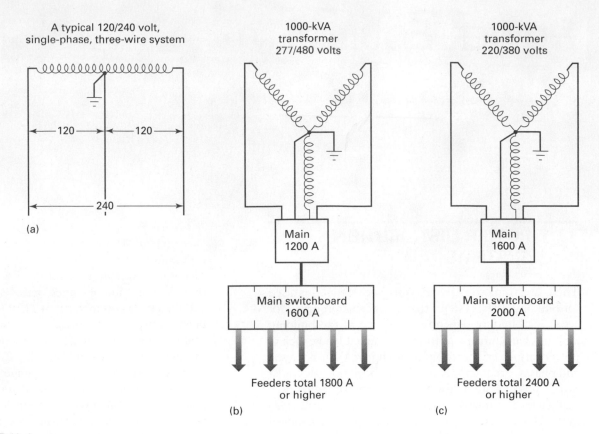

A typical 120/240 volt, single-phase, three-wire system

120 ← 120 → 120

← 240 →

(a)

1000-kVA transformer 277/480 volts

Main 1200 A

Main switchboard 1600 A

Feeders total 1800 A or higher

(b)

1000-kVA transformer 220/380 volts

Main 1600 A

Main switchboard 2000 A

Feeders total 2400 A or higher

(c)

■ **FIGURE 11–1**

(a) One-line diagram of a 120/240-V single-phase system. Similar systems of the same voltage class include 115/230 V and 125/250 V. (b) One-line diagram of a 1000-kVA, 277/480-V three-phase four-wire power distribution system showing its size (the capacity of the system components). (c) 1000-kVA system identical to (b), but at 220/380 V it requires about 25 percent higher ampacity of its distribution components and thus a higher capital investment. More convincingly, when a 277/480-V system is compared with a 120/208-V system, the lower-voltage system will require nearly 150 percent more ampacity.

11.1.2 Loads Greater Than 100 kVA

Loads greater than 100 kVA are usually served by three-phase systems. An exception is multiunit apartments or small shopping centers, where it is entirely appropriate to use a single-phase system, even if the total load far exceeds 100 kVA. As the load increases, systems at higher voltages are favored, owing to economy of equipment and to reduction of wire sizes.

Popular systems used in the United States are 120-, 120/240-, 240-V single-phase systems and 120/208-, 240-, 277/480-, and 480-V three-phase systems. The 220/380-V systems are quite common in countries where the utilization voltage for appliances is 220 V rather than 120 V. In such a system, 220-V single-phase lines will serve small appliances, and 380-V three-phase lines will serve larger loads. However, the 220/380-V system will eventually be replaced by a 277/480-V system, which is considerably more economical. For example, a 1000-kVA, 380-V three-phase system requires a

1600-A main service, whereas a 480-V three-phase system requires only a 1200-A main service. This 25 percent reduction in current affects the distribution equipment and feeders all the way down the line and, thus, the overall system cost. (See Figures 11–1(b) and 11–1(c).)

Medium-voltage systems such as 2400/4160 V and high-voltage systems such as 7200/12,470 V are also used for buildings in which the total load exceeds the service capacity of low-voltage systems.

11.1.3 Common Distribution Systems

Figure 11–2 provides a summary of power systems appropriate for various load capacities. For extremely large buildings with over 1 million square feet in gross floor area, or for campus-type facilities, power distribution may consist of many small systems, all interconnected to increase the system's flexibility and reliability. The figure shows five primary and secondary distribution systems,

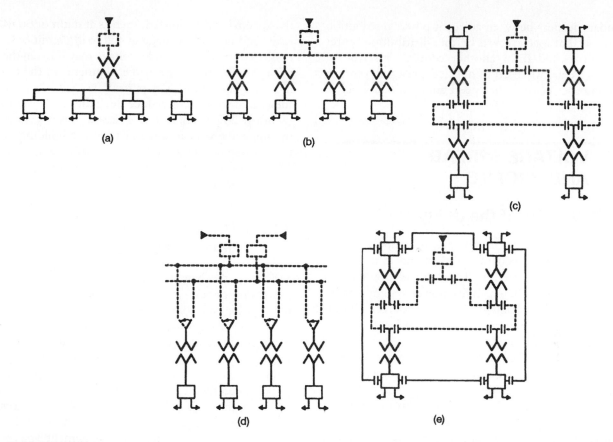

■ **FIGURE 11–2**

Typical power distribution systems. (a) Simple radial system. (b) Distributed radial system. (c) Secondary loop system. (d) Primary and secondary loop system. (e) Primary selective system.

ranging from simple secondary radial to the more complex primary and secondary selectives. The term *primary* refers to the portion of a distribution system before a transformer, and *secondary* indicates the portion after a transformer.

Simple Secondary Radial

This system, shown in Figure 11–2(a), is the simplest of all. It usually contains a single step-down transformer that transforms the primary voltage (say, 4160 V) to a secondary voltage (say, 120–240 V) for distribution to the loads. Typical applications of simple secondary radial systems are small office buildings, stores, and large residences. The transformers may be external to the building, or within the building, depending on need.

Distributed Radial

This system, shown in Figure 11–2(b), consists of multiples of simple secondary radial systems. Typical applications are shopping centers, apartment complexes, large department stores, schools, and institutional buildings.

Primary Loop

This system, shown in Figure 11–2(c), consists of several simple secondary radial distribution systems having the primary feeders connected in a loop. In case a section of the primary feeder is faulty, that section of the primary feeder may be isolated and power may be fed from another direction. This system is normally used in industrial plants or campus-type facilities where the cost of dual primary feeders can be avoided.

Primary Selective

This system, shown in Figure 11–2(d), consists of dual primary feeders to each transformer substation. This is most commonly used in high-rise buildings, hospitals, and research facilities where reliable power is essential. The second primary power source may be either from another power source or from a group of standby generators.

Primary Loop with Secondary Tie

This system, shown in Figure 11–2(e), consists of a primary loop as in Figure 11–2(c) and interconnected secondary feeders between adjacent power distribution centers so that loads served by one distribution center may be back-fed from another center. This type of system is frequently used in hospitals, research centers, and computer centers that need an additional level of reliability.

11.2 VOLTAGE SPREAD AND PROFILE

11.2.1 Voltage of the Utility

Most utility power systems regulate their supply voltage automatically at the generating plant and at the distribution substation to within 1 percent of the nominal voltage; however, when a utility power system is overloaded, its supply voltage may drop. A 5 percent fluctuation is quite common and is within the legal limits of utility regulations.

Under unusual circumstances, when a utility network is extremely overloaded, the utility company may resort to lowering its distribution voltage by 5 to 10 percent of its normal voltage—for example, from 4160 V to 3750 V. This will correspondingly lower the secondary voltage in the building by a similar percentage and more or less reduce the building demand load by the same magnitude; however, when the voltage drop is excessive, the building equipment may suffer detrimental effects, such as an overheated motor, lowered energy efficiency, or the malfunction or tripping of sensitive electrical or electronic instruments.

11.2.2 Voltage Spread

Voltage spread is the difference between the maximum and minimum voltages of the system between no-load and full-load conditions. Voltage spread accounts for the voltage drop through the transformers, distribution equipment, feeders, and branch circuits. For example, if the voltage drop through the transformer of a 240-V system is 6 V between no load and full load, and the voltage drops through the feeder and branch circuits supplying the load are 2 and 4 V, respectively, then the voltage spread at the load will be $(6 + 2 + 4) = 12$ V. The total voltage spread is $(100 \times 12/240)$, or 5 percent.

11.2.3 Voltage Profile

The cumulative effect of utility voltage fluctuations and the voltage drop within the building distribution system may subject the load equipment to an overvoltage when the system is lightly loaded, such as at night or on weekends, and an undervoltage at full-load conditions. Excessive overvoltage or undervoltage may shorten the life of or even burn out the load equipment. In the case of incandescent lamps, a 5 percent voltage drop below their rated voltage will reduce the light output (lumens) by nearly 20 percent. For this reason, it is imperative to minimize the voltage spread within the building distribution system. Figure 11–3 gives a graphic illustration of voltages at various points of a distribution system under extreme conditions.

11.2.4 System and Equipment Voltage Ratings

Even with the best designed systems, the voltage of a power distribution system will unavoidably drop between its supplying end (source) and receiving end (load), owing to the impedances inherent in the distribution components. For this reason, equipment or appliance voltage ratings are usually about 5 percent lower than system voltage ratings. For example, single-phase electrical appliances are usually rated for 110–115 V for use on the 120–125-V system in the United States. Similarly, 460-V motors are used on 480-V three-phase systems. Table 11–1 lists the differences between system and equipment voltages of common power distribution systems.

11.3 GROUNDING

11.3.1 Grounded System

When an electrical system is connected to the earth, either intentionally or accidentally, it is said to be *grounded*. In general, all building electrical systems are intentionally grounded at the point where the voltage to ground is the lowest. There are several reasons for grounding:

1. Grounding protects the system and equipment from overvoltage due to accidental contact with higher-voltage sources, such as the primary voltage side of the distribution system, which may exceed several hundred thousand volts.
2. When lightning strikes a building and its electrical system, the electrical wiring and insulation on equipment may break down, and lightning current will flash over to seek the ground. If the system is properly grounded, the lightning current will follow a direct path to the ground, bypassing the feeders and equipment.
3. Grounding protects people from heavy electrical shock. If the system and equipment are grounded

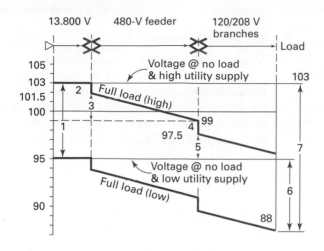

As Percent of Initial Voltage

1 Utility voltage spread
= 103-95
= 8%
2 Distribution transformer voltage drop
= 103-101.5
= 1.5%
3 Feeder voltage drop
= 101.5-99
= 2.5%
4 Building transformer voltage drop
= 99-97.5
= 1.5%
5 Branch circuit voltage drop
= 97.5-96%
= 1.5%
6 Utilization voltage spread (no load)
= 96-88
= 9%
7 Total system spread (full load)
= 103-88
= 15%

■ **FIGURE 11–3**

Voltage profile of a typical electrical distribution system. The profile shows extreme conditions. A high-quality utility service and a well-designed distribution system should keep all drops and spreads below 50 percent of the figures shown.

TABLE 11–1

Voltage rating of common distribution systems showing the system versus equipment voltages and the maximum equipment rating versus the system capacity. In general, the maximum rating of a single piece of inductive load (motor) should not exceed 15 to 20% of the system capacity if the system also serves other loads.

Distribution Systems	Nominal System Voltage, V	Nominal Equipment Voltage, V	Application Limits	
			Max. Inductive Equipment Rating,* kVA	System Capacity, kVA
1 φ, 3 W	120/240	115/230	20	100
3 φ, 4 W	120/208	115/2.00	150	750
	220/380	210/365	500	3000
	277/480	265/460	500	3000
	2400/4160	2300/4000	5000	15,000
	7200/12,470	6900/12,000	10,000	50,000
	19,920/34,500	N/A	N/A	No maximum
3 φ, 3 W	240	230	150	750
	480	460	500	3000
	600	575	500	3000
	4160	4000	5000	15,000
	13,800	13,200	10,000	50,000
	34,500	N/A	N/A	No maximum

*Refers to a single piece of inductive load, such as a motor that can be connected on this system without causing excessive voltage dip. This is because of the extremely but momentarily large inrush of current during start-up.

and the electrical circuit is accidentally shorted to the equipment, the circuit protection device (fuses or circuit breakers) should trip and cut off the circuit. If the system is not grounded, then any accidental grounding of the wire through the equipment may pass through a person who happens to be touching the equipment.

4. A grounded system is more economical than an ungrounded one. The grounded side of a circuit must not be switched. Thus on a single-phase circuit, only a single-pole switch is required, whereas a double-pole switch must be used on an ungrounded single-phase system.

11.3.2 Ungrounded Systems

While grounding is preferred for general building systems, it does have one major disadvantage: A grounded system can easily be "tripped," or put out of service, whenever its ungrounded side accidentally touches ground. This is a problem especially with critical loads, such as lighting and power for hospital surgical rooms. For this reason, the electrical power supply to the surgical rooms is ungrounded; however, ungrounded power systems should be isolated from the main system through individual isolation transformers and equipped with a sensitive ground detection system to recognize any accidental grounding. The detection system allows the power system to continue operation until the fault (accidental grounding) is corrected.

11.3.3 Grounding Methods

A common practice is to use an underground water main as the grounding electrode, because the water main is extensively and solidly buried in the ground; however, experience indicates that grounding an electrical system in this manner may cause corrosion of the water main. Because of this problem, water utilities discourage such grounding by installing nonconductive couplings at the main or by using plastic piping materials. A less effective but still acceptable alternative is to use copper rods or plates driven into the ground as the grounding electrode. A more reliable, but costlier, method is to install a grounding grid under the building or a grounding loop around the building as a system ground. There are two grounding requirements.

System Grounding

The entire distribution system should be grounded at the neutral, or lowest-voltage point of the system. Figure 11–4 indicates the locations of grounded systems of some interior distribution systems. The voltage between an ungrounded line with the ground (earth) or the grounded line (usually neutral) is the voltage to ground.

Equipment Grounding

All electrically conductive enclosures, such as the frame of a motor or the enclosure of an electrical appliance, should be grounded by a separate color-coded (green) conductor. Although metal raceways can be used instead as the ground, a separate equipment-grounding conductor is preferred. (See Figure 11–5(a) and (b).)

11.3.4 Polarization

Since one side of all building interior distribution systems is grounded, the enclosure of any electrical equipment, such as the shell of a lamp socket, the metal case of an electrical tool, or the steel enclosure of a clothes-washing machine, as in Figure 11–5(b), should be connected to the grounded side of the electrical circuit to safeguard the person who operates the equipment. This requirement is satisfied by the installation of polarized receptacles with equipment-grounding terminals. The ungrounded side of the receptacle is smaller than the grounded sides; thus, it can receive only the smaller blade from the equipment plug.

11.3.5 Ground Fault Circuit Interrupter

For added safety, certain receptacles and circuits that supply loads for swimming pools, kitchens, garages, bathrooms, laundries, or outdoor equipment should be of the type that can trip or cut out the circuit if an imbalance of current flow is sensed between the grounded and ungrounded sides of the circuit. Such an imbalance indicates that current is leaking through the ungrounded side of the circuit; thus, the situation should be corrected by a ground fault circuit interrupter (GFCI). The requirements for using GFCI are given in Article 210 of the NEC. Figure 11–5 illustrates the configuration and operating principle of a GFCI receptacle.

11.4 SHORT-CIRCUIT AND INTERRUPTING CAPACITY

11.4.1 Short Circuit

Electrical circuits are designed to allow only a limited amount of current to flow through; this is done to prevent overheating of the wires due to I^2R loss. The limitation on the current is largely due to the resistance or

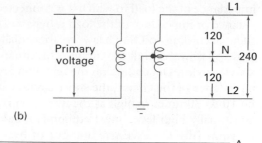

(a)

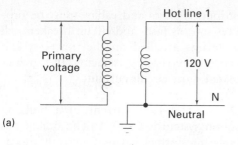

(b)

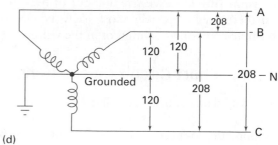

(c)

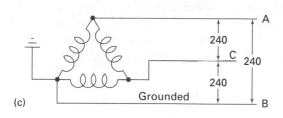

(d)

■ **FIGURE 11–4**

Typical grounded system.

(a) 120-V single-phase two-wire system. Line to ground = 120 V

(b) 120/240-V single-phase three-wire system. Line to ground = 120 V.

(c) 240-V three-phase three-wire system. Line to ground = 240 V.

(d) 120/208-V three-phase four-wire system. Line to ground = 120 V.

(e) 120/240-V three-phase four-wire system, midpoint of one phase to ground. Line to ground = 120 V and 208 V on the high leg.

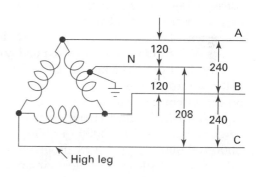

High leg

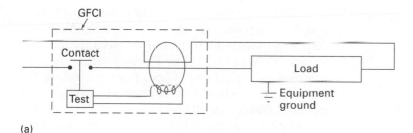

(a)

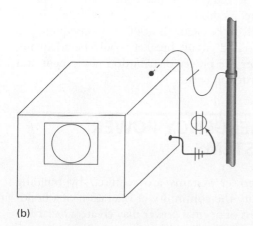

(b)

■ **FIGURE 11–5**

(a) Schematic diagram of the operating principle of a GFCI. Any unbalanced current flow between the two wires indicates a ground current leakage, which trips open the contact.

(b) The metal enclosure of a domestic washing machine or similar equipment is grounded either to the piping system or to the ground wire of the equipment-grounding system.

impedance of the load to which it is connected. If the resistance or impedance of the load is bypassed, or *shorted*, then, according to Ohm's law, an abnormally high current will flow through the circuit. This situation is called a *short circuit*. Depending on the remaining resistance or impedance of the circuit, the short-circuit current could be 10 to 30 times as high as the normal current. At this abnormally high level, most equipment and wiring will be ruined by the excessive amount of heat generated. Furthermore, there will most likely be fire of combustible components within or in the vicinity.

11.4.2 Short-Circuit Calculations

To avoid damage from excessive heat and the magnetic force created by a short circuit, all electrical circuits and equipment connected to the system must have an *interrupting rating*, or *interrupting capacity*, equal to or greater than the calculated *short-circuit capacity* of the system. The calculation of the short-circuit capacity of a power system is very involved and complex. Conservatively, it can be as high as 20 to 30 times the normal full-load current of the system. For example, if a power system is designed to carry a full-load current of 2000 A, then the short-circuit capacity could be in the neighborhood of 20×2000 (40,000) A, or even 30×2000 (60,000) A. Computer programs are required to calculate a realistic level of short-circuit currents.

In its simplified form, the short-circuit current can be conservatively approximated by the following equations.

$$I_S = E/Z \qquad (11\text{--}1a)$$

or

$$I_S = 100 \times I/(Z_p + Z_t) \qquad (11\text{--}1b)$$

where I_S = short-circuit current, A
  E = voltage, V
  Z = total impedance, ohms
  I = full-load current of transformer, A
  Z_p = impedance of the primary, %
  Z_t = impedance of the transformer, %

Conservatively, the impedance of the primary on a large utility substation or a network may be considered negligible (near zero); thus, a short circuit near the secondary side of the transformer (or the system mains) is totally determined by the internal impedance of the transformer (or mains). For example, the short circuit (I_S) of a large system connected to a transformer rated with 5 percent impedance could be $100 \times I/5$, or $20I$.

11.4.3 Interrupting Capacity

To prevent the system and equipment from being destroyed by heat, arcing, and fire caused by a short circuit, all distribution wires and cables must be protected by devices such as fuses and circuit breakers. All equipment or devices installed on the power system must also carry an interrupting capacity (IC) greater than the calculated short-circuit current.

Example 11.2 The full-load current of a building power distribution system is 1200 A. The building is served by a single transformer having 5 percent impedance. The utility power service supplying the transformer is from a nearby substation with practically unlimited power. Determine the available short-circuit current at the main switchboard.

Answer From Eq. (11–1),

$$I_S = 100 \times I/(Z_p + Z_t)$$

where I = 1200 A
  Z_p = 0 (impedance of the substation is considered
    to be near zero because the substation has an
    unlimited power supply)
  Z_t = 5 (transformer's impedance, as a percent of
    its own rating)

Thus, $I_S = 100 \times 1200/(0 + 5) = 24,000$ A. In other words, even though the normal full-load current is only 1200 A, the system may instantly produce as much as 24,000 A during a fault.

Example 11.3 For the system in Example 11.2, determine the minimum IC rating of the main circuit breaker.

Answer On the basis of available commercial products, IC ratings of circuit breakers with 1200 A full-load current are made to withstand between 20,000 and 50,000 A. Naturally, higher-IC-rated circuit breakers cost more than the lower-IC-rated circuit breakers. For this particular power distribution system, a breaker rated at over 24,000 A—say 30,000 A—would be a safe choice.

On the other hand, if the short-circuit current was correctly engineered (rather than conservatively approximated by the method above), the true short-circuit current is likely to be below 20,000 A. Consequently, a 20,000-A IC-rated circuit breaker would be adequate. This will lower the power distribution equipment and related construction costs.

11.5 EMERGENCY POWER SYSTEMS

Emergency power systems are required by building codes to ensure the continuity of operation of a building when a loss of normal power may create a hazard to life, a fire hazard, or a loss of property or business.

11.5.1 Alternative Energy Power Supplies

Three types of emergency power supply are in common use:

1. *Tap ahead of the main switch.* When the main switch is disconnected, the emergency branch will still be active. This is the least reliable alternative emergency power system, since if the power is lost from the utility, both the normal branch and the emergency branch are out of commission. This alternative is good only for residences and small businesses. (See Figure 11–6.)
2. *On-site generator.* One or more electrical generators installed on-site or in the building automatically start up to provide power to the essential loads. This is the most reliable alternative, but it is more costly to own and maintain.

3. *Separate power source.* The emergency load is automatically transferred to a separate power source either from the same utility or from another power source. This alternative is not always available other than in extremely large buildings.

11.5.2 Types of Loads

Emergency power systems are required in the event of power outage for exit lights, lighting to prevent panic in assembly areas, fire alarm, safety communications, fire suppression, elevators for firefighter use, and other functions whose interruption would endanger life or health. For a simple building, exit lights and a fire alarm may be the only loads requiring emergency power. In this case, batteries with sufficient run time will suffice. For buildings with fire pumps or other larger emergency power needs, a generator will be required.

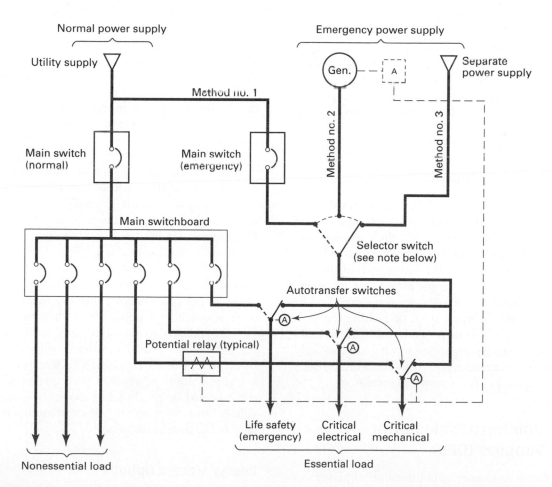

■ FIGURE 11–6

One-line diagram of a power system showing three methods for emergency power supply. Method 1 is used for small systems only. (*Note:* The selector switch is not required if the alternative methods are mechanically and electrically interlocked.)

The generator can not only serve emergency loads but also provide standby power. There are two categories of standby power. Legally required standby power systems provide power to loads whose interruption would create a hazard or hamper rescue or fire-fighting operations. These might include smoke control systems, critical heating and refrigeration, sewage disposal, and others at the discretion of the authority having jurisdiction.

Systems to provide standby power beyond legal requirements are termed *optional standby systems* and are generally used to operate computers, communications equipment, and processes for which continuity is important for the owner.

A single generator system can be used to power all three of the emergency and standby power systems if attention is paid to proper design of circuitry and transfer switches according to code.

11.5.3 Emergency Power Capacities

The capacity of essential loads (emergency and critical) varies with the design and occupancy of a building. Obviously, a single-story office building does not require anywhere near the essential load of a high-rise building, a hospital, a laboratory, or a special industrial plant. In general, the essential load for offices may range from 10 percent to 20 percent of the total connected load and may even be as high as 40 percent for super-high-rise buildings or hospitals.

11.5.4 Power Generator Sets

A *generator* is equipment that generates electrical power. A *generator set* is a complete unit, including a prime mover that drives the generator and the necessary start-up, speed, and power controls. Automatic sensors initiate the start-up of the prime mover within seconds after the loss of normal power. The prime mover may be either a gasoline or a diesel engine or a steam turbine. Standby power is required for many types of buildings, such as hospitals and hotels. It is also essential for high-rise buildings (over seven stories) regardless of what kinds of building they are—for example, high-rise offices or apartment buildings. Figure 11–7(a) shows a generator set.

11.5.5 Uninterruptible Power Supplies (UPS)

Some applications require an uninterruptible power supply (UPS). Examples include computers, health-care monitoring equipment, network servers, control systems, and laboratory analytical equipment. Having an emergency or standby generator will maintain power during a utility outage, but the system requires 5 to 10 seconds for automatic starting and switching. These types of loads cannot tolerate even a momentary discontinuity in power supply without serious consequences. Even the momentary outages experienced in utility power supply, which may last only a fraction of a second will be problematic. For such applications a UPS system should be used if the consequences of malfunction are severe enough to warrant the additional expense.

UPS systems typically consist of rectifiers to change utility power from AC to DC, batteries to store enough power to coast through the outage event, and inverters to convert the DC back to AC for use in critical loads. The UPS system must have sufficient capacity to operate the critical load and may be as small as a desk-mounted unit for a computer workstation or large enough to power an entire data center. (See Figure 11–8.)

For critical computer operations (e.g., at financial institutions and in military defense), continuous operation may be required, and the UPS system is generally modular, with one or more extra modules installed than are required to carry the load. Modular design adds to the complexity of the system, owing to the need to install components in parallel to allow individual modules to coordinate the voltage and frequency of their outputs. With a single extra module, one module can fail or be taken out of service for maintenance without jeopardizing the load. With two extra modules, failure of a single module can be tolerated while another is down for maintenance. For applications requiring very high reliability, dual systems comprising multiple modules may be employed with separate distribution feeders to the load. These types of systems generally employ high-speed electronic switches to transfer from one to the other. High-speed switches are designed to transfer in milliseconds, so fast that the load is unaffected by the switching.

Obviously, the cost of such provisions is high, and the consequences of failure may not warrant the extra expense. In this case a maintenance bypass may be installed to allow switching to utility power when the unit is down.

In addition to providing uninterruptible power, the UPS system provides power conditioning by regulating voltage and frequency. This feature is important if available utility power is of poor quality, or problems are expected in the building power system owing to voltage drop from starting of large motors or interference fed back into the system.

Energy Storage Options

Battery capacity for UPS systems is sized for a selected period of effective backup. Fifteen minutes of battery time is a commonly used value. This period will allow

(a)

(b)

(c)

■ **FIGURE 11-7**

(a) Typical diesel engine-driven generator or generator set. (b) Indoor installation, requiring extensive ventilation provisions. (c) Outdoor installation in weatherproof housing. Note double-wall above-ground fuel tank and redundant fuel pumps in foreground.

■ **FIGURE 11–8**
Desk-mounted UPS used to protect workstation computers. (Courtesy of WTA.)

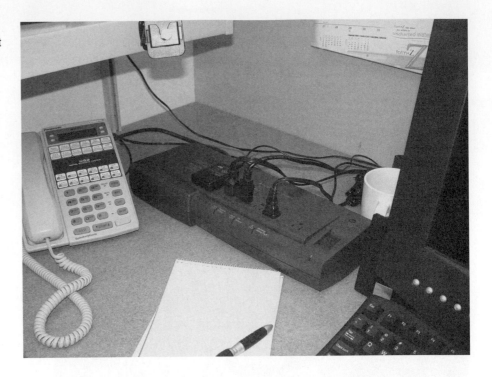

orderly shutdown of equipment if there are no provisions for standby power (generators), or if the generators do not start properly. If high reliability is required, multiple, redundant generators will be required.

Battery technology for UPS systems varies depending on size of application. Small UPS systems for desktop or dedicated equipment applications are usually dry-cell nicad batteries. (See Figure 11–9.) Larger UPS systems serving multiple loads can be dry- or wet-cell. (See Figure 11–10.) Wet-cell lead–acid batteries, vented or sealed, are used in the largest applications, and newer installations generally use sealed batteries to minimize fumes and hydrogen hazards associated with older installations. (See Figure 11–11.) Regardless of sealing and venting provisions, wet-cell batteries contain sulfuric acid, and battery rooms must be equipped with provisions for spill control, acid neutralization, safety showers, and eyewash stations.

Flywheel systems can be used as an alternative to batteries, to avoid the maintenance and potential hazard and environmental problems inherent in large battery installations. These systems use motors to convert utility power to kinetic energy stored in a rotating mass. The flywheel's energy is used to run a generator to serve critical loads. Motor-generator technology was used in early UPS systems, but the storage time was limited to seconds rather than minutes. Advances in flywheel technology have extended storage time to be competitive with battery systems.

■ **FIGURE 11–9**
UPS for installation in a small computer room with integral dry-cell batteries. (Courtesy: Liebert Corp.)

11.6 POWER EQUIPMENT

There are various levels of equipment in an electrical power system, starting with the power service, transformers, and distribution equipment and ending with the load-end protection devices.

■ **FIGURE 11–10**
Large UPS module, generally installed in remote electrical
room. (Courtesy: Liebert Corp.)

11.6.1 Service Entrance

Utility power may enter a building through an overhead
service drop for small systems or underground duct
banks for large systems. The service entrance power may
be at the building utilization voltage, such as 120/240 V
or 120/208 V or at a higher voltage, which is then
reduced to the utilization voltage through step-down
transformers.

11.6.2 Switchboards and Panelboards

A *switchboard* is an assembly of switches and circuit protection devices from which power is distributed. The
switchboard serves as the main distribution center of a
small system or as a portion of the distribution center of
a large system. (See Figures 11–12(a) and 11–12(b).)

A *panelboard* is an assembly of switches and circuit
protection devices as the final serving point of the
power distribution system. Figure 11–13(c) shows the
interior of a small panelboard. Depending on the load
to which it is connected, a panelboard may be identified
as a lighting, power, heating, specialty, or combination
panelboard. Figures 11–13(a) and (b) show several
types of panelboards. There is no clear demarcation between a switchboard and a panelboard, although a
switchboard is always referred to as the distribution
equipment closest to the supply end of the system.

■ **FIGURE 11–11**
Large array of wet-cell batteries,
typically sized for 15 to 20 minutes
of backup power capacity. Wet-cell
batteries contain acid and need
special provisions for safety.

(a)

(b)

■ **FIGURE 11–12**
(a) Typical unit substation showing the primary cable
section, a liquid-filled and fan-cooled transformer, control
and metering sections, and main secondary switches.
(b) Several lineups of motor control centers. This
arrangement saves space and eliminates the maze of
interconnecting control wires.

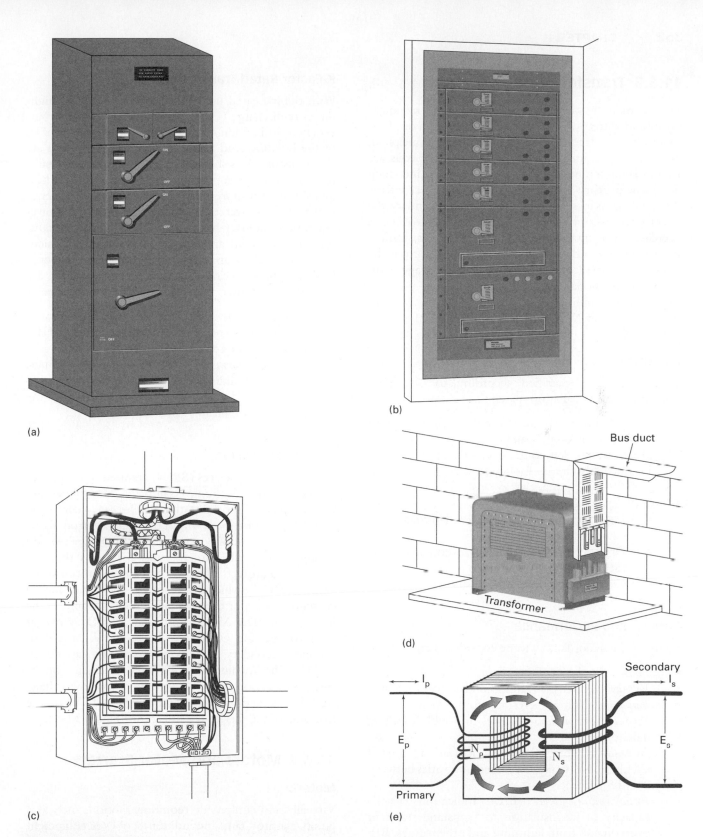

■ FIGURE 11–13

(a) Freestanding switchboard. (b) Wall-mounted power distribution panelboard. (c) Interior view of a lighting panelboard, showing 20 single-pole circuit breakers, the main feeder from the top of the panel, and branch circuit wires (black) from the circuit breakers to various loads through four feeder conduits. The white wires are neutral conductors for these branch circuits. (d) Small pad-mounted dry-type transformer with bus duct connection. (e) Operating principle of a transformer: $E_s/E_p = N_s/N_p$, where N_s and N_p denote the number of turns of secondary and primary coils in the transformer, respectively. (Reproduced with permission from Cutler-Hammer/Westinghouse.)

11.6.3 Transformers

Transformers are power transmission equipment primarily intended to convert a system's voltage from one level to another. All transformers operate on the principle of magnetic induction: Primary and secondary coils are wound on a common silicon steel core. The incoming side is the primary and the outgoing side the secondary. If the voltage is increased from primary to secondary, the transformer is a *step-up* transformer; if the voltage is decreased, the transformer is a *step-down* transformer. With magnetic induction, voltage is induced from the primary to the secondary. The resulting voltage is directly proportional to the ratio of the primary-to-secondary winding turns; that is, $E_s = (N_s / N_p) \times E_p$ (i.e., $E_s = \frac{1}{2} E_p$, if N_p is twice N_s). (See Figure 11–13(e).)

Classification

Transformers are classified according to their size, application, and insulation material.

- *Size* From a fraction of a kilovolt-ampere to thousands of kilovolt-amperes.
- *Application* Self-contained, substation service, or unit substation; for linear load or nonlinear load.
- *Insulation* Liquid or dry.
  - Liquid types—mineral oil, nonflammable, low flammable, etc. (See Figure 11–12(a).)
  - Dry types—ventilated, nonventilated, etc. (See Figure 11–13(d).)
  - Combination of liquid and dry types.

Ratings and Performance

Some of the major factors to be considered are:

- *Phase* Single- or three-phase.
- *Voltages* Primary (in) and secondary (out) voltages.
- *Capacity* Volt-amperes, kilovolt-amperes.
- *Winding connections* Single- or double-winding, delta or wye, etc.
- *Impedance* Internal impedance on a selected kVA base. Normally, the impedance varies between 2 and 5 percent.
- *Basic impulse level (BIL)* Indication of the capacity of its insulation to withstand transient overvoltages from lightning and other surges. BIL values are generally about six times line-to-line voltage for dry-type transformers and eight times or higher for liquid-filled transformers.
- *Other features* Voltage taps, temperature rise, sound level, manual or automatic load tap changing, grounding requirements, etc.

K-Factor Rated Transformers

With the increased use of fluorescent and other high-intensity-discharge (HID) lights and personal computers (PCs) in buildings, the voltage and current profiles of the building load are no longer sine waves but are rather distorted, as shown in Figure 11–14. The resulting effects to the supply transformers are greater heat generation and more noisy and more radio-frequency radiation. To overcome these problems the power equipment industry has designed K-factor-rated transformers for nonlinear loads. The definition and calculation of *K-factor* are outside the scope of this book; however, the following K-factor-rated transformers are now commonly used with certain types of buildings:

- *K-4 transformers* Used in offices or similar buildings having 50 percent nonlinear load such as HID lighting and electronic equipment.
- *K-13 transformers* Used in research buildings, computer buildings, and similar buildings or portions of buildings where nearly 100 percent of the loads are nonlinear.

Surge Suppression

Surge protection devices (SPDs) are used in systems to protect sensitive electronic equipment from voltage and current spikes of short duration. Spikes can occur owing to induction from loads within the building, from lightning, or from utility switching. SPD devices are arranged in parallel circuits with the loads they are designed to protect. The SPD consists of a metal-oxide varistor (MOV), which has very high impedance and no current flow under the normal voltage available to the load served. The MOV's impedance breaks down at high voltage and essentially shorts excess current to ground. Secondary surge arrestors (SSA) can be installed at the building's service entry. Transient voltage surge suppressors (TVSS) are rated for low-voltage (600 V or less) distribution and may be installed on the load side of the service entrance. (See Figure 11–15.)

11.6.4 Motors and Motor Starters

Motors

Virtually any equipment requiring motion, such as a pump, elevator, fan, air conditioner, or even equipment as small as an electric clock, needs one or more motors. Motors are classified according to the following characteristics:

- *Size* Fractional or integral horsepower, etc.
- *System and equipment voltage* 120, 208, 240, 277, 380, 480, 600, 2300, 4160 volts, etc.

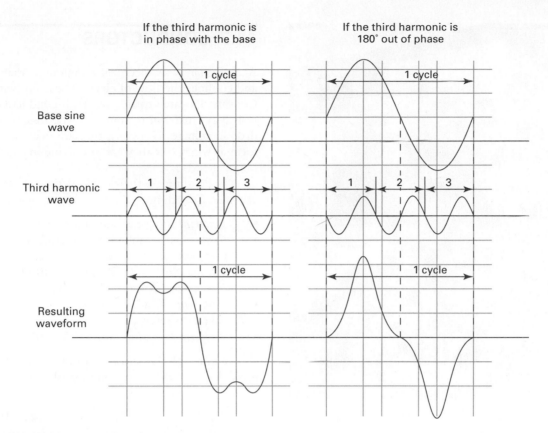

If the third harmonic is
in phase with the base

If the third harmonic is
180° out of phase

Base sine
wave

Third harmonic
wave

Resulting
waveform

■ FIGURE 11–14

Effect of the third harmonic wave on the composite voltage waveform of a power distribution system having high nonlinear third-harmonic loads. A diminishing effect will also be caused by other harmonics, such as the fifth and ninth. The effect of harmonics on current waves and their RMS values can be similarly demonstrated.

- *Number of poles* Two-pole (3600 rpm), four-pole (1800 rpm), six-pole (1200 rpm), etc. (based on 60Hz) synchronous speed.
- *Phase* Single-, double-, three-phase, etc.
- *Operating principle* Universal, split phase, induction (squirrel cage, wound rotor), synchronous, etc.
- *Construction* Open drip proof, totally enclosed fan cooled (TEFC), watertight, explosion-proof, etc.
- *Starting characteristics* High torque, low starting current, etc.

Most motors used in building equipment are of the squirrel-cage induction type. Owing to inductive reactance in the motor winding, induction motors always have a lagging power factor that can range from 70 to 80 percent at full load to as low as 10 to 20 percent during start-up. Consequently, the starting current of a motor may be as high as 10 times its full-load current.

The size of a motor is rated in horsepower (hp). One horsepower (1 hp) is equivalent to 746 W, or 0.75 kW. The full-load current of a motor varies with the design of

the motor. It may be calculated approximately from the basic power formula given in Chapter 10.

Operating on the principle of slippage, an induction motor has a normal speed that is slightly slower than its synchronous speed. For example, a two-pole motor normally has a synchronous speed of 3600 rpm (60 Hz × 60 sec/min) but has a rated speed of 3450–3500 rpm running on a 60-Hz system. Similarly, a four-pole motor designed for operating on 50-Hz power will have a rated speed of 1440–1480 rpm and a synchronous speed of 1500 rpm.

Motor Starters

When a motor starts up, its current is many times higher than its normal full-load current for several seconds. The persistence of the starting current depends on how fast the equipment can be brought up to full speed, which in turn depends on the inertia of the load. Ordinary-duty manual on-off switches are not capable of withstanding the momentary inflow of current. Thus, motor circuit–rated switches must be used. Automatic starters

■ FIGURE 11–15
TVSS module, generally installed adjacent to a branch circuit panel.

are required for large motors of 1 hp or higher. These starters allow a large inflow of current, as well as protection against continous overload. Motor starters may be classified according to the following properties:

- *Operating principle* Electromagnetic, solid state, variable speed drive, etc.
- *Protection devices* With or without a disconnect switch, with or without short-circuit protection.
- *Starting circuitry* Across-the-line, reduced voltage (autotransformer type), reduced inrush (delta-wye, part-winding types).
- *Protection circuitry* Overcurrent, overvoltage, undervoltage, reverse phasing, etc.
- *Construction* General service, weathertight, waterproof, explosion-proof, etc.

Figure 11–16 illustrates two of many motor starters that are used to reduce the inrush current during the start-up of a motor load. Motor starters may be individually mounted or preassembled as a motor control center (Figure 11–12(b)) to facilitate in interlocking control wiring in large systems.

11.7 CONDUCTORS

A *conductor* is an electrical component that conducts and confines the flow of electrical current within itself. Conductors are made of high-conductivity (low-resistivity) material to minimize the loss of power and drop in voltage. Normally, they are made in cylindrical form as wires, but they are also made in square or rectangular sections.

Depending on their construction, conductors are classified according to the following characteristics:

- *Material* Copper, aluminum, etc.
- *Form* Wire, cable, bus, bus-duct, etc.
- *Composition* Solid, stranded, etc.
- *Voltage class* 100, 300, 600, 5000 V, etc.
- *Insulation* Rubber, thermoplastic, asbestos, etc.
- *Covering* Lead, aluminum, nonmetallic, cross-linked polymer, etc.
- *Temperature rating* 60°C, 75°C, 90°C, 250°C, etc.

Figure 11–17 shows various kinds of wire and cable. Some of the more common 600-V general building wires used in raceways are the following:

- *THHN* Heat-resistant, thermoplastic, 90°C for wet and dry locations; used mostly for branch circuits.
- *THWN* Heat- and moisture-resistant, 75°C; used mostly for branch circuits.
- *USE* Underground service entrance cable, 75°C, heat- and moisture-resistant insulation with nonmetallic covering.
- *XHHW* Heat- and moisture-resistant, cross-linked synthetic polymer, 75°C for wet and dry locations; used mostly for larger feeders.

In addition to wires installed in raceways, certain wires and cables are permitted to be installed without raceways. The following such wires are used in buildings:

- *Mineral-insulated (MI)* These are metal-sheathed wires with a temperature rating of 90° to 250°C. These MI cables may be directly buried in concrete, pavement, walls, or ceilings.
- *Nonmetallic (NM and NMC)* This sheathed cable is a factory assembly of two or more insulated conductors having an outer sheath of moisture-resistant, flame-retardant, nonmetallic material. It is used mostly in residential dwellings and other types of buildings not exceeding three stories in height. Such cable is commonly called Romex.
- *Armored cable (AC)* This cable contains two or more insulated conductors in a metallic enclosure. It can be used in exposed or concealed applications, mostly for small sizes. It is sometimes called BX.

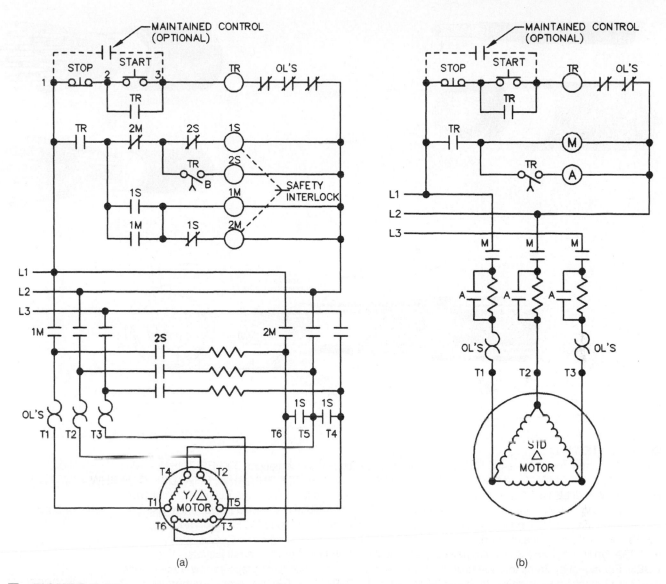

(a)   (b)

■ FIGURE 11–16

Reduced voltage and reduced inrush motor-starting methods. (a) Wye-delta (y/Δ) starter in which the motor is connected in wye during start-up (stator windings at 57.7% of line voltage) and connected in delta after an adjustable time delay (stator windings at 100% of line voltage). This method requires a special six-lead motor. (b) Primary resistor starter utilizes standard three-lead induction motor but requires a large bank of resistors to reduce the line voltage during start-up.

■   *Flat conductor cable (FC)*   Flat cable is an assembly of parallel conductors formed integrally with an insulating material web specifically designed for field installation in a surface metal raceway. It can be used only as branch circuit wiring for lighting and small appliances. It was formerly also approved for use under carpet tiles; however, this method of installation is no longer approved by the National Electrical Code (NEC).

11.7.1 Wire Sizes

American Wire Gauge

Conductors are numbered according to the American wire gauge (AWG) from No. 36 to No. 0000 (#4/0). The numbers are retrogressive (i.e., a smaller number denotes a larger size). A #4/0 solid (nonstranded) wire should have a diameter of 0.5 in. Each smaller size will have a reduction of diameter in the ratio of 1.123.

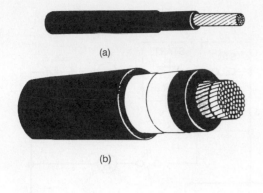

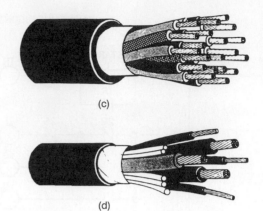

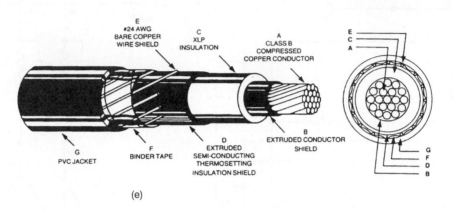

■ FIGURE 11–17

(a) Single-conductor types using PVC/nylon, neoprene, rubber-type insulations. TYPES: THHN-THWN, XHHW, NM, AMERFLEX, TW, THW UF AMERTHERM, UF and NMC AMERTHERM, RHH-RHW, RHW, USE-RHH-RHW, USE-RHW, RHH, RHW, STYLE RR, SE STYLE U. USES: Homes or factories for low-voltage power distribution.

(b) Single- and multiconductor thermosetting insulated. Unshielded and shielded power cable using XLP, EPR, or butyl rubber insulations; PVC and neoprene jackets; metallic tape or wire shields; 5-, 8-, and 15-kV ratings. USES: General power distribution.

(c) IMSA types 19–1 and 20–1 with polyethylene insulation and PVC or PE overall jackets. USES: For signal systems in underground conduit or aerial cable use when supported by a messenger.

(d) Thermoplastic and thermosetting insulations using PVC, PE, XLP, EPR/neoprene, rubber/neoprene, and EPR CSPE Insulations; and PVC, neoprene, or CSPE overall jackets. Thermosetting insulated and jacketed types suitable for fossil fuel and nuclear power plant use. USES: For general and supervisory usage.

(e) 5-kV copper cable for use in aerial, direct burial, conduit, and underground duct installation. (Source: Figures 11–17(a)–(d), reproduced with permission from American Insulated Wire Corp. Figure 11–17(e) reproduced with permission from Southwire Company.)

In other words, the diameter of a solid #3/0 wire should be 0.5/1.123, or 0.405 in. The actual diameter of stranded wire is, of course, larger than solid wire of the same AWG.

Circular Mil and Square Mil

Conductors larger than #4/0 are described in circular or square mils. A mil is defined as 1/1000 of an inch, and a circular mil is the area of a circle 1 mil in diameter. Similarly, a square-mil is the area of a square with sides

1/1000 in length. A thousand CM is also expressed as one KCM (or kcmil), where the first K stands for 1000. For example, a conductor of 300,000 CM is 300 KCM. Large conductors come in rectangular shapes or bars called *bus bars*.

11.7.2 Current-Carrying Capacity

The current that can be safely carried by a conductor depends on the size of the conductor as well as the type of insulation, method of installation, number of wires within a raceway, and surrounding temperature. The allowable current-carrying capacities, or ampacities, of various types and sizes of wires are given in the NEC. Table 11–2 is a condensed table of commonly used cables. (See the NEC for properties of other types of cable.) The allowable ampacity of conductors is reduced at ambient temperatures higher than 88°F. (See Table 11–2 for the relevant correction factors.) The allowable ampacity of conductors also is reduced when more than three current-carrying conductors are installed in the raceway. (See Table 11–3 for the relevant correction factors.)

11.7.3 Dimension of Conductors

The NEC provides data on bare and covered conductors for the purpose of sizing raceways. A condensed listing of the dimensions of rubber- and thermoplastic-covered conductors is given in Table 11–4. The use of the table will be demonstrated in Section 11.9.

Example 11.4 What is the code-allowed maximum ampacity of a #8 AWG copper conductor in a conduit with covering rated for 75°C?

Answer From Table 11–2, the answer is 50 A.

Example 11.5 What is the allowed maximum ampacity if there are nine #8 AWG copper conductors in the same conduit?

Answer From Table 11–3 the correction factor is 0.70; thus, $50 \times 70\% = 35$ A.

Example 11.6 If the ambient temperature of the installation is 100°F, what is the allowed ampacity in example 11.5?

Answer From Table 11–2, the temperature correction factor is 0.88; thus, $35 \times 0.88 = 30.8$ A.

Example 11.7 What are the diameter and area of a #1/0-type THHN conductor?

Answer From Table 11–4, the diameter is 0.491 in. and the area is 0.1893 sq in.

11.8 WIRING METHODS

There are over 20 NEC approved wiring methods. Those approved for use in buildings are generally wires and cables installed within raceways, with the exception of types NM (Romex), AC (BX), and FCC

TABLE 11–2
NEC-allowed maximum ampacities of most commonly used single-insulated conductors (based on 30°C/86°F ambient temperature and not more than three conductors in raceway)

Size	In Raceways				In Free Air			
	Copper		Aluminum		Copper		Aluminum	
AWG and MCM	75°C	90°C	75°C	90°C	75°C	90°C	75°C	90°C
14	20	25			30	35		
12	25	30	20	25	35	40	30	35
10	35	40	30	35	50	55	40	40
8	50	55	40	45	70	80	55	60
6	65	75	50	60	95	105	75	80
4	85	95	65	75	125	140	100	110
3	100	110	75	85	145	165	115	130
2	115	130	90	100	170	190	135	150
1	130	150	100	115	195	220	155	175
1/0	150	170	120	135	230	260	180	205
2/0	175	195	135	150	265	300	210	235
3/0	200	225	155	175	310	350	240	275
4/0	230	260	180	205	360	405	280	315
250	255	290	205	230	405	455	315	355
500	380	430	310	350	620	700	485	545

Ambient	Ampacity Correction Factor							
88–95°F (31–35°C)	0.94	0.96	0.94	0.96	0.94	0.96	0.94	0.96
97–104°F (36–40°C)	0.88	0.91	0.88	0.91	0.88	0.91	0.88	0.91

TABLE 11-3
Correction factors for more than three current-carrying conductors in raceway or cable

Conductors	4–6	7–9	10–20	21–30	31–40
Factor	0.80	0.70	0.50	0.45	0.40

(Reprinted with permission from NFPA 70-2008, *National Electric Code®*, Copyright © 2007, National Fire Protection Association, Quincy, MA 02169. This reprint material is not the complete and official position of the NFPA on the reference subject, which is represented only by the standard in its entirety.)

(flat wires), which can be installed without raceways. The following are commonly used wiring methods:

1. *Electrical metallic tubing* (EMT) is commonly referred to as thin-wall conduit. It is made in ½-in. to 4-in. sizes, using a crimp type of coupling for quick installation. EMT is the wiring method most commonly used for all building applications, except in locations where wiring must be watertight, or explosion-proof.

2. *Rigid conduit* is similar to EMT, except that it uses threaded couplings. It may be used for all applications, including explosive and wet locations. It is more expensive than EMT; therefore, it is used only for 4-in. and larger sizes or where EMT is not permitted.

3. *Wireways* are used to enclose a large number of wires. They are usually 3 in. to 8 in. in size, contain tens or hundreds of wires, and should be installed where they are accessible. Surface metal raceways, such as "wiremold" and "plugmold," are examples of smaller wireways. (See Figure 11–18.)

4. *Bus ducts* are used for feeding large power distribution systems. They come with a feeder or plug-in design. (See Figure 11–19.)

5. *Underfloor ducts* are raceways cast into a poured floor slab to supply electrical wiring to the center of large rooms. Single, double, and triple ducts may be allocated for different electrical systems. (See Figure 11–20.)

TABLE 11-4
Dimensions of several rubber- and thermoplastic-covered conductors (Refer to NEC for a complete listing of conductors as well as other properties)

Size, AWG MCM	Types RFH-2, RH, RHH		Types TF, THW, TW		Types TFN, THHN, THWN	
	Approximate Diameter, in.	Approximate Area, sq in.	Approximate Diameter, in.	Approximate Area, sq in.	Approximate Diameter, in.	Approximate Area sq in.
14	.204*	.0327	.131	.0135	.105	.0087
12	.221*	.0384	.148	.0172	.122	.0117
10	.242	.0460	.168	.0222	.153	.0184
8	.328	.0845	.245	.0471	.218	.0373
6	.397	.1238	.323	.1819	.257	.0519
4	.452	.1605	.372	.1087	.328	.0845
3	.481	.1817	.401	.1263	.356	.0995
2	.513	.2067	.433	.1473	.388	.1182
1	.588	.2715	.508	.2027	.450	.1590
1/0	.629	.3107	.549	.2367	.491	.1893
2/0	.675	.3578	.595	.2781	.537	.2265
3/0	.727	.4151	.647	.3288	.588	.2715
4/0	.785	.4840	.705	.3904	.646	.3278

*The dimensions of RHH and RHW.

■ **FIGURE 11–18**
Section of surface wireway showing divided compartments, plug-in receptacles, and wires. (Reproduced with permission from The Wiremold Co.)

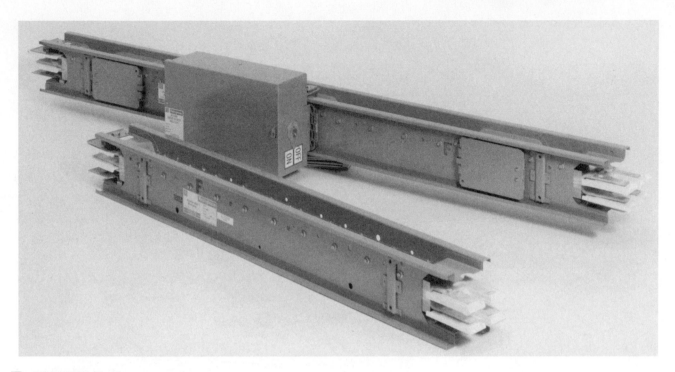

■ **FIGURE 11–19**
Sections of bus ducts. The upper one is the plug-in type ready to receive plug-in switches or circuit breakers. The lower one is a feeder bus duct that does not provide for a plug-in. (Reproduced with permission from Cutler-Hammer/Westinghouse.)

6. A *cellular floor* is a combined structural floor and electrical raceway system for a modern office or research building where electrical power is required throughout the floor area. It is a very flexible system. A cellular floor system may use one or more structural cells as electrical cells, depending on the electrical system used. The cost of having the system installed is usually higher than that for a raceway system; however, on a life-cycle basis, a cellular floor is both economical and necessary for modern office buildings. (See Figure 11–21.)

7. A *raised floor* installed above the structural floor creates space between the two floors to run wiring with or without additional raceways. Raised floors are commonly used for computer mainframe space and spaces with a concentration of electrical equipment. (See Figure 11–22.) A raised floor is flexible but costly.

With the increased popularity of telecommunications, data transmission, and local area networks, most modern office buildings have some form of underfloor raceways.

11.9 INSTALLATION OF WIRES IN RACEWAYS

11.9.1 Partial Fill of Raceways

The installation of wires (or cables) in raceways is strictly regulated. Generally, no more than 40 percent of the cross-sectional area of the raceway can be filled with wires or cables. The limitation is necessary for two key reasons:

- *To prevent excessive heat buildup* All wires have resistances and impedances that create a power loss which turns into heat and, if unabated, may cause the breakdown of the insulation material or even a fire.

- *To permit the physical installation of the wires* Wires in conduits must be pulled into the conduits by special tools. A clear space must be provided for the wires to be pulled in easily, without damage.

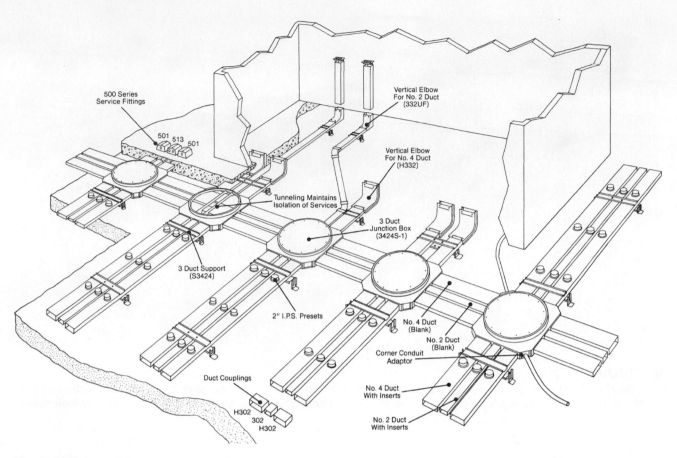

■ FIGURE 11–20

Single-level, three-cell underfloor duct system. Normally, the smaller cells are for power and data, and the larger cell is for communication wiring. For the proper class of wiring, data and communication wiring may be installed in the same cell. (Courtesy: General Electric Company, Nela Park, Cleveland, OH.)

Rules governing the number and size of wires that can be installed in a raceway are given in detail in the National Electrical Code. Tables 11–2, 11–3, 11–4, and 11–6 are condensed from the NEC tables for some cable types.

11.9.2 Accessible Boxes

- *Pull boxes* When the raceway (conduit) is too long or contains too many bends, pull boxes must be installed at locations to facilitate the pulling of conductors into the raceways.
- *Junction boxes* When the conductors need to be spliced or connected to another conductor, the splices must be made within a box.
- *Outlet boxes* Every wiring device must be installed within an outlet box.

All pull, junction, and outlet boxes must be accessible for inspection and services. NEC defines "accessible" as applied to wiring methods means capable of being removed or exposed without damaging the building structure or finish, or not permanently obstructed by the structure or finish of the building.

Example 11.8 How many No. 12-type THWN wires can be installed in a 1-in. conduit?

Answer From Table 11–5, the maximum number is 19.

Example 11.9 What is the allowable ampacity of three 90°C-rated, 600-V, No. 10 copper conductors in a conduit?

Answer From Table 11–2, the maximum allowed is 40 A.

Example 11.10 What is the allowable ampacity if eight No. 10 copper conductors are installed in one conduit?

Answer From Table 11–2, the maximum ampacity of 90°C-rated conductor is 40 A for not more than three conductors in the conduit.

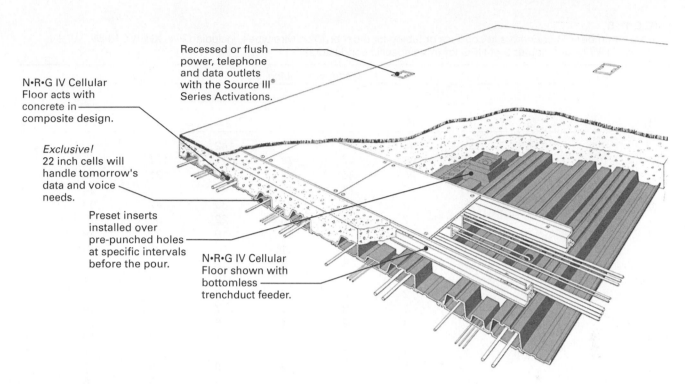

■ **FIGURE 11–21**

Cutaway view of an underfloor electrical distribution system consisting of a trench duct above the cellular floor cells. The trench duct is for main distribution, and the cells are for wiring to floor outlets. (Reproduced with permission from Walker Systems, Inc., a Wiremold Company.)

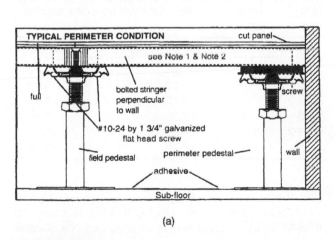

(a)

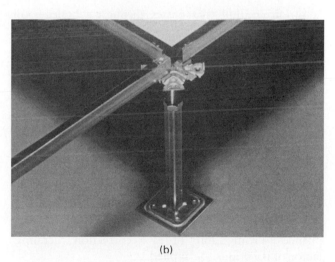

(b)

■ **FIGURE 11–22**

(a) Cutaway section of a raised floor with adjustable pedestal supports. The panels are normally 2-ft × 2-ft modules.
(b) Anchoring details of a pedestal. (Reproduced with permission from Tate Access Floors, Inc.)

Answer From Table 11–3, the correction factor is 0.7. Thus, the allowable ampacity is 0.7 × 40, or 28 A.

Example 11.11 For the same installation as in Example 11.9, but with the wires exposed to an ambient temperature of 100°F, what is the allowable ampacity?

Answer From Table 11–2, the correction factor is 0.91; thus, the allowable ampacity is further reduced to 0.91 × 28, or 25.5 A.

Example 11.12 Two sets of 120/208-V, three-phase, four-wire distribution system feeders are installed in a

TABLE 11-5
Maximum number of conductors in conduits or tubing for most building wire types, including TW, XHHW, RHW, TWHN, RHH, TW, THW (See Chapter 9 of NEC for more specific conductors)

AWG and MCM	Conduit or Tubing, in.									
	½	¾	1	1¼	1½	2	2½	3	3½	4
14	9	15	25	44	60	99	142			
12	7	12	19	35	47	78	111	171		
10	5	9	15	26	36	60	85	131	176	
8	2	4	7	12	17	28	40	62	84	108
6	1	3	5	9	13	21	30	47	63	81
4	1	2	4	7	9	16	22	35	47	60
3	1	1	3	6	8	13	19	29	39	51
2	1	1	3	5	7	11	16	25	33	43
1		1	1	3	5	8	12	18	25	32
1/0		1	1	3	4	7	10	15	21	27
2/0		1	1	2	3	6	8	13	17	22
3/0		1	1	1	3	5	7	11	14	18
4/0			1	1	2	4	6	9	12	15
250			1	1	1	3	4	7	10	12
300			1	1	1	3	4	6	8	11
350			1	1	1	2	3	5	7	9
500				1	1	1	3	5	6	8

common conduit that passes through a boiler room with a maximum ambient temperature of 102°F. The demand current of feeder 1 is calculated to be 100 A, and that for feeder 2 is 50 A. Determine the feeder sizes, based on 90°C copper wires (cables), and select the common conduit size. Assume the selected feeders are type THHN copper.

Answer There are four wires in each set of feeders, or eight for feeders 1 and 2. Theoretically, the neutral conductors may not carry any current if the load is balanced between phases A, B, and C; however, recent design practices have been to treat the neutral conductor as a current-carrying conductor, owing to the third harmonics of inductive loads such as PCs and electronic appliances. From Table 11–3, a correction (derating) factor of 0.70 must be applied.

The ambient temperature in the boiler room is 102°F; thus, a correction (derating) factor of 0.91 must be applied for the 90°C-rated wires (cables).

The overall derating factor for ampacity is $0.70 \times 0.91 = 0.637$; thus, feeder 1 must be selected for 100 A/0.637 = 157 A, and feeder 2 must be selected for 50A/0.637 = 78.5 A.

From Table 11–2, feeder 1 must be a minimum size of 1/0 AWG, which is rated for 170 A under normal conditions, and feeder 2 must be a minimum size of No. 4 AWG.

TABLE 11-6
Dimensions and internal areas of electrical metallic tubing (EMT) and conduit (See NEC for other conduit sizes and interior dimensions)

Size, in.	Internal Diam., in.	Area, sq in.
1-1/2	1.61	2.04
2	2.07	3.36
2-1/2	2.73	5.86
3	3.36	8.85
3-1/2	3.84	11.55
4	4.33	14.76

From Table 11–4, No. 1/0 THHN cable has 0.1893 sq in. of cross-sectional area, and No. 4 cable has 0.0845 sq in. The total cross-sectional area of all the cables is

$$(4) \times 0.1893 + (4) \times 0.0845 = 1.160 \text{ sq in}$$

Given the maximum 40 percent fill rule, the conduit must have a minimum cross-sectional area of 1.160/40 percent, or 2.9 sq in. From Table 11–6, a 1-½ inch conduit has a cross-sectional area of 2.04 sq in. It is smaller than the required 2.9 sq in. Thus, the next larger size, 2-in. conduit having a cross-sectional area of 3.36 sq in., must be used.

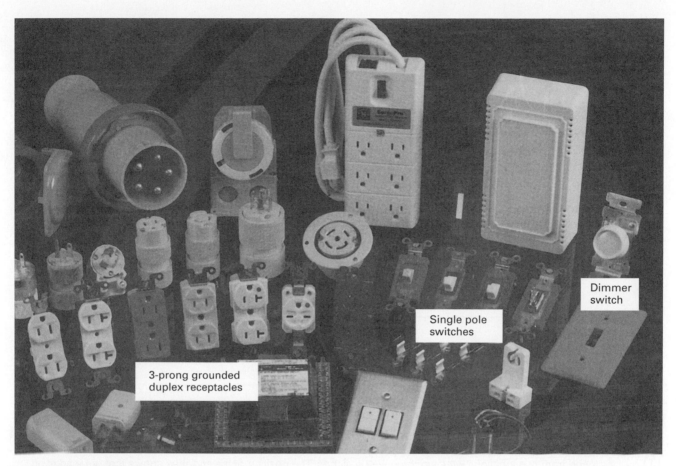

■ **FIGURE 11–23**

Typical receptacles and switches. (Courtesy: General Electric Company, Nela Park, Cleveland, OH.)

11.10 WIRING DEVICES

Various wiring devices—from switches, receptacles, and overcurrent protection devices to contactors and dimmers—are used in electrical systems. All wiring devices must be installed in code-approved boxes, regardless of the wiring system. The devices most commonly used are described in the rest of this section; some are shown in Figure 11–23.

11.10.1 Switches

A switch is a device for making, breaking, modulating, or changing the connections in an electrical circuit. Switches are classified according to the following criteria:

1. *NEC rating.* General service, isolating, motor duty, etc.
2. *Method of engaging contact.* Sliding, snap, liquid (mercury), etc.
3. *Voltage rating.* 125, 250, 277, 480, 600, 5000 V, etc.

4. *Number of breaks.* Single break, double break, etc.
5. *Number of poles.* 1, 2, 3, 4 poles, etc.
6. *Number of closed positions.* Single throw, double throw, etc.
7. *Method of operation.* Manual, magnetic, motor-operated, etc.
8. *Speed of operation.* Slow make/slow break, quick make/quick break, etc.
9. *Enclosure.* Open, enclosed, weathertight, watertight, explosion-proof, etc.
10. *Control function.* Single-acting, three-way, four-way, etc.
11. *Method of protection.* Nonfused, fused, circuit breaker, combination, etc.
12. *Performance of contacts.* Maintained contact, momentary contact, etc.
13. *Duty.* Light-duty, heavy-duty, load-interrupting-duty, etc.
14. *Other functions.* Dimming or voltage control, photoelectric, time clock, electrically or mechanically held, auxiliary controlled—pressure, temperature, flow, infrared, motion, proximity-sensitive, etc.

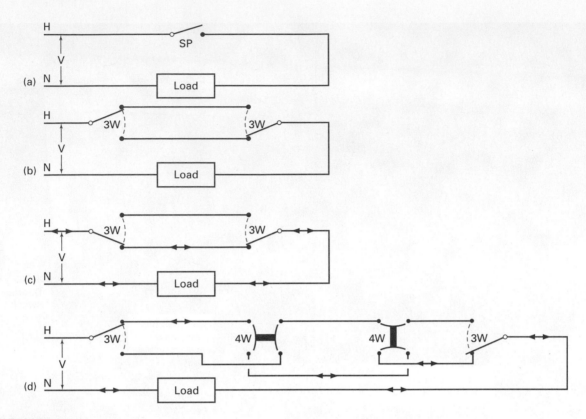

■ **FIGURE 11–24**
Schematic wiring diagrams.
(a) Single-pole (two-way) switched circuit.
(b) Load switched from two locations with two three-way switches. The position of the switches shows that the load is off.
(c) Same circuit as (b) with the load on. Change the position of any one of the three-way switches, and the load will be off.
(d) Load is controlled at four locations with two three-way and two four-way switches. The position of the switch shows that the load is on: Change the position of any switch, and the load will be off.

Light switches are normally single-pole, single-throw switches. When lights need to be switched from more than one location, three-way and four-way switches are used. The operating principles of three-way and four-way switches are illustrated in Figure 11–24. As a rule, the first and last switches must be three-way switches, and the in-between switches must be four-way switches.

11.10.2 Remote-Control Low-Voltage Switching

When switching of lights and appliance loads is desired at multiple locations, a remote-control low-voltage switching system will provide flexibility and economy. With this system, all control wires are reduced to 24 V or less and thus need not be installed in raceways (conduits). The loads, whether 120, 240, or 277 V, are operated by one or more momentary contact–type electromagnetic relays. A typical remote-control low-voltage wiring system is shown in Figure 11–25.

11.10.3 Receptacles

A *receptacle* is a wiring device installed within an outlet box for the connection of electrical apparatus through an attachment plug. (See Figure 11–23.) Receptacles may be classified according to the following characteristics:

1. *Number of receptacles in the assembly.* Single, double (duplex), triple (triplex), etc.
2. *Current rating.* 5, 15, 20, 30, 50, 60, 100 A, etc.
3. *Voltage rating.* 125, 250, 277, 480, 600 V, etc.
4. *Number of poles and wires.* Two-pole, two-wire; two-pole, three-wire grounding; three-pole, three-wire; three-pole, four-wire grounding, etc.
5. *Shape of the blades.* Straight blades, locking blades (which can lock the plug in place by twisting).
6. *Configuration of the blades.* Parallel, angled, flanged, polarized, etc.

(a)

(b)

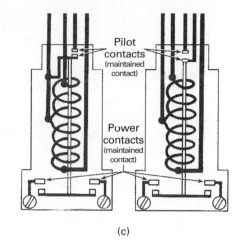
(c)

(d)

■ **FIGURE 11–25**

Remote-control low-voltage switching.
(a) Low-voltage switch.
(b) Low-voltage relay.
(c) Sectional view of low-voltage relay, illustrating the pilot and maintained contacts and the double-acting electromagnetic coils.
(d) Ganged master control for low-voltage switches.
(e) Typical wiring diagram of a remote-control low-voltage switching system.
(Courtesy: General Electric Company, Nela Park, Cleveland, OH.)

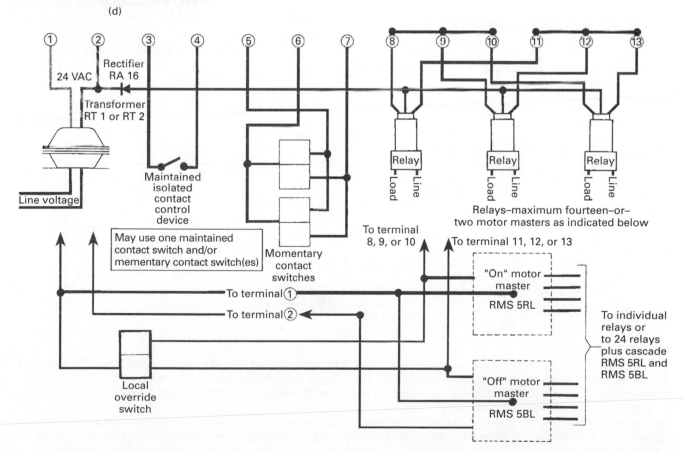
(e)

Receptacles commonly used for plug-in types of apparatus such as typewriters, portable lights, and televisions are also known as *convenience outlets* and are normally rated for 125 V and 15 A. They may be two-wire with nonpolarized parallel blades, or three-wire with polarized blades. Since all 120-V electrical systems in a building are grounded, a two-wire parallel-blade nonpolarized receptacle is not able to distinguish the polarity of the wires and, hence, which side of the plug is grounded. Logically, there is a 50–50 chance that the equipment is plugged in on the wrong side and will subject a person to the line voltage if the ungrounded side of the wire touches the equipment enclosure. Fortunately, the 125-V system is not deadly if a person accidentally touches it, although an electrical shock will definitely be felt. For this reason, the NEC no longer approves nonpolarized and ungrounded receptacles in new installations. All two-wire ungrounded receptacles will eventually be replaced.

The National Electrical Manufacturers' Association (NEMA) has issued standard configurations of various wiring devices, including plug-in receptacles based on ampacity rating, blade configuration, and number of poles. Some NEMA standards/receptacles are shown in Figure 11–23.

11.10.4 Contactors and Relays

Contactors and relays are remote-control power-transmitting devices. The former are normally used to carry line voltage power, and the latter are normally used to carry line or low-voltage power. The construction of these devices is the same and includes an electromagnetic coil and a set of electrical contacts. When the coil is energized, the magnetic force that is created causes the contacts, and thus the electrical circuit, to open or to close. Figure 11–26 illustrates the basic wiring diagrams of a contactor and a relay. The wiring diagram of magnetic motor starters shown in Figure 11–25 is, in effect, a combination magnetic contactor with overload relays. Contactors may be classified according to the following properties:

1. *Voltage.* 24, 48, 125, 250, 600, 5000 V, etc.
2. *Number of contacts (poles).* 2, 3, 4, 6, 12 poles, etc.
3. *Type of contacts.* Button, mercury, knife blade, etc.
4. *Type of operations.* Maintained, momentary contact, etc.
5. *Current rating.* 15, 30, 50, 60, 100, 200, 400, 800, 1200 A, etc.
6. *Duty.* Inductive, noninductive loads, etc.

Figure 11–27 shows a control scheme using momentary-contact stop-start push buttons. The start push button will close the control circuit through the coil (*M*), which closes the normally open (NO) main contact, allowing power to pass through L1 and L2. The auxiliary contact (*M*) wired across the start push button is closed to keep the control circuit energized even when the push button is released. Figure 11–27(b) is similar to Figure 11–27(a) except that the control switch is a maintained-contact switch, and the main

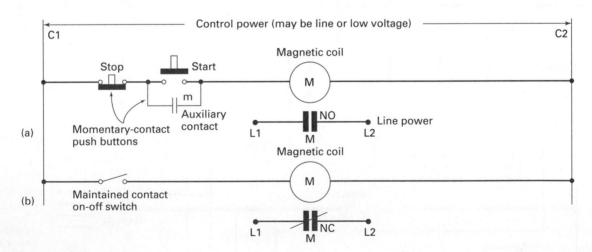

■ **FIGURE 11–26**

(a) The elementary diagram shows a control scheme using momentary-contact stop-start buttons (switches). The start push button will close the control circuit through the coil (M), which closes the normally open (NO) main contact, allowing power to pass through L1 and L2. The auxiliary contact (m) wired across the start push button is closed to keep the control circuit energized even when the push button is released.

(b) The elementary diagram shows a control scheme using a maintained-contact switch, and the main contact is normally closed (NC). This means that when the control circuit is closed, the main contact (M) of the power circuit will be opened (deenergized).

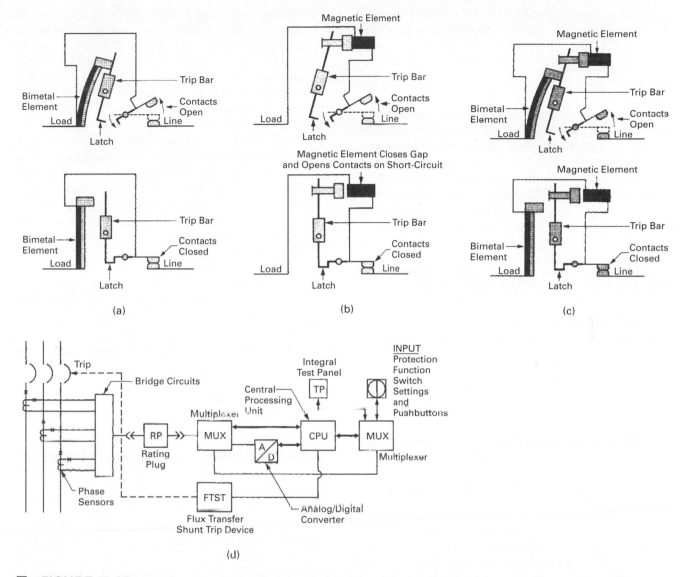

(a)   (b)   (c)

(d)

■ **FIGURE 11-27**

Operating principles of circuit breakers. (a) The thermal element of the tripping mechanism is a bimetal that is heated by the load current. On a sustained overload, temperature will build up on the bimetal. Owing to the difference in coefficient of thermal expansion of the two metals, the bimetal will deflect, causing the operating mechanism to trip (open). The speed of action of the thermal element is inversely proportional to the magnitude of the current. (b) The magnetic trip action is achieved through the use of an electromagnet whose winding is in series with the load current. When a short circuit (fault) occurs, the magnetic field caused by the abnormally high current flow will trip (open) the contacts. The magnetic trip is nearly instantaneous to quickly isolate the electrical system and equipment from the fault. (c) Most circuit breakers are designed to include both thermal and magnetic protection devices. (d) Electronic circuit breaker. A digital trip unit normally starts with analog inputs, multiplexer, and analog/digital converter feeding into a microprocessor or central processing unit (CPU). Input data are continuously scanned and updated with appropriate action taken. The microprocessor can be reprogrammed to fit the load changes. (Reproduced with permission from Cutler-Hammer/Westinghouse.)

contact is normally closed (NC). This means that when the control circuit is energized, the power circuit will be deenergized.

11.10.5 Dimmers

Dimmers are operating devices that reduce the input to, and thus the output from, an apparatus. Although most dimmers are used to control the intensity of light output, they are also used to control the speed of a fan, drill, etc. Dimmers usually operate on the principle of adjustable resistance, adjustable voltage with autotransformer, or solid-state elements to control the shape of the wave or duration of the current through a load.

11.11 PROTECTIVE DEVICES

Electrical circuits that include feeders, distribution equipment, branch circuits, and the load equipment must be protected from exceeding their rated capacity, which may occur as a result of many circumstances. Some examples are:

- *Overcurrent,* due to mechanical overload or internal or external electrical faults.
- *Overvoltage,* due to short-circuiting between primary and secondary wiring or due to a lightning strike.
- *Reverse in polarity of a three-phase system,* due to a change of power service.

The most common method used to prevent the damage caused by overloading conditions is the installation of protective devices at strategic locations (e.g., on switchboards, on panelboards, at the beginning of a feeder, in a branch circuit, or at the equipment). These devices are divided into three general types: relays, circuit breakers, and fuses. Relays are normally used by utility companies to protect their primary distribution system or large primary equipment within a network. For building systems and equipment, circuit breakers and fuses are usually used.

11.11.1 Circuit Breakers

Classification of Circuit Breakers

A circuit breaker (CB) is defined in NEMA standards as a device designed to open and close a circuit by nonautomatic means, and to open the circuit automatically on a predetermined overcurrent, without damage when properly applied within its rating. There are three types of circuit breakers.

1. *Molded-case circuit breakers (MCCB).* The current-carrying parts, mechanisms, and trip devices are completely contained within a molded case of insulating material. MCCBs are available in small and medium frame sizes from 30 A to 800 A, and with trip sizes from 15 A to 800 A.
2. *Low-voltage-power circuit breakers (LVPCB).* These CBs are also known as *air breakers.* They are primarily used in drawout switchgear construction. Drawout circuit breakers are mounted on rails that can be pulled out like a drawer for ease of maintenance. LVPCBs have replaceable contacts and are designed to be maintained in the field. LVPCBs are available in medium and large frame sizes from 600 A to 4000 A. LVPCBs are rated from 600 V, whereas MVPCBs are rated for up to 72.5 kV, and HVPCBs for over 72.5 kV.
3. *Insulated-case circuit breakers (ICCB).* These have the construction characteristics of both MCCBs and LVPCBs and are used primarily in fixed mounted switchboards but are also available in drawout configurations. The frame sizes range from 600 A to 4000 A.

Construction and Features of Circuit Breakers

Circuit breakers are also classified by other construction and operating features according to the following:

1. *Arc quenching (extinguishing) media.* Air or oil.
2. *Operating principle.* Thermal, magnetic, thermal magnetic, solid state (electronic), etc.
3. *Voltage class.* 125, 250, 600 V, 5, 12, 15, 35 kV, etc.
4. *Frame size.* 30, 50, 100, 225, 400, 600, 800, 1200, 2000, 4000 A, etc.
5. *Trip ratings.* 15, 20, 30, 50, 90, 100 A, and up to the frame size ratings.
6. *Interrupting capacity.* 5000, 10,000, 15,000, 20,000, 30,000, 50,000, 75,000 A, and up.
7. *Operating methods.* Manual, remote operating, etc.
8. *Other features.* Overvoltage, undervoltage, auxiliary contacts, reverse current, reverse phase, etc.

Operating Principles of Circuit Breakers

There are two types of circuit-breaking (tripping) components within a CB.

1. *Bimetal/electromagnetic type.* Consists of a bimetal element, which responds to temperature buildup inside the CB; and an electromagnet, which responds to the magnetic force caused by an abnormally high current flow. The bimetal provides thermal protection, and the magnet provides short-circuit protection. The operating principles of these

elements are illustrated in Figure 11–27(a), (b), and (c).

2. *Solid-state (electronic) type.* Consists of analog or digital devices to sense the electrical characteristics or circumstances of the circuit and process the data through a central processing unit (CPU) with pre-programmed actions. The analog type senses the peak current, whereas the digital type senses the rms current of the current flow. The rms value is a more realistic representation of the AC current. The operating principle of a digital-type solid-state CB is shown in Figure 11–27(d).

Advantages of Circuit Breakers

The following are advantages of circuit breakers over other types of protection devices, namely fuses:

- Easily resettable when the system is tripped
- More compact
- Adaptable for remote control and electrical inter-lock with other equipment
- Can serve as a disconnect switch (although should not be used as an operating switch)

Figure 11–28(a) and 11–28(b) shows several types of circuit breakers—a single-pole molded-case branch circuit breaker (MCCB) and a high-interrupting capacity (LVPCB) breaker with solid-state sensing and controls.

11.11.2 Fuses

A fuse is an electrical protective device that melts upon sensing an abnormal current and opens the circuit in which it is installed. It is a self-destructive device.

Classification of Fuses

There are many types of fuses, classified as follows (see Figure 11–29):

1. *Voltage class.* 12, 24, 125, 250, 600, 5000 V, and higher voltages.
2. *Current rating.* From a fraction of an ampere to 6000 A.
3. *Construction.* Nonrenewable or renewable, single or dual elements, etc.
4. *Principle of operations.* Fast clearing, time delay, current limiting, etc.
5. *Short-circuit interrupting capacity.* 5000 to 200,000 A.
6. *Fusible material.* Lead, tin, copper, silver, etc.

Operating Principle of Fuses

The operating principle of a single-element fuse is very simple. The fusible link is made of a lead–tin–antimony eutectic alloy that has a single-temperature melting point without softening prior to reaching its melting point. The fusible element is precisely made, having bottlenecks (narrow sections) that melt upon overheating due to a higher current flow. (See Figure 11–30(a).)

The dual-element fuses as illustrated in Figure 11–30(b) are time-delay fuses. Each consists of two fusible elements. Under normal operating conditions, the time-delay element will break loose when the fusible material holding the S-connector is melted. Because a heat sink is attached to the S-connector, the melting is intentionally delayed to avoid nuisance tripping of the connected load, such as a motor load. A motor load has a high inrush during start-up. If there is no time delay to compensate for the high inrush current, then the motor load will be cleared before it can be started. Figure 11–30(b) demonstrates the action of a dual- element fuse under various conditions. The second fusible element, as in the single-element fuse, melts when it senses an abnormally high short-circuit current.

Advantages of Fuses

Two major advantages of fuses are that they are fast-acting and self-coordinating in a properly designed distribution system. If the same class of fuse is used at all levels of protection, the lower-level fuses (naturally, smaller in ampere ratings) will clear first and prevent interruption of the upper-level fuses. Figure 11–31(a) illustrates the normal trip time of a molded-case circuit breaker, which is over one cycle. Figure 11–31(b) is the clearing time of a fuse, which is normally less than half a cycle. Figure 11–31(c) illustrates the natural coordination of fuses in a distribution system in which the feeder fuse has a shorter clearing time than the main fuses.

11.11.3 Application of Protective Devices

Through improved mass-production technology, small low-voltage circuit breakers for single- and double-pole breakers up to 200 A are now as economical as the corresponding fused switches. Thus, circuit breakers are universally used for lighting and receptacle panels; however, for large-ampacity loads—and especially systems with 40,000 A or higher or available short-circuit current—fused switches are definitely more economical. A combination of fuses and circuit breakers is often used for systems having an available short-circuit current in excess of 50,000 A.

Coordination is extremely important among different levels of protective devices, so that upon overload or fault, only the protective device next to the load or fault will open, while the others remain in service.

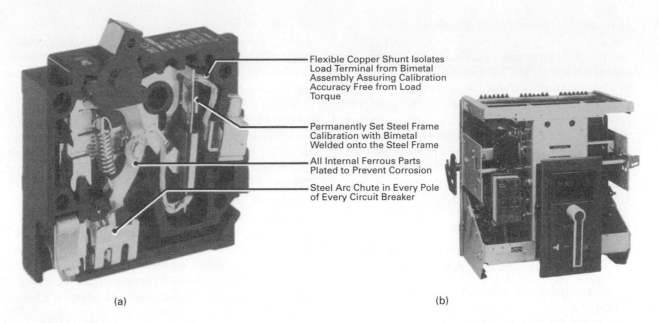

Flexible Copper Shunt Isolates Load Terminal from Bimetal Assembly Assuring Calibration Accuracy Free from Load Torque

Permanently Set Steel Frame Calibration with Bimetal Welded onto the Steel Frame

All Internal Ferrous Parts Plated to Prevent Corrosion

Steel Arc Chute in Every Pole of Every Circuit Breaker

(a)

(b)

■ **FIGURE 11–28**

Typical construction of circuit breakers. (a) Internal mechanism of a single-pole, molded-case, thermal-magnetic circuit breaker. Nominal rating from 15 A to 100 A, with an interrupting capacity not to exceed 10,000 A. (b) Large drawout low-voltage-power circuit breaker (LVPCB). Nominal rating from 600 A to 4000 A, 480 V to 5000 V, with interrupting capacity up to 200,000 A. (Courtesy: (a) Reproduced with permission from Challenger Electrical Equipment Corp. (b) Reproduced with permission from Cutler-Hammer/Westinghouse.)

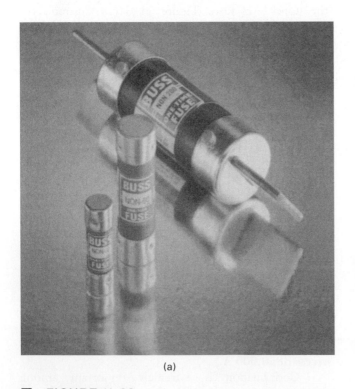

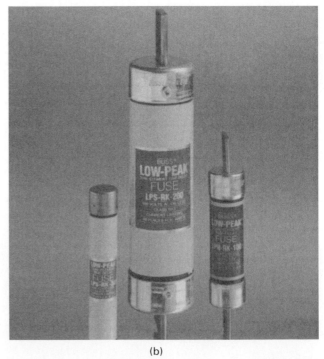

(a)

(b)

■ **FIGURE 11–29**

Several cartridge fuses. (a) General-purpose, single- or dual-element type. (b) High-current-limiting dual-element type. (Courtesy: Bussmann Division, Cooper Industries, St. Louis, MO.)

1. Cutaway view of a typical single-element fuse.

2. Under a sustained overload, a section of the link melts and an arc is established.

3. The open single-element fuse after opening a circuit overload.

4. When the fuse is subjected to a short-circuit current, several sections of the fuse link melt almost instantly.

5. The open single-element fuse after opening a shorted circuit.

(a)

Overload section consists of spring loaded "S" connector held in place by fusing alloy

Short-circuit sections are connected through the heat absorber

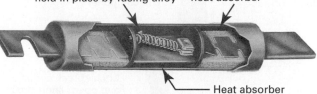

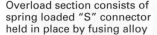

Heat absorber

1. The dual-element fuse has distinct and separate overload and short-circuit elements.

2. Under a sustained overload, the trigger spring fractures the calibrated fusing alloy and releases the connector.

3. The open dual-element fuse after opening a circuit overload.

4. A short-circuit current causes the restricted portions of the short-circuit elements to melt and causes arcing to burn back the resulting gaps until the arcs are suppressed by the arc-quenching material and increased arc resistance.

5. The open dual-element fuse after opening a short circuit.

(b)

■ **FIGURE 11–30**

(a) Sequence of operation of a single-element fuse. (b) Sequence of operation of a dual-element (time-delay) fuse.
(Courtesy: Bussmann Division, Cooper Industries, St. Louis, MO.)

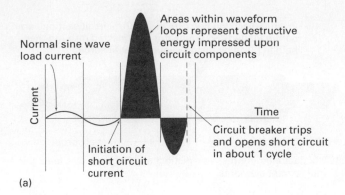

(a)

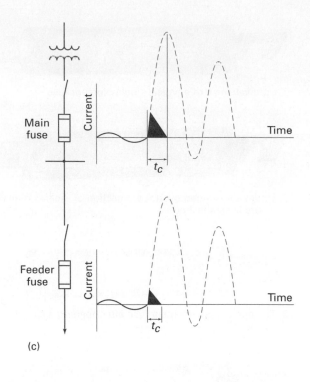

(c)

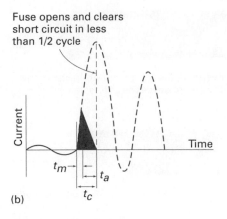

(b)

■ **FIGURE 11–31**

(a) Typical operation of a molded-case circuit breaker, which opens a short circuit in about one cycle (1/60 of a second on 60 Hz power system). (b) Typical operation of a current-limiting fuse, which opens a short circuit in less than half a cycle (t_m = melting time, t_a = arcing time, and $t_m + t_a = t_c$ = clearing time). (c) Smaller fuses have a faster clearing time than larger fuses of the same class and thus will open first to selectively protect the distribution system from nuisance tripping. (Courtesy: Bussmann Division, Cooper industries, St. Louis, MO.)

Without proper coordination, more than one device may open, unnecessarily interrupting the services of other loads. Figure 11–32 illustrates the coordination between circuit breakers and relays. The same time–current graphs are used to coordinate fuses or fuses and circuit breakers.

QUESTIONS

11.1 What electrical distribution systems are normally used to serve residences? Small business buildings with load about 100 kVA?

11.2 What electrical distribution systems are normally selected to serve a combination of single-phase and three-phase loads?

11.3 What is meant by voltage spread of any electrical distribution system? Voltage drop?

11.4 What is the recommended maximum kVA rating of a single motor-driven piece of equipment on a

100-kVA, 120/240-V, single-phase, three-wire system? Why is the recommended maximum considerably lower than the system rating?

11.5 What is the voltage to ground of a 120/208-V, three-phase, four-wire system?

11.6 What is the voltage to ground of a 480-V, three-phase, three-wire system?

11.7 Why should the interior electrical distribution system be grounded?

11.8 Must all electrical systems be grounded?

11.9 Should the interrupting capacity (IC) of the switchboard be rated higher or lower than the available short-circuit current (I_s)?

11.10 Name a few NEC-classified wires for building interior wiring systems and temperature ratings.

11.11 What on-floor power system would you recommend for a high-rise office building designed for flexibility and capability of using PC computers and electronic equipment?

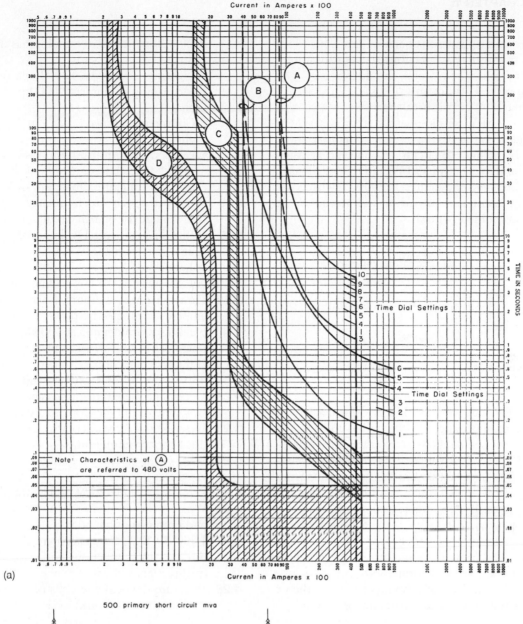

(a)

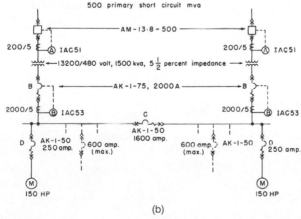

(b)

■ FIGURE 11–32

Coordination between protective devices. (a) The time–current curves of relay A and circuit breakers (B), (C), and (D) are selected so that (D) will trip first, followed by (C), (B), and (A). (b) With proper clearance between the tripping times, system integrity and reliability are maintained, and there is less chance of nuisance tripping or system interruption. (Courtesy: General Electric Company, Nela Park, Cleveland, OH.)

11.12 If you wish to control a bank of lights in a room from two different locations, how many three-way and four-way switches should you use?

11.13 Name several advantages of fuses. Name several advantages of circuit breakers.

11.14 A 120/240-V, single-phase system is a system with its center tap as the neutral point. (See drawing.) If the load of circuit 1 connected between line 1 and neutral is 1500 W, and that of circuit 2 between line 2 and neutral is 1000 W, what is the current in amperes carried by each circuit? What is the combined neutral?

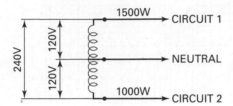

11.15 If the full load current of a transformer having 5 percent impedance is 500 A, what could be the expected short-circuit current at the main switchboard? (*Note:* For simplicity, a conservative approach is to assume that the utility power network serving this transformer is sufficiently large to have practically zero impedance during a fault.)

11.16 According to NEC, what is the allowable (maximum) current that can be carried by a single #3/0, 90°C rated conductor in the conduit? What if there are six conductors in the same conduit? What size conduit should be used for the six conductors (including the neutrals)?

11.17 A 20,000-sq-ft school building has the following estimated demand loads:
- Lighting: 20,000 W, 120-V one-phase
- Mechanical equipment: 100-hp, 480-V, three-phase
- Convenience power: 1 W/sq ft, 120-V, one-phase
- What power distribution system or systems would you choose? (*Note:* 1 hp = 746 W, 0.75 kW, or approximately 1 kVA @ 0.75 PF.)

11.18 What is the standard frequency and the utilization voltage for branch circuits in Europe?

11.19 If the line-to-neutral voltage of a three-phase four-wire system is 220 V, what is the line-to-line voltage? Similarly, for a 2400-V line-to-neutral voltage, what is the line-to-line voltage?

11.20 Determine the maximum short-circuit current in amperes of an electric power system having one main transformer rated at 300 kVA, 480 V, three-phase. The transformer impedance is 3 percent.

11.21 If the system secondary voltage is 240 V in lieu of 480 V, what is the maximum short-circuit value?

11.22 Determine the interrupting capacity rating of the two systems in Questions 11.20 and 11.21.

11.23 Describe the three alternative power sources that can be used for buildings.

11.24 What is the difference between K-4- and K-13-rated transformers?

11.25 Why are large electrical cables stranded?

11.26 What is the difference between standby power and uninterruptible power?

11.27 How do K-type transformers, UPS, and TVSS differ in their functions with regard to power-quality management?

11.28 What are the pros and cons of placing a generator indoors versus outdoors?

COMMUNICATIONS, LIFE SAFETY, AND SECURITY SYSTEMS*

12

COMMUNICATIONS, LIFE SAFETY, AND SECURITY SYSTEMS are part of the numerous telecommunication systems that use electrical power to generate, process, store, or transmit information. Thus telecommunication systems are also considered information transport systems.

The scope of telecommunication systems in buildings is growing at a very rapid pace. What was considered a luxury or state-of-the-art technology only a few years ago is now a minimum requirement in newer buildings. For example, a major office building built to operate in the 21st century should be equipped with data, communications, security, audio visual and life safety systems, in addition to the traditional telephone, sound, signal, building automation, and fire alarm systems. Provisions should be made for raceways and/or cabling if some of these systems will not be installed initially.

With advancement in technology, architects and building systems design engineers should be fully cognizant of all available telecommunication systems, even though they may be expert in one or more of these systems. This chapter introduces only the scope and operating principles of some major telecommunication systems frequently required in modern buildings. Appendix A includes acronyms and abbreviations of basic terms with which a professional should be familiar. Some of these terms are discussed in the body of the chapter, and the others are explained in Appendix A.

12.1 FORM AND FUNCTION OF INFORMATION SYSTEMS

According to Davis and Davidson,[†] all information can be expressed by one or more of the following forms and functions:

1. *Forms*
   - Data   numbers, quantitative values
   - Text   written language
   - Sound   voice, signal, or music
   - Image   video, motion, or still

2. *Functions*
   - Generation   creating information into various forms
   - Processing   editing, calculating, analyzing, synthesizing, expanding, interpolating, or extrapolating information
   - Storage   keeping, filing, or memorizing information for retrieval
   - Transmission   sending or receiving information

As illustrated in Figure 12–1, a basic telephone system is an information system that transmits sound, and only sound (one form and one function). A more sophisticated telephone system can encompass the transmission and storage of data, text, sound, and images (four forms and one function). The interfacing of a telephone system with computer technology adds processing capability (four forms and two functions). Finally, the networking of computer, telephone, and telecommunication systems completes all four forms and four functions. As science and technology continue to advance, the time will come when the intelligence level of information systems approaches or even surpasses that of humans.

12.2 COMMON CHARACTERISTICS OF TELECOMMUNICATION SYSTEMS

Several characteristics common to all systems are helpful in planning building systems:

1. *Telecommunication systems operate on 30 V or less, either AC or DC.* These systems are frequently

*Steve Brohammer, RCDD, and Timothy D. Ruiz, RCDD are the contributing authors of this chapter. See Acknowledgments.
[†]Stan Davis and Bill Davidson, *20/20 Vision*, Simon & Schuster, New York, 1991.

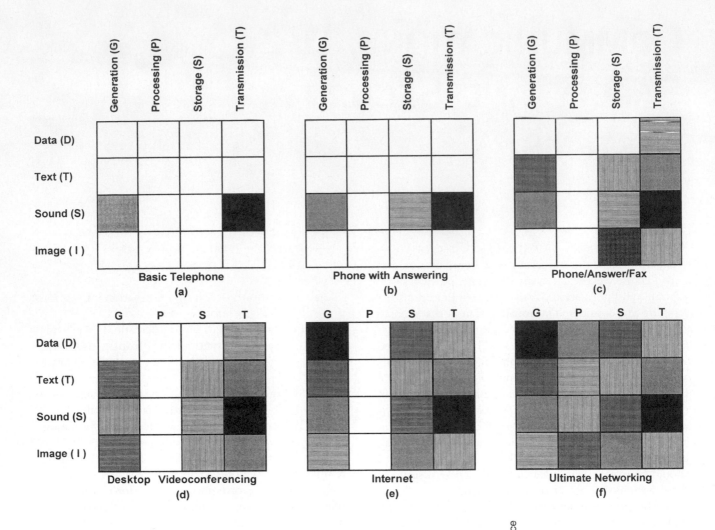

Basic Telephone
(a)

Phone with Answering
(b)

Phone/Answer/Fax
(c)

Desktop Videoconferencing
(d)

Internet
(e)

Ultimate Networking
(f)

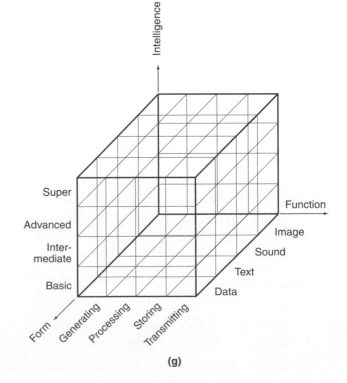

■ **FIGURE 12–1**

All information systems can be expressed by their forms and functions.

(a) A telephone system generates and transmits sound.

(b) A phone with an answering machine generates, transmits, and stores sound.

(c) A phone with a fax also stores text and transmits data, text, and images.

(d) A desktop videoconferencing system with an answering machine and a fax adds images to its form and function.

(e) A computer–phone interface adds processing capability.

(f) The networking of computer and telecommunications can perform all forms and functions.

(g) The ultimate information systems will increase in intelligence until they may supersede that of human beings.

described as low-voltage systems. At this voltage level, wiring may be installed without raceways, although raceways or cable trays are often used to facilitate installation, to provide protection, and to prevent tampering.

2. *Telecommunication systems feature limitless functions and options.* It is desirable to integrate these systems with the building systems design.

12.3 CLASSIFICATION OF TELECOMMUNICATION SYSTEMS

There are two major groups of telecommunication systems:

1. *Communication systems*

- Audio — public address, intercom, music, radio, etc.
- Video — TV, CATV, MATV, SATV, etc.
- Telephone — public private exchanges (PAX, PBX), voice over IP (VOIP), etc.
- Data — modem, local area network, wide-area network, wireless, etc.
- Signals — time, program, signals, etc.
- Multimedia — combination of audio/video/ telephone and data, such as video-conferencing and distance learning systems, etc.

2. *Building operational systems*

- Safety — fire alarm, sprinkler alarm, emergency evacuation, etc.
- Security — access control, video surveillance intrusion detection, etc.
- Automation — BAS, BMS, BMAS, etc.
- Specialty — electronic signage, sound masking and systems unique for special building occupancies such as hospitals, defense, retail stores, food services, and theatrical.

12.4 COMPONENTS AND WIRING

12.4.1 Basic Components

Most telecommunication systems operate on DC circuitry, because DC devices are more sensitive than AC. The current drawn can be at the level of milliamperes (1/1000 A) or microamperes (1/1,000,000 A). The following components are common to most systems.

Power Supply Unit

The power supply unit consists of electromagnetic transformers to transform a utility AC system from 110/220 V to less than 30 V. The AC circuit is then rectified into DC by a solid-state device known as diode or silicon-controlled-rectifier (SCR). Figure 12–2(a) illustrates the voltage transformation and rectifying circuitry of most telecommunication systems.

Sensing and Signaling Devices

Most telecommunication systems and/or its components detect, control, or amplify variations in energy, such as sound, light, motion, temperature, color, infrared, ultraviolet, heat, microwave, and other forms of low-level energy, and convert these forms of energy into electrical energy to operate signaling devices, such as speakers, telephones, clocks, and lights. This energy is converted and controlled with semiconductors or solid-state technology.

A semiconductor is usually defined as a material that has an electrical resistance between that of a conductor, such as metal, and an insulator, such as plastic. Semiconductors are, in general, crystalline materials containing varying degrees of impurities. Silicon and germanium are the most popular semiconductors. Others include copper oxide, selenium, and cadmium sulfide. Semiconductors conduct electricity because of the presence of free electrons in them. Semiconductors may carry negative charges (*n* type), positive charges (*p* type), or both (*p-n* junction type). These types form the various semiconductor devices and circuits, such as thyristors (silicon unilateral switches) and rectifiers (SCRs).

Control Devices

Control devices are manufactured to provide on-off, variable-output (voltage or current), maintained or momentary-contact, mechanical or electronic, direct or remote-controlled switching functions. Associated with switching are the necessary relays, circuit boards, and signal and sensing devices.

12.4.2 Wiring for Telecommunication Systems

Line voltage (110 V or higher) wiring for telecommunication systems follows the same guidelines as other wiring systems for power, lighting, and equipment discussed in Chapter 13; however, most telecommunication systems operate on low voltage and with limited power. Thus, the wiring system will not be a life or fire hazard. Low-voltage wiring is, therefore, normally installed without a traditional raceway but typically will utilize open cabling supported by J-hooks, bridle rings, cable tray, or similar methods.

■ FIGURE 12–2

(a) A typical full-wave rectifier circuit consists of a transformer and two diodes. The output is a pulsating DC varying voltage having the same frequency as the AC power supply. (Courtesy: Radio Shack.) (b) The upper diagram illustrates how a telephone voice signal is sampled at four equal intervals and converted into four 8-bit digital signals to represent the original analog input in millivolts. The lower diagram illustrates the conversion in reverse at the receiving end. (Source: Stephen Bigelow, *Understanding Telephone Electronics*, Prentice Hall Computer Publishing, Indianapolis, IN.) (c) Typical unshielded twisted pair (UTP) cable. Generally, a 2-in. to 6-in. twist is used. For high-speed data transmission, a tighter twist less than 1 in. is required. (d) Typical shielded twisted pair (STP) cable with metallic shield and a separate ground wiring. (e) Typical coaxial cable. (f) Illustrated at the left is a typical 4-pair UTP cable, and at the right is a typical 4-pair STP cable. Multipair UTP and STP cables are made in sets of up to 24 or more pairs. In all the illustrations of cable, the sizes are enlarged for clarification. In general, the conductors of UTP, STP, and coaxial cables are made with AWG No. 20 to No. 24 copper wires. The external diameter of the cable depends on the construction of the cable, as well as its shielding, insulation, and jacket thickness. A four-pair shielded cable may have an overall diameter of 0.25 in. (Reproduced with permission from Berk-Tek, Inc.)

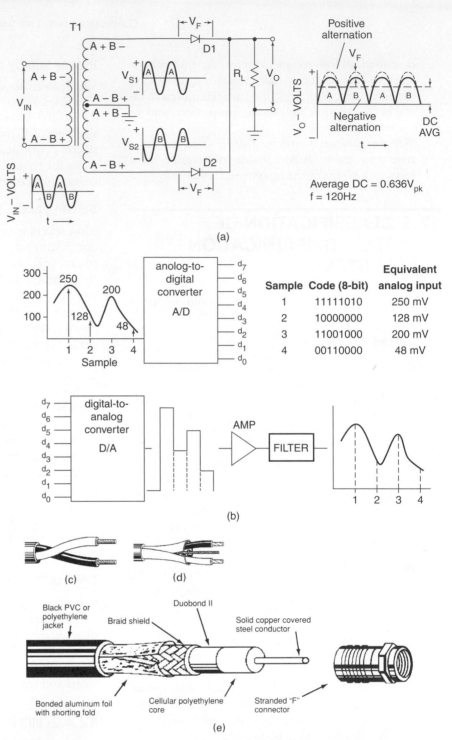

Average DC = $0.636V_{pk}$
f = 120Hz

Sample	Code (8-bit)	Equivalent analog input
1	11111010	250 mV
2	10000000	128 mV
3	11001000	200 mV
4	00110000	48 mV

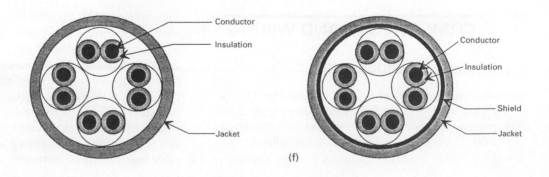

12.4.3 Basic Wiring for Low-Voltage and High-Frequency Systems

One of the characteristics of low-voltage systems is low signal strength. This is particularly true of systems such as voice (telephone), sound (music), radio, and video (TV). The power levels of these systems are usually measured in milliwatts or microwatts; thus the systems are very sensitive to external disturbances, such as electromagnetic interference (EMI). The transmission wiring is evaluated according to the following characteristics.

Characteristic Impedance

Most telecommunication systems operate at much higher frequencies than do power systems. Power systems operate at 60 to 400 Hz, whereas telecommunication systems operate in the kilohertz (10^3), megahertz (10^6) or gigahertz (10^9) ranges. At these high frequencies, it is extremely important to match the impedance of the transmission wiring to that of the terminal equipment to minimize the loss or attenuation of the signal. The term *characteristic impedance* is used to express the performance characteristics of the wires. It relates more to frequency than length. Typical impedances of low-voltage wires for high-frequency telecommunication systems are rated between 50 and 300 ohms.

Transmission Capacity

The transmission capacity of a medium is measured differently for different types of signals. *Analog signals* are continuous waves. Telephone and TV signals are typically of this variety. The transmission of an analog signal is measured by the bandwidth of the signal, in hertz. The *bandwidth* may be defined as the highest frequency that can be transmitted in a given medium without excessive attenuation. Typically, the upper limit is specified as the point at which the signal's strength has dropped by 3 dB. The wider the bandwidth, the higher the frequency a medium can support.

The bandwidth of a wiring system is limited not only by the construction of the wires but also by mechanical components (connectors, jacks, patch panels, etc.), by imperfections in wiring, and by effects of the environment (heat, moisture, proximity to power and other signals, etc.).

Digital signals are separate on-off pulses. Data, telephone, and computer signals are transmitted in digital format. The speed of transmission of digital signals is measured in bits, megabits, or gigabits per second. It should be obvious that a 100-megabyte-per-second (Mbps) wiring system can carry as much information as a 10 Mbps system. The one drawback of digital transmission is the increased bandwidth required to transmit

digital pulses compared with the bandwidth required for equivalent analog signals. A major advantage of digital transmission is the fidelity of the regenerated signal, which is relatively independent of the distance over which the signal is transmitted. (See reference 8.)

Analog and digital signals can be converted from one to the other. For example, the telephone network converts analog signals (the voice from the telephone handset) into digital signals for long-distance transmission and then converts the signal back to analog form on the other end. For a telephone system containing signals up to 4000 Hz (4 kHz), scanning (or sampling) is made at twice the frequency (8000 times per second). With a 7-bit code that contains the values from 0 to 127, the result is a very realistic reproduction of the voice waveform. With 8000 samples per second and a 7 bps digital rate, the speed of data transmission is 7×8000, or 56,000 bits per second. For high-fidelity speech or music, the sampling rate is usually four times faster, or 32,000 samples per second and 224,000 bps. (See Figure 12–2(b) for a graphic illustration of analog and digital signals.)

There is no precise equivalence between the digital and the analog transmission capacity of a medium. An approximate equivalency exists between 10,000 bps of digital capacity and 10 MHz analog capacity.

Basic Wiring Systems

There are three choices for wiring low-voltage and telecommunication systems: twisted pair, coaxial cable, and fiber-optic cable.

Twisted pair (TP) *wire* is the most commonly used type of wire and is constructed as shown in Figure 12–2(c). TP consists of a pair of insulated wires, generally copper, twisted around each other, with the length of the twist—the *lay*—selected to reduce interference from external electrical and magnetic fields. TP is normally constructed with an outer jacket for physical protection. Additional protection against external fields is afforded by installing an overall shield (a metallic tape or braid), as shown in Figure 12–2(d). These two types of TP wiring are referred to as unshielded twisted pair (UTP) and shielded twisted pair (STP), respectively.

UTP and STP have been used for many years in such applications as voice communications (telephone and public address systems), control systems (low-voltage lighting, building management, fire alarm systems, etc.), low-speed digital signals in direct digital control (DDC) building management systems, and short-haul computer communications. The recent proliferation of computers and the demand for high-speed digital transmission have resulted in significant improvement in UTP and STP performance (bandwidth), permitting their use in applications requiring rates as high as 10 Gigabits per second

(Gbps). According to the Telecommunications Industry Association (TIA)/Electronic Industries Alliance (EIA), UTP cables are further divided into six classifications—categories 1 through 7 including enhanced category (5E), corresponding to increasing performance and permitting the selection of wiring appropriate to the application and the budget. As of this writing, the category 7 committee is currently reviewing performance standards.

Coaxial cable is constructed as shown in Figure 12–2(e). It consists of two concentric conductors sharing the same axis (hence the name), separated by an insulating material and surrounded by an overall protective jacket. (Harsh environments may necessitate two or more layers of protective jackets.) The center conductor is almost always solid wire, while the outer conductor is braided or metallized tape. Although the outer conductor has an appearance similar to the shield in STP, it is in fact a current carrier and affords no shielding properties.

RGB Cable

This type of cable consists of four or five coaxial cables in a single jacket. It is typically used for high-resolution analog video signals such as computer graphics to an LCD data projector or a CRT-type data projector. The video signal is broken down into its component parts (red, green, blue, hoz. sync., and vert. sync.), with each coaxial cable carrying a single component.

12.4.4 Fiber Optics

Fiber optics is a technology that uses light to transmit information. The transmitting medium is constructed of thin filaments (strands) of glass through which light beams are transmitted. For illumination, light may be in any part of the visible spectrum. For data transmission applications, light must be of a single wavelength to be totally reflected (refracted) within the fiber.

A fiber-optic system consists of many components, including the light source, transmitter, receiver, repeater, regenerator, optical amplifier, and cable, as well as accessories. To transmit light at low loss and for flexibility, *optic fiber cables* are made of fine fibers about 100 to 200 microns in diameter. The cable consists of a glass fiber core coated with a thin layer of glass or plastic as cladding and covered with one or more layers of material for dielectric isolation and physical protection. The light travels through the core, while the cladding keeps the light contained within the core because of its different refractive index. (See Figure 12–3(f).)

Since optic fiber must be very thin (or small in diameter) to keep the light within the angle of total reflection, the *light source* must be very tiny. The best sources are light-emitting diodes (LEDs), short-wavelength

laser compact disc (CDs), vertical cavity surface-emitting laser (VCSELs), and laser diodes (LDs). In addition, lasers at near-infrared wavelengths can keep loss at a minimum in long-distance transmissions. The popular laser and LED light sources are generated from gallium (Ga), aluminum (Al), and arsenic (As) compounds. A wavelength of 750 to 900 nm (1 nm $= 10^{-12}$ m) is used for low-cost, short-distance applications, in which the typical light power loss is about 1.5 to 2.5 dB/km. A 130-nm wavelength is used for medium-distance applications, in which the typical light power loss is about 0.35 to 0.5dB/km. A wavelength of 1500 nm is used for long-distance applications, in which the typical light power loss is about 0.2 to 0.3 dB/km.

The *transmitter* is used to convert electrical signals into light signals. The electrical signals may be in digital format (e.g., computer data) or analog format (e.g., TV and radio). Light transmission can accommodate either type of signal. In general, digital signals can tolerate distortion, whereas analog signals are subject to disturbance. The operating speed of a fiber-optic transmitter is measured by its bandwidth for analog signals and data rate for digital signals.

The *receiver* does the opposite of a transmitter, converting light signals into electrical signals. *Repeaters, regenerators*, and *amplifiers* are used to reinforce a signal or to minimize attenuation in long-distance applications.

Finally, *connectors* and *splicers* are used to make connections and splices in fiber-optic cables. A secure and square connection with low dimensional offset is extremely important. The most common connector designs are the ST, SC, and the FDDI types, with the SC type adopted by TIA/EIA as the standard. (See Figure 12–4.) A new class of small form factor (SFF) fiber-optic connectors have been introduced with the goal of reducing the size of the connector while maintaining or reducing the cost. The SFF connectors are well suited for high-density applications. There is no standard for choosing the SFF connector type utilized for different applications. Several different SFF designs have been developed by different manufacturers. These types include LC, MT-RJ, MU, and E2000.

Light Transmission within Transparent Fiber

As an electromagnetic wave, light travels in a straight line. It can be deflected, however, by reflection from a surface or by refraction at the boundary or interface between two transparent media, such as between air and glass or between glass and plastic. The degree of deflection or bending depends on the refractive indexes of the two media and the angle at which the light strikes the interface (boundary) between two media. See Figure 12–3(a). If light within the medium with the higher index hits the

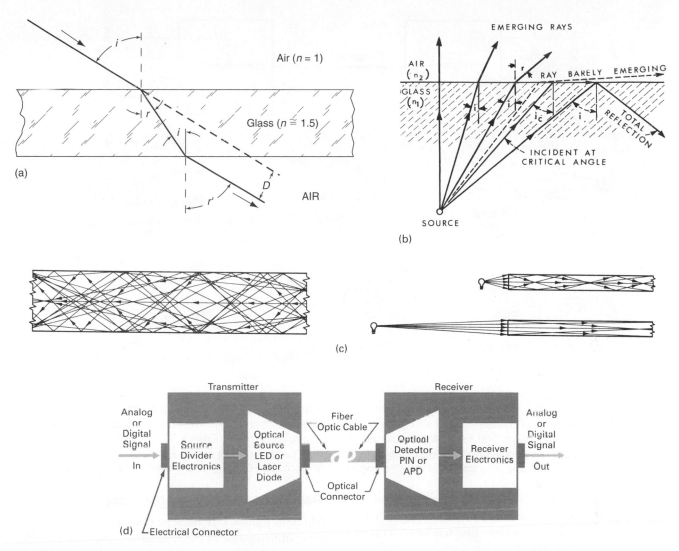

■ FIGURE 12–3

(a) Refraction of a light ray at a plane surface causes bending of the incident ray and displacement (*D*) of the emergent ray. Snell's law of refraction is expressed as

$$n_1 \sin i = n_2 \sin r$$

where n_1 = index of refraction of the first medium
   i = angle of incidence (angle the incident ray forms with the normal to the surface)
   n_2 = index of refraction of the second medium
   r = angle of refraction

(b) As the angle of incidence increases, the light ray bends more and more. The angle at which time the reflected ray is parallel to the surface is called the *critical angle* (i_c). A light ray incident at any angle greater than the critical angle is totally refracted (or reflected) back to the medium. (c) With angles of incidence greater than the critical angle between the glass core and its glass (or plastic) cladding, a light ray is totally reflected, but scattered. This phenomenon is called *pulse dispersion*. (Top illustration.) Light incident with a lower acceptance angle (bottom illustration) has a low pulse dispersion and thus a higher speed of transmission. (Source: (a)-(c) Reproduced with permission from IESNA.) (d) This simple schematic diagram shows an optical transmitter and receiver connected by a length of optical cable in a point-to-point link. The transmitter converts electronic signal voltage into optical power, which is launched into the wire by an LED, a laser diode, or a laser. At the photodetector point, either a positive-intrinsic-negative (PIN) or avalanche photodiode (APD) captures the light pulses for conversion back into electrical current. (Source: (d) Reproduced with permission from Belden Wire & Cable Company.)

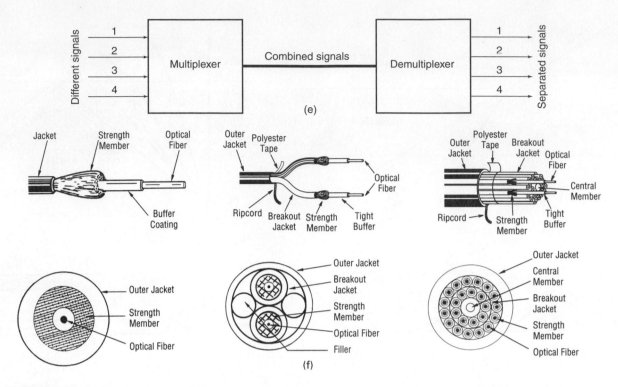

■ **FIGURE 12–3** *(Continued)*
(e) Several signals can be combined or multiplexed into a single signal through a device called a *multiplexer*. The combined signal is then transmitted through a single transmission line, which greatly increases the capacity of transmission. A demultiplexer at the receiving end separates the signals into their original forms. (f) From left to right: Construction of fiber-optic cables. Left: single fiber, tight buffer; center: two fibers, tight buffer; right: 24 fibers, tight buffer. The fiber size varies from 50 to 100 microns. (Source: (e) and (f) Reproduced with permission from Belden Wire & Cable Company.)

boundary with the medium with the lower index at an angle smaller than a certain critical angle, the light beam will be totally refracted within the medium. This is known as *total reflection* (it is no longer described as refraction). (See Figure 12–3(b).)

Conversion of Electromagnetic Waves

Since light and electrical energy are all part of the electromagnetic spectrum, they differ only in wavelength and frequency. Light and electrical energy can, therefore, be converted from one form to another. This unique property becomes a great asset in modern communication technology when electrical signals in either analog or digital form are converted into light signals for low-loss transmission through an optic fiber and then are converted back to electrical signals for processing. (See Figure 12–3(d).) Furthermore, many signals may be transmitted simultaneously within the same fiber, a process known as *multiplexing*. (See Figure 12–3(e).) In practice, a fiber-optic cable can transmit thousands of signals. With the proper light source (an LED), cable construction, and channels (pulse dispersion), a multimode fiber-optic cable may transmit several millions of digital signals simultaneously.

Fiber-Optic Multiplexing Methods

Multiple signals can be sent over a single fiber-optic cable utilizing one of three methods: time division multiplexing (TDM), wave division multiplexing (WDM), and dense wave division multiplexing (DWDM). Synchronous optical networking (SONET) is a good example of TDM over fiber. (See Table 12–1, optic cable level versus line rate.) Utilizing TDM, speeds up to 10 Gbps (10 Gig) are possible.

TDM equipment breaks up the available bandwidth into "time slots." These time slots are then rotated through the various signals at a given rate. For example, suppose that you are multiplexing four separate signals (A, B, C, D). A TDM sequence might appear as follows; A-B-C-D-A-B-C-D-A-B-C-D-A-B. . .

WDM takes multiplexing even further, allowing up to 28 TDM signals to be multiplexed together. A WDM takes each signal but does not sequence the signals; instead, WDM uses different colors of light for each signal to be transmitted. These signals are on 6-nm centers and are multiplexed through a prism. This technology allows the signals to be transmitted independently of one another. DWDM combines multiple optical signals so that they can be amplified as a group and

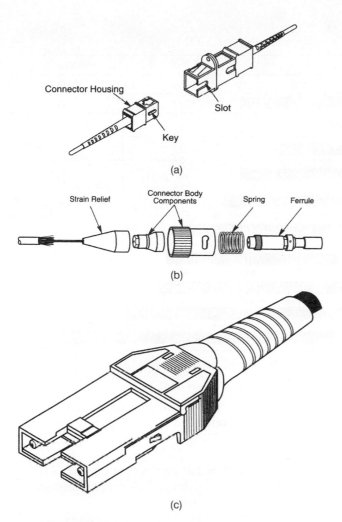

Connector Housing

Slot

Key

(a)

Strain Relief

Connector Body Components

Spring

Ferrule

(b)

(c)

■ **FIGURE 12–4**

During earlier years of fiber cable development, each cable manufacturer often developed its own design for connecting or splicing cables. This led to the problem of interfacing between equipment and systems. To eliminate the proliferation of connector designs, the International Standards Organization (ISO) and the Telecommunications Industry Association (TIA) have endorsed the square connector type SC, the round connector type ST, and the fiber distributed data interface (FDDI) connector as preferred standards. Configurations of these connectors are shown in (a), (b), and (c). (Source: (a) Courtesy: Tyco Electronics Corporation; (b) and (c) Courtesy of Corning Cable Systems, Hickory, NC.)

TABLE 12–1
Optic cable level versus line rate

OC Level	Line Rate, Mbps
OC–1	51.8
OC–3	155.52
OC–9	466.56
OC–12	622.08
OC–18	933.12
OC–24	1244.16
OC–36	1866.24
OC–48	2488.32
OC–192	9953.28

Advantages and Application of Fiber Optics

The primary advantages of fiber-optics technology are its high transmission capacity and low loss characteristics. In addition, electromagnetic interference and moisture do not affect fiber optics. Some typical applications of fiber optics are the following:

- Long-distance telephone lines, on land and at sea
- Local and wide-area networks
- Cable television between microwave receivers and head-end equipment
- Transmission of a signal where moisture may be a problem.
- Transmission of signals where electromagnetic interference is a problem.
- High-security applications, such as financial, military, and intelligence systems

12.4.5 Selection of Wiring Systems

The selection of wires and cables among twisted pairs, coaxial cables, optic fiber, and combinations thereof depends on the application, the performance of the medium in terms of bandwidth, transmission speed, attenuation, impedance, shielding, etc., and, of course, the cost of installation of the wires and cables. Optic fiber is best as regards bandwidth and speed, but it is the most costly of the three media; however, with the narrowing cost differential fiber has become a viable solution. Optic fiber is now being installed in backbone wiring in buildings and in local and wide-area networks. For design and installation guidelines, the EIA and TIA standards listed in the reference section of this chapter are the best resources.

Table 12–2 lists the basic applications of the popular transmission media and the ranges of speed at which they may be used. As with the design of any building system, the future needs of the building and trends in technology must be carefully weighed in the selection of wiring systems for all telecommunication systems.

transported over a single fiber. Each signal can be carried at a different rate and in a different format. Signals operating over a DWDM infrastructure can achieve capacities of over 40 Gbps.

TABLE 12–2
Applications of telecommunication cables

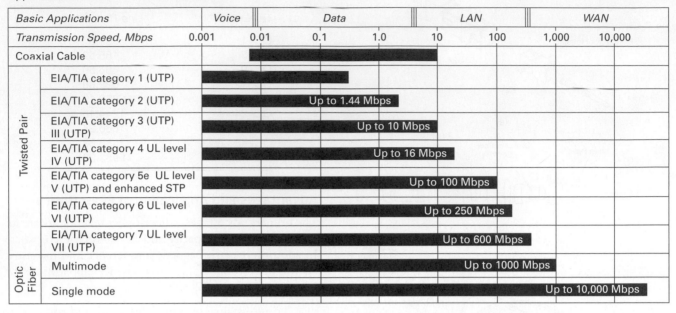

Basic Applications		Voice	Data		LAN	WAN
Transmission Speed, Mbps	0.001	0.01	0.1	1.0	10	100	1,000	10,000
Coaxial Cable							
Twisted Pair — EIA/TIA category 1 (UTP)							
EIA/TIA category 2 (UTP)			Up to 1.44 Mbps				
EIA/TIA category 3 (UTP) III (UTP)				Up to 10 Mbps			
EIA/TIA category 4 UL level IV (UTP)				Up to 16 Mbps			
EIA/TIA category 5e UL level V (UTP) and enhanced STP					Up to 100 Mbps		
EIA/TIA category 6 UL level VI (UTP)					Up to 250 Mbps		
EIA/TIA category 7 UL level VII (UTP)					Up to 600 Mbps		
Optic Fiber — Multimode						Up to 1000 Mbps	
Single mode							Up to 10,000 Mbps

12.5 TELECOMMUNICATION SYSTEMS

A telecommunication system is defined as any electrical system that transmits, emits, or receives signals, images, sound, or information of any nature by wire, radio, video, or some other form of energy within the electromagnetic spectrum. Thus, telephone, radio, microwave, radar, security, fire alarm intercom, public address, CCTV/video surveillance, broadcasting TV, CATV, and SATV are all telecommunications systems. Technology is changing so rapidly that a new system may become obsolete within a few months. Thus, this section discusses only the general concept, without dealing with specific design data or guidelines.

12.5.1 The Electromagnetic Spectrum

The electromagnetic spectrum is a graphic representation of radiant energy in an orderly arrangement according to its wavelength or frequency. Radiant energy within the spectrum differs in wavelength and frequency, but its speed of transmission is constant. The speed of radiant energy in vacuum is 299,793 kilometers per second (km/s), or 186,282 miles per second (mi/s). The speed of radiant energy in air is slightly slower (299,724 km/s). In round numbers, the speed may be considered to be 300,000,000 m/s (3×10^8 m/s). The following equation expresses the relationship among speed, frequency, and wavelength:

$$s = f \times G \qquad (12\text{–}1)$$

where s = speed, 3×10^8 meters/second (m/s)
  f = frequency, cycles per second (cps or hertz)
  G = wavelength, meters (m)

Example 12.1 The standard frequency of electrical power in the United States is 60 Hz. What is the wavelength of the electrical power?
  Answer The wavelength of 60-Hz power is $3 \times 10^8 / 60 = 5 \times 10^6$ m.

Example 12.2 If the radio frequency of an FM station is 100 MHz, what is the wavelength?
  Answer The wavelength of the FM radio station is $(3 \times 10^8)/(100 - 10^6) = 3$ m.

12.5.2 Standard Radio and Video Frequencies

The region of radio and video frequencies within the electromagnetic spectrum extends from 3 kHz to 300 GHz. (One GHz is 1 billion Hz.) The boundary of each subdivision of a region is gradual; there are no specific demarcation lines.

To minimize interference between transmitted signals, the applications of frequencies are strictly controlled by governments. The regulatory agencies in the

TABLE 12–3
Typical applications of FCC/NTIA frequency standards[a]

Frequency Bands[b]	Frequency	Applications
POWER BANDS (extremely low to infralow frequencies)		
Standard frequency, Hz	50–60	Public utilities
Intermediate frequency, Hz	400	Aircraft power
High frequency, Hz	3000	High-frequency lighting
RADIO BANDS (medium to ultrahigh frequencies)		
Amplitude modulation (AM), kHz	530–1600	Off-air radio
Amateur, kHz	1800–1900	Ham radio
Telephone, MHz	1.6–1.8	Wireless telephony (short range)
Amateur, MHz	3.5–4.0	Ham radio
High frequency, MHz	5–40	International radio
Frequency modulation (FM), MHz	88–108	Off-air radio
Telephone, MHz	400–500	Mobile telephone system
Telephone, MHz	824–894	Cellular telephone system
Products, MHz	902–928	Wireless, applications (short range)
TELEVISION BANDS (very high to ultrahigh frequencies)		
VHF (low band), MHz	54–88	Channels 2–6
VHF (midband), MHz	121–169	Channels 14–22
VHF (high band), MHz	174–216	Channels 7–13
VHF (superband), MHz	217–295	Channels 23–36
UHF (hyperband), MHz	301–451	Channels 37–62
SATELLITE BANDS (ultrahigh to superhigh frequencies)		
Ultrahigh frequency, MHz	950–1750	Satellite TV
Superhigh frequency, low band, GHz	3.7–4.2	C-Band satellite
Superhigh, midband, GHz	11.7–12.2	KU-Band satellite
Superhigh, midband, GHz	12.2–12.7	Direct broadcast system (DBS)
MICROWAVE BANDS (ultrahigh to tremendously high frequencies)		
Superhigh frequency, GHz	2–10	Long-distance transmission
Extremely high frequency, GHz	10–50	Radar
Tremendously high frequency, GHz	50–500	Long-distance transmission

[a]Includes only those applications commonly related to audio and video reception and distribution systems in buildings.
[b]Classification of Frequency Bands

ELF	below 300 Hz	Extremely low frequency	VHF	30–300 MHz	Very high frequency
ILF	300–3000 Hz	Infralow frequency	UHF	300–3000 MHz	Ultrahigh frequency
VLF	3–30 kHz	Very low frequency	SHF	3–30 GHz	Superhigh frequency
LF	30–300 kHz	Low frequency	EHF	30–300 GHz	Extremely high frequency
MF	300–3000 kHz	Medium frequency	THF	300–3000 GHz	Tremendously high frequency
HF	3–30 MHz	High frequency			

United States are the National Telecommunications and Information Administration (NTIA) of the U.S. Department of Commerce and the Federal Communications Commission (FCC). Table 12–3 lists some typical applications of FCC/NTIA standards to power, radio, video, and radar signals. The governments of all developed and developing countries also control their transmission frequencies. In general, the frequencies allocated by other countries are comparable to those used in the United States, forming the basis for international standards.

12.5.3 Off-Air/Over-the-Air (OTA) Systems

Off-air systems refer to radio and TV systems that receive their electromagnetic signals "off air" (on the air),

cable networks, or signals generated locally. This section discusses systems that receive and transmit off air only.

High-definition audio and video transmission has introduced challenges in receiving those signals in both antenna and signal processing equipment. While analog off-air signals are still being generated, the mandatory cutoff date is February 17, 2009, which will end the analog television broadcasting era.

Exterior Antennas

An antenna is a device for receiving or transmitting off-air signals. Although most off-air receiving equipment—e.g., radio or TV—has built-in antennas, heavy concrete or steel building walls and roofs attenuate these signals to the extent that they are often distorted or noisy. In most buildings, exterior antennas are necessary to ensure good reception of the signal.

An antenna on or near a building has a major impact on the aesthetics of the building. The designer must be keenly aware of the size and location of the antenna during the design process, so that these elements may best be coordinated with the overall design, rather than be added as an afterthought. This is particularly important for buildings such as hotels and offices, which are increasingly dependent on satellite data transmission.

The ideal location of an exterior antenna is on the roof of a building or in an open space on the ground. Antennas should be mounted in such a manner that there is no major obstruction between the antenna and the transmitted signals.

External antennas come in all shapes and sizes. Some popular models are the yagi type (straight bars) for VHF frequencies, loop type for FM/UHF frequencies, and dish type for satellite frequencies. Figure 12–5(a) illustrates a typical radio and VHF antenna. Higher-frequency UHF signals are more directional. Their behavior is close to that of visible light, which travels in a straight line but can be reflected by a reflector or focused by a parabolic-shaped dish. Figure 12–5(b) illustrates a typical dish-type antenna.

If outlets are limited in number and in distance, as in a residence, the strength (gain) of a signal from an outdoor antenna should be sufficient without further amplification. On the other hand, larger buildings having tens or hundreds of outlets will require additional head-end equipment to filter, boost, convert, amplify, modulate, and combine the various frequencies into a master antenna system (MATV). When the signals are combined, the distribution from the head-end equipment to the individual outlets can be accomplished by a single coaxial cable looping between outlets, rather than with multiconductor cables. The economic benefit is obvious.

Telecommunication Satellites

Most telecommunication satellites are of the geostationary type, in an orbit 22,300 miles directly above the equator. The satellite maintains the same relative position and attitude with respect to the earth by the use of small station-keeping jets. The electrical power required to operate these jets is obtained from batteries recharged by photovoltaic solar collectors.

The satellite itself is an assembly of many receivers and transmitters, with associated amplification. Each satellite is designed to cover certain areas on the earth. The signal is weaker at the edge of an area, and as the site of the receiving station moves farther from the equator, the size of the receiving station is increased to compensate. In other words, a satellite dish is made considerably larger in Alaska than in Florida in order to receive signals

from the same transmitter. The angle of the receiving station is lower the farther it is from the equator.

The speed of the signal is the speed of light. This means that there is a minimum known delay in the signal of at least the time it takes to travel the 22,300 miles between the receiver/transmitter and the satellite, or 0.24 second. The farther from the equator the receiver/transmitter is, the longer is the distance to cover, and the greater the reduction in signal strength and the lower the angle of the receiver/transmitter. This lower angle makes it more probable that obstructions will be encountered. Processing time at the satellite (to turn the signal around) usually makes the transmission time approximately 0.4 second, or 400 ms, or greater.

Satellite Receiver/Transmitter for Building Services

A satellite receiver/transmitter installed for building services is referred to as a VSAT, which stands for *very small aperture transceiver*. A transceiver is a combination transmitter and receiver. The dish portion of the device is a section of a parabola. The focal point concentrates the incoming signal at the receiver for amplification at the horn. The alignment of the dish is critical for reception and transmission of the signal to the satellite.

The angle and azimuth necessary to acquire the satellite signal are known from any location that can "see" the satellite. The viewing angle must be clear. Although it is possible to acquire a signal through deciduous trees, this situation should be avoided, since obstruction will attenuate the signals.

A VSAT ranges from less than 1 m up to 10 m in diameter. The size required depends on enviromental conditions, the altitude and longitude of the building and the satellites. With the advance of technology, antenna dishes have diminished drastically in size in recent years.

Smaller antenna terminals with more powerful amplifiers are being developed. The diameter ranges from about 1 ft for ultra SAT (USAT) to less than 1 ft for tiny SAT (TSAT). Associated with the reduction in size are an increase in frequency and transmission speed and a narrowing of the beam pattern.

Signal Distribution

Each of the signals (frequencies) received is processed by a separate controller that is connected to the OTA and VSAT antenna by coaxial cable. The allowable distance between the antenna and the controller is limited and varies with the total configuration. Figure 12–5(c) is a schematic diagram of an OTA signal distribution system in a large building.

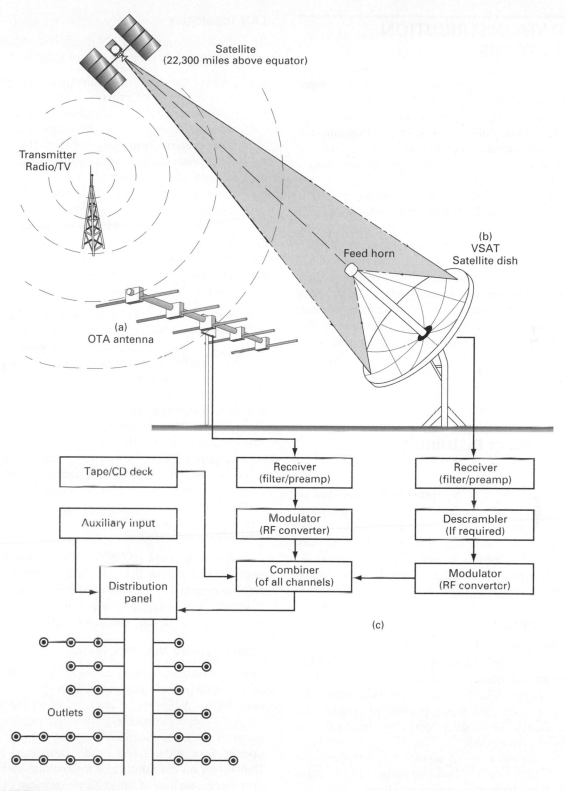

■ **FIGURE 12–5**

(a) Typical radio and TV antenna (54–451 MHz).

(b) Typical VSAT satellite dish (0.9 to 12.7 GHz).

(c) Typical OTA system distribution wiring diagram. The signals are filtered, amplified, modulated, and combined prior to distribution.

12.6 DATA DISTRIBUTION SYSTEMS

Data distribution through networking is no longer limited to major facilities, such as computer centers, hospitals, hotels, department stores, and corporate headquarters. Most buildings or group of buildings are equipped with central computers, remote terminals, or multiple units of PCs and should consider installing a data distribution system.

The biggest advantages of networking are that it allows the sharing of files, programs, and equipment and that it saves money. Networking also makes it easy to set up an electronic mail (E-mail) system or establish a shared connection to the Internet. With the proliferation of the Internet over the last several years, it is penetrating our lives from almost every angle and provides access to a wealth of information. With data sharing, data need be entered only once, thus eliminating the wasted effort of reentry of the data at each station. Data networking has become as common as telephone cabling. Most commercial and educational buildings are cabled with enhanced category 5, (5E) or, when budgets permit, category 6.

12.6.1 Types of Distribution Systems

When different pieces of equipment are connected to distribute data or files to each other, the equipment is said to be connected as a network. The data to be distributed are usually in a digital format, although they may be in analog format as well. Networking may be confined to within a building or may be extended to thousands of miles away.

Local Area Network

A *local area network* (LAN) is a system that connects computers and peripheral equipment within a building or within several nearby buildings for the purpose of sharing resources. A LAN system is practical within several thousand feet. Thus it is applicable to campus-type facilities, such as universities, corporate headquarters, and industrial complexes.

LANs are made up of servers, workstations, and PCs. The server needs a faster processor and a larger hard disk than the workstation does. Each workstation that is to be connected to the LAN requires a network interface card (NIC). The card is the interface between the network electronics and the workstations. There are many network adapter cards, such as ATM, Ethernet, (10-base-T, 100-base-T, 1000-base-T), ARCnet, Token Ring, FDDI, and Zero-Slot LANs.

LAN Topologies

The physical layout of a LAN is called its *topology*. The most common LAN topologies are as follows:

- *Point-to-point* is the simplest format, connecting two computers together.
- *Star topology* is an extension of point-to-point topology to multiple workstations. This topology is the preferred topology of Ethernet LAN.
- *Bus topology* has one cable connecting all the workstations and the server.
- *Ring topology* is similar to a bus network, except that the ends are tied together to form a ring. This is the topology of IBM Token Ring LANs.
- *Switchboard topology* is an arrangement in which each workstation is considered its own network segment. This increases speed by dedicating bandwidth rather than sharing resources.

Star topology consists of home-run cabling from each workstation back to a central point of connectivity, or hub. The hub maintains whichever logical topology is appropriate for the network protocol. Star topology is the basis for the structured cabling system in buildings defined in the TIA/EIA, 568-B (2001): "Commercial Building Telecommunications Wiring Standard."

In a ring network, the cabling between nodes forms a ring. In bus network, the cabling between nodes forms a bus. Figure 12–6(a) through (c) illustrates the physical topologies of data distribution systems. Figure 12–6(d) is a block diagram of a combination data, telephone, and A/V system.

Wide-Area Network

A *wide-area network* (WAN) is defined as a data communication network that uses telecommunication circuits to extend a LAN beyond the building or campus it serves. WANs typically transmit over phone lines, microwave towers, and satellites without limitation.

Virtual Private Network (VPN)

One increasingly common variation of a WAN is known as a *virtual private network* (VPN). In a VPN, each node of the WAN has a connection to the Internet. Then, by secure protocols, a "free" connection is made between sites. Applications running on the network operate as if a dedicated circuit connected the sites. Depending on the number of remote sites and connection speeds, savings of up to 75 percent can be realized by utilizing VPN technology.

Wireless Communications

Wireless communications consists of many different technologies, including cellular, satellite, infrared, mobile radio, and wireless networks.

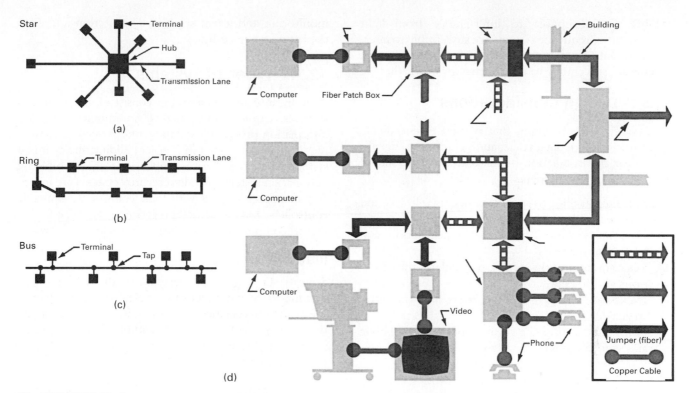

■ FIGURE 12–6

(a) Star LAN is arranged around a single hub. The hub can be simply a passive coupler or an active controller. This is the preferred arrangement for all LAN installations. (b) Ring LAN is linked point to point. A bit pattern (token) is circulated to each node (workstation or peripheral equipment), to control the data transmission. (c) Bus LAN uses carrier-sensing multiple access with collision detection to control data transmission. (d) LAN wiring diagram incorporating telephone, computer, and A/V links with multiple- and single-mode fibers as risers (backbone) and twisted pair or coaxial wires for equipment connections. (Reproduced with permission from Belden Wire & Cable Company.)

Wireless networks when applied to telecommunication can be further subdivided into types including WLAN, WPAN, WMAN, and WWANs. Data and A/V signals may be transmitted through a wireless network, which usually operates on infrared or radio frequencies. In a *wireless local-area network* (WLAN) small networks of three or more wireless devices are connected to a network. Wireless LANs provide all the functionality of wired LANs but without the physical constraints of the wire itself. The typical range of a WLAN is 100 to 500 ft, with transfer rates up to 54 Mbps.

A wireless personal-area network (WPAN) typically covers a small area surrounding a user's work space and provides the ability to synchronize computers, transfer files, and gain access to local peripherals. The typical range of a WPAN is 20 to 35 ft with transfer speeds up to 54 Mbps. In a wireless metropolitan-area network (WMAN) a wireless LAN–LAN bridge connects LANs in separate buildings using microwave communication. Distances range from less than a mile up to 20 miles with speeds up to 360 Mbps; however, a clear line of sight is needed between the receiving and transmitting devices. In some systems, the bridge meets or exceeds traditional telecommunication wireline standards, guaranteeing 99.999% availability.

A wireless wide-area network (WWAN) generally uses cellular phone insfrastructure to enable notebooks and handheld devices to access the Internet. Generally, a WWAN is operated by public carriers and uses open standards such as GSM, CDMA, AMPS, and TDMA. The typical range of the WWAN is several miles with transfer speeds from 5 kbps to 20 kbps. The major advantage of a wireless system is flexibility. Disadvantages of this type of system are usually higher cost initially and slower throughput, but the technology allows systems to be installed in hard-to-wire places and eliminates the need to trench or bore to lay cables. Many systems operate at unregulated frequencies of 2.4 GHz or 5.3 GHz and do not require FCC licensing, eliminating costly registration delays.

Another disadvantage of wireless LANs is the possibility of external interference with the signal. Infrared systems are susceptible to interference from light sources, radio frequency systems are susceptible to interference from electromagnetic sources, and microwave systems are susceptible to interference from weather.

12.6.2 Design Considerations

Structured wiring is wiring that is used to coordinate the transmission of data, voice, audio, and video signals. A plan for the installation of structured wiring should cover seven design elements:

1. Entrance facilities (i.e., conduits, manholes, and armored cables)
2. Equipment rooms, including construction and space requirements
3. Telecommunication rooms, particularly their locations and sizes
4. Backbone wiring, including risers and the infrastructure
5. Horizontal wiring (i.e., wiring between backbone and workstations)
6. Work area
7. Administration

Specific recommendations for each design element are given in EIA/TIA Standard No. 568-B. When the recommendations are followed faithfully, they will ease network administration and enhance computerized tracking and documentation, as well as minimize attenuation of the signal. The wiring recommended for backbone and horizontal distribution includes 100-ohm UTP, and 62.5/125- or 50/125-micrometer optical fibers.

Wiring within a building for a LAN is usually made with twisted pair or fiber-optic cables. The selection depends on the type of data to be transmitted. For high-speed transmission and broad bandwidth, optic fiber is preferred at a higher cost. (See Table 12–2 for the applications of telecommunication cables.)

Cables are best connected and terminated at one or more telecommunication rooms distributed strategically throughout the building, preferably not more than 500 ft apart.

Wiring for (digital) data distribution systems may be combined with (analog/digital) A/V and voice systems when the proper interface devices are provided. (See Figure 12–6(d) for a typical wiring diagram of a combined data, telephone, and A/V system.)

12.7 SECURITY SYSTEMS

A security system installed in a building safeguards property and people. Security systems range from security guards at building entries or exits to sophisticated monitoring and alarm systems. This section addresses the latter systems only.

12.7.1 Types of Systems

Basically, a security system can consist of various access controls, annunciations, alarms, communications, and information-processing components. A security system may itself be converged with other building management systems, such as fire alarm, public address, and building automation systems. Modern security systems are invariably computer systems with logic chips, programmable controllers, and/or central processing units. The fundamental components of a security system are as follows.

Intrusion Detection

Security starts at the property line with fences or walls that may or may not incorporate electronic surveillance. Electronic surveillance usually operates on infrared (active and passive), acoustical (audible and ultrasonic), microwave (beam pattern and field effect), or vibration detection principles.

Access Controls

An access control system identifies the person seeking to enter or leave a building. The techniques include the following:

- *I.D. cards*, which may incorporate magnetic strips, proximity-tuned passive circuits, coded pattern capacitors, infrared optical marks, or mechanical (Hollerith) coded holes arranged in a pattern.
- *Biometric identifications*, which make use of several unique physiological characteristics of a person, such as fingerprints, a retinal scan, hand geometry, a signature, or a voiceprint. Of these characteristics, fingerprint, hand geometry, and eye retina identification offer the most secure features and are thus desirable for high-security applications.
- *Video surveillance* Cameras not only provide visual identification of persons but also broaden the ability to "watch" large facilities from a single location. The use of cameras as part of a security system has proven to increase the efficiency of security staff and, in cases where recording features are added, to provide a valuable means of obtaining evidence.

Detection within the Building

Detection devices are installed in corridors, stairs, elevators, and critical spaces. Among these devices are:

- *Motion detectors*, which utilize infrared, microwave, or capacitance principles.

- *Photoelectric detectors*, which generate a light beam between the photocell and a receiver. If the beam is interrupted, a signal will be initiated.
- *Door position status contacts*, which may be either normally closed or normally open. When the positions of the contacts are disturbed, the circuit will be activated. Typical applications of contact devices are doors and windows.
- *Video surveillances*, camera can be programmed to record only when motion is detected, 24/7, or upon human activity detection and/or recognition.

Annunciation

When the detection devices are activated, the security system should announce the event at strategic locations in the building. Commonly used devices include:

- *Computer alarm monitoring station.*
- *Alarms*—bells, horns, buzzers, etc.
- *Annunciators* (i.e., lights and lighted panels with locations identified).
- *Monitors,* and digital video recorders (DVR), as part of the video surveillance system.
- *Speakers*, interconnected to the building's public address or intercom system.
- *Digitized voice messages.*
- *Wireless radio system*, utilizing the building's radio communication system, and can send a text message to a wireless phone or to a text pager.

Recording Information

All security-related activities should be recorded through a computer-based printer or file to register the sequence of events.

12.7.2 Closed-Circuit Television

Closed-circuit television (CCTV) is a wired, self-contained television system. Although it is widely used in security applications for video surveillance, it is also used in industrial process controls, business promotion, sports training, traffic control, experimentation, data filing, etc.

A basic CCTV system consists of an electronic camera that converts an optical image into analog signals, a transmission medium (usually coaxial cables or optic fiber), and a monitor to convert the signal back to an image. The signal is stored on tape or digital video recorders (DVRS) through a video recorder, multiplexer, or video printer. The system may be expanded into a multiple camera, monitor, and recorder system. A CCTV system may have one or more of the following features:

- *Mode* Black-and-white or color; continuous, sequential-switched, or time-lapse monitoring; or

multiplexed recording, which allows numerous cameras to be recorded on a single analog or digital recorder.

- *Cameras* Fixed-position or pan/tilt/zoom (PTZ); sensitivity to normal or infrared spectrum; freeze action; remote controls.
- *Monitors* Single or multiple units; full or split screen; various sizes of screen; various resolutions (300 to 1200 lines per screen).
- *Recorders* Video recorders; time-lapse recorders; sequential switches; time-date generators or digital computer-based recording.

Figure 12–7(a) shows the various CCTV components, and Figure 12–7(b) shows the connection diagram of a system with a multicamera, time-lapse recorder, sequential switching, and a date generator. In practice, CCTV systems usually converge with other systems into a combined security system. (See Figure 12–7(c).)

12.8 TELEPHONE SYSTEMS

Since the invention of the telephone in 1878 by Alexander Graham Bell telephony has become the primary means of communication in modern societies. The level of development of a nation or a region can often be judged by the number of telephones installed per capita.

Telephone services are usually provided by public utilities, which may be owned by private enterprise. With the exponential development of new telephone technology, such as cellular telephone systems, private branch exchanges (PBX), wireless telephones, pagers, Voice over Internet Protocol (VoIP), and facsimile (FAX or fax), the planning of telephone systems in a building has become more complex. Often, the advice of the public utility or an independent telecommunication consultant is necessary if a new facility is to have state-of-the-art capabilities.

The fundamental principle of a telephone system is very simple. The system starts with individual telephone sets, a central switching or exchange facility, a DC power supply, and distribution wiring in between. The traditional analog telephone set consists of a transmitter, a receiver, a ringing circuit, and switching and coding devices. The transmitter is in effect a microphone, and the receiver is a speaker. When a person talks into the transmitter, the acoustical (sound) energy compresses a carbon-granule-filled diaphragm, causing a change in its electrical resistance, which in turn varies the electrical current (at the milliampere level) through the loop circuit to the receiving party's telephone. The receiver then converts the electrical current back into acoustical energy. Figure 12–8(a) illustrates the construction of a

(a)

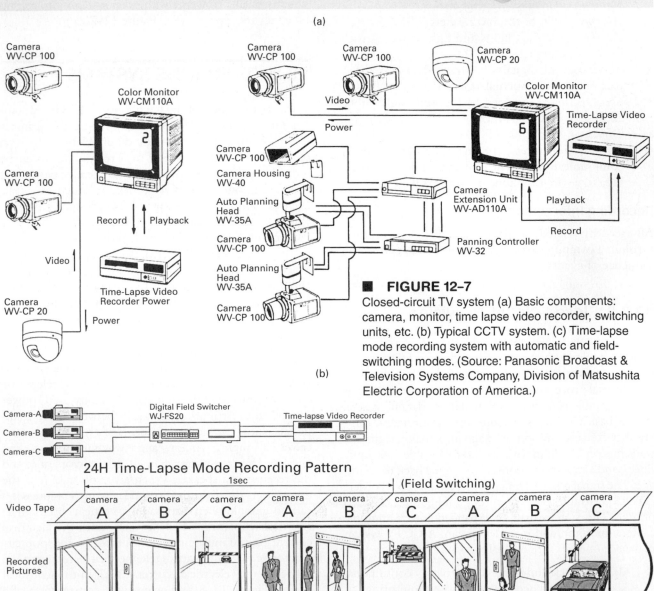

Camera
WV-CP 100

Color Monitor
WV-CM110A

Camera
WV-CP 100

Record ↕ Playback

Video

Time-Lapse Video
Recorder Power

Camera
WV-CP 20

Power

Camera
WV-CP 100

Camera
WV-CP 100

Camera
WV-CP 20

Video

Power

Camera
WV-CP 100
Camera Housing
WV-40

Auto Planning
Head
WV-35A

Camera
WV-CP 100

Auto Planning
Head
WV-35A

Camera
WV-CP 100

Color Monitor
WV-CM110A

Time-Lapse Video
Recorder

Camera
Extension Unit
WV-AD110A

Playback

Record

Panning Controller
WV-32

(b)

■ FIGURE 12–7

Closed-circuit TV system (a) Basic components:
camera, monitor, time lapse video recorder, switching
units, etc. (b) Typical CCTV system. (c) Time-lapse
mode recording system with automatic and field-
switching modes. (Source: Panasonic Broadcast &
Television Systems Company, Division of Matsushita
Electric Corporation of America.)

Camera-A

Camera-B

Camera-C

Digital Field Switcher
WJ-FS20

Time-lapse Video Recorder

24H Time-Lapse Mode Recording Pattern

1sec   (Field Switching)

Video Tape

| camera A | camera B | camera C | camera A | camera B | camera C | camera A | camera B | camera C |

Recorded
Pictures

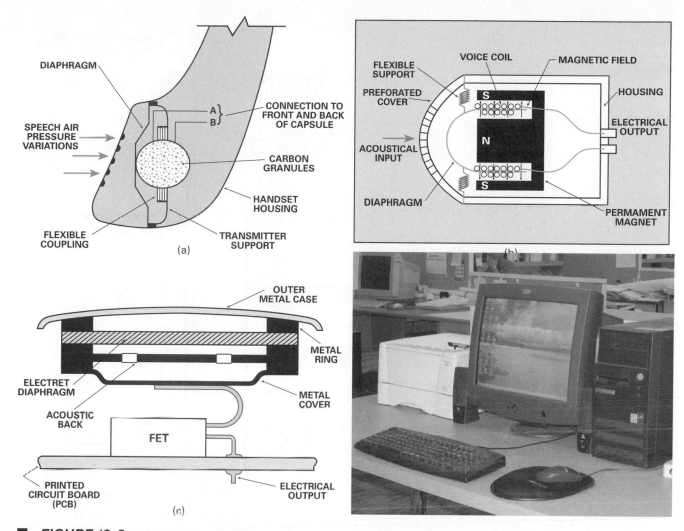

■   FIGURE 12–8

Basic construction and features of three types of transmitters (microphones) for a telephone set. The traditional carbon-granule type will soon be replaced by the newer electrodynamic and electret types. (a) Carbon-granule type. (b) Electrodynamic type. (c) Electret type. (Source: Stephen J. Bigelow, *Understanding Telephone Electronics*, Sams Publishing, Indianapolis, IN.) (d) Computer-based PBX system. The system can be fully automatic without anyone in attendance. If employed, an operator is usually assigned other duties when not in demand. (Figure continues, page 394.)

traditional telephone transmitter. Presently, transmitter/receivers are increasingly the electrodynamic type, utilizing the principle of mutual inductance between electricity and magnetism. (See Figure 12–8(b).) The electret type utilizes the property of electrostatic charge on a coated dielectric diaphragm. Vibration of the diaphragm will change the electric voltage and current in the loop circuit. With amplification, the signal is transmitted to the receiving telephone. (See Figure 12–8(c).)

For long-distance transmission, the analog voice signals are converted into digital signals so that they may be integrated with other digital signals utilizing multiplexing technology. The quality of new telephone systems is improving steadily, to the point that long-distance conversation over thousands of miles away is just as clear as if it were next door. With the emergence of new-IP based telephone systems or VoIP, it is not unusual to achieve full CD-quality voice communication.

Some cellular telephone systems operate on narrow radio frequency bands. (See Table 12–3 for their operating frequencies.) The signals are currently transmitted from closely spaced transmitter towers to cover a defined area. The technology of some systems will likely migrate to using satellites and eliminate the costly local transmission towers.

■ **FIGURE 12–8** *(Continued)*
(e) Typical telephone wiring distribution (riser)
diagram. (Courtesy: William Tao &
Associates.)

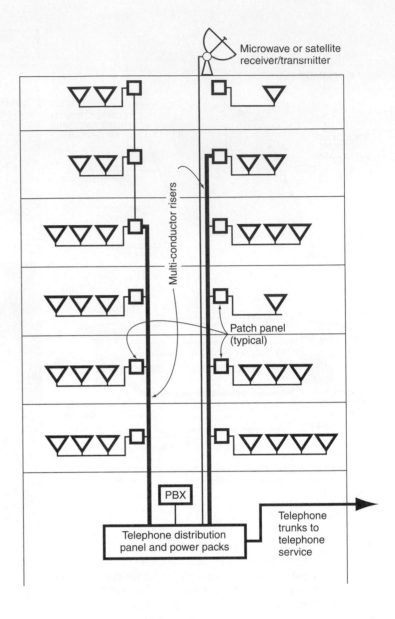

12.8.1 Types of Systems

Telephone services are provided directly by the telephone company. There are two types of dialing systems: pulse (a rotary dial) and tone (a push-button keypad). The latter technology can interface with digital signals. A touch-tone telephone has 12 push buttons (0 to 9 plus, and 1). When pushed, each button generates two tones, one low-frequency tone between 697 and 941 Hz and one high-frequency tone between 1209 and 1477 Hz. This feature, known as dual-tone multifrequency (DTMF), is designed to prevent false signals.

A *private branch exchange* (PBX) is a private telephone system serving one building or a group of buildings. A PBX system connects to the public telephone system through a switchboard or automatic switching equipment. The switchboard may or may not require an operator. The abbreviation PBX is used interchangeably with PABX, which stands for *private automatic branch exchange*, since current PBX systems are all of the automatic type. Most PBX systems have direct outward dialing, a feature whereby one dials "9" to connect to the public telephone trunk line. A PBX system can be interfaced with computer systems. Figure 12–8(d) illustrates a PC-based PBX system. Some typical features of a PBX system are the following:

- Ancillary device connection
- Automated attendant
- Automatic call distribution
- Automatic redial
- Automatic trunk-to-trunk traffic
- Call forwarding
- Call transfer
- Call holding
- Call privacy

- Conference calls
- Code restriction
- Customized messaging
- Delay announcement
- Direct inward dialing (DID)
- Direct outward dialing (DOD)
- Speed dialing
- Intercom
- Fax interface
- Hot line
- Key/multifunction registration
- Night answer
- Off-hook voice announcement
- Off-premise extension
- PC programming
- Route advance block
- System/station speed dial
- System data up/download
- Security code (password)
- Toll restrict
- Trunk-to-trunk transfer
- Voice mail integration
- Voice recording service
- Centrex compatibility
- All call

Centrex is a business telephone service offered by the local telephone company. It is basically single-line telephone service with special features, such as intercom, call forwarding, call transfer, and least-cost routing. The advantages of Centrex over a PBX system are Centrex's lower initial investment, equipment space savings, continuous updating with new technology, and lower level of maintenance; however, the special features are limited, and the service may not be available in many areas.

Facsimile (FAX or fax) is the technology that allows written material, data, and images to be transmitted through the switched telephone system and printed out at the receiving end. There are five internationally accepted equipment groups: 1, 2, 3, 3E (enhanced), and 4. Groups 3 and 3E are the most popular types, with transmission speeds between 9600 and 14,400 bps. Group 4 is designed for digital lines transmitting at speeds between 56,000 and 64,000 bps.

12.8.2 Design Considerations

Wiring for telephone systems need not be installed in raceways; however, for security and aesthetic reasons, all wiring within finished spaces should be concealed. Receptacles should be provided within the telephone closets for equipment. Telephone trunk lines should be terminated in telephone panels or in telephone closets for splicing and distribution. Figure 12–8(e) illustrates the distribution diagram of a telephone system. For preliminary purposes, one linear foot of wall or closet space is allowed for each 3000 sq ft of finished floor space.

12.9 FIRE ALARM SYSTEMS

A fire alarm system is not limited to initiating an alarm during a fire; it may also serve to identify the location of the fire, transmit audio or visual messages, activate fire-extinguishing systems, and interface with building management systems. A more sophisticated fire alarm system is in effect a fire management system. This section introduces various fire alarm systems and their electrical wiring requirements. See Chapter 9, which discusses planning, detection, and signaling devices commonly used in fire alarm systems.

12.9.1 Code Requirements

The requirements of a fire alarm system and the locations of devices are governed by the National Fire Protection Code (NFPA 72), Americans with Disabilities Act (ADA), and local building codes, which vary with the occupancy, location, size, and height of the building. In general, a fire alarm system is required in public assembly buildings, businesses, educational institutions, hotels, factories, and hospitals. Theaters and entertainment facilities are exempted, as a false alarm may cause panic.

12.9.2 Types of Systems

A fire alarm system is usually a combination of several basic systems. Among these are the following.

Central Station versus Local System

The alarm signals are transmitted to a central monitor remote from the building. With modern communication technology, the central station may be miles away from the building, at the fire department, or even in another city to which the signal is transmitted through telephone or modem lines. Local systems are wired completely within the building. A key factor in a reliable fire alarm system is the reliability of the electrical power supply. On-site generators are preferred. Fire alarm system battery backup is required.

Manual or Automatic System

A fire alarm system may consist simply of manual alarm stations connected to sound bells, speakers, or horns; however, for building use groups that involve the public and for residential buildings over three stories in height, automatic detection devices are usually mandatory. In automatic systems, signals are initiated by automatic detection devices, as well as from manual alarm stations.

Coded or Noncoded System

A fire alarm system may provide a general alarm signal continuously until the system is turned off manually. This is known as a *noncoded system*. By contrast, if the signal is intermittent in duration or frequency, then the system is said to be *coded*. A coded system is usually designed to produce three rounds of signals to identify the location of the fire or the initiating device. For example, a signal with one short and one long sound may mean that the fire is at the first-floor east wing, whereas a signal with two short and two long sounds may mean that the fire is at the second-floor west wing. A coded system is a necessity for multiwing or multisectional buildings. The coding may have to be subzoned for large buildings when it becomes too complex.

Supervised or Nonsupervised System

With a nonsupervised system, accidental grounding or breaking of the wiring or contacts will undetectably disable the system until the problem is discovered. In the meantime, the system is inoperative and the building is unprotected. In a supervised system, the signal notifies the building management of the problem until it is fixed. Supervised or double-supervised systems are preferred in large buildings. Nonsupervised systems are typically not recommended and do not adhere to the NFPA standards and most local codes.

Single or Zoned System

In a single-zone system, all alarms are activated at once; in contrast, a zoned system may divide the alarms into two or more zones according to the locations of the alarm devices. For example, a building with east, center, and west wings may be divided into 3 zones, 1 for each wing. If the building is five stories high, then it may be divided into 15 zones, 1 for each floor of each wing. Caution must be exercised not to overdivide the zones, as the signal may become too complicated to be recognized.

Single-Stage or Two-Stage Systems

With a single-stage system, all alarms within each zone will be energized to signal that the building should be evacuated. Two-stage systems may provide a preliminary warning or alert to the occupants during the first stage and a signal to evacuate the building only when the second stage is energized.

General Alarm or Presignal System

In buildings where a general alarm signaling a minor fire may cause panic and the building is under full-time supervision by qualified personnel, a presignal system may be desirable. With this system, detection devices or manual alarm stations will send the signal only to limited locations so that management can determine whether and when a general alarm is to be activated. Schools and stadiums are good candidates for such a system. Note that the use of a presignal system in hospitals is prohibited by the NFPA, unless specifically approved by code officials.

Voice Communication System

Modern fire-signaling systems can also include a public address system to provide prerecorded or live voice instructions to the occupants and, in later stages, by the fire department. The public address system may also be supplemented by two-way communication devices, such as telephones, intercoms, and radio-frequency modules. Speakers may be used to generate a tone in lieu of bells or horns. Voice communication systems are required in all high-rise buildings and public assembly use groups.

Addressable or Nonaddressable System

A system is addressable when each fire detection or signaling device is assigned a unique coded frequency (address). The address may be adjustable to allow reassignment in the field. With this capability, the fire protection plan (zoning and staging) may easily be changed to fit a building's operational program. Addressability is especially useful for a tenant-occupied building when the tenants are frequently changing. Nonaddressable systems do not have this flexibility, but are, of course, initially more economical. A system may combine addressable and nonaddressable devices through an interfacing device. Figure 12–9 illustrates a typical multizone fire alarm system.

Standalone or Integrated System

The fire alarm system may be a standalone system or may be integrated with other building systems, such as an automatic sprinkler, building management, life safety, and vertical transportation systems. The advantages of integrating are many, including space and initial cost savings, flexibility, and better coordination. The primary disadvantages are the complexity of the system and the cost associated with proprietary systems and components after installation.

Thus a fire alarm system combination may be a system such as the following:

- Manual, single-zone, single-stage system
- Manual and automatic, multizone, noncoded, single-stage system
- Automatic, multizone, coded, two-stage with presignal and voice communication system

Naturally, a more sophisticated system offers better protection and requires a greater capital investment and

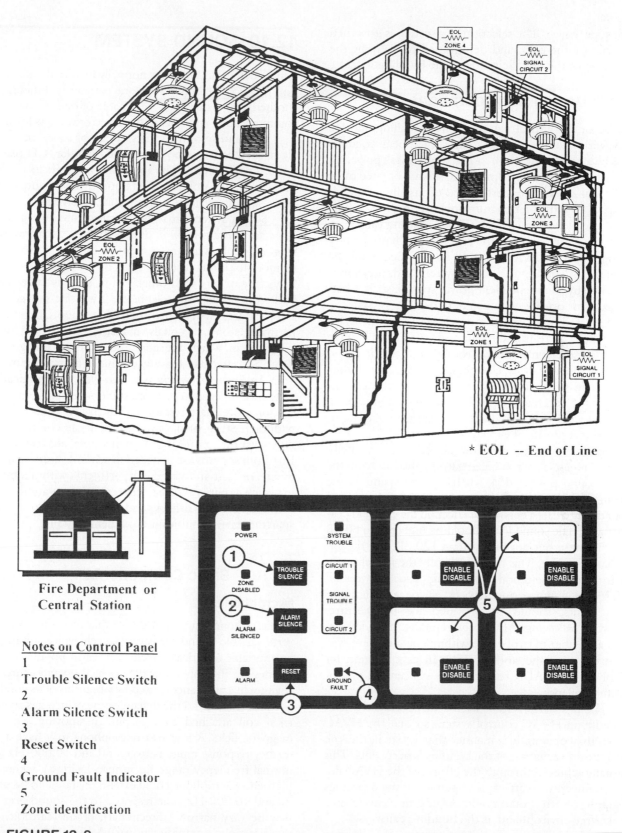

* EOL -- End of Line

Fire Department or Central Station

Notes on Control Panel

1
Trouble Silence Switch
2
Alarm Silence Switch
3
Reset Switch
4
Ground Fault Indicator
5
Zone identification

■ **FIGURE 12–9**

Small multizone fire alarm system showing the location of the control panel, signal, and alarm devices. This is a four-zone system. Additional zones can be accommodated. (Reproduced with permission from Edwards Co.)

more supervision. The selection of a system is governed primarily by the code and determined partially by the evaluation of the specific needs of the building.

12.9.3 Design Considerations

Because of the critical nature of fire signalling and life safety, wiring for fire alarm systems should be in conduits or raceways separate from other, noncritical power or low-voltage systems. In some cases, the use of rated cable can be utilized to save cost in lieu of installing conduit and wire. In return-air plenum applications, it is important to use plenum-rated low-smoke-emitting cable.

Fire alarm systems are required to have battery backup. Electrical power should be supplied from the emergency section of the power system and, preferably, backed up by an independent power system, such as battery packs or on-site engine generators.

Fire and smoke detection devices should be located where there is a likely hazard, such as in storage rooms, electrical–mechanical rooms, and public egress areas (e.g., in entrances, elevators, or lobbies, near stairways). The fire command center, where the control panel or the command console is located, should be in a secure locale, constructed of fire walls and, preferably, with a direct means of egress.

Audible fire annunciation devices, such as bells, horns, chimes, and tone generators, should have the proper sound power (dB), so that all occupants in the building can hear the signals. On the other hand, too loud a sound may create a deadening effect on the ears, and possibly panic. The sound level should be higher than 60 dB at any location but not higher than 120 dB near the annunciation device. This range is usually the guideline for determining the selection and spacing of such devices.

Visual annunciating devices, such as flashing or strobe lights, should be provided for the hearing-impaired, following ADA requirements. In addition, when a number of strobes are within a field of view, synchronization of the storbes offers an easy solution for complying with ADA requirements concerning photosensitive epilepsy.

The fire alarm system should be coordinated with the building HVAC control system so that the HVAC system may operate in a manner that assists in the safe and speedy evacuation of the building's occupants. This is usually achieved through the efforts of the HVAC design engineer, who attempts to sequence the quantities of supply, return, exhaust, and outside air to maintain a smoke-free environment at the building exitways.

In designing the fire alarm system, the designer should study the feasibility and economics of establishing an integrated fire alarm and building management system that is most appropriate for the building.

12.10 SOUND SYSTEM

Sound systems in buildings normally involve the distribution of voices or music from one part of the building to another. The system may be capable of one- or two-way communication and offer either wired or wireless features.

A sound system consists of one or more input devices, such as a microphone, tape recorder, CD player, radio, or telephone. The input is converted to electric signals that are amplified and transmitted to output devices, such as speakers or telephone receivers.

Sound systems may be classified into two groups. A *public address system* is designed to broadcast a voice message, such as an announcement or speech, from one or more locations to other parts of the building. It also may carry or mix with other inputs, such as radio, tape, background music, or telephone. It is a one-way system. Public address systems are installed in schools, hotels, department stores, hospitals, industrial plants, and other buildings in which the administration or management have a need to make announcements or to transmit music.

An *intercom system* is a two-way communication system between two or more stations. It may consist of one or more master stations, which can communicate with each other, or slave units, which can communicate only with the master unit to which they are connected. Intercom systems are installed in schools, homes, apartments, offices, and specific departments within a building, such as between the sales area and the warehouse of a store, between offices, and between the lobby and apartments in a multiunit apartment building.

12.10.1 Basic Components

Microphones

A microphone is an input device that reacts to and converts variable sound pressure (speech, voice, or music) into variable electrical current. The most popular microphones are the condenser type, which operates through changes in capacitance caused by vibrations of its conductive diaphragm; and the dynamic type, which has an electrical coil attached to a diaphragm moving within a magnetic field. A low-cost microphone may have a frequency response range between 50 and 10,000 Hz (the normal frequency range for human hearing), whereas a high-end microphone could extend the range to between 20 and 20,000 Hz. The response to the sound pressure may be very narrow (directional), broad (cardioid), or nondirectional (omnidirectional). A microphone must be selected so as to match the impedance with other components of the system. Figure 12–10(a) illustrates the internal construction of a dynamic microphone.

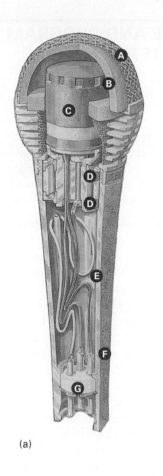

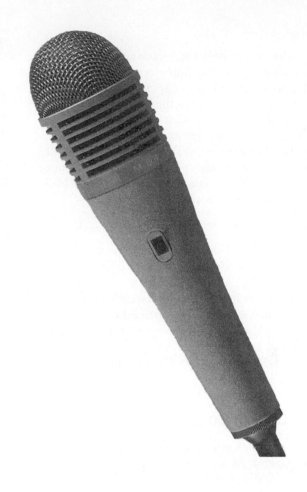

(a)

■ **FIGURE 12–10**
(a) Cutaway and exterior
views of a dynamic
microphone. (Courtesy:
Paso, Inc.) (b) Typical ceiling
speaker assembly, including
speaker with matching
transformer and ceiling baffle
(top); typical trumpet speaker
(bottom). (Reproduced with
permission from Atlas/
Soundolier.) (Figure
continues, page 401.)

(b)

Speakers

A speaker is an output device that converts electrical signals into sound signals through electromagnetic voice coils and a driver that vibrates a speaker cone. Speakers are designed to respond to either the full frequency range or tuned frequencies—that is, bass (a woofer, for low frequencies) or treble (a tweeter, for high frequencies). Speakers are installed in or on ceiling or wall surfaces. Most speakers are provided with matching transformer and volume controls. Figure 12–10(b) illustrates several types of speakers.

Amplifiers

An amplifier receives a sound-generated electrical signal and boosts the signal for transmission to output devices. Amplifiers for public address systems are rated in watts of sound power with a frequency response from 20 to 30,000 Hz. They may be self-contained or may be designed to include switches, a radio tuner, tape or compact disks, and a telephone interface. Figure 12–10(c) illustrates a desktop model of a public address system and its block diagram.

12.10.2 Design Considerations

In designing a building sound system, one should determine the location of a control center where messages are normally initiated and the locations, sizes, and ambient noise levels of rooms to which the messages are to be transmitted. Figure 12–10(d) is a one-line diagram of a typical public address system.

Next, one must determine the approximate number of speakers required for each space. In general, ceiling speakers (usually 8 in. in diameter) can cover 250 sq ft of floor area at an 8-ft ceiling height and up to 600 sq ft at a 12–ft ceiling height. A trumpet (horn) type of speaker may cover 8000 sq ft in large, quiet open spaces; the area of coverage is reduced in noisy spaces.

For planning the power required for different speakers is as follows:

- 8-in. ceiling or single-projection wall speakers 1 W
- 12–in. ceiling or double-projection wall speakers 2 W
- Column speakers 10–30 W
- Trumpet speakers 10–30 W
- Signal (ring) generators 2–4 W

The amplifier is sized for the total power of the speakers, plus an allowance for future expansion. To achieve good sound reproduction and long life, it is good practice not to load the amplifier to its maximum rating.

Sound systems are wired primarily with shielded or nonshielded twisted pair cables. In conduits or wireways, shielded twisted pair (STP) is preferred.

12.11 TIME AND PROGRAM SYSTEMS

Time and program systems are necessary in schools, sports facilities, and certain industries where synchronized time is important. These systems consist of a control center (master clock, code generator, switches, etc.), individual clocks (secondary or slave clocks), program signaling devices (bells, buzzers, or annunciators). These systems can be either wired or wireless. There are various types of devices for synchronizing time.

- *Synchronous motor-driven type.* Secondary clocks and signaling devices are wired to the control center. The secondary clock runs by a synchronous motor, but its indication of time is monitored and adjusted by the master clock.
- *Impulse-driven type.* Secondary clocks and signaling devices are wired directly to the control center. The secondary clock is driven by an impulse signal of the master clock and advances once every minute in unison with the master clock. During a power failure, the control center will accumulate missed minute impulses for up to 24 hours by its standby battery power. Upon restoration of normal power, rapid correction pulses will be sent to advance the secondary clocks to the correct time.
- *Carrier-frequency type.* Individual clocks are plugged into the standard 120-V power without special wiring, but the indication of time is corrected by a high-frequency signal (between 3000 and 8000 Hz) carried (i.e., superimposed) through the 120-V power wiring system. With the carrier-frequency system, dedicated wiring between the control center and the individual clocks (and signaling devices) is eliminated.
- *Radio frequency type.* The individual clocks are plugged into a standard 120-V power outlet or rely on a battery and do not require special wiring. The time is corrected by a radio-frequency generator. The frequency may be tuned to the National Institute of Science and Technology (NIST) universal time for extreme accuracy in special research applications.

Figure 12–11 illustrates several components and the wiring diagram of a typical time and program system.

Videoconferencing

Videoconferencing systems are becoming more common in business and education. Videoconferencing allows virtual face-to-face meetings and sharing of data without the need for excessive travel, overnight lodging, and other related expenses. A typical videoconference

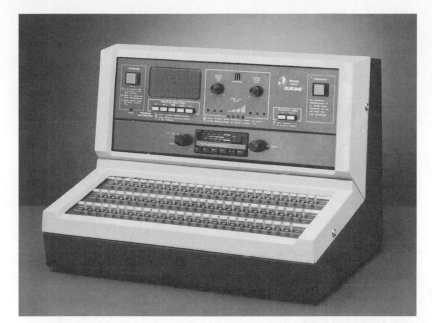

■ **FIGURE 12–10** *(Continued)*
(c) Desktop model of legacy PA/Tutorcom
system control panel with telephone interface, and a block diagram of system components. (Reproduced with permission from
Dukane Corp.) (d) One-line diagram of a
typical PA system.

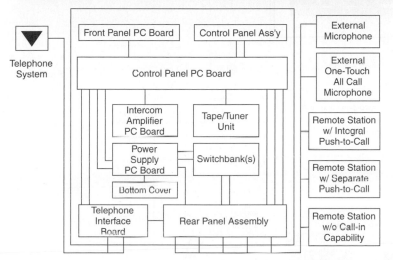

(c)

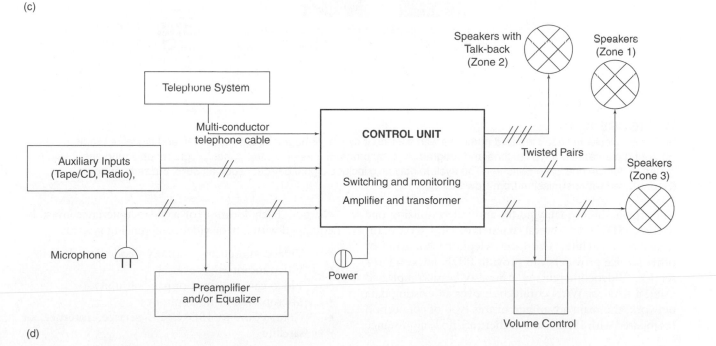

(d)

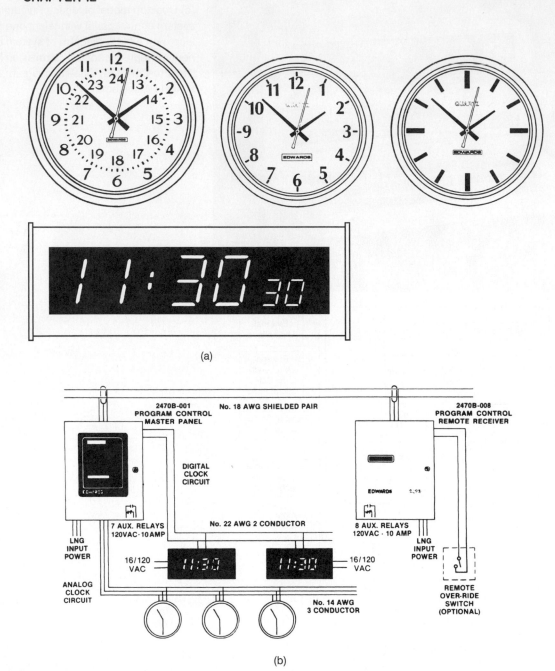

(a)

(b)

■ **FIGURE 12–11**

(a) Typical clocks include standard numerals with a 24-hour dial, a 12-hour dial, a graphic dial, and digital readouts.
(b) Wiring diagram for a typical time and program system: program master control panel, programmable modules, backup batteries, relays, and zone switches. The system may also include signal devices, such as bells, buzzers, or horns. (Reproduced with permission from Edwards Co.)

will provide high-quality audio and video utilizing one to three ISDN lines, several channels of a T-1 service, Internet or a satellite. The local telephone company or other service providers can provide ISDN lines and T-1 service. Videoconferencing can also be accomplished using a LAN or WAN connection over an existing data network; the main advantage of this type of connection (compared with ISDN) is that there are no long-distance

charges. Each location of a videoconference must be equipped with a basic videoconferencing system:

- Videoconferencing CODEC (code decoder)
- Video monitor
- Video camera (fixed or pan/tilt unit)
- Microphone or microphones
- Connection to ISDN, T-1 service, Internet, or satellite

These components can be grouped on a mobile cart, integrated into a single piece of equipment, or installed in a dedicated room.

One important consideration for a videoconferencing system is the refresh rate it is capable of providing; 15 FPS (frames per second) to 30 FPS is full-motion video similar to the quality from a television signal; 15 FPS is acceptable but will provide a jagged and jerky picture when movement occurs.

Videoconferencing systems are standards-based systems, which allow most manufacturers' equipment to work together. The current standards are H.320, which governs the protocols used for ISDN and other public switched network interfaces. H.323 governs the protocols used for videoconferencing over a LAN and WAN connection.

Audio/Visual (A/V) Systems

Audio/Visual (multimedia) systems are typically found in corporate boardrooms, conference rooms, auditoriums, classrooms, and training centers. These systems can consist of a multitude of audio, video, and control equipment.

1. *Typical equipment.*
   - LCD data projector—used to project video or data onto a large screen for viewing by a large group.
   - Document camera—video camera and light table that allows a document or three-dimensional object to be displayed on an LCD projector.
   - Slide to video converter—allows standard 35 mm slides to be displayed through an LCD projector.
   - *Interactive presentation board*—allows storage and play back of interactive marker board output/input with an attached projector and computer.
2. *Projection screen.* A manual or motorized projector screen. The use of a tab-tensioned screen provides a very flat projection surface, which is extremely important for viewing computer images such as spreadsheets or video presentations.
3. *Sound reinforcement system.* Consisting of ceiling or wall-mounted speakers and wired or wireless microphones on podiums, tabletops, lapels, headsets, or microphone stands to reinforce speech or program material (audio from VCR tape DVD, CD, etc.) Additionally, a mixer, equalizer, and amplifier are required.
4. *Controls system.* Consists of wired or wireless tabletop or podium-mounted push-button touchscreen panels. The control system consolidates and automates numerous devices into a single easy-to-operate panel. For example a single button on the control panel could dim the lights, close the blackout shades, turn on the OCD projector, and lower the projection screen. The operator can then select the source to display (such as the VCR, CD, DVD, computer) and then control the transport functions of the media player.

These types of control panels are fully customizable to meet specific end-user requirements, allowing flexibility and user-definable options.

12.12 MISCELLANEOUS AND SPECIALTY SYSTEMS

Because of space considerations, many commonly used telecommunication systems have not been covered in this chapter. Nonetheless, they should not be overlooked in building design. For example, a doorbell system, although very simple, is necessary in many kinds of buildings. In addition, facilities such as offices, apartment complexes, hotels, hospitals, department stores, schools, and entertainment and sporting facilities often have many other unique systems that must be addressed during the initial planning phase of the project. Some of these systems are mentioned here without explanation. They serve as a reminder to the designer that telecommunication systems are becoming more and more prevalent in the building design process. The following list mentions some of these facilities and the systems often found in them:

- *Hospitals and health-care facilities* Nurse's call; doctor's paging; patient-monitoring systems in intensive care units; bedside monitoring and control panel; telemedicine diagnoses; remote imaging readout; videoconferencing and multimedia conference and educational centers.
- *Hotels and motels* Voice mail and wake-up-call systems; satellite and CATV, front-of-the-house and back-of-the-house management; card-access systems, central reservation and accounting point of sale, high-speed Internet access.
- *Sporting and convention facilities* Central lighting controls, scoreboards, CCTV, and video recording, wireless communication, special open-space fire and smoke detection systems (infrared beam); high-quality sound reinforcement system.
- *Residences and apartments* Door intercom; intrusion and security; wired and wireless sound; TV and satellite antenna; whole-house audio; home theater systems; high-speed broadband access; fully automated home appliances and controls.
- *Department and retail stores* Shoplifting detection systems; general surveillance and CCTV;

point of sales; master distribution of TV and satellite signals, high-speed Internet access; staff-locating systems.

QUESTIONS

12.1 What are the forms and functions of information systems?

12.2 What is the difference between analog and digital signals? Can they be converted from one to the other?

12.3 What are the forms and functions that can be provided by radio? TV? A computer with a modem?

12.4 Human hearing frequency extends beyond 10,000 Hz. What is the highest frequency that can normally be transmitted by a conventional telephone system?

12.5 What wires are commonly used for telecommunication systems?

12.6 What is fiber optics? What is an optic fiber?

12.7 Name two primary advantages of an optic fiber over twisted pair or coaxial cables.

12.8 What is meant by critical angle in an optic fiber?

12.9 The index of refraction of glass is 1.5 ($n_1 = 1.5$), and the index of refraction of air is 1.0 ($n_2 = 1.0$). What is the critical angle when light passes from glass to air?

12.10 What are the metal compounds commonly used to generate laser in fiber-optic systems? What are the range of wavelengths in nanometers (nm)?

12.11 What is the applicable transmission speed range in Mbps for a coaxial cable? A single-mode optic fiber?

12.12 What are the frequency bands assigned for FM radio in the United States? VHF/TV channels (2 to 6)? Cellular telephones? Satellite UHF?

12.13 What is a geostationary-type telecommunication satellite? What is its orbit (miles above the earth)?

12.14 What is a LAN? A WAN?

12.15 What are the disadvantages of wireless LAN?

12.16 What techniques are commonly used in controlling access?

12.17 What principles are commonly applied in detecting intrusions?

12.18 What detection devices are commonly used within a building?

12.19 What are the two types of telephone dialing systems?

12.20 What are PBX, DOD, and DID in a telephone system?

12.21 What is a coded, a supervised, a zoned, and an addressable fire alarm system?

12.22 Why are flashing or strobe lights required in fire alarm systems?

12.23 What is a public address (PA) system? What are its basic components?

12.24 What size ceiling- or wall-mounted speaker is normally used in buildings, and what is the approximate area of coverage?

12.25 What types of wires are normally used for sound systems? Can optic fiber be used?

12.26 What is a time and program system? Where is it usually installed?

12.27 How are clocks synchronized in the time-clock system?

12.28 Name as many as you can of the telecommunication systems that should be considered in a large residence, a hospital, a hotel, and a department store.

12.29 Low-voltage cabling typically does not require traditional EMT conduits. What methods are commonly used to support low-voltage cable?

12.30 When is conduit (EMT) necessary when low-voltage cable is installed?

12.31 RGB cable is typically used for high-resolution video signals. What are the individual video components carried by an RGB cable?

12.32 What are the two most common types of data projectors used to display computer graphics?

12.33 Name the card installed in a PC workstation to connect it to a LAN or WAN.

12.34 What is the most common type of topography used for a LAN?

REFERENCES

NFPA-70, National Electrical Code, National Fire Protection Association, Quincy, MA.

EIA/TIA 586, Commercial Building Wiring Standard.

Hecht, Jeff, *Understanding Fiber-Optics*. Indianapolis: Prentice Hall Computer Publishing, 1993.

Newton, Harry, *Newton's Telecom Dictionary*, 7th ed. New York: Flatiron Publishing, 2000.

Bigelow, Stephen, *Understanding Telephone Electronics*. Indianapolis: Prentice Hall Computer Publishing, 1997.

Paynter, Robert T., *Introductory Electronic Devices and Circuits*. Englewood Cliffs, NJ: Prentice Hall, 1997.

Lighting Handbook. New York: Illuminating Engineering Society of North America, 2000.

BICSI, *Telecommunications Distribution Methods Manual*, 10th ed. Tampa, FL, 2003.

Brohammer, Steve, and Tim Ruiz, "Bridging the Future," *Consulting Specifying Engineer*, July 2002.

ELECTRICAL DESIGN AND WIRING

13

ELECTRICAL SYSTEM DESIGN IS AN INTEGRAL PART OF THE overall building design process. Nearly all mechanical equipment, such as air conditioners, pumps, and fans, as well as building equipment, such as elevators and appliances, is electrically powered. Even gas or oil-fired equipment, such as boilers and heaters, requires electric power. The selection of an electrical power system is often influenced by which mechanical system is chosen. For example, if a building containing many apartments uses individual room air conditioners, the electrical distribution system will probably be a 120/240-V, single-phase system, whereas if the apartments are served by a central chiller and air-handling system, the electrical distribution system will probably be a 120/208-V, three-phase, four-wire system.

With the explosion of communication systems in buildings, the demand for electrical power has increased dramatically in recent years. For example, the demand for convenience power (plug-in types of equipment) in an office building has increased severalfold. What were acceptable standards a few years ago are now inadequate for use with data-processing and communication equipment in all types of buildings, including residences.

Generally speaking, electrical systems do not require much building space, compared with mechanical systems; however, most electrical operating devices are normally exposed in occupied spaces; therefore, their location, configuration, and aesthetics must be precisely coordinated with architectural and interior design. Unfortunately, more often than not, these electrical operating devices, such as switches, receptacles, controls, and alarms, are installed indiscriminately, without regard to their location, size, shape, or color.

13.1 ELECTRICAL DESIGN PROCEDURE

There are five basic steps in electrical system design:

1. Analyze building needs.
2. Determine electrical loads.
3. Select electrical systems.
4. Coordinate with other design decisions.
5. Prepare electrical plans and specifications.

In practice, some of these steps may be temporarily bypassed in order to keep pace with the project's progress. For example, the power needs for a building are determined by its lighting, mechanical, and building equipment loads, from which the electrical power system is selected. However, before these steps can be completed sequentially, the utility company might wish to know the types of services in advance; the architect would like to know the sizes and locations of the electrical equipment rooms; the structural engineer would like to know the weight of all major equipment; and the HVAC engineer would like to know any system options for evaluating the cost of HVAC equipment. Under these circumstances, reasonable assumptions must be made based on experience or statistical data, some of which are given in this book. The assumptions will have to be assessed, modified, and finalized at each phase of the design process.

13.2 ANALYSIS OF BUILDING NEEDS

The first step in electrical system design is to identify the needs of the building established by the architectural program. Following are the major factors affecting electrical systems:

1. *Occupation factors.* Type of building occupancy, number of occupants, present and future electrical appliances to be installed or anticipated in the building, etc.
2. *Cost factors.* Whether the building has an austere budget, is of average quality, or is a high-image building, etc.
3. *Architectural factors.* Size of the building, number of floors, floor-to-floor height, building footprints, elevations, etc.
4. *Building environments.* Whether the building is heated, whether it is air-conditioned, whether the systems are central or unitary, etc.
5. *Illumination criteria.* Lighting level and the predominant type of light sources to be used.
6. *Other mechanical systems.* Need for electricity for cold water, hot water, sewage disposal, fire protection systems, etc.

7. *Building equipment.* Vertical transportation systems, food preparation, recreational equipment, processing equipment such as computers, and production equipment requiring electrical power, etc.

8. *Auxiliary systems.* Systems such as building management, time clock, fire alarm, telecommunication, radio and TV antenna, public address, security, data, etc. Specialty systems may be needed in hospitals, hotels, and industrial buildings. (The impact of an auxiliary system on a building's power distribution system is negligible.)

9. *Future needs.* More than any other components of the building systems, electrical power load in buildings has been known to grow consistently year after year. The question is not whether to build in some spare capacity for growth but rather how much to build in and for how long. While the structure may last hundreds of years, the electrical power system may need to be upgraded every 20 to 30 years, owing to advances in technology. The designer must make an independent judgment or consult with the architect or building management for specific future loads as well as for general growth. In general, a minimum of 25 percent spare capacity should be provided at the power main and at one or more power distribution centers.

13.3 DETERMINATION OF ELECTRICAL LOADS

The consumption of electrical power in buildings—particularly office buildings—has risen dramatically, as noted above. For example, the demand for convenience power for portable or plug-in equipment in office buildings increased from an average of 1 to 3 W/sq ft (on net floor area basis) in the 1980s to an average of 2 to 4 W/sq ft since 1990 for convenience power in intelligent-type office buildings, which have a higher concentration of personal computers, telecommunication equipment, and fax and copy machines. Similarly, the number of electrical appliances in homes has also increased steadily. Personal computers, microwave ovens, food processors, multiple units of television sets, videotape CD and DVD players are just a few appliances that are becoming a higher standard of living rather than a luxury.

The increase in the electrical load of a building also has a direct effect on the building's air-conditioning (cooling) system. Heat generated from electrical loads in an environmentally controlled space must be removed through ventilation or the air-conditioning system for the space to be maintained at a comfortable level. In general, all electrical power input (watts) will become internal heat gain (Btuh) for the building. Following the fundamental law of energy conversion, 1 W of electrical power used for 1 hour will generate 3.4 Btu of heat. It must be understood, however, that the total heat gain of all equipment should be based on the net *demand load* rather than the gross *connected load* of all loads. The difference between demand and connected loads is explained in later sections.

Electrical loads for buildings can be analyzed according to a number of different categories.

13.3.1 Lighting

Lighting design is usually the coordinated effort of the architect, interior designer, lighting designer, and electrical engineer. Lighting accounts for one of the larger electrical loads in most buildings. In general, lighting fixtures are designed for 120-V, single-phase power; however, they may also be designed for use on 208-, 240-, and 277-V, single-phase power systems.

A continuous improvement in light-source technology has increased the efficiency of converting electrical energy to lighting energy, which in turn has reduced the need for electrical power for lighting in buildings. It was common to require 3 to 5 W/sq ft for lighting in office buildings in the 1980s, but only about 2 W/sq ft was needed in the early 1990s, and 1.5 W/sq ft in 2000. (See Table 13–1.)

For planning electrical power, the average power required by various types of indoor areas or activities is given in Table 13–1. The figures shown are only approximations and should not be used for lighting design purposes. (See Chapter 17 for lighting design calculations.)

13.3.2 Mechanical Equipment

Electrical power required for mechanical equipment varies widely with the building, type of climate, architectural design, size of the building, type of mechanical systems, and intended method of operation of these systems. For planning, preliminary data on power are obtained from the design mechanical engineers and are updated periodically as the design progresses.

The mechanical system includes the HVAC, plumbing, and fire protection systems. Equipment such as chillers, boilers, pumps, and fans usually requires a large power capacity and is more economically designed for higher voltages, such as 208-, 240-, and 480-V, three-phase power; however, residential equipment is normally designed for 120- or 240-V, single-phase power.

TABLE 13-1

Unit (lighting) power density (UPD)—W/ft² (W/m²) for common and specific areas

Common Spaces for All Building Occupancies	UPD	Building Occupancy or Specific Space Functions	UPD
Office—enclosed	1.5 (16.1)	Exhibit space	3.3 (35.5)
Office—open plan	1.3 (14.0)	Religious building	2.2 (21.6)
Conference/multipurpose	1.5 (16.1)	Museum (general exhibition)	1.6 (10.8)
Classroom (except penitentiary)	1.6 (17.2)	Professional sports (court area only)	4.3 (57.0)
Library		Athletic facility (playing area)	1.9 (20.5)
• file/cataloging	1.4 (15.1)	Professional sports televised (court +	
• stacks	1.9 (20.4)	perimeter playing area)	3.8 (40.9)
• reading area	1.8 (19.4)	Gymnasium (playing area)	1.9 (20.4)
Lobby (hotels)	1.7 (18.3)	Auditorium (seating area)	1.7 (18.3)
Lobby (motion picture)	.8 (8.61)	Airport (concourse)	0.7 (7.5)
Lobby (peforming arts)	1.2 (12.91)	Airport (ticket counter)	1.8 (19.4)
Lobby (except theaters & hotels)	1.8 (19.4)	Hospital	
Lobby (auditorium, theater)	(1.2/.8) (8.6)	• emergency	2.8 (34.0)
Atrium (first 3 floors)	1.3 (14.0)	• nurse station	1.8 (22.6)
Atrium (each additional floor)	0.2 (2.2)	• corridor	1.6 (10.8)
Food preparation	2.2 (23.7)	• examination	1.6
Restroom	1 (10.8)	• pharmacy	2.3
Corridor/transition spaces (except hospital, manufacturing)	.7 (7.5)	Hotel/Motel (guest room)	2.5 (26.9)
Stairs—active	.9 (9.7)	Multi-Family (private living space)	2 (21.5)
Storage—active	1.1 (11.8)	Dormitory (living space)	1.9 (20.4)
Storage—inactive	0.3 (3)	Bank (banking area)	2.4
Warehouse (fine material)	1.6 (17.2)	Restaurant	Varies
Warehouse (medium/bulky)	1.1 (1.8)	Retail (general sales area)	2.1 (22.6)*
Garage (parking area) .2-pedestrian, .1-attendant	0.3 (3.2)	• Mall concourse	1.8
Elec./mech.space	1.3 (8.6)		

Note: UPD values given in this table are for load-estimating purpose only. It shall not be used for lighting design or for Electrical Code compliance. Refer to ASHRAE / IESNA Standard 90.1-2001 and NEC-2002 for specific requirements for compliance.

*Refer to Ashrae 90.1 2001 for Additional UPD Credits.

13.3.3 Building Equipment

Building equipment includes vertical transportation equipment (elevators and escalators), food service equipment, and household, recreational, and miscellaneous operational equipment. The power requirement for this type of equipment varies widely in capacity and operating characteristics. The electrical power required for building equipment is usually gathered by the architect, the user, or various specialty consultants.

13.3.4 Auxiliary Systems

Normally, auxiliary systems, such as those systems covered in Chapter 12, do not require a large power capacity; thus, they are usually designed for 120- or 240-V, single-phase power. Depending on the type of occupancy—residential, commercial, institutional, or industrial—each building may require one or more of the following systems:

- Building management systems
- Security system
- Time-clock systems
- Fire alarm systems
- Telecommunication systems
- Radio and TV antenna systems
- Public address systems
- Data and networking
- Specialty systems

For power-plannings, usually one or two 20- to 30-A, single-phase circuits are sufficient for each auxiliary system in most buildings.

13.3.5 Convenience Power

Convenience power is power provided for plug-in equipment such as household appliances, personal computers, office equipment, laboratory instruments, service equipment, portable lights, and audio and video equipment. With the proliferation of electrical appliances in homes, offices, and schools, the demand for on-floor power for plug-in equipment has increased steeply.

The total appliance load for typical offices and homes has already exceeded that required for lighting and becomes a dominant load of the building electrical systems. The design of convenience power circuits must, therefore, be carefully analyzed with regard to capacity and appropriate location. Table 13–2 provides typical power ratings of household appliances. Table 13–3 provides the suggested convenience power to be allowed for various interior spaces or building occupancies.

13.3.6 Connected and Demand Loads

Permanently wired electrical equipment is a *fixed load*. Plug-in electrical equipment is a *convenience load*. All electrical equipment has a nameplate current rating in amperes (A), or working power in watts (W) or (kW), or in apparent power in (VA) or (kVA). Any of these power ratings, but in consistent units, may be used to calculate the total loads of the system or the entire building. The conversion between these units was given in Chapter 10.

TABLE 13–2
Typical power rating of household appliances

Appliances	Range of Power (VA)
Window air conditioner	
• 1/2 ton, 115 V	500–700
• 3/4 ton, 115 V	900–1100
• 1 ton, 115 V	1200–1400
• 2 ton, 230 V	2400–2800
Refrigerators, 115 V	300–1000
Freezers, 115 V	300–800
Washing machines, 115 V	800–1200
Dryers, 115/230 V	3000–5000
Gas dryers, 115 V	200–300
Water heaters, 115/230 V	3000–6000
Toaster, 115 V	500–1000
Oven, 115/230 V	2000–5000
Microwave oven, 115 V	300–1000
Combination oven, 115/230 V	1000–5000
Television, 115 V	200–1000

TABLE 13–3
Typical power allowance for convenience (portable or fixed loads)

Type of Space	Power Allowance[a]
Offices	2–4 W/sq ft (22–44 W/m²)
Classrooms—general (no PCs)	2–3 W/sq ft (22–33 W/m²)
Classrooms—specific (with PCs)	Per load
Meeting rooms	3–5 W/sq ft (33–55 W/m²)
Residential spaces:	–
• Kitchen	(2)–(6) 20-A circuits
• Dining and family	(2) 20-A circuits
• Bedrooms, not air-conditioned	(1) 20-A/room
• Bedrooms, centrally air-conditioned	(1) 20 A/2 rooms
• Laundry[b]	(2) 20-A circuits
• Exterior	(1) 20 A/exposure

[a]In net occupied spaces.
[b]Provide 30- to 50-A, 220-V circuit for electric clothes dryers.

Connected Load

The connected load of an electrical power system is the algebraic sum of all electrical loads connected to the system. It does not take into account how and when these loads are being used.

Equation 13–1 expresses the simple relations between connected loads (CL) of various load groups, such as lighting, equipment, and convenience power groups, and the gross connected load (GCL) of a power system or a building:

$$GCL = CL_1 + CL_2 + CL_3 + CL_4 + \ldots + CL_n$$
$$= \Sigma CL_i \qquad (13\text{–}1)$$

Demand Load

The demand load of an electrical power system indicates the net load that would probably be used at the same time for each load group. When all connected loads are used at the same time, the demand load is equal to the connected load; however, in most buildings, the demand load is always lower than the connected load. For example, to estimate the load of plug-in receptacles (convenience outlets), NEC recommends using 1.5 A or 180 VA for each duplex outlet. If the building contains 1000 duplex outlets, then the connected load is 180,000 VA, but in reality, the demand may be at most 50 percent of the connected load, or 90,000 VA. In this example, the demand factor is 50 percent. The simple relationship between connected load (CL), demand load (DL), demand factor (DF), and gross demand load (GDL) is expressed in Eqs. 13–2 and 13–3.

$$DL = CL \times DF \qquad (13\text{–}2)$$

and

$$GDL = \Sigma DL \qquad (13\text{–}3)$$

Diversity Coefficient

The diversity coefficient accounts for the diversity of demand between different groups of loads. For example, when there is a very high appliance and lighting load in a space, the heat generated from this load may cause the heating system to be turned off and reduce the demand load of the heating system. Another typical example is the diversification between heating and cooling systems when only one system is in use at a time.

The diversity coefficient is also called the diversity factor (DF); however, to present possible confusion with demand factor (also DF), the authors prefer to use the term *diversity coefficient* (DC), and it is used as a divisor rather than a multiplier. The diversity coefficient is time-consuming to calculate. In this book, DC = 1.0 is used for systems that do not have obvious load diversification, and DC = 1.2 is used for large systems or for systems with diversified load groups. When the diversification of loads is greater than the suggested range of diversity coefficients, a detailed analysis of connected loads under all operating conditions will have to be made hour by hour or minute by minute, differentiating between day and night, weekday and weekend, summer and winter, etc. Often, diversity coefficient is neglected by designers; this can result in oversized systems. The relationship between the net demand load (NDL), gross demand load (GDL), and diversity coefficient (DC) is expressed in Eq. 13–4:

$$NDL = \frac{GDL}{DC} \qquad (13\text{–}4)$$

Example 13.1 A building electrical system is calculated to have the following connected loads by load groups. The estimated demand factors are shown in the table. What is the net demand load of the system, assuming that the diversity coefficient is 1.1?

Demand Load Calculations

Load Group	Connected Load, kW	DF	Demand Load, kW
Lighting	125	0.9	112.5
Receptacles	85	0.2	42.5
Mechanical equipment	200	0.8	160.0
Building equipment	150	0.6	90.0
Connected load (CL) 560			
Gross demand load (GDL) .... 405 kW			
Net demand load (NDL), (405/1.1)...................... 368 kW			
(use 370)			

13.4 SYSTEM SELECTION AND TYPICAL EQUIPMENT RATINGS

Some power distribution systems commonly used in buildings, including their limitations, were briefly introduced in Chapter 11. In actual design applications, the designer must also weigh the various other factors in alternative systems before making the final selection. These factors include cost of investment, cost of maintenance, system reliability, serviceability, space availability, and the impact on the other building systems that require electrical power. Often, the phases and voltages of some mechanical equipment are standardized without options. In such cases, the power system or a portion of the power system must be so selected as to be compatible with the equipment standard.

13.4.1 Three-Phase versus Single-Phase Systems

On the basis of load analysis, electrical power systems are selected to complement the load characteristics. For example, if the loads in the building are predominantly 120-V, single-phase, then the system should be either a 120/240-V, single-phase, three-wire or a 120/208-V, three-phase, four-wire system. One of the advantages of a three-phase, four-wire system not mentioned earlier is its inherent economy. It allows one common neutral wire to serve up to three single-phase loads; however, with the increased use of inductive loads, such as computers in offices, there is a tendency to overheat the common neutral wire owing to the third harmonics. In such a case, individual neutral wires for each circuit are recommended.

13.4.2 Common Voltage Ratings

The electrical power system for a large building can be a combination of single-phase and three-phase, low-voltage systems supplied by one or more high-voltage systems. Figure 13–1 is a simplified one-line diagram of a larger power distribution system in which the primary power, 13,800-V (13.8 kV), three-phase three-wire, passes through step-down transformers to three different low-voltage systems: 120/208-V, three-phase, four-wire; 277/480-V, three-phase, four-wire; and 480-V, three-phase, three-wire systems.

Naturally, for smaller buildings, a single power system will suffice; however, in some applications, it is desirable to separate the large power load from the lighting and appliance loads to minimize voltage fluctuation of the system owing to the on-off nature of the larger loads. The design electrical engineer must select the system early and inform the architect and the mechanical engineers of any options, so that equipment selected by them will match properly. In general, the following equipment greatly affects the selection of the system during the preliminary design period.

Elevators

Elevators usually range from 15 to 200 hp, depending on their speed and load capacity. They are usually equipped with three-phase motors whose voltage is compatible with that of the building power system.

Food Service Equipment

Food service equipment of 30 kW or less can be either single-phase or three-phase; equipment of higher capacity is usually three-phase. In all cases, the equipment voltage must match that of the building system. For example, if the building power system is 208 V and the equipment is rated for 240 V, then the equipment output will probably be reduced by 15 percent in capacity unless a local transformer is installed. The reverse is true when the system voltage is higher than the equipment voltage.

Mechanical Equipment

While large pieces of mechanical equipment are always designed for three-phase power, smaller equipment, such as fans, coils, and unitary equipment, may be made for either single-phase or three-phase power. The selection of such equipment must be a joint decision of the mechanical and electrical engineers during the preliminary design phase.

Household Appliances

Although there are other options, most household appliances are made for a 120/240-V, single-phase system. If a 120/208-V, three-phase, four-wire system is selected for a large residence or an apartment building, then either specially designed appliances must be used or local load-size transformers must be installed at the equipment to boost the system voltage to match that of the equipment.

13.5 COORDINATION WITH OTHER DESIGN DECISIONS

13.5.1 Interfacing of Building Systems

The simplified one-line diagram in Figure 13–1 clearly illustrates the complex interfacing required of all building systems. For example, in a high-rise building consisting of 100 apartments served by a central cooling and heating system, the main electrical power system should be a 480-V, three-phase, three-wire system for the central equipment and a 120/208-V, three-phase, four-wire system for lighting and appliance loads. On the other hand, if each apartment has its own electrical meter for all connected loads—including individual unitary heating and cooling units—then the electrical power system would probably be multiples of small 120/240-V systems, one for each apartment—even though the building is large. Usually, the selection of the mechanical system affects the selection of the electrical system.

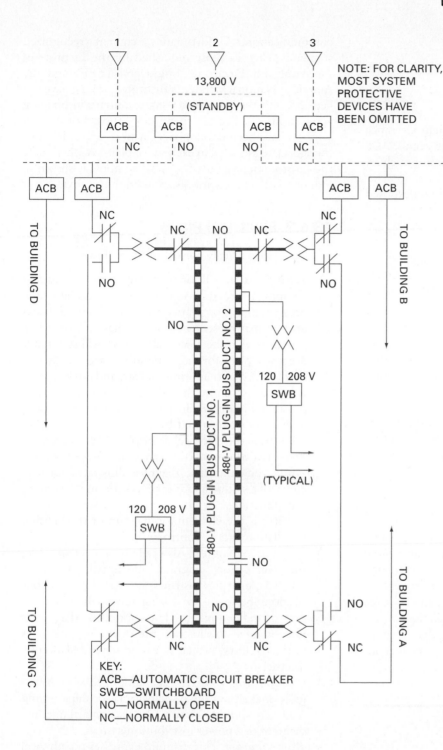

■ FIGURE 13–1

A simplified one-line diagram of a multiple-building power distribution system consisting of a high-rise main building and four low-rise buildings—A, B, C, and D. The main building has two plug-in-type bus ducts for plug-in connection to 480-V, three-phase, four-wire step-down transformers supplying power to 120/208-V distribution switchboards serving building loads.

KEY:
ACB—AUTOMATIC CIRCUIT BREAKER
SWB—SWITCHBOARD
NO—NORMALLY OPEN
NC—NORMALLY CLOSED

13.5.2 Space Planning

Space planning is not an exact science, yet it is vital to the orderly progression of the building design process. Once the electrical or mechanical space is allocated within the building, it is difficult to change its size or location, as doing so will undoubtedly affect the work in progress of many other disciplines or systems.

In the design of a bathroom, the size of plumbing fixtures is fairly standardized; by contrast, the size of electrical or mechanical equipment varies with the capacity, features, and manufacturers of the equipment. Educated guesses and experience of the designers play important roles in this connection. A more detailed discussion is beyond the scope of this book. The following are general guidelines.

Accessibility

All equipment and devices must be accessible for inspection, service, and replacement.

Safety

Electrical equipment is hazardous. Sufficient clearance must be provided on all sides that require access. As a rule, the NEC requires the following minimum clearances in front of all accessible sides of the equipment. The larger values are for equipment containing exposed live components:

- Up to 150 V: 3 ft
- 151 to 600 V: 3 to 4 ft
- 601 to 2500 V: 3 to 5 ft
- 2501 to 9000 V: 4 to 6 ft

Common Access Spaces

To minimize the loss of useful building space due to access requirements, double-loaded, center-aisle design is preferred. The common aisle space between two lineups of motor control centers shown in Figure 11-12(b) in Chapter 11 is a typical example. As passageways, corridors are frequently utilized for wall-mounted switchboards and panelboards. The NEC describes other safety requirements that are referred to in the design of building electrical systems.

Integration of Electrical and Structural Elements

The electrification of structural floors is a popular method for distributing on-floor power. The system may be an underfloor duct system (Figure 11–20) or a cellular floor system (Figure 11–21). It is most applicable for offices, especially offices with an open landscape design. Although the initial cost of installation of underfloor distribution systems is greater than that of other systems, statistics have proven their overall economy, with short payback periods.

13.6 DRAWING UP OF ELECTRICAL PLANS AND SPECIFICATIONS

13.6.1 Graphic Symbols

Graphic symbols are used to indicate various aspects of electrical design, including equipment, devices, wiring, and raceways. Without these symbols, electrical design would be difficult to illustrate. Standardized symbols are intended for use as a common language for communication. Unfortunately, current recognized standards, such as those published by the Institute of Electrical and Electronic Engineers (IEEE) and the American National Standards Institute (ANSI), have not kept up with the growth of electric systems in building applications. As a result, many nonstandard or custom symbols have been developed. A symbol schedule must, therefore, be included in each set of plans. A schedule of the most commonly used graphic symbols is shown in Figure 13–2. These symbols are used in this text.

13.6.2 Electrical Plans

Electrical plans usually consist of the following:

1. *Floor plans.* In a floor plan, electrical devices and equipment are superimposed on an architectural background. For clarity, the architectural background may be screened or may have a lighter line weight so that the electrical features will stand out. Electrical plans for major buildings are usually further divided into lighting, power, and auxiliary system plans to show the wiring design as clearly as possible. Figure 13–3(a) is a perspective of a room with a light controlled by one switch, and Figure 13–3(b) is the electrical floor plan of the room.

2. *Schematic diagram.* Also called an *elementary diagram,* the schematic diagram illustrates the circuitry of a system and is the basis for understanding the functions of an electrical system. Figure 13–3(c) is the elementary diagram of the lighting installation and its controls.

3. *Connection diagram.* Also called a *wiring diagram,* the connection diagram, illustrated in Figure 13–3(d), provides instructions regarding the connections between the wiring terminals of various devices and equipment. The connection diagram is not intended to illustrate the operating principles of the circuitry; rather, it is used for installation by electricians.

4. *One-line diagram.* This is a simplified system diagram that shows the principal relationships among major equipment. Figure 13–1 is a typical one-line diagram of a power distribution system.

5. *Riser diagram.* This diagram expresses the physical relationship between different pieces of equipment or devices and is frequently used to show the vertical relationship between floors. (See Figure 13–4.)

13.6.3 Specifications

Specifications are the written portion of a design document. They are used to supplement the drawings in the document. Drawings and specifications are used jointly,

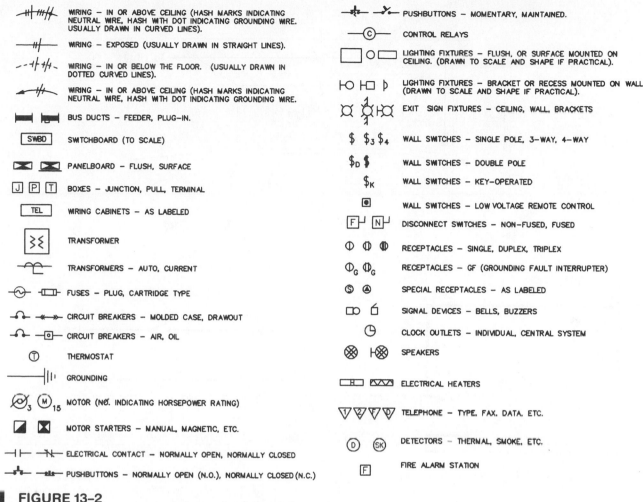

■ FIGURE 13–2

Graphic symbols commonly used for building electrical plans.

and what is called for in one is considered to be called for in both.

13.7 NATIONAL ELECTRICAL CODE

The National Electrical Code (NEC) prepared by the National Fire Protection Association (NFPA) was originally intended to assist electrical design professionals and the electrical contracting industry to prevent fire in buildings resulting from poor engineering and construction practice. It is normally revised every 3 years by a panel of experts. In as much as the NEC has been adopted as a part of the national and state building codes in the United States by reference as a whole, it is, in effect, considered a part of the law.

The design, installation, and maintenance of an electrical system in compliance with the NEC will result in an installation that is essentially efficient, convenient, and safe; however, it should be remembered that the requirements of NEC establish only a minimum standard of safety. In practice, engineering design must often exceed the minimum requirements for reasons of improved system reliability, desired higher level of life safety, optimized equipment performance, and benefits of life-cycle savings.

The NEC has also been widely adopted by many foreign countries throughout the world with or without minor variations or exceptions owing to local customs, variance in products construction, or differences in living standards and units of measure.

Building officials usually inspect the electrical installation of a building for compliance with NEC and other codes prior to issuing an occupancy permit before a new building can be occupied. Thus, designers and contractors must be thoroughly familiar and comply with all applicable sections of the NEC in all electrical installations.

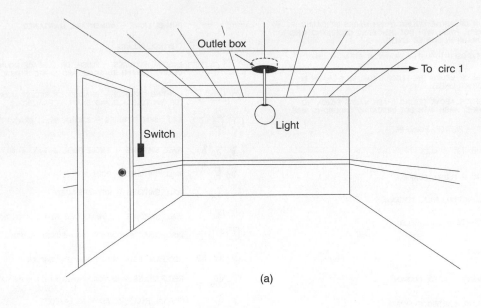

(a)

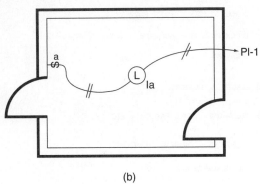

(b)

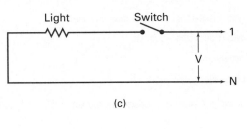

(c)

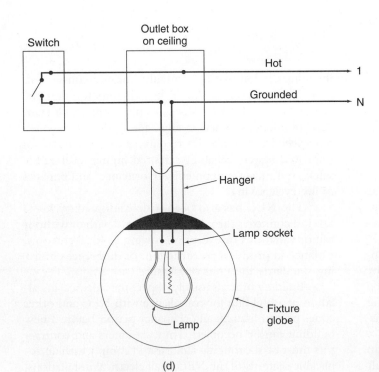

(d)

■ **FIGURE 13–3**

Various forms of electrical plans and diagrams of a simple lighting installation. (a) Perspective of a room with one light controlled by one switch. (b) Electrical floor plan of the room in (a). (c) Schematic diagram of the circuitry in the room. (d) Connection diagram of the circuitry.

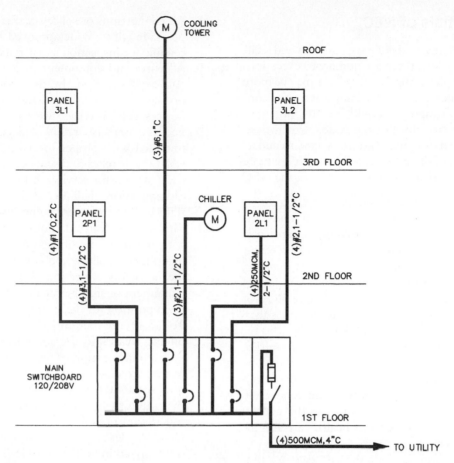

■ FIGURE 13–4

Typical riser diagram of a power distribution system, showing feeders between main switchboard, panels, and major motor loads.

Hazards often occur because a wiring system is overloaded beyond the NEC recommended values, commonly in older systems designed when the use of electricity was only a fraction of what it is today. Electrical system design should allow for growth; 20 to 30 percent spare capacity is the norm.

13.7.1 Scope and Arrangement of NEC

The NEC covers methods of connection to electrical power and the installation of electrical conductors and equipment in public and private facilities. The code is divided into nine chapters, as follows:

- *Chapter 1* General guidelines for compliance with the code.
- *Chapter 2* Wiring and protection: methods for sizing feeders and branch circuits, overcurrent protection devices, and grounding requirements.

- *Chapter 3* Wiring methods and materials: types of conductors and methods of installing conductors.
- *Chapter 4* Equipment for general use: construction requirements for lighting fixtures, appliances, electrical heaters, motors, generators, and other equipment using or producing electrical power.
- *Chapter 5* Special occupancies: requirements for hazardous locations, places of assembly (such as auditoriums and arenas), manufacturing, health care, and other facilities.
- *Chapter 6* Special equipment: requirements for elevators, low-voltage systems, swimming pools, etc.
- *Chapter 7* Special conditions: requirements of emergency systems, standby power, and optical fiber cables.
- *Chapter 8* Communication systems: requirements for radio and TV antennas and equipment.
- *Chapter 9* Tables and examples. Data pertaining to conductors and raceways, and examples for calculating loads and capacities for dwellings.

13.7.2 Essentials of NEC

The NEC contains hundreds of pages of rules and technical data. For each rule, there are numerous exceptions to cover various applications. In fact, there may be more exceptions than rules. Accordingly, any attempt to condense the code to simplify it could lead to misunderstanding and even violations of the code. Nonetheless, it is of great value to designers and users alike to understand the essentials of the NEC in order to achieve more useful and safer electrical systems. In general, the NEC rules may be summarized as follows:

1. Electrical power systems shall have a minimum capacity for various types of building occupancies. Naturally, this requirement varies for different countries, depending on local needs. For example, the electrical service for a single-family dwelling in the United States should not be less than 60 A for a 120/240-V, three-wire service and not less than 100 A for initial loads of 10 kVA or more. This requirement is less stringent in most other countries.

2. Alternating current systems of 50 to 600 V shall be grounded wherever possible so that the maximum voltage to ground on the ungrounded conductors does not exceed 150 V. Exceptions to this general rule include special systems in health-care and industrial facilities.

3. Only wires listed in the code shall be used in buildings. Application of the wires should not exceed the temperature and voltage ratings of their insulation for approved dry or wet locations.

4. The feeder capacity and the number of branch circuits shall not be lower than either the actual demand load or the calculated demand load given under Articles 200, 210, 215, and 220, whichever is greater.

5. There are numerous code-approved insulated wires and cables. Most of these shall be installed in protective raceways, such as conduits, wireways, or cable trays. For practical purposes, all commercial, institutional, and industrial buildings are designed to use wires in raceways, primarily in EMT or rigid conduits (heavy walls).

6. The allowable ampacities of insulated conductors are given in NEC Article 310, including the derating factors for ambient temperatures and when there are more than three conductors in the raceway. Unless otherwise stated, all references are based on copper conductors.

7. Nonmetallic sheathed cables, types NM and NMC, may be used without raceways in one-family or multiple-family dwellings and other structures, except when any dwelling or structure exceeds three floors above grade.

8. All connections or splices of wires and cables must be made within code-approved boxes that are accessible for inspection and service.

9. All wires and equipment shall be protected from abnormal situations by the proper installation of overcurrent, overvoltage, and overload protection devices, such as relays, fuses, or circuit breakers.

10. Feeders, branch circuits, and equipment shall be provided with proper means of disconnection or isolation switches to disengage them from the electrical system for service and repair. The means of disconnection shall be readily accessible and in sight of the equipment being disconnected.

11. Motors and electrical equipment shall be provided with means of disconnection, branch circuit protection, controllers, and overload protection. The means of disconnection shall be in sight of the equipment or shall be capable of being locked in an open position for the safety of service personnel.

12. To avoid the buildup of heat and to facilitate pulling wires into the conduit, wires shall occupy only a small portion of the conduit. In general, the overall cross-sectional area of all conductors shall not exceed 40 percent of the cross-sectional area of the conduit when there are three or more conductors in the conduit.

13. General rules for wiring of low-voltage systems are summarized in Section 13.11.

13.8 BRANCH CIRCUITS

According to the NEC, a *branch circuit* is that portion of a wiring system extending beyond the final overcurrent device protecting the circuit and the outlet or the load. A branch circuit may be classified as an appliance or as a general-purpose, individual, multiwire, or motor circuit. Every branch circuit shall be protected from overcurrent by overcurrent devices, which shall be located at the power supply end of the circuit.

There is no limit on the size of motor branch circuits, provided that the conductors are properly protected in accordance with NEC Article 430. In general, overcurrent devices with higher ampacity ratings are allowed in motor branch circuits than are allowed in general-purpose branch circuits, so as to accommodate the extremely high current that builds up during the start-up of a motor due to its inductive reactance or lower power factor. The ampacity ratings of general-purpose branch circuits are limited to 20 A for lighting and 120-V appliances, 30 A for lighting with heavy-duty lamp holders and high-wattage appliances, and 50 A for cooking and laundry equipment in dwellings.

13.8.1 Branch Circuiting Design

Branch circuiting is the basis for the wiring of individual loads, and in the final design of the power distribution system. Branch circuiting work is mostly tedious and time-consuming. On the other hand, such work can be greatly simplified if the following rules are observed:

1. No wire smaller than No. 14 AWG shall be used for dwelling applications.
2. No wire smaller than No. 12 shall be used for commercial, industrial, and institutional applications.
3. For two-wire (120-V) circuits, the continuous load per circuit shall be limited to 1200 W for 15-A circuits and 1500 W for 20-A circuits.
4. For heavy-duty lamp circuits, the load per circuit shall not exceed 2000 W for No. 10 wires, 2500 W for No. 8, and 3000 W for No. 6.
5. As a rule, the NEC code requires that the rating of any one portable appliance shall not exceed 80 percent of the branch circuit rating; the connected load of inductive lighting, such as fluorescent or HID fixtures, shall not exceed 70 percent of the branch circuit rating; and the total load of fixed wired appliances shall not exceed 50 percent of the branch circuit rating if lighting or portable appliances are also connected to the same circuit.
6. Where the run from a panelboard to the first outlet of a lighting branch circuit exceeds 75 ft, the wire shall be at least one size larger than that determined by any of the preceding considerations.
7. No run longer than 100 ft between a panelboard and the first outlet of a lighting branch circuit shall be made, unless the intended load is so small that the voltage drop can be restricted to 2 percent between the panelboard and any outlet on that circuit.
8. Where the run from a panelboard to the first outlet of a convenience outlet circuit exceeds 100 ft, No. 10 wires shall be used.
9. No convenience outlet shall be supplied by the same branch circuit that supplies ceiling or show-window lighting outlets.
10. The maximum number of convenience outlets included in one circuit shall be based on the following:
    - Outlets supplying specific appliances and other loads based on the ampere rating of appliances
    - Outlets supplying heavy-duty lamp holders—use 5 A per receptacle
    - Other outlets (general purpose)—use 1.5 A per receptacle or 3 A per duplex receptacle.
11. Fifteen to 25 percent of spare circuits shall be provided in each general-purpose panel, in addition to spare circuits for known future loads.

12. Motor branch circuits shall comply with NEC Article 430. In general, branch circuit conductors shall be rated as follows:
    - Single-motor circuit: at least 125 percent of full-load rating of motor
    - Multiple-motor circuit: at least 125 percent of the highest-rated motor, plus the full-load rating of all other motors

13.8.2 Branch Circuiting Layout

Once the branch circuit design is formulated, it is shown on the electrical floor plans indicating the circuiting number, wire size (other than the basic No. 12 or No. 14 AWG), home runs, interconnection between outlets or boxes, points of control, and special notation.

Branch Circuiting with NM Cables

NM cables are used mostly for dwellings and building interior finishes. NM cables come with two conductors (black and white) or three conductors (black, white, and red). Each cable contains a bare grounding conductor within the nonmetallic protective jacket. Since NM cables are limited in their variety, the indication of complete branch circuit wiring on floor plans is quite straightforward and, in fact, can be eliminated in plans of simple dwellings. A typical electrical floor plan with NM wiring is presented in Figure 13–14. In general:

- Use minimum No. 14 AWG for 15-A circuits and No. 12 for 20-A circuits.
- Use two-wire cable for single-circuit, single-switch, and 120-V receptacles.
- Use three-wire cable for two circuits, three way switches, and 240-V appliances.
- Use the white wire for the grounded side of the power system only.
- Use the bare wire for grounding the conductive housing of the lighting fixtures, wiring devices, equipment, and boxes.

Branch Circuiting with Wiring in Conduit

Laying out branch circuits in conduits requires careful planning. When properly done, it may reduce complications in the field with regard to rerouting or enlarging some incorrectly sized conduits, and the cost of construction. The goal in laying out conduit is to minimize the length and size of the conduits. Following are some simple rules:

- EMT shall be used primarily for dry locations, although it may be used in wet locations with corrosion-resistant supporting components.
- EMT is limited to 4 in., and no conduit smaller than ½ in. shall be used.

- All wiring splices shall be made at outlet boxes, gutters of panelboards, or other junction boxes. All boxes must be accessible.
- Conductors for different voltage systems may be installed in the same conduit, provided that all conductors are rated for the highest voltage enclosed.
- In general, conductors for telephones, signals, radio, and TV systems may not occupy the same conduit or enclosure that lighting and power systems occupy, except when otherwise permitted under NEC Articles 800 through 820. See Section 13.11 for wiring of low-voltage systems.
- Conductors of a three-phase circuit shall be installed in the same conduit to reduce the induction effect (induced heat). The neutral conductor of a circuit shall also be installed in the same conduit whenever possible.
- No conduit shall have more than four equivalent 90° bends between boxes, including the bends at the boxes.
- It is good practice to feed the branch circuit first at the lighting fixture outlet box and then at the switchboxes. This will help in future alterations and in troubleshooting.
- Separate convenience outlet and appliance circuits from lighting circuits, except for incidental loads.
- Avoid using conduits larger than 1 in. for general-purpose branch circuits. When necessary, split the wires into two or more 1-in. or smaller conduits. This practice is, of course, limited to branch circuit wiring. It is not intended for equipment and distribution feeders. In such applications, 4-in. or multiples of 4-in. conduits are the practical limits.
- Avoid having more than nine conductors in a single conduit, other than with low-voltage communication-type systems.

Typical wiring plans for branch circuiting in conduit is illustrated in Figures 13–5 and 13–6.

13.9 TABLES AND SCHEDULES

In the design of any building systems or subsystems, not all pertinent information can be expressed by drawings. Tabulated data and combination of graphics with tabulated data are effective ways to provide the design information of a project. In electrical system design, the following schedules are usually required.

13.9.1 Panelboard Schedules

A panelboard schedule is the most important schedule of power system design. It serves to summarize the individual branch circuit loads from which the overall load system is calculated. A complete panelboard schedule used for large projects is shown in Figure 13–7. It includes four subschedules:

- *Branch circuit connections* Including the circuit number, load (VA, W, kW, or kVA), phase to which the load is to be connected (*x* or *y* for single-phase system or A, B, C for three-phase system), type of overcurrent protection devices (OCP) and current rating (A), number of poles (P).
- *Demand-load calculations* Based on the load types, the estimated demand factor (DF), and the estimated diversity coefficient (DC).
- *Power supply data* Describing how the power is fed.
- *Branch data number, poles, and rating of poles.*
- *Specifications* Capacity rating, short-circuit rating, OCP ratings, physical dimensions, and construction features.

It is important to have the actual demand load of each phase be reasonably balanced so that each feeder feeding the panel will share the total load equally or nearly equally. This is easier said than done because the use pattern of each circuit is different and for most the time hard to predict. As shown in Figure 13–7, the total connected loads are 11.8, 10.2, and 10.3 kVA for phases A, B, and C, respectively. They are about 10 percent unbalanced as circuited, but the actual demand load of each phase may vary further. In practice, the electrical contractor will be asked to interchange one or more of the branch circuits after the use pattern of the connected loads by the users is established.

13.9.2 Feeder Schedule

A feeder schedule usually contains the feeder designation, type of cable insulation, size in AWG or MCM, method of installation (with or without raceways), overcurrent protection, connected load in amperes, and demand load in amperes.

13.9.3 Mechanical Equipment Data Schedule

Most modern buildings contain electrically powered equipment, such as HVAC equipment, plumbing and fire protection equipment, elevators, and food preparation equipment. This equipment is usually motor-driven and thus constitutes inductive loads. Wiring design for inductive loads differs from that for lighting loads in that inductive loads have large inrush (starting) current and thereby require different design criteria and protection. Figure 13–8 shows a typical mechanical equipment data schedule. It consists of two parts: basic description of the load and the wiring data.

(North side)

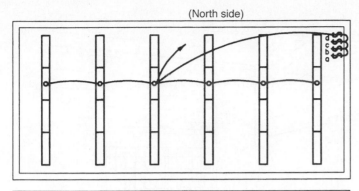

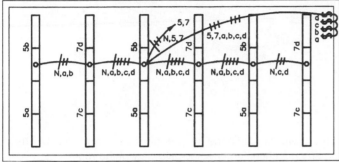

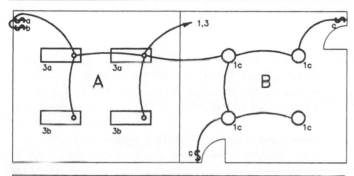

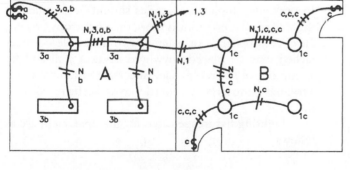

■ FIGURE 13–5

(top) The room is an architectural design studio with wall-to-wall windows on the south exposure. It is desired that these lights be controlled by four switches and the first two fixtures on each row be switched separately from the remaining fixtures in that row. This is to utilize the available daylight near the south side of the room. (second from top) Wiring in conduits is shown in the wiring plan with the function of each wire. (third from top) Room A has four fluorescent fixtures. These fixtures shall be connected to circuit 3. The upper two fixtures shall be controlled by switch *a* and the lower two by switch *b*. Room B has four incandescent fixtures. These fixtures shall be connected to circuit 1 and controlled by switch *c*. One switch shall be located at each door location. (bottom) Branch circuits 1 and 3 shall share the same neutral on a 120/208-V, three-phase, four-wire system. The number of wires in a conduit is shown by hash marks across the conduit. The long hashes represent the neutral wire, and the short hashes represent the circuit or switch wires. For learning purposes, the function of each wire is identified alongside the marks. In practice, these identifications are omitted.

13.9.4 Lighting Fixture Schedule

Lighting can consist of up to 50 percent of the total connected load in a building. Figure 13–9 shows a typical lighting fixture schedule.

13.9.5 Control and Automation Schedules

A control and automation schedule usually contains the name of the equipment to be controlled and the required control functions in a matrix format. The schedule is used in conjunction with the technical specifications and control diagrams. The schedule normally contains the following:

- Operating mode: manual, programmed, or automatic start or stop for equipment.
- Control functions and indications: temperatures, pressure, or flow for various media, such as air, water, or fluid, and for equipment, such as fans, pumps, and valves.

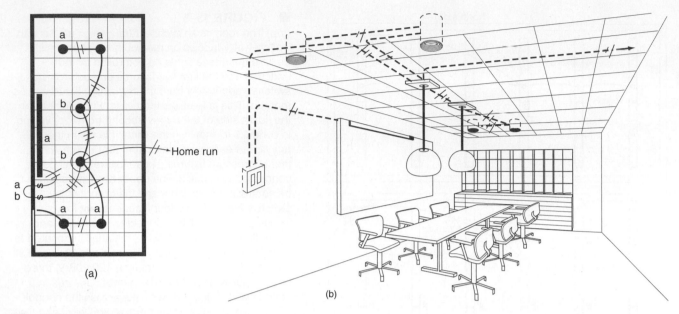

(a)

(b)

■ FIGURE 13-6

(a) Electrical branch circuit wiring plan showing the conduit layout, switching, and the number of wires in conduits. (b) The conduit and wiring of the branch circuit plan are superimposed on an architectural perspective view of the room. The perspective clearly illustrates how conduits are running from outlet to outlet.

- Control point adjustments of the media and equipment.
- Interfacing of building systems between HVAC, fire protection, plumbing, security, transportation, lighting, and standby power systems.

13.10 POWER WIRING DESIGN PROBLEM

Design the electrical system for a small apartment with load data given on page 000 and the equipment layout shown in Figure 13–10.

A wiring layout is a major part of an electrical system design. It provides the information for the electrician to rough in all concealed wires, boxes, and raceways before the building's interior finishes are installed. It consists of instructions for lighting, receptacles, switches, and any equipment which requires electrical power or controls. The procedure for wiring layout can best be illustrated with a practical example.

Figure 13–10 shows the architectural floor plan of a 1000-sq-ft apartment, including the intended furniture layout, which is essential in electrical wiring planning. Should a furniture layout not be available, as is often the case for buildings during the planning phase, then reasonable assumptions and allowances have to be made by the engineer in consultation with the architect and the owner.

On the basis of the furniture layout, the lighting, receptacles, and switch positions can be located. Methods for lighting design and layout will be covered in Chapters 14 through 17.

Convenience receptacles are located as needed to coordinate with the furniture layout. The NEC and other building codes have specific requirements for locating convenience receptacles in residential dwellings. In principle, duplex receptacles should be installed in the wall so that no point is more than 6 ft from a receptacle. An appliance or a lamp with a flexible cord attached may be placed anywhere along a wall and be within 6 ft of the receptacle. In certain instances, when a room is quite large, receptacles may also be installed in the floor.

Lighting fixtures selected for this apartment are as follows:

Type-A 100-W incandescent, wall-mounted.

Type-B 60-W incandescent, ceiling recess-mounted.

Type-C 50-W fluorescent including ballast, 4 ft/ fixture, ceiling-mounted.

Type-D 100-W 2-lamp fluorescent, 4ft/fixture.

Type-E 150-W total of 9 incandescent lamps, pendent mounted.

Type-F 150-W total of 3 incandescent lamps, ceiling-mounted.

Panelboard Schedule
Type T -For 3P/4W System

Project	MICDS/SLMO (GYM)
Project No.	WT9914A
Designer	STA
Date	May-99
Revised Date	

LOAD	LOAD KVA			OCP		CKT	PH. BALANCE LOAD KVA			CKT	OCP		LOAD KVA			LOAD
	Light	Recept	O/M	A	P	No.	A	B	C	No.	P	A	Light	Recept	O/M	
Sound Board			1.6	20	1	1	1.8			2	1	20		0.2		Roof Top Recpt.
S.E. Ctr. Ct. FL Recpt		1.6		20	1	3		1.6		4	1	20				Spare
S.E. Ctr. Ct. FL Recpt		1.6		20	1	5			1.6	6	1	20				Spare
S.E. Bat C. FL Recpt		1.6		20	1	7	3.1			8	1	20			1.5	Rm.128, Hand Dryer
S.E. Motorized Grill			0.5	20	1	9		2.0		10	1	20			1.5	Rm.128, Hand Dryer
S.E. Backstop			0.4	20	1	11			1.9	12	1	20			1.5	Rm.122, Hand Dryer
S.E. Curtain			0.4	20	1	13	1.9			14	1	20			1.5	Rm.122, Hand Dryer
S.E. Backstop			0.4	20	1	15		1.9		16	1	20			1.5	Rm.122, Hand Dryer
S.E. Curtain			0.4	20	1	17			1.9	18	1	20			1.5	Rm.121, Hand Dryer
S.E. Batting Cage			0.4	20	1	19	1.9			20	1	20			1.5	Rm.121, Hand Dryer
S.E. Exterior Recpt.		1.6		20	1	21		3.1		22	1	20			1.5	Rm.129, Hand Dryer
S.E. EWC			0.2	20	1	23			1.7	24	1	20			1.5	Rm.130, Hand Dryer
S.E. Recpt.		1.6		20	1	25	3.1			26	1	20			1.5	Rm.130, Hand Dryer
S.E. Recpt.		1.6		20	1	27		1.6		28	1	20				Spare
S.E. Recpt.		1.6		20	1	29			1.6	30	1	20				Spare
S.E. Recpt.		1.6		20	1	31			1.6	32	1	20				Spare
Spare				20	1	33				34	1	20				Spare
Spare				20	1	35				36	1	20				Spare
Space				-	1	37				38	1	-				Space
Space				-	1	39				40	1	-				Space
Space				-	1	41				42	1	-				Space
Subtotal		12.8	4.3											0.2	15.0	Subtotal
Total (KVA)		13.0	19.3				11.8	10.2	10.3							
Total (KVA)													REMARKS:			
Total (KVA)																

DEMAND LOAD				
LOAD	CL (KVA)	DF (PU)	DL (KVA)	DL (KW)
Lighting	0.1	1.0	0.1	0.1
Receptacle	10.0	1.0	10.0	8.5
Receptacle	33.6	0.5	16.8	14.3
Motor		0.8		
Other	20.1	0.8	16.1	13.7
Pnl. 1NH5	21.04	1.0	21.0	17.9
a Present Total DL			64.0	54.4
b Spare at 25%			16.0	10.0
c Gross Overall DL			80.0	68.0
d Est. DC	1.00			
e Net Overall DL			80.0	68.0
f Estimated PF		.85		

Current Demand Calculation

1 PH $1000 \times \dfrac{\text{KVA} =}{\text{V}} = \_\_\_\_$ A

3 PH $580 \times \dfrac{80.0 \text{ KVA} =}{208 \text{ V}} = 223$ A

POWER SUPPLY DATA				
F	From		T-1	1NH1
E	Type	Phase	350 kcmil	1
E	&	Neutral	350 kcmil	-
D	Size	Ground	4	6
E	Length, Ft.			
R	Voltage Drop %			
	Raceway Quan/Size		3"	1 1/2"
OCP	CB	Frame/Trip	-	125
	FS	Size/Fuse	-	-

BRANCHES			
Quan	Poles	Rating A.	Total Poles
Total Poles			

SPECIFICATION			
Available Fault, Symmertical Amp			
IC Rating Symmertical Amp			10,000
Branch	Type (CB) (FS)		CB
Devices	Model		
	Special Gutters	Special Features	
X Ground Bus			
Contactor			
Split Bus			
Shunt Trip			
Flush Trim			
X Surface Trim			

Dimensions	Depth	Width	Height
	6"	20"	

M	Feed From	Top	
A	Lugs	Quantity	
I	Per PH	Size Range	
N	Size & Type	300A MCB & SFL	
	System V-PH-W	277/480V-3PH-4W	
	Panel Location	Electric Room 120	
	Transformer T-1 = 75 KVA		

LEGEND:	OCP	- Over Current Protection	CL	- Connected Load
	O/M	- Other or Motor Loads	DF	- Demand Factor
	IC	- Interruption Capacity	DL	- Demand Load
	V-PH-W	- Voltage-Phase-Wires	DC	- Diversity Coefficient

If Multiple Section, see Panel Sect.	PANEL DESIGNATION
	1NL1
	SOUTHEAST

Form E1-a

P:\Proj\9914a\Data\Panels.xls

■ **FIGURE 13–7**

Actual project panelboard schedule with design data filled in. The schedule is designed for use with a three-phase, four-wire power distribution system. Other schedules are designed for one-phase, three-wire or three-phase four-wire systems. (Courtesy: William Tao & Associates, St. Louis, MO.)

Equipment Data Schedule

EQUIPMENT	MARK	HP (KW)	VOLT PHASE SPEED	EQUIPMENT LOCATION	ACCESS-ORIES I	TYPE II	SIZE	PANEL DESIG.	BR. CKT. DEVICE TYPE	BR. CKT. DEVICE SIZE	CONDUCTORS PHASE	CONDUCTORS GRD	CONDUIT SIZE	NOTES
CHILLER	CH-1	295MCA (162)	480-3-1	LEVEL 1, AREA C 1071 CHILLER ROOM	-	PWCP	-	MSB2	CB	350A-3P	(2) 2/0	(2) 3	(2) 2"	5
CHILLER	CH-2	295MCA (162)	480-3-1	LEVEL 1, AREA C 1071 CHILLER ROOM	-	PWCP	-	MSB2	CB	350A-3P	(2) 2/0	(2) 3	(2) 2"	5
CHILLER (FUTURE)	CH-3	295MCA (162)	480-3-1	LEVEL 1, AREA C 1071 CHILLER ROOM	-	-	-	MSB1	CB	350A-3P	-	-	-	
HOT WATER BOILER	B-1	7.5+1/4	480-3-1	LEVEL 1, AREA C 1013 BOILER ROOM	-	PWCP	-	MCC-1EH1	FS	30/17.5/3	12	12	3/4"	4,5
HOT WATER BOILER	B-2	5+1/4	480-3-1	LEVEL 1, AREA C 1013 BOILER ROOM	-	PWCP	-	MCC-1EH1	FS	30/15/3	12	14	3/4"	4,5
HOT WATER BOILER	B-3	5+1/4	480-3-1	LEVEL 1, AREA C 1013 BOILER ROOM	-	PWCP	-	MCC-1EH1	FS	30/15/3	12	14	3/4"	4,5
STEAM BOILER	B-4	2 @ 2	480-3-1	LEVEL 1, AREA C 1013 BOILER ROOM	-	PWCP	-	MCC-1EH1	FS	30/10/3	12	14	3/4"	4,5
HOT WATER BOILER (FUTURE)	B-5	7.5+2+1/4	480-3-1	LEVEL 1, AREA C 1013 BOILER ROOM	-	-	-	MCC-1EH1	FS	30/-/3	-	-	-	4
BOILER FEED WATER UNIT	BFU-1	2 @ 3/4 EA.	480-3-1	LEVEL 1, AREA C 1013 BOILER ROOM	-	PWCP	-	MCC-1NH1	FS	30/3.5/3	12	14	3/4"	5
CHILLER LOOP PUMP	CHP-1	10	480-3-1	LEVEL 1, AREA C 1071 CHILLER ROOM	HOA-P	COMB	1	MCC-1NH1	FS	30/17.5/3	12	12	3/4"	
CHILLER LOOP PUMP	CHP-2	10	480-3-1	LEVEL 1, AREA C 1071 CHILLER ROOM	HOA-P	COMB	1	MCC-1NH1	FS	30/17.5/3	12	12	3/4"	
CHILLER LOOP PUMP (FUTURE)	CHP-3	10	480-3-1	LEVEL 1, AREA C 1071 CHILLER ROOM	HOA-P	COMB	1	MCC-1NH1	FS	30/-/3	-	-	-	
BLDG. CHILLED WATER PUMP	CP-1	25	480-3-1	LEVEL 1, AREA C 1071 CHILLER ROOM	-	AFC	-	MCC-1EH1	FS	60/45/3	8	10	3/4"	4,5

Column group headers: NAMEPLATE DESIGNATION (EQUIPMENT, MARK) · MOTOR OR EQUIPMENT DATA (HP (KW), VOLT PHASE SPEED, EQUIPMENT LOCATION) · CONTROLLER DATA (ACCESSORIES I, TYPE II, SIZE, PANEL DESIG.) · BRANCH CIRCUIT DATA (BR. CKT. DEVICE: TYPE/SIZE, CONDUCTORS: PHASE/GRD, CONDUIT SIZE, NOTES).

■ FIGURE 13–8

Equipment data schedule for mechanical equipment wiring, for an actual project. (Courtesy: William Tao & Associates, St. Louis, MO.)

LIGHTING FIXTURE SCHEDULE

TYPE	GENERAL DESCRIPTION AND APPLICATION		LAMP & POWER DATA					SPECIFICATION DESCRIPTION	
(I)	GENERAL DESCRIPTION	TYPICAL APPLICATIONS	TYPE (II)	QUAN x W/LAMP	LAMP CODE	SUPPLY VOLT(S)	W/FIX. (III)	MANUFACTURER AND CATALOG NUMBER SERIES	NOTES (IV)
BAB	BRACKET MTD. DIE CAST ALUM. WALL HOOD WITH OPAL GLASS GLOBE	CLOSETS MH 7'-6" RISER RM MH 7'-6"	P	2x7	F7TT/27K	277	25	DAYBRITE #FFW14PL-CPC-277	14, 15
BAB-1	BRACKET MTD. DIE CAST ALUM. WALL HOOD WITH OPAL GLASS GLOBE	ELEV. PITS MH 3'-C"	P	2x7	F7TT/27K	120	25	DAYBRITE #FFW14PL-CPC-120	14, 15
BAG	BRACKET MOUNTED 4'-0" FLUOR. DIRECT CORNER COVE WITH SEALED PRISMATIC LENS, HINGED & GASKETED DOOR FRAME & ONE PIECE 20 GAUGE STEEL BODY	MECH ROOM- MOUNT AS DIRECTED BY ARCH.	F	2x32	FO32/41K	277	70	DURAY #CL240-OCTRON 277V	4, 7, 8,14
BAG-1	BRACKET MTD. 4'-0" WALL FLUOR. WITH PLIABLE WRAPAROUND OPAL ACRYLIC LENS & 20 GA. STEEL BODY	TOILET ROOM MOUNT AS DIRECTED BY ARCH.	F	2x32	F032/41K	277	70	ALKCO #6240-277V OCTRON WHITE FINISH	3, 4, 8, 14
BAU	BRACKET MTD. 4" x 4-3/4", 72'-0" LONG EXTRUDED ALUMINUM RECTANGULAR ILLUMINATED RAIL WITH (2) 90 DEGREE CORNERS & PRISMATIC LENS.	BRIDGE SEE DRAWINGS 4.101 & 4.103	F	18 x32	F032/41K	277	360	PEERLESS #P771588KC	3, 4, 6, 8, 17, 18
BBG	BRACKET MTD. 4'-0" FLUOR UP & DOWN LIGHT WITH ALUM. FACIA	STAIR MH 7'-6"	F	2x32	FO32/41K	277	70	MARK #2/40-277-4-4-PT	3, 4, 8, 14
BCP	ADJUSTABLE DIE CAST ALUM. FLOODLIGHT WITH HINGED DOOR/ TEMPERED GLASS LENS, C-CLAMP HANGER & 3'-0" CORD & NEMA L5-20P PLUG	PLATFORM - 133 MOUNT AS DIRECTED BY ARCH.	Q	1x300	Q300T2- 1/2 / CL	120	300	DAYBRITE #QF300 W/ YOLK	BASED ON G.E. LAMP CODE
BLN	BRACKET MTD. FLUORESCENT DOWNLIGHT 6"W x 8" H x 12" L EXTRUDED ALUM. ENCLOSURE WITH POLISHED FINISH, 1" x 1" BLADE, 3" LONG MTG. BRACKETS, 0 DEGREE BALLASTS & DAMP LABEL	LIGHT RAIL STATION GRD. LEVEL	F	6 x 32	FO32/41K	277	192	PEERLESS #B-277-12-B-101-OC	
BLQ	BRACKET MTD.4" DIA. TUBULAR 38 FT. LONG ROTATABLE FLUOR. DOWNLIGHT WITH BAFFLE	9TH FLOOR LOUNGE	F F	2x25 8x32	F025/41K F032/41K	277	345	GARDCO #M7-0187-1 CUSTOM R40 RWK-132/25 OCT-38 PAT-SC-277-RL	3, 4, 8, 14, 17, 18

FIGURE 13–9
Typical lighting fixture schedule identifying the fixture type as labeled on drawings; general description of the fixture, lamp, and power data; and model or specifications. (Courtesy: William Tao & Associates, St. Louis, MO.)

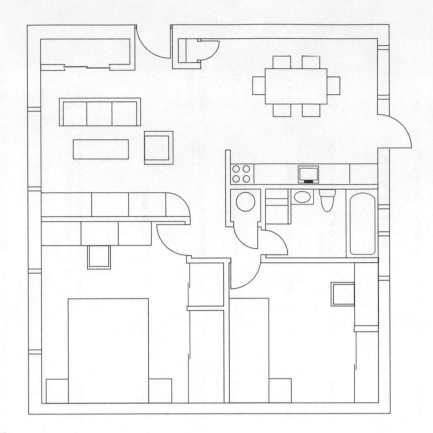

■ FIGURE 13–10

Architectural floor plan with furniture layout of a small apartment, as described in Section 13.10.

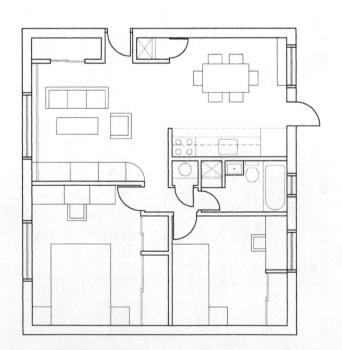

■ FIGURE 13–11

Electrical background floor plan with furniture layout.

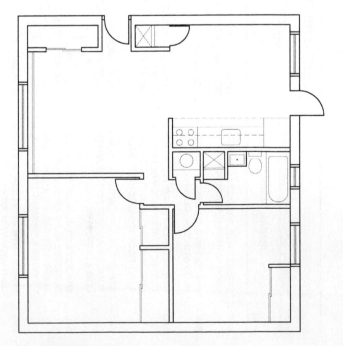

■ FIGURE 13–12

Electrical background floor plan without furniture layout.

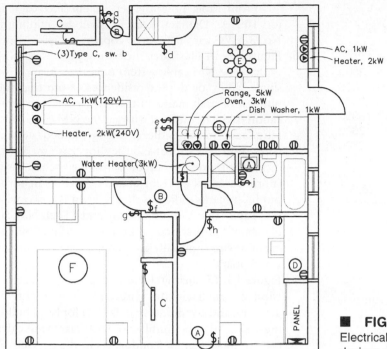

■ FIGURE 13–13
Electrical floor plan including lighting fixtures and electrical devices prior to wiring.

The mechanical and household equipment load as determined by the owner, architect, and mechanical engineer are as follows:

Cooling: Window air conditioners, two, 1 kW each, 120-V, single phase.

Heating: Electrical heaters, two, 2 kW each, 120/240-V, single-phase in the living room only. No heating or cooling is required in other spaces.

Water: Hot-water heater, 3-kW, 240-V, single-phase.

Cooking equipment: Range, 8-kW, 240-V, single-phase. Oven, 3-kW, 240-V, single-phase. Microwave oven, 1 kW, 120-V, single-phase.

Kitchen appliances: Dishwasher, refrigerator, etc., 2-kW total, 120-V, single-phase.

Other appliances: Standard convenience outlets (5 A/outlet).

Auxiliary system: Low-voltage systems, such as doorbell, security, and TV antenna, are considered convenience loads.

Using the preceding information, work out the following problems:

1. Determine the present and future total connected loads by load groups for summer and winter operations.
2. Determine the present and future demand loads by load groups for summer and winter operations.
3. Determine the future net demand load for winter allowing a diversity coefficient of 1.1.
4. What is the calculated future net demand current in amperes?
5. What should be the size of the service entrance conductor? (Use 75°C rated copper cable.)
6. Complete the electrical wiring plan, using NM cables without raceways.
7. Complete the electrical wiring plan, using wires in raceways (conduits).

Answers to Design Problems

1. The present and future connected loads are summarized in Table A:

TABLE A

Load Group	Connected Load, W	
	Summer	Winter
Lighting[a]	3000	3000
HVAC	2000	4000
Water heater	3000	3000
Cooking equipment	12,000	12,000
Kitchen appliances	2000	2000
Convenience[b]	2500	2500
Present connected	24,500	26,500
Future spare (25%)	6100	6600
Future connected	30,600	33,100

[a]NEC table 220-3(b), 3 VA/sq ft (3 W/sq ft @ 1.0 PF)
[b]NEC 220-3, 14 outlets @ 180 VA/outlet, or 2520 VA total, use 2500 W.

TABLE B

| | Demand Load, W | |
Load Group	Summer	Winter
Lighting[c]	3000	3000
HVAC	2000	4000
Water heater	3000	3000
Cooking equipment[d]	8000	8000
Kitchen appliances	2000	2000
Convenience	2500	2500
Present demand	20,500	22,500
Future spare (25%)	5200	5700
Future demand	25,900	28,200

[c]NEC table 220-11, 100% DF
[d]NEC 220-19, note 3, use 8 kW for 12 kW or less.

2. The present and future demand loads are summarized in Table B.
3. The future net demand load for winter with a diversity coefficient of 1.1 is

$$\text{Net future demand} = 28,200 \text{ W} \times 1/1.1$$

$$= 25,640 \text{ W}$$

The future demand load is an indication of the real demand that could be experienced by the electrical utility company, from which it could plan its distribution systems.

4. The future demand current is calculated from future demand load (28,200 W) without a diversity coefficient. (See NEC Article 220.) A power factor (PF) should be calculated or estimated from the characteristics of the winter loads. Since nearly all winter loads are of the resistive type (incandescent lamps, resistance heaters, and food service equipment), a power factor of 95 percent is reasonable. Accordingly, from Equation 10–19:

$$P = E \times I \times \text{PF}, \text{ or } I = \frac{P}{E \times \text{PF}}$$

$$I = \frac{28,200}{240 \times 0.95} = 124 \text{ A}$$

5. Since the calculated 124 A future demand load current exceeds 100 A, the service entrance cable and the main distribution panel will have to be selected for the next standard size, which is 200 A. The cost of a 200-A service is considerably higher than that of a 100-A service. As a designer, you now face the decision of replacing some of the electrical equipment with other energy sources or reducing the future spare load in order to reduce the future demand load to below 100 A.

From Table 11-2, the service entrance cable is for 100 A. Demand current shall be minimum No. 3 AWG for 75°C rated copper.

6. Figure 13–14 illustrates the wiring plan using NM cables. The NM cable system is used exclusively for single-family or duplex residences. Note that the number of hash marks in the cable runs is an indication of the number of wires in the NM cable. The long hash marks indicate the grounded neutral wires (normally white) and the short hash marks indicate the number of circuit wires or switched wires (normally black or color-coded). A wire run without hash marks indicates that the run is a two-wire NM cable (one hot and one ground). Each NM cable should also contain an uninsulated equipment ground wire (normally bare copper), not shown as a hash mark.
7. Figure 13–15 illustrates the wiring plan of the electrical design using the wire and conduit system. The wire and conduit system is used for large buildings. Similarly, the number of hash marks indicates the number of single conductors in the conduit. A separate equipment ground wire (green) is not indicated as a hash mark.

13.11 WIRING OF LOW-VOLTAGE SYSTEMS

Extra-low-voltage systems, such as life safety, security, and telecommunication systems (covered in Chapter 12), are referred to as low-voltage systems by NEC. These systems operate either under 50 V or with limited power capacity. The acceptable practice for installing these systems in buildings or dwellings is open wiring. Raceways, such as conduits, or wireways are also used when the wiring penetrates through floors or studs up in wall cavities, or when they are needed for extra security or fire protection.

Wiring methods for these systems are governed by NEC Article 725 (Remote Controls and Signaling Systems), Article 760 (Fire Protection Systems), and Article 800 (Communication Circuits). The readers should refer to NEC for more specific requirements. Wiring for optical fiber (OF) cables is governed by NEC Article 770. Their installation is also governed by other industry standards, which will not be addressed in this section.

13.11.1 Types of Cables

There are hundreds of specialty wires and cables developed and produced for low-voltage systems. The cable designations usually follow the NEC designations and

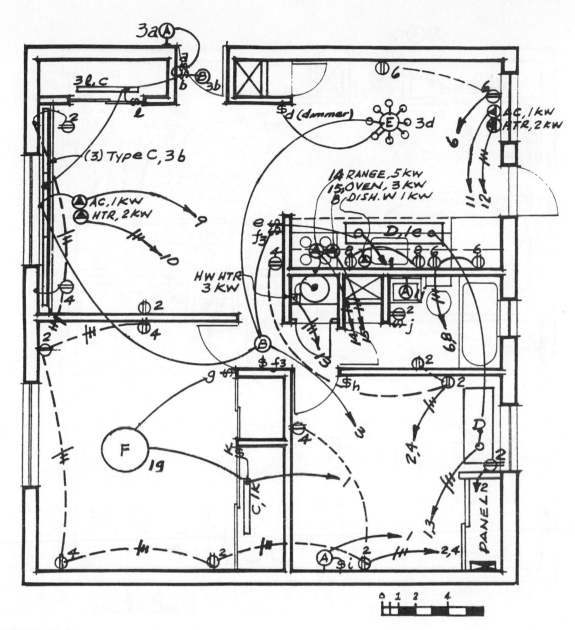

■ FIGURE 13–14
Completed wiring plan, with NM cables without conduits.

consist of code letters (acronyms). Some frequently used code letters are:

- P—plenum, R—riser, G—general, X—limited use
- MP—multipurpose, CM—communications, FP—fire protection, TC—tray cable
- PL—power-limited,   NPL—non-power-limited, LP—limited plenum, LR—limited riser, UC—under carpet
- CL1—Class 1 (basically an extension of building power)
- CL2—Class 2, CL3—Class 3

For example:

MPP stands for a multipurpose cable rated for use in plenum space.

FPLP stands for fire protection cable rated for use in limited plenum space.

NPLFR stands for non-power-limited fire alarm riser cable.

CL2X stands for Class 2 cable, limited use.

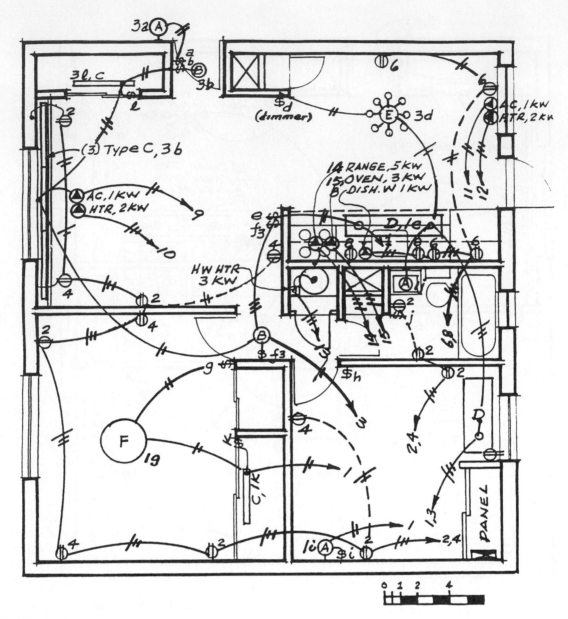

FIGURE 13–15
Completed wiring plan, with wires in conduits.

13.11.2 Cable Substitution and Hierarchy

In lieu of the specific cable designated for every system, an alternative cable may be used, provided that it is of a higher hierarchy, as listed by the NEC. Figure 13–16 illustrates the substitution and hierarchy of NEC-listed cables. The chart starts with the higher-listed cables at the top and descends to the lower-listed cables. For example, MPP, the highest-listed cable, may be used to substitute for CMP CL3P, and so on all the way down. On the other hand, MPG is not allowed to substitute for MPP, and so on.

13.11.3 Planning Low-Voltage Systems

1. Like other systems, large low-voltage systems may require distribution panels, terminal boxes, or pull boxes. These should be coordinated with the power distribution systems.
2. A riser diagram depicting the destination of risers is a must, as is a block diagram illustrating the interrelationship of equipment and sensing, announcing, and operating devices.
3. Low-voltage equipment, although usually compact and lightweight, requires workspace for operation and service. Adequate space from front, sides, and rear must be provided as required.

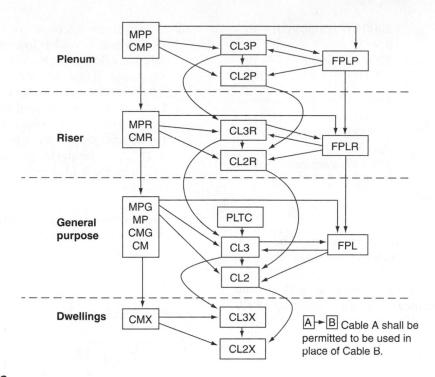

■ FIGURE 13–16

NEC-approved cable substitution and hierarchy for low-voltage systems other than optic fiber (OF) systems. (Reprinted with permission from NFPA 70-2008, National Electrical Code ®, Copyright © 2007, National Fire Protection Association, Quincy, MA 02169. This reprinted material is not the complete and official position of the NFPA on the referenced subject, which is represented only by the standard in its entirety. National Electrical Code and NEC are registered trademarks of the National Fire Protection Association, Quincy, MA.)

4. There are four guidelines to follow when choosing cables:
   - Application and environment determine the type of cable that can be used.
   - A cable substituted for another cable must be rated higher than the NEC code requires.
   - The NEC is a general guideline that can be adopted wholly or in part. Always check with the local, state, county, city, or municipal approved code. Contact the local authorities for verification of the code.
   - The local authority or fire marshal has the final authority to approve or disapprove any installation of cable on the basis of the NEC or a local code.

QUESTIONS

13.1 Although electrical equipment does not require as much space as most mechanical equipment, the design of electrical systems and the layout of electrical components demand even closer interfacing with the architectural elements in space. Why?

13.2 What are the basic steps in electrical design? Must they be followed in a logical order?

13.3 Unit power density (UPD) is a suggested allowance or budget of electrical power for lighting in a building. It is given in watts per square foot of floor area. The UPD values are convenient for estimating the amount of electrical power required for spaces or buildings during the preliminary planning phase; however, UPD is not intended for final lighting design, which varies widely in layout and in the selection of light sources. As practice, estimate the electrical power required for lighting of an office building having the floor area shown in the following table:

Area or Activities	Description	Area, sq ft
General office	Large, with no partitions	1000
Private offices	Small, enclosed	500
Accounting offices	Small, enclosed	200
Storage	Active and fine visual task	300
Transient spaces	Corridors, stairs, etc.	400
	Total gross floor area	2400

13.4 Determine the gross and net demand loads of a power distribution for a large building complex with the following design data:

Lighting	2000 kW connected, with 0.9 DF
Convenience power	1500 kW connected, with 0.1 DF
Mechanical systems	3000 kW connected, with 0.5 DF
Elevators	1000 kW connected, with 0.5 DF
Building systems	1000 kW connected, with 0.8 DF
Estimated diversity coefficient	1.2

13.5 For safety, adequate clearance must be maintained around the front and any side with access to the interior of all electrical equipment. What is the minimum clearance for a 120-V wall-mounted panelboard? A free-standing 480-V switchboard with rear access? A 4160-V controller which is not totally enclosed?

13.6 Name several methods for providing convenience electrical power to the middle of a space not accessible from wall outlets.

13.7 What is the difference between a schematic or elementary diagram and a connection or wiring diagram?

13.8 A room has eight fluorescent fixtures. Each fixture uses three fluorescent lamps totaling 150 W including the ballast. It is desired to switch on and off the center lamp of each fixture for a low level of illumination and the other two lamps by another switch for a medium level of illumination. Switching on all three lamps will provide the highest level. The switching shall be accessible from either of the two entries to the room. The room configuration, lighting layout, and switching locations are shown in the following floor plan. Determine the number of branch circuits required, and prepare the branch circuit wiring plan.

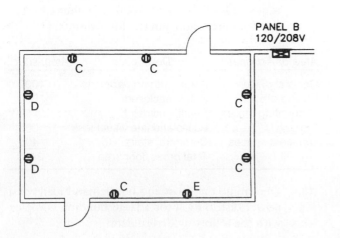

13.9 Complete the electrical floor plan for the wiring of convenience outlets in a room as shown in the following floor plan. Section 13.8.1 provides the suggested loads for general-purpose outlets, which is 1.5 A per outlet, or 3 A per duplex outlet. If an outlet is identified as serving equipment, then the load for that outlet shall be based on the nameplate rating of the equipment, for example, a TV or a computer. For the purposes of this problem, the load at each receptacle is identified as follows:

C—general-purpose receptacle	3 A
D—special equipment	5 A
E—special equipment	10 A

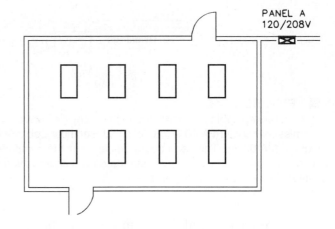

13.10 Why is the system capacity for future loads important? What is the general practice?

13.11 Name the advantages of a three-phase system over a single-phase system.

13.12 If a piece of new equipment that will be installed in a building has a different voltage rating or phase connections, what would you do?

13.13 What would you do if equipment rated for 50 Hz is connected to a 60-Hz system?

13.14 Should the National Electrical Code (NEC) be followed to the letter?

13.15 What is the minimum size wire for branch circuits in residential buildings and for institutional buildings?

13.16 If a branch circuit is rated for 30 A, what is the maximum rating of portable equipment that can be connected to this circuit?

13.17 Is EMT conduit allowed for in-the-floor installations? In wet locations?

13.18 Should all three phases of a circuit be installed in the same conduit or in different conduits?

LIGHT AND LIGHTING

14

LIGHT IS RADIANT ENERGY THAT CAN ORIGINATE FROM the sun or be converted from other forms of energy. Lighting is the utilization of either natural or converted light energy to provide a desired visual environment for working and living.

This chapter covers the properties of light, human vision and responses, color, and fundamentals of lighting techniques applicable to the built environment.

14.1 LIGHT AND THE ENERGY SPECTRUM

14.1.1 The Electromagnetic Energy Spectrum

Natural light is a form of radiant energy that originates from the sun. The human eye is sensitive to only a very narrow portion of the broad electromagnetic energy spectrum that natural light occupies. This visible portion of the electromagnetic spectrum known as the visible spectrum of visible light, extends from 380 to 780 nanometers (nm) in wavelength. One nanometer is defined as one billionth of a meter, or 1 nm $= 10^{-9}$ m.

Figure 14–1 illustrates the electromagnetic energy spectrum, which extends from very short wavelengths of about 10^{-6} to 10^{-5} nm, known as cosmic rays, to very long wavelengths of 5×10^{16} nm, familiar to us as electrical power, such as 60-Hz AC.

All forms of radiant energy are transmitted at the same speed in a vacuum—186,282 miles per second, or 3×10^8 meters per second. Light and other forms of energy travel somewhat slower in air, water, and other media. All forms of radiant energy differ in wavelength, and thus in frequency. Equation (12–1) in Chapter 12 expressed the relation between the frequency and wavelength of radiant energy. That equation is repeated here:

$$s = f \times G \qquad (14\text{–}1)$$

where s = speed of electromagnetic energy, 3×10^8 m/s

f = frequency, cycles per second, Hz
G = wavelength, m

Figure 14–1 also shows the correspondence between wavelength and frequency. For example, at 380 nm, the lowest wavelength of the visible spectrum, the corresponding frequency is 7.8×10^{14} Hz, and the frequency at 760 nm is 3.9×10^{14} Hz. It is more convenient to use the smaller values of wavelength expressed in nanometers than the larger values of frequency expressed in exponentials when dealing with the visible energy spectrum; however, it is more convenient to use frequency when dealing with electrical power. Figure 14–1 indicates the convention or general rules for specifying different forms of electromagnetic energy. Frequency is used to the right of the dividing line between infrared and radar, and wavelength is used to the left of the dividing line.

14.1.2 Light Converted from Other Energy

Light can be converted from other forms of energy; for example, the burning of oil converts chemical energy into heat and light energy. The most efficient means of converting energy into light is through electrical energy, which is also part of the electromagnetic spectrum, but with a considerably lower frequency than that of light—for example, 60 cycles per second. With modern technology, the conversion efficiency of electrical energy to lighting energy has advanced steadily, and all modern light sources used in buildings are, without exception, electric light sources.

14.2 PHYSICS OF LIGHT

14.2.1 Photometric Units

Although the International System (SI) of units is accepted worldwide as a standard for calculations of illumination, the traditional units of quantities (foot, pound, second) have not been phased out in the United States.

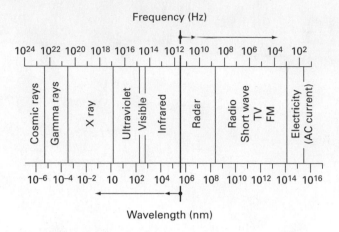

■ **FIGURE 14–1**

The radiant energy (electromagnetic) spectrum (Source: Helms and Belcher: *Lighting for Energy-Efficient Luminous Environments.*)

Accordingly, photometric quantities introduced in this section will be given in both sets of units. The basic photometric quantities are given in Table 14–1.

14.2.2 Energy of Light (*Q*)

As stated in the opening section, visible light is a part of the electromagnetic energy spectrum. It originates from the sun, but it also can be converted from other forms of energy, such as heat, electrical, and chemical energy, however, light generated in buildings is nearly all converted from electrical energy. Because light energy quickly degrades into heat energy, it cannot be stored. Thus, to maintain a certain lighting level in a space, electrical power must be supplied continuously. By definition,

$$Q = \int \Phi \, dt \qquad (14\text{–}2a)$$

or $Q = F \times t$ (when light power is constant)   (14–2b)

where Q = luminous (light) energy (lm − sec) or (lm − hr)
   Φ = luminous flux lumen (lm)
   t = time (s)
   F = luminous power (lm)

Table 14–2 shows the values of various solar energy properties in terms of power density or intensity. Those terms will be defined later.

14.2.3 Power of Light (*F*)

Power is the rate of consuming energy. In lighting, power is the luminous flux emitted by a light source in a unit of time, or

$$F = \Phi = dQ/dt \qquad (14\text{–}3)$$

where F = luminous power (lm)

A *lumen* is defined as the luminous power emitted within a unit solid angle (one steradian) by a point source having a uniform luminous intensity of 1 candela (cd).

If the conversion of electrical energy to light energy results in no loss, then one electrical watt can be converted to 683 lm of a single-wavelength green light. If the conversion is into a band of different wavelengths, as white light, the conversion produces only about 200 lm/W.

Chapter 15 discusses the luminious conversion efficiency, or efficacy, of various light sources, which varies widely from 10 to 180 lm/W.

14.2.4 Intensity of Light (*I*)

The intensity of light or *luminous intensity* (*I*) is defined as the flux density in a given direction, or the ability of a light source to produce illumination (illuminance) in

TABLE 14–1
Basic photometric quantities

Photometric Terms			Units and Abbreviations	
Scientific	(Layperson's)	Symbol	Conventional Units	SI Units
Luminous energy	(Light energy)	Q	lumen-hours (lm-hr)	lumen-hours (lm-hr)
Luminous flux	(Light power)	F	lumen (lm)	lumen (lm)
Luminous intensity	(Light intensity)	I	candela (cd)	candela (cd)
Illuminance (Light power density)	(Illumination level)	E	foot-candle (fc)	lux (lx)
Luminance	(Brightness)	L	foot-lambert (fL)(lm/ft²) (cd/3.14 ft²) (cd/452 in².)	candela/m² (cd/m²)
Exitance (reflected)	(Reflected power)	M	candela/in.² (cd/in.²)	candela/m² (cd/m²)
Exitance (transmitted)	(Transmitted power)	M	candela/in.² (cd/in.²)	candela/m² (cd/m²)
Contrast	(Contrast)	C	per unit	per unit

TABLE 14–2
Facts about solar energy (values in various units are representative values)

	Conventional Units	SI Units
Solar power density (maximum)	330 Btuh/ft^2	3800 KJ/m^2
Solar heat gain through clear glass (vertical) (40° north latitude)	250 Btuh/ft^2	2800 KJ/m^2
Daylight (sunlight + skylight)	10,000 foot-candles	108,000 lux
Daylight (overcast sky)	1000 foot-candles	10,800 lux
Sky luminance near horizon (clear day)	1500 candela/ft^2	16,000 candela/m^2
Sky luminance near horizon (overcast day)	500 candela/ft^2	5400 candela/m^2
Sky luminance overhead (clear day)	500 candela/ft^2	5400 candela/m^2
Sky luminance overhead (overcast day)	1500 candela/ft^2	16,000 candela/m^2

a given direction, measured in candela (cd). One candela is mathematically equal to one lumen per steradian of solid angle. Thus,

$$I = d\Phi/d\omega \qquad (14\text{–}4)$$

where I = luminous intensity (lm/sr) or (cd)
Φ = luminous flux (lm)
ω = solid angle (steradian, sr)
($\omega = dA/d^2$, where d is the radius of an imaginary sphere)

In laypersons' terms, intensity used to be explained as the luminous intensity produced by a standard candle, or one candlepower (cp). A *standard candle* was at one time called an *international candle* and defined as a source candle that produced luminous power equal to that of a blackbody at the freezing point of platinum (2045 K). Because a sphere contains 4π or 12.57 sr, then one candlepower (cp) would be equivalent to 12.57 lm. The term *candlepower* is deprecated.

14.2.5 Light Power Density: Illuminance (*E*)

Determination of Illuminance (*E*) from Luminous Power (*F*)

Illuminance is the density of luminous flux incident on a surface. It is analogous to Btu/hr/area in thermodynamics or W/area in electrical power. By definition:

$$E = d\Phi/dA \qquad (14\text{–}5a)$$

or $E = F/A$ (when luminous flux is constant) (14–5b)

where E = illuminance (fc) when area is in square feet (ft^2), (lx) when area is in square meters (m^2)
F = total lumens incident on the surface (lm)
A = area (ft^2) or (m^2)

The terms *illumination* or *illumination level* are still used often by many in the lighting industry, as they are more distinguishable from the word *luminance*, which means brightness to a layperson. Figure 14–2 illustrates

a lighting source with 1 cd intensity producing 1 lm/sq ft of illuminance of 1 fc or 1 lm/m^2 of illuminance of 1 lx. Table 14–3 indicates typical illuminance values encountered in nature and in buildings.

Determination of Illuminance (*E*) from Intensity (*I*): The Inverse Square Law

Because intensity (I) is defined as $d\Phi/d\omega$, or $d\Phi = I \, d\omega$, and $d\omega$ is dA/d^2, or $dA = d^2 d\omega$, then we can also express illuminance (E) as follows:

$$E = d\Phi/dA = I \, d\omega/d^2 d\omega$$
$$= I/d^2 \qquad (14\text{–}5c)$$

Equation (14–5c) is also known as the *inverse square law*, and states that the illuminance on a surface is directly proportional to the intensity (I) and inversely proportional to the square of the distance between the source and the surface, providing the surface is perpendicular (normal) to the direction of the source.

Figure 14–3(a) illustrates the inverse square law. For the same light source (intensity), the same luminous flux covers 1 ft^2 at 1 ft; 4 ft^2 at 2 ft; and 9 ft^2 at 3 ft.

The Cosine Law

When the surface receiving the luminous flux is not perpendicular to the light source, then the flux will cover a larger area, in relation to the cosine of the angle, as illustrated in Figures 14–3(b) and 14–3(c). Therefore, we can modify Eq. (14–5c) as follows:

$$E = (I/d^2) \times \cos\theta$$

$$\text{or } E = I\cos\theta/d^2 \qquad (14\text{–}6a)$$

$$\text{or } E = I\cos^3\theta/h^2 \qquad (14\text{–}6b)$$

where θ = the angle between the direction of luminous flux and the normal to the plane
d = the distance between the light source and the surface, at point E (ft), or (m)
h = the distance between the light source and the surface, at the normal to the surface (ft) or (m)

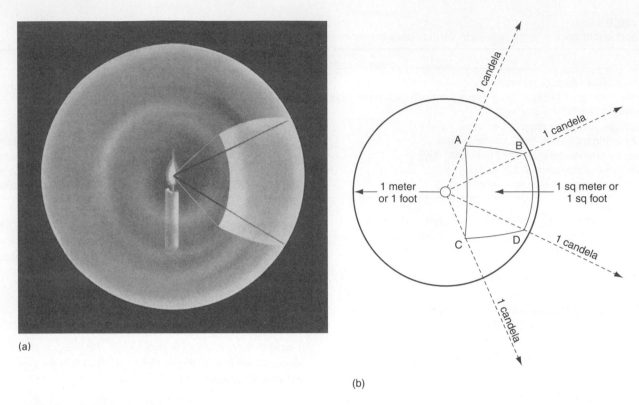

(a)

(b)

■ **FIGURE 14–2**

(a) One-candela source at the center of a clear sphere. (b) The illuminance at any point on the sphere is 1 lx (one lumen per square meter) when the radius is 1 m, or 1 fc (one lumen per square foot) when the radius is 1 ft. The solid angle subtended by area *ABCD* is one steradian. The flux density is therefore one lumen per steradian, which corresponds to a luminous intensity of one candela, as originally assumed. The sphere has a total area of 12.57 (4π) sq m or sq ft, and there is a luminous flux of 1 lm falling on each square meter or square foot. Thus, the source provides a total of 12.57 lm.

TABLE 14–3
Illuminance encountered in nature and in buildings

Surroundings	Nominal Level, fc*
Starlight	0.0002
Moonlight	0.02
Street light	1.5–10.0
Daylight:	
In shade (outdoors)	100–1000
In direct sunlight	5000–10,000
Transient space (corridor, etc.)	5–10
Office lighting	50–70
Classrooms (kindergarten to college)	20–70
Drafting rooms	70–150
Sports facilities (schools to professional facilities)	30–300

*Multiply by 10.76 to obtain lux.

14.2.6 Luminance (*L*)

Luminance is a quantity that can be difficult to grasp. It relates directly to the perceived brightness of a real or imaginary surface, or the visual appearance of a surface

produced by the illuminance of a surface. Brightness is the *subjective* evaluation of a surface, whereas luminance is the *objective* measured characteristic of a surface. In fact, illuminance is invisible. We do not see illuminance (power of light), rather, we see the luminance (brightness) of a surface or the difference in luminance between surfaces (contrast). Luminance can be defined as the luminous intensity (*I*) of a surface in a given direction (θ) per unit projected area (A_θ) from that direction, or

$$L = dI_\theta/dA_\theta \qquad (14\text{–}7a)$$

where L = luminance (cd/in².) or (cd/m²)

$$\text{or } L = d^2\phi/d\omega\, dA_\theta \qquad (14\text{–}7b)$$

where L = luminance (lm/ft²) or (fL)

$$\text{and } 1\text{ cd/in².} = 452\text{fL} = 1550\text{ cd/m²} \qquad (14\text{–}7c)$$

As a frame of reference, the approximate luminance values of various light sources are given in Table 14–4 where one will notice the great difference in luminance values between the sun at 1,540,000,000 cd/m² and the soft overcast sky at only 2000 cd/m².

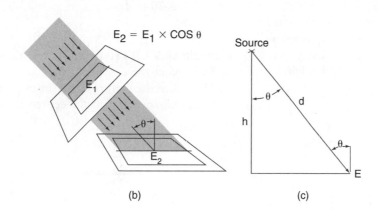

(a)

$$E_2 = E_1 \times \cos \theta$$

E_1

E_2

(b)

Source

θ

d

h

θ

E

(c)

TABLE 14–4
Approximate luminance values for various light sources

Light Source	Typical Luminance, fL	Typical Luminance, cd/m²
Sun (as observed from Earth)	450,000,000	1,540,000,000
Moon (as observed from Earth)	2400	8000
Snow in sunlight	9000	31,000
Overcast sky	600	2000
Candle flame	2900	10,000
Filament lamp (60 W inside frosted)	8800	30,000
40-W fluorescent (cool white) lamp	5000	17,000

14.2.7 Exitance (*M*)

Exitance is defined as the total luminous flux leaving a surface regardless of its direction. By direction,

$$M = d\phi/dA \qquad (14\text{–}8)$$

where M = exitance (lm/ft²) or (lm/m²)

If the surface is a reflective surface, such as a painted wall, then the quantity is the reflected exitance; and if the surface is a translucent surface, such as an opal white globe, then it is the transmitted exitance.

As an approximation, exitance may be calculated from the reflectance or transmittance value of the surface as follows:

$$M = \rho \times E \qquad (14\text{–}8a)$$

and

$$M = \tau \times E \qquad (14\text{–}8b)$$

where M = exitance (lm/ft²) or (lm/m²)
ρ = reflectance of a surface, per unit
τ = transmittance of a surface, per unit
E = illuminance (fc) or (lx)

14.2.8 Contrast (*C*)

Contrast is not a physical property of light; rather it is a phenomenon of our visual response to the light energy transmitted or reflected from surfaces or objects. For the human eye to see details, there must be a difference in luminous intensity (brightness) or in wavelength (color).

Simply stated, we do not see foot-candles or lux, but our eyes sense the luminance (brightness) or difference in luminance (contrast) or difference in wavelength (color) or the combination thereof. Luminance may be generated from an object such as the sun or a lightbulb; it may be reflected from an object, such as the wall, an apple, or the paper of a book; or it may be transmitted from an object, such as a stained glass window or a television screen.

If two objects of the same color and texture are situated side by side, we will not be able to differentiate them unless one is more reflective than the other. In fact, we can read black print on white paper with ease because there is considerable luminance contrast between the white paper and black print. If the black print is printed on dark gray paper, then the print will not be so easily readable. This phenomenon is called *contrast*. By definition,

$$C = |L_t - L_b|/L_b \qquad (14\text{–}9)$$

where C = contrast
L_t = luminance of the task, candela per unit area
L_b = luminance of the background, candela per unit area

(*Note:* The absolute value $|L_t - L_b|$ is used when L_t is darker than L_b, and C is always less than unity.)

Plate 2(a) provides an excellent illustration of typed text over a gradually changing background. The difference in readability or visual acuity of the text is obvious. Contrast is even more important in the case of a fast-moving object, such as a baseball thrown at a speed exceeding 100 mi/hr. Plate 2(b) shows a baseball on a white background compared with one on a dark background. There are true stories about great professional players losing sight of routine fly balls in some indoor stadiums. The cause is very simple: there is no contrast between the white ball and the white stadium roof.

When the foreground and background are different colors, visual acuity improves even at the same luminance level. Plate 3 illustrates this effect in its view of a tennis ball (normally yellow) against several colored backgrounds. As shown in the so-called color wheel in Plate 11(a), the further the color difference, the better the visual acuity becomes.

Example 14.1 If the lighting power reaching a surface of 10 sq ft is 900 lm, what is the average illuminance in foot-candles or lux?
From Equation (14–5b):

$$E = \frac{F}{A} = 900/10 = 90 \text{ fc}$$

or

$$E = \text{fc} \times 10.76 = 90 \times 10.76 = 960 \text{ lx}$$

Example 14.2 A lighting fixture has an intensity of 9000 cd directly below the fixture. What is the illuminance on a table 10 ft below? 20 ft below?
From Equation (14–5c):

$$\text{At 10 ft: } E = \frac{I}{d^2} = \frac{9000}{10 \times 10} = 90 \text{ fc}$$

$$\text{At 20 ft: } E = \frac{9000}{20 \times 20} = 22.5 \text{ fc}$$

Example 14.3 If a spotlight with a luminous intensity of 5000 cd at the center is aimed at a painting on the wall 5 ft from the light (h), and the angle Φ is 45°, what is the illuminance level at the center of the painting?

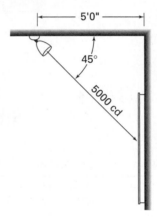

From Equation (14–5):

$$E = \frac{5000 \times (0.707)^3}{5^2} = 71 \text{ fc}$$

or

$$E = 70 \times 10.76 = 764 \text{ lx}$$

Example 14.4 A room 10 ft × 20 ft is illuminated with eight fixtures. Each fixture has 3000 lumens of light output (power). If 70 percent of the light power can be utilized at the desktop level, what is the average illuminance at the desktop?
From Equation (14–5b):

$$E = \frac{(8)(3000)(70\%)}{10 \times 20} = \frac{16{,}800 \text{ lm}}{200 \text{ ft}^2} = 84 \text{ fc}$$

Example 14.5 If the luminance L of an object is 1000 cd/m² with a surrounding background luminance of 50 cd/m², what is the luminance contrast? If the luminances of the object and its background are reversed, what is the luminance contrast?[1]

From Eq. (14–9), $C = |L_t - L_b|/L_b = (1000 - 50)/50 = 19$. If the luminances are reversed, $C = (50 - 1000)/1000 = 1(-)0.951 = 0.95$.

Example 14.6 The illuminance on a wall is measured to be 500 lx. If the reflectance value of the wall surface is 70%, or 0.70, what is the approximate value of exitance?

From Eq. (14–8a), the exitance M is $0.70 \times 500 = 350$ lm/sq ft.

Example 14.7 The illuminance at the outside surface (plane) of a diffused white glass window is 5000 lx, and the transmittance of the glass is 50% or 0.5, what is the exitance of the window?

From Equation (14–8b), the exitance M is $0.50 \times 5000 = 2500$ lm/sq ft.

14.3 VISION AND THE VISIBLE SPECTRUM

The human eye is so constructed that when light enters the cornea and lens, light energy is focused on the retina and transferred to the brain by optic nerve cells. The brain then translates the information back to the eyes, forming the optical image. The impressions from both eyes integrate the information into three-dimensional images, a very complex process indeed. A cross-sectional view of the human eye is shown in Figure 14–4.

Human beings see in two ways: by differences in color and by contrasts in luminance (brightness). Details of color will be covered in later sections; here, we introduce some fundamental concepts relating to these properties. We can see red objects on a blue background, but not easily on a background of the same red color. There must be a *color difference* for our eyes to see. Similarly, we can read black print on white paper, but not easily on dark gray paper, and not at all on black paper. Thus, there must also be a *luminance contrast* for our eyes to see.

---

[1]Luminance contrast varies considerably, depending on whether the background is darker or brighter. One alternative convention is to express the equation as $C = (L_g - L_e)/L_g$, where L_g is greater luminance and L_e is lesser luminance. With this convention, the luminance contrast will always be smaller than unity.

14.3.1 Visual Comfort

Although high luminance contrast improves visual acuity, too high a contrast may cause eyestrain after prolonged viewing. It is often more desirable to print black text on slightly colored paper, such as tan, ivory, or light blue. Obviously, visual acuity is diminished when black text is printed on dark-colored paper, such as saturated red or blue, or other colors of low reflectance value.

14.3.2 Glare

When the luminance of a light source or an object is so high (so bright) that it begins to interfere with vision, it is called *glare*. When glare is so strong that it causes physiological discomfort, it is called *discomfort glare*. When glare actually affects the ability to see, it is called *disability glare*. Glare that originates from a light source is called *direct glare*, and glare reflected from a surface is called *reflected* or *indirect glare*. Direct glare can be avoided by relocating the light source away from the line of sight. If practical, indirect glare can be minimized by replacing the reflecting surface with nonglare (matte) or low-reflectance (dark) surfaces.

14.3.3 Modeling

Modeling is the ability of a lighting system to reveal the three-dimensional image of an object. Without modeling, an object would appear to be flat. Modeling is extremely important in sports. Plate 8 illustrates the effect of lighting on the appearance of a three-dimensional tennis ball.

14.3.4 The Visible Spectrum

As illustrated in Plate 1, the visible spectrum extends from 380 to 780 nm in wavelength. Different wavelengths of light produce different sensations of color. Wavelengths of colored light are not physically divided but shift gradually from one color to the other.

Representative Colors

The common recognizable colors of light are blue, blue-green (cyan), green, green-red (yellow), orange, red, and red-purple (violet or magenta). Violet has the shortest wavelengths, from 380 to 430 nm; and red has the longest wavelengths, from 630 to 780 nm. The wavelengths of other colors are in between these.

White Light and the Rainbow

White light is a mixture of all colors in the visible spectrum. A perfectly white light is the result of mixing all

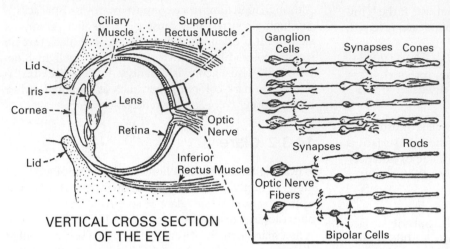

Light waves enter the eye through the cornea, which roughly focuses the pattern of waves on the foveal pit in the retina. The waves are fine-focused as they pass through the lens. The iris acts as a diaphragm that expands or contracts the pupil (opening in the iris to the lens), controlling the amount of light that is permitted to enter the eye.

The rods and cones are the ultimate receivers for individual parts of the image. They transform the received optical image pattern from radiant energy into chemical energy, which energizes millions of nerve endings. The optical pattern then becomes a series of electrical impulses traveling within a very special group of nerves that connect to the optic nerve. The optic nerves (from both eyes) combine and transmit the selective impulses to the brain, where they are interpreted.

■ **FIGURE 14–4**
How the eye works. (Courtesy: General Electric Company, Nela Park, Cleveland, OH.)

colors of equal energy level. Daylight at noon is close to being perfect, but daylight varies greatly throughout the day and is affected by the weather. (See Plate 9.) Thus, there really is no perfect white light; rather, there are different white lights—bluish, greenish, yellowish, or reddish. The fact that white light is a mixture of colored lights can be demonstrated by using a glass prism, as illustrated in Plate 4. When white light enters the prism, it bends by refraction. Since light of various wavelengths bends differently, the result is the splitting of white light into its component colors. This phenomenon occurs in nature when water vapor refracts sunlight to create a rainbow after heavy rain.

Ultraviolet and Infrared

The shortest wavelength that is still easily visible is about 380 nm and is called violet. Wavelengths shorter than violet, in the region between 200 and 380 nm, are somewhat visible. This region is called ultraviolet, meaning "beyond violet." The longest wavelength that is only somewhat visible is in the region between 780 nm and beyond. This region is called infrared, meaning "below red."

Primary Colors

Of all the colors, red, green, and blue are the most dominant. Because they are the basis for forming all the other colors, these three colors are called *primary colors*

of light. In contrast, magenta, yellow, and cyan are the primary color of pigments or paints. Mixing all three primary lights at equal energy levels produces white light, as is illustrated in Plate 5.

Secondary Colors

When two of the three primary colors of light are mixed, they produce secondary colors of light: magenta (red + blue), cyan (green + blue), and yellow (green + red). Adding the primary color with its complementary secondary color will produce white light. For example, adding green light to magenta (red + blue) produces white, or adding red light to cyan (green + blue) produces white also.

Subtractive Primaries

In pigments or paints (colorants), a primary color is defined as one that subtracts or absorbs a primary color of light and reflects or transmits the other two. So, as noted above, the primary color in pigments (also called subtractive primaries) are magenta, cyan, and yellow, which correspond to the secondary colors of light. The subtractive nature of pigment is easily demonstrated by placing magenta, cyan, and yellow pigment filters over a white light source. Each of the filters absorbs or subtracts one of the primary colors from the light. Where two filters overlap, only the opposite primary color can pass through. For example, the yellow filter absorbs blue (transmitting red and green); the magenta filter absorbs green (transmitting red and blue). Together, the two filters transmit only red (see Plates 5 and 6).

When a white light passes through all three subtractive primary filters, all colors are absorbed, and the result is no light, or black.

Reflection of Light

Light can be reflected from opaque or translucent materials. In fact, we see most objects in our visual field by reflected light. As illustrated in Plate 6(b), a red surface is a material that reflects red light and absorbs all other colors of light.

14.4 COLOR

14.4.1 Color Specifications

The human eye is sensitive to the visible part of the spectrum. (See Plate 1.) Color is a matter of visual perception, which varies with a person's subjective interpretation. A red apple may be described in common terms as red, deep red, bright red, rosy red, or even apple red; however, these terms are only a general description of a color. They are inadequate to define a color precisely. This section introduces some basic concepts used to quantify the color of a light or an object. In this regard, the references at the end of the chapter are valuable.

Three Basic Characteristics of Color

The color of a light or an object can be described by the following three characteristics:

- *Hue* is the basic color, such as red, yellow, green, or blue; and the mixture of these colors, such as red-yellow, blue-green, or red-blue. The basic color hues are illustrated in Plate 11(a), commonly known as a *color wheel*.
- *Value* is the shade of color, such as light or dark red. In pigment or paint applications, the lighter color is the result of mixing the hue color with white, and the darker color is the result of mixing the hue color with black. In lighting applications, the value of light is often disregarded, since black is simply the absence of light.
- *Chroma* is the intensity or degree of color saturation—that is, whether the color is vivid or dull. In lighting applications, it is the result of the dilution of spectrum (saturated) light with white lights.

Color Specifications

There are many systems for specifying color. Two systems commonly used for lighting are the Munsell system and the CIE color system.

- *Munsell system* This is a method of notation for describing the color of an object or a surface, such as a wall or carpet. The Munsell system uses two color charts. The first is a color wheel containing 20 basic hues of saturated colors. Each color is assigned an alphanumeric label, such as 5B for blue or 10G for the mixture of green and blue green. See Plate 11(a). The second chart is a set of color chips of the basic hues, but with different values and chromas. The value (vertical) scale is a lightness–darkness scale ranging from 1 for black to 10 for white. The value scale is useful in lighting calculations, and the value is approximately equal to the square root of the percentage reflectance of the color. For example, a surface with a Munsell value of 6 should have a reflectance value of approximately 36 percent. The chroma (horizontal) scale is a purity or saturation scale, with the pure spectrum color as 10 and the fully diluted color as 1. See Plate 11(c). The green coffee cup shown in Plate 11(b) can be described in Munsell notation as 5G5/10 green.
- *CIE Color System* The Commission Internationale de l'Éclairage (CIE) adopted a color notation system for lighting applications. The system consists of a color diagram of the spectrum known as the *chromaticity diagram;* see Plate 12(d). All colors can be found on this diagram, whether they are emitted, transmitted, or reflected. The horseshoe-shaped diagram is plotted on the x (red) and y (green) axes. The coordinate scales, from 0 to 1, are fractions of the red and green colors of the total primary colors, including blue. The z (blue) values can be determined by subtracting x and y from 1. Plate 7 shows a standard CIE chromaticity diagram. All saturated colors are located on the perimeter of the diagram, which is known as the *spectrum locus*. All colors fade into white at the center of the diagram, which is called *equal-energy white* (0 percent saturated). Any color can be expressed in terms of its x, y-coordinates. For example, a saturated yellow is identified at $x = 0.44$ and $y = 0.56$, and a diluted yellow as $x = 0.38$ and $y = 0.46$.

14.4.2 Color Temperature

A theoretically black object, or *Planckian radiator*, which absorbs all radiant energy incident upon it, is called a *blackbody*. Such a body will become dull red when heated to 800 kelvin (K), bright red at 2000 K, yellow at 3000 K, white at 5000 K, near equal-energy white at 6500 K, pale blue at 8000 K, and brilliant blue at 60,000 K. Thus, color temperature is frequently used to express the color of light sources. Representative values of light sources expressed in color temperatures are shown in Table 14–5.

The blackbody or Planckian radiator temperature locus can be found on the CIE chromaticity diagram (Plate 12). Plotted on the same diagram are the color temperatures of several fluorescent and incandescent lamps, along with daylight sources of illumination. The color temperature rating of a lamp is a convenient way to describe the lamp's approximate color characteristics. It has nothing to do with the operating temperature or with the color-rendering quality of the lamp.

14.4.3 Color Rendering

Mixing of Colors

White light may be obtained by mixing various amounts of the three primary colors. In fact, it can also be produced by the proper mixing of intermediate or unsaturated colors, since all unsaturated colors are mixtures of saturated colors. The color of an object is perceived as the composition of the light reflected or transmitted from the object. Thus, colored objects such as paints or textiles may appear to be different colors under different light sources—even different "white" light sources.

In general, standard white fluorescent lamps are rich in blue and lacking in yellow, whereas incandescent lamps are rich in yellow and red, but are lacking in blue. Neither light source renders the true color of an object. The spectrum distribution characteristics of various light sources are given in Chapter 15.

The CIE chromaticity diagram is a very useful tool in lighting design because it predicts color effects in merchandising or on a stage when there are overlapping light sources. As an illustration (see Plate 7), a white light (W) may be produced by mixing equal power (lumens) of light source A ($x = 0.35$, $y = 0.52$) and light

source B ($x = 0.26$, $y = 0.14$). The same white light may also be produced by mixing light source C ($x = 0.18$, $y = 0.45$) with light source D ($x = 0.46$, $y = 0.20$), or by any other combination of light sources in proper proportion.

Matching Colors

When matching the color of two objects, one should view the objects under the light source that will be used. One may have an experience similar to that of matching wall paints or drapery materials in a store under one kind of light, only to find that they are not the same color under light at home. A practical way to match colors is to observe the materials under two or more different light sources to see whether they match under all sources.

Psychology of Color

Psychologically, people tend to associate the longer-wavelength colors (red through orange) as warm, exciting, and dynamic and the shorter-wavelength colors (blue through green) as cool, peaceful, and calm. Wavelengths in the middle (cyan, yellow, and tan) are neutral. In general, cooler colors are preferred in a hot climate or environment, and warmer colors are preferred in a cold climate or environment.

Color Harmony and Contrast

In general, colors adjacent in hue on the color wheel (see Plate 11) are similar. For example, red and orange (Munsell 5R and 10R) are next to each other. Colors separated by five or more positions, such as red and yellow (Munsell 5R and 5Y) or red and purple (Munsell 5R and 5P), are contrasting colors. Colors on the opposite side of the color wheel, such as red and blue green (Munsell 5R and 5BG), are complementary. Whether the colors in a space should be similar, contrasting, or complementary depends largely on applications and individual preferences.

Color Rendering Index

One way to quantify the color-rendering quality of light sources is the color-rendering index (CRI) tested for the light sources. The CRI is a measure of the color shifts when standard color samples are illuminated by the light source, as compared with a reference (standard) light source. This procedure is explained in more detail in Chapter 15.

14.4.4 Color Selection

Color preference varies among individuals and is often affected by one's ethnic background, education, age,

TABLE 14–5
Color temperatures of various light sources

Light Source	Color Temperature, K
Candle flame	2000
Gas-filled incandescent lamp	3000
Warm white fluorescent lamp	3500
Daylight incandescent lamp	4000
Cool white fluorescent lamp	4500
Daylight photoflood lamp	5000
Daylight fluorescent lamp	6500
Skylight (varies with time)	5500–28,000

gender, etc. A color scheme for a building and its interior also depends on the building occupancy, spatial relations, geographical locations, climate, user's and designer's preferences, etc. Certain fundamentals, however, should be followed in selecting light sources to work with colored surfaces:

1. In general, warmer white light sources—e.g., incandescent, warm fluorescent, or high-pressure sodium lamps—should be used to enhance warm colors such as red, and orange. Cooler white light sources—e.g., cool white fluorescent lamps—should be used to enhance cool colors such as green, blue, and purple.

2. Warm white light is preferred for sources used at low light levels; conversely, cool sources are preferred for high light levels. For example, in an elegant restaurant where a low lighting level is designed for a relaxed atmosphere, warm light sources, such as incandescent or warm white fluorescent lamps, are the proper choice. On the other hand, in a fast-food restaurant a cooler light source is preferred to speed customer turnover.

3. Cooler white light sources should be used for high lighting levels. For example, cool white fluorescent lamps should be used in research laboratories, where the high lighting level is required for close observation.

4. When color rendering is important, light sources with a high color-rendering index (CRI) should be selected.

The purpose of lighting in a building is not just to provide lighting at the proper level but also to provide color harmony resulting from the coordinated efforts of the architect, interior designer, and lighting engineer.

14.5 LIGHT CONTROLS

The lighting in a space may be too bright or may come from the wrong direction, causing visual discomfort or inefficient utilization. Or the light may be the wrong color, causing poor color discrimination. For all these reasons, light must be controlled.

14.5.1 Means of Controlling Light

Light travels in clean air without bending or notable loss until it is intercepted by another medium, which will either reflect, absorb, transmit, refract, diffuse, or polarize the light. These characteristics of varying materials are utilized as a method of controlling light to achieve better lighting. Six means of control normally employed in illumination design are illustrated in Figure 14–5.

Reflection

Light is reflected from the surface of a material. If the surface is shiny, or specular, such as the surface of a mirror, then reflected light will follow the law of reflection: the angle of reflection is equal to the angle of incidence, as illustrated in Figure 14–6(a). Even a transparent material reflects some light from its surface, Figure 14–6(b).

Diffusion

When the surface is matte—that is, not shiny—then the reflected light will be diffused. The following are some types of diffusing characteristics.

- *Spread reflection*, in which light is reflected from a rough or textured surface and still follows the law of reflection but is somewhat spread out.
- *Specular diffusion*, in which light is reflected from a surface such as porcelain enamel and is generally diffused, but a narrow component follows the law of reflection.
- *Perfect diffusion*, in which light is reflected from a surface such as a matte-finished paper. This type of surface is also known as a *Lambertian surface*, and the reflected intensity follows the cosine relation with (I) maximum at 90° to the surface, and zero reflection at 0° to the surface. Figure 14–5 (2) illustrates this characteristic. A Lambertian surface will have a constant luminance irrespective of the angle of incident light or the direction of viewing angle.

Transmission

When the material is transparent (clear glass), spread (etched glass), or totally diffused (white glass), light will pass through it in a controlled mode.

Absorption

Light is absorbed when it is directed to an opaque material or passed through a transparent or translucent material. There will be a loss of light in either case. The amount of light absorbed is the balance of the incident light that is reflected or transmitted.

Refraction

The direction of light changes at the interface between two different materials, such as air and glass; this property is called refraction. Refraction is the most effective means of controlling light and is commonly used by lighting designers. Snell's law of refraction is

$$n_1 \times \sin i = n_2 \times \sin r \qquad (14\text{–}10)$$

where n_1 = index of refraction of the first medium

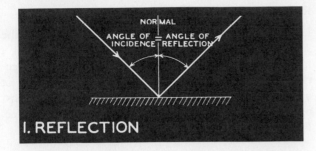
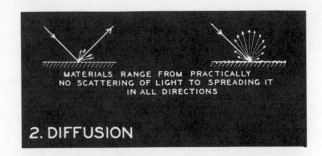
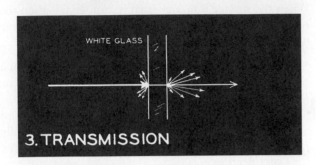
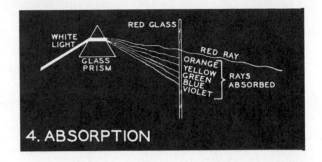
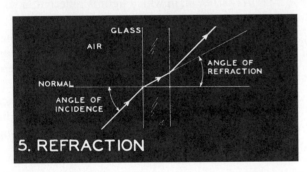
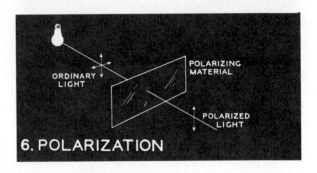

■ FIGURE 14–5
Six means of controlling light. (Courtesy: General Electric Company, Nela Park, Cleveland, OH.)

i = angle the incident light ray forms with the normal to the surface
n_2 = index of refraction of the second medium
r = angle the refracted light ray forms with the normal to the surface

When the first medium is air ($n_1 = 1$), Equation 14–10 can be simplified to

$$\sin r = \sin i \times n_1/n_2 = \sin i/n_2 \quad (14\text{–}11)$$

Figure 14–7 illustrates the refraction of a light ray from air to glass and back to air. If the angle of the light ray in the glass is such that the sine of the refracted angle is equal to unity, light will be trapped (confined) within the glass. The angle of incidence is then called the *critical angle* (θ_c) between glass and air. This phenomenon is the basic principle of fiber optics.

Polarization

Light travels at high speed with waves vibrating in all planes at right angles to the direction of travel. *Polarization* is the phenomenon wherein the waves vibrate in only one plane. A polarizing material (filter) is called a *polarizer*. When light passes through two polarizers in tandem, but with their optical axes oriented at 90°, the light will be totally polarized. One of the more popular uses of polarizers is in polarized sunglasses, which are recognized as the most effective means of reducing glare from the sun. If we position two polarized sunglasses at right angles to each other, we find that the sunlight is nearly 100 percent filtered. Polarization is used in lighting controls in the form of multilayered polarizing lenses (diffusers). Figure 14–8 illustrates the principle of polarization.

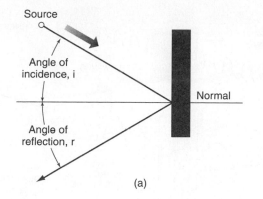

Source

Angle of
incidence, i

Normal

Angle of
reflection, r

(a)

■ **FIGURE 14–6**

(a) The law of reflection states that the angle of incidence, i, is equal to the angle of reflection, r, from a specular surface. (b) Reflection from a transparent medium is from the front as well as the rear. (c) Reflection from a glass mirror with the rear surface coated is primarily from the rear, but a small portion is also reflected from the clear front surface. The images reflected are offset by the index of refraction of the glass medium. (Reproduced with permission from IESNA.)

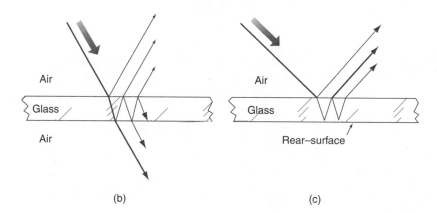

Air

Glass

Air

(b)

Air

Glass

Rear–surface

(c)

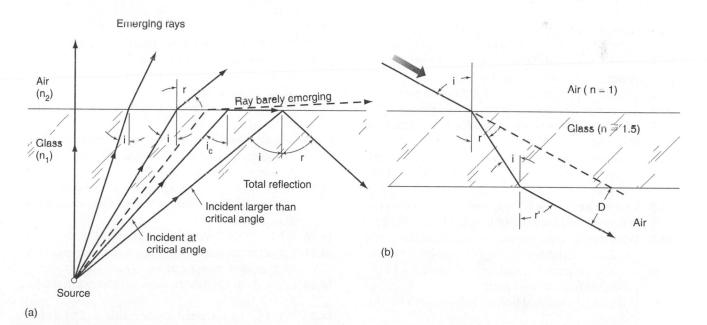

Emerging rays

Air
(n_2)

Ray barely emerging

Glass
(n_1)

i i i_c i r

Total reflection

Incident larger than
critical angle

Incident at
critical angle

Source

(a)

Air ($n = 1$)

Glass ($n = 1.5$)

i

r

i

r'

D

Air

(b)

■ **FIGURE 14–7**

(a) Total reflection occurs when sin $r = 1$. The critical angle at which light is totally refracted (reflected) varies with the medium. The critical angle between glass ($n = 1.5$) and air ($n = 1$) is 41.8°. (b) Refraction of light rays at a plane surface causes bending of the incident rays and displacement of the emerging rays. A ray passing from a lighter to a denser medium bends toward the normal, while a ray passing from a denser to a lighter medium bends away from the normal. (Reproduced with permission from IESNA.)

■ **FIGURE 14–8**

A light wave traveling in space may be compared to random waves traveling along a rope when one end of it is shaken. The waves vibrate in all planes at right angles to the direction of travel. The first polarizer (A) allows only the vertical vibrations to go through, thus polarizing the waves. The second polarizer (B) stops the vertical waves, so none of the polarized waves can go beyond, resulting in total darkness. (Reproduced with permission from IESNA.)

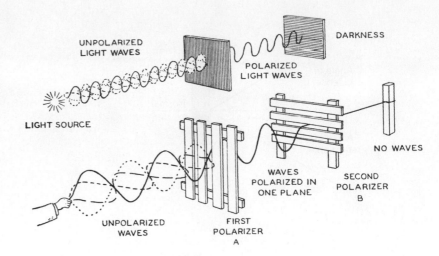

14.5.2 Application of Light Control Techniques

Any combination of the six methods of light control can be incorporated into lighting design. Popular geometric forms for lighting controls are illustrated in Figures 14–9 through 14–16.

QUESTIONS

14.1 Natural light is radiant energy from the sun. The visible spectrum ranges from ( ) to ( ) nanometers.

14.2 Light generated from electrical light sources (lamps) is radiant energy. (True) (False).

14.3 The three primary colors of light are ( ).

14.4 The three subtractive primaries of light are ( ).

14.5 What unit of measure is used for illuminance, or illumination level?

14.6 Daylight under sunlight is about ( ) foot-candles.

14.7 Daylight under an overcast sky is about ( ) foot-candles.

14.8 Color temperature is a way to define the heat generated by various types of lamps. (True) (False).

14.9 What is the approximate color temperature of a gas-filled incandescent lamp? A fluorescent lamp?

14.10 What is the conventional unit of measure of light power? What is the SI unit?

14.11 What are the normal human responses to various colors?

14.12 What is the meaning of luminous intensity?

14.13 Color difference also improves visual acuity even if the object and its background are at the same luminance level or zero contrast. (True) (False)

14.14 A higher luminance contrast is desired for viewing a fast-moving object, such as a baseball, but is not important for a stationary task. (True) (False).

14.15 Does a high luminance contrast provide more visual comfort?

14.16 What are glare, direct glare, reflected glare, discomfort glare, and disability glare?

14.17 A ray of sunlight reaching the earth has a balanced energy level of all wavelengths. (True) (False)

14.18 What wavelength is most sensitive to human eyes?

14.19 White light is a mixture of colored lights. (True) (False)

14.20 What are the basic characteristics of color?

14.21 If 64 lm reach a surface of 10 sq ft, what is the illumination level in foot-candles and in lux?

14.22 What is the recommended illuminance level for office occupancies?

14.23 When are the inverse-square law and the cosine law used?

14.24 A light source having a candlepower of 20 cd illuminates a surface 2 ft away. What will be the illuminance on the source?

14.25 If the light source in Question 14.24 is at a 60° angle with the normal to the surface, what is the illuminance?

14.26 What are the means of light control?

14.27 Light travels in a straight line in a single medium, such as air, water, or glass. (True) (False)

14.28 Light always bends the same way between media. (True) (False)

14.29 What is the refracted angle inside a glass if light enters from air into the glass at 40° from the normal axis and the index of refraction of glass is 1.4?

14.30 Refraction is the most effective means of light control. (True) (False)

14.31 The reflected light beams from a parabolic reflector are parallel to each other. (True) (False)

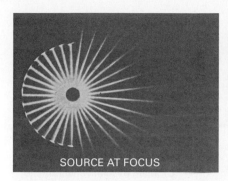
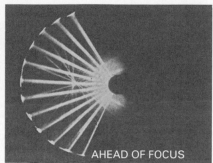
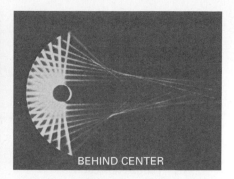

SOURCE AT FOCUS   AHEAD OF FOCUS   BEHIND CENTER

■ **FIGURE 14–9**

The circular reflector is used to reflect light to produce either spread out or concentrated light patterns, depending on the position of the lamp in relation to the center of the circle. (Courtesy: General Electric Company, Nela Park, Cleveland, OH.)

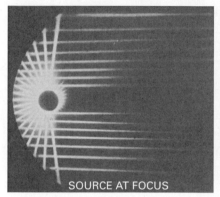
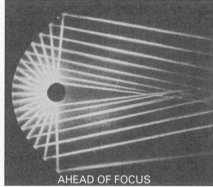

SOURCE AT FOCUS   AHEAD OF FOCUS   BEHIND FOCUS

■ **FIGURE 14–10**

A parabola has one focal point. When the lamp is located at the focus, all reflected light beams will be parallel to each other. Parabolic reflectors are used primarily for spotlight applications. Varying the position of the lamp changes the distribution from a very narrow spotlight to a very wide floodlight. (Courtesy: General Electric Company, Nela Park, Cleveland, OH.)

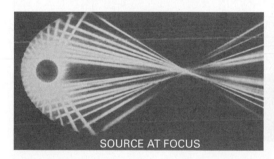
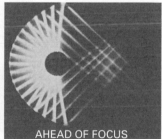
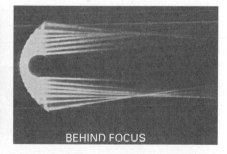

SOURCE AT FOCUS   AHEAD OF FOCUS   BEHIND FOCUS

■ **FIGURE 14–11**

An elliptical reflector is used primarily to focus the light beam through the second focal point, allowing all light beams to pass through a pinhole. (Courtesy: General Electric Company, Nela Park, Cleveland, OH.)

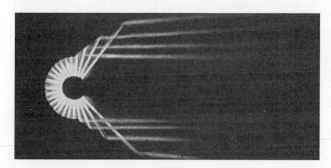
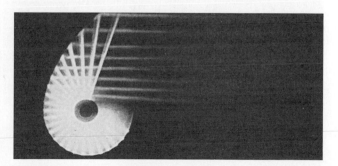

■ **FIGURE 14–12**

Reflection with compound circular-parabolic reflectors. (Courtesy: General Electric Company, Nela Park, Cleveland, OH.)

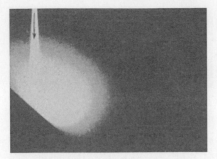 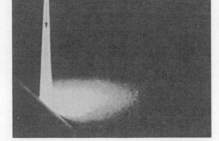 

Diffuse reflection	Spread reflection	Diffuse-specular reflection
• matte paint • limestone	• aluminum paint	• glossy enamel
• plaster • terra cotta	• oxidized aluminum	• glossy paper

■ **FIGURE 14–13**

Characteristics of diffuse materials. (Courtesy: General Electric Company, Nela Park, Cleveland, OH.)

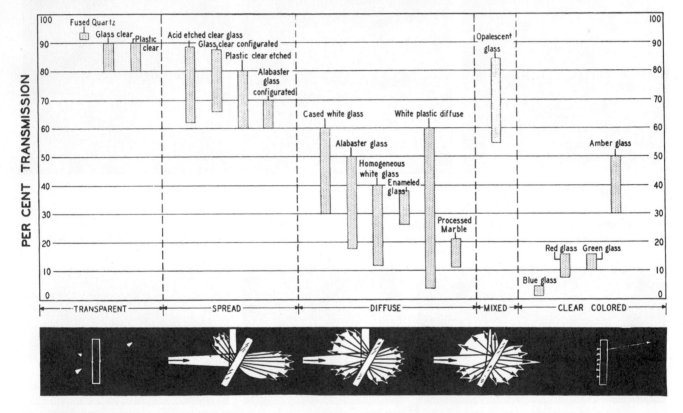

■ **FIGURE 14–14**

Characteristics of transmission materials. (Courtesy: General Electric Company, Nela Park, Cleveland, OH.)

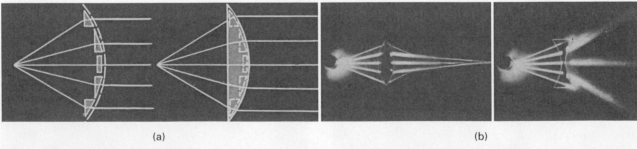

(a)

The refraction of light through a series of tiny prisms and a single convex lens is identical.

(b)

The refraction of light through double convex and concave lens.

■ **FIGURE 14–15**

Refraction of light through lenses and prisms. (a) Refraction of light through a series of tiny prisms and refraction through a single convex lens is identical. (b) Refraction of light through double convex and concave lens. (Courtesy: General Electric Company, Nela Park, Cleveland, OH.)

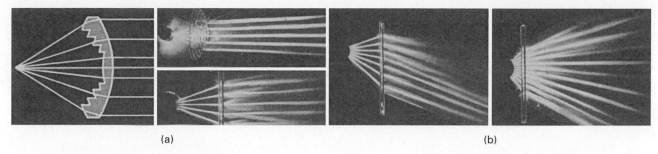

(a)

Construction of convex Fresnel lens
and prismatic lens plate.

(b)

Lens plates may provide a symmetrical
light pattern.

■ **FIGURE 14–16**

Refraction of light through a Fresnel lens and plates. (a) Construction of convex Fresnel lens and prismatic lens plate.
(b) Lens plates may provide a symmetrical light pattern. (Courtesy: General Electric Company, Nela Park, Cleveland, OH.)

14.32 What is the primary use of an elliptical reflector?

14.33 What is a Fresnel lens?

14.34 Describe the light control principles used for the following light sources or lighting fixtures:
   a. Fixture with a diffusing glass globe.
   b. Metal can with an open bottom. The inside surfaces of the fixture are painted black.
   c. Ceiling-mounted fixture with translucent plastic diffusers at the bottom.
   d. Ceiling-mounted fixture with a Fresnel (prismatic) control lens.
   e. Suspended fixture with opaque bottom and open top (indirect distribution).

14.35 A book is printed in black ink on buff paper. The luminance measured on the paper is 60 cd/m² and on the print is 4 cd/m². What is the contrast of the reading task (print and its background)?

14.36 A book is 0.80 sq ft in area and the average lighting power, directly and indirectly, reaching a 4-sq-ft desk is 75 lm. What is the illuminance on the book? On the desk?

14.37 A lighting system consists of 10 spotlights, of which one is located directly above a task, and the other 9 are aimed elsewhere. The reflected illuminance from these 9 fixtures is calculated to be 3 fc on the task. What is the illuminance on the task if the candlepower at the center of the spotlight is 2500 cd, and the spotlight is 10 ft above the task?

14.38 What is the luminous exitance of the task if its reflectance is 25 percent?

14.39 Why do approaching automobile headlights appear so bright at night?

14.40 How would you describe your visual sensation when the approaching headlights are on low beam? On high beam?

14.41 A lamp rated for 1000 lm of light power is installed within a lighting fixture (luminaire). Can all the lumens be utilized to produce useful illumination in the space?

14.42 When a lamp or several lamps are installed in a lighting fixture or luminaire, not all the lamp light power (lumens) may be emitted from the luminaire. If 80 percent of the total lumens are emitted from a luminaire, then the luminaire is said to be 80 percent efficient. Does this mean that 80 percent of the total lamp light power will be utilized on the task anywhere in the room?

REFERENCES

Illuminating Engineering Society of North America (IESNA), *Lighting Handbook: Reference and Application*, 8th ed. New York, 1994.

General Electric Lighting Company, *Fundamentals of Light and Lighting*. Nela Park, OH: 1960.

General Electric Lighting Business Group, *Light and Color*. Nela Park, OH: 1995.

General Electric Lighting Company, *Specifying Light and Color*. Nela Park, OH: 1995.

Minolta Corporation, *Precise Color Communication*. Ramsey, NJ: 1995.

IESNA, *IES Education Series (Introductory)*. New York: 1995.

IESNA, *IES Education Series (Intermediate)*. New York: 1995.

Murdoch, Joseph B., *Illumination Engineering*. New York: Macmillan, 1985.

Helms, Ronald H., and M. Clay Belcher, *Lighting for Energy-Efficient Luminous Environments*. Englewood Cliffs, NJ: Prentice Hall, 1991.

Egan, M. David, *Concepts in Architectural Lighting*. New York: McGraw Hill, 1983.

LIGHTING EQUIPMENT AND SYSTEMS*

15

LIGHT MAY ORIGINATE IN MANY WAYS—FROM SOLAR energy (daylight), from combustion and chemical reactions, and from the conversion of electrical energy. Of all light sources, daylight is the most plentiful, and free of charge; however, daylight is not available at night and fluctuates widely during the day, sometimes being too bright for visual comfort or too hot to stay under for very long. Still, when properly controlled, e.g., with shading and air conditioning, it is the most economical of all sources of light.

In buildings, electric lighting has become the sole source of light at night and a supplemental source during the day. Frequently, electric lighting must be used even if there is plenty of daylight—for example, during a visual presentation in a classroom or conference room, and during many sporting events, when daylight from low angles may cause glare that blinds the players or the spectators.

Figure 15–1 illustrates the chronological development of light sources and the relationship between them. All modern light sources applicable to the interior lighting of buildings are electric lights; combustion-type light sources, such as gaslights, are limited to decorative use only.

The fundamental component of lighting equipment is the light source, commonly called a *lamp*. The assembly that holds a lamp or lamps together to provide lighting is the luminaire, commonly called a lighting fixture. Although discouraged in academic circles, the term *lighting fixture* is more accepted than *luminaire* by the general public and among professionals in the design and construction industries. Thus, the two terms are used interchangeably in this book, as are *light sources* and *lamps*.

Lighting fixtures must be designed for a particular type and size of lamp and are usually not suitable for other types. One reason for this is the difference in configuration of the lamp holder (socket). Another reason is

the amount of heat generated, which affects the material of the fixture, the ventilation, and the physical clearance between the lamp and the surrounding surfaces. Thus, lighting fixtures and lamps must be compatible and within the rated power (wattage) limit of the fixture.

Lighting fixtures are important elements of interior design, since they are often prominently displayed in the space. In addition to lighting performance, other features such as physical size, texture, shape, and color must be considered when selecting a fixture. To create a successful lighting design, the architect, interior designer, and lighting engineer must be knowledgeable about both light sources and the performance of various fixtures. This chapter presents the fundamentals of lamps, fixtures, and the performance characteristics of lighting systems.

Lighting accounts for about 25 percent of all electrical energy consumed in the United States. For each kilowatt-hour of electrical energy consumed, a corresponding amount of pollutant is being released into the atmosphere. The impact of energy consumption on the global environment is addressed in Chapter 1. Lighting design professionals have a responsibility to minimize energy consumption by using efficient light sources and equipment while creating an environment that is aesthetically pleasing and conducive to higher human performance.

15.1 ELECTRICAL LIGHT SOURCES

Although thousands of lamps are made for diverse applications, they can be grouped into four major classes, based on their operating principles.

Incandescent lamps convert electrical energy into heat at a temperature that causes the filament of the lamp to become incandescent (red or white hot). The process closely resembles the heating of a blackbody, discussed in Chapter 14.

Fluorescent lamps contain mercury vapor. When proper voltage is applied, an electric arc is produced between the opposing electrodes, generating some

*David A Krailo, LC. is contributing author of this chapter. See Acknowledgments.

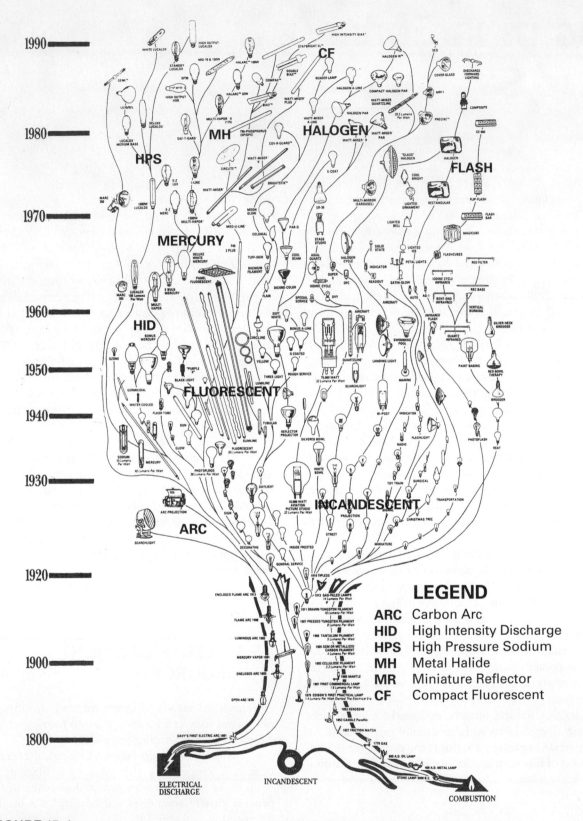

■ FIGURE 15–1

Tree of light: historic development of light sources. (Courtesy: General Electric Company, Nela Park, Cleveland, OH.)

visible, but mostly invisible, ultraviolet radiation. The ultraviolet radiation excites the phosphor coating on the inside of the bulb, which then emits visible light.

High-intensity discharge (HID) *lamps* produce high-intensity light within an inner arc tube contained in an outer bulb. The metallic gas within the arc tube may be mercury, sodium, or a combination of other metallic vapors. The outer bulb may be clear or coated with phosphor. Mercury vapor, metal halide, and high-pressure sodium lamps are all classified as HID light sources.

Miscellaneous lamps include a wide variety of lamps operating on different principles. Although they have limited application in buildings, future breakthroughs in technology and production may provide a new dimension to the world of architecture and environmental design. Some promising new types are the following:

1. *Short-arc lamps,* or compact-arc lamps, such as the xenon family of lamps, produce light in small arc tubes and are the closest thing to a true point source of high luminance. They are used primarily as searchlights, in projectors, and in optical instruments.
2. *Low-pressure sodium lamps* (LPSs) are monochromatic lamps in the yellow region of the spectrum (589 to 589.6 nm). The efficacy of an LPS lamp is as high as 180 lumens/watt, (lpw), but because of its color, this lamp has limited applications. Typical applications are along highways and in storage yards.
3. *Electroluminescent lamps* emit light by the direct excitation of phosphor by an alternating current. Therefore, they can be made in any shape, size, and form. Electroluminescent lamps can produce different colors by mixing phosphors. Although these lamps are extremely efficient at about 200 lpw, their use is limited to signs and decorative applications.
4. *Electrodeless lamps* are gaseous lamps excited by electromagnetic or microwave energy without the use of electrodes. These lamps have a promising future in building lighting applications through the use of specially designed fixtures or the use of "light pipes" and will be discussed later in the chapter.
5. *Light-emitting diodes* (LEDs) are based on the principle that electrons move in layers within atoms. When electrons fall from one layer to another, photon energy (light) is released. LEDs are illuminated by the movement of electrons in layers of semiconductor material. With no filament to fail, these light sources have extremely long life. As available output and efficiency increase, LEDs are being incorporated into architectural lighting products.

15.2 FACTORS TO CONSIDER IN SELECTING LIGHT SOURCES AND EQUIPMENT

There are many factors to consider in selecting lamps of all varieties.

15.2.1 Light Output

Light output, expressed in lumens, is defined as follows:

- *Initial lumens* Rated light output when the lamp is new, typically measured after 100 hours of operation.
- *Mean lumens* Lumen output of a light source after the source has been used. Mean lumen values for fluorescent and HID lamps are typically measured at 40 percent of their rated lives. Most high-pressure sodium and mercury lamps are measured at 50 percent of their rated lives. All measurements are made on ANSI reference ballasts. Mean lumens are not typically measured for incandescent and tungsten halogen lamps.
- *Beam lumens* The lumen output of a directional light source measured within the beam angle, which is the angle between the two directions for which the intensity is 50 percent of the center-beam candlepower.
- *Field lumens* The lumen output of a directional light source measured within the field angle, which is the angle between the two directions for which the intensity is 10 percent of the center-beam candlepower.

15.2.2 Intensity

Light intensity is expressed in candelas (cd) at various angles from the lamp or fixture. The data are usually provided by manufacturers in the form of candlepower distribution curves.

15.2.3 Luminous Efficacy

Luminous efficacy, or simply efficacy is defined as the light output per unit of electrical power (watts) input, or lumens/watt (lpw). Theoretically, 1 W of electrical power can be converted to 683 lm of monochromic green light, or about 200 lm of white light of equal energy level among all visual spectrum wavelengths. With this as reference, the 10- to 25-lpw efficacy of incandescent lamps is far short of ideal. The efficacy of a lamp should include the power consumed by its accessories, that is:

- *Lamp efficacy*, in lumens/watt, for lamps only, or
- *System efficacy*, in lumens/watt, of the lamps and accessories (e.g., ballast), for electrical discharge (fluorescent and HID) lamps.

15.2.4 Luminaire Efficiency

Luminaire efficiency is the ratio of the total light output of the lamps in lumens versus the total light input of the luminaire. The value is expressed as a percentage. Luminaire efficiency is a good measure for comparing luminaires of similar candlepower distribution characteristics, but it is not necessarily a measure of how well the light is being utilized. For example, if one must illuminate a painting on a wall, a bare-bulb fixture, which is 100 percent efficient, suspended in front of the painting will not be as good as a directional luminaire that is only 60 percent efficient. Chapter 16 will introduce the concept of the coefficient of utilization (CU) in illumination calculations.

15.2.5 Rated Lamp Life

The rated life of a lamp is defined as the time elapsed when 50 percent of a group of lamps remain burning. Rated life closely follows the mortality curve of most statistics for a large number of subjects. Figure 15–2 shows the mortality curve of incandescent lamps.

Example 15.1   If the rated life of one type of incandescent lamp is 750 hours, what would be the expected percentage of survival for a large number of lamps installed in a building after 500 hours of use?

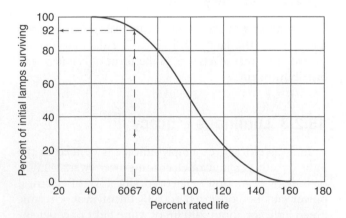

■ **FIGURE 15–2**
Range of typical mortality or life expectancy curves for incandescent lamps.

Solution. The hours of use of the lamp are ⅔, or 67 percent, of the rated life. As shown in Figure 15–2, 92 percent of the lamps would probably survive.

15.2.6 Lumen Depreciation

Light output depreciates with time. The loss of light, known as *lumen depreciation*, may vary from 10 to 40 percent of a lamp's initial light output. This characteristic must be taken into consideration in illumination design. Typical lumen depreciation characteristics for various lamps are listed in Table 15–1.

15.2.7 Color Temperature

Color temperature is the perceived color of a light source as judged by the human's eye. It is expressed in degree Kelvin (K). Chapter 14 provides the background for using this concept to describe the color of various types of white, bluish-white, yellowish-white, or reddish-white light sources that are closely following the Black Body Locus. Color temperature cannot be used to express the perceived color of other light sources, for example, saturated red, yellow, green, blue, etc. Table 15–2 lists the typical color temperature ranges of common white light sources.

15.2.8 Color-Rendering Index

The concept of the color-rendering index (CRI) was introduced in Chapter 14. CRI is useful for measuring how well a lamp renders color, compared with a reference light source of the same color temperature range. An incandescent lamp with color temperature at 3000 K has been selected by the CIE as the reference source (CRI = 100) for comparison with all other lamps having color temperature below 5000 K. This selection of CRI = 100 does not mean that this particular incandescent lamp can render the true colors of every object or material. Such a lamp is selected only because it has a smooth profile of spectral power distribution (SPD) without extremely narrow energy bands.

15.2.9 Flicker and Stroboscopic Effect

Theoretically, the cyclic flow of a 60-Hz current through a lamp can produce light fluctuations 120 times per second. This is called *flicker*. Because of the light retention characteristic of incandescent filaments and the phosphor coatings of discharge lamps, flicker is not normally perceivable except in noncoated HID lamps. When a rapidly moving object is observed under a clear HID lamp, the object may appear to be at a standstill or moving at lower harmonic frequencies.

TABLE 15–1
Mean lumens for various lamp types as percent of initial lumens

Lamp Type	Mean Lumens as % of Initial Lumens[a]	Lamp Life (hrs)
Fluorescent		
F40T12/halophosphor	85	20,000
F40T12/rare-earth triphosphor	90	20,000
F96T12/halophosphor	88	12,000
F96T12/rare-earth triphosphor	90	12,000
F96T12/HO	75	12,000
F96T12/VHO	70	10,000
T8/rare-earth triphosphor	90	20,000
T5/rare-earth triphosphor	95	20,000
Incandescent @70% life	87	750–2000 Avg.: 1000
Tungsten halogen @70% life	95	1000–4000 Avg.: 2500
Metal halide		
M400/U	70	20,000 vertical
MP100/U/Med	75	15,000 vertical
MS400/HOR	65	20,000
High-pressure sodium		
LU100 @50% life	84	24,000+
LU400 @50% life	90	24,000+

[a]Lumen depreciation @ 40% of lamp life unless otherwise noted. Multiply rated lumens by this percentage to calculate mean lumen or "design lumen."

TABLE 15–2
Color temperature of typical light sources

Light Source	Color Temperature Range, K
Incandescent,	
60 W	2500–2700
100 W	2700–2900
500 W	2900–3100
Halogen, tungsten	3000–3200
Fluorescent,	
Warm white	2000–3000
Cool white	4000–5000
Daylight	6000–6500
Mercury,	
Clear	5500–5800
Improved	4400–4500
Metal halide,	
Clear	3700–3800
Coated	3200–4000
High-pressure sodium,	
Normal	2000–2100
Color improved	3000–4000
Low-pressure sodium	1700–1800
Photoflood lamps	3200
Photoflash lamps	5500

This is called the *stroboscopic effect*. It can be eliminated or minimized by the use of lead-lag ballasts for multiple lamp fixtures and the wiring of multiple fixtures on alternate three-phase electrical circuits.

When the frequency of the electrical power system is lower than 60 Hz, flicker and the stroboscopic effect will be exaggerated. Conversely, at higher frequencies, such as 400 or 3000 Hz, flicker will not be perceived.

15.2.10 Brightness

Physically, small light sources of high intensity, such as incandescent lamps, are excellent for light control, but they can be too bright for visual comfort. The location and aiming of the fixtures must therefore be carefully selected. Uncomfortable glare can be minimized by methods discussed in Chapter 14.

15.2.11 Intensity Control

Light intensity can be controlled by multilevel switching or by dimming. Incandescent lamps can be dimmed easily by the use of autotransformers or solid-state dimmers. Fluorescent and HID lamps can be dimmed by special ballasts and circuitry, but at a considerably higher cost. Other means of intensity control include using fixture components such as lenses and louvers to reduce light source brightness in certain directions. (See Chapter 17 for the effect of dimming on typical residential or public spaces.)

15.2.12 Accessories

Among the many accessories that should be considered before selecting light sources are ballasts, starters, and dimmers.

15.3 INCANDESCENT LIGHT SOURCES

Edison's first successful lamp, using carbon filament in a vacuum, produced 1.4 lpw. Since then, incandescent lamps have improved dramatically, using a tungsten filament in a bulb filled with inert gas. Incandescent lamps are easily controlled, both in intensity and in direction; however, they are the least energy-efficient lamps. Following are some characteristics of incandescent lamps.

15.3.1 Shape and Base

Incandescent lamps are available from a fraction of a watt to several thousand watts. They come with a variety of shapes and bases. Figures 15–3(a) and (b) illustrate these features. Listed here are selected American National Standard Institute (ANSI) lamp shape designations of the most popular lamp types:

A – Standard*	G – Globe	S – Straight inside
B – Decorative	GT – Globe tubular	T – Tubular
C – Conical	PAR – Parabolic	T/C – Tubular, circular
E – Elliptical	P – Pear shaped	TU – Tubular, U-shaped
ER – Elliptical reflector	PS – Pear, straight neck	MR – Multifaceted reflector†
F – Flame		R – Reflector
M.O.L. – Maximum overall length		L.C.L. – Light center length

*Arbitrary spherical with narrow neck.
†Commonly referred to as minireflector.

15.3.2 Size Designation

The physical size of a lamp is described numerically in ⅛-in. increments of the widest part of the bulb. For example, if the diameter of a tubular lamp is 1 in., then the lamp will carry the designation "T8." Similarly, a pear-shaped lamp that is 2⅝ in. at its widest part is designated to be P21. (Two inches is 16, and ⅝ in. is 5; thus, 16 + 5 = 21.)

The *maximum overall length* (M.O.L.) of a lamp is measured from the bottom of its base to the tip of the bulb, whereas *light center length* (L.C.L.) is measured from the bottom of its base to the optical center of the lamp. These are critical dimensions for lighting fixture manufacturers when designing new luminaires.

15.3.3 Rated Life

Incandescent lamps are short-lived. General-service types are normally rated for 750 hours as a bare lamp. (See Figure 15–2 for the range of a typical mortality curve for lamps.) Halogen and quartz lamps have a longer rated life, from 1000 to 4000 hours; however, they are still inferior from a life standpoint compared with fluorescent and HID lamps, which have rated lives of 10,000 to 40,000 hours, depending lamp design and power rating.

15.3.4 Efficacy

As defined earlier, efficacy is luminous efficiency expressed in lumens per watt (lpw). Incandescent lamps are among the poorest lamps in terms of efficacy. They range from 10 lpw for 40-W lamps to 25 lpw for 1500-W lamps. Compared with the 80- to 120-lpw efficacy of fluorescent and HID lamps, the 10- to 25-lpw efficacy of incandescent lamps renders them only about one-fifth as efficient. In other words, an incandescent lamp requires about five times as much electrical power as a fluorescent or HID lamp to produce the same amount of illumination. The use of incandescent lamps should thus be limited to applications that do not require high illuminance except where the task to be illuminated is small or where it is difficult to reach.

15.3.5 Depreciation of Light Output

Standard incandescent lamps have a moderate rate of depreciation of their light output. At the rated life of the lamp, the lumen output is about 82 percent of its initial output. (See Figure 15–4.) Halogen lamps, on the other hand, are extremely good at maintaining their light output, having practically no depreciation at all.

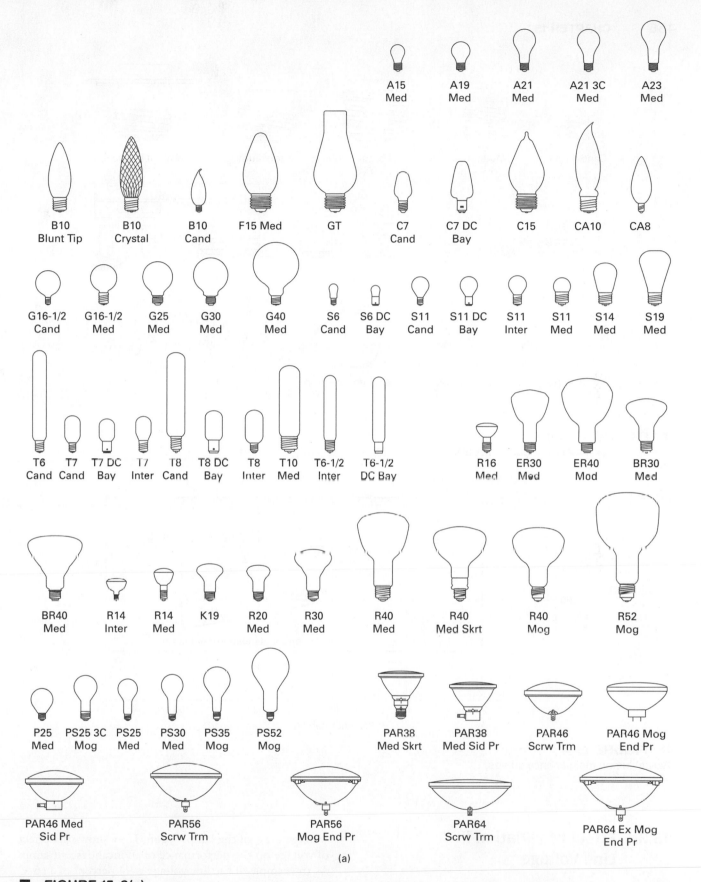

■ **FIGURE 15–3(a)**

Typical incandescent lamp bulb sizes and shapes. See Section 15.3.1, "Shape and Base," for letter designations. (Courtesy: Osram Sylvania, Inc., Danvers, MA.)

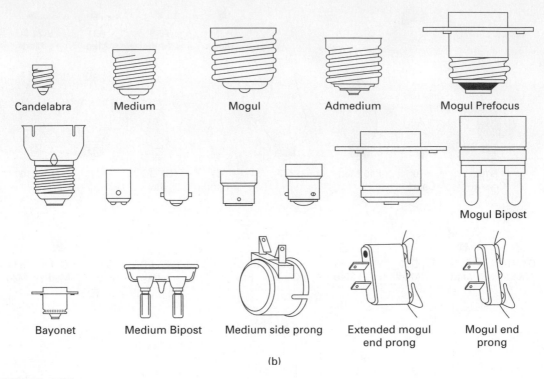

Candelabra   Medium   Mogul   Admedium   Mogul Prefocus

Mogul Bipost

Bayonet   Medium Bipost   Medium side prong   Extended mogul end prong   Mogul end prong

(b)

■ **FIGURE 15–3(b)**
Typical incandescent lamp bases. (Courtesy: General Electric Company, Nela Park, Cleveland, OH.)

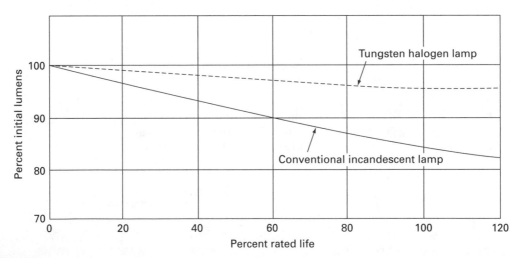

■ **FIGURE 15–4**
Typical lumen maintenance curves.

15.3.6 Effect of Variation in Line Voltage

Incandescent lamps are normally rated at 115 V. If the actual line voltage is higher or lower than the rated voltage, it will greatly affect the life, lumen output, and wattage draw of the lamp. Figure 15–5 shows the effect of voltage on the performance of an incandescent lamp. As an example, if the line voltage is 94 percent of 115 V, or 108 V, then the light output will drop to 80 percent of its rating, the power will drop to 90 percent, and the life of the lamp will increase to 220 percent.

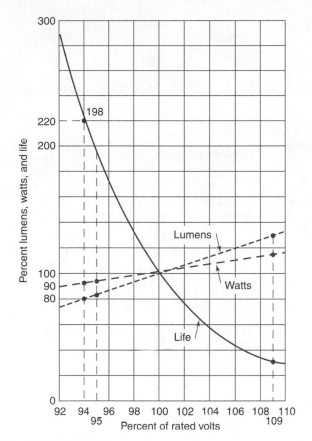

■ FIGURE 15-5

Effect of voltage on performance of incandescent lamp. (Answers to Example 15.1 are shown on the dotted line to the left.)

Example 15.2 If the voltage at an incandescent lamp base rated for 120 V is only 114 V, what will be (a) the lumen output of the lamp, (b) the actual power drawn, and (c) the expected life of the lamp?

Assume that the lamp is an A19, 100-W, general-service lamp rated for 120 V and with a rated output of 1750 lm and rated life of 750 hours.

Answer The actual 114 V at the lamp is lower than its 120 V rated voltage, or $(114/120) \times 100 = 95\%$ of its rated voltage.

From Figure 15-5:

a. The light output will be only 82%, or $1750 \times 82\% = 1435$ lm.
b. The power consumption will be 95%, or $100\,W \times 95\% = 95$ W.
c. The actual life will be about 198%, or $750 \times 1.98 = 1485$ hours.

In other words, the lamp only will produce 82% of its rated lumens but will consume only 95% of the rated power, extending its life by 98%.

15.3.7 Color Rendering and Color Preference

Incandescent lamps have excellent color rendering and color preference. Indeed, they are considered the best reference light source at their given color temperature.

15.3.8 Major Types of Incandescent Lamps

All incandescent lamps use tungsten as their filament (see Figure 15-6). Tungsten filaments can be designed to operate at temperatures ranging from 3800 to 5000°F. Higher operating temperatures will produce a whiter spectrum and higher energy conversion efficiency (efficacy) but a reduced life for the filament. The extremes of incandescent lamp design can best be illustrated by the performance of photoflood lamps. Whereas an incandescent lamp normally has an efficacy between 15 and 25 lpw, a 500-W R40 photoflood lamp can produce 38,000 lm, or 76 lpw, but has only 4 hours of rated life. Conversely, lamps can be designed to last essentially forever, but at extremely low efficacy.

There are numerous types of incandescent lamps. Following are some that are commonly used in building systems.

General-Service Type

These lamps are the standard for use in buildings. They are designed to have relatively good efficacy (15–25 lpw), a moderate rated life (750–1000 hours), and good color-rendering characteristics (95–100 CRI).

Rough and Vibration Services

These lamps are used in locations subject to rough handling or vibration, such as on machinery or on trains or automobiles. They are specially designed with heavy tungsten filaments and supports. Efficacy is between 10 and 14 lpw.

Extended-Life Service

These lamps are designed with heavy filaments to operate at considerably lower than 3800°F and have a rated life of from 2500 to 10,000 hours. It should be expected that the longer the rated life, the lower the efficacy. A number of products claim to have 10,000 hours or more of lamp life. Most of these are constructed of heavier filaments with more filament supports, but they operate at low temperatures and thus at extremely low light efficacy generally about 7 to 10 lpw.

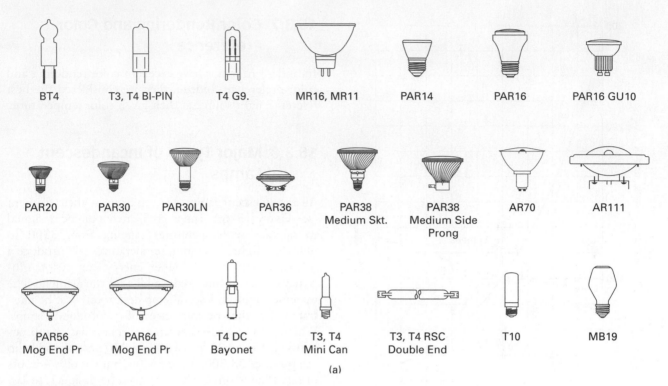

BT4   T3, T4 BI-PIN   T4 G9.   MR16, MR11   PAR14   PAR16   PAR16 GU10

PAR20   PAR30   PAR30LN   PAR36   PAR38
Medium Skt.   PAR38
Medium Side
Prong   AR70   AR111

PAR56
Mog End Pr   PAR64
Mog End Pr   T4 DC
Bayonet   T3, T4
Mini Can   T3, T4 RSC
Double End   T10   MB19

(a)

■ **FIGURE 15–6(a)**
Tungstern–halogen lamp bulb shapes. (Courtesy: Osram Sylavania, Inc., Danvers, MA.)

Dichroic Reflector Lamps

These lamps transmit color selectively through a molecular layer of chemical coating, allowing only the desired wavelength of color to pass through. They are used to reduce the infrared wavelength, which causes heat. Typical applications of dichroic lamps are in retail merchandising displays and on art paintings, where infrared heat in the light beam is substantially reduced. Life and efficacy are similar to those of standard incandescent lamps. Because most of the infrared energy is passed back through the dichroic reflector, one must take care in selecting fixtures that can manage this additional heat coming from the back of the lamp.

Krypton Lamps

These lamps are filled with the gas krypton and are designed for long life. Krypton gas is a heavier fill gas than argon, which is typically used in general-service incandescent lamps. The heavier gas retards the evaporation of the lamp filament and allows for longer lamp life. Although krypton lamps are not more efficient than standard incandescent types, their longer life makes them excellent for applications where lamp replacement is difficult or expensive. One of the most common applications for krypton lamps is in traffic signals.

Tungsten–Halogen Lamps

PAR lamps with tungsten–halogen capsules are available in several sizes, including PAR14, PAR16, PAR20, PAR30, PAR38. Each of these lamps has a medium screw base and serves as a direct replacement for a standard incandescent R- or PR-shaped lamp. They are fully dimmable, offer the whiter light of halogen, and are available in various beam angles from 9° spot to 55° wide flood. The newest versions of these halogen PAR lamps offer improved reflector and lens designs to provide smoother beam patterns that significantly reduce stray light and striations and makes them ideal for retail and display lighting applications. The molding of precisely calculated shapes into the lamp's reflector and/or the arrangement of the lenticules on the lens surface into specialized patterns has improved the performance of these unique optical systems over older lamp designs. Figure 15–6(a) shows the various halogen lamp bulb shapes that are available. Figure 15–6(b) shows the principle of the use of halogen technology applied to a PAR lamp.

Infrared Technology   Infrared (IR) technology is also being used to increase the efficiency of tungsten–halogen PAR lamps. All incandescent and halogen lamps produce a significant amout of IR energy, most of which is given

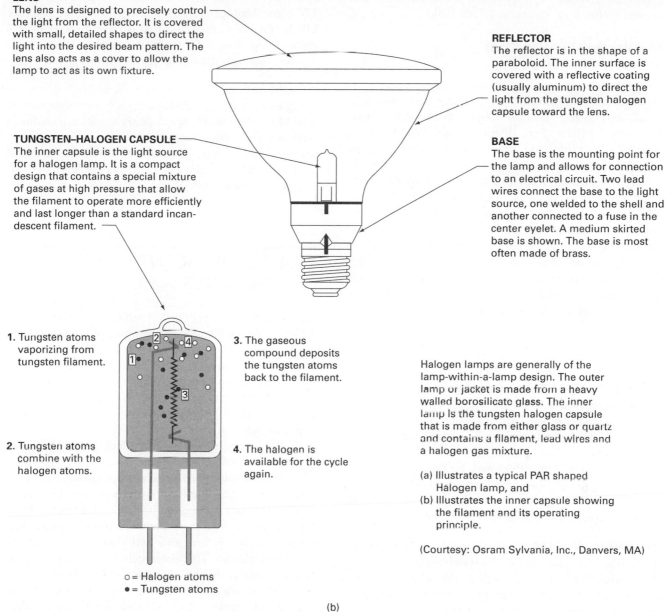

LENS
The lens is designed to precisely control the light from the reflector. It is covered with small, detailed shapes to direct the light into the desired beam pattern. The lens also acts as a cover to allow the lamp to act as its own fixture.

REFLECTOR
The reflector is in the shape of a paraboloid. The inner surface is covered with a reflective coating (usually aluminum) to direct the light from the tungsten halogen capsule toward the lens.

TUNGSTEN–HALOGEN CAPSULE
The inner capsule is the light source for a halogen lamp. It is a compact design that contains a special mixture of gases at high pressure that allow the filament to operate more efficiently and last longer than a standard incandescent filament.

BASE
The base is the mounting point for the lamp and allows for connection to an electrical circuit. Two lead wires connect the base to the light source, one welded to the shell and another connected to a fuse in the center eyelet. A medium skirted base is shown. The base is most often made of brass.

1. Tungsten atoms vaporizing from tungsten filament.

3. The gaseous compound deposits the tungsten atoms back to the filament.

2. Tungsten atoms combine with the halogen atoms.

4. The halogen is available for the cycle again.

Halogen lamps are generally of the lamp-within-a-lamp design. The outer lamp or jacket is made from a heavy walled borosilicate glass. The inner lamp is the tungsten halogen capsule that is made from either glass or quartz and contains a filament, lead wires and a halogen gas mixture.

(a) Illustrates a typical PAR shaped Halogen lamp, and
(b) Illustrates the inner capsule showing the filament and its operating principle.

(Courtesy: Osram Sylvania, Inc., Danvers, MA)

o = Halogen atoms
● = Tungsten atoms

(b)

■ FIGURE 15–6(b)

Halogen lamps are generally of the lamp-within-a-lamp design. The outer lamp or jacket is made from a heavy-walled borosilicate glass. The inner lamp is the tungsten–halogen capsule that is made from either glass or quartz and contains a filament, lead wires, and a halogen gas mixture. (Top) A typical PAR-shaped halogen lamp. (Bottom) The inner capsule showing the filament and its operating principle. (Courtesy: Osram Sylvania, Inc., Danvers, MA.)

off as excess heat. Newer versions of halogen PAR as well as MR16 lamps have a multilayered, IR-conserving coating on the inner halogen capsule. This coating allows the visible light to pass through it while reflecting heat back to the filament. Because the heat is reflected back to the filament within the lamp capsule, less energy is required to maintain the filament at its optimum operating temperature. This allows halogen lamps with IR-coated capsules to provide improved performance and offer energy savings compared with standard halogen capsule lamps.

Figure 15–6(b) shows the principle of the use of halogen capsule technology applied to a PAR lamp.

Multifaceted Pressed-Glass Reflector (MR) Lamps

These lamps are basically compact halogen lamps in wattages ranging from 20 to 60 W that are housed in a small, aluminized reflector. They are available in two basic sizes. MR16 (2-in. diameter) and MR11 (1⅜-in. diameter). Multifaceted pressed-glass reflector lamps are compact, easy to control, and are available in a wide variety of beam angles, from 8° (narrow spot) to 60° (very wide flood). Unlike conventional PAR lamps, which control the beam angle with a lens, MR lamps utilize the reflector to control the shape of the beam. Most versions are designed to operate at 12 V and are typically used on a transformer that may be integral with the lighting fixture or externally mounted. MR lamps are popular in interior display and accent lighting applications such as in retail, residential, museum, and office spaces.

Standard Lamps   Standard MR16 and MR11 lamps utilize a dichroic reflector, which allows the visible light to leave the front of the lamp while the infrared heat energy passes back through the reflector. Because a small percentage of the light passes out the back of the lamp along with much of the heat, these lamps can create a rainbow effect when used in gimbal-ring-type lighting fixtures. This effect may be desirable in some display lighting applications. These standard MR16 and MR11 lamps may exhibit some color shift over their life as the dichroic coating degrades ove time.

Aluminized MR16 Lamps   This family of lamps has an aluminized reflector that does not deteriorate over time. They offer highly stable color and crisp white light output throughout their service life. Unlike dichroic MR lamps, which are designed to send much of the heat out the back of the lamp, aluminized MR lamps direct most of the lamp's heat forward. This can help reduce premature socket and transformer failure and makes these lamps ideal for recessed fixtures and track lights with integral transformers. Because the reflector is completely opaque, there is no backlight or rainbow effect when mounted in gimbal ring fixtures. Some of these types are offered with a cover glass over the face of the lamp.

Titanium-Coated MR16 Lamps   Unlike standard dichroic MR lamps, in which the reflector coating can degrade and cause color shift, these lamps use a special hardened titanium oxide dichroic reflector coating that ensures constant color throughout the lamp's life. As with other dichroic-coated MR lamps, much of the heat generated is sent out the back of the lamp. The lamps incorporate a UV-control halogen capsule that significantly reduces the UV-B and UV-C radiation that is most associated with color fading. Some of these lamps are also available with a cover glass over the face of the lamp.

Infrared MR Lamps   The same IR technology that is being used in halogen PAR lamps is being incorporated into MR16 lamps. These lamps are available in reduced wattages and offer energy savings over conventional-wattage MR16 lamps. They also incorporate a hardened dichroic reflector and a UV-control capsule that significantly reduces the UV-B and UV-C radiation from the lamps.

15.4 FLUORESCENT LIGHT SOURCES

Fluorescent lamps were developed in France in the early 1930s by André Claude, the inventor of the neon lamp, and were initially manufactured in the United States in 1939. During the second half of the twentieth century, fluorescent lamps improved manyfold in performance, efficacy (see Figure 15–7(b)), color, life expectancy, and cost. They have become the chief light source in the world of lighting.

A fluorescent lamp contains electrodes at both ends of a tube that is filled with mercury vapor. When an electric voltage is impressed between the electrodes, ultraviolet energy is generated and converted to visible energy by the phosphor coating on the inside of the bulb. Based on the construction of their electrodes, fluorescent lamps may be divided into two classes—cold-cathode and hot-cathode. The construction of a typical hot-cathode lamp is shown in Figure 15–7(a). Most fluorescent lamps used for general lighting are of the hot-cathode variety. Cold-cathode lamps have larger electrodes operating at a lower temperature than the hot-cathode lamps and thus will last longer, up to 50,000 or more hours of rated life. They can be made in longer lengths and usually have higher initial cost. Examples of cold-cathode fluorescent lamps are those used in decorative signs similar to neon tubes, in architectural coves, and where lamp replacement is more difficult.

The phosphor coating on the inside of a fluorescent lamp is a mixture of many chemicals that emit visible light when excited by the ultraviolet energy (at 253.7 nm) generated by the mercury vapor. Different phosphors emit different colors. The basic phosphor in white fluorescent lamps is calcium halophosphate, which emits light in the range of 350–750 nm with a peak energy at 610 nm. Other phosphors commonly used are

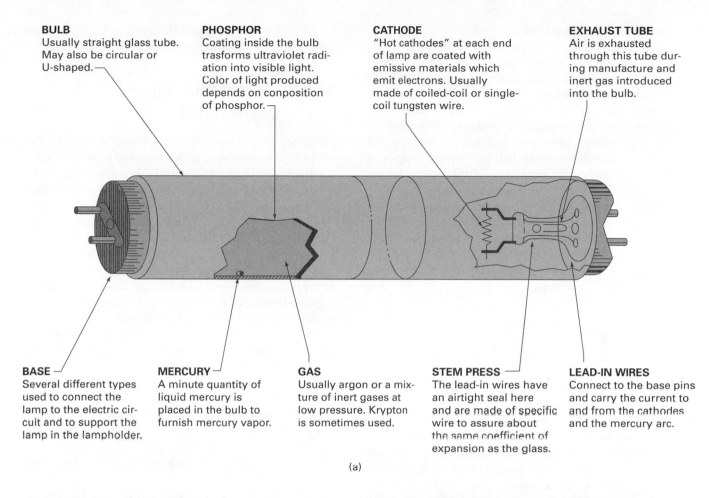

BULB
Usually straight glass tube. May also be circular or U-shaped.

PHOSPHOR
Coating inside the bulb trasforms ultraviolet radiation into visible light. Color of light produced depends on conposition of phosphor.

CATHODE
"Hot cathodes" at each end of lamp are coated with emissive materials which emit electrons. Usually made of coiled-coil or single-coil tungsten wire.

EXHAUST TUBE
Air is exhausted through this tube during manufacture and inert gas introduced into the bulb.

BASE
Several different types used to connect the lamp to the electric circuit and to support the lamp in the lampholder.

MERCURY
A minute quantity of liquid mercury is placed in the bulb to furnish mercury vapor.

GAS
Usually argon or a mixture of inert gases at low pressure. Krypton is sometimes used.

STEM PRESS
The lead-in wires have an airtight seal here and are made of specific wire to assure about the same coefficient of expansion as the glass.

LEAD-IN WIRES
Connect to the base pins and carry the current to and from the cathodes and the mercury arc.

(a)

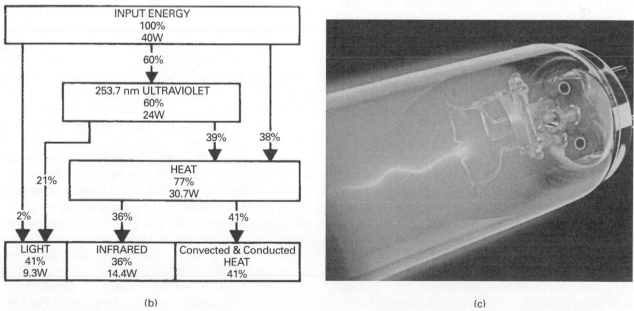

(b)

(c)

■ **FIGURE 15–7**

(a) Basic construction of a hot-cathode lamp. (Courtesy: Osram Sylvania Inc., Danvers, MA.) (b) Energy distribution of a typical 40-W cool white fluorescent lamp. (c) Electric arc across the electrodes. (Courtesy: General Electric Company, Nela Park, Cleveland, OH.)

cadmium borate (pink), calcium silicate (orange), calcium tungstate (blue), T5 and zinc silicate (green).

The mixing of phosphors in different proportions creates hundreds of white fluorescent lamps, such as cool white, warm white, white, daylight, and deluxe white; they vary slightly with each manufacturer. Rare-earth triphosphors in red, blue, and green represent the latest technology. Fluorescent lamps using triphosphors are more efficient than standard fluorescent lamps and offer excellent color rendition. Triphosphor lamps, especially T8, have become the standard for energy-efficient general lighting applications. To select the most appropriate light source, it is imperative that the lighting designer be knowledgeable about the SPD characteristics of these variations.

The following subsections present some characteristics of fluorescent lamps.

15.4.1 Shape and Size

Fluorescent lamps are generally tubular and are of varying lengths and diameters. The bulb is normally designated by the letter (T), indicating tubular, followed by a number indicating the maximum diameter in eighths of an inch. Thus T12 indicates a tubular bulb $1\frac{2}{8}$ in. in diameter, and T5 indicates a tubular bulb $\frac{5}{8}$ in. in diameter. Nontubular fluorescent lamps include U-shaped, circular (TC), compact fluorescent (CFL), and helicoid or twist lamps. The designations of common fluorescent lamp shapes and bases are shown in Figure 15–8.

15.4.2 Temperature Effect on Operation

The light output of a fluorescent lamp is very much dependent on the concentration of mercury vapor, which in turn depends on temperature. The effect of temperature on mercury vapor pressure manifests itself as variations in light output and color. Most fluorescent lamps, which are intended primarily for indoor use, have been designed so that their light output and luminous efficacy will reach optimum values at a bulb wall temperature of 38°C (100.4°F) and an ambient temperature of 25°C (77°F). In well-designed luminaires, this temperature is typically reached when the lamps are operated at rated power under usual indoor temperature. The effect of ambient temperature on the output of fluorescent lamps is considerable, and when the ambient temperature is near or below freezing, the lamp may cease to operate. Under such operating conditions, the lamp should be fully enclosed within a glass or plastic envelope, and ballasts specially designed for low-temperature operations should be used. The new generation of T5 and T5/HO

lamps for general lighting is designed to have peak light output at an ambient temperature of 35°C (95°F).

15.4.3 Efficacy and Lumen Depreciation

In fluorescent lamps, three main energy conversions occur in the process of generating light. Initially, electrical energy is converted into kinetic energy by accelerating charged particles. This kinetic energy is then converted to electromagnetic radiation (particularly UV) during particle collision. Lamp phosphors then convert this UV radiation to visible energy. During each of these conversions some energy is lost. In fact, only about 25 percent of the total electrical power input is converted to visible energy. One of the advantages of fluorescent lamps is their high efficacy. Depending on the type and rating, fluorescent lamps have a nominal efficacy from 50 to 104 lpw. Figure 15–7(b) illustrates the distribution of energy of a typical 40-W cool white rapid-start lamp. The lumen output of fluorescent lamps also depreciates substantially with time. For standard rapid-start lamps, the depreciation is about 20 percent, and for VHO (very high output) lamps, the depreciation is as high as 40 percent at the end of the rated life of the lamp. The use of triphosphors in T8 and T5 fluorescent lamps has improved their lumen maintenance characteristics. Lumen depreciation in these types can be as low as 5–10 percent. See Figure 15–9(c), comparing the depreciation characteristics of various fluorescent lamps.

Newer and more energy-efficient fluorescent lamps are filled with a mixture of two gases, argon and krypton, rather than only argon. These lamps are rated at 34 W rather than 40 W. The lamp has a transparent conductive coating to lower the starting voltage.

The introduction of T8 fluorescent lamps and instant-start electronic ballasts has revolutionized the lighting industry. The 4-ft T8 lamps utilize triphosphors and are designed specifically for use on electronic ballasts to optimize their efficacy. T8 lamps operated on electronic ballasts have become the standard for energy efficiency and good color rendition. The lamps have efficacies of up to 104 lpw and offer other distinct advantages. The instant-start circuit offers additional energy savings over conventional rapid-start systems, because no power is required to heat the lamps' cathodes.

These T8 lamps, using triphosphor technology, also offer excellent lumen maintenance in addition to superior energy efficiency. They are available in a wide range of color temperatures from 2700 K to 6500 K to suit a variety of applications. Although these lamps are available in conventional lengths from 18 in. to 8 ft, they do require ballasts that are different from those

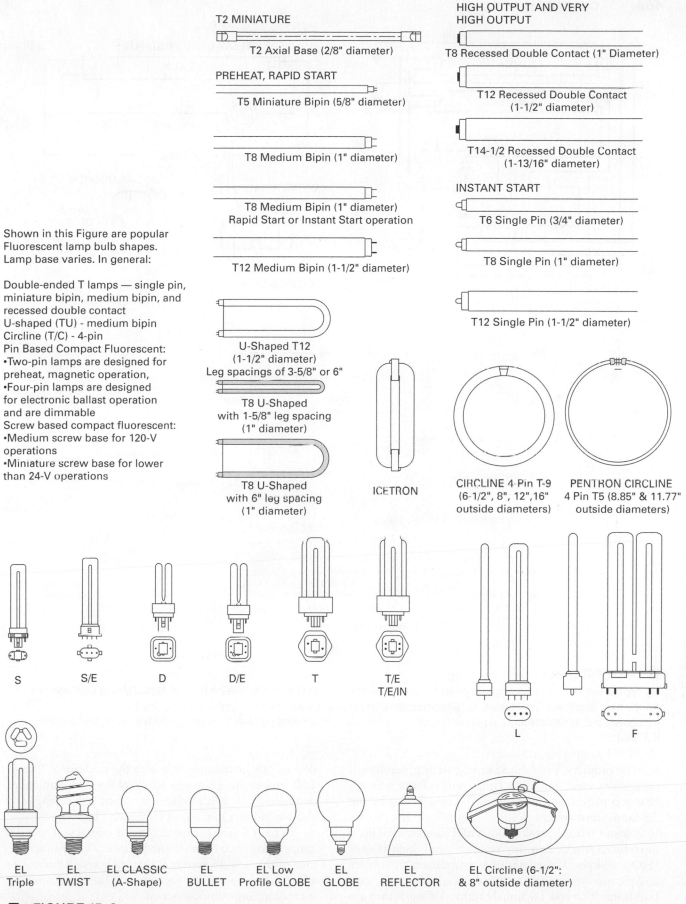

Shown in this Figure are popular Fluorescent lamp bulb shapes. Lamp base varies. In general:

Double-ended T lamps — single pin, miniature bipin, medium bipin, and recessed double contact
U-shaped (TU) - medium bipin
Circline (T/C) - 4-pin
Pin Based Compact Fluorescent:
•Two-pin lamps are designed for preheat, magnetic operation,
•Four-pin lamps are designed for electronic ballast operation and are dimmable
Screw based compact fluorescent:
•Medium screw base for 120-V operations
•Miniature screw base for lower than 24-V operations

T2 MINIATURE
T2 Axial Base (2/8" diameter)

PREHEAT, RAPID START
T5 Miniature Bipin (5/8" diameter)

T8 Medium Bipin (1" diameter)

T8 Medium Bipin (1" diameter)
Rapid Start or Instant Start operation

T12 Medium Bipin (1-1/2" diameter)

U-Shaped T12
(1-1/2" diameter)
Leg spacings of 3-5/8" or 6"

T8 U-Shaped
with 1-5/8" leg spacing
(1" diameter)

T8 U-Shaped
with 6" leg spacing
(1" diameter)

ICETRON

HIGH OUTPUT AND VERY HIGH OUTPUT
T8 Recessed Double Contact (1" Diameter)

T12 Recessed Double Contact
(1-1/2" diameter)

T14-1/2 Recessed Double Contact
(1-13/16" diameter)

INSTANT START
T6 Single Pin (3/4" diameter)

T8 Single Pin (1" diameter)

T12 Single Pin (1-1/2" diameter)

CIRCLINE 4-Pin T-9
(6-1/2", 8", 12",16"
outside diameters)

PENTRON CIRCLINE
4 Pin T5 (8.85" & 11.77"
outside diameters)

S

S/E

D

D/E

T

T/E
T/E/IN

L

F

EL
Triple

EL
TWIST

EL CLASSIC
(A-Shape)

EL
BULLET

EL Low
Profile GLOBE

EL
GLOBE

EL
REFLECTOR

EL Circline (6-1/2":
& 8" outside diameter)

■ **FIGURE 15–8**

Shape and base of fluorescent lamps. (Courtesy: Osram Sylvania, Inc., Danvers, MA.)

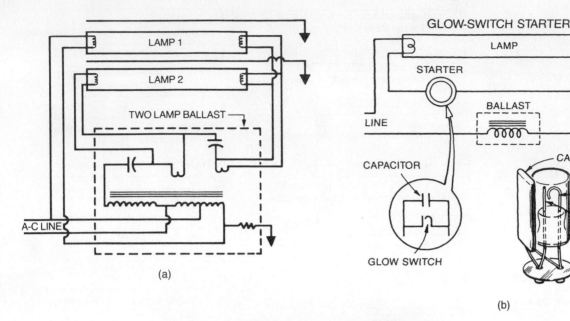

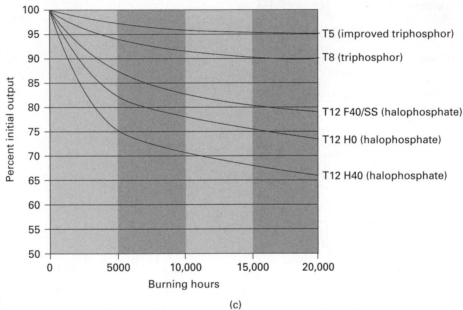

■ FIGURE 15–9

(a) Typical two-lamp series-sequence rapid-start circuit. (Reproduced with permission from Philips Lighting Company.)

(b) Preheat circuit with a glow switch. (Reproduced with permission from Philips Lighting Company.)

(c) Lumen depreciation based on operating on suitable ballast at 3 hours per start. (Reproduced with permission from IESNA.)

used by ordinary T12 lamps. Energy-saving retrofit opportunities exist, but both lamps and ballasts must be changed to realize the efficiency gains afforded by the T8 fluorescent system.

More recently, two families of T5 lamps have been introduced that, include medium- and high-output (HO) versions. These lamps have miniature bipin bases and feature metric lengths that are slightly shorter than traditional T12 and T8 lamp lengths. Although they are not directly interchangeable with the traditional T8 and T12 lamps, the T5 lamps are ideal for new luminaires designed to fit easily into U.S. standard grid ceilings and also for indirect luminaires for general lighting.

The T5 lamps have the same efficacy as the T8 lamps being used in the United States. The main advantage of the new T5 system is that their smaller diameter affords the opportunity for reducing luminaire size and increasing luminaire efficiency.

Although the high-output T5 lamps are not as efficient as the medium- output T5 lamps, they offer higher lumen packages than have been available in T12 or T8 lamps. The 4-ft, 54-W T5/HO lamps have a peak light output rating of 5000 lm. They are ideal for indirect lighting applications. Both medium- and high-output versions are designed for peak light output at 35°C (95°F) as opposed to the 25°C (77°F) peak light output of T12 and T8 fluorescent lamps. Both systems utilize rate-earth phosphors that offer excellent color rendition (82 CRI) and high lumen maintenance (93 percent). Like the domestic T8 lamps, the T5 and T5/HO lamps were specifically designed to operate on electronic ballasts and offer another energy-efficient option for lighting system designers and users.

15.4.4 Rated Life

Fluorescent lamps have an excellent rated life. The popular types of fluorescent lamps are rated between 5000 and 30,000 hours. Because some of the emissive coating is lost from the electrodes during each start, the number of starts (on/off) influences lamp life. Therefore, the life of the lamp is affected by the burning cycle, as measured by hours per start. The rated average life of fluorescent lamps is usually based on 3 hours of operation per start. At 3 hours per start, rapid-start lamps can have a rated life exceeding 24,000 hours. New electronic ballasts have been introduced for T8 and T5 lamps that feature programmed rapid start. This starting method makes lamp life less dependent on starting frequency and makes them ideal for use with occupancy sensors. This is about 30 times the rated life of general-service types of incandescent lamps.

15.4.5 Color Temperatures

The color produced by fluorescent lamps depends on the blend of phosphors. The apparent color of the light generated by fluorescent lamps has many variations, ranging from warm (rich in red) to cool (rich in blue). Color varies among manufacturers. The color temperature can be varied between 2500 K and 6500 K by changing the concentration of the phosphor components. A representative color temperature rating of fluorescent lamps is given in Table 15–3, and the relative positions of these colors on the CIE chromaticity diagram are shown in Plate 12(d).

15.4.6 Color Rendering

With the rapid improvement of fluorescent technology, the color-rendering indices of some fluorescent lamps are nearly as good as those of incandescent lamps. An

TABLE 15–3
Apparent color (color temperature) of selected fluorescent lamps

Lamp Designation	Color Temperature Range, K
Triphosphor (warm)	3000
Halophosphor (warm white)	3000
Triphosphor (white)	3500
Triphosphor (cool white)	4100
Triphosphor (cool)	5000
Triphosphor (daylight)	6500

increasingly popular approach to blending phosphors for fluorescent lamps is the triphosphor system. Three highly efficient narrow-band rare-earth activated phosphors have been developed that show emission peaks in the blue, green, and red regions of the visible spectrum. When a mixture of these premium color phosphors is used in lamps, the result is both high color rendering and good efficiency. Whereas standard fluorescent lamps have ratings from 57 to 80, specially formulated triphosphor lamps have a CRI from 75 to 85 and also deliver higher lumen per watt performance. Plate 13 illustrates the color-rendering effect of several fluorescent lamps.

15.4.7 Ballasts

Fluorescent lamps have so-called negative electrical resistance. That is, once an arc is struck across the lamp, the ionized mercury vapor becomes increasingly more conductive, and thus more current will flow until a lamp burns out. For this reason, fluorescent as well as HID lamps must be operated on a ballast, which serves both as a transformer to boost the voltage at the lamp terminals and as a choke to limit the maximum flow of current. There are four basic types of ballast:

1. *Magnetic type.* This is the conventional electromagnetic core-and-coil type of ballast operating at 60 Hz with secondary voltage between 200 and 700 V, depending on the length of the lamp. A capacitor is provided in high-power-factor (HPF) models, and a thermal cutout is provided in P-rated models.
2. *Hybrid magnetic type.* This is basically a magnetic ballast, except that it has a built-in electronic switching device to save energy by disengaging the cathode current after the lamp or lamps are started.
3. *Hybrid electronic type.* This is a combination electronic and electromagnetic type of ballast that consists of an input electromagnetic interference (EMI) filter, a rectifier to convert standard-frequency (50–60 Hz) AC into DC, an inverter (oscillator) to

convert DC to high-frequency AC (20–50 kHz), a capacitor to correct the power factor, and an output transformer to provide the operating voltage and current limitation to the lamps.

4. *Electronic type.* This is the newest type of ballast with all-electronic components. In addition to the rectifier and inverter, it incorporates a DC power preconditioner that provides power factor correction and a constant DC source to power the inverter. Models with integrated circuit and feedback controls can have additional control functions, such as manual and automatic dimming, detection of occupancy, and interfacing with the energy management system. Electronic ballasts are quiet in operation, low in weight, and cool in temperature, and they have the highest ballast efficiency factor (BEF). As their cost becomes more competitive they will undoubtedly displace the electromagnetic type of ballast for all major lamp sizes in the 30- to 200-W range. Small linear preheat fluorescent lamps probably will continue to use the electromagnetic type for some years to come. Figure 15–10(a) is a block diagram of an electronic ballast. Electronic ballasts should meet FCC standards on radio frequency interference (RFI) or EMI, IEEE/ANSI requirements on harmonics, and UL requirements on safety. These requirements are too detailed to be discussed here.

15.4.8 Ballast Factor

The ballast factor (BF) is the ratio of the light output produced by lamps operating on a commercial ballast to the light output of the same lamps operating on a standard reference ballast in the testing laboratory. The rated light output measurement of a fluorescent lamp is always made on a reference ballast. If a ballast has a BF of 0.95, it means that the actual light output of the lamps connected with this commercial ballast will be only 95 percent of the manufacturer's published lumen output. In general, ballasts certified by the Certified Ballast Manufacturers (CBM) or Underwriters Laboratories (UL) must have a minimum BF of 0.85. Today, electronic ballasts are available with BFs from 0.71 to 1.20. The lighting designer should be aware of BF and should select the ballast that is appropriate for the application in terms of light output and power consumption. The BF per unit can be expressed as in Eq. (15–1):

$$\text{BF} = F_a/F \qquad (15\text{–}1)$$

where F_a = actual light output of the lamps using the commercial ballast (lm)

F = published light output listed in the lamp manufacturer's catalog (as tested with a reference ballast in the lab)

The actual light output F_a of the lamps used in the calculation of illuminance is

$$F_a = F \times \text{BF} \qquad (15\text{–}2)$$

15.4.9 Ballast Efficiency Factor

The ballast efficiency factor (BEF) is the ratio of the ballast factor to the power input to the specific ballast-and-lamp combination. The National Appliance Energy

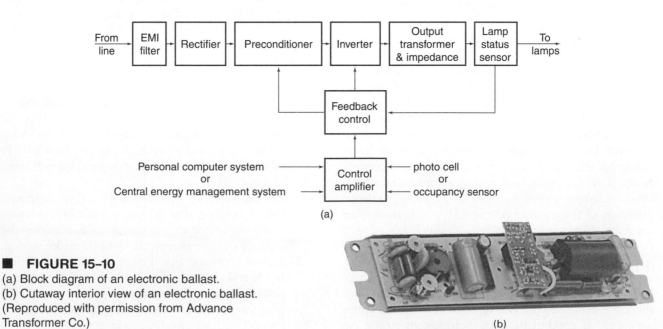

(a)

■ **FIGURE 15–10**
(a) Block diagram of an electronic ballast.
(b) Cutaway interior view of an electronic ballast.
(Reproduced with permission from Advance Transformer Co.)

(b)

Conservation Act (NAECA) establishes the minimum energy efficiency standards for ballasts as

$$BEF = BF \times 100/P \qquad (15\text{--}3)$$

where BF = ballast factor

P = power input to the ballast-and-lamp combination (W)

For example, NAECA specifies that the BEF shall not be lower than 1.805 for a single 40-W, 4-ft T12 lamp ballast and 1.060 for a two-lamp ballast combination. Unfortunately, no correlation can be established between these BEF values.

15.4.10 Lamp Ballast Efficacy Factor (LBEF)

A more useful factor is the LBEF. It can be used for evaluating any lamp–ballast combination. In lumens/watt (lpw):

$$LBEF = F_a/P = F \times BF/P \qquad (15\text{--}4)$$

where F_a = actual light output of a lamp–ballast combination (lm)

P = total lamp/ballast power (W)

Example 15.3 A 4 ft T8, 32-W, RS lamp is rated for 2950 lm. If the lamp is wired with an electronic ballast having a BF = 0.87 and a power input of 31 W, then,

$$LBEF = F \times BF/P = (2950 \times 0.87)/31 = 82.8 \, lpw$$

This value can be compared with any other lamp–ballast combinations. A higher LBEF always represents a more efficient combination for lamps of the same size or different sizes.

15.4.11 Dimming of Fluorescent Lamps

Like incandescent lamps, fluorescent lamps can also be dimmed. Dimming a fluorescent lamp differs from dimming an incandescent lamp in two ways. First, fluorescent dimmers do not provide dimming to zero light output. Second, when fluorescent lamps are dimmed, the color temperature does not vary substantially over the dimming range. Dimming is achieved by reducing the effective lamp current. Although early magnetic dimming ballasts achieved dimming by lowering the primary voltage to the ballast transformer, the vast majority of available fluorescent dimming ballasts are of the soild-state (electronic) type. Electronic dimming ballasts are generally more efficient, quieter, and less bulky than their autotransformer predecessors. Furthermore, they can substantially reduce lamp flicker.

15.4.12 High-Frequency Operation

With the advent of highly efficient electronic ballasts, the fluorescent lighting industry is rapidly progressing toward high-frequency operation of fluorescent lamps. A fluorescent lamp's efficacy increases with its frequency. As shown in Figure 15–11, the efficacy steadily increases to 111 percent from the standard utility frequency (50–60 Hz) to about 100 kHz. There is no gain from 20 to 100 kHz. An aircraft's electrical system is generally designed for 400 Hz in order to reduce the weight of power equipment. At 400 Hz, fluorescent lights will gain about 6 percent more efficacy.

15.4.13 Types of Fluorescent Lamps

Fluorescent lamps may be divided into two varieties: hot-cathode and cold-cathode. The major difference between the two is in the construction and operating temperature of their cathodes. Because most fluorescent lamps operate under the hot-cathode principle, their identification as such is often omitted. The following subsections briefly describe commercially available fluorescent lamps.

Preheat Lamps

A manual or automatic starter switch is required in preheat lamps. In preheat circuits, the lamp electrodes are heated before application of the high voltage across the lamp. Lamps designed for such operation have bipin bases to facilitate electrode heating. Many preheat-starting compact fluorescent lamps have the starting devices built into the lamp base. Ballasts are available to operate some preheat lamps without the use of starters. See Figure 15–9(b) for the preheat circuit with a glow switch.

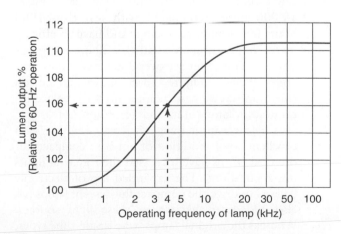

■ **FIGURE 15–11**
Percent efficacy of fluorescent lamps versus operating frequency.

Instant-Start Lamps

These lamps depend solely on the application of a high voltage across the lamp without a starter. The ballast provides a high enough voltage to strike the arc instantly. Because no preheating of electrodes is required, instant-start lamps need only a single contact at each end. Thus, a single pin is used on most instant-start lamps. These are commonly called *slimline* lamps. A few instant-start lamps use bipin bases with the pins connected internally.

Rapid-Start Lamps

These lamps are designed for rapid-start operation and typically have low-resistance cathodes. Normally, the cathodes are heated continuously by the application of cathode voltage while the lamps are in operation. Ballasts for rapid-start lamps have separate windings to provide continuous heating voltage for the lamp electrodes, as shown in the circuit diagram in Figure 15–9(a). The lamp starts in less than 1 second, nearly instantaneously. Following are a number of variations:

- Straight-tube rapid-start
- U-tube rapid-start
- Circline rapid-start
- High-output rapid-start
- Very high output rapid-start

Compact Lamps

Two kinds of compact fluorescent lamps can be used to replace low-efficacy incandescent lamps:

- *Compact fluorescent* (CFL) *lamps* with external ballasts are available from 5 W up to 70 W. Owing to their compact size, they are popular for illuminating general spaces that previously were lit by incandescent fixtures. CFL lamps may last as long as 10,000 hours, compared with less than 1000 hours for incandescent lamps and have an efficacy up to 75 lpw compared with 15–20 lpw for incandescent. Thus, they are over 300 percent more energy efficient. Figure 15–8 shows an Edison (medium screw) base adapter used to replace incandescent lamps in incandescent luminaires.

    The compact size, efficiency, long life, and excellent color rendition of pin-base compact fluorescent lamps have increased their general lighting applications. The expanded use of electronic ballasts to drive the compact fluorescent lamps has also added to the use of these light sources. Whereas the original lamps were single-tube types, newer CFLs are available in double- and triple-tube varieties that offer higher lumen packages while remaining compact in size. Compact fluores-

cent lamps are now available with lumen packages up to 5200 lm. Figure 15–8 shows the bulb shapes and base types for most of the CFL lamps available today.

    Owing to their efficiency and compact size, these light sources are being used in such general lighting applications as retail lighting, office lighting, and lighting for hallways and corridors. Owing to their color consistency and instant-restart characteristics, the higher-wattage CFL lamps are also being used instead of low-wattage metal halide lighting in many applications.

    Many of the new, higher-wattage lamps also incorporate amalgam technology. The use of amalgam as a mercury vapor pressure control allows these higher-wattage lamps to have a more consistent light output over a wider range of operating temperatures. This increases their lumen output performance in higher-temperature applications such as multilamp-recessed downlights and low- to medium-bay industrial fixtures.

- *Integrally ballasted lamps* with medium screw base and integral ballasts have also seen dramatic growth as direct replacement for standard incandescent lamps. They are now available in wattages from 15 to 23 W with lumen packages up to 1450 lm. Their long life and high efficiency make them an ideal alternative for standard incandescent lamps in several applications. Several new lamp styles are also available, such as triple tube, twist, and globe shapes for general lighting. There are also reflector shapes that can be used in recessed downlighting applications. Figure 15–8 shows several of the currently available integrally ballasted compact fluorescent lamps.

Specialty Lamps

There are numerous varieties of specialty fluorescent lamps:

- *Black light lamps* produce energy in the near-ultraviolet range.
- *UV lamps* produce ultraviolet energy below 320 nm for use as germicides.
- *Plant-growth lamps* are designed especially to stimulate photosynthesis.
- *Cold-cathode lamps* are phosphor-coated lamps filled with mercury vapor and argon gas that operate at from 700 to 1000 V. The cathodes are thimblelike in shape and can emit sufficient electrons at a much lower temperature than hot-cathode lamps. Cold-cathode lamps are generally small in diameter (T4 to T8) and can be bent into various shapes up to 20 ft in length. They have low efficacy

(not much above half that of cathode lamps) and a long rated life (over 40,000 hours). With different gas fills, they can also produce different colors. Cold-cathode lamps are used primarily for decorative and sign lighting applications. Cold-cathode lamps are often used in lieu of neon lights in the interiors of buildings.

- *Neon lamps* are noncoated cold-cathode lamps operating at extremely high voltages (exceeding 5000 V). The lamp tube is small in diameter and can easily be bent into any shape. Different gases generate different colors, e.g., neon emits red, and argon and mercury together emit blue and, combined with a blue-absorbing glass tube, will emit green. Other combinations of gas fills and glass will result in various colors. Neon lights are used primarily for signs and decorative applications.
- *Subminiature lamps* constitute a family of tiny fluorescent lamps with a 5–7-mm (T2 ½) diameter with ratings from 1 to 3 W. They are used principally for backlighting liquid crystal display signs and for lighting instruments. The hot-cathode versions from 6 to 13 W may also be used for display accent, and as undercabinet lighting.
- *Reflector lamps* have an internal reflector to cover up a portion of the bulb; thus, they reflect the light to the open-aperture portion of the lamp at a higher intensity than general-service lamps. Reflector lamps may be considered as a lighting fixture–lamp with built-in reflector. They are useful for display and in cove-lighting applications.

15.5 HIGH-INTENSITY-DISCHARGE LIGHT SOURCES

High-intensity-discharge (HID) lamps are a family of lamps that incorporate a high-pressure arc tube within the lamp envelope. The tube is filled with metallic gas, such as mercury, argon, or sodium. When the gas is fully vaporized, owing to the flow of electrical current, the arc tube will have a high internal pressure of around 2 to 4 atmospheres (200 to 400 kilopascals). The concentrated light power generated in the arc tube has considerably high intensity (candela/sq in.), from which the name of the lamp is derived. In the mercury type of HID lamp, the mercury gas generates a broader spectrum consisting of five principal visible spectral lines (404, 435, 546, 577, and 579 nm) in lieu of the single ultraviolet wavelength at 257 nm that low-pressure fluorescent lamps produce. The envelope (bulb) of HID lamps may be clear or coated with phosphor to improve color rendition. Figure 15–12 shows the available bulb shapes and base configurations of typical HID light sources. The ANSI designation of lamp shape and sizes is similar to that used for other types of light sources. In addition, some designations specific for the HID lamps are BT—bulbous tubular, ED—ellipsoidal, and ET—ellipsoidal tubular. Lamps with screw bases have one lead-in wire soldered or welded to the center contact and the other soldered or welded to the upper rim of the base shell.

15.5.1 Types of HID Lamps

There are three major classes of HID lamps.

- *Mercury vapor lamps* contain only mercury vapor and produce a predominantly bluish white light (5500–5800 K). Mercury vapor lamps are used mostly for industrial and outdoor applications, and they have an efficacy between 40 and 70 lpw. Figure 15–13(a) shows the typical construction of a mercury vapor lamp.
- *Metal halide lamps* contain mercury vapor and other halides to improve both their efficacy and their color-rendering characteristic. Typical halides include scandium, sodium oxide, dysprosium, indium oxide, and other rare earth iodides. These lamps have a color temperature around 3200 K to 4000 K, compared with 6000 K, for mercury lamps. Because they have an extremely high lamp efficacy—up to 120 lpw, metal halide lamps are applicable to most indoor lighting needs. Figure 15–13(b) shows the construction of a metal halide lamp.

  The high efficiency and very good color-rendering properties of metal halide lamps make them an excellent choice for applications such as sports lighting, parking lot lighting, and in many retail applications. Metal halide lighting systems are now becoming more popular than high-pressure sodium systems for exterior applications where white light is desired.

  The introduction of pulse-start ballast systems and protected lamps that can be operated in unshielded luminaires has expanded the application of metal halide lamps. Metal halide lamps are now available in wattages as low as 39 W, and there are several types from 50 W to 150 W that incorporate medium screw bases.

  The latest development in metal halide technology is the use of ceramic arc tubes. These arc tubes have a more precise shape and size, which allows the lamps to have much better color consistency lamp to lamp, as well as over lamp life. Ceramic-arc-tube metal halide lamps are available

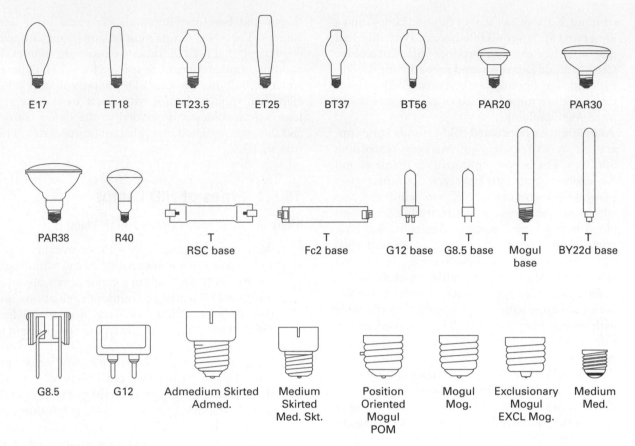

■ FIGURE 15–12

Shapes and bases of popular HID light sources. (Courtesy: Osram Sylvania Inc., Danvers, MA.)

in a variety of shapes and sizes, including PAR30 and PAR38 directional types. These new ceramic metal halide lamps are being used in such applications as retail lighting and general interior lighting where higher-lumen packages and excellent color rendition is desired. They are available in color temperatures from 2900 K to 4200 K and offer color-rendering indices from the mid 70s to the low 90s.

■ *High-pressure sodium (HPS) lamps* contain xenon as a starting gas and an amalgam of sodium and mercury that is partially vaporized when the lamp attains its operating temperature. HPS lamps are constructed with two envelopes: an inner envelope (arc tube) made with material that has electrical resistance to sodium and an outer envelope designed to protect the arc tube. Some HPS lamps are also available with diffuse coatings on the outer bulb. Figure 15–13 shows the construction of HPS lamps.

Although HPS lamps radiate energy across the entire visible spectrum, yellow-orange wavelengths from 550 to 600 nm predominate. See Plate 12(a) for a typical spectral distribution. HPS lamps may be described as golden white. Their color temperatures vary from 2000 to 4000 K. Increasing the sodium pressure increases the percentage of red radiation and thus improves color rendition, but at the sacrifice of life and efficacy. Figure 15–13(d) shows HPS lamps in sizes from 35 and 100 W. These lamps have a CRI of 80 as compared with standard HPS lamps, which have a CRI of 20. Plate 12(b) and (c) compares the color rendering of a high-CRI HPS lamp with that of an incandescent lamp.

HPS lamps have the highest luminous efficancy of all HID lamps, ranging from 60 to 140 lpw. Most general-lighting HPS lamps also carry a life rating of 24,000+ hours. These two characteristics have made HPS lamps a popular choice for street and roadway lighting, when high color rendition has not been desired. One of the drawbacks of high-pressure sodium lamps is that they tend to cycle on and off at the end of lamp life. This has been a maintenance nuisance in many street lighting

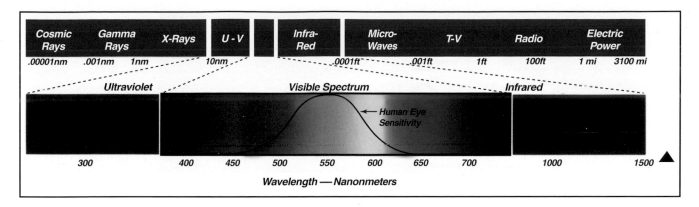

Plate 1 The electromagnetic spectrum. (Courtesy: General Electric Company, Nela Park, Cleveland, OH.)

(a)

(b)

Plate 2 (a) The statement reads, "Contrast improves visual acuity and recognition." Without contrast, visual acuity is lost. Too much contrast may create visual discomfort. (b) Visual acuity is critical in high-speed sports. A white baseball in front of a dark background is easily visible, compared with a baseball on a white background. A baseball against a dark blue background shows up even more clearly because there is luminance contrast as well as color difference.

Plate 3 Visual acuity is even more critical in tennis, where a ball is being hit at close range in random directions. The tennis court surface and the backdrops (vertical surfaces) should have high luminance contrast and color differences with the yellow or white tennis balls.

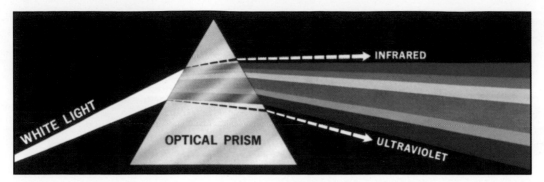

Plate 4 A prism can bend white light into its component colors. The effect of the prism is to bend shorter wavelengths more than longer wavelengths, separating them into distinctly identifiable bands of color. The colors can be recombined into a beam of white light by accurately orienting a properly designed second prism to intercept the dispersed colored light rays. Note that the prism is useful for separating infrared and ultraviolet rays, as well as visible rays. (Courtesy: General Electric Company, Nela Park, Cleveland, OH.)

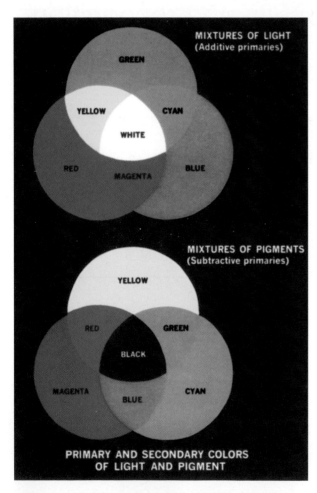

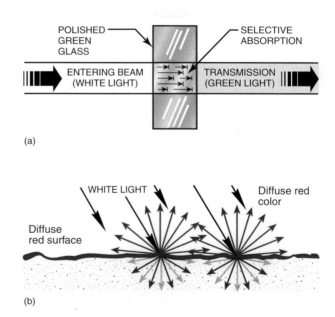

Plate 5 The primary colors of light are blue, green, and red. These are called *additive primaries*. Adding equal energy levels of the additive primaries produces white light. The primary colors of pigments are magenta, yellow, and cyan. These are called *subtractive primaries*. Adding equal amounts of subtractive primaries produces a black pigment. (Courtesy: General Electric Company, Nela Park, Cleveland, OH.)

Plate 6 (a) White light passing through a green glass becomes green light because the green glass absorbs all color wavelengths except green. (b) A red painted or naturally red surface reflects only red wavelengths and absorbs all other wavelengths. (Courtesy: General Electric Company, Nela Park, Cleveland, OH.)

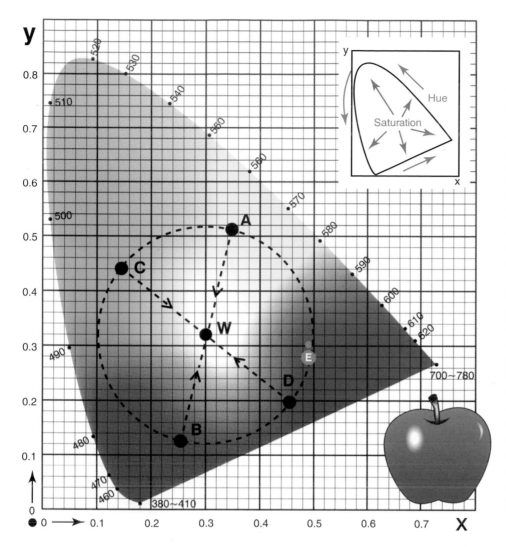

Plate 7 The CIE (x, y) chromaticity diagram is a diagram of the hue (color) and chroma (saturation) of spectrum colors on a selected neutral value (lightness-darkness plane). The diagram is plotted on the x- and y-coordinates with saturated colors at the perimeter. The white light at point W may be produced by mixing equal lumens of light sources A and B, or light sources C and D, or the combination of any other sources opposite point W. The specifications of the light sources represented at points A, B, C, D, and W can be read directly from the x- and y-coordinates. For example, light source D is a light with CIE chromaticity of $x = 0.46$ and $y = 0.2$. Point E on the chromaticity diagram represents the color of the apple with color values of $x = 0.49$ and $y = 0.30$. (Reproduced with permission from Minolta Corporation, Instrument Systems Division.)

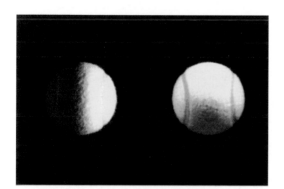

Plate 8 Light orientation is important to bring out the modeling (the 3-dimensional image) of an object. Shown at the left is a tennis ball illuminated from the right side only. The ball appears to be semispherical instead of a full sphere. When light comes from multiple directions with luminous differences between them, the 3-dimensional form of the ball is revealed, as shown in the photo on the right.

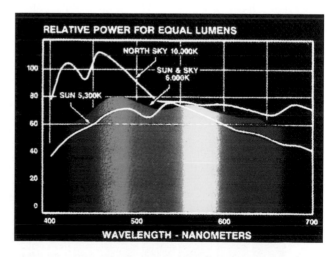

Plate 9 The distribution of light energy varies with the time of day. Shown are three variations: sun and sky at 6000K, sun at 5300K, and north sky at 10,000K. (Courtesy: General Electric Company, Nela Park, Cleveland, OH.)

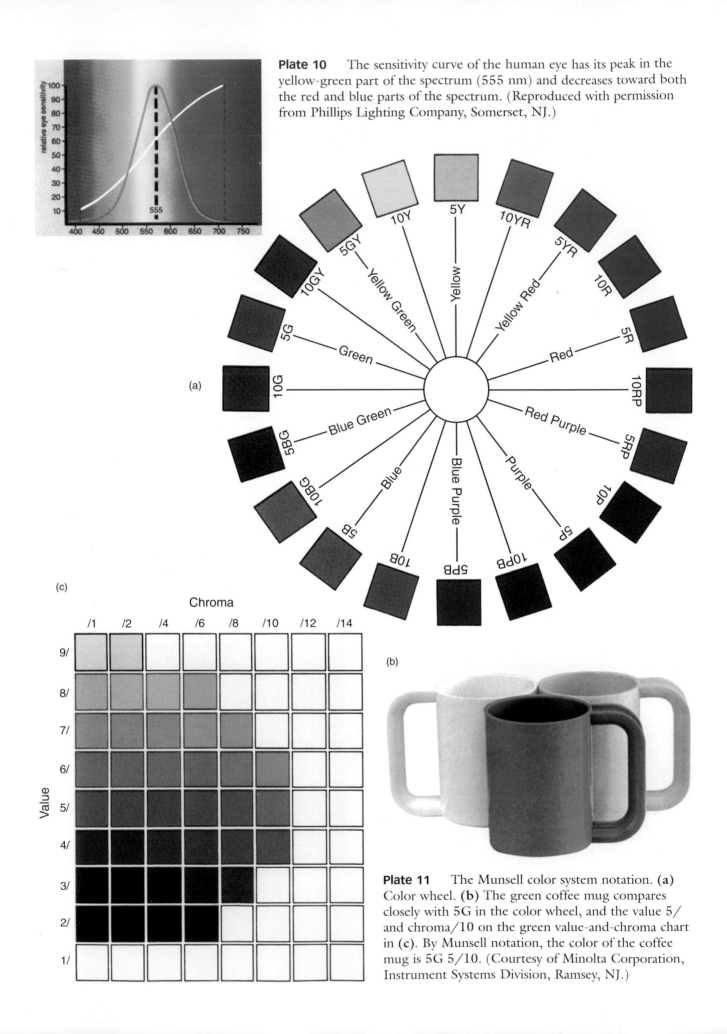

Plate 10 The sensitivity curve of the human eye has its peak in the yellow-green part of the spectrum (555 nm) and decreases toward both the red and blue parts of the spectrum. (Reproduced with permission from Phillips Lighting Company, Somerset, NJ.)

Plate 11 The Munsell color system notation. (a) Color wheel. (b) The green coffee mug compares closely with 5G in the color wheel, and the value 5/ and chroma/10 on the green value-and-chroma chart in (c). By Munsell notation, the color of the coffee mug is 5G 5/10. (Courtesy of Minolta Corporation, Instrument Systems Division, Ramsey, NJ.)

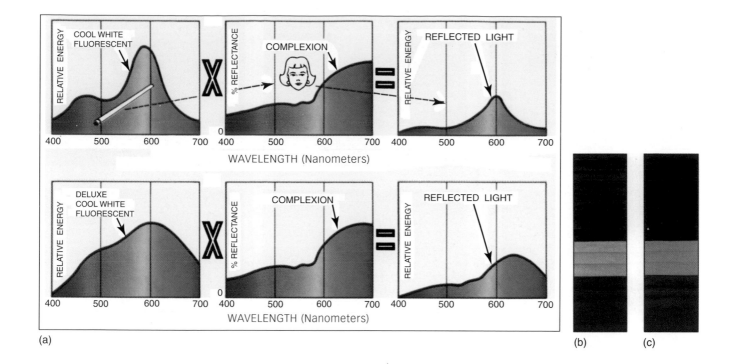

(a)

(b)   (c)

Plate 12   (a) How a light source affects color appearance. The illustration shows how spectral energy distributions (SEDs) of Cool White (CW) and Deluxe Cool White (CWX) lamps are modified by the reflectance characteristics of a typical human complexion. Under the deluxe lamp, much more red is reflected to the eye. The result is a healthier, more natural appearance. This is true even though CW and CWX lamps have approximately the same whiteness. The Deluxe Warm White lamp, which deemphasizes blues somewhat, is even more flattering to complexions. (b) The color of a multiple-color fabric as illuminated by a 150-watt incandescent flood lamp. (c) The color of the same as in (b) but illuminated by a 50-watt, high CRI, high-pressure sodium lamp by Phillips. (d) The position of various "white" light sources are indicated on the CIE chromacity diagram. They range from yellowish white to reddish white, but none is "pure (equal energy)" white. (Courtesy: General Electric Company, Nela Park, Cleveland, OH.)

(d)

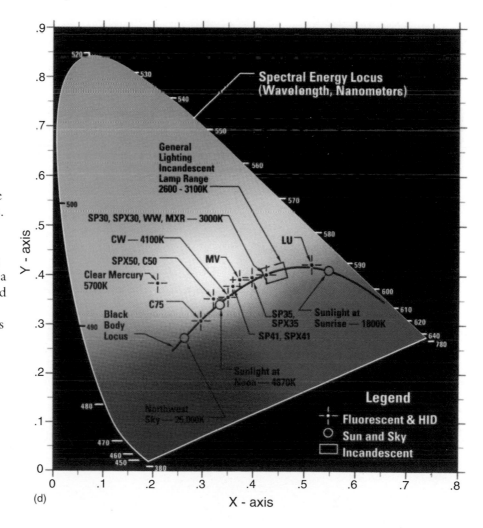

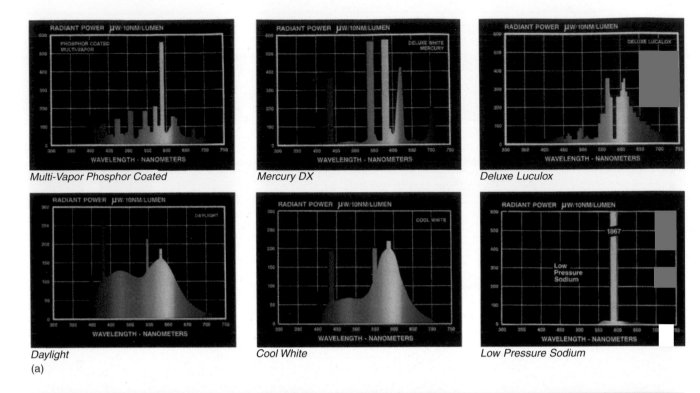

Multi-Vapor Phosphor Coated

Mercury DX

Deluxe Luculox

Daylight

Cool White

Low Pressure Sodium

(a)

(b)

Plate 13 (a) Spectral energy distribution (SED) of fluorescent and LPS lamps. [Note: Multi-vapor is metal halide, and Luculox is HPS of General Electric lamps.] (b) Effect of the light source on the complexion of a model. The pictures are intended, within the limits of modern high-speed printing, to give a good indication of the differences between SP and SPX lamps, at various chromaticities. [Note: SP and SPX are designations of General Electric. The number indicates the color temperature rating.] (Courtesy: General Electric Company, Nela Park, Cleveland, OH.)

Plate 14 Lighting systems and their effect on office environments. (**a**) *Indirect system*: High luminance on the ceiling; soft luminance contrast on furniture; low glare on the video display screen. (**b**) *Parabolic diffusers*: Uniform luminance; wall luminance varies; video display screen has more spill light. (**c**) *Prismatic lens*: Walls and desk are evenly illuminated; video display screen is washed out. (**d**) *High-cutoff louvers*: Walls are dark with pronounced scalloping; poor vertical illumination; less horizontal uniformity; video display screen has less spill light. (**e**) *Directional light produces shadows*. The degree of shadow (soft, harsh, strong) depends on the luminaire's orientation and distribution. (**f**) Any high luminance will appear on the video display screen and will follow the law of reflection. (Courtesy of Peerless Lighting Corp., Berkeley, CA.)

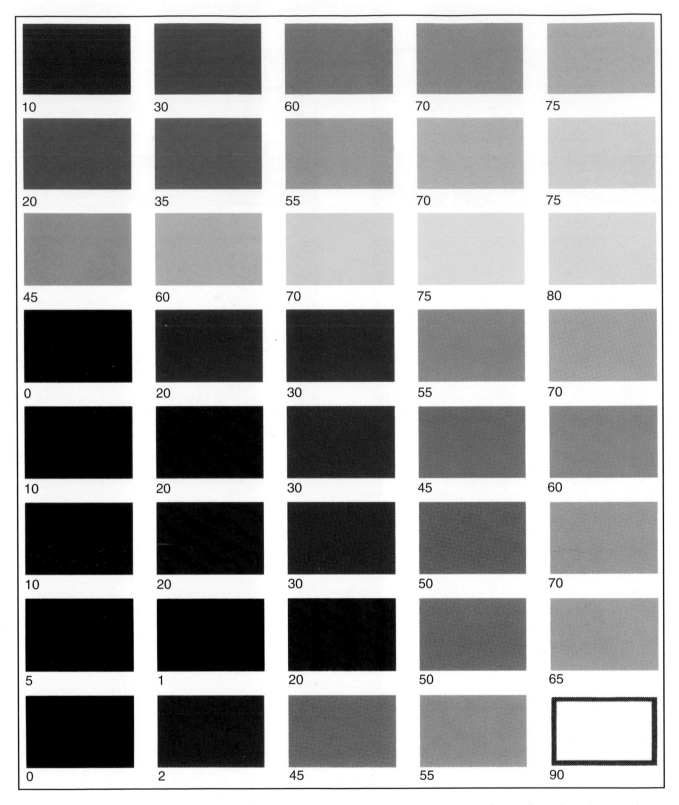

Plate 15 Color chips showing base reflectance values of representative room surfaces. The value given under each color sample is the light reflectance value (LRV) of the surface expressed as a percentage. For example, the saturated red in the first row of the chart has an LRV of 10, or 10% of the incident light flux; the light pink in the first row has an LRV of 75 (75%). The base reflectance of white is between 80 and 95, because there are many shades of white ranging from off-white, to warm white, to high reflectance white, etc. In design applications, 80% LRV is normally selected for white surfaces to allow for aging and discoloration.

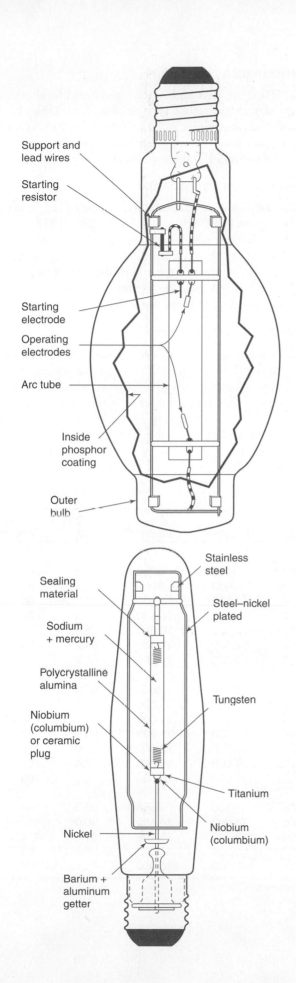

Support and
lead wires

Starting
resistor

Starting
electrode

Operating
electrodes

Arc tube

Inside
phosphor
coating

Outer
bulb

Sealing
material

Sodium
+ mercury

Polycrystalline
alumina

Niobium
(columbium)
or ceramic
plug

Nickel

Barium +
aluminum
getter

Stainless
steel

Steel–nickel
plated

Tungsten

Titanium

Niobium
(columbium)

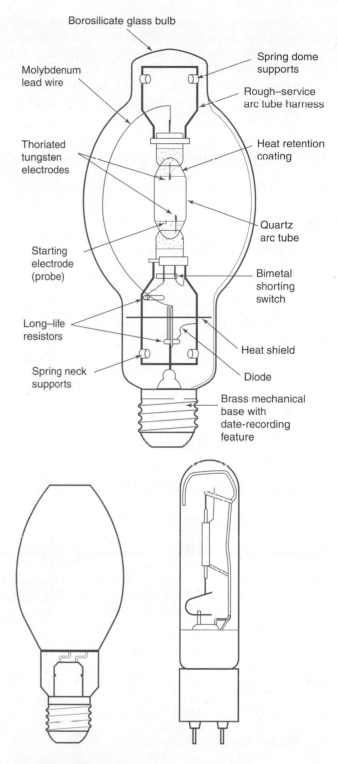

Borosilicate glass bulb

Molybdenum
lead wire

Thoriated
tungsten
electrodes

Starting
electrode
(probe)

Long–life
resistors

Spring neck
supports

Spring dome
supports

Rough–service
arc tube harness

Heat retention
coating

Quartz
arc tube

Bimetal
shorting
switch

Heat shield

Diode

Brass mechanical
base with
date-recording
feature

■ **FIGURE 15–13**

Typical construction of HID lamps. (top left) Mercury vapor
(Courtesy: Osram Sylvania, Inc., Danvers, MA.)
(top right) Metal halide (Courtesy: Osram Sylvania, Inc.,
Danvers, MA.)
(bottom left) High-pressure sodium (Courtesy: General
Electric Company, Nela Park, Cleveland, OH.)
(bottom center and right) Super white HPS (Reproduced
with permission from Philips Lighting Company.)

applications. In response to this application issue, new HPS lamps have been introduced that will not cycle at the end of lamp life. They are available from 50 W to 400 W and can be used for new projects or retrofitted into existing installations without a ballast change. These noncycling HPS lamps carry a life rating of 30,000 hours.

15.5.2 Characteristics of HID Lamps

HIDs and fluorescents are all electric discharge lamps. Following are some common characteristics of HID lamps.

Configuration and Designation

The configurations of mercury vapor and metal halide lamps are similar, but HPS lamps are more cylindrical. See Figure 15–12 for typical shapes. The ANSI designation of mercury vapor lamps starts with "H," MH lamps with "M," and HPS lamps with "S." Other numerals and letters in the designation denote the bulb size and wattage of the lamp. For example: M400/C/U stands for metal halide, 400 W, coated and designed for universal burning.

Performance

All HID lamps have high efficacy and long life, although they vary considerably in type and size. Table 15–4 shows the dimensions, rated life, initial lumens, efficacy (lpw), and lamp lumen depreciation (see next subsection) of typical HID lamps.

Lamp Lumen Depreciation

HID lamps have a relatively higher rate of lamp lumen depreciation (LLD) than fluorescent lamps; MH lamps have the highest rate, down to 60 percent at the end of their rated life. Economics will justify the replacement of these lamps prior to the end of their rated life, however. Table 15–4 indicates the initial and mean lumens of the various lamp types and sizes.

Starting Characteristics

HID lamps are not instant-start lamps. They all require a warm-up time to reach their rated light output. Metal halide lamps require the longest time to start, and once turned off, they take a longer time to restart (restrike). In general, the starting time of HID lamps is from 2 to 5 minutes, and the restarting time is from 1 to 15 minutes. There are also lamp circuits in which lamps can

TABLE 15–4
Typical characteristics of selected HID lamps[a]

Watts	Finish	ANSI Designation	Length, in.	Rated Hours	Initial Lumens	Mean Lumens[b]	LLD, %[c]	Lumens per Watt[d]
Mercury Lamps								
100	Phos.	H38JA–100/DX	7½	24,000	4100	3300	80.5	41
175	Phos.	H39KC–175/DX	8⁵⁄₁₆	24,000	7800	6800	87.2	45
250	Phos.	H37KC–250/DX	8⁵⁄₁₆	24,000	12,500	10,000	80.0	50
400	Clear	H33CD–400	11½	24,000	21,000	18,900	90.0	53
400	Phos.	H33GL–400/DX	11½	24,000	22,000	16,600	75.5	55
1000	Clear	H36GV–1000	15⅛	24,000	57,500	48,400	84.2	58
1000	Phos.	H36GV–1000/DX	15⅛	24,000	60,500	48,500	80.2	61
Metal Halide Lamps								
175	Clear	M57PE–175	8⁵⁄₁₆	10,000	13,600	8600	63.2	78
250	Clear	M58PG–250	8⁵⁄₁₆	10,000	20,800	13,500	64.9	83
400	Clear	M59PJ–400	11½	20,000	36,000	24,000	66.7	90
400	Phos.	M59PK–400	11½	20,000	35,000	21,000	60.0	88
1000	Clear	M47PA–1000	15⅛	12,000	110,000	88,000	80.0	110
1000	Phos.	M47PB–1000	15⅛	12,000	107,000	85,000	79.4	107
1500	Clear	M48PC–1500	15⅛	3000	155,000	126,000	81.3	103
High-Pressure Sodium (HPS) Lamps								
100	Clear	S54SB–100	7¾	24,000	9500	8550	90.0	95
150	Clear	S55SC–150	7¾	24,000	16,000	14,400	90.0	107
250	Clear	S50VA–250	9¾	24,000	28,500	25,600	89.8	114
400	Clear	S51WA–400	9¾	24,000	50,000	45,000	90.0	125
1000	Clear	S52XB–1000	15¹⁄₁₆	24,000	140,000	126,000	90.0	140

[a]Performance data are based on median value of three leading lamp manufacturers, (2000).
[b]Mean lumen is the lamp output at 40% of rated lamp life.
[c]LLD is lamp lumen depreciation factor.
[d]Lumens per watt (lpw) is calculated from initial lumen/lamp power.

gain full brightness in only a few seconds at considerably higher cost. Thus, the designer should be knowledgeable about the operation of the space where HID lamps are to be used. HID lamps are not suitable for frequent on-and-off operations. Their rated life is based on a 10-hour operating cycle. When they are used in public spaces, such as an arena or a stadium, an auxiliary lighting system, such as fluorescent or incandescent lighting, should be provided to maintain minimum lighting during a power interruption.

Operation

For continuous operation (24 hours/day and 7 days/week), the lamps should be turned off a minimum of 10 minutes at least once a week. This will reduce the chance that the arc tube will rupture at the end of lamp life.

Equipment Operating Factor

The lumen output of HID lamps depends not only on the ballast but also on the lamp operating position and other factors. The equipment operating factor (EOF) is defined as the ratio of the flux of an HID lamp– ballast–luminaire combination operating in a specific position to the lamp–luminaire combination operating with a reference ballast in the vertical position. The EOF includes the effect of the lamp position (or tilt) factor and the ballast factor. For determining initial illuminance in engineering calculations, the EOF should be applied to the manufacturer's published light output. For example, if the EOF of a 400-W metal halide lamp–ballast–luminaire combination is given as 0.91, then the actual light output of this combination, using a 400-W M59PK–400 metal halide lamp, is 36,000 lm multiplied by 0.91, or only 33,000 lm. See Chapter 16 for the use of lamp operating factor (LOF) instead of EOF.

15.6 MISCELLANEOUS LIGHT SOURCES

Many other light sources are used in lighting systems. The following are of particular interest.

15.6.1 Short-Arc Lamps or Compact-Arc Lamps

Short-arc lamps characteristically provide a source of very high luminance. They are primarily used in searchlights, projectors, display systems, and optical instruments, and for simulation of solar radiation. To facilitate lamp starting, short-arc lamps contain argon, xenon, or another rare gas at a pressure of several kilopascals. These lamps produce high-intensity light from a small bulb, thus closely resembling a point source, which is important for critical light beam controls. The lamps are made with rated wattages from 5 to 32,000 W and are available for operation in vertical or horizontal positions. Short-arc lamps are under considerable pressure during operation and therefore must be operated in an enclosure at all times. In addition, precaution must be taken to ensure protection from the powerful UV radiation emitted from these lamps.

15.6.2 Low-Pressure Sodium Lamps

Low-pressure sodium (LPS) lamps produce a monochromic yellow-wavelength light at 589 nm. (See Plate 1 for the spectral energy distribution of typical LPS lamps.) The lamp is filled with sodium vapor and small amounts of argon, xenon, or helium as the starting gas. The arc tube temperature, which is critical, is optimal at 260°C. Deviating from this temperature will result in a great reduction in efficacy. For this reason, the outer bulb of the lamp is kept in a state of high vacuum, to retard heat transfer. When operating at the desired arc tube temperature, the lamp can produce up to 200 lpw of light power, the highest of all light sources. LPS is not suitable for interior lighting applications in buildings. The principal applications of LPS lamps are highway exchanges, tunnels, storage yards, and wherever color rendition is unimportant or a strong dominant color (yellow) is desired. For example, using LPS lamps in tunnels provides a message of caution to motorists. On the other hand, LPS lamps are not recommended for parking lots, because their poor color-rendering characteristics (−15 to 0 CRI) make automobiles difficult to recognize (and thus difficult to find).

15.6.3 Electrodeless Lamps

Electrodeless lamps include several new generations of lamps, and they promise to be the lamps of the future. They have a good color-rendering property, high efficacy, and up to 100,000 or more operating hours. Lamp life is not affected by starting frequency because there are no electrodes to deteriorate and fail. They are ideal for difficult to reach places. Several types of electrodeless lamps are:

- *Magnetic induction* This type uses magnetic-induction technology instead of an electrode at each end to power the discharge. Figure 15–14 shows a typical construction of such a lamp. It utilizes an inductively coupled plasma driven in a closed-loop discharge tube within which RF power is evenly distributed along the discharge path. The electric field that initiates and maintains the plasma inside the tube is created not by electrodes but by an RF magnetic field concentrated within ferromagnetic ring cores. This configuration allows a low-profile geometry that avoids excessive thermal stress near the excitation area, which is typical of RF lamps with internal RF drives. The system design is optimized for high-frequency, high-lumen output and maximum reliability.

- *Electromagnetic*   This type utilizes the principles of electromagnetism to excite the gas in the lamp. It consists of a magnetic core-and-coil assembly at the lamp's center, but external to the lamp envelope. The construction and operating principle of electromagnetic electrodeless lamps are shown in Figure 15–14(a) and (b).

- *Microwave.*   This type utilizes a concentrated microwave generator to direct microwaves to a glass bulb filled with sulfur gas. Depending on the microwave power, the light generated can be of very high intensity. Figure 15–14(c) and (d) illustrates the construction and operating principle of the microwave lamp. Because of its concentrated light power within a small bulb, this lamp is ideal for use with a light tube—a tubular luminaire with light originating from one end. By controlling refraction and reflection in the tube, light can be uniformly emitted throughout the entire length of the tube, which may stretch up to several hundred feet. This eliminates the need for multiple luminaires in the space.

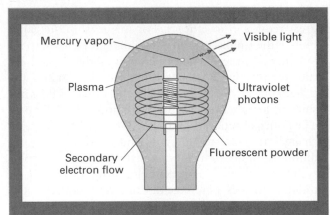

(a)

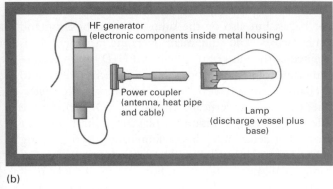

(b)

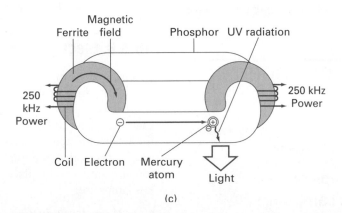

(c)

■ **FIGURE 15–14**

(a) and (b) The principle of light generation of an induction lamp, which has no direct electrical power connection but, rather, a power coupler, that generates high frequency in the megahertz range. The high frequency induces an alternating magnetic field that excites the mercury gas in the lamp, which generates light energy in the same manner as an HID lamp. (Reproduced with permission from Philips Lighting Company.) (c) Ferromagnetic rings at a low frequency of 250 kHz surround a closed-loop phosphor-coated glass tube. The alternating magnetic force excites the mercury atoms within the tube, generating visible energy much like a fluorescent lamp does except without the need for electrodes. This low-frequency design minimizes electromagnetic interference problems and ballast complexity. (Courtesy: Osram Sylvania, Inc., Danvers, MA)

15.6.4 Electroluminescent Lamps

An electroluminescent lamp is a thin, flat area source in which light is produced by a phosphor excited by a pulsating electric field. Green, blue, amber, yellow, or white light may be produced by choice of phosphor. Green phosphor has the highest luminance. Lamps are available in rigid ceramic or flexible plastic and are easily fabricated into simple solid or complex multisegmented shapes. Their thin profile, light weight, solid-state reliability, and negligible heat generation make them an ideal low-level light source. The life of electroluminescent lamps is long, and their power consumption is low. Electroluminescent lamps are used in decorative lighting, instrument panels, switches, and emergency lighting and signs, and for backlighting liquid crystal displays (LCDs).

15.6.5 Light-Emitting Diodes

A light-emitting diode is a semiconductor *p-n* junction lamp that consists of several layers of semiconductor material. When the diode is forward biased, light is generated in the thin active layer. Unlike incandescent lamps, which produce light as a result of blackbody radiation (heating of a filament) and which produce a continuous spectrum, an LED converts electricity directly to visible light and emits an almost monochromatic light of a particular color. The color of the light depends on the material used. Two material systems—AlInGaP and InGaN—are used for creating high-brightness LEDs in all colors from red through blue. A white LED is a blue LED that uses a yellow phosphor to adjust the color characteristics. Figure 15–15 shows the detailed structure of an LED chip.

The typical size of an LED is 0.25 mm², which allows the modules to be made very small and thin. The semiconductor is mounted in a package for easy electrical contact and environmental protection. There are two basic types of LED packages. THT (through-hole technology) and SMT (surface mount technology). Figure 15–15(b) shows these two types of LED packages.

The highest luminous intensity is produced with a DC voltage source. The necessary forward voltage depends on the color of the LED, varying between 2 V and 4 V at a forward current of up to 70 mA. LEDs are measured and categorized by on-axis intensity. Varying the LED drive current changes the intensity. The generated light increases directly with increased current rather than voltage.

The efficiency of LEDs has improved a great deal in the last few years and has already reached levels of 20 lpw for white LEDs and up to 60 lpw for others, depending on the color. Efficacies of up to 100 lpw have been reported for yellow LEDs.

One attractive feature of LEDs is their long service life. Unlike other light sources whose rated life is typically taken when 50 percent of the lamps have failed, the rated life of LEDs is taken when the brightness drops to 50 percent of its initial value. The life of LEDs can also be affected by temperature and current in a relationship similar to that for the light output.

Recent developments in materials have allowed LEDs to be used for applications such as traffic signals, pedestrian crosswalk signs, and exit signs. Today, LEDs are being incorporated into lighting modules for use in many applications such as sign lighting, accent lighting, low-level walkway and way-finding lighting, as well as architectural highlighting.

15.6.6 Nuclear Light Sources

Nuclear light sources are self-contained and thus require no power supply. These sources consist of a sealed glass tube or bulb internally coated with a phosphor and filled with tritium gas. Low-energy beta particles (electrons) from the tritium, an isotope of hydrogen, strike the phosphor, which in turn emits light of a color characteristic of the type of phosphor used. Thus the mechanism of light production is very similar to that in a conventional television tube. The higher the ratio of the quantity of the tritium to the phosphor area, the greater the luminance. Highest brightness is obtained from green and yellow phosphors. Their luminance can be as high as 7 cd/m². Even a person whose eyes are not dark-adapted can easily see them, and some forms are visible at considerable distances. Nuclear light sources typically provide illumination for instrument panels, controls, and clocks.

15.6.7 Carbon-Arc Lamps

Carbon-arc lamps were the first commercially practical electric light sources. They were used for many years in applications where extremely high luminance, high correlated color temperature, or high color rendering was necessary, such as motion picture projection lamphouses, theatrical follow-spots, searchlights, and "supplemental daylight" for film production.

Carbon arcs are operated in lamphouses, which shield the outside from stray radiation. These lamphouses may incorporate optical components, such as lenses, reflectors, and filters, for eliminating radiation from undesired parts of the spectrum. Arcs for projection of motion pictures generally operate on direct current to prevent disturbing stroboscopic effects on the projector shutter. Both motor generators and rectifiers are used. Flame arcs are widely used on both direct and alternating current. Low-current arcs have a negative

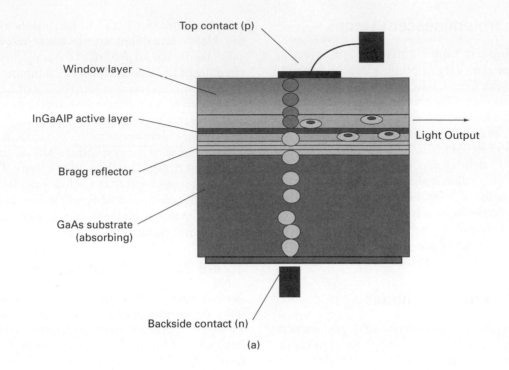

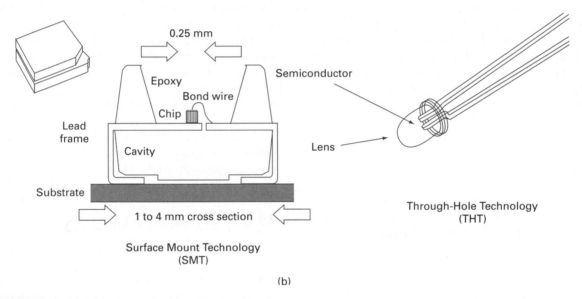

■ **FIGURE 15–15**

(a) Construction of an LED chip. When the top contact (*p*) and bottom contact (*n*) are forward biased, light is generated in the then-active layer. The top layer is used for light transmission and physical protection. (b) The two types of LEDs: (left) Surface mount technology (SMT) with the LED chip seated in the cavity surrounded by a plastic reflector. The cavity is filled with transparent epoxy resin to protect the chip. (right) Through-hole technology (THT) with the LED chip seated within a reflector–lens assembly. (Courtesy: Osram Sylvania, Inc., Danvers, MA.)

volt-ampere characteristic and therefore must be operated from circuits which include ballast resistances or reactances, in either the generating or the rectifying equipment or as separate units in the arc circuits. High-intensity carbon arcs have positive volt-ampere characteristics and may be operated without ballast.

15.7 GENERAL COMPARISON OF LIGHT SOURCES

With the basic knowledge covered to this point, the reader should be able to make appropriate selections of light sources for various applications. Following is a summary of the factors that enter into these decisions:

1. *Efficacy.* Incandescent lamps—including the tungsten-halogen types—are the least energy-efficient light sources compared with fluorescent or HID lamps, which are three to five times as efficient.
2. *Spectral power distribution.* No light source can produce equal power across the entire visible spectrum. Even daylight at 7500 K is richer in blue and poorer in red. Incandescent lamps are rich in the orange and red portion of the spectrum while fluorescent lamps are rich in blue and green. Low-pressure sodium lamps generate only a very narrow band of yellow spectral light and lack all other colors. Typical spectrum power distributions of selected light sources are shown in Plate 13(a).
3. *Color rendering.* This is a very important aspect of lighting design. Plate 13(b) illustrates how a model appears under various light sources. The final color appearance is the combination of the spectrum power on the life of the light source and the reflectivity of the model, as illustrated in Plate 12(a). Many new light sources have excellent color-rendering qualities as well as high luminous efficacy. See Plates 12(b) and 12(c). The HPS lamp used in the illustration has a CRI of 80; the color rendition compares favorably with that of an incandescent lamp with a CRI of 100. Similar results can be achieved using metal halide lamps.
4. *Color temperature.* The apparent color of white light sources, whether they are warm (yellowish, reddish) or cool (bluish, greenish), can be expressed by the blackbody temperature along the Planckian locus in the CIE chromaticity diagram. The characteristics of representative white light sources including that of incandescent, fluorescent lamps, sunlight, and skylight are identified in Plate 12(d). It should be noted that owing to the manufacturing process metal halide lamps may have color temperature shifts from 2000 k to 6000 k during their rated life.

5. *Rated life.* The lives of different light sources vary widely. In general, fluorescent and HID lamps last 10 to 20 times as long as incandescent lamps. This means that during the average life of a fluorescent or HID lamp, an incandescent lamp may have to be replaced 20 times.
6. *Cost of operation.* The energy savings and cost of maintenance are considerably lower for fluorescent and HID lamps than for incandescent lamps. The use of incandescent light sources should be limited to special-task illumination rather than general space illumination in most interior spaces, including highly decorative commercial spaces.
7. *Lamp start-up.* Although many lamp sources including fluorescent do not achieve full lumen output instantly, it should be noted that HID sources may take 2 to 10 minutes to strike, making them unacceptable for emergency lighting or switched applications, as in theaters, churches, or classrooms. Fluorescent lamps are temperature sensitive.

The following tables are very useful, as they provide side-by-side comparisons of commonly used light sources.

- Table 15–5 gives general characteristics of commonly used light sources.
- Table 15 6 gives a general comparison of types of lamps, efficacy, life, and applications. For specific values of selected light sources, such as initial lumens, efficacy (lpw), lumen maintenance, life (hrs), CRI, starting and warm-up time (minutes), and dimming range (percentage light output), the reader should refer to Figure 6 3 of the ninth edition of *IESHB*.

15.8 LUMINAIRES

A luminaire, commonly called a lighting fixture or fixture, is a complete lighting unit that contains one or more lamps, structural supports, accessories, auxiliaries, luminous control elements, and electrical components.

15.8.1 Classification of Luminaires

Luminaires can be classified in a number of ways according to their construction, physical configuration, photometric distribution, type of light sources, and electrical characteristics.

1. *Classification by light source.* Type of lamps contained: incandescent, halogen, fluorescent, mercury, metal halide, high-pressure sodium, etc. In some special applications, the light source within a

TABLE 15-5
General characteristics of commonly used light sources*

Light Source	Wattage Range	Efficacy, lpw	Life	Lumen Maintenance	Starting Time	Color Rendition	Ballast Required	Dimming Capability	Optical Control
Incandescent filament	10 to 1500	Very low	Very low to low	Fair to good	Very good	Very good	No	Very good	Good
Tungsten–halogen	10 to 2000	Very low to low	Very low to low	Good to very good	Very good	Very good	No	Good	Very good
Low-pressure discharge									
Standard fluorescent	15 to 40	Low to good	Fair to very good	Fair to good	Good to very good	Low to very good	Yes	Good	Poor
Slimline fluorescent	20 to 75	Fair to good	Fair to good	Fair to good	Very good	Fair to very good	Yes	Low	Poor
High-output fluorescent	35 to 110	Fair to good	Fair to good	Fair to good	Very good	Fair to very good	Yes	Good	Poor
Very high output fluorescent	38 to 215	Fair to good	Fair to good	Fair to good	Very good	Fair to very good	Yes	Good	Poor
Energy-saving fluorescent (T12)	30 to 185	Fair to good	Fair to good	Fair to good	Very good	Low to very good	Yes	Low	Poor
High-efficacy fluorescent	18 to 40	Good	Good	Good	Very good	Good to very good	Yes	Fair	Poor
Compact fluorescent	5 to 40	Good	Fair to good	Good	Good to very good	Good to very good	Yes	Very low	Fair
High-intensity discharge									
Mercury	40 to 1000	Low to Fair	Good to very good	Very low to fair	Low	Very low to fair	Yes	Fair	Poor
Self-ballasted mercury	100 to 1500	Very low	Fair to very good	Low to fair	Fair	Low to fair	No	Very low	Poor
Metal halide	32 to 1500	Good	Low to Fair	Very low to fair	Good	Low	Yes	Low	Good
High-pressure sodium	35 to 1000	Fair to good	Fair to very good	Fair to good	Fair	Low to good	Yes	Low	Good
Miscellaneous									
Low-pressure sodium	10 to 180	Fair to very good	Fair to good	Good to very good	Fair	Very low	Yes	Very low	Poor
Cold-cathode	10 to 150	Low	Very good	Fair to good	Very good	Low to very good	Yes	Good	Poor

*See manufacturers' catalogs for specific data.
(Reproduced with permission from IESNA.)

TABLE 15-6

General comparison of different types of lamps, efficacy, life, and applications

Category	Type	Typical Efficacy, lpw	Rated Life, hours	Characteristic Features	Typical Application
Incandescent Lamps	General Service and Reflector	10 to 25	750 to 2000	Easy to install, easy to use; many different versions; instant start; low cost; reflector lamps allow concentrated light beams	General lighting in the home; decorative lighting; localize lighting; accent and decorative lighting (reflector lamps)
	Halogen	15 to 36	1000 to 4000	Compact; high light output; white light; easy to install; long life compared with normal incandescent lamps	Accent lighting, floodlighting
Fluorescent Lamps	Tubular	50 to 104	12,000 to 30,000	Wide choice of light colors; high lighting levels possible; economical in use	All kinds of commercial and public buildings; streetlighting; home lighting
	Screw in base	50 to 80	9000 to 10,000	Energy–effective; direct replacement for incandescent lamps	Most applications where incandescent lamps were used before
	Compact Fluorescent	30 to 80	10,000 to 40,000	Compact; long life; energy–effective	To create a pleasant atmosphere in social areas, local lighting; signs, security, orientation lighting and general lighting
Gas-Discharge Lamps	Self-ballasted mercury	20 to 25	12,000 to 16,000	Long life; good color rendering; easy to install; low efficacy but better than incandescent lamps	Direct replacement for incandescent lamps; small industrial and public light projects; plant irradiation
	High-pressure mercury	50 to 60	12,000 to 24,000	High efficacy; long life; reasonable color quality	Residential area lighting; factory lighting; landscape lighting
	Metal halide	80 to 120	10,000 to 20,000	Very high efficacy combined with excellent color rendering; long life	Floodlighting; especially for color TV; industrial lighting; road lighting, plant irradiation; retail lighting
	High-pressure sodium	60 to 140	10,000 to 30,000	Very high efficacy; extremely long life; good color rendering	Public lighting; floodlighting, industrial lighting; plant irradiation; roadway lighting.
	Low-pressure sodium	130 to 200	14,000 to 18,000	Extremely high efficacy; long life; high visual acuity; poor color rendering; monochromatic light	Many different application areas; wherever energy/cost effectiveness is important and color is not critical, such as tunnel lighting

Note: Ballast factor (BF) must be applied to fluorescent and gas-discharge lamps to obtain the net lamp-ballast efficacy.
(Reproduced with permission from Philips Lighting Company.)

luminaire may be xenon, electroluminescent, carbon arc, laser (light amplification by stimulated emission of radiation), LED (light-emitting diodes), etc.

2. *Classification by photometric distribution.* For indoor luminaires, a method specified by the International Commission on Illumination (CIE) is commonly used. CIE classifies indoor luminaires into five distribution types, namely: direct, semidirect, direct–indirect, semi-indirect, and indirect. The IESNA formally includes one additional photometric distribution class, the general diffuse type. The authors believe the additional classification is meaningful, so it is included for better understanding of its applications.

Figure 15–16 illustrates the six types.

3. *Classification by application.* Indoor or outdoor use or special applications:
   - *Indoor* General-purpose, industrial, emergency, exit signs, portable, tabletop, floor standing, light track, wall washers, residential applications, etc.
   - *Outdoor* roadway, highway, pathway, area flood, sports, sign, security, landscape, etc.
   - *Special applications* Swimming pool, mining, aviation, theater, medical, food processing, clean room, etc.

4. *According to construction.* Waterproof, weatherproof, dust-tight, explosion-proof, insect-tight, corrosion-resistant, etc.

5. *According to the method of mounting.* Recessed in ceiling or on the ceiling surface, suspended from

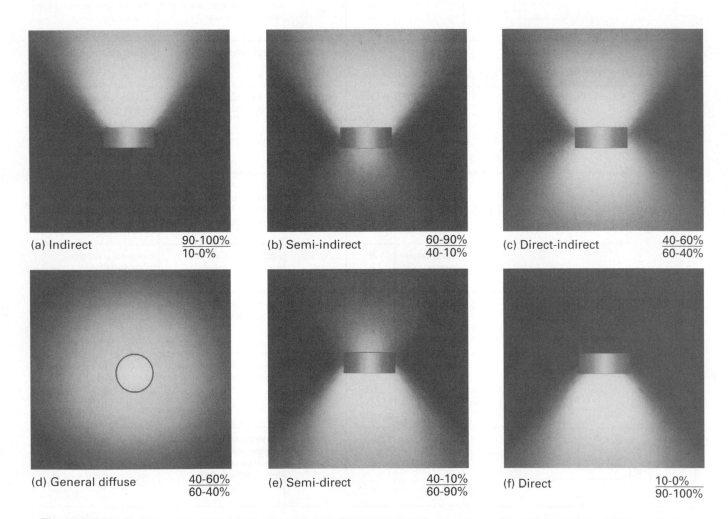

(a) Indirect — 90-100% / 10-0%

(b) Semi-indirect — 60-90% / 40-10%

(c) Direct-indirect — 40-60% / 60-40%

(d) General diffuse — 40-60% / 60-40%

(e) Semi-direct — 40-10% / 60-90%

(f) Direct — 10-0% / 90-100%

■ **FIGURE 15–16**
Indoor luminaires are classified by the CIE into five types as indirect, semi-indirect, direct–indirect, semidirect and direct based on the percentage of light flux distribution above and below the horizontal axis. In addition, the IESNA further classifies luminaires with a nearly balanced light flux distribution at all angles as general diffuse type. (Courtesy: William Tao & Associates, Inc., Consulting Engineers, St. Louis, MO.)

the ceiling, mounted on brackets from the wall, recessed in the wall or concrete, etc.

6. *According to its light flux control medium:* Shielded by housing or diffused by means of prismatic lenses translucent diffusers, parabolic louvers, etc. Refer to Chapter 14 for the means of light controls. Some fixtures rely on the lamp itself to control light flux, such as MR16 and PAR lamp fixtures.

7. *According to the electrical characteristics:*
   - *Power system* voltage and frequency, such as 120, 277, or 480 Volt, 60 Hz; 230 V, 50 Hz; 24 V 3000 Hz, etc. Many fixtures with an electronic ballast are rated for more than one voltage.
   - *Lamp envelope size and maximum wattage rating:* A21, 100-W, general-service incandescent; T8, 32-W, fluoresecent; etc. The maximum wattage of the lamps is given to prevent overheating of the luminaire, a potential fire hazard. Although large-wattage lamps may fit, the maximum rated wattage should not be exceeded.

8. *Classification by other features or components:* Electromagnetic (EM) shielding, a built-in battery as a standby power source, an extra lamp to be wired to an emergency power supply, cutoff angle at which angle light flux is reduced to zero, etc. The cutoff angle is especially important for spotlighting or sports lighting applications to provide accurate light control. Cutoff is also important for office lighting and for glare control when viewing computer screens and other tasks.

15.8.2 Examples of Luminaire Specifications

Example 15.4   A luminaire may be described as follows: The type RA luminaire (fixture) shall be recessed lay-in ceiling-mounted fluorescent fixture for direct distribution. It shall contain three 32-W, T8 fluorescent lamps rated for 3500 K color temperature, and not less than 85 Color Rendering Index (CRI). The fixture shall have a 24-cell parabolic louver. The photometric distribution shall be a batwing pattern and shall have minimum coefficient of utilization of 75 percent based on 80–50–20 ceiling-wall-floor effective reflectances and room cavity ratio (RCR) of 2.0. The fixture housing shall be fabricated of 24-gauge steel with baked-on white enamel finish. The ballast shall be UL listed, Class P rated, and thermally protected.

Example 15.5   A luminaire may be described as follows: The type SB fixture shall be a suspension-mounted fixture with a 12-in.-diameter opal glass globe. The fixture shall be designed for use with one 150-W general-service incandescent lamp. The fixture shall be provided with a 36-in.-long stem and ceiling canopy and outlet box. The canopy shall be made of cast aluminum with satin finish.

15.9 LUMINAIRES: PHOTOMETRY

Photometric data (Figures 15–17 and 15–18) are the most important pieces of information required to determine the performance of a luminaire. They are the basic tool used in calculations of illumination. Photometric data are normally prepared by an independent testing laboratory or the laboratory of a qualified fixture manufacturer. They should contain the following information (when applicable):

1. *Description of the luminaire.*
   - Manufacturer's name, luminaire type, and catalog number of luminaire.
   - Type of mounting.
   - Reflectances of light-reflecting surfaces.
   - Drawing showing luminaire shape, dimensions, and luminaire photometric center.
   - Ballast data: number of lamps to be used, power factor, voltage, frequency, lamp current where applicable, and catalog number of ballasts.
   - Ballast factor (average for the test luminaire where multiple ballasts are involved).
   - Input power (watts).

2. *Test lamp description.*
   - Number of lamps in the luminaire, type of lamps, bulb designations, and length where applicable.
   - Rated power (watts)
   - Rated lumen output.
   - Color.

3. *Photometric characteristics.*
   - Luminous intensity distribution curves in various planes.
   - Luminous intensity values, tabulated (for example, for indoor fluorescent luminaires, values for five planes at 5° increments are usually provided).
   - Tabulations of zonal lumens and percentages of bare-lamp lumens.
   - Tabulation of average luminance data where applicable.

4. *Luminaire efficiency.* That is, the total lumens output versus the lamp lumens. Note that this is distinct from lamp efficacy.

5. *Coefficient of utilization (CU).* This number expresses the portion of lamp lumens that can be utilized on a working surface normally selected to be 30 in. above the floor for horizontal illumination (although any height can be used). The CU may be

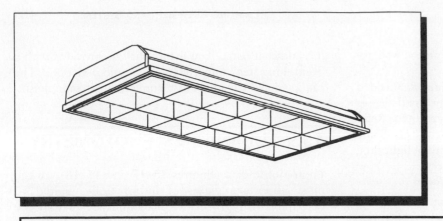

2' X 4'
3-Lamp
T12 or T8
18 or 24 Cell

PHOTOMETRIC DATA

CATALOG # 2P3G340-36SL **LAMPS =** F40 ES **INPUT WATTS =** 118 **LER =** FP-41
TEST #15698 **S/MH=** 1.6 **BALLAST =** ESB **BALLAST FACTOR =** .88

COMPARATIVE YEARLY LIGHTING ENERGY COST PER 1000 LUMENS = **$5.85** BASED ON 3000 HRS. AND $.08 PER KWH.

CANDLEPOWER

Angle	End	45	Cross
0	2210	2210	2210
5	2210	2212	2221
10	2172	2217	2260
15	2108	2206	2272
20	2038	2169	2241
25	1948	2102	2172
30	1852	2013	2205
35	1731	1917	2360
40	1593	1906	2127
45	1436	1809	1719
50	1255	1495	1135
55	1039	1022	706
60	777	575	417
65	343	255	172
70	54	66	49
75	22	20	22
80	10	9	9
85	4	4	3

MAINTAINED ILLUMINATION TABLE- Square Feet/Fixture*

- 80-50-20 Reflectances (Ceiling-Wall-Floor)
- LLF = 0.73 2650 Lumens/Lamp very clean
- Room width divided by room height = 5 or more, 2 or 1

Fixture Size & # of Lamps	Room Width / Room Height =	Approx. Area (sq. ft.) per Fixture				
		10 ft-c	30 ft-c	50 ft-c	70 ft-c	100 ft-c
2' X 4' 3 Lamp	5	–	143	86	61	43
	2	–	102	61	44	31
	1	–	76	45	32	–

*Observe Fixture S/MH Requirements for Specific Applications

AVERAGE LUMINANCE CD/SQ.M WITH 2650 LUMEN LAMPS

ANGLE	END	45°	CROSS
45	3244	4087	3884
55	2894	2846	1966
65	1297	964	650
75	136	123	136
85	73	73	55

LLF = .73 LLF = LIGHT LOSS FACTOR LLF = LDD X LLD X BF LDD = VERY CLEAN 0.94 CLEAN 0.90
LLD = 0.88 @ 40% RATED LAMP LIFE BF = 0.88 ESB BALLAST & 34W LAMP (RELAMP AT 70% LAMP LIFE)

TYPICAL V.C.P.'s

Room Size	Mounting Height Lengthwise		Crosswise	
	8.5	10	8.5	10
30x30	87	83	90	86
40x40	92	88	94	90
60x30	92	88	94	91
60x60	94	91	96	93
100x100	97	95	98	96

COEFFICIENT OF UTILIZATION

pfc	20							
pcc	80			70			50	
pw	70	50	30	70	50	30	50	30
RCR								
0	81	81	81	80	80	80	77	77
1	77	73	71	75	72	70	69	68
2	70	66	61	68	65	61	63	59
3	65	58	54	64	57	54	56	52
4	59	53	47	58	52	46	51	46
5	56	47	41	54	46	41	46	40
6	52	42	38	50	42	36	40	36
7	47	39	34	46	39	33	38	33
8	45	35	29	44	34	29	34	29
9	41	33	28	40	33	27	32	27
10	39	29	25	38	29	25	28	25

LIGHT DISTRIBUTION

DEGREES	LUMENS	% LAMP	% FIXTURE
0-30	1795	22.6	32.7
0-40	3041	38.3	55.5
0-60	5167	65.0	94.2
0-90	5483	69.0	100.0

PHOTOMETRIC DATA

CATALOG # 2P3GS332-36SL-1/3-EB **LAMPS =** F32 T8 **INPUT WATTS =** 88 **LER =** FP-67
TEST #15694 **S/MH=** 1.7 **BALLAST =** ELECTRONIC **BALLAST FACTOR =** .93

COMPARATIVE YEARLY LIGHTING ENERGY COST PER 1000 LUMENS = **$3.58** BASED ON 3000 HRS. AND $.08 PER KWH.

CANDLEPOWER

Angle	End	45	Cross
0	2520	2520	2520
5	2523	2527	2541
10	2476	2543	2608
15	2405	2544	2630
20	2323	2504	2587
25	2226	2429	2498
30	2107	2310	2426
35	1961	2188	2857
40	1801	2146	2909
45	1628	2238	2106
50	1424	1907	1211
55	1180	1205	743
60	880	636	460
65	375	276	200
70	65	75	56
75	26	23	26
80	11	10	11
85	4	4	3

MAINTAINED ILLUMINATION TABLE- Square Feet/Fixture*

- 80-50-20 Reflectances (Ceiling-Wall-Floor)
- LLF = 0.77 2900 Lumens/Lamp very clean
- Room width divided by room height = 5 or more, 2 or 1

Fixture Size & # of Lamps	Room Width / Room Height =	Approx. Area (sq. ft.) per Fixture				
		10 ft-c	30 ft-c	50 ft-c	70 ft-c	100 ft-c
2' X 4' 3 Lamp	5	–	–	105	75	53
	2	–	126	76	54	38
	1	–	93	56	40	–

*Observe Fixture S/MH Requirements for Specific Applications

AVERAGE LUMINANCE CD/SQ.M WITH 2900 LUMEN LAMPS

ANGLE	END	45°	CROSS
45	3678	5056	4758
55	3287	3356	2069
65	1418	1043	756
75	160	142	160
85	73	73	55

LLF = .77 LLF = LIGHT LOSS FACTOR LLF = LDD X LLD X BF LDD = VERY CLEAN 0.94 CLEAN 0.90
LLD = 0.88 @ 40% RATED LAMP LIFE BF = .93 ELECTRONIC BALLAST & T8 LAMP (RELAMP AT 70% LAMP LIFE)

TYPICAL V.C.P.'s

Room Size	Mounting Height Lengthwise		Crosswise	
	8.5	10	8.5	10
30x30	86	81	89	84
40x40	90	86	92	88
60x30	91	87	93	89
60x60	93	90	95	92
100x100	96	94	97	95

COEFFICIENT OF UTILIZATION

pfc	20							
pcc	80			70			50	
pw	70	50	30	70	50	30	50	30
RCR								
0	86	86	86	84	84	84	81	81
1	81	79	76	80	77	75	73	71
2	75	69	66	73	68	65	67	63
3	69	63	57	68	61	56	59	56
4	64	56	51	63	55	50	54	48
5	58	51	45	57	50	44	47	44
6	55	46	40	53	45	40	44	39
7	51	41	35	50	40	34	40	34
8	46	38	32	46	38	32	36	32
9	44	34	28	44	34	28	34	28
10	40	32	27	40	32	27	30	26

LIGHT DISTRIBUTION

DEGREES	LUMENS	% LAMP	% FIXTURE
0-30	2061	23.7	32.3
0-40	3508	40.3	55.0
0-60	6015	69.1	94.4
0-90	6374	73.3	100.0

■ FIGURE 15–17

Comprehensive photometric data sheet of a fluorescent fixture. This report contains photometrics of two different lamp–ballast configurations for the same fixture. The upper one uses three 40-W, T12, rapid-start lamps with electromagnetic ballast having a 0.88 ballast factor. The lower one uses three 32-W, T8 rapid-start lamps with electronic ballast having a 0.93 ballast factor. Note the substantial energy cost savings of the fixture using T8 lamps with electronic ballast over the T12 lamps with electromagnetic ballasts. (Reproduced with permission from Thomas Industries.)

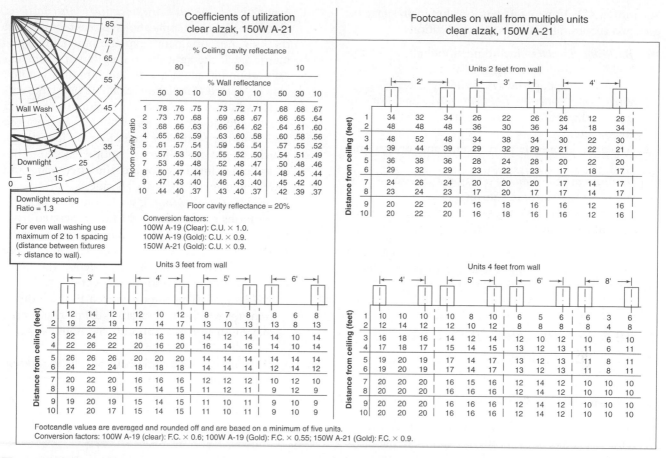

■ **FIGURE 15–18**
Typical photometric report for an incandescent luminaire. (Courtesy: Lightolier.)

considered the "net efficiency" of the selected luminaire for a particular installation. It depends on the room's physical characteristics, such as:

- Reflectance of the room surfaces: ceiling, wall, floor, or the average of the floor and the desktop reflectances together.
- Room dimensions and mounting: ceiling height, luminaire mounting height, and configuration and size of the room. The terms *room ratio* and *room index* are used to represent the overall effect of the room and luminaire relations.

The procedure for determining the CU of a design application is given in Chapter 16.

6. *Maximum spacing-to-mounting height ratio*. This concerns the uniformity of illumination on the selected working plane.

7. *Visual comfort probability (VCP)*. This is a rating of the light system expressed as the percentage of people who will find the system comfortable. The VCP is limited to office and school applications at selected positions in the room. It is not normally calculated.

8. *Light loss factor (LLF)*. This is a modification factor expressing the depreciation of light output of the lighting system, including the luminaires and room conditions. It depends on the construction of the luminaire, the type of lamp used, and the cleanliness of the surroundings. Methods for determining the LLF are given in Chapter 16.

15.10 OUTDOOR LUMINAIRES

Outdoor luminaires are luminaires that are specially designed for outdoor applications, although any indoor luminaires can also be used outdoors if it is properly shielded from the weather; however, the specifier/designer should always be aware of any factors that affect the longevity or performance of the luminaire—for example, the light output of fluorescent lamps can be drastically reduced in cold weather conditions. See

Section 15.4.2. In general, outdoor luminaires may be grouped according to five major applications:

1. *Building exteriors*—including floodlighting of the building facade, signs, special features, entries, exits, landscape, pathways, walkways, etc.
2. *Area and roadway lighting*—including parking lots, parks, outdoor storage areas, small or side streets, etc.
3. *Sports lighting*—for recreational and professional sports, etc.
4. *Transportation lighting*—major through streets, highways, and airport runways, etc.
5. *Special-purpose lighting*—monuments, harbors, searchlights, advertising signs, etc.

Our discussion addresses the first three application groups only, as transportation and special-purpose lighting are not normally associated with building design.

15.10.1 Construction of Outdoor Luminaires

In general, outdoor luminaires have the following characteristics:

1. *Water-resistant.* Whether the luminaire is totally enclosed or is open at the bottom, the luminaire must be water-resistant. The electrical wiring and components must be protected from rain or snow. The lamp, if exposed to rain or snow, must be of a kind that can withstand thermal shock. Such fixtures are typically noted by manufacturer as either UL listed for damp or wet locations. Fixtures for hose-down applications often carry an IP number, which is a more stringent European rating system.
2. *Corrosion resistant.* All components must be constructed of noncorrosive materials, such as aluminum and copper alloys, plastics, or stainless steel. If a surface is made of steel subject to corrosion, then the surfaces exposed to weather must be treated or coated to withstand the weather conditions.
3. *Insectproof.* If the luminaire is enclosed with glass or plastic lens, the assembly must be gasketed to prevent insects from entering the assembly.

15.10.2 Photometrics of Outdoor Luminaires

In addition to the standard CIE/IES classifications, namely, indirect, semi-indirect, direct–indirect (general diffuse), semidirect, and direct, there are other requirements so that the luminaires can fulfill the intended illumination goals effectively and economically.

1. *Symmetrical versus asymmetrical distributions.* Outdoor lighting applications often require that the light flux originate from the luminaire asymmetrically. If the light flux is distributed symmetrically around its physical axis, much of the light flux may be wasted. An obvious example of asymmetrical distribution is a wall-mounted luminaire whose flux should be directed forward or away from the wall. Another example is street lighting luminaires, which should have internal optical controls that emit higher intensity toward the street side and lower intensity on the sidewalk side.
2. *Flood versus spotlighting.* Usually, outdoor lighting applications require that light flux generated from luminaires be confined within a certain angle, which may vary between less than 10° up to more than 100°. The following terms define the luminaires and the distribution of the light flux.
   - *Field angle*   The angle between the two directions for which the intensity is 10 percent of the maximum intensity through the nominal beam centerline. (See Figure 15–19.)
   - *Beam angle*   The angle between the two directions for which the intensity is 50 percent of the

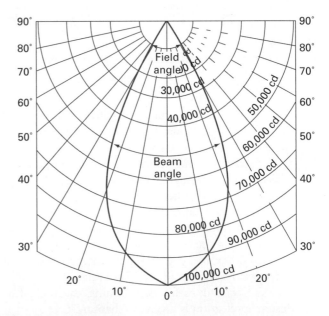

■ **FIGURE 15–19**
As illustrated in this intensity distribution curve on polar coordinates, the maximum intensity is 100,000 cd, the field angle is 60°, the angle included between the intersecting points of the curve at 10,000 cd (10% of 100,000 cd). The beam angle is 48°, the angle included between the intersecting points of the curve at 50,000 cd (50% of 100,000 cd). With this beam angle, the luminaire is a floodlight and is classified as NEMA Type 4. (Reproduced with permission from IESNA.)

maximum intensity through the nominal beam centerline. (See Figure 15–19.)

- *Spotlight* A luminaire with a narrow beam angle. Strictly speaking, the beam angle should be within 10°, although any luminaire with a beam angle of 30° or less is usually perceived as a spotlight.
- *Floodlight* A luminaire with a medium to wide beam angle from 10° to more than 100°, although narrow-angle floodlights can be perceived as spotlights. For area and sports lighting applications, the National Electrical Manufacturers Association (NEMA) has classified floodlights into seven beam types, from Type 1 (field angle 10° to 18°) to Type 7 (field angle more than 130°), with Types 2 through 6 in between these values. It should be pointed out that even with the use of widest field angle flood luminaires, there still will be unevenness on a large surface to be illuminated unless the field angles between adjacent flood lights are properly overlapped.

3. *Glare and light pollution.* Because most of the luminaires are installed outdoors, the high luminance of the light source contrasted against a dark background will make the luminaire appear extremely bright or glary.

Outdoor lighting for streets, highways, parking lots, building exteriors, airports, and sport facilities is the primary cause of the modern phenomenon known as *sky glow*. It is also known as *light pollution*. Guidelines for responsive exterior lighting design will be covered in Chapter 17. One of the simple ways to minimize this problem is to properly select luminaires, which is addressed in the next section.

4. *Types of luminaires.* Exterior luminaires may be classified as follows:

- *According to applications* Building or area floodlights, pathway, landscape, roadway, street, parking lots (exterior) parking garage (interior), underwater, etc.
- *According to design* Pole mounted; post-top mounted; wall surface, recess, or bracket mounted; bollards (low post); floor-recess mounted, etc.
- *According to construction* Weathertight, waterproof, explosion proof, etc.
- *According to electrical voltage* Low-voltage (usually below 24 V), 120 V, 277 V, etc.
- *According to optical controls* general diffuse, shielded or nonshielded, sharp cutoff, etc. The sharp cutoff design, sometimes referred to as "shoebox" design, eliminates light flux from undesirable directions for glare control and minimizes the problem of sky glow. This topic will be further discussed in Chapter 17.

5. *Photometric data of exterior luminaires.* In addition to the standard candela distribution curves plotted on polar coordinates, the performance data of outdoor lighting luminaires are usually given in a combination of isocandela and lumen distribution tables. Photometric data of a typical outdoor luminaire are shown in Figure 15–20.

6. *Typical outdoor luminaires.* There are hundreds, if not thousands, of manufactured products applicable to or exclusively designed for outdoor applications. Figure 15–21 shows many of the typical products, ranging from traditional to ultramodern. Their selection should be compatible with the architectural style of the building and its surroundings but, more importantly, should be energy-effective, minimize direct and reflected glare, be environmentally friendly, and meet the required lighting tasks.

15.11 GENERAL COMPARISON OF LIGHTING SYSTEMS

For each lighting system, thousands of luminaires are commercially available, and for each luminaire, there are many variations of lamp type, size, and control medium, such as lens, louvers, baffles, etc. Therefore, it would be meaningless to illustrate just a few of these combinations; however, we present the following overview of lighting systems:

- A *direct system* is a lighting system that is based on the use of direct-distribution luminaires. It is the most effective system for horizontal illumination; however, there is the potential for both direct and reflected glare, the latter from visual situations, such as viewing a visual display screen. There are many ways to overcome the glare—for example, by using large low-brightness ceilings or luminaires with sharp angle cutoffs.
- A *semidirect system* has small upward components. These will soften the high contrast between the ceiling and the luminaires and will improve the spatial brightness relations.
- A *general diffuse system* uses luminaires with light flux more or less uniformly distributed in all directions. This system is likely to achieve more of a modeling effect, with soft shadows, but may have more direct glare. The fixtures are pendant-mounted. If they are too close to the ceiling, the system will behave like a semidirect system.
- A *direct–indirect system* is similar to a general diffuse system, except that the light flux within 10° to 30° below the horizontal plane is reduced or

Source: Metal halide
Watts: 1,000

Lumens: 110,000
NEMA type: 4

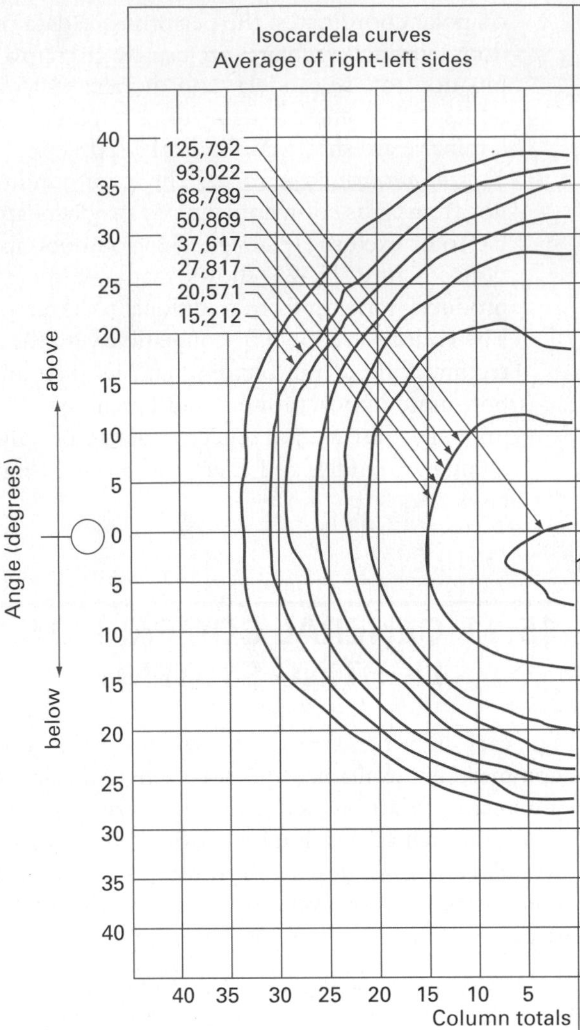

Isocandela curves — Average of right-left sides

Isocandela curve values: 125,792 / 93,022 / 68,789 / 50,869 / 37,617 / 27,817 / 20,571 / 15,212

Lumen distribution — Average of right-left sides

Angle (degrees)	0	5	10	15	20	25	30	35	40	Row total
(top)	0	0	0	0	0	0	0	0	0	0
40 above	151	141	117	101	88	79	73	63	42	855
35	310	241	190	146	105	83	73	66	52	1266
30	452	432	370	245	152	98	74	66	59	1948
25	507	509	492	409	237	131	61	66	61	2493
20	630	596	593	518	364	170	96	67	58	3092
15	719	668	591	567	454	219	113	69	57	3457
10	743	731	660	599	494	262	126	73	57	3745
5	922	868	836	673	544	275	129	75	56	4378
0	1,036	978	833	669	513	238	118	70	55	4510
5 below	943	845	734	623	449	192	102	67	56	4011
10	718	697	581	522	284	145	86	67	57	3237
15	655	584	465	280	154	104	76	68	60	2446
20	286	238	202	128	98	75	69	66	61	1223
25	120	103	86	79	69	68	67	64	52	708
30	68	64	61	61	65	67	65	57	39	547
35	61	61	61	61	65	63	55	40	25	492
40	0	0	0	0	0	0	0	0	0	0
Column totals	8,076	7,853	7,100	5,681	4,135	2,269	1,403	1,044	847	38,408

Angle to left and right (degrees): 40 35 30 25 20 15 10 5 | 0 5 10 15 20 25 30 35 40

■ **FIGURE 15–20**

Photometric data of a typical outdoor or sports luminaire indicating the isocandela curves on the left side and lumen depreciation values on the right side. The last column at the extreme right indicates the accumulated lumens within each horizontal row. These values should be doubled to obtain the total accumulated lumens for both sides of the luminaire. (Reproduced with permission from IESNA.)

shielded. This feature will greatly reduce direct glare, particularly on video display screens.

- An *indirect system* has more light flux directed above the horizontal plane, so direct glare is reduced. In turn, however, the ceiling brightness may become too great for comfort. These systems should not be installed unless there is adequate ceiling height and the fixtures are suspended within the recommendations of the fixture manufacturer. An indirect system is ideal for a video

display space. If the overall illuminance level is not adequate for tasks such as close reading, the system may have to be supplemented with localized track lighting.

With any of the preceding systems, the color and reflectance of wall surfaces is of great importance. Plate 15 is a chart of colored surfaces from which the reflectance value can be selected. As illustrated in Plate 14, the effects of color and reflectance of surfaces are quite

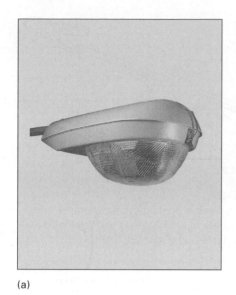

(a)

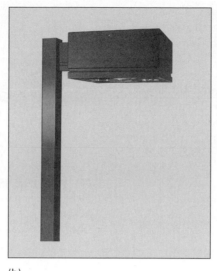

(b)

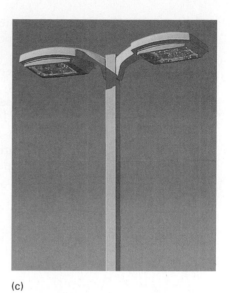

(c)

(d)

(e)

(f)

■ **FIGURE 15–21**

Typical commercially luminaires for open field or roadways.

(a) Cobra-head-style luminaire used for roadway and area lighting. (Courtesy: Cooper Lighting.)

(b) Standard Shoebox-shape area light for roadway and area lighting with semicutoff optics. (Courtesy: Cooper Lighting.)

(c) Architectural roadway and area light with low profile, reduced wind load housing, and segmented full cutoff optics. (Courtesy: Gardco Lighting.)

(d) Contemporary indirect area light. (Courtesy: Louis Poulsen Lighting Incorporated.)

(e) Traditional acorn-style area light with refractive glass lens and general diffuse distribution. (Courtesy: Holophane Lighting.)

(f) Traditional acorn-style area light with internal louvers for IESNA cutoff, dark-sky-compliant distribution. (Courtesy: Holophane Lighting.)

different for different lighting systems. The proper selection of color will enhance the appearance of the spatial relations. The reflectance values are important for lighting designers and will be discussed in Chapters 16 and 17.

As a matter of general interest, several lighting system concepts are shown in Figures 15–22 and 15–23. These include the light pipe, fiber-optic cables, and track lighting. Although track lighting is not a new concept, it has become increasingly popular.

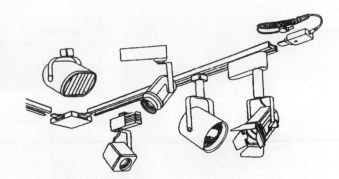

■ **FIGURE 15–22**
Light track is a very flexible lighting system. It can accommodate a variety of light sources and luminaires. (Reproduced with permission from IESNA.)

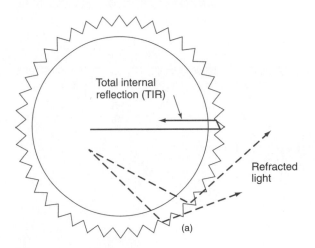

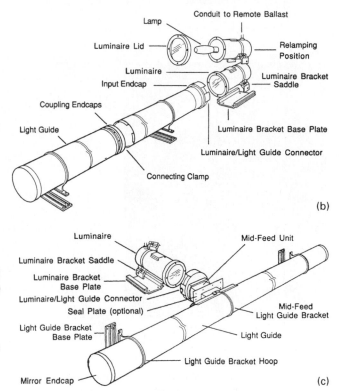

■ **FIGURE 15–23**
Light pipe consists of a concentrated light source and a prism light guide (PLG). When light enters the interface between the transparent material and its prismatic boundary interface, light will either refract or totally internal reflect (TIR) depending on the angle of incidence. This property is illustrated in (a); (b) and (c) illustrate the construction of some TIR units. (Courtesy: TIR Systems LTD, Vancouver, B.C. Canada.)

QUESTIONS

15.1 Daylight is plentiful and economical; however, it ( ) widely and its ( ) shifts during the day.

15.2 Fluorescent lamps were first introduced in the (1930s) (1940s) (1950s).

15.3 Light output depreciates with time. The light output of general-service incandescent lamps is about ( ) percent at their rated life.

15.4 The rated life of lamps is based on the time elapsed when only ( ) percent of a group of lamps still remain burning.

15.5 Which light source has a higher color temperature, one that is described as "warm" or one that is described as "cool"?

15.6 The CRI for a 3000-K incandescent lamp is given a rating of 100. Thus it can reproduce the true color of any object. (True) (False)

15.7 To reduce the stroboscopic effect from fluorescent and HID lamps, it is desirable to wire lighting fixtures on AC circuits or to use lead-lag ballasts for multiple lamp fixtures. (True) (False)

15.8 What lamp base is used on a T8 general-service fluorescent lamp?

15.9 What does PAR stand for in an incandescent reflector lamp?

15.10 What is the maximum bulb diameter, in inches, of (a) a PAR38 halogen lamp? (b) a T5 fluorescent lamp?

15.11 If the voltage at the lamp socket is 6 percent over the rated voltage of an incandescent lamp, what will be the expected percent increase in light output? What will be the percent decrease in the life of the lamp?

15.12 All other parameters being equal, is a higher-wattage incandescent lamp more or less efficient than a lower-wattage lamp?

15.13 The rated life of an HID lamp is determined by operating the lamp at ( ) hours per start.

15.14 Tungsten–halogen lamps contain a halogen gas (iodine, fluorine, etc.), which reacts with the tungsten filament to form a halogen–tungsten compound and redeposits the tungsten back on to the filament, prolonging the life of the lamp. (True) (False)

15.15 The rated life of a fluorescent lamp is determined by operating the lamp at ( ) hours per start.

15.16 The color temperature of fluorescent lamps is fixed at 3500 K. (True) (False)

15.17 The use of krypton gas in an incandescent lamp results in what benefit?
a. Increased efficacy.
b. Increased lamp life.

15.18 An electromagnetic ballast for fluorescent lamps may consume up to ( ) percent of the rated lamp power. If a lamp is rated for 40 W, the ballast may consume ( ) W; thus, the total power consumed by the lamp and ballast is ( ) W.

15.19 Which of the following statements applies to PAR lamps that utilize halogen capsules?
a. They are a direct replacement for incandescent PAR lamps.
b. They have better beam control than incandescent PAR lamps.
c. They are fully dimmable.
d. They run hotter than incandescent PAR lamps.
e. They produce a whiter light than incandescent PAR lamps.
f. They cannot replace standard R-shaped incandescent lamps.

15.20 The light output of a standard 40-W fluorescent lamp is about ( ) percent of its input energy.

15.21 Name the three major types of high-intensity discharge lamps.

15.22 What are the characteristics of LPS lamps?

15.23 How can infrared technology improve the efficacy of tungsten–halogen lamps?

15.24 The efficacy of HPS lamps is as high as ( ) lumens/watt, excluding the power loss of the ballast.

15.25 All HID lamps require a ballast. (True) (False)

15.26 All incandescent lamps have the same color temperature. (True) (False)

15.27 The rated lumen output that appears in a manufacturer's catalog for a fluorescent lamp assumes the lamp is operated on a ballast having a ballast factor of ( ).

15.28 The ballast factor is always smaller than 1. (True) (False)

15.29 Where are MR16 and MR11 lamps typically used?

15.30 List five ways in which aluminized MR16 lamps differ from dichroic MR16 lamps.

15.31 What are the primary benefits of titanium-coated MR16 lamps over other MR16 types?

15.32 High-pressure sodium lamps are monochromic. (True) (False)

15.33 What is the typical efficacy of a 1000-W, clear MH lamp? (85) (100) (150) lpw.

15.34 What is the depreciated lamp lumen output of a 400-W, clear HPS lamp at 70 percent of its rated life? (65 percent) (70 percent) (73 percent)

15.35 Name two electrodeless lamps.

15.36 What is the difference between the IES and CIE lighting system classifications?

15.37 What are the two most valuable pieces of data on the performance of a luminaire in the luminaire's photometric report?

15.38 What is a light pipe, and what are its applications?

15.39 In addition to the type of light source, installation, and maintenance, what are the important elements in selecting a lighting fixture for interior space?

15.40 Initial lumens of a light source are typically measured after how many hours of operation?

15.41 Lighting accounts for approximately what percentage of all electrical energy consumed in the United States?

15.42 Define *luminous efficacy*.

15.43 Theoretically, 1 W of electric power can be converted to approximately how many lumens? (a) 600, (b) 60, (c) 6

15.44 What is the approximate lamp life of
a. An incandescent lamp including halogen and quartz?
b. A fluorescent and HID lamp?

15.45 What is the approximate luminous efficacy range of
  a. Incandescent lamps?
  b. Fluorescent lamps?
  c. High-intensity discharge lamps?

15.46 For incandescent lamps, how is the rated life related to efficacy?

15.47 The use of rare-earth triphosphor in fluorescent lamps offers what two benefits?

15.48 Explain briefly the light-generating process of a fluorescent lamp.

15.49 For most fluorescent lamps, what are the optimal bulb-wall temperature and the optimal ambient temperature?

15.50 What are the three most common lamp-color temperatures for fluorescent lamps?

15.51 Rank the luminous efficacy of HID lamps (mercury, metal halide, HPS) in the order of increased efficacy.

15.52 Rank the following from low to high in luminous efficacy and in color temperature:
  a. Mercury vapor lamp.
  b. Metal halide lamp.
  c. High-pressure sodium lamp.

15.53 Which type of lamp has a better lumen depreciation factor—fluorescent or metal halide?

15.54 Which of the following light sources is (are) monochromatic—fluorescent, mercury vapor, metal halide, high-pressure sodium, low-pressure sodium, and LEDs?

15.55 Which of the following is (are) not appropriate for describing a low-pressure sodium lamp?
  a. High luminous efficacy.
  b. Poor color rendition.
  c. Used for interior lighting.

15.56 What is meant by triphosphor?

15.57 What are two basic types of fluorescent lamps and where are they typically used?

15.58 The newer T5 fluorescent lamps are more efficient than T8 fluorescent lamps. (True) (False)

15.59 Ambient temperature has a great effect on the light output of a fluorescent lamp. (True) (False)

15.60 The newer T5 fluorescent lamps are designed to have a peak light output at higher ambient temperature than other general-lighting fluorescent lamps. (True) (False)

15.61 The newer T5 fluorescent lamps offer luminaire manufacturers what two design opportunities?

15.62 What is the best application for a fluorescent lighting system that utilizes programmed rapid-start electronic ballasts and why?

15.63 What two characteristics of higher-wattage compact fluorescent lamps make them an attractive alternative to low-wattage metal halide lamps?

15.64 What are two benefits of electrodeless fluorescent lamps over conventional fluorescent lamps that have cathodes?

15.65 Name three benefits of ceramic metal halide lamps over those that use quartz arc tubes.

15.66 LEDs produce light (a) in a continuous spectrum or (b) with a nearly monochromatic spectrum.

15.67 The rated life of a LED is based on 50% _____
  a. survival.
  b. of initial light output.

15.68 For a directional light source such as a PAR lamp or a MR16 lamp, field lumens is the output of the lamp taken within ( ) percent of the center-beam candlepower, whereas beam lumens is the output of the lamp taken within ( ) percent of the center-beam candlepower.

CALCULATIONS OF ILLUMINATION*

16

THIS CHAPTER COVERS THE PROCEDURES FOR analyzing the visual quality requirements and for determining the quantity requirements of a specific space or visual task.

The procedure involves the following sequential steps:

- Analysis of the visual environment
- Determination of illumination requirements
- Tentative equipment selection
- Preliminary engineering calculations
- Final equipment selection
- Final engineering calculations
- Final equipment layout

In fact, like any other design process, lighting design is a trial-and-error process. Several tentative selections and equipment layouts may have to be made prior to achieving an optimum solution. Experienced designers can often shorten this process by selecting the proper equipment and layout from the start, bypassing much of the repetitions. Nevertheless, the procedure still has to be followed.

The final layout of the lighting equipment, more appropriately called the *lighting system design*, is as much an art as a science. Lighting system design will be discussed in detail in Chapter 17.

16.1 QUANTITY AND QUALITY OF ILLUMINATION

Strictly speaking, quantity and quality of illumination are inseparable in that quantitative needs are closely related to quality. High-quality lighting can often compensate for a lack of quantity; conversely, poor lighting often requires higher lighting quantity to overcome the loss of visual performance. For example, a light source that is aimed at a visual task, such as a book, but is not

shielded from the normal viewing angle of the reader, will create glare or vailing reflection, which washes out the reading task. At higher illuminance levels, the glare may cause discomfort or even temporary blindness. In this case, a higher quantity of illumination could even be counterproductive.

Quality of lighting can be quantified, yielding numerical values that can be compared. However, the calculation processes are usually very complex; thus the terms specifying quality will be introduced only briefly here:

1. *Visual comfort* may be evaluated in terms of the *discomfort glare rating* (DGR), a numerical assessment of the *index of sensation* (M) of all light sources within the field of an observer, such as luminaires (lighting fixtures), windows, and skylights. The index M is a function of the size of the light sources, their luminance, and their locations relative to the field of view. The values may vary from 20 to 600, with a lower number being more visually comfortable.

   Visual comfort can also be expressed in terms of *visual comfort probability* (VCP), which converts the DGR values into a probable comfort level for a group of observers. The VCP values are given in percentage (%) from 1 to 99 with higher values being more comfortable. A VCP value of 70 percent is considered the borderline between comfort and discomfort.

   Unified glare rating (UGR) is a more recently developed glare rating system based on similar input values of luminance, size, and position factors as criteria. Final values range from 5 to 30, with higher numbers indicating higher discomfort.

2. *Equivalent sphere illumination* (ESI) evaluates the relative effectiveness of the illuminance. It is calculated in ESI foot-candles (lux). ESI is used primarily for evaluating large office and classroom lighting systems using sophisticated mathematics or computer calculations. The procedures are complex and beyond the scope of this text.

3. *Uniformity of illuminance*—although uniformity is not necessarily a desirable criterion for many lighting applications, such as in residential and entertainment

*Steve T. Andert P.E. is contributing author of this chapter. See Acknowledgment.

facilities uniformity is important for most working spaces, such as in general offices, laboratories, and classrooms. Uniformity of illuminance is especially important in sports lighting applications and is critical for high-speed sports such as tennis, baseball, and ice hockey in which players must react quickly to the moving target of a ball or an ice puck at close range. The reader is directed to the IES publication PR-6-2003 for specific recommendations of uniformity criteria and methods of evaluation.

4. *Color Rendering Index* (CRI) is a quality factor to be taken into account in nearly all lighting design considerations. The values of CRI range from near 0 to 100, as discussed in Chapter 14 and 15. A higher CRI is desired, although it is often compromised owing to economical and maintenance considerations.

16.2 EVALUATING THE VISUAL ENVIRONMENT

Before the quantitative requirements for a visual task or a space can be determined, the design issues must be systematically analyzed to identify each issue's importance or relevance. The Illuminating Engineering Society of North America in its ninth edition handbook (*IESHB*) identified a series of issues that must be evaluated in the design of a lighting system. The recommendation is to rate each issue as very important, important, somewhat important, or not important or not applicable.

The design issues and the recommended illuminance categories are:

1. *Design issues.*
   - Appearance of space and luminaires
   - Color appearance and color contrast
   - Daylighting integration and control
   - Direct glare
   - Flicker and strobe
   - Light distribution on surfaces
   - Light distribution on task plane (uniformity)
   - Luminances of room surfaces
   - Modeling of faces or objects
   - Points of interest
   - Reflected glare (veiling reflection)
   - Shadows
   - Source and task geometry
   - Sparkle or desirable reflected highlights
   - Surface characteristics
   - System control and flexibility
2. *Illuminance selection.*
   - *Horizontal illuminance:* its importance and the recommended illuminance category.
   - *Vertical illuminance:* its importance and the recommended illuminance category.
3. *Evaluation* of these design issues and luminance category will enable the designer to narrow down the options to one or more design solutions.
   - Preliminary and final selection of lamp sources in terms of appearance, color, direct glare, reflected glare, flicker, sparkle, modeling, shadow, system controls, and switching flexibility.
   - Preliminary and final selection of luminaires in terms of appearance, direct glare, reflected glare, light flux distribution, controls.
   - Preliminary and final layout of the luminaires in terms of flux distribution, daylighting integration and controls, modeling of faces or objects, points of interest, direct and reflected glare, geometry.

The evaluation process addresses all aspects of lighting design issues and leads to the determination of the illuminance level considered most appropriate for the space or task. A complete listing of commonly encountered locations and tasks are given in the IESNA Lighting Design Guide in Chapter 10 of the ninth edition of the *IESHB*. A condensed version of this design guide is shown in Tables 16–1A, B, and C that is considered adequate for the intent of this book. The user will find the complete listing of all tasks and locations at the end of Chapter 10 of the *IESHB*.

The user should also be reminded that the design of a lighting system is not limited to the determination of lighting quality and quantity. Economic factors including initial construction and maintenance costs, and environmental factors including energy consumption and ecological impact on the environment are all part of the final design process.

16.3 ILLUMINANCE CATEGORIES AND RECOMMENDED ILLUMINANCE LEVELS

The IESNA established nine illuminance categories in 1979 with a range of illuminance values for categories from A thru I. Each category covers a range of illuminance values. These categories have been reduced to seven by consolidating the last three categories into one and the illuminance range for each category into a single value. The reasons for such consolidations are given in the handbook. The authors, with years of experience in building design, believe there is merit to giving the designer flexibility in selecting an illuminance value slightly below or above a single value. The

TABLE 16–1A
Design guide for selected interior lighting tasks and locations

Tasks & Locations	Appearance of Space and Luminaires	Color Appearance (and Color Contract)	Day Lighting Integration and Control	Direct Glare	Flicker (and Strobe)	Light Distribution on Surface	Distribution on Task Plane (Uniformity)	Luminances of Room Surfaces	Modeling of Faces or Objects	Point(s) of Interest	Reflected Glare	Shadows	Source/Task/Eye Geometry	Sparkle/Desirable Reflected Highlights	Surface Characteristics	System Control and Flexibility	Special Considerations	Notes on Special Considerations	Illuminance (Horizontal)	Category (see Table 16-2)	Illuminance (Vertical)	Category (see Table 16-2)	Notes on Illuminance	Additional Notes
Letter designation of tasks and locations of design issues: V - Very important / I - Important / S - Somewhat important																								
Auditorium																								
Assembly	S	I	I	S	S		S	S								V			I	(C)				
Social activities	S	I	I	S	S		S	I								V			S	(B)	S	(A)		
Churches & synagogues																								
Congregational areas	I	V	I	V	V	S	I	S	I	S	I	I	I	S	I	I			I	(C)	I	(A)		
Leadership area	V	V	V	V	V	I	I	I	V	I	I	V	V	V	V	V			I	(D)	V	(D)		
Highlighted items	I	I	S	I	I	S	S	S	I	V	S	S	S	I	S	S	S				V	(D)		
Merchandising																								
Sales area	I	I	I	V	S	S	I	S	I	S	I	S	I						I	(D)				
Circulation	I	I		S	S	S	S		S		S								S	(C)				
General display	I	V	I	I	S	I	I	S	I	S	I		S	I		S			I	(E)	I	(C)		
Feature display	V	V	I	V	S	I	S		V	I	I	S	I	I		S			I	(F)	I	(D)		
Show windows	V	V	V	I	S	I	S		V	I	V	S	I	I		S			S	(G)	I	(E)		
Museums																								
Exhibit cases	I	V	V	V	S	V	V	I	V	I	V	V	V	I	I	I	V	(7)	V	(D)	V	(B)		
Three-dimensional objects	I	V	V	V	S	V	V	I	V	I	V	V	V	S	I	I	V	(7)	V	(D)	V	(B)		
General areas	V	S	V	S	S	S	S	S	I	V		S		I	S	S	S	(7)	I	(C)	I	(A)		
Offices																								
Filing	S	S		I	S	I	I	S	S		I	I	I		I				I	(E)	V	(C)		
Intensive VDT	I	I	I	V	I	I	I	V	I		V	I	V		I		S	(14/15)	I	(D)	V	(B)		
Private office	I	I	V	V	I	I	I	V	I		V	I	I		I				I	(E)	I	(B)		
Residences																								
General lighting	S	S	S	S		S	S	S	S	S		S	S	S					S	(B)				
Relaxation, conversing	S	V	S	S		S		S	V	S	S		S	S	V				S	(A)	S	(A)		
Dining	S	V	S	V		S	S	S	V	S	S		S	S	S	S			V	(B)				
Kitchen counter (cutting)		V		V		I	I	I			V	S	V		I				V	(E)	V	(C)		
Educational																								
Classrooms (reading)	S	S	I	I	I	I	I	I	S	I	I	S	I		I	S				(16)		(16)		
Art room		V	V	I	S	I	I	S	V		S	I	V	I					V	E	V	D		
Laboratories (science)		V	V	V	I		V	I	I	I	V	V	V		S				I	E	I	D		

Notes:
(7) Degradation factors important to consider.
(14) Lighting should be flexible to accommodate changes in office furniture layout.
(15) Acoustical aspects of luminaires need to be considered.
(16) See Table 16–3 for reading tasks

Based on the recommended procedure of the Illuminating Engineering Society of North America. (Refer to the *IESHB*, 9th ed. or latest, for a detailed listing of tasks and locations.)

TABLE 16–1B
Design guide for selected outdoor lighting tasks and locations

Tasks & Locations	Appearance of Space and Luminaires	Color Appearance (and Color Contract)	Direct Glare	Light Distribution on Surface	Light Pollution/Trespass	Modeling of Faces or Objects	Peripheral Detection	Point(s) of Interest	Reflected Glare	Shadows	Source/Task/Eye Geometry	Sparkle/Desirable Reflected Highlights	Surface Characteristics	Special Considerations	Notes on Special Considerations	Illuminance (Horizontal)	Category (see Table 16-2)	Illuminance (Vertical)	Category (see Table 16-2)	Notes on Illuminance	Special Notes
Building exteriors																					
Entrance (active)	V	V	V	I	V	V	V	V	V	V	V	I	V			V	(B)	V	(A)		
Entrance (inactive)	S	I	I	S	V	I	I	I	S	I	I	S	I			I	(A)	I	(A)		
Prominent structure	V	I	I	V	V	I	S	V	V	V	V	S	V			I	(B)	V	(A)		
Buildings, floodlighted																					
Light surface	V	I	I	V	V	I		I			I	I	S	V	(2)			I	(A)		
Medium light	V	I	I	V	V	I		I			I	I	S	V	(2)			I	(B)		
Medium dark	V	I	I	V	V	I		I			I	I	S	V	(2)			I	(B)		
Dark	V	I	I	V	V	I		I			I	I	S	V	(2)			I	(C)		
Poster board																					
(dark surroundings)																					
Light surface	V	I	I	V	I	V		V	V	S	V	S	V	S	(2)	S	(A)	V	(C)		
Dark surface	V	I	I	V	I	V		V	V	S	V	S	V	S	(2)	S	(A)	V	(D)		
Loading and unloading																					
Platforms			I		I		I			S						I	(C)	S	(A)		
Freight car interior			I		I		I			S						I	(B)	S	(A)		
Parking areas	I	V	I	V	V	V	I	V	V	V			S			I		V		(9)	
Retail space—outdoor																					
Fast-food restaurant	V	I	V	I	V	I	I	V	V	I	V	I	V			I	(C)	V	(A)		
Convenience store	V	I	V	I	V	I	I	V	V	I	V	I	V			I	(A)	V	(A)		
Pedestrian mall	V	I	V	I	V	I	I	V	V	I	V	I	V			I	(A)	V	(A)		
Roadways	V	S	V	V	V	I	V	S	V	V	V		S			I		V		(9)	
Walkways	I	V	V	I	V	V	V	I	V	V	V		S	V	(3)	I		V		(9)	

Notes:

(2) Lighting must not interfere with visibility of pedestrians, motorists, or boaters.

(3) Hazards such as stairs or areas adjacent to bodies of water should be clearly identified and lighted for safety.

(9) The recommended illuminance values generally are below 10 lx; however, for safety and security reasons, many municipal codes require considerably higher illuminance values. The design shall comply with the code requirements. See Chapter 17 for design considerations.

Based on the recommended procedure of the Illuminating Engineering Society of North America. (Refer to the *IESHB*, 9th ed. or latest, for a detailed listing of tasks and locations.)

TABLE 16–1C

Design guide of selected sports and recreational tasks and locations

Tasks & Locations — Letter designation of tasks and locations of design issues: V – Very important; I – Important; S – Somewhat important	Color Appearance (and Color Contract)	Day Lighting Integration and Control	Direct Glare	Flicker (and Strobe)	Distribution on Task Plane (Uniformity)	Light Pollution/Trespass	Luminaire Noise	Modeling of Faces and Objects	Reflected Glare	Shadows	Special Considerations	Notes on Special Considerations	Illuminance (Horizontal)	Category (see Table 16–2)	Illuminance (Vertical)	Category (see Table 16–2)	Notes on Illuminance	Special Notes
Baseball																		
Professional (outdoor)	I		V	V	V	S		V		V			V		I		(2)	
Recreational	S		V	S	V	S		S		S			V		S		(2)	
Basketball																		
Indoor	I	I	V	I	V		I	I	V	V			V		I			
Outdoor	S		V	I	V	S		I		I			V		S			
Football																		
Professional (indoor)	I	I	V	V	V		I	V	I	V			V		S			
Professional (outdoor)	I		V	V	V	S		V		V			V		I			
Recreational (outdoor)	S		V	S	V	S		S		S			V		S			
Hockey/skating, ice																		
Indoor	S	I	V	I	V		I	S	V	I			V		I			(4)
Outdoor	S		V	I	V	I		S	V	I			V		I			
Water sports																		
Indoor (on deck)	S	I	V		I		I	I	V	S			V		I			
Outdoor (on deck)	S		V		I	S	S	I	V	S			V		I			
Tennis																		
Indoor	S	I	V	V	V		V	V	I	I			V		V			
Outdoor	S		V	V	V	S	S	V	I	I			V		V			
Volleyball																		
Indoor	S	I	V	V	V		V	V	I	I			I		V			
Outdoor	S		V	V	V	S	S	V	I	I			I		V			

Notes:

(2) Infield values.

(4) Supplementary lighting may be necessary to minimize harsh shadows next to the dashboard if the light sources are positioned behind the hockey rink dashboard boundary lines.

Based on the recommended procedure of the Illuminating Engineering Society of North America. (Refer to the *IESHB*, 9th ed. or latest, for a detailed listing of tasks and locations.)

primary reason is that the illuminance values between the adjacent categories are too far apart. For example, current recommended illuminance values of categories D, E, and F are 300, 500 and 1000 lx respectively. The values are nearly doubled for each of the next categories. In practice, aesthetics, equipment availability, spacing, and cost factors frequently demand that designers fine-tune a design for optimum solutions. For these reasons, Table 16–2, Determination of illuminance categories, contains two columns—one lists the single value of current recommended practice, and the other lists a range of values for designers to evaluate, with the midrange of each category corresponding to the current IES recommended value. One will notice that the upper value of a category is close or equal to the lower value of the next higher category, thus making the selection of illuminance values more continuous rather than increasing in large steps. For example, the higher range of category B (75 lx) is equal to the lower range of category C (75 lx), and the higher range of category C (150 lx) is close to the lower range of category D (200 lx).

It should be noted that owing to both uncertainty in photometric measurements and uncertainty in space conditions, such as room surface reflectance or ceiling obstructions or configurations, variances of ±10 percent between the measured and calculated illuminance values are to be expected in actual installations.

16.3.1 Guidelines for Selecting the Illuminance Categories

The most common visual tasks encountered in lighting design are among the D, E, and F categories, in which visual tasks can be described in terms of luminance contrast and task size. The criteria for determining those two factors are given in Figures 10–10 and 10–11 of the *IESHB*, 9th ed., and briefly described here.

1. *Criteria for determining illuminance contrast.* As defined in Eq. (14–1) of Chapter 14, contrast of a visual task against its immediate background is the absolute difference between their luminance values divided by the background luminance. Equation (14–1) is repeated here:

$$C = |L_t - L_b|/L_b \qquad (14\text{–}1)$$

Because luminance (L) is proportional to reflectance (ρ) under the same lighting (illumination) conditions, Eq. (14–1) can also be expressed as a ratio of reflectance,

$$C = |\rho_t - \rho_b|/\rho_b \qquad (14\text{–}1a)$$

where C = contrast, per unit (pu)*

(*) Use either ρ_t or ρ_b as the denominator, whichever is greater.

Therefore, C should be always smaller than unity, or 1.

2. *Borderline between high and low contrast.* It is suggested that a value of 0.3 be considered as the borderline between high and low contrast. Accordingly:

   - If $C \le 0.3$, the task is of low contrast, and at even lower contrast, the higher illuminance range of the same category is recommended.
   - If $C > 0.3$, the task is of high contrast, and at even higher contrast, the lower illuminance range of the same category is recommended.

3. *Criteria for determining task size.* The size of a visual task is not determined by its physical size but rather by the visual angle extended between the task and the viewer. The visual angle in a three-dimensional (3-D) space is called a *solid angle* and is measured in steradians (sr). One steradian is equal to $2\pi (1 - \cos\theta)$, where θ is the half-cone angle of a 3-D visual cone in degrees. The following examples serve to illustrate the relationship between real visual tasks and their solid-angle dimensions:

   - 8-point typeface (*) read from 19 in. (50 cm) 3.1×10^{-6}
     (*average solid angle of total printed area for numerical digits)
   - 12-point typeface read from 19 in. 6.9×10^{-6}
   - 3×3 in. (7.5 cm $\times$ 7.5 cm) viewed from 100 ft (30 m) 6.3×10^{-6}
   - 12×12 in. (30 cm $\times$ 30 cm) viewed from 100 ft 1.0×10^{-4}
   - AWG No. 20 wire (0.81 mm diameter) viewed from 15 in. (40 cm) 3.5×10^{-6}
   - 0.04 drilled hole (1.02 mm diameter) viewed from 15 in. 5.6×10^{-5}

4. *Borderline between large- and small-size visual tasks.* It is suggested that a solid angle of 4.0×10^{-6} be considered as the borderline between large and small visual tasks. Accordingly,
   - If the visual task size is $\le 4.0 \times 10^{-6}$, the task is considered small; and if the task is smaller, the higher illuminance value of the same category is recommended.
   - If the visual task size is $>$ than 4.0×10^{-6}, the task is considered large; and if the task is larger, the lower illuminance value of the same category is recommended.

TABLE 16–2
Determination of illuminance categories[a]

For orientation and simple visual tasks

These visual tasks are found in public spaces where reading and
visual inspection are only occasionally performed. Select higher categories
are recommended for special tasks located within those spaces

Category	Type of Activities	Target and Range of Illuminances[d]	
		Target (lx)[b]	Range (lx)[c]
A	Public spaces	30	20–30–50
B	Simple orientation for short visits	50	30–50–75
C	Working spaces where simple visual tasks are performed	100	75–100–150

For common visual tasks

These visual tasks are found in commercial, industrial, and residential
applications. Higher levels are recommended for visual tasks with
critical elements of low contrast and small size.

D	Performance of visual tasks of high contrast or large size	300	200–300–500
E	Performance of visual tasks of high contrast and small size or low contrast and large size	500	300–500–750
F	Performance of visual tasks of low contrast or small size	1000	750–1000–1500

For special visual tasks

These visual tasks are very specialized, including those with very small or
very low contrast critical elements. Recommended illuminance levels
should be achieved with supplementary task lighting.

G	Performance of visual tasks of low contrast and small size	3000–10,000	3000–10,000

[a]To account for both uncertainty in photometric measurements and uncertainty in space reflections, measured illuminances should be with $\pm$ 10% of the recommended values. It should be noted, however, that the final illuminance may deviate from these recommended values owing to other light design criteria.
[b]Target illuminance is the maintained illuminance values as recommended in the *IESHB,* 9th ed.
[c]Range of illuminance is the maintained illuminance values as recommended in the *IESHB,* 8th ed., which allows the designer to choose among low, and medium, high levels within the illuminance range. See suggested procedures for selecting the illuminance value within the recommended range of each category
[d]Values indicated are in lux. Divide by 10.76 for equivalent foot candles.

(The tabulated illuminance values for each of the categories are merged from the *IESHB,* 8th and 9th eds.)

16.3.2 Additional Factors Affecting the Selection of Illuminance Values

Two additional factors affect the selection of illuminance values: age and speed.

1. *Age factor.* It has been well documented that the visual acuity and sensitivity of older people, say, 60 years and older, are greatly reduced. Thus, in a lighting environment designed for older people, such as in retirement homes, or in mixed-use facilities where older persons are frequently present,

higher illuminance value of the same category, or the next-higher category, should be considered. Because of the higher illuminance level, associated direct and reflected glare for the light sources and visual tasks should be carefully evaluated.

2. *Speed factor.* The speed factor can easily be understood. In a high-speed sport, with the ball or playing object approaching the player or players at high speed, say 100 mph (161 km/h) at close range, a high-illuminance and low-glare system is necessary. In sports such as baseball and tennis, the balls are actually

relatively large (about 3 in. in diameter, or 6.3×10^{-6} sr at 100 ft). However, when the ball is approaching a player only a few feet away, the solid angle is no longer a factor; rather, it is the time of response that governs the visual need. Table 16–1C provides the general guidelines for evaluating sports facilities; however, it does not provide recommended illuminance categories. The recommended illuminance category for various playing levels of all sports, including college and professional levels, are published in IESNA *Manual RP-6-01: Recommended Practice for Sports and Recreational Area Lighting*, which should be referred to for specific guidelines. For quick reference, the recommended illuminance category of several selected sports are tabulated in Table 16–3. It should be noted that the recommended illuminance categories for most sports are still given in horizontal illumination even though vertical illumination is equally important for most sports. This is because horizontal illuminance is easier to measure in the field, and if the uniformity in the field is maintained as recommended, vertical illuminance will be 50 percent or better with modern sports luminaires.

16.3.3 Examples

Example 16.1 Select the illuminance level for a conferring/meeting room of a retirement home.

- A retirement home is one form of residential housing. According to Table 16–1A, most design issues are generally in the somewhat important (S) area, and the recommended horizontal illuminance values are in the A and B categories.
- From Table 16–3, the recommended illuminance category for a general conferring/meeting room is B, or 50 lx for vertical and D or 300 lx for horizontal. Because this room is used for retired persons, the high values of the B and D range are recommended. Thus, the vertical illuminance should be 75 lx in lieu of 50 lx, and the horizontal illuminance should be 500 lx in lieu of 300 lx.

Example 16.2 Select the illuminance level for a kindergarten room.

- Kindergarten rooms are generally for reading large prints, graphic displays, and larger-size visual tasks.
- According to Table 16–1A most of the design issues are in the very important (V), and important (I) categories. From Table 16–3, for reading of

mixed materials, the recommended horizontal illuminance category is E, or 500 lx; however, since this room is for young children, and most of the reading tasks are for short duration and casual, the lower illuminance value of the E category, or 300 lx, is recommended.

Example 16.3 Select the illuminance value for two college baseball fields—one for intercollegiate competition and another for intramural play.

- From Table 16–1C, direct glare, flicker, uniformity, modeling of face and objects, shadows, and vertical illuminance are very important for professional level (higher than intercollegiate level), whereas flicker, light pollution, modeling, and shadow issues are less important for the recreational (intramural) level.
- From Table 16–3, Class I level, 1500 lx (150 fc) for the infield, and 1000 lx (100 fc) for the outfield, are recommended for this intercollegiate field. Refer to *IESNA RP-6-01* for recommended illuminance of Class II, if desired.
- From Table 16–3, Class IV, 300 lx (30 fc) for the infield, and 200 lx (20 fc) for the outfield, are recommended for the intramural field. Refer to *IESNA RP-6-01* for recommended illuminance of Class III, if desired.

Example 16.4 Select the illuminance level for a computer or word-processing laboratory.

- The type of task would normally be high contrast and small size. From Table 16–2 illuminance category E (500 lx) or F (1000 lx) would be the proper selection. Final selection between categories E and F may be affected by the selected lighting systems.
- Of all the design issues, direct glare from the light sources and reflected glare (image) of the luminaires on the VDT terminals are both very important. Thus, an indirect lighting system (category E) or a narrow distribution direct lighting system (category F) are probably the systems to consider.

Example 16.5 Select the illuminance level for a medium-size hotel lobby.

- A hotel lobby includes various spaces, such as passageways (category A), seating (category B), casual reading (categories C and D), snack bar (categories C and D), check-in counter (category D or E), and cashier (category E or F).

TABLE 16–3

Recommended illuminance values of common task/areas

Tasks & Locations	V	H	Tasks & Locations	V	H	Tasks & Locations	V	H
						Sports & Recreation (1) & (2)		
Interiors			Interiors					
Auditoriums			Reading			Indoor Sports (in fc; x 10 for lux)		
Assembly	–	C	Photocopies	–	D	Basketball		
Social activity	A	B	CRT screens	A	A	Class I	(2)	125
Banks			#3 pencil and softer leads	–	E	Class II	(2)	80
Lobby—general	A	C	#4 pencil and harder leads	–	F	Class III	(2)	50
Writing area	A	D	Ballpoint pen	–	D	Bowling		125
Teller's station	A	E	Reading mixed material	–	E	Approach	–	80
Conference Rooms			Schools			Target	120	–
Conferring/meeting	B	D	Classrooms (see Reading)			Ice Hockey		
Critical seeing (see Reading)			Science laboratories	D	E	Class I	(2)	125
Video	D	E	Shops (see IES handbook)			Class II	(2)	100
Drafting			Stairways and Corridors	–	C	Class III	(2)	75
CAD	A	C	Residential Spaces			Tennis		
Mylar	D	F	General lighting	–	B	Class I	(2)	125
Exhibition/Convention			Entertainment	A	A	Class II	(2)	75
General	A	C	Passage areas	–	B	Class III	(2)	60
Display	D	F	Specific visual tasks			Outdoor Sports (in fc; x 10 for lux)		
Libraries			Dining	–	B	Baseball	(2)	
Reading areas (see Reading)			Grooming	B	D	Class I (Infield/ outfield)	(2)	150/100
Book stacks—active	D	–	Kitchen general	B	D	Class II (Infield/ outfield)	(2)	100/70
Card files	B	D	Kitchen counter	C	E	Class IV (Infield/ outfield)		30/20
Audio/visual areas	–	D	Kitchen range	C	E	Football		
Merchandising Spaces			Kitchen sink	C	E	→ Class I	(2)	100
Circulation	–	C	Laundry	A	D	Class II	(2)	50
Merchandise display	C	E	Music study (piano)	B	D	Class III	(2)	30
Offices			Advanced scores		E	Golf driving range		
Accounting (see Reading)			Reading (services)	C	E	At tee	–	20
Audio/visual area						Down the range	10	–
Conference (see Conference Rooms)						Softball		
Drafting (see Drafting)						Class II (Infield/ outfield)	(2)	100/70
General and private offices	B	E	Exteriors			Class IV (Infield/ outfield)	(2)	30/20
Libraries (see Library)			Building entrance (active)	A	B	Swimming		
Lobbies, lounges and reception areas	A	C	Prominent structure	A	B	Class I	–	75
			Floodlighted (light)	A	–	Class II		50
Mail sorting	A	E	Floodlighted (dark)	C	–	Class III	–	30
Offset printing and duplicating area	A	D				Skiing		
Spaces with VDTs (1)						Class IV	0.5	–

Notes: (1) Classes I, II, III, and IV are for professional, national, college, regional, club, high school, and recreational levels, respectively, with considerable overlap between these levels.
(2) See Section 16.3.2, explanation for vertical illuminance values.

- No single lighting system will satisfy all the spaces and tasks. In all cases, color and appearance are very important. In addition, a hotel lobby often has paintings and decorative objects, which are points of interest requiring special modeling. Occasional high contrast in illuminance with soft shadows may even create visual interest. A liberal use of point light sources for special modeling, well-shielded fluorescent light sources for the working counters, and large-area and low-brightness light sources, including skylights over the seating areas, are just a few of the possible design solutions. (See Chapter 17 for layout and design considerations.)

16.4 BASIS FOR ILLUMINATION CALCULATIONS

Illumination calculations involved in lighting design are based on the principle of luminous flux transfer from the light source or sources to a surface. Normally, the transfer is through clean air and is assumed to have no loss.

Many quantitative and qualitative values of a lighting system can be calculated; however, the most important one is the lighting power density or illuminance and, to a lesser degree, the luminance (brightness) and luminance ratio (contrast) of surfaces.

16.4.1 Calculating Light Power Density: Illuminance

There are two methods to calculating illuminance:

1. *Lumen method.* By the definition given in Eq. (14–5b) of Chapter 14, light power density or illuminance (E) on a surface is the number of lumens per unit area on a surface illuminated by the light source or sources,

   or $$E = F/A \qquad (14\text{–}5b)$$

   If the surface is in units of square feet, then the illuminance is given in foot-candles (fc). If the surface is in units of square meters, then the illuminance is given in lux (lx).

2. *Point method.* By the definition given in Eq. (14–5c), illuminance (E) on a surface is also equal to the light intensity (I) of the light source divided by the square of the distance from the surface,

   or $$E = I/d^2 \qquad (14\text{–}5c)$$

If the distance is measured in feet, then the illuminance is in foot-candles. If the distance is measured in meters, then the illuminance is in lux.

It should be noted that the exact dimensional equivalent of 1 fc is 10.76 lx, although for lighting design purposes, designers often equate 1 fc to 10 lx for approximation. These values are not mathematically equivalent.

For example, if a flashlight produces 100 lm of light power, all of which falls uniformly on a surface 2 sq ft in total area, then the illumination level or illuminance is $100/2 = 50$ fc. If the 100 lm falls on a surface 2 m² in total area, then the illuminance is $100/2 = 50$ lx; however, in the first example, the exact equivalent of illuminance of 50 fc is $50 \times 10.76 = 538$ lx.

When the preceding definitions are applied in actual lighting design, Eqs. (14–2) and (14–3) must be modified by including several correction factors or coefficients, since not all luminous flux coming from the light source or sources can be aimed totally on the surface. Furthermore, the light flux of electric light sources (lamps) will depreciate with age and must be accounted for. Those correction factors are introduced and elaborated on in Sections 16.5 and 16.6.

16.4.2 Applicability of the Two Methods

Depending on the relative size of the light source or sources and the receiving surface or surfaces, either one or both of these two methods may be applicable. In general, if the light source is truly small (as a point) or very small (as a MR lamp) aiming at a surface, then the point method will yield more accurate results. If the light source is large (as a fluorescent lamp), or is diffused (as luminaires with diffused lens), then the lumens method will be the only practical method to use manually. In practice, most sophisticated computer programs for illuminance calculations are usually based on the point method by tracing the interreflection of light rays among the room surfaces; however, it is not practical to perform such tedious calculations manually. The point method is used almost exclusively for sports lighting design, especially for outdoor athletic fields.

16.4.3 Calculating Luminance and Luminance Contrast

Sometimes it is necessary to determine the surface brightness of an illuminated surface for glare control, to calculate the luminance contrast of a task against its background to study visual acuity, or to measure the luminance difference of adjacent surfaces for visual comfort. The methods for calculating these values are given in Sections 14.2.7 and 14.2.8 of Chapter 14. The applicable equations can be used directly.

16.5 THE ZONAL CAVITY METHOD

The zonal cavity method is an application of the Lumen's Method ($E = F/A$) to determine the horizontal illuminance on a working plane in an interior space. If all surfaces in a room are painted black, then the illuminance on a horizontal working surface can easily be calculated by knowing the intensity of the direct component of the luminaire or luminaries that are reaching the working surface. However, in reality much of the luminous flux reaching the working surface also originates from the room surfaces through inter-reflection. To account for these inter-reflected components, one would need to perform hundreds or thousands of tedious calculations to achieve reasonable accuracy.

The zonal cavity method utilizes the ray tracing technique by introducing a simple modification factor known as the Coefficient of Utilization (CU) factor for various room conditions and luminaire types. In addition, it also includes a light loss factor (LLF) to account for light losses due to depreciation. Thus, Equation (14–2) can now be written as follows:

$$E = (F/A) \times CU \times LLF \qquad (16\text{–}1)$$

where E = maintained illuminance selected as most appropriate for the task or locations as determined in Section 16.2, fc or lx.

F = total number of lamps times the rated lumens of each lamp as published by the lamp manufacturer, lm.

CU = coefficient of utilization, as a percentage of the total lamp lumens that can be utilized in the room based on the room size, room configuration, room surface reflectance and the performance characteristics of the luminaries, per unit or percent/100.

LLF = light loss factor as a percentage of the rated lamp lumens. Attainable in the field with regard to character of the room, electrical power, luminaire construction, and other maintenance factors, per unit or percent/100.

Each of these terms is explained in the following sections.

16.5.1 Light Loss Factor (LLF)

The light loss factor is a correction factor representing the components of light loss of a lighting system initially as well as during normal operations. These components may be divided into two groups, with each group containing several factors.

1. *Nonrecoverable light loss factors.* The following conditions are inherent in the installed system and cannot be corrected through routine maintenance.
   - *Voltage factor (VF)* All lamps are rated for a standard electrical voltage, such as 115, 120, 125, 208, 230, or 277 V; however, the actual voltage at the lamp socket, even with a well-designed electrical distribution system, will fluctuate during the course of a day. Some older systems may even be operating permanently below their rated voltage. Table 16–4 lists the correction factors for various types of lamps. For normal installation in new buildings, VF = 1 may be assumed.
   - *Temperature factor (TF)* Lamps and luminaires are normally tested at 25°C (77°F). Variations in ambient temperature normally encountered in

TABLE 16–4

Voltage factor (VF) for various types of lamps, percent[a]

Lamp Type	Percent Deviation from Lamp Rated Voltage					
	−5	−4	−3	−2	0	+2
Fluorescent (magnetic ballast)	95	96	97	98	100	102
Fluorescent (electronic ballast)[b]	–	–	–	–	100	–
Mercury (nonregulated ballast)	88	90	92	95	100	105
Incandescent (general service)	83	86	89	94	100	106
Incandescent (halogen)	80	84	88	92	100	108
Mercury (constant wattage)	97	98	99	99	100	102
Metal halide (reactor ballast)	91	93	95	97	100	103
High-pressure sodium	–	–	–	–	100	–

[a]Light output rounded to the nearest 0.5%
[b]Obtain photometric data from fixture or ballast manufacturer for the true lumens output of the lamp ballast combinations.

TABLE 16–5

Typical ballast factor (BF) of electrical discharge lamps

Lamp–Ballast Combination	Typical BF
Fluorescent (30 W and larger) with CBM-certified magnetic ballasts[a]	0.95
Fluorescent (below 30 watts) with CBM-certified magnetic ballasts[a]	0.85
Fluorescent with electronic ballasts (verify with ballast–lamp manufacturer)	0.8 to 1.2
HID lamps (mercury, metal halide, HPS)	EOF[b]

[a]Certified Ballast Manufacturers (CBM).
[b]Luminaire manufacturers often provide a combined factor known as the *equipment operating factor* (EOF) for HID luminaires. This factor includes TF, BF, and PF.

building interiors have little effect on the lumen output of incandescent or HID lamps. There is more of a temperature effect on air-handling-type fluorescent luminaires and outdoor-mounted luminaires. For example, open fluorescent luminaires operating in freezing weather may produce only about 80% of their rated light output. Designs should include the appropriate TF value obtained from the luminaire manufacturer. Normally, for indoor applications, TF = 1 may be assumed.

- *Ballast factor (BF)*   Lamps are tested in the laboratories with a standard reference ballast. The ballast factor is defined as the lumen output of the lamp–ballast combination divided by the rated lamp lumens under the lamp's laboratory testing conditions. Most magnetic ballasts will give a BF < 1.0, and most electronic ballasts will have a BF > 1.0. Typical BF values of selected lamp–ballast combinations are given in Table 16–5. The actual BF value for a specific ballast should be obtained from the product manufacturer.

Example 16.6
A fluorescent luminaire is designed to use a lamp–ballast combination with lamps rated at 3000 lm and a CBM-certified magnetic ballast. The lamp is rated for operating at 120 V, but the voltage at the lamp is only 115 V. Determine the initial lamp light output.

- The actual 115 V at the lamp socket is 4 percent lower than the lamp rated voltage at 120 V. Therefore, from Table 16–4, VF = 0.96.
- From Table 16–5, the CBM-certified ballast has a BF = 0.95.
- The combined light output multiplier is 0.96 × 0.95 = 0.91. (This value represents the initial lamp operating condition of this particular installation. It is also called the *lamp operation factor* (LOF).

- Thus, the initial lamp light output is 3000 × 0.91 = 2730 lm.
- *Lamp position or tilt factor (PF)*   HID lamps are normally designed for burning in a vertical position. Lumen output will be lower when a lamp is mounted at different angles. The manufacturer's latest catalog data should be consulted for the specific lamp-position factor; otherwise, PF = 1.
- *Luminaire surface depreciation factor (LSDF)* Luminaire reflecting and transmitting surfaces will deteriorate with time, some more than others. It is difficult to predict this deterioration, which depends largely on manufacturers' quality control and the building environment. Normally, glass, porcelain, and processed aluminum have negligible depreciation and can be restored to original reflectance; painted surfaces will deteriorate with time; acrylic plastic lasts longer than polystyrene. Although the LSDF may have a drastic impact on the maintained illuminance of an installed lighting system, no reasonable factor can be given. The luminaire should be refinished or replaced as necessary. It is the designer's responsibility to do a life-cycle analysis and to compare alternative products of varying quality. For design purposes, LSDF = 1.

The position factor (PF) is used for HID lamps that are not mounted vertically. The temperature factor (TF) is used for air-return types of luminaires and for cold-weather exterior application only, and the luminaire surface deterioration factor (LSDF) is important only for poor air-quality locations. Therefore, the applicable nonrecoverable light loss factor (LLF) can be greatly simplified as follows:

$$\text{LOF} = \text{VF} \times \text{BF} \qquad (16\text{–}2)$$

where LOF = lamp operating factor

The LOF modifies the rated lumen output of a lamp, and the ballast factor (BF) is applicable to fluorescent and HID lamps only.

2. *Recoverable LLF.* Recoverable factors are those that can be changed by regular maintenance, such as cleaning, relamping, and painting of room surfaces.
- *Lamp lumen depreciation factor (LLD)* Lamp light output depreciates gradually and continuously until the lamp burns out. The LLD is the fraction of the initial lumens produced at a specific time during the life of the lamp. The general characteristics of different lamp types are thoroughly discussed in Chapter 15. LLDs vary widely, even among different sizes of the same

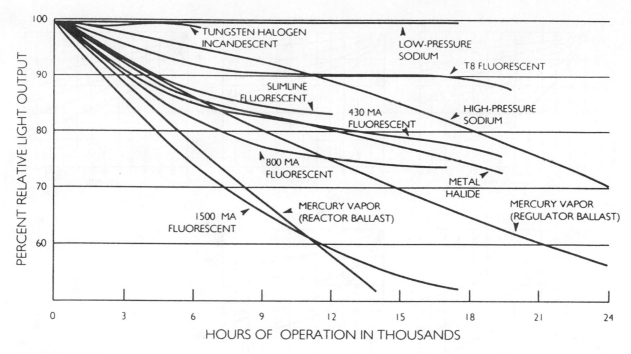

■ FIGURE 16–1

Lamp lumen depreciation (LLD) of typical light sources. LLD will vary between lamp sizes and models. Refer to manufacturers' publications for applicable values. (Courtesy of The National Lighting Bureau.)

type. The designer should refer to manufacturers' catalogs for information on specific types and sizes. For general information, Figure 16–1 shows the LLD for some selected lamp types and sizes. The performance data may be used for comparisons.

Example 16.7

Determine the light output as a percentage of its rated lumens for a T8 fluorescent lamp at 15,000 operating hours (75 percent of its 20,000-hour rated life).

Answer

From Figure 16–1, the lumen output of a T8 fluorescent lamp at 15,000 operating hours is about 90 percent.

- *Luminaire dirt depreciation factor (LDD)* The effect of dirt accumulation on the surface of a luminaire depends on the construction of the luminaire and the air quality in the building in which the luminaire is installed. For example, if the luminaire is installed in an air-conditioned building in a very clean (VC) atmospheric environment, and the luminaire top and bottom is open (maintenance category I), then very little dirt should accumulate on the luminaire. Figure 16–2 shows six sets of graphs indicating the luminaire maintenance categories (I to V), five atmospheric conditions (very clean to very dirty), and LDD versus the number of months between cleanings.

Example 16.8

Determine the LDD of a lighting installation using a luminaire with both top and bottom closed, installed in a very clean (VC) environment, with a 12-month cleaning cycle.

Answer

From Figure 16–3, the VC curve of category V and a 12-month cleaning cycle gives an LDD of 93 percent.

- *Room surface depreciation factor (RSD)* This factor assumes that the room surfaces deteriorate with time. For example, it is usually assumed that a white ceiling or white acoustical tile has 80 percent reflectance; however, a white ceiling will age to a less reflective yellowish color, say 70 percent. Naturally, the illuminance value of the lighting system will drop as well. In such cases it may be advisable to consider using a lower reflectance value for the ceiling initially, say, 70 percent in lieu of 80 percent. Then the effect of depreciation of wall surfaces may be neglected. For design purposes, RSD may be considered as 1.0.

- *Lamp burnout factor (LBO)* Lamp burnout in a multiple lamp installation lowers illumination level than desired. LBO is defined as the ratio of the number of lamps that remain burning to the total installed that is permissible for continuing to perform the visual task intended. This is often

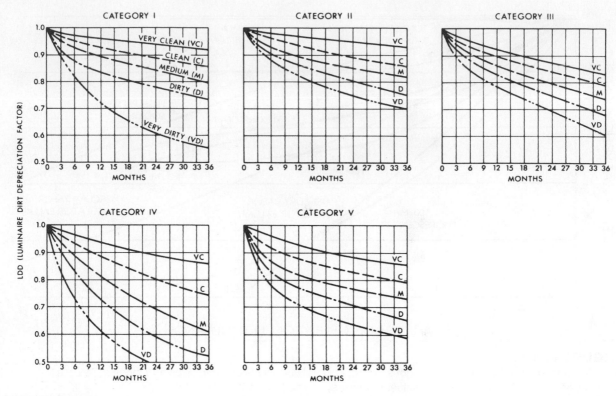

■ **FIGURE 16–2**

Luminaire dirt depreciation (LDD) factor for five luminaire categories. (Reproduced with permission from IESNA.)

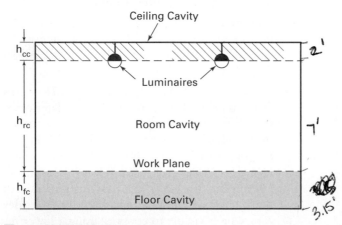

■ **FIGURE 16–3**

Terms of zonal cavity method.

the practice in large industrial plant operations when the luminaires are located in high bay spaces where relamping is difficult. In such spaces it is usually the practice to replace all lamps at once after a predetermined number of hours of use, say, 80 percent of the rated life of the lamps. This practice is called *group relamping*. Lighting

design should make sure that the luminous flux coverage of adjacent luminaires overlaps. Therefore, the burnout of a few lamps will not render any area too dark. Group relamping should be avoided for normal working spaces such as offices, schools, and commercial spaces. For normal design of this type of working space, LBO shall be taken as 1.0.

Of the four recoverable LLF factors, the room surface depreciation factor (RSD) is often compensated for by selecting a lower ceiling reflectance value, and the lamp burnout factor (LBO) is used only for group relamping–type operations. Therefore, the applicable recoverable LLF can be simplified as follows:

$$LLF = LLD \times LDD \qquad (16\text{–}3)$$

where LLF = the applicable portion of the recoverable LLF.

LLD = lamp lumen depreciation factor. See Figure 16–1 or lamp manufacturer's catalog.

LDD = luminaire dirt depreciation factor. (See Figure 16–2.)

16.5.2 Coefficient of Utilization (CU)

The *coefficient of utilization* is the factor for determining the portion of the luminous flux of all lamps (per unit or percent) that reaches the work plane directly and indirectly through inter-reflection in calculating the average illuminance by the zonal cavity method; that is,

$$\text{CU} = \text{flux reaching the working plane} / \text{flux of all lamps} \quad \text{- - - - - - - - - - -}(16\text{-}7)$$

where CU = coefficient of utilization, per unit, or percent/100

The method divides a room into three zones or cavities—ceiling, room, and floor cavities, as illustrated in Figure 16–3.

There are four basic steps to follow in determining the CU of a lighting design or a lighting system:

- *Step 1*   Calculate the cavity ratios of each zone.
- *Step 2*   Select the basic reflectance of each zone.
- *Step 3*   Determine the effective reflectance of each zone.
- *Step 4*   Determine the coefficient of utilization of this lighting system for the space or room.

1. *Step 1: Calculate the cavity ratios.* From the room dimensions and luminaire mounting, as illustrated in Figure 16–3 calculate the cavity ratios (CR) as follows.

$$\text{CR} = 2.5 \times (\text{perimeter area of the cavity} / \text{floor area of the room})$$

$$= 2.5 \times (\text{perimeter area} \times \text{cavity height}) / \text{floor area of the room}$$

$$= 2.5 \times (\text{perimeter/floor area of the room}) \times \text{cavity height}$$

$$= 2.5 \times \text{PAR} \times h \qquad (16\text{-}4)$$

where h = cavity height, *ft* (m)

$\boxed{\text{PAR} = \textit{ratio of perimeter to floor area}}$
$$= 2 \times (L + W)/(L \times W) \text{ for}$$
rectangular rooms (16–5a)
$$= 4/D \text{ for circular rooms} \quad (16\text{-}5b)$$
$$= 3.27/D \text{ for semicircular rooms}$$
$$\qquad\qquad (16\text{-}5c)$$

in which L = length of a rectangular room, ft (m)
W = width of a rectangular room, ft (m)
D = diameter of a circular room, ft (m)

Also,

$$\text{CCR} = 2.5 \, \text{PAR} \times h_{cc} \qquad (16\text{-}6a)$$

$$\text{RCR} = 2.5 \, \text{PAR} \times h_{rc} \qquad (16\text{-}6b)$$

$$\text{FCR} = 2.5 \, \text{PAR} \times h_{fc} \qquad (16\text{-}6c)$$

in which h_{cc} = ceiling cavity height, ft (m)
h_{rc} = room cavity height, ft (m)
h_{fc} = floor cavity height, ft (m)

Example 16.9

A room is 20 ft by 25 ft with h_{cc}, h_{rc}, and h_{fc}, 2.0, 7.0, and 2.5 ft, respectively. Find the cavity ratios.
From Equation 16–5a,
PAR = 2 × (20 + 25)/20 × 25 = 0.18
From Equation 16–6a,
CCR = 2.5 × 0.18 × 2.0 = 0.9
From Equation 16–6b,
RCR = 2.5 × 0.18 × 7.0 = 3.15
From Equation 16–6c,
FCR = 2.5 × 0.18 × 2.5 = 1.13

2. *Step 2: Select the base reflectances.* The surface reflectances of the ceiling, walls, and floor greatly affect the coefficient of utilization of a lighting installation. Although reflectance values may not be known during the initial design of the space, the designer must make a reasonable assumption or an educated guess. In general, unless otherwise specified:
   - As regards ceiling reflectance (R_c), assume a white ceiling having 70 to 80 percent base reflectance, unless otherwise given.
   - For wall reflectance (R_w), assume 50 percent base reflectance for medium- to light-colored walls, 20–30 percent for dark wood paneling, and 60–70 percent for white walls.
   - With respect to floor reflectance (R_f), normally use 20 percent base reflectance for the combination of furniture and floor. Use 10 percent for dark floor finishes and 30 percent for light floor finishes.

3. *Step 3: Determine the effective reflectances.*
   - *Effective ceiling reflectance* (ρ_{cc}) As illustrated in Figure 16–3 when lighting fixtures are mounted a certain distance below the ceiling, the upper portion of the wall between the ceiling and the wall is, in effect, an extension of the ceiling. Thus, the effective reflectance of the ceiling is the combined reflectances of both the ceiling and the upper wall. For surface- or recess-mounted lighting installation (CCR = 0), ρ_{cc} is, of course, the same as the base ceiling reflectance. Table 16–6 gives effective ceiling or floor reflectances calculated from given base ceiling and floor reflectances.
   - *Effective wall reflectance* (ρ_w) The effective wall reflectance is numerically equal to the base wall reflectance ($P_w = R_w$).
   - *Effective floor reflectance* (ρ_{fc}) The effective floor reflectance for CU tables published by the lighting fixture manufacturers is standardized at

TABLE 16-6

Selection of effective ceiling or floor reflectance from base ceiling or floor reflectance*

Cavity Ratio	Percent Base Reflectance 80										Percent Base Reflectance 70										Percent Base Reflectance 60										Percent Base Reflectance 50									
Percent Wall Reflectance	90	80	70	60	50	40	30	20	10	0	90	80	70	60	50	40	30	20	10	0	90	80	70	60	50	40	30	20	10	0	90	80	70	60	50	40	30	20	10	0
0.2	79	78	78	77	77	76	76	75	74	72	70	69	68	68	67	67	66	66	65	64	60	59	59	59	58	58	57	56	55	53	50	50	49	49	48	48	47	46	46	44
0.4	79	77	76	75	74	73	73	72	70	68	69	68	67	66	65	64	63	62	61	58	60	59	59	58	57	55	54	53	52	50	50	49	48	48	47	46	45	45	44	42
0.6	78	76	75	73	71	70	68	66	65	63	69	67	65	64	63	61	59	58	57	54	60	58	57	56	55	53	51	51	50	46	50	49	47	46	45	44	43	42	41	38
0.8	78	75	73	71	69	67	65	63	61	57	68	66	64	62	60	58	56	55	53	50	59	57	56	55	54	51	48	47	46	43	50	48	47	45	44	42	40	39	38	36
1.0	77	74	72	69	67	65	62	60	57	55	68	65	62	60	58	55	53	52	50	47	59	57	55	53	51	48	45	44	43	41	50	48	46	44	43	41	38	37	36	34
1.2	76	73	70	67	64	61	58	55	53	51	67	64	61	59	57	54	50	48	46	44	59	56	54	51	49	46	42	40	38	37	50	47	45	43	41	39	36	35	34	29
1.4	76	72	68	65	62	59	55	53	50	48	67	63	60	58	55	51	47	45	44	41	59	56	53	49	47	44	41	39	38	36	50	47	45	42	40	38	35	34	32	27
1.6	75	71	67	63	60	57	53	50	47	44	67	62	59	58	53	47	45	43	41	38	59	55	52	48	45	42	39	37	35	33	50	47	44	41	39	36	33	32	30	26
1.8	75	70	68	62	58	54	50	47	44	41	66	61	58	54	51	46	42	40	38	35	58	55	51	47	44	40	38	35	33	31	50	46	43	40	38	35	31	30	28	25
2.0	74	69	64	60	56	52	48	45	41	38	66	60	56	52	49	45	40	38	36	33	58	54	50	46	43	39	37	33	31	29	50	46	43	40	37	34	30	28	26	24
2.2	74	68	63	58	54	49	45	42	38	35	66	60	55	51	48	43	38	36	34	32	58	53	49	45	42	37	34	31	29	28	50	46	42	38	36	33	29	27	24	22
2.4	73	67	61	56	52	47	43	40	36	33	65	80	54	50	46	41	37	35	32	30	58	53	48	44	41	36	32	30	27	26	50	46	42	37	35	31	27	25	23	21
2.6	73	68	60	55	50	45	41	38	34	31	65	59	54	49	45	40	35	33	30	28	58	53	48	43	39	35	31	28	26	24	50	46	41	37	34	30	26	23	21	20
2.8	73	65	59	53	48	43	39	38	32	29	65	59	53	48	43	38	33	30	28	26	58	53	47	43	38	34	29	27	24	22	50	46	41	36	33	29	25	22	20	19
3.0	72	65	58	52	47	42	37	34	30	27	64	58	52	47	42	37	29	29	27	24	57	52	46	42	37	32	28	25	23	20	50	45	40	36	32	28	24	21	19	17
3.2	72	65	57	51	45	40	35	33	28	25	64	58	51	46	40	36	31	28	25	23	57	51	45	41	36	31	27	23	22	18	50	44	39	35	31	27	23	20	18	16
3.4	71	64	56	49	44	39	34	32	27	24	64	57	50	45	39	35	29	27	24	22	57	51	45	40	35	30	26	23	20	17	50	44	39	35	30	26	22	19	17	15
3.6	71	63	54	48	43	38	32	30	25	23	63	56	49	44	38	33	28	25	22	20	57	50	44	39	34	29	25	22	19	16	50	44	39	34	29	25	21	18	16	14
3.8	70	62	53	47	41	36	31	28	24	22	63	56	49	43	37	32	27	24	21	19	57	50	43	38	33	29	24	21	19	15	50	44	38	34	29	25	21	17	15	13
4.0	70	61	53	46	40	35	30	26	22	20	63	55	48	42	36	31	26	23	20	17	57	49	42	37	32	28	23	20	18	14	50	44	38	33	28	24	20	17	15	12
4.2	69	60	52	45	39	34	29	25	21	18	62	55	47	41	35	30	25	22	19	16	56	49	42	37	32	27	22	19	17	14	50	43	37	32	28	24	20	17	14	12
4.4	69	60	51	44	38	33	28	24	20	17	62	54	46	40	34	29	24	21	18	15	56	49	42	36	31	27	22	19	16	13	50	43	37	32	27	23	19	16	13	11
4.6	69	59	50	43	37	32	27	23	19	15	62	53	45	39	33	28	24	21	17	14	56	49	41	35	30	26	21	18	16	13	50	43	36	31	26	22	18	15	13	10
4.8	68	58	49	42	36	31	26	22	18	14	62	53	45	38	32	27	23	20	16	12	56	49	41	34	29	25	21	18	15	12	50	43	36	31	26	22	18	15	12	09
5.0	68	58	48	41	35	30	25	21	18	14	61	52	44	36	31	26	19	19	16	11	56	48	40	34	28	24	20	17	14	11	50	42	35	30	25	21	17	14	12	09
6.0	66	55	44	38	31	27	22	19	15	10	60	51	41	35	28	24	19	16	13	09	55	45	37	31	25	21	17	14	11	07	50	42	34	29	23	19	15	13	10	06
7.0	64	53	41	35	28	24	19	16	12	07	58	48	38	32	26	22	17	14	11	06	54	43	35	30	24	20	15	12	09	05	49	41	32	27	21	17	14	11	08	05
8.0	62	50	38	32	25	21	17	14	11	05	57	46	35	29	23	19	15	13	10	05	54	33	33	28	22	18	14	10	08	04	49	40	30	25	19	16	12	10	07	03
9.0	61	49	36	30	23	19	15	13	10	04	56	45	33	27	21	18	14	12	09	04	52	40	31	26	20	16	12	09	07	03	48	39	29	24	18	15	11	09	07	03
10.0	59	46	33	27	21	18	14	11	08	03	55	43	31	25	19	16	12	10	07	03	51	39	29	24	18	15	11	09	07	02	47	37	27	22	17	14	10	08	06	02

*Values in this table are based on a length to width ratio of 1.6.
†Ceiling, or floor cavity.

Percent Base[†] Reflectance / Percent Wall Reflectance / Cavity Ratio

Percent Base Reflectance = 40

Cavity Ratio	90	80	70	60	50	40	30	20	10	0
0.2	40	40	39	39	38	38	37	36	36	36
0.4	41	40	39	39	38	37	36	35	34	34
0.6	41	40	39	38	37	36	34	33	32	31
0.8	41	40	38	37	36	35	33	32	31	29
1.0	42	40	38	37	35	33	32	31	29	27
1.2	42	40	38	36	34	32	30	29	27	25
1.4	42	39	37	35	33	31	29	27	25	23
1.6	42	39	37	35	32	30	27	25	23	22
1.8	42	39	36	34	31	29	26	24	22	21
2.0	42	39	36	34	31	28	25	23	21	19
2.2	42	39	36	33	30	27	24	22	19	18
2.4	43	39	35	33	29	27	24	21	18	17
2.6	43	39	35	32	29	26	23	20	17	15
2.8	43	39	35	32	28	25	22	18	16	14
3.0	43	39	35	31	27	24	21	18	16	13
3.2	43	39	35	31	27	23	20	17	15	13
3.4	43	39	34	30	26	23	20	17	14	12
3.6	44	39	34	30	26	22	19	16	14	11
3.8	44	38	33	29	25	22	18	16	13	10
4.0	44	38	33	29	25	21	18	15	12	10
4.2	44	38	33	29	24	21	17	15	12	10
4.4	44	38	33	28	24	20	17	14	11	09
4.6	44	38	32	28	23	19	16	14	11	08
4.8	44	38	32	27	22	19	16	13	10	08
5.0	45	38	31	27	22	19	15	13	10	07
6.0	44	37	30	25	20	17	13	11	08	05
7.0	44	36	29	24	19	16	12	10	07	04
8.0	44	35	28	23	18	15	11	09	06	03
9.0	44	35	26	21	16	13	10	08	05	02
10.0	43	34	25	20	15	12	08	07	05	02

Percent Base Reflectance = 30

Cavity Ratio	90	80	70	60	50	40	30	20	10	0
0.2	31	31	30	30	29	29	29	28	23	27
0.4	31	31	30	30	29	28	28	27	25	25
0.6	32	31	30	29	28	27	26	26	25	23
0.8	32	31	30	29	28	26	25	25	23	22
1.0	33	32	30	29	27	25	24	23	22	20
1.2	33	32	30	28	27	25	23	22	21	19
1.4	34	32	30	28	26	24	22	21	19	18
1.6	34	33	29	27	25	23	22	20	17	17
1.8	35	33	29	27	25	23	21	19	17	16
2.0	35	33	29	26	24	22	20	18	15	14
2.2	36	32	29	26	24	22	19	17	15	13
2.4	36	32	29	26	24	22	19	16	14	12
2.6	36	32	29	25	23	21	18	16	14	11
2.8	37	33	29	25	23	21	17	15	13	11
3.0	37	33	29	25	22	20	15	15	12	10
3.2	37	33	29	25	22	19	16	14	12	10
3.4	37	33	29	24	22	19	16	14	11	09
3.6	38	33	28	24	21	18	15	13	10	08
3.8	38	33	28	24	21	18	15	13	10	08
4.0	38	33	28	24	21	18	14	12	09	07
4.2	38	33	28	24	20	17	14	12	09	07
4.4	39	33	28	24	20	17	14	11	09	06
4.6	39	33	28	24	20	17	13	10	08	06
4.8	39	33	28	24	20	17	13	10	08	05
5.0	39	33	28	24	19	16	13	10	08	05
6.0	39	33	27	23	18	15	11	09	06	04
7.0	40	33	26	22	17	14	10	08	05	03
8.0	40	33	26	21	16	13	09	07	04	02
9.0	40	33	25	20	15	12	09	07	04	02
10.0	40	32	24	19	14	11	08	06	03	01

Percent Base Reflectance = 20

Cavity Ratio	90	80	70	60	50	40	30	20	10	0
0.2	21	20	20	20	20	20	19	19	19	17
0.4	22	21	20	20	20	19	19	18	18	16
0.6	23	21	21	20	19	19	18	18	17	15
0.8	24	22	21	20	19	19	18	17	16	14
1.0	25	23	22	20	19	18	17	16	15	13
1.2	25	23	22	20	19	17	17	16	14	12
1.4	26	24	22	20	18	17	16	15	13	12
1.6	26	24	22	20	18	17	15	15	13	11
1.8	27	25	23	20	18	17	15	14	12	10
2.0	28	25	23	20	18	16	15	13	11	09
2.2	28	25	23	20	18	16	14	12	10	09
2.4	29	26	23	20	18	16	14	12	10	08
2.6	29	26	23	20	18	16	13	11	09	07
2.8	30	27	23	20	16	15	13	11	09	06
3.0	30	27	23	20	17	15	13	11	09	07
3.2	31	27	23	20	17	15	12	11	09	06
3.4	31	27	23	20	17	15	12	10	08	06
3.6	32	27	23	20	17	15	12	10	08	05
3.8	32	28	23	20	17	14	12	10	07	05
4.0	33	28	23	20	17	14	11	09	07	05
4.2	33	28	23	20	17	14	11	09	07	04
4.4	34	28	24	20	17	14	11	09	07	04
4.6	34	29	24	20	17	14	11	09	07	04
4.8	35	29	24	20	16	13	10	08	06	04
5.0	35	29	24	20	16	13	10	08	06	04
6.0	36	30	24	20	16	13	10	08	05	02
7.0	36	30	24	20	15	12	09	07	04	02
8.0	37	30	23	19	15	12	08	06	03	01
9.0	37	29	23	19	14	11	08	06	03	01
10.0	37	29	22	18	13	10	07	05	03	01

Percent Base Reflectance = 10

Cavity Ratio	90	80	70	60	50	40	30	20	10	0
0.2	11	11	11	10	10	10	10	09	09	09
0.4	12	12	11	11	11	10	10	09	09	08
0.6	13	13	12	11	11	11	10	09	08	08
0.8	15	14	13	12	11	11	10	09	08	07
1.0	16	14	13	12	12	11	10	09	08	07
1.2	17	15	14	13	12	11	10	09	07	06
1.4	18	16	14	13	12	11	10	09	07	06
1.6	19	17	15	14	12	11	09	08	07	06
1.8	19	17	15	14	13	11	09	08	06	05
2.0	20	18	16	14	13	11	09	08	06	05
2.2	21	19	16	14	13	11	09	07	06	05
2.4	22	19	17	15	13	11	09	07	06	05
2.6	23	20	17	15	13	11	09	07	06	04
2.8	23	20	18	16	13	11	09	07	05	03
3.0	24	21	18	16	13	11	09	07	05	03
3.2	25	21	18	16	13	11	09	07	05	03
3.4	26	22	18	16	13	11	09	07	05	03
3.6	26	23	19	16	13	11	09	06	04	03
3.8	27	23	19	17	13	11	09	06	04	02
4.0	27	23	20	17	14	11	09	06	04	02
4.2	28	24	20	17	14	11	09	06	04	02
4.4	29	24	20	17	14	11	08	06	04	02
4.6	29	25	20	17	14	11	08	06	04	02
4.8	29	25	20	17	14	11	08	06	04	02
5.0	30	25	20	17	14	11	08	06	04	02
6.0	31	26	21	18	14	11	08	06	03	01
7.0	32	27	21	17	13	11	08	06	03	01
8.0	33	27	21	17	13	10	07	05	03	01
9.0	34	28	21	17	13	10	07	05	02	01
10.0	34	28	21	17	12	10	07	05	02	01

* Values in this table are based on a length to width ratio of 1.6.
† Ceiling, or floor cavity.

(Reproduced with permission from IESNA.)

TABLE 16-7

Multiplier for other than 20 percent effective floor cavity reflectance

% Effective Ceiling Cavity Reflectance, p_{cc}	80				70				50			30			10		
% Wall Reflectance, p_w	70	50	30	10	70	50	30	10	50	30	10	50	30	10	50	30	10

For 30 Percent Effective Floor-Cavity Reflectance (20 Percent = 1.00)

Room cavity ratio	70	50	30	10	70	50	30	10	50	30	10	50	30	10	50	30	10
1	1.092	1.082	1.075	1.068	1.077	1.070	1.064	1.059	1.049	1.044	1.040	1.028	1.026	1.023	1.012	1.010	1.008
2	1.079	1.066	1.055	1.047	1.068	1.057	1.048	1.039	1.041	1.033	1.027	1.026	1.021	1.017	1.013	1.010	1.006
3	1.070	1.054	1.042	1.033	1.061	1.048	1.037	1.028	1.034	1.027	1.020	1.024	1.017	1.012	1.014	1.009	1.005
4	1.062	1.045	1.033	1.024	1.055	1.040	1.029	1.021	1.030	1.022	1.015	1.022	1.015	1.010	1.014	1.009	1.004
5	1.056	1.038	1.026	1.018	1.050	1.034	1.024	1.015	1.027	1.018	1.012	1.020	1.013	1.008	1.014	1.009	1.004
6	1.052	1.033	1.021	1.014	1.047	1.030	1.020	1.012	1.024	1.015	1.009	1.019	1.012	1.005	1.014	1.008	1.003
7	1.047	1.029	1.018	1.011	1.043	1.026	1.017	1.009	1.022	1.013	1.007	1.108	1.010	1.004	1.013	1.007	1.003
8	1.044	1.026	1.015	1.009	1.040	1.024	1.015	1.007	1.020	1.011	1.006	1.017	1.009	1.004	1.013	1.007	1.003
9	1.040	1.024	1.014	1.007	1.037	1.022	1.014	1.006	1.019	1.011	1.005	1.016	1.009	1.004	1.013	1.007	1.002
10	1.037	1.022	1.012	1.006	1.034	1.020	1.012	1.005	1.017	1.010	1.004	1.015	1.009	1.003	1.013	1.007	1.002

For 10 Percent Effective Floor Cavity Reflectance (20 Percent = 1.00)

Room cavity ratio	70	50	30	10	70	50	30	10	50	30	10	50	30	10	50	30	10
1	.923	.929	.935	.940	.933	.939	.943	.948	.956	.960	.963	.973	.976	.979	.989	.991	.993
2	.931	.942	.950	.958	.940	.949	.957	.963	.962	.968	.974	.976	.980	.985	.988	.991	.995
3	.939	.951	.961	.969	.945	.957	.966	.973	.967	.975	.981	.978	.983	.988	.988	.992	.996
4	.944	.958	.969	.978	.950	.963	.973	.980	.972	.980	.986	.980	.986	.991	.987	.992	.996
5	.949	.964	.976	.983	.954	.968	.978	.985	.975	.983	.989	.981	.988	.993	.987	.992	.997
6	.953	.969	.980	.986	.958	.972	.982	.989	.977	.985	.992	.982	.989	.995	.987	.993	.997
7	.957	.973	.983	.991	.961	.975	.985	.991	.979	.987	.994	.983	.990	.996	.987	.993	.998
8	.960	.976	.986	.993	.963	.977	.987	.993	.981	.988	.995	.984	.991	.997	.987	.994	.998
9	.963	.978	.987	.994	.965	.979	.989	.994	.983	.990	.996	.985	.992	.998	.988	.994	.999
10	.965	.980	.989	.995	.967	.981	.990	.995	.984	.991	.997	.986	.993	.998	.988	.994	.999

For 0 Percent Effective Floor Cavity Reflectance (20 Percent = 1.00)

Room cavity ratio	70	50	30	10	70	50	30	10	50	30	10	50	30	10	50	30	10
1	.859	.870	.879	.886	.873	.884	.893	.901	.916	.923	.929	.948	.954	.960	.979	.983	.987
2	.871	.887	.903	.919	.886	.902	.916	.928	.926	.938	.949	.954	.963	.971	.978	.983	.991
3	.882	.904	.915	.942	.898	.918	.934	.947	.936	.950	.964	.958	.969	.979	.976	.984	.993
4	.893	.919	.941	.958	.908	.930	.948	.961	.945	.961	.974	.961	.974	.984	.975	.985	.994
5	.903	.931	.953	.969	.914	.939	.958	.970	.951	.967	.980	.964	.977	.988	.975	.985	.995
6	.911	.940	.961	.976	.920	.945	.965	.977	.955	.972	.985	.966	.979	.991	.975	.986	.996
7	.917	.947	.967	.981	.924	.950	.970	.982	.959	.975	.988	.968	.981	.993	.975	.987	.997
8	.922	.953	.971	.985	.929	.955	.975	.986	.963	.978	.991	.970	.983	.995	.976	.988	.998
9	.928	.958	.975	.988	.933	.959	.980	.989	.966	.980	.993	.971	.985	.996	.976	.988	.998
10	.933	.962	.979	.991	.937	.963	.983	.992	.969	.982	.995	.973	.987	.997	.977	.989	.999

(Reproduced with permission from IESNA.)

20 percent. Effective floor reflectances other than 20 percent should be modified by a multiplier. The impact of the effective floor reflectance is less significant than that of the effective ceiling reflectance. In design practice, an effective floor reflectance in the 10–30 percent range is often selected. Table 16–7 provides the multipliers required from the values given for a 20 percent floor in a photometric report. The multiplier varies from 1.1 for 30 percent effective floor reflectance for large spaces to 1.0 for zero percent reflectance for small spaces. For extremely light-colored floors, the multiplier may be extrapolated from these data.

4. *Step 4: Determine the CU.* Use the manufacturer's photometric data and the previously calculated values RCR, ρ_{cc}, ρ_w, and ρ_{fc} to determine CU.

Photometrics normally include data on the CU of the luminaire. The latter data are presented in a tabular format so that the designer can choose between the room cavity ratio (RCR), effective ceiling reflectances (80, 70, and 50 percent), wall reflectances (70, 50, and 30 percent), and effective floor reflectance (20 percent) only. A typical CU table for a fluorescent luminaire is shown in Figure 16–5.

Example 16.10 Determine the specific CU for a ceiling mounted luminaire with its CU data shown in Figure 16–4 installed in a room with an effective floor reflectance (ρ_f) of 20 percent, an effective ceiling reflectance (ρ_c) of 70 percent, and a wall reflectance (ρ_w) of 30 percent. The room configuration is such that its RCR is 2.5.

COEFFICIENT OF UTILIZATION

pfc pcc pw	(20) 80 70	 50	 30	(70) 70	 50	 (30)	50 50	 30
RCR								
0	81	81	81	80	80	80	77	77
1	77	73	71	75	72	70	69	68
(2)	70	66	61	68	65	(61)	63	59
(3)	65	58	54	64	57	(54)	56	52
4	59	53	47	58	52	46	51	46
5	56	47	41	54	46	41	46	40
6	52	42	38	50	42	36	40	36
7	47	39	34	46	39	33	38	33
8	45	35	29	44	34	29	34	29
9	41	33	28	40	33	27	32	27
10	39	29	25	38	29	25	28	25

■ **FIGURE 16–4**
Typical data on the CU of a luminaire, provided by luminaire manufacturers.

ρ_{fc} ρ_{cc} ρ_w	20 80 70	20 80 50	20 80 30	20 70 70	20 70 50	20 70 30	**20** **52** **30**	20 50 30
RCR								
0	81	81	81	80	80	80		77
1	77	73	71	75	72	70		68
2	70	66	61	68	65	61		59
(3)	65	58	54	64	57	(54)	(52.2)	52
4	59	53	47	58	52	46		46
5	56	47	41	54	46	41		40
6	52	42	38	50	42	36		36
7	47	39	34	46	39	33		33
8	45	35	29	44	34	29		29
9	41	33	28	40	33	27		27
10	39	29	25	38	29	25		25

Example 16.11 → (at RCR 3, ρ_w 30 column)
Example 16.10 (vertical label in the 52/30 column)

■ **FIGURE 16–5**
Exercise in visual interpolation from tabulated data.

Answer From Figure 16–4, the CU for 70 percent ceiling, 30 percent wall, and RCR = 2 is 61 percent; with RCR = 3, the CU is 54 percent. Thus, by interpolation, with RCR = 2.5, the CU is 57.5 percent.

Example 16.11 Using the same luminaire as in Example 16.10, calculate the CU of a lighting design for a small workshop. Use the following data:

- Room dimensions: length, 20 ft; width, 15-ft ceiling height, 11 ft
- Room finishes: base ceiling = 80 percent; wall = 30 percent; base floor = 30 percent
- Work plane: 3.0 ft above floor (bench top)
- Luminaire: Suspended 3.0 ft below ceiling

Calculations:

- From Equation 16–6c:
  FCR = 2.5 × 0.23 × 3.0 = 1.7
- From Table 16–6:
  Effective ceiling reflectance (ρ_{cc}) for 80 percent base ceiling, 30 percent wall, and CCR of 1.7 is between 53 and 50, or 51.5 percent (use 52).
- From Table 16–6:
  Effective floor reflectance (ρ_{fc}) for 30 percent base floor, 30 percent wall, and FCR of 1.7 is 21.5 percent (use 20 percent as an approximation).
- From Equation (16–6b):

$$RCR = 2.5 \times (70/300) \times (11 - 3 - 3)$$
$$= 2.5 \times 0.233 \times 5 = 2.9 \text{ (use 3.0)}$$

- From Figure 16–5, determine the CU, based on the following data:

  $\rho_{cc} = 52$ percent
  $\rho_w = 30$ percent (same as base reflectance)
  $\rho_{fc} = 20$ percent
  RCR = 3.0

Thus, CU = 52.2 (use 52)
Figure 16–5 shows how the preceding calculation is arrived at using the manufacturer's data.

16.5.3 Preliminary Method for Determining the Coefficient of Utilization

The previous example demonstrates the tedious and time-consuming nature of the manual procedure for calculating CU. To save time and effort one may use a computer program. For preliminary design purposes, the designer may bypass some of the time-consuming steps by choosing effective reflectances at the outset, rather than base reflectances. With this approach, CU can be found directly from the photometric data, although it will normally be higher than if it is calculated by the full zonal cavity method.

Example 16.12 For the same installation as in Example 16.10, assume the effective reflectances are 80, 30, and 20 percent for the ceiling, walls, and floors respectively. Then from Figure 16–4, with RCR = 3, the CU is found to be 54 percent in one step. There is about 4 percent variance or (52–54)/52 between the two methods. For most applications, the variance will not be noticed. A more conservative selection of the effective reflectances is recommended.

16.5.4 Coefficient of Utilization of Generic Luminaires

The CU values of every luminaire should be provided by the manufacturer based on certified laboratory tests, preferably by an independent testing laboratory. Experience indicates that the CU of a higher-quality product may be considerably greater than that of a lower-quality product of the same appearance, owing to a number of factors, such as the effective reflectance of luminaire surfaces, the positioning of the lamps, the geometry of the housing, the efficacy of the diffusing media, and the power loss of the ballast. This variation may result in a need for more or less luminaires to be installed in an identical space. It is, therefore, extremely important to use the specific laboratory testing data of the selected luminaire prior to final design calculations.

However, for preliminary design and evaluation, using the CU values of generic luminaire types published in the *IESNA Handbook* can provide valuable comparisons between different types of luminaires. These are shown in Figure 16–6.

Two luminaires may appear to be similar but have widely different CU values. For example, with effective ceiling reflectance at 70 percent, wall at 30 percent, and RCR = 1 (a large room), luminaire CU for type 7 is 1.03 (> 1.0) and only 0.91 for type 8. Type 7 is definitely more effective for this room configuration; however, for the same luminaires in another room with RCR = 10 (a very small room), CU for type 7 is only 0.45; CU for type 8 is 0.53. Therefore, type 8 is a more effective selection.

16.6 APPLICATION OF THE ZONAL CAVITY METHOD

Having defined the light loss factor (LLF) and the coefficient of utilization (CU), we can further modify Eq. (16–1) as follows:

$$E = (F \times CU \times LLF)/A \qquad (16\text{–}1)$$
$$= (N \times LPF \times RLL \times LOF) \times CU \times LLF/A \qquad (16\text{–}1a)$$

where E = maintained illuminance of a lighting system (fc) or (lx).
F = total lamp lumens required
= (no. of fixtures (N) $\times$ no. of lamps / fixture (LPF) $\times$ rated lamp lumens (RLL))

(*Note:* The word *fixture* is used here in lieu of *luminaire* to minimize the confusion with lumen or light, or lamp.)

LOF = lamp operating factor = VF $\times$ BF (16–2)
LLF = applicable portion of recoverable
LLF = LLD $\times$ LDD (16–3)
RLL = rated lamp lumens

We can rewrite equation (16–1a) in the following way to determine the number of fixtures (N) required for the desired maintained illuminance of a specific room and selected fixtures:

$$N = (E \times A)/LPF \times RLL \times LOF \times CU \times LLF \qquad (16\text{–}1b)$$

Often, a designer may have to compare the initial selection with many alternatives to determine an optimum selection having the best overall merit by comparing the performance, aesthetics, initial cost, and life-cycle cost of the choices.

The zonal cavity method is the method used for manual calculations as well as the algorithm for

Coefficients of Utilization for 20 Per Cent Effective Floor Cavity Reflectance ($\rho_{FC} = 20$)

Typical Luminaire	Maint. Cat.	SC	RCR	ρCC→80 ρW 50	30	10	ρCC 70 ρW 50	30	10	ρCC 50 ρW 50	30	10	ρCC 30 ρW 50	30	10	ρCC 10 ρW 50	30	10	ρCC 0	WDRC
1. Pendant diffusing sphere with incandescent lamp (35½%↑, 45%↓)	V	1.5	0	.87	.87	.87	.81	.81	.81	.70	.70	.70	.59	.59	.59	.49	.49	.49	.45	
	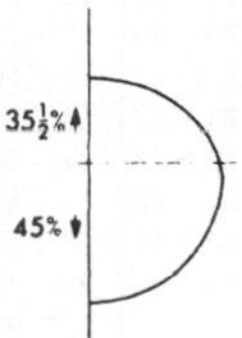		1	.71	.66	.62	.65	.61	.58	.55	.52	.49	.46	.44	.42	.38	.36	.34	.30	.368
			2	.60	.53	.48	.55	.50	.45	.47	.42	.38	.39	.35	.32	.31	.29	.26	.23	.279
			3	.52	.44	.38	.48	.41	.36	.40	.35	.31	.33	.29	.26	.27	.24	.21	.18	.227
			4	.45	.37	.32	.42	.35	.29	.35	.30	.25	.29	.25	.21	.23	.20	.17	.14	.192
			5	.40	.32	.27	.37	.30	.25	.31	.25	.21	.26	.21	.18	.21	.17	.14	.12	.166
			6	.35	.28	.23	.33	.26	.21	.28	.22	.18	.23	.19	.15	.19	.15	.12	.10	.146
			7	.32	.25	.19	.29	.23	.18	.25	.20	.16	.21	.16	.13	.17	.13	.11	.09	.130
			8	.29	.22	.17	.27	.20	.16	.23	.17	.14	.19	.15	.12	.15	.12	.09	.07	.117
			9	.26	.19	.15	.24	.18	.14	.21	.16	.12	.17	.13	.10	.14	.11	.08	.07	.107
			10	.24	.17	.13	.22	.16	.12	.19	.14	.11	.16	.12	.09	.13	.10	.08	.06	.098
2. Concentric ring unit with incandescent silvered-bowl lamp (83%↑, 3½%↓)	II	N.A.	0	.83	.83	.83	.72	.72	.72	.50	.50	.50	.30	.30	.30	.12	.12	.12	.03	
	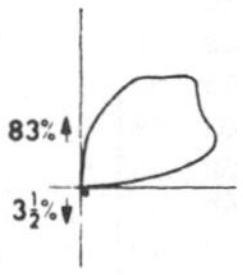		1	.72	.69	.66	.62	.60	.57	.43	.42	.40	.26	.25	.25	.10	.10	.10	.03	.018
			2	.63	.58	.54	.54	.50	.47	.38	.35	.33	.23	.22	.20	.09	.09	.08	.02	.015
			3	.55	.49	.45	.47	.43	.39	.33	.30	.28	.20	.19	.17	.08	.07	.07	.02	.013
			4	.48	.42	.37	.42	.37	.33	.29	.26	.23	.18	.16	.15	.07	.06	.06	.02	.012
			5	.43	.36	.32	.37	.32	.28	.26	.23	.20	.16	.14	.12	.06	.05	.05	.01	.011
			6	.38	.32	.27	.33	.28	.24	.23	.20	.17	.14	.12	.11	.06	.05	.04	.01	.010
			7	.34	.28	.23	.30	.24	.21	.21	.17	.15	.13	.11	.09	.05	.04	.04	.01	.009
			8	.31	.25	.20	.27	.21	.18	.19	.15	.13	.12	.10	.08	.05	.04	.03	.01	.008
			9	.28	.22	.18	.24	.19	.16	.17	.14	.11	.10	.09	.07	.04	.03	.03	.01	.008
			10	.25	.20	.16	.22	.17	.14	.16	.12	.10	.10	.08	.06	.04	.03	.03	.01	.007
3. Porcelain-enameled ventilated standard dome with incandescent lamp (0%↑, 83½%↓)	IV	1.3	0	.99	.99	.99	.97	.97	.97	.93	.93	.93	.89	.89	.89	.85	.85	.85	.83	
	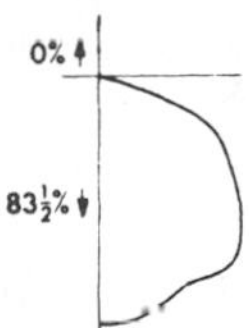		1	.87	.84	.81	.85	.82	.79	.82	.79	.77	.79	.76	.74	.76	.74	.72	.71	.323
			2	.76	.70	.65	.74	.69	.65	.71	.67	.63	.69	.65	.62	.66	.63	.60	.59	.311
			3	.66	.59	.54	.65	.59	.53	.62	.57	.53	.60	.56	.52	.58	.54	.51	.49	.288
			4	.58	.51	.45	.57	.50	.45	.55	.49	.44	.53	.48	.44	.51	.47	.43	.41	.264
			5	.52	.44	.39	.51	.44	.38	.49	.43	.38	.47	.42	.37	.46	.41	.37	.35	.241
			6	.46	.39	.33	.46	.38	.33	.44	.38	.33	.43	.37	.33	.41	.36	.32	.31	.221
			7	.42	.34	.29	.41	.34	.29	.40	.33	.29	.39	.33	.29	.38	.32	.28	.27	.203
			8	.38	.31	.26	.37	.31	.26	.36	.30	.26	.35	.30	.25	.34	.29	.25	.24	.187
			9	.35	.28	.23	.34	.28	.23	.33	.27	.23	.33	.27	.23	.32	.26	.23	.21	.173
			10	.32	.25	.21	.32	.25	.21	.31	.25	.21	.30	.24	.21	.29	.24	.20	.19	.161
4. Recessed baffled downlight, 140 mm (5½") diameter aperture—150-PAR/FL lamp (0%↑)	IV	0.5	0	.82	.82	.82	.80	.80	.80	.76	.76	.76	.73	.73	.73	.70	.70	.70	.69	
			1	.78	.77	.75	.76	.75	.74	.74	.73	.72	.71	.70	.70	.69	.68	.68	.67	.051
			2	.74	.72	.71	.73	.71	.70	.71	.70	.68	.69	.68	.67	.67	.66	.66	.65	.050
			3	.71	.69	.67	.71	.68	.67	.69	.67	.66	.67	.66	.65	.66	.65	.64	.63	.049
			4	.69	.66	.64	.68	.66	.64	.67	.65	.63	.66	.64	.63	.64	.63	.62	.61	.048
			5	.67	.64	.62	.66	.63	.62	.65	.63	.61	.64	.62	.61	.63	.61	.60	.59	.047
			6	.64	.62	.60	.64	.61	.60	.63	.61	.59	.62	.60	.59	.61	.60	.59	.58	.045
			7	.63	.60	.58	.62	.60	.58	.61	.59	.57	.61	.59	.57	.60	.58	.57	.56	.044
			8	.61	.58	.56	.60	.58	.56	.60	.58	.56	.59	.57	.56	.59	.57	.56	.55	.043
			9	.59	.56	.55	.59	.56	.55	.58	.56	.54	.58	.56	.54	.57	.55	.54	.54	.042
			10	.58	.55	.53	.57	.55	.53	.57	.55	.53	.56	.54	.53	.56	.54	.53	.52	.041
5. Recessed baffled downlight, 140 mm (5½") diameter aperture—75ER30 lamp (0%↑, 85%↓)	IV	0.5	0	1.01	1.01	1.01	.99	.99	.99	.95	.95	.95	.91	.91	.91	.87	.87	.87	.85	
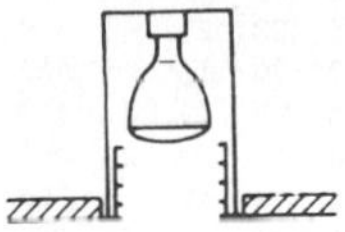	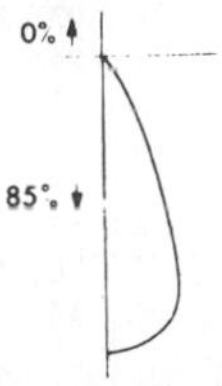		1	.96	.94	.93	.94	.93	.91	.91	.89	.88	.88	.86	.85	.85	.84	.83	.82	.085
			2	.91	.88	.86	.90	.87	.85	.87	.85	.83	.84	.83	.81	.82	.81	.80	.79	.084
			3	.87	.83	.81	.86	.83	.80	.83	.81	.79	.81	.79	.78	.80	.78	.77	.75	.082
			4	.83	.79	.76	.82	.79	.76	.80	.77	.75	.79	.76	.74	.77	.75	.73	.72	.080
			5	.79	.76	.73	.79	.75	.72	.77	.74	.72	.76	.73	.71	.75	.72	.71	.70	.078
			6	.76	.72	.70	.76	.72	.69	.74	.71	.69	.73	.71	.68	.72	.70	.68	.67	.076
			7	.73	.69	.67	.73	.69	.67	.72	.69	.66	.71	.68	.66	.70	.68	.66	.65	.073
			8	.71	.67	.64	.70	.67	.64	.69	.66	.64	.69	.66	.64	.68	.65	.63	.62	.071
			9	.68	.64	.62	.68	.64	.62	.67	.64	.62	.66	.63	.61	.66	.63	.61	.60	.069
			10	.66	.62	.60	.66	.62	.60	.65	.62	.59	.64	.61	.59	.64	.61	.59	.58	.067
6. EAR-38 lamp above 51 mm (2") diameter aperture (increase efficiency to 54½% for 76 mm (3") diameter aperture) (0%↑, 43½%↓)	IV	0.7	0	.52	.52	.52	.51	.51	.51	.48	.48	.48	.46	.46	.46	.45	.45	.45	.44	
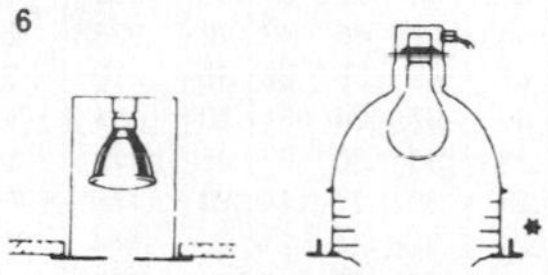	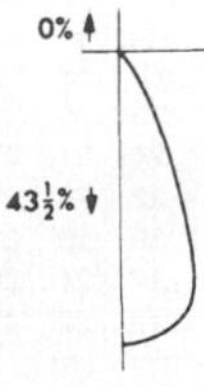		1	.49	.48	.47	.48	.47	.46	.46	.45	.45	.44	.44	.43	.43	.43	.42	.41	.055
			2	.46	.44	.43	.45	.44	.43	.44	.43	.42	.43	.42	.41	.41	.41	.40	.39	.054
			3	.43	.41	.40	.43	.41	.40	.42	.40	.39	.41	.39	.38	.40	.39	.38	.37	.053
			4	.41	.39	.37	.41	.39	.37	.40	.38	.37	.39	.37	.36	.38	.37	.36	.35	.052
			5	.39	.37	.35	.39	.37	.35	.38	.36	.35	.37	.36	.34	.36	.35	.34	.34	.051
			6	.37	.35	.33	.37	.35	.33	.36	.34	.33	.35	.34	.33	.35	.34	.32	.32	.049
			7	.35	.33	.31	.35	.33	.31	.34	.33	.31	.34	.32	.31	.33	.32	.31	.30	.048
			8	.34	.31	.30	.33	.31	.30	.33	.31	.30	.32	.31	.29	.32	.31	.29	.29	.046
			9	.32	.30	.28	.32	.30	.28	.31	.30	.28	.31	.29	.28	.31	.29	.28	.28	.045
			10	.31	.28	.27	.31	.28	.27	.30	.28	.27	.30	.28	.27	.30	.28	.27	.26	.043

■ **FIGURE 16–6**

Coefficient of utilization (CU) of generic luminaires.
(Reproduced with permission from IESNA.)

Coefficient of utilization table — Figure 16-6 (Continued). Columns: ρcc (80, 70, 50, 30, 10, 0) with ρw (50, 30, 10) sub-columns; RCR; WDRC. Coefficients of Utilization for 20 Per Cent Effective Floor Cavity Reflectance (ρFC = 20).

7 — R-40 flood without shielding (Maint. Cat. IV; SC 0.8; 0% ↑, 100% ↓)

RCR	80/50	80/30	80/10	70/50	70/30	70/10	50/50	50/30	50/10	30/50	30/30	30/10	10/50	10/30	10/10	0	WDRC
0	1.19	1.19	1.19	1.16	1.16	1.16	1.11	1.11	1.11	1.06	1.06	1.06	1.02	1.02	1.02	1.00	
1	1.08	1.05	1.03	1.06	1.03	1.01	1.02	1.00	.98	.98	.97	.95	.95	.93	.92	.90	.241
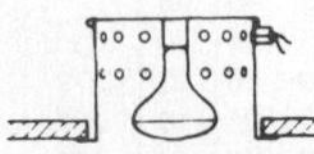 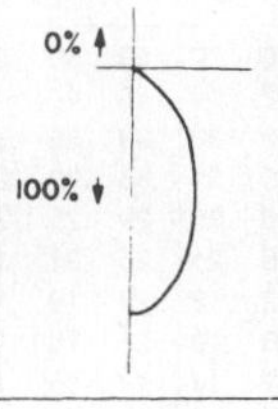 2	.99	.94	.89	.97	.92	.88	.93	.90	.86	.90	.87	.84	.88	.85	.83	.81	.238
3	.90	.84	.79	.88	.83	.78	.86	.81	.77	.83	.79	.76	.81	.77	.74	.73	.227
4	.82	.75	.70	.81	.75	.70	.79	.73	.69	.77	.72	.68	.75	.71	.67	.66	.215
5	.76	.68	.63	.75	.68	.63	.73	.67	.62	.71	.66	.62	.69	.65	.61	.59	.202
6	.70	.62	.57	.69	.62	.57	.67	.61	.57	.66	.60	.56	.64	.60	.56	.54	.191
7	.65	.57	.52	.64	.57	.52	.62	.56	.52	.61	.56	.52	.60	.55	.51	.50	.180
8	.60	.53	.48	.59	.53	.48	.58	.52	.48	.57	.52	.47	.56	.51	.47	.46	.169
9	.56	.49	.44	.55	.49	.44	.54	.48	.44	.53	.48	.44	.52	.47	.44	.42	.160
10	.52	.46	.41	.52	.45	.41	.51	.45	.41	.50	.45	.41	.49	.44	.41	.39	.152

8 — R-40 flood with specular anodized reflector skirt; 45° cutoff (Maint. Cat. IV; SC 0.7; 0% ↑, 85% ↑)

RCR	80/50	80/30	80/10	70/50	70/30	70/10	50/50	50/30	50/10	30/50	30/30	30/10	10/50	10/30	10/10	0	WDRC
0	1.01	1.01	1.01	.99	.99	.99	.94	.94	.94	.90	.90	.90	.87	.87	.87	.85	
1	.95	.93	.91	.93	.91	.89	.89	.88	.87	.86	.85	.84	.83	.82	.82	.80	.115
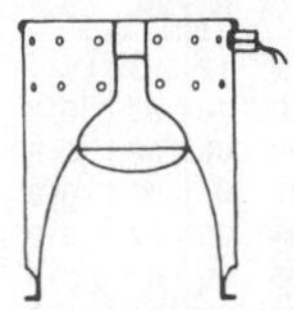 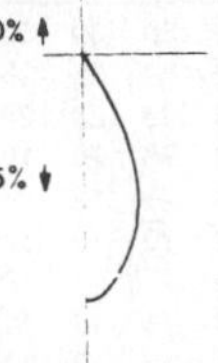 2	.89	.86	.83	.87	.84	.82	.85	.82	.80	.82	.80	.79	.80	.78	.77	.76	.115
3	.83	.80	.77	.82	.79	.76	.80	.77	.75	.78	.76	.74	.76	.74	.72	.71	.113
4	.79	.74	.71	.78	.74	.71	.76	.73	.70	.74	.71	.69	.73	.70	.68	.67	.110
5	.74	.70	.67	.74	.69	.66	.72	.68	.66	.71	.68	.65	.69	.67	.65	.63	.107
6	.70	.66	.62	.70	.65	.62	.68	.65	.62	.67	.64	.61	.66	.63	.61	.60	.104
7	.67	.62	.59	.66	.62	.59	.65	.61	.58	.64	.61	.58	.63	.60	.58	.57	.100
8	.63	.59	.56	.63	.58	.55	.62	.58	.55	.61	.58	.55	.60	.57	.55	.54	.097
9	.60	.56	.53	.60	.56	.53	.59	.55	.52	.58	.55	.52	.58	.54	.52	.51	.094
10	.57	.53	.50	.57	.53	.50	.56	.52	.50	.56	.52	.50	.55	.52	.49	.48	.091

9 — 2-lamp prismatic wraparound—see note 7 (Maint. Cat. V; SC 1.5/1.2; 11½% ↑, 58½% ↓)

RCR	80/50	80/30	80/10	70/50	70/30	70/10	50/50	50/30	50/10	30/50	30/30	30/10	10/50	10/30	10/10	0	WDRC
0	.81	.81	.81	.78	.78	.78	.72	.72	.72	.66	.66	.66	.61	.61	.61	.59	
1	.71	.68	.66	.68	.66	.63	.63	.61	.59	.58	.57	.56	.54	.53	.52	.50	.223
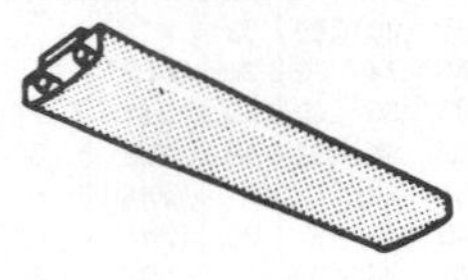 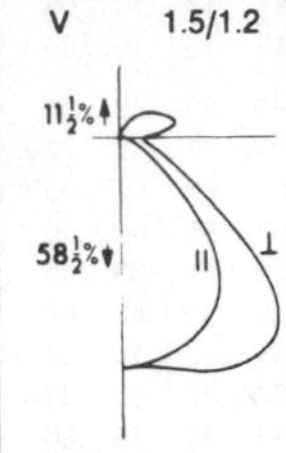 2	.63	.58	.55	.60	.56	.53	.56	.53	.50	.52	.50	.47	.48	.46	.45	.43	.201
3	.56	.50	.46	.54	.49	.45	.50	.46	.43	.47	.43	.41	.43	.41	.39	.37	.183
4	.50	.44	.40	.48	.43	.39	.45	.40	.37	.42	.38	.35	.39	.36	.34	.32	.167
5	.45	.39	.34	.43	.38	.34	.40	.36	.32	.38	.34	.31	.35	.32	.30	.28	.153
6	.40	.34	.30	.39	.34	.30	.37	.32	.28	.34	.30	.27	.32	.29	.26	.25	.142
7	.37	.31	.27	.35	.30	.26	.33	.29	.25	.31	.27	.24	.30	.26	.23	.22	.131
8	.33	.28	.24	.32	.27	.23	.30	.26	.23	.29	.25	.22	.27	.24	.21	.20	.122
9	.31	.25	.21	.30	.25	.21	.28	.24	.20	.26	.23	.20	.25	.22	.19	.18	.114
10	.28	.23	.19	.27	.22	.19	.26	.21	.18	.24	.21	.18	.23	.20	.17	.16	.107

10 — Prismatic bottom and sides, open top, 4-lamp suspended unit—see note 7 (Maint. Cat. VI; SC 1.4/1.2; 33% ↑, 50% ↓)

RCR	80/50	80/30	80/10	70/50	70/30	70/10	50/50	50/30	50/10	30/50	30/30	30/10	10/50	10/30	10/10	0	WDRC
0	.91	.91	.91	.85	.85	.85	.74	.74	.74	.64	.64	.64	.54	.54	.54	.50	
1	.80	.77	.74	.75	.72	.70	.65	.63	.61	.57	.55	.54	.49	.47	.47	.43	.179
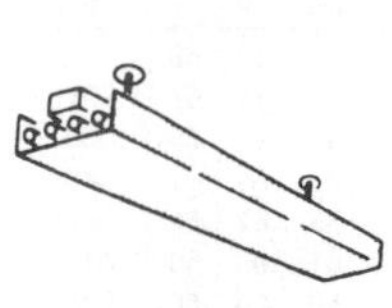 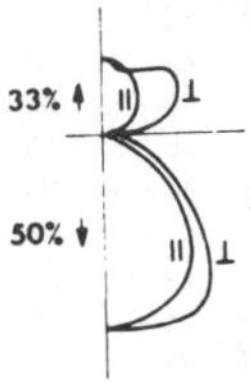 2	.70	.65	.61	.66	.62	.58	.58	.54	.52	.50	.48	.46	.43	.42	.40	.37	.166
3	.62	.56	.51	.58	.53	.49	.51	.47	.44	.45	.42	.39	.39	.37	.35	.32	.153
4	.55	.49	.44	.52	.46	.42	.46	.41	.38	.40	.37	.34	.35	.32	.30	.27	.140
5	.50	.43	.38	.47	.41	.36	.41	.37	.33	.36	.33	.30	.32	.29	.26	.24	.129
6	.45	.38	.33	.42	.36	.32	.37	.33	.29	.33	.29	.26	.29	.26	.23	.21	.119
7	.40	.34	.29	.38	.32	.28	.34	.29	.26	.30	.26	.23	.26	.23	.21	.19	.111
8	.37	.30	.26	.35	.29	.25	.31	.26	.23	.28	.24	.21	.24	.21	.19	.17	.103
9	.34	.27	.23	.32	.26	.22	.29	.24	.21	.25	.22	.19	.22	.19	.17	.15	.096
10	.31	.25	.21	.29	.24	.20	.26	.22	.19	.23	.20	.17	.21	.18	.15	.14	.090

11 — 2-lamp diffuse wraparound—see note 7 (Maint. Cat. V; SC 1.3; 8% ↑, 37½% ↓)

RCR	80/50	80/30	80/10	70/50	70/30	70/10	50/50	50/30	50/10	30/50	30/30	30/10	10/50	10/30	10/10	0	WDRC
0	.52	.52	.52	.50	.50	.50	.46	.46	.46	.43	.43	.43	.39	.39	.39	.38	
1	.44	.42	.40	.42	.40	.39	.39	.37	.36	.36	.35	.33	.33	.32	.31	.30	.201
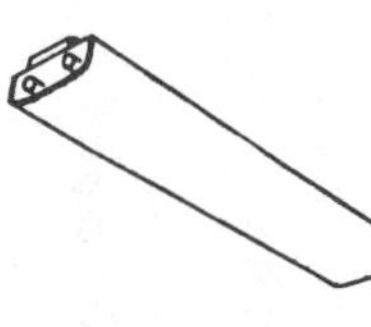 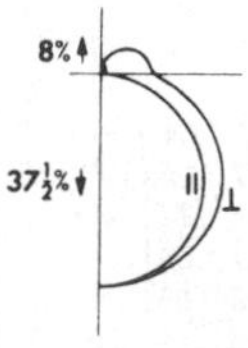 2	.38	.35	.32	.37	.33	.31	.34	.31	.29	.31	.29	.27	.28	.27	.25	.24	.171
3	.33	.29	.26	.32	.28	.25	.29	.26	.24	.27	.25	.22	.25	.23	.21	.20	.149
4	.29	.25	.22	.28	.24	.21	.26	.23	.20	.24	.21	.19	.22	.20	.18	.17	.132
5	.26	.22	.19	.25	.21	.18	.23	.20	.17	.21	.18	.16	.20	.17	.15	.14	.117
6	.23	.19	.16	.22	.18	.16	.21	.17	.15	.19	.16	.14	.18	.15	.13	.12	.106
7	.21	.17	.14	.20	.16	.14	.19	.15	.13	.17	.15	.12	.16	.14	.12	.11	.096
8	.19	.15	.12	.18	.15	.12	.17	.14	.12	.16	.13	.11	.15	.12	.11	.10	.088
9	.17	.14	.11	.17	.13	.11	.16	.13	.10	.15	.12	.10	.14	.11	.09	.09	.081
10	.16	.12	.10	.15	.12	.10	.14	.11	.09	.14	.11	.09	.13	.10	.09	.08	.075

12 — Fluorescent unit dropped diffuser, 4-lamp 610 mm (2') wide—see note 7 (Maint. Cat. V; SC 1.2; 1% ↑, 60½% ↓)

RCR	80/50	80/30	80/10	70/50	70/30	70/10	50/50	50/30	50/10	30/50	30/30	30/10	10/50	10/30	10/10	0	WDRC
0	.73	.73	.73	.71	.71	.71	.68	.68	.68	.65	.65	.65	.62	.62	.62	.60	
1	.63	.60	.58	.62	.59	.57	.59	.57	.55	.56	.55	.53	.54	.53	.51	.50	.259
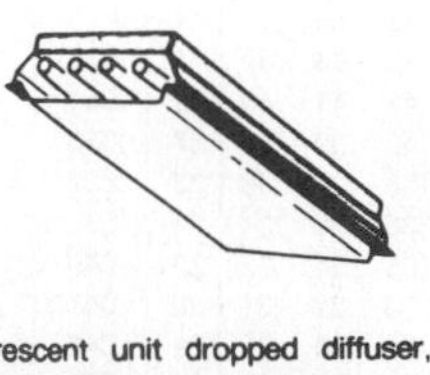 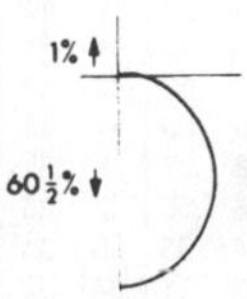 2	.55	.51	.47	.54	.50	.46	.51	.48	.45	.49	.46	.44	.47	.45	.43	.42	.236
3	.48	.43	.39	.47	.42	.39	.45	.41	.38	.43	.40	.37	.42	.39	.36	.35	.212
4	.43	.37	.33	.42	.37	.33	.40	.36	.32	.39	.35	.32	.37	.34	.31	.30	.191
5	.38	.33	.29	.37	.32	.28	.36	.31	.28	.35	.31	.28	.33	.30	.27	.26	.173
6	.34	.29	.25	.34	.29	.25	.33	.28	.24	.31	.27	.24	.30	.27	.24	.23	.158
7	.31	.26	.22	.31	.26	.22	.30	.25	.22	.29	.25	.21	.28	.24	.21	.20	.144
8	.28	.23	.20	.28	.23	.20	.27	.23	.19	.26	.22	.19	.25	.22	.19	.18	.133
9	.26	.21	.18	.26	.21	.18	.25	.21	.17	.24	.20	.17	.24	.20	.17	.16	.123
10	.24	.19	.16	.24	.19	.16	.23	.19	.16	.22	.19	.16	.22	.18	.16	.15	.115

* Also, reflector downlight with baffles and inside frosted lamp.

■ **FIGURE 16-6** (Continued)
Coefficient of utilization (CU) of generic luminaires.
(Reproduced with permission from IESNA.)

Typical Luminaire	Maint. Cat.	SC	RCR ↓	80 ρw=50	80 30	80 10	70 50	70 30	70 10	50 50	50 30	50 10	30 50	30 30	30 10	10 50	10 30	10 10	0	WDRC	RCR ↓
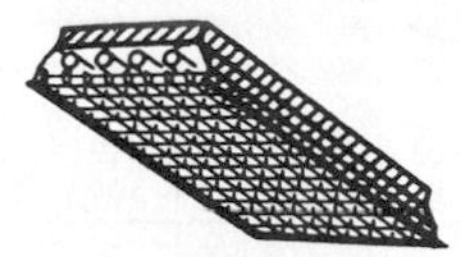 **13** 4-lamp, 610 mm (2') wide troffer with 45° white metal louver—see note 7	IV	0.9	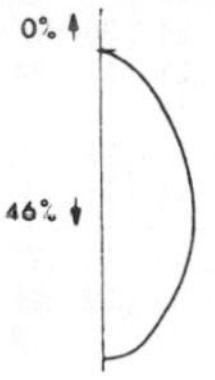 0	.55	.55	.55	.54	.54	.54	.51	.51	.51	.49	.49	.49	.47	.47	.47	.46		0
			1	.49	.48	.46	.48	.47	.46	.46	.45	.44	.45	.44	.43	.43	.42	.42	.41	.137	1
			2	.44	.42	.40	.43	.41	.39	.42	.40	.38	.40	.39	.37	.39	.38	.37	.36	.131	2
			3	.40	.37	.34	.39	.36	.34	.38	.36	.33	.37	.35	.33	.36	.34	.32	.32	.122	3
			4	.36	.33	.30	.36	.33	.30	.35	.32	.30	.34	.31	.29	.33	.31	.29	.28	.113	4
			5	.33	.30	.27	.33	.29	.27	.32	.29	.27	.31	.28	.26	.30	.28	.26	.25	.104	5
			6	.30	.27	.24	.30	.27	.24	.29	.26	.24	.29	.26	.24	.28	.25	.24	.23	.097	6
			7	.28	.25	.22	.28	.24	.22	.27	.24	.22	.26	.24	.22	.26	.23	.22	.21	.090	7
			8	.26	.23	.20	.26	.22	.20	.25	.22	.20	.25	.22	.20	.24	.22	.20	.19	.085	8
			9	.24	.21	.19	.24	.21	.19	.23	.20	.18	.23	.20	.18	.23	.20	.18	.18	.079	9
			10	.23	.19	.17	.22	.19	.17	.22	.19	.17	.22	.19	.17	.21	.19	.17	.16	.075	10
14 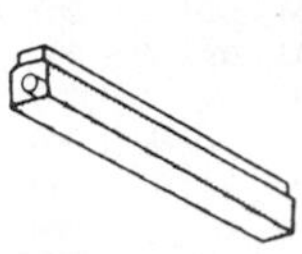 Bilateral batwing distribution—one-lamp, surface mounted fluorescent with prismatic wraparound lens	V	N.A.	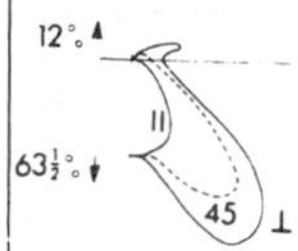 0	.87	.87	.87	.84	.84	.84	.77	.77	.77	.72	.72	.72	.66	.66	.66	.64		0
			1	.75	.72	.69	.72	.69	.66	.67	.64	.62	.62	.60	.58	.57	.56	.54	.52	.296	1
			2	.65	.60	.56	.63	.58	.54	.58	.54	.51	.54	.51	.48	.50	.47	.45	.43	.261	2
			3	.57	.51	.46	.55	.49	.45	.51	.46	.42	.47	.43	.40	.44	.41	.38	.36	.232	3
			4	.50	.44	.39	.48	.42	.38	.45	.40	.36	.42	.38	.34	.39	.35	.32	.30	.209	4
			5	.45	.38	.33	.43	.37	.32	.40	.35	.31	.37	.33	.29	.35	.31	.28	.26	.189	5
			6	.40	.33	.28	.39	.32	.28	.36	.31	.26	.34	.29	.25	.31	.27	.24	.22	.172	6
			7	.36	.29	.25	.35	.29	.24	.32	.27	.23	.30	.26	.22	.28	.24	.21	.19	.158	7
			8	.33	.26	.22	.31	.25	.21	.29	.24	.20	.28	.23	.20	.26	.22	.19	.17	.146	8
			9	.30	.23	.19	.29	.23	.19	.27	.22	.18	.25	.21	.17	.24	.20	.17	.15	.135	9
			10	.27	.21	.17	.26	.21	.17	.25	.20	.16	.23	.19	.16	.22	.18	.15	.13	.126	10
15  Radial batwing distribution—4-lamp, 610 mm (2') wide fluorescent unit with flat prismatic lens—see note 7	V	1.7	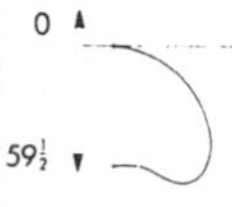 0	.71	.71	.71	.69	.69	.69	.66	.66	.66	.63	.63	.63	.61	.61	.61	.60		0
			1	.62	.59	.57	.60	.58	.56	.58	.56	.54	.55	.54	.52	.53	.52	.51	.50	.251	1
			2	.53	.49	.46	.52	.48	.45	.50	.47	.44	.48	.45	.43	.46	.44	.42	.41	.237	2
			3	.46	.41	.37	.45	.41	.37	.44	.40	.36	.42	.39	.36	.40	.38	.35	.34	.216	3
			4	.41	.35	.31	.40	.35	.31	.38	.34	.30	.37	.33	.30	.36	.32	.30	.28	.196	4
			5	.36	.30	.26	.35	.30	.26	.34	.29	.26	.33	.29	.26	.32	.28	.25	.24	.178	5
			6	.32	.27	.23	.32	.26	.23	.31	.26	.22	.29	.25	.22	.29	.25	.22	.21	.162	6
			7	.29	.24	.20	.28	.23	.20	.28	.23	.19	.27	.22	.19	.26	.22	.19	.18	.149	7
			8	.26	.21	.17	.26	.21	.17	.25	.20	.17	.24	.20	.17	.24	.20	.17	.16	.137	8
			9	.24	.19	.15	.24	.19	.15	.23	.18	.15	.22	.18	.15	.22	.18	.15	.14	.127	9
			10	.22	.17	.14	.22	.17	.14	.21	.17	.14	.20	.16	.14	.20	.16	.14	.12	.118	10
16 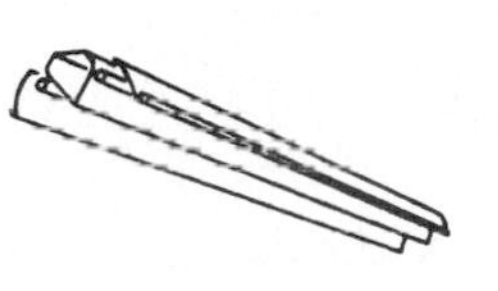 Diffuse aluminum reflector with 35°CW shielding	II	1.5/1.3	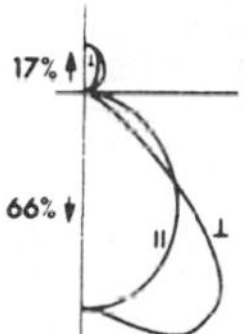 0	.95	.95	.95	.91	.91	.91	.83	.83	.83	.76	.76	.76	.69	.69	.69	.66		0
			1	.85	.82	.79	.81	.79	.76	.75	.73	.71	.69	.67	.66	.63	.62	.61	.59	.197	1
			2	.75	.71	.67	.72	.68	.65	.67	.63	.61	.62	.59	.57	.57	.55	.53	.51	.194	2
			3	.67	.61	.57	.65	.59	.55	.60	.56	.52	.55	.52	.49	.51	.49	.46	.44	.184	3
			4	.60	.54	.49	.58	.52	.48	.54	.49	.45	.50	.46	.43	.46	.43	.41	.39	.173	4
			5	.54	.47	.43	.52	.46	.42	.49	.43	.40	.45	.41	.38	.42	.39	.36	.34	.162	5
			6	.49	.42	.37	.47	.41	.37	.44	.39	.35	.41	.37	.33	.38	.35	.32	.30	.151	6
			7	.44	.38	.33	.43	.37	.32	.40	.35	.31	.38	.33	.30	.35	.31	.28	.27	.141	7
			8	.40	.34	.29	.39	.33	.29	.37	.31	.28	.34	.30	.27	.32	.28	.26	.24	.132	8
			9	.37	.31	.26	.36	.30	.26	.34	.29	.25	.32	.27	.24	.30	.26	.23	.21	.124	9
			10	.34	.28	.24	.33	.27	.23	.31	.26	.23	.29	.25	.22	.28	.24	.21	.19	.117	10
17  Porcelain-enameled reflector with 30°CW × 30°LW shielding	II	1.0	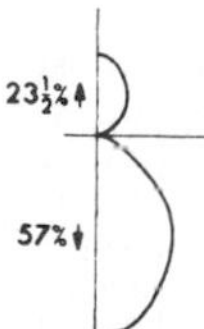 0	.91	.91	.91	.86	.86	.86	.77	.77	.77	.68	.68	.68	.61	.61	.61	.57		0
			1	.80	.77	.75	.76	.74	.71	.69	.67	.65	.62	.60	.59	.55	.54	.53	.50	.182	1
			2	.71	.67	.63	.68	.64	.60	.61	.58	.55	.55	.53	.51	.50	.48	.46	.43	.174	2
			3	.63	.58	.53	.60	.55	.51	.55	.51	.47	.50	.46	.44	.45	.42	.40	.38	.163	3
			4	.57	.51	.46	.54	.49	.44	.49	.45	.41	.45	.41	.38	.41	.38	.35	.33	.151	4
			5	.51	.45	.40	.49	.43	.39	.45	.40	.36	.41	.37	.34	.37	.34	.31	.29	.140	5
			6	.46	.40	.35	.44	.38	.34	.41	.36	.32	.37	.33	.30	.34	.30	.28	.26	.130	6
			7	.42	.36	.31	.40	.35	.30	.37	.32	.29	.34	.30	.27	.31	.28	.25	.23	.121	7
			8	.38	.32	.28	.37	.31	.27	.34	.29	.26	.31	.27	.24	.29	.25	.23	.21	.113	8
			9	.35	.29	.25	.34	.28	.25	.31	.27	.23	.29	.25	.22	.27	.23	.21	.19	.106	9
			10	.33	.27	.23	.31	.26	.22	.29	.24	.21	.27	.23	.20	.25	.21	.19	.17	.099	10
18 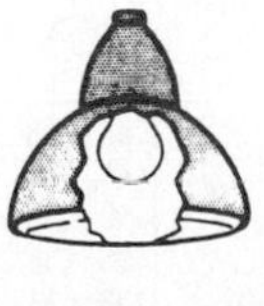 Enclosed reflector with an incandescent lamp	V	1.4	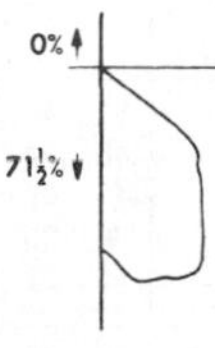 0	.85	.85	.85	.83	.83	.83	.80	.80	.80	.76	.76	.76	.73	.73	.73	.72		0
			1	.77	.75	.73	.76	.74	.72	.73	.71	.69	.70	.69	.67	.67	.66	.65	.64	.189	1
			2	.70	.66	.63	.68	.65	.62	.66	.63	.60	.64	.61	.59	.61	.60	.58	.56	.190	2
			3	.63	.58	.54	.62	.57	.54	.60	.56	.53	.58	.54	.52	.56	.53	.51	.50	.183	3
			4	.56	.51	.47	.56	.51	.47	.54	.50	.46	.52	.49	.46	.51	.48	.45	.44	.174	4
			5	.51	.46	.42	.50	.45	.41	.49	.44	.41	.48	.44	.40	.46	.43	.40	.39	.164	5
			6	.46	.41	.37	.46	.41	.37	.45	.40	.36	.43	.39	.36	.42	.39	.36	.34	.155	6
			7	.42	.37	.33	.42	.37	.33	.41	.36	.33	.40	.36	.32	.39	.35	.32	.31	.146	7
			8	.39	.33	.30	.38	.33	.29	.37	.33	.29	.37	.32	.29	.36	.32	.29	.28	.137	8
			9	.36	.30	.27	.35	.30	.27	.35	.30	.27	.34	.30	.26	.33	.29	.26	.25	.129	9
			10	.33	.28	.24	.33	.28	.24	.32	.27	.24	.31	.27	.24	.31	.27	.24	.23	.122	10

Coefficients of Utilization for 20 Per Cent Effective Floor Cavity Reflectance (ρFC = 20)

* Also, reflector downlight with baffles and inside frosted lamp.

■ FIGURE 16–6 *(Continued)*
Coefficient of utilization (CU) of generic luminaires.
(Reproduced with permission from IESNA.)

Coefficient of utilization (CU) of generic luminaires table.

Typical Luminaire	Typical Intensity Distribution and Per Cent Lamp Lumens		ρcc →	80			70			50			30			10			0		ρcc →
			ρw →	50	30	10	50	30	10	50	30	10	50	30	10	50	30	10	0	WDRC	ρw →
	Maint. Cat.	SC	RCR ↓	Coefficients of Utilization for 20 Per Cent Effective Floor Cavity Reflectance (ρFC = 20)																	RCR ↓
 19	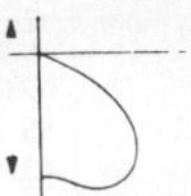 IV	1.7	0	.67	.67	.67	.65	.65	.65	.62	.62	.62	.60	.60	.60	.57	.57	.57	.56		
			1	.60	.58	.56	.58	.57	.55	.56	.55	.53	.54	.53	.52	.52	.51	.50	.49	.177	1
			2	.53	.49	.46	.52	.48	.46	.50	.47	.45	.48	.46	.44	.46	.44	.43	.42	.179	2
	0° ▲		3	.46	.42	.39	.46	.42	.38	.44	.41	.38	.42	.40	.37	.41	.39	.37	.35	.172	3
			4	.41	.36	.33	.40	.36	.33	.39	.35	.32	.38	.34	.32	.37	.34	.31	.30	.161	4
			5	.37	.32	.28	.36	.31	.28	.35	.31	.28	.34	.30	.27	.33	.30	.27	.26	.150	5
	56° ▼		6	.33	.28	.24	.32	.28	.24	.31	.27	.24	.30	.27	.24	.30	.26	.24	.23	.139	6
			7	.30	.25	.21	.29	.25	.21	.28	.24	.21	.28	.24	.21	.27	.23	.21	.20	.129	7
Wide spread, recessed, small open bottom reflector with low wattage diffuse HID lamp			8	.27	.22	.19	.26	.22	.19	.26	.22	.19	.25	.21	.19	.24	.21	.19	.17	.120	8
			9	.25	.20	.17	.24	.20	.17	.24	.20	.17	.23	.19	.17	.22	.19	.17	.16	.112	9
			10	.22	.18	.15	.22	.18	.15	.22	.18	.15	.21	.18	.15	.21	.17	.15	.14	.105	10
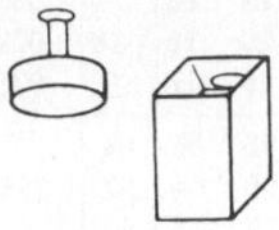 20	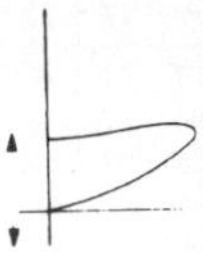 VI	N.A.	0	.74	.74	.74	.63	.63	.63	.43	.43	.43	.25	.25	.25	.08	.08	.08	.00		
			1	.64	.62	.59	.55	.53	.51	.38	.36	.35	.22	.21	.20	.07	.07	.07	.00	.000	1
			2	.56	.52	.48	.48	.45	.42	.33	.31	.29	.19	.18	.17	.06	.06	.06	.00	.000	2
			3	.49	.44	.40	.42	.38	.35	.29	.26	.24	.17	.15	.14	.05	.05	.05	.00	.000	3
			4	.43	.38	.34	.37	.33	.29	.26	.23	.20	.15	.13	.12	.05	.04	.04	.00	.000	4
	78° ▲		5	.38	.33	.28	.33	.28	.25	.23	.20	.17	.13	.12	.10	.04	.04	.03	.00	.000	5
	0° ▼		6	.34	.28	.24	.29	.25	.21	.20	.17	.15	.12	.10	.09	.04	.03	.03	.00	.000	6
			7	.31	.25	.21	.26	.22	.18	.18	.15	.13	.11	.09	.08	.03	.03	.03	.00	.000	7
Open top, indirect, reflector type unit with HID lamp (mult. by 0.9 for lens top)			8	.28	.22	.18	.24	.19	.16	.16	.13	.11	.10	.08	.07	.03	.03	.02	.00	.000	8
			9	.25	.20	.16	.21	.17	.14	.15	.12	.10	.09	.07	.06	.03	.02	.02	.00	.000	9
			10	.23	.17	.14	.20	.15	.12	.14	.11	.09	.08	.06	.05	.03	.02	.02	.00	.000	10

Typical Luminaires		ρcc →	80			70			50			30			10			0
		ρw →	50	30	10	50	30	10	50	30	10	50	30	10	50	30	10	0
		RCR ↓	Coefficients of utilization for 20 Per Cent Effective Floor Cavity Reflectance, ρFC															
21		1	.42	.40	.39	.36	.35	.33	.25	.24	.23	Coves are not recommended for lighting areas having low reflectances.						
		2	.37	.34	.32	.32	.29	.27	.22	.20	.19							
		3	.32	.29	.26	.28	.25	.23	.19	.17	.16							
		4	.29	.25	.22	.25	.22	.19	.17	.15	.13							
		5	.25	.21	.18	.22	.19	.16	.15	.13	.11							
		6	.23	.19	.16	.20	.16	.14	.14	.12	.10							
		7	.20	.17	.14	.17	.14	.12	.12	.10	.09							
		8	.18	.15	.12	.16	.13	.10	.11	.09	.08							
Single row fluorescent lamp cove without reflector, mult. by 0.93 for 2 rows and by 0.85 for 3 rows.		9	.17	.13	.10	.15	.11	.09	.10	.08	.07							
		10	.15	.12	.09	.13	.10	.08	.09	.07	.06							
22 ρcc from below ~65%		1				.60	.58	.56	.58	.56	.54							
		2				.53	.49	.45	.51	.47	.43							
		3				.47	.42	.37	.45	.41	.36							
		4				.41	.36	.32	.39	.35	.31							
		5				.37	.31	.27	.35	.30	.26							
		6				.33	.27	.23	.31	.26	.23							
Diffusing plastic or glass		7				.29	.24	.20	.28	.23	.20							
1) Ceiling efficiency ~60%; diffuser transmittance ~50%; diffuser reflectance ~40%. Cavity with minimum obstructions and painted with 80% reflectance paint—use ρc = 70.		8				.26	.21	.18	.25	.20	.17							
		9				.23	.19	.15	.23	.18	.15							
2) For lower reflectance paint or obstructions—use ρc = 50.		10				.21	.17	.13	.21	.16	.13							
23 ρcc from below ~60%		1				.71	.68	.66	.67	.66	.65	.65	.64	.62				
		2				.63	.60	.57	.61	.58	.55	.59	.56	.54				
		3				.57	.53	.49	.55	.52	.48	.54	.50	.47				
		4				.52	.47	.43	.50	.45	.42	.48	.44	.42				
		5				.46	.41	.37	.44	.40	.37	.43	.40	.36				
		6				.42	.37	.33	.41	.36	.32	.40	.35	.32				
Prismatic plastic or glass.		7				.38	.32	.29	.37	.31	.28	.36	.31	.28				
1) Ceiling efficiency ~67%; prismatic transmittance ~72%; prismatic reflectance ~18%. Cavity with minimum obstructions and painted with 80% reflectance paint—use ρc = 70.		8				.34	.28	.25	.33	.28	.25	.32	.28	.25				
		9				.30	.25	.22	.30	.25	.21	.29	.25	.21				
2) For lower reflectance paint or obstructions—use ρc = 50.		10				.27	.23	.19	.27	.22	.19	.26	.22	.19				
24 ρcc from below ~45%		1							.51	.49	.48				.47	.46	.45	
		2							.46	.44	.42				.43	.42	.40	
		3							.42	.39	.37				.39	.38	.36	
		4							.38	.35	.33				.36	.34	.32	
		5							.35	.32	.29				.33	.31	.29	
		6							.32	.29	.26				.30	.28	.26	
Louvered ceiling.		7							.29	.26	.23				.28	.25	.23	
1) Ceiling efficiency ~50%; 45° shielding opaque louvers of 80% reflectance. Cavity with minimum obstructions and painted with 80% reflectance paint—use ρc = 50.		8							.27	.23	.21				.26	.23	.21	
		9							.24	.21	.19				.24	.21	.19	
2) For other conditions refer to Fig. 6–18.		10							.22	.19	.17				.22	.19	.17	

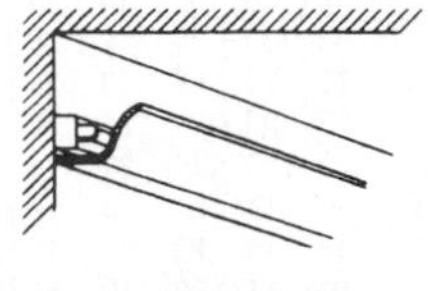

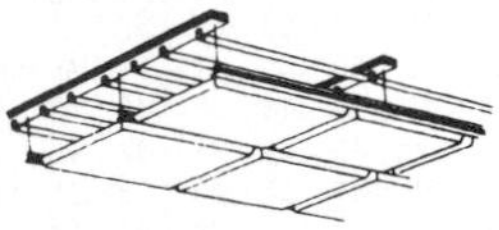

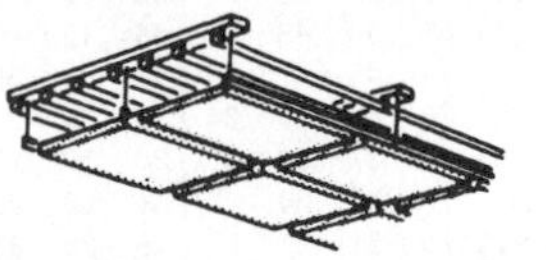

■ FIGURE 16–6 *(Continued)*
Coefficient of utilization (CU) of generic luminaires.
(Reproduced with permission from IESNA.)

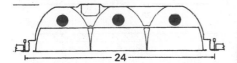

Typical CU for a 3-lamp, T8 fluorescent, with 3-in. parabolic louvers

pcc→	80			70			50		
pw→	70	50	30	70	50	30	50	30	10
0	0.87	0.87	0.87	0.85	0.85	0.85	0.81	0.81	0.81
1	0.81	0.78	0.76	0.79	0.77	0.74	0.74	0.72	0.70
2	0.75	0.70	0.66	0.73	0.69	0.65	0.66	0.63	0.61
3	0.69	0.63	0.58	0.68	0.62	0.57	0.60	0.56	0.52
4	0.64	0.56	0.51	0.62	0.55	0.50	0.54	0.49	0.46
5	0.59	0.51	0.45	0.58	0.50	0.44	0.48	0.44	0.40
6	0.55	0.46	0.40	0.53	0.45	0.40	0.44	0.39	0.35
7	0.51	0.42	0.36	0.50	0.41	0.36	0.40	0.35	0.31
8	0.47	0.38	0.32	0.46	0.38	0.32	0.37	0.32	0.28
9	0.44	0.35	0.29	0.43	0.35	0.29	0.34	0.29	0.25
10	0.41	0.32	0.27	0.40	0.32	0.27	0.31	0.26	0.23

■ **FIGURE 16–6** *(Continued)*
Coefficient of utilization (CU) of generic luminaires.
(Reproduced with permission from IESNA.)

computer programming. Figure 16–7 is a typical lighting design form for calculating the average illuminance of a space using the zonal cavity method. The form includes blank spaces for entering all design data and variables and for calculating the required number of luminaires to satisfy the design conditions. The following is a summary of the step-by-step design procedure.

1. Determine the average maintained illuminance for the space resulting from the analysis of design issues (Section 16.2, "Evaluating the Visual Environment").
2. Make a preliminary selection of light source and luminaires (from Chapter 15, "Lighting Equipment and Systems," from Figure 16–6, or from manufacturers' laboratory reports).
3. Determine and select the room characteristics: dimensions, reflectance, fixtures to be used and the mounting height of these fixtures, work plane height, etc.
4. Calculate the LLF and CU to determine the required number of fixtures, using the design form shown in Figure 16–7 or a computer program to expedite the calculation process.
5. Compare with other selections and choose the optimum one.
6. Make drawings to illustrate the desired layout or alternative layouts. (The procedure for laying out of a lighting system will be discussed in Chapter 17, "Lighting Design.")

16.6.1 Illustration of the Zonal Cavity Method

The following design problems serve to illustrate the use of the zonal cavity method to calculate an intended design for a classroom lighting system, and a dinning room.

1. *Design Problem No. 1.* Given the fluorescent fixture shown in Figure 16–8, calculate the number of fixtures required for a classroom in a university.

   The design criteria and room data are as follows:

Room dimensions	40 ft × 22 ft
Ceiling height	8 to 11 ft, avg. 9.5 ft
Reflectances	$R_c = 80\%$, $R_w = 40\%$ avg., $R_f = 20\%$
Lamps	2 × 40 W, T-12, fluorescent, 3200 lumens/lamp
LLF	70%

Answer (see Figure 16–9):

a. Select the illuminance, $E = 50$ fc.
b. Calculate P, A, PAR, CCR, RCR, and FCR.
c. Determine $\rho_{cc} = R_c$; ρ_w; R_w; $\rho_{fc} = R_f$.
d. Determine CU (see calculation form).
e. Estimate LLF = 0.70.
f. Calculate number of fixtures, $N = 17.5$ (use 18).

2. *Design Problem No. 2.* Determine the lamp size for a single incandescent lamp fixture with a diffused glass sphere (luminaire type 1) to be installed at the center of a circular dining room. Information on the room is as follows:

Room dimensions	12-ft diameter, 9-ft ceiling
Illumination maintained	15 fc
Room finishes	$R_c = 80\%$, $R_w = 70\%$, $R_f = 20\%$
Luminaire mounting	7 ft above floor ($h_{cc} = 2$)

AVERAGE ILLUMINANCE CALCULATION FORM

FOR ROOM [_____]

PROJECT: _____

PROJECT NO: _____

CALCULATION BY: _____

DATE: _____ PAGE: _____

ILLUMINANCE CRITERIA	IES ILLUMINANCE CATEGORY			
	MAINTAINED ILLUMINANCE, FC, (LUX)			
FIXTURE DATA	MFR/MODEL			
	TYPE DISTRIBUTION			
	NO. OF LAMPS PER FIXTURE			
	RATED LAMP LUMEN & WATTS/LAMP			
	LUMENS PER FIXTURE (LPF)			

ROOM DIMENSIONS	h		W, width		L, length	
ROOM CHARACTERS	h_{cc}		R_C		R_{w1}	
	h_{rc}		R_w		R_{w2}	
	h_{fc}		R_f		R_{w3}	

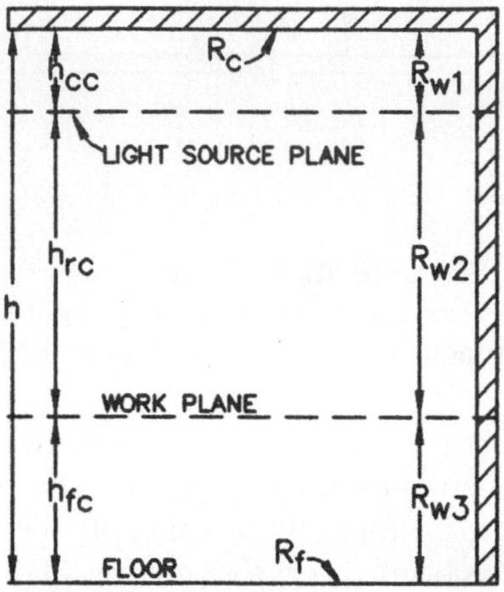

(Diagram: Room cross-section showing R_C, h_{cc}, R_{w1}, LIGHT SOURCE PLANE, h_{rc}, R_{w2}, h, WORK PLANE, h_{fc}, R_{w3}, FLOOR, R_f)

P	PERIMETER, FT(M):	
A	AREA, SF(SM):	
PAR	PERIMETER/AREA RATIO (P ÷ A)	
CCR	2.5 x PAR x h_{cc}	
RCR	2.5 x PAR x h_{rc}	
FCR	2.5 x PAR x h_{fc}	
ρ_{cc}	FROM R_C & R_{w1} & CCR	
ρ_w	SAME AS R_w OR R_{w2}	
ρ_{fc}	FROM R_f & R_{w3} & FCR	
CU	FROM CU TABLE OF FIXTURE MFGR. INTERPOLATING BETWEEN RCR AND ρ_{cc}, ρ_w, ρ_{fc}	
LOF	BF – BALLAST FACTOR		
	VF – VOLTAGE FACTOR		
	OTHER		
LLF	LLD–LAMP LUMEN DEPREC.		
	LDD–LUMINAIRE DIRT DEPREC.		
	OTHER		

FLOOR OR CEILING PLAN

(USE SEPARATE DRAWINGS FOR ADDITIONAL LAYOUTS)

CALCULATIONS

MAINTAINED ILLUMINANCE

$$E = \frac{N \times (LPF \times LOF) \times CU \times LLF}{A}$$

INITIAL ILLUMINANCE

$$E_i = E \div LLF$$

CALCULATION & REMARKS:

* N – NUMBER OF FIXTURES

■ **FIGURE 16–7**

Typical average illuminance calculation form using the zonal cavity method. (Courtesy: William Tao & Associates, St. Louis, MO.)

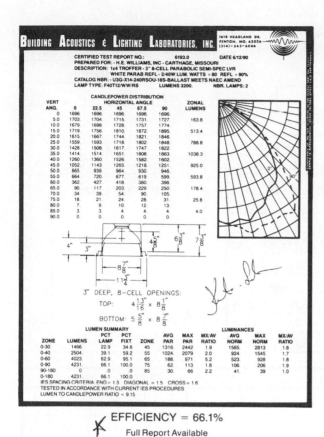

ZONAL CAVITY COEFFICIENTS OF UTILIZATION
EFFECTIVE FLOOR CAVITY REFLECTANCE = .20

CEILING	.80				.70				.50			.30		
WALL RCR	.70	.50	.30	.10	.70	.50	.30	.10	.50	.30	.10	.50	.30	.10
0	.79	.79	.79	.79	.77	.77	.77	.77	.73	.73	.73	.70	.70	.70
1	.74	.72	.70	.68	.72	.70	.68	.67	.68	.66	.65	.65	.64	.63
2	.69	.65	.62	.59	.68	.64	.61	.58	.62	.59	.57	.60	.58	.56
3	.64	.59	.55	.51	.63	.58	.54	.51	.56	.53	.50	.54	.52	.49
4	.60	.53	.49	.45	.58	.52	.48	.45	.51	.47	.44	.49	.46	.44
5	.55	.48	.43	.39	.54	.47	.43	.39	.46	.42	.39	.45	.41	.38
6	.51	.43	.38	.35	.50	.43	.38	.34	.42	.37	.34	.41	.37	.34
7	.47	.39	.34	.30	.46	.39	.34	.30	.38	.33	.30	.37	.33	.30
8	.43	.35	.30	.26	.42	.35	.30	.26	.34	.29	.26	.33	.29	.26
9	.40	.31	.26	.23	.39	.31	.26	.23	.30	.26	.23	.30	.25	.23
10	.37	.28	.23	.20	.36	.28	.23	.20	.27	.23	.20	.27	.23	.20

VISUAL COMFORT PROBABILITY
Reflectances = 80/50/20
Work Plane Illumination = 100 fc.

Room	Lengthwise				Crosswise			
	Mounting Height							
W	L	8.5	10	13	16	8.5	10	13	16
20	20	80	75	71	75	85	81	70	73
30	30	85	80	74	69	81	76	70	66
30	60	88	84	79	74	84	80	74	68
60	30	89	85	79	72	86	82	76	71
60	60	90	87	82	76	88	84	79	72

Calculated in accordance with RQQ2 1972

RCR

$$^\dagger RCR = \frac{H \times 5 \times (L+W)}{L \times W}$$

H = DISTANCE FROM LUMINAIRE TO ILLUMINATED PLANE (TASK)
L = AREA LENGTH
W = AREA WIDTH

$$^+ RCR = 2.5 \times PAR \times h_{re}^{\wedge}$$

L = Area length
W = Area width

EFFICIENCY = 66.1%
Full Report Available

■ FIGURE 16–8
Photometric report of a fluorescent fixture. (Reproduced with permission from H.E. Williams, Inc.)

Answer (see Figure 16–10):
a. Enter maintenance (illuminance: 15 fc)
b. Enter fixture and room data
c. Calculate perimeter, $P = \pi D = 37.7$ ft
d. Calculate area, $A = \pi r^2 = 113$ sq ft
e. Calculate: PAR, CCR, RCR, FCR
f. Determine ρ_{cc} from Table 16–6 with $R_c = 0.80$, $R_w = 0.70$, and CCR = 1.65; ρ_{cc} is interpolated to be = 0.66.
g. Determine CU from photometric data of type 1 (Figure 16–6)
   ρ_{cc} of 0.66 (between 70 percent and 50 percent), RCR of 3.7 (between 3 and 4), and wall reflectance of 70 percent extrapolated beyond 50 percent; the interpolated and extrapolated value of CU is 0.47 (see calculation form).
h. Estimate LLF to be 0.7 (LLD = 0.7, LDD = 1.0).
i. Calculate lumens per fixture required = 5152 lumens

The lumen output of a 300-W, PS-25 general-service type of incandescent lamp, found in the *IESNA Handbook* or in the lamp manufacturers' data, is 6100 lm. The output is about 20 percent higher than required. However, it would be appropriate to use a dimmer to vary the illuminance under different operating modes and to extend the life of the lamp.

If the wall is painted or covered with 30 percent wall covering, the ρ_{cc} will be reduced to 0.43 and the CU reduced to about 0.3. As a result, the lamp has to be enlarged to 500 W, an increase of 66 percent (CU = 0.47 versus 0.30). The impact of wall reflectance on illumination design, particularly in low-RCR spaces, is demonstrated.

16.6.2 Initial Illuminance of a Lighting System

Although a lighting system must be designed to maintain a minimum illuminance level with use, it is also important for the designer or the owner to know the initial illuminance level, realizing that it will eventually depreciate to the maintained level. This is especially important during the commissioning of a new system, when

AVERAGE ILLUMINANCE CALCULATION FORM

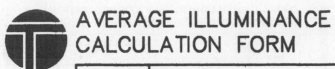

FOR ROOM	CLASSROOM

PROJECT: __UNIVERSITY__

PROJECT NO: __WORK PROB. 16-1__

CALCULATION BY: __WKT__

DATE: __2/17__   PAGE: __1__

ILLUMINANCE CRITERIA	IES ILLUMINANCE CATEGORY			E
	MAINTAINED ILLUMINANCE, FC, (LUX)			50
FIXTURE DATA	MFR/MODEL	WILLIAMS/SERIES U5G		
	TYPE DISTRIBUTION	DIRECT		
	NO. OF LAMPS PER FIXTURE		2	
	RATED LAMP LUMEN & WATTS/LAMP		3200/40	
	LUMENS PER FIXTURE (LPF)		6400 Lms	

ROOM DIMENSIONS	h	9.5	W, width	22	L, length	40
ROOM CHARACTERS	h_{cc}	0	R_c	0.7	R_{w1}	0.4
	h_{rc}	7	R_w	0.4	R_{w2}	0.4
	h_{fc}	2.5	R_f	0.2	R_{w3}	0.4

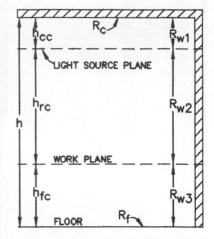

P	PERIMETER, FT(M):		124
A	AREA, SF(SM):		880
PAR	PERIMETER/AREA RATIO (P ÷ A)		0.14
CCR	2.5 x PAR x h_{cc}		0
RCR	2.5 x PAR x h_{rc}		2.45
FCR	2.5 x PAR x h_{fc}		0.88
ρ_{cc}	FROM R_c & R_{w1}		0.70
ρ_w	SAME AS R_w OR R_{w2}		0.40
ρ_{fc}	FROM R_f & R_{w3}		0.20
CU	FROM CU TABLE OF FIXTURE MFGR. INTERPOLATING BETWEEN RCR AND ρ_{cc}, ρ_w, ρ_{fc}		0.59
LOF	BF – BALLAST FACTOR	0.95	0.95
	VF – VOLTAGE FACTOR	1.0	
	OTHER	–	
LLF	LLD–LAMP LUMEN DEPREC.	0.80	0.70
	LDD–LUMINAIRE DIRT DEPREC.	0.87	
	OTHER	–	

FLOOR OR CEILING PLAN

40'

22'

(USE SEPARATE DRAWINGS FOR ADDITIONAL LAYOUTS)

CALCULATIONS

MAINTAINED ILLUMINANCE

$$E = \frac{N \times (LPF \times LOF) \times CU \times LLF}{A}$$

INITIAL ILLUMINANCE

$$E_i = E \div LLF$$

CALCULATION & REMARKS:

$E = \dfrac{(N \times 6400 \times 0.95) \times 0.59 \times 0.70}{880} = 50$ fc

N = 17.5 FIXTURES (USE 18) @ 2 ROWS

THE DESIGNED ILLUMINANCE FOR 18 FIXTURES:

E = 18 x 6400 x 0.59 x 0.70 / 880 = 54 fc

* N – NUMBER OF FIXTURES

■ **FIGURE 16–9**

Calculation form for Design Problem No. 1.

AVERAGE ILLUMINANCE CALCULATION FORM

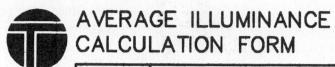

PROJECT: RESIDENCE

PROJECT NO: WORK PROB. 16-2

CALCULATION BY: WKT

DATE: 2/17   **PAGE:** 1

FOR ROOM	DINING ROOM

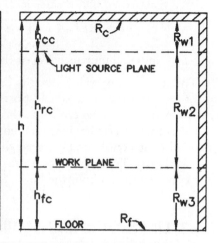

ILLUMINANCE CRITERIA	IES ILLUMINANCE CATEGORY					C
	MAINTAINED ILLUMINANCE, FC, (LUX)					15
FIXTURE DATA	MFR/MODEL	TYPE 1 FIGURE 16-4				
	TYPE DISTRIBUTION	GENERAL DIFFUSE				
	NO. OF LAMPS PER FIXTURE					1
	RATED LAMP LUMEN & WATTS/LAMP					TBD
	LUMENS PER FIXTURE (LPF)					TBD
ROOM DIMENSIONS	h	9	W, width	—	L, length	—
ROOM CHARACTERS	h_{cc}	2	R_c	0.8	R_{w1}	0.7
	h_{rc}	4.5	R_w	0.7	R_{w2}	0.7
	h_{fc}	2.5	R_f	0.2	R_{w3}	0.7

P	PERIMETER, FT(M):	37.7
A	AREA, SF(SM):	113
PAR	PERIMETER/AREA RATIO (P ÷ A)	0.33
CCR	2.5 x PAR x h_{cc}	1.65
RCR	2.5 x PAR x h_{rc}	3.7
FCR	2.5 x PAR x h_{fc}	2.1
ρ_{cc}	FROM R_c & R_{w1} & CCR	0.66
ρ_w	SAME AS R_w OR R_{w2}	0.70
ρ_{fc}	FROM R_f & R_{w3} & FCR	0.20
CU	FROM CU TABLE OF FIXTURE MFGR. INTERPOLATING BETWEEN RCR AND ρ_{cc}, ρ_w, ρ_{fc}	0.47

LOF	BF — BALLAST FACTOR	1.0	1.0
	VF — VOLTAGE FACTOR	1.0	
	OTHER	—	
LLF	LLD—LAMP LUMEN DEPREC.	0.7	0.7
	LDD—LUMINAIRE DIRT DEPREC.	1.0	
	OTHER	—	

FLOOR OR CEILING PLAN

D = 12 ft

(USE SEPARATE DRAWINGS FOR ADDITIONAL LAYOUTS)

CALCULATIONS

MAINTAINED ILLUMINANCE

$$E = \frac{N \times (LPF \times LOF) \times CU \times LLF}{A}$$

INITIAL ILLUMINANCE

$$E_i = E \div LLF$$

CALCULATION & REMARKS:

$$E = \frac{(1 \times 1.0 \times LPF) \times 0.47 \times 0.7}{113} = 15 \text{ fc}$$

LPF = 5152 Lumens

USE (1) PS-50 500W, GENERAL/SERVICE LAMP RATED FOR 6100 LUMENS

* N — NUMBER OF FIXTURES

■ **FIGURE 16-10**

Calculation form for Design Problem No. 2.

the building owner and tenant will need to accept the building without waiting for several months or longer to know the actual maintained illuminance level.

Once the initial illuminance level is measured or confirmed, the maintained illuminance level can be simulated. The initial illuminance is, of course, simply the calculated maintained illuminance level with LLF = 1,

or $E_i = (N \times LPF \times RLL \times LOF) \times CU/A$ (16–4)

where E_i = initial illuminance (fc) or (lx)

The initial illuminance value is extremely important for high-level or professional sports facilities where the illuminance level is so critical to the performance of the players that the facility management may choose to replace all lamps considerably ahead of the rated lamp life to keep the illuminance level on the playing area nearly constant throughout the life of the facility.

16.6.3 Limitations and Applications of the Zonal Cavity Method

- The illuminance calculated by the zonal cavity method is a representative average value only if the luminaires are installed to meet manufacturers' recommended minimum mounting height and spacing. Even so, some variations are to be expected. Most likely the illuminance will be higher at the center of the space and lower near the walls.
- The calculated illuminance is valid only under the conditions assumed for the calculation. It is entirely possible that the illuminance of a room with dark walnut paneling (15–20 percent reflectance) would be doubled if the wall was refinished in high-reflectance colors. This, of course, depends on the distribution characteristics of the luminaires. The effect of wall reflectance values is more pronounced for wide-distribution, direct or indirect lighting systems, including the diffuse (or direct-indirect) type. Ceiling reflectance is more critical for indirect systems and has little effect on narrow (spot) types of direct (down) lights. Calculated illuminance is for an assumed working plane, such as a desktop. Although design calculations normally select 30 in. as the working plane, this method will work well with the work plane at other heights.
- The zonal cavity method may also be applied to determine the illuminance and luminance values on vertical surfaces by using the wall reflected radiation coefficient (WRRC), which can be found in the *IESNA Lighting Handbook*. Methods also are available for calculating illuminance in irregular-shaped spaces, illuminance affected by

low partitions, and other applications. Refer to the *IESNA Lighting Education Manual* (ED-150) for these applications.

16.7 POINT METHOD

The point method, also referred to as the point-by-point method, is based on the definition that the illuminance on a surface perpendicular to the light beam incident on it is inversely proportional to the square of the distance from the light source to the surface (Eq. 14–5c). If the surface is not perpendicular to the light beam, then a cosine factor must be included (Eq. 14–6a). The latter equation can be rearranged as follows.

16.7.1 Initial Illuminance

The component of the initial illuminance on a horizontal plane is

$$E_{ih} = \frac{I_a \times \cos \theta}{D^2} = \frac{I_a \times \cos \beta}{D^2}$$
$$= \frac{I_a \times \cos^3 \theta}{H^2} \qquad (16-8)$$

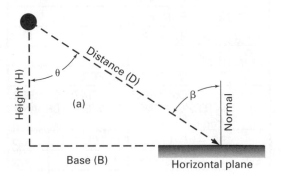

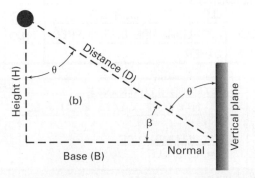

■ FIGURE 16–11
Trigonometric relationships applicable to the inverse-square law. Light incident on (a) horizontal plane and (b) vertical plane.

Refer to Figure 16–11. The component of the initial illuminance on a vertical plane is

$$E_{iv} = \frac{I_a \times \sin\theta}{D^2} = \frac{I_a \times \cos\beta}{D^2}$$

$$= \frac{I_a \times \cos^2\theta \times \sin\theta}{H^2} \quad (16\text{–}9)$$

The intensity (I) in any direction from a luminaire is normally presented in a polar diagram in the photometric report provided by the manufacturer. The intensity values in candela (cd) are the actual intensity (I_a) of the luminaire including temperature factor (TF), ballast factor (BF), and position factor (PF) except for the voltage factor (VF). Thus, an LOF factor need not be included in the initial illuminance calculations when the values from the photometric report are used.

16.7.2 Maintained Illuminance

The component of the maintained illuminance on a horizontal plane is

$$E_{mh} = E_{ih} \times \text{LLF} = \frac{I \times \cos^3\theta}{H^2} \times \text{LLF} \quad (16\text{–}10)$$

The component of the maintained illuminance on a vertical plane is

$$E_{mv} = E_{iv} \times \text{LLF}$$

$$= \frac{I \times \cos^2\theta \times \sin\theta}{H^2} \times \text{LLF} \quad (16\text{–}11)$$

16.7.3 Typical Photometric Report

Photometric reports for point sources, such as downlights, wall washers, and floodlights, usually include a candlepower distribution curve in polar coordinates. In addition, the manufacturer may provide a handy chart depicting the illuminance levels at various locations (planes) away from the luminaire or group of luminaires. Figure 16–12 illustrates the construction and photometric report of a typical point-source luminaire.

Example 16.14 Using the photometric report shown in Figure 16–12, calculate the illuminance on the wall at 60°, 45°, and 30° using zero as the nadir (the point directly below the luminaire). The luminaire is installed 4 ft from the wall. The lamp is 150 WR-40 and is rated for 1900 lm.

Answer From Equation 16–9, for the illuminance on a vertical plane,

$$E_{iv} = (I \times \sin\theta)/D^2 \quad (16\text{–}2)$$

The calculations of illuminance can best be presented in tabular format, accompanied by a diagram (Figure 16–13).

The foregoing example illustrates the steps used for determining the illuminance on a surface by the point method. To avoid this tedious work, the manufacturer often provides precalculated illuminance levels in tabulated format, as shown in Figure 16–12. For sports

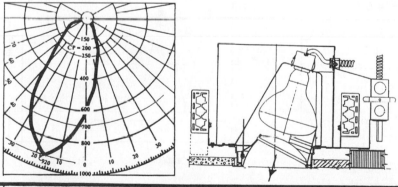

Illumination on Vertical Plane

Units 3 Feet From Wall (D)

DD	Spacing 3'			4'			5'			6'			DD
1'	10	8	10	8	5	8	8	3	8	8	2	8	1'
2'	14	14	14	11	10	11	10	6	10	9	4	9	2'
3'	16	15	16	13	10	13	12	7	12	11	5	11	3'
4'	18	17	18	15	11	15	13	8	13	13	5	13	4'
5'	16	16	16	13	11	13	12	8	12	11	6	11	5'
6'	14	14	14	11	10	11	9	7	9	8	5	8	6'
7'	11	11	11	9	8	9	7	6	7	7	5	7	7'
8'	9	9	9	7	7	7	6	5	6	5	4	5	8'
9'	8	8	8	6	6	6	5	5	5	4	4	4	9'

Illumination on Vertical Plane

■ **FIGURE 16–12**

Typical photometric report of a luminaire with an asymmetrical candlepower distribution pattern. The report includes a precalculated illuminance table for luminaires shown at the indicated locations. (Courtesy: Cooper Lighting.)

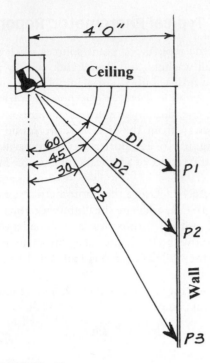

FIGURE 16–13
Section through the room showing location of the fixture in front of the wall.

ρ	θ	$\sin\theta$	$D = 4/\sin\theta$	I	$E = I \times \sin\theta/D^2$
ρ_1	60	0.87	$D_1 = 4.6$	$I_1 = 150$	$E_1 = \dfrac{150 \times .87}{(4.6)^2} = 6.2$
ρ_2	45	0.71	$D_2 = 5.6$	$I_2 = 250$	$E_2 = \dfrac{250 \times 0.71}{(5.6)} = 5.7$
ρ_3	30	0.50	$D_3 = 8.0$	$I_3 = 700$	$E_3 = \dfrac{700 \times 0.5}{8} = 5.5$

lighting applications, the precalculated illuminance levels may be given in the form of isolux (equal-illuminance) contours.

The illuminances calculated in the example are initial values. An LLF is applied to determine the maintained values.

The point method is commonly used for the following:

- Manual calculation of the illuminance at a point on a surface from a single luminaire.
- Manual calculation of the illuminance at a point on a surface from multiple luminaires, provided that the calculations have taken into consideration the intricate angles of incidence between the luminaires and the point.

- Computerized calculation of the illuminance of a point on any surface from multiple sources at different angles, including the reflected flux from the interior surfaces. The algorithm of the calculations is complex and includes thousands of calculations; thus it can be done only by a computer. Computer programs are available from the Illuminating Engineering Society of North America, consulting engineers, lighting designers, and most of the leading lighting fixture manufacturers.
- Computerized calculations of sports lighting. The point method is the only method applicable to sports lighting design, which demands accurate prediction of illuminances on the horizontal and vertical planes, as well as the uniformity ratios. The design of sports lighting is very specialized.

Design Problem No. 3   The following design is for a commercial carpet showroom that displays carpets on the floor. From IES categories, it is determined that the medium value of category D is proper (i.e., 30 fc). The preliminary design determined to use two type-A indirect distribution luminaires (Figure 16–6, luminaire type 2) and six type-B direct distribution luminaires (Figure 16–6, luminaire type 8), is shown in Figures 16–14(a) and (b).

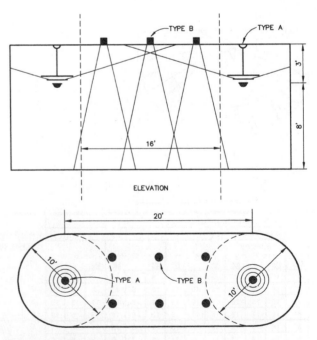

FIGURE 16–14
(a) Section of a showroom.
(b) Ceiling plan of the show-room.

Determine the lamp lumens per fixture required, based on the following information:

1. Room dimensions: Rectangular room with semicircular ends, as shown in Figure 16–14(b).
2. Reflectances:
   - Ceiling, 80 percent
   - Wall, 50 percent average
   - Floor, 20 percent without carpets in place.
3. This work problem is designed to illustrate a number of variables encountered in lighting design applications:
   - The work plane is the floor, not the standard 30-in. level at desktops.
   - Although it is assumed that the floor reflectance is 20 percent, the actual floor reflectance could vary considerably, depending on the reflectance of the carpets displayed in various parts of the room.
   - The room is not rectangular but rather is a composite of a rectangular and a semicircular shape.
   - Two types of luminaires are to be used, each covering a different part of the room.
4. There are a number of approaches to tackling this design. One suggested way is to calculate the illuminance produced by the type-A and type-B luminaires separately and then superimpose the illuminance produced by both.

16.8 COMPUTER CALCULATIONS AND COMPUTER-AIDED DESIGN

Computerized calculations and design are valuable tools essential in today's lighting design practices. Mastering this technology will simplify the design process, generating more accurate solutions while creating more effective visual presentations and additional outputs not feasible with manual methods. Nevertheless, computer software for producing visual presentations such as an accurate illuminance map over a large work plane with accurate luminance and color prediction is still in their infancy. Much still needs to be improved before the outputs can fully simulate what is viewed by human eyes.

The manual calculation methods previously introduced in this chapter are the building blocks for computer programs. In this section we introduce several popular computer programs that are either available commercially or in the public domain.

Although today's computer programs are generally user friendly, it is imperative that the user be familiar with the fundamental physics of light and lighting as discussed in Chapter 14, the properties of light sources and equipment discussed in Chapter 15, as well as the luminance and illuminance calculation methods given in this chapter. The "garbage in–garbage out" principle holds true in software applications of lighting programs just as with any other program.

16.8.1 Overview of Computer Technology

Let us look at the example of calculating the illuminance on the work plane in a room with only a few ceiling-mounted down lights. To use the point-by-point method to calculate a single point on the work plane, one would need to input the luminous intensity (candela) values of every down light in the room at the specific angles and distances away from this selected point. This procedure would need to be repeated for multiple points as needed on a work plane to determine average illuminance and uniformity values for the entire room. Furthermore, calculations for the reflected components from room surfaces that contribute to the illuminance on the work plane would need to be included. This process would require thousands of calculations to achieve accurate results and would be impractical to do manually, however, such large-scale calculations are an elementary process for computer programs, and results are output in just a few seconds.

Modern computers are compact, economical, and capable of operating at speeds up to many gigahertz, which makes computerized lighting calculations and modeling affordable and practical and opens the door to more sophisticated studies and modeling as an alternative or supplement to real-world mockups.

16.8.2 Lighting Calculation Programs

There are many lighting programs available today to aid lighting designers in calculating quantitative values such as illuminance, luminance, and uniformity, as well as qualitative issues such as luminance contrast and visual comfort projection for both interior and exterior electric lighting applications. In addition, some computer programs can include the solar azimuth intensity and energy levels for day-lighting calculation and design.

One of the most useful applications of computer technology in lighting design is the increased use of 3-dimensional computer-aided design (CAD) software to model perspective views simulating various lighting design options and to generate photolike renderings. Most modeling programs are based on ray-tracing or radiosity algorithms or a combination of both to calculate light levels.

- A ray-tracing algorithm is a way of computing an image based on tracing paths of light rays from the eye back to the luminaires. This method results in determining the direct lighting contribution to the work plane but lacks accuracy in calculating the contribution from reflected components, which should be significant for all distribution classes of lighting except the direct distribution class.

- The radiosity algorithm was originally invented to calculate the radiated heat transfer from one surface to another. Because lighting is a form of energy, this algorithm was reapplied to identify the intensity of light on a surface. With this method, a surface is divided into small meshes, and light energy is redistributed through interreflectance from surface to surface, resulting in a more realistic representation of the light energy distribution.

Each of these methods has its advantages and disadvantages with respect to accuracy, rendering quality, computation time, and computer memory requirements. A combination of ray-tracing and radiosity algorithms can result in higher-quality images and a greater variety of output formats.

The following list is a sampling of lighting design and calculation programs currently available. Some of these programs are available commercially, and others are provided by institutions or government agencies available in the public domain.

- *Advance Graphic Interface* A comprehensive lighting calculation program developed by Lighting Analyst, used for interior, exterior, and day-lighting applications and capable of complex geometry.
- *DOE2* An energy modeling program with daylighting capabilities developed by the U.S. Department of Energy under public domain and may be obtained by writing. The program is a comprehensive energy program that includes lighting program.
- *Light Pro* A Windows-based graphics program developed by Hubble Lighting, Inc., used for interior and exterior electric lighting calculations.
- *Light Scape* A comprehensive lighting calculation program developed by AutoDesk Software Company capable of producing light-accurate renderings.
- *Luxicon* A Windows-based graphics program developed by Cooper Lighting, Inc., used for interior and exterior electric lighting.
- *Radiance* A comprehensive lighting calculation program developed by Lawrence Berkeley Laboratories, capable of producing light-accurate renderings. The program may be obtained from the laboratory or may be downloaded from the Internet.
- *Visual* A Windows-based graphics program developed by Lithonia Lighting Company used for interior and exterior electric lighting calculations.

16.8.3 Applying Computer-Aided Design Software

When performing computer modeling it is important to identify the purpose of the model and the project-specific characteristics that will directly affect this purpose. The purpose may be simply to identify the lighting level associated with a specific design, or to identify the uniformity and visual comfort, or to compare one lighting system with another system, or to determine the impact of various room dimensions and finishes on a specific lighting design.

Computer modeling can be carried to any level of detail. To create a more realistic image of the lighting system in a space, one must input the correct room shape, surface texture, surface reflectance, and furniture or major objects in the space. This can be very time-consuming and can overload the computer's computing power and memory. Experienced operators can often simplify the input data to eliminate insignificant details. For example, perforated acoustical ceiling tiles may have a reflectance rating of 80 percent, but in reality they will have a slightly lower reflectance because the tiny perforations trap and thus absorb part of the incident light. Instead of inputting all the perforations, the operator may just input a smooth white acoustical tile with a reduced reflectance, say, 75 percent, and achieve the same net results. As with any computer program, there is a margin of error inherent in any computer modeling.

Normally, modeling a room by entering only the major dimensions of length, width, and height as a simple rectangular box together with the photometric data of the lighting equipment is sufficient for most illuminance modeling projects. A typical computer-generated output with illuminance values on uniform grid points is shown in Figure 16–15.

When more sophisticated modeling of a space in perspective is desired, advanced modeling software is available to generate color-scale images of illuminance levels and realistic renderings. Figure 16–16 illustrates the modeling of a curved interior space illustrating the visual image of the space including both the interior electric lighting system and day lighting.

Although computer modeling technology is advancing rapidly, current programs still require intelligent input and adjustments by experienced operators to produce satisfactory results. To accurately evaluate the data generated by the program and to draw conclusions, the designer will have to interpret the data outputted based on professional experience and the specific needs of the project.

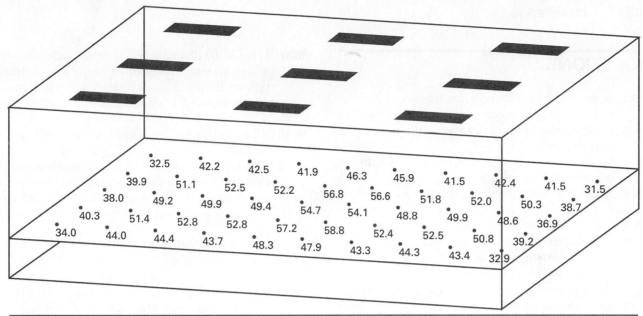

Summary: Horizontal illuminance on the desktop plane							
Project: A typical classroom							
Room	**CalcType**	**Units**	**Avg**	**Max**	**Min**	**Avg/Min**	**Max/Min**
Room No. 1	Illuminance	Fc	46.44	57.2	31.5	1.47	1.82

■ **FIGURE 16–15**

A typical computer–generated output of illuminance values on uniform grid points.

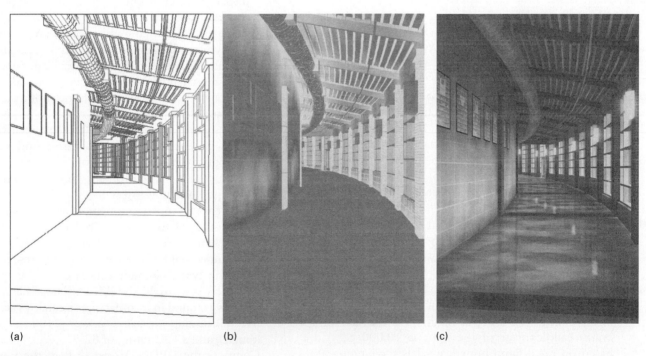

(a)　　　　　　　　　　　(b)　　　　　　　　　　　(c)

■ **FIGURE 16–16**

The stages and output formats for advanced lighting calculations. This model includes both artificial lighting and daylight contributions. (a) 3-D wire frame model with surface characteristics defined forms the basis of the model. (b) Radiosity solution based on manufacturer's photometric files (IES format) and summer solstice solar data. Illuminance levels are represented in a color spectrum gradient with the blue end of the spectrum representing the lowest light level and the red end representing the highest light level. For this model the spectrum limits are set with minimum at 5 fc and maximum at 112 fc. (c) Based on calculated light levels and material definitions a light-accurate rendering is generated. Images Courtesy William Tao and Associates, Inc. (*Note:* See Plate 13(a) of the images in full color.)

QUESTIONS

16.1 Name some illumination quality factors that can be expressed as numerical values.

16.2 What are the factors determining the selection of illuminance?

16.3 Name the steps recommended by IESNA for the selection of illuminance.

16.4 Select the maintained illuminance for a sewing factory.

16.5 What is the recommended illuminance for professional football in an indoor stadium?

16.6 State the reasons why initial illuminance has important design applications.

16.7 What is the lumen output for a general-service incandescent lamp if the voltage at the lamp socket is 5 percent below its rated voltage?

16.8 What are two most important factors that make up the lamp operating factor (LOF)?

16.9 Is the ballast factor always lower than 1.0?

16.10 What is the ceiling cavity ratio (CCR) for a ceiling recess-mounted fluorescent fixture? What is the effective ceiling reflectance if the base ceiling reflectance is 75 percent?

16.11 What is the simplified approach in zonal cavity method calculations? What variance can be expected?

16.12 What is the maintenance category for a luminaire with open top and bottom and with 65 percent uplight?

16.13 Determine the coefficient of utilization of a type-2 indirect distribution incandescent luminaire installed in a room with RCR = 2, ρ_{cc} = 80 percent, ρ_w = 50 percent, ρ_{fc} = 20 percent.

16.14 For the luminaire in question 16.13, but installed in a room with RCR = 10, ρ_{cc} = 30 percent, ρ_w = 30 percent, and ρ_{fc} = 20 percent, what is the CU?

16.15 If the luminaire is installed in an air-conditioned office building in a suburban environment, what is the appropriate LDD factor with a 12-month cleaning cycle?

16.16 A spotlight is aimed at a bulletin board on a wall. The light is 6 ft in front of the board. The center of the board is 12 ft below the light. If the maximum candlepower of the light is 10,000 cd, based on rated lamp lumen, what is the illuminance at the center of the board?

16.17 What is the illuminance 5 ft below the ceiling for a wall-washing installation, using the lighting fixtures illustrated in Figure 16–12? The fixtures are mounted 3 ft from the wall and spaced on 4-ft centers.

16.18 What are the principal applications of the point method?

16.19 As a lighting designer, you are commissioned to calculate two alternative lighting schemes for the merchandise display room of a store. The two preliminary schemes are:

Scheme 1 Using lighting system A1 comprised of type 1 luminaires on 4-ft hangers as illustrated in Figure 16–6. Use 150 W incandescent lamps rated for 2850 initial lumens, LLF = 70%

Scheme 2 Using lighting system B1 comprised of type 5 luminaires as illustrated in Figure 16–6. Use 90 W PAR38 lamps rated for 1200 initial lumens, LLF = 70%

The 30 × 50 ft room has 70% ceiling, 30% wall and 20% floor with 12 ft ceiling height. Typical sales display counter height is 4′0″ above floor (selected work plane).

a. Calculate the required number of luminaires of each system for providing general ambient space illumination of 20 fc. (Note: Task lighting for merchandise display will be provided by separate track and showcase lighting, not a part of this problem).

b. Evaluate these two systems as to their relative quality in terms of aesthetics, direct and reflected glares, modeling effect on the merchandise and on the customers, room surface luminance, and uniformity.

c. As an energy conservation measure, calculate the number of fixtures required for the same two schemes described above using EL-T, triple-tube, self-ballasted compact fluorescent lamps, in lieu of incandescent sources. Designate these systems A2 and B2. The lamp characteristics as found in manufacturer's catalogs are: 34 W including the ballast power, 2100 initial lumens, and a CRI rating of 82.

d. Compute the UPD in W per sq ft of the four lighting systems—A1, A2, B1, and B2.

LIGHTING DESIGN

17

LIGHT CAN AFFECT HUMAN BEHAVIOR, ENHANCE OR degrade a designed environment, and improve or hinder human productivity. Light has a profound effect on how we organize and live our daily lives.

Light can alter the character of a space, making it appear more spacious by lighting its walls, or more intimate by creating a barrier. Light can have an impact on people's moods by the combination of color, brightness, and contrast, or simply by its presence. Light can originate from the ceiling, from the walls, or even from the floor. Light can be a prominent feature in space or can be totally hidden through reflection or refraction techniques. Light is the most versatile of all design elements.

Lighting design is an art as well as a science. The scientific aspects of lighting and lighting calculations have been introduced in Chapters 14, 15, and 16. This chapter deals with design issues including the artistic aspect of lighting design.

17.1 DESIGN CONSIDERATIONS

17.1.1 Visual Performance

A lighting design is a plan to achieve the visual performance desired for the visual task or tasks for which the associated lighting system is to be used. Determining clearly the types of visual tasks to be performed in a space is fundamental to good design. The designer should seek this information from the user or key decision makers on the project. Also, it is often useful to survey a user's existing facility. Once these tasks are identified, levels of visual performance can be established that are in keeping with the user's needs and expectations.

17.1.2 Selection of Illuminance

Chapter 16 gives calculations of illumination for visual task categories and the criteria for selecting illuminance, taking into account visual display, the age of the observers, the speed and accuracy required for the task, and the reflectance of the task and its background. Further refinements should take into account other influencing factors, such as the duration of the task, the interfacing of adjacent tasks in the space, and other spatial, architectural, structural, and mechanical relationships.

Specific luminance ratios for various applications such as offices, educational facilities, institutions, industrial areas, and residences can be found in the *IESNA Lighting Handbook*. In general, a higher luminance ratio (contrast) is desired between the task and its immediate background for better visibility and visual acuity. (See Figure 17–1.) However, a low luminance ratio is actually more desirable for visual comfort. Good lighting design tries to balance these conflicting criteria and arrive at appropriate solutions.

17.1.3 Visual Comfort

Visual comfort is achieved when there is no prolonged visual sensation due to excessively high luminances within the visual field. One measure of visual comfort is the *visual comfort probability* (VCP) of a lighted space due to either daylight or an interior lighting system. A VCP of 70 or higher is considered acceptable for visual comfort. Reflected glare can cause a loss of visibility. A lighting system with low reflected glare can be designed with the *equivalent spherical illumination* (ESI) method, which requires specialized computer software. In general, reflected glare can be minimized by providing a proper surface texture for the task and by controlling the orientation of the luminous flux.

17.1.4 Architectural Lighting Needs

Lighting is called on in many situations to enhance architectural form or detail. Early in the design phase of a project, the lighting specifier should try to determine what architectural attributes may require lighting reinforcement. As the overall lighting design evolves, consideration of these architectural lighting needs should take place concurrently, fostering a more integrated solution. Indirect light coves for ceiling illumination, recessed wall slots to accent vertical surfaces, and concealment of a highlighting source within a column capital are a few common examples.

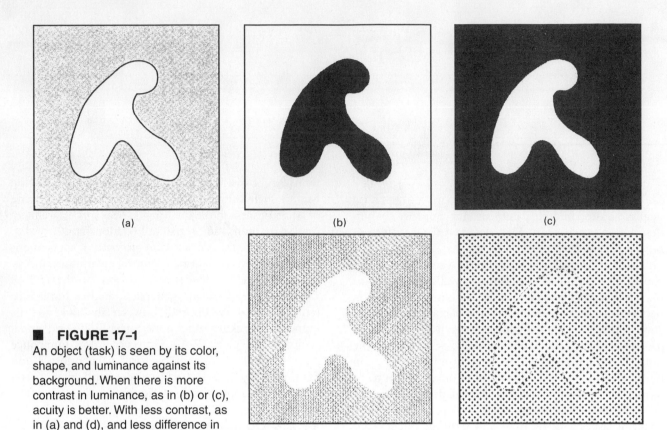

FIGURE 17–1

An object (task) is seen by its color, shape, and luminance against its background. When there is more contrast in luminance, as in (b) or (c), acuity is better. With less contrast, as in (a) and (d), and less difference in texture (e), acuity is poorer.

In many cases, an actual luminaire may become the architectural lighting embellishment. Themed or decorative luminaires in the form of chandeliers and sconces are often featured as focal points of architectural lighting in interior spaces.

17.1.5 Light Reinforcement of Spatial Impressions

It has been clearly demonstrated through both research and practice that light plays an identifiable role in shaping human impressions. Although lighting's support of the performance of visual tasks is of prime importance, its ability to either enhance or detract from the users' experience cannot be overlooked.

Examples of how the distribution, intensity, and color of the light can reinforce users' impression of a space include these:

- To create an impression of spaciousness, provide uniform wall lighting.
- To create an impression of relaxation, create nonuniform light distributions, provide relatively low intensity, and emphasize peripheral light placement.
- To create an impression of visual clarity, provide bright, uniform light.

A conscious awareness of these factors will help the lighting designer determine how and where to distribute light in order to reinforce the desired impressions. This use of light as a subjective influence can significantly enhance the user's experience of space.

17.1.6 Color

For a better understanding of color, chromaticity, color rendering, and the use of color, the lighting designer should refer to Section 5 of the reference volume of the *IESNA Lighting Handbook*. The terms *chromaticity* and *color rendering* are often misused. To put it simply, *chromaticity* refers to the color appearance of a light source, such as its color temperature, and *color rendering* refers to the ability of the light source to render colors of surfaces and objects as one would expect them to appear at the same color temperature. Both factors are important in finding the proper light source for a lighting design.

17.1.7 Interactions with Interior Design

Lighting design is strongly affected by architecture and interior design. Chapter 16 shows the lighting designer how to determine the number of fixtures of a given kind to produce the desired illumination levels in a given space. Knowledgeable lighting designers understand the procedure. Proactive lighting designers will share the implications of their understanding with their architectural and interiors colleagues to produce more efficient overall building solutions.

The zonal cavity method relies on luminaire manufacturers' tables of CU, or coefficient of utilization. The higher the CU, the fewer luminaires are required. High CU means efficient lighting. Cutting CU from 40 to 80 percent would reduce the number of fixtures by half, reducing both the initial cost of lighting systems and ongoing costs of operation, including energy, relamping, and air conditioning. Considering the factors that affect CU gives insights into design of efficient lighting systems.

CU is affected by room geometry and reflectance of surfaces in the room. Open office planning is an option to closed offices. Closed offices have more wall area to absorb light. Light absorbed by wall surfaces does not reach task planes, so closed offices will need 10 to 30 percent more fixtures to produce the same illumination levels. Selection of finishes will also affect system efficiency. Medium colors (40 to 60 percent reflectance) will reduce CU in closed offices by another 25 to 35 percent compared with light colors (> 80 percent reflectance).

17.1.8 Energy

Lighting consumes approximately 30 percent of the energy used in buildings, so energy performance is a key design issue. Fluorescent and other high-efficiency light sources are used almost exclusively in new buildings, owing in large part to economics and to prescriptions of the National Energy Policy Act. New types of fluorescent lamps and fixtures can now be used in almost every situation that previously may have required incandescent light sources, such as recessed down lights and applications requiring dimming. As technology is improved, even highly efficient HID light sources may be considered in dimming applications.

Lighting controls can also be employed to avoid unnecessary energy costs. Occupancy sensors can be used to shut off lights in unoccupied spaces. Daylight sensors can be used to dim or step lighting in spaces where there is a significant daylight contribution.

The lighting system interacts with other systems. Heat from lights is removed by the air-conditioning system, which uses energy in mechanical equipment. The amount of energy used by air-conditioning systems can be 20 to 30 percent of the energy used by the lighting. On the other hand, lighting can help heat perimeter spaces during cold weather; however, electricity used for lighting is an expensive use of energy for heating compared with gas or oil. Energy-efficient lighting design is always beneficial and most effective in warm climates and large buildings with minimal perimeter space.

17.1.9 Economics

The lighting designer must be aware of the economics surrounding the project and the budget within which the lighting must be developed. Failure to recognize this early in the process may result in wasted effort and unfulfilled expectations. The responsibility for making sure the lighting budget is met should be clearly defined by the client. If this task falls to the lighting designer, then it will be extremely important to monitor equipment cost estimates frequently during the design's development in order to anticipate problems.

17.1.10 Maintenance and Operation

The long-term maintenance and operation of a lighting system should not be neglected during the design process. Care must be taken to select fixtures for ease of open cleaning, relamping, and ballast replacement. Otherwise, the fixtures may be damaged during relamping, or maintenance may not be performed. Also, the designer should consider access to the fixture. Fixtures over stairs and fixed seating should be avoided, for instance, since ladder or mechanical lift access is difficult in such areas.

17.1.11 Selection of Lighting Systems

The IESNA defines six general classifications of lighting systems, which are illustrated in Figure 15–16.

>   Direct
>   Semidirect
>   General diffuse
>   Direct–indirect
>   Semi-indirect
>   Indirect

Selecting the system and associated luminaire types that are appropriate for a specific project will be based on the various visual needs of the project. Which system or systems best fulfill the visual performance requirements of the tasks occurring in each space? Which will appropriately enhance the important architectural features? Which will provide light distribution and intensity that will reinforce the desired impressions for each room? Which will provide the appropriate spatial modeling? (See Figure 17–2.)

(a)   (b)

■ FIGURE 17–2
Light orientation is important in lighting design. (a) Harsh shadows are produced on the objects with unidirectional lighting. (b) Shadows are softened with diffused lighting. (Courtesy: General Electric Company, Nela Park, Cleveland, OH.)

Indirect versus Direct Lighting Systems

Indirect and semi-indirect lighting systems offer higher-quality lighting, usually with less quantity of light. Recessed direct systems are the workhorse for commercial and institutional buildings and often result in glare, shadows, veiling reflections, and reflected images in computer screens. Indirect or semi-indirect lighting is an alternative that produces better visibility of tasks at lower levels of illumination. Indirect lighting is theoretically less efficient than direct lighting owing to light lost in reflection from room surfaces; however, indirect light is generally more uniform, eliminates glare, results in fewer shadows, and can be designed at lower light levels to produce a better environment at lower energy cost.

Initial costs for indirect fixtures and their installation have historically been higher than for direct systems. The disparity in cost is likely to be eliminated with new light sources designed specifically for indirect systems. Direct lighting systems generally use 48-in. T8 lamps, which produce approximately 2850 lm per lamp. Newer 48-in. T5 high-output lamps are available, which produce 5000 lm. These lamps are much brighter than conventional T8 lamps and cannot be used in direct systems owing to glare but are ideal for indirect systems, which hide the light source from direct view. With almost 75 percent more light output per lamp, T5 indirect systems can use fewer fixtures for the same performance and be very competitive with conventional direct systems.

Although one lighting system may serve as the predominant choice in most areas, it is likely that a range of luminaires will be required to address specific needs or special conditions. Wall washing for presentation surfaces, accent lights for artwork and signage, theatrical lighting for auditorium stages, and high-intensity spotlighting for indoor plants are just a few frequently encountered conditions in which the basic overall lighting system may not be sufficient.

Architecturally integrated lighting and decorative luminaires for a project should be selected collaboratively with the architect, the interior designer, and, if appropriate, the end user. Although these systems may add some functional light to a room, their primary role is to create spatial atmosphere and reinforcement.

Task Lighting versus Uniform Ceiling Lighting

Lighting designers can use the task lighting concept to supply adequate light in specific locations with higher lighting requirements than surrounding areas. The result is less overall lighting equipment, less power, and less energy usage than would be used in uniformly illuminated spaces. Task lighting has the potential for lower installed cost, lower utility bills, and a more visually interesting design than conventional solutions.

The most common application of the task lighting concept is to use furniture-mounted fixtures to illuminate work surfaces at a level suitable for conducting visual tasks requiring high acuity. Surrounding areas rely on ceiling fixtures to provide lower-level illumination. A general lighting system may be designed for 20 to 30 fc with supplemental task lights providing 80 fc or more on selected work surfaces.

Ceiling systems can also be designed with task lighting in mind. Workstations are generally laid out in an organized, modular fashion with primary and secondary circulation paths. Ceiling lighting systems are usually designed with the goal of uniform illumination across the entire floor, which overilluminates circulation areas. Light fixtures need not be uniformly spaced but, rather, can be clustered to provide high lighting levels over workstations and to use spill light for circulation paths.

Clustering ceiling lighting over groups of workstations will use fewer fixtures, saving initial cost. In addition, switching can be designed to coordinate with work groups so that lights on can be restricted to spaces occupied during off-hour operation at night and on weekends. Motion sensors may also be considered to operate specific lighting groups in response to actual occupancy needs.

Using the ceiling cluster concept in conjunction with furniture-mounted task lights can create three levels of lighting. High illumination can be achieved at specific work surfaces, medium levels in workstation areas, and lower levels in circulation space.

The cluster concept requires coordination among the several parties in the design team. Typically, the architect and the electrical engineer design the ceiling grid and ceiling lighting. The space planner and/or interior designer position the workstations and circulation space. The interior designer selects workstation components, including furniture-mounted task lights.

All parties must work together to achieve an integrated solution that satisfies not only the initial occupancy but also the inevitable changes in work clusters that necessitate changes in the arrangement of ceiling fixtures and circuiting. Lay-in fixtures and modular wiring are most appropriate to allow flexibility in the ceiling lighting system to accommodate changes on the floor.

17.2 LIGHTING DESIGN DEVELOPMENT

Once the basic lighting design has been formulated and a lighting system selected, it is necessary to apply the design concept throughout the entire space or building. Lighting layouts and arrangements must satisfy a variety of requirements in order to fulfill the design concept. They must deliver the determined levels of illuminance necessary to support the performance of the space's visual tasks. They must work harmoniously with the architecture and interior design so as to present an integrated, overall concept; and they must create a luminous environment that appropriately reinforces the impressions intended.

The most difficult of the designer's charges is satisfying all these needs in a single, unified layout. More often than not, it will be necessary to examine a series of alternative arrangements for the lighting system's various luminaire types in order to arrive at a desirable solution.

17.2.1 Lighting Layout

The number of luminaires for a desired level of visual performance within a space is determined using the calculation methods outlined in Chapter 16. Whether the goal is to provide uniform illumination throughout an entire space or within only a portion of it, a lighting layout can be developed to meet the objective. Two basic criteria are used to ensure uniform illumination across a defined work place or task surface—the principle of general area layout, and the ratio of spacing to mounting height.

The working area to be illuminated is divided into as many unit areas as the number of luminaires to be installed. For example, if calculations show that a space requires eight luminaires, then the space is divided into eight unit areas, and one luminaire is located at the center of each unit area. The divided areas should be as symmetrical to the luminaire distribution pattern as is practical. Figure 17–3(a) and (b) illustrates the principle involved. If the objective is to illuminate only a portion of the space, then the working area to be illuminated is still divided into unit areas accordingly. Figures 17–3(c) and (d) illustrates this principle. Often, the pattern must be compromised to fit the ceiling grid.

The general layout principle applies to all luminaires with symmetrical flux distribution. There are, however, luminaires designed for asymmetrical distribution where the luminous flux (lumens) is stronger (higher) in one direction than the other, such as wall washers or adjustable accent lights. In such cases, the location of luminaires is governed by their photometric distribution.

For uniform illumination, the spacing of the luminaire must not exceed the spacing-to-mounting height ratio (SMHR) recommended by the manufacturer. The problem associated with a design that exceeds the recommended SMHR is illustrated in Figures 17–3(e) and (f), showing a relatively dim area (lower illuminance) between the luminaires. Although most manufacturers furnish the SMHR as part of the technical information in their catalogs, it can be visually estimated by not exceeding the angles between the maximum and 50 percent values of the candlepower distribution curve provided in the photometric data. The SMHR is greater (more spread out) for semi-indirect and totally indirect luminaires. Narrow-beam direct distribution luminaires are normally selected to highlight a localized area, and selecting them to light a large space uniformly is inappropriate.

Table 17–1 provides general guidelines for determining the maximum SMHR of different classes of luminaires. For example, the SMHR of a general diffuse class of luminaire should not exceed 1.0, and that of an indirect distribution luminaire can be approximately 1.25. In design applications, if the lumen output and class of distribution of luminaires are properly selected, the actual SMHR for most highly demanding visual performance spaces, such as classrooms, offices, and retail

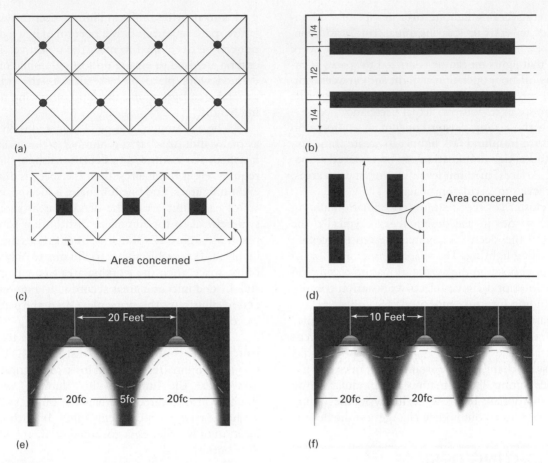

■ **FIGURE 17–3**

Layout of luminaires for uniform lighting. (a), (b) The entire space is divided into unit areas for total uniformity. (c), (d) The luminaires are located in the area where uniformity is desired. (e), (f) Insufficient luminaire spacing will create dark (shadow) areas.

stores, is normally lower than the maximum SMHR required. Whenever the manufacturer's photometric test report for the specific SMHR is available, the designer should consult it.

17.2.2 Lighting Expressions

The arrangement or organization of a lighting system's layout will be a factor in an occupant's perception of a space. According to William Lam, the expression of a lighting system may be described as any of the following:

1. *Neutral.* The lighting system is deemphasized, and the lighting elements do not draw the special attention of the occupants or visitors to the space.
2. *Expressive.* The lighting system is designed to harmonize with, supplement, or enhance the architectural expressions in the space.
3. *Dominant.* The lighting system dominates the space, overpowering most other elements.

4. *Confused (disorganized).* The lighting system is disorderly, either in its configuration or in its relation to other elements in the space. In this case, lighting becomes a liability of the entire design.

For example, a small square luminaire installed at the center of the ceiling of a square room is obviously neutral, a large square could be expressive, and a square ⅔ of ceiling dimensions would be dominant. These designs are illustrated in Figure 17–4.

Although the classification of expressions is somewhat subjective, it is not difficult to reach consensus. No line can be drawn between these four expressions (neutral, expressive, dominant, and confused). A design slightly different in dimensions, form, or location could easily be switched from one expression to the other; however, it is not difficult to agree on the expressions identified in Figure 17–5.

Figures 17–6, 17–7, and 17–8 illustrate lighting layouts for various space configurations. Some layouts can be easily identified as representing one expression

TABLE 17–1
Spacing-mounting height of luminaires

Mounting Height of Luminaires (above floor) except for Indirect and Semi-Indirect Luminaires, use Ceiling Height (above floor)	Suspension Distance For Indirect and Semi-Indirect Luminaires	Indirect	Semi-Indirect	General Diffusing	Semi-Direct	Direct	Semi-Concentrating Direct	Concentrating Direct	Distance from Walls All Types of Luminaires
8	1-3	9.5	9.5	8	7	7	6.5	5	
9	1.5-3	10.5	10.5	9	8	8	7	5.5	⅓ Spacing
10	2-3	12	12	10	9	9	8	6	Distance if
11	2-3	13	13	11	10	10	9	6.5	desks or
12	2.5-4	14.5	14.5	12	11	11	9.5	7	work benches
13	3-4	15.5	15.5	13	12	12	10.5	8	are against
14	3-4	17	17	14	12.5	12.5	11	8.5	walls, other-
15	3-4	18	18	15	13.5	13.5	12	9	wise ½.
16	4-5	19	19	16	14.5	14.5	13	9.5	
18	4-5	22	22	18	16	16	14.5	11	
20 or more	4-6	24	24	20	18	18	16	12	

(All dimensions in feet) — MAXIMUM* SPACING DISTANCE BETWEEN LUMINAIRES

*The actual spacing is usually less than these maximum distances to suit bay or room dimensions or to provide adequate illumination. At an established mounting height, it is often necessary to reduce the spacing between luminaires or rows of them to provide specified footcandles. In such systems and particularly in small rooms, the utilization is reduced because the luminaires are closer to the walls. For example, in an 80-30-10 room with a room ratio of 0.6 and a flux ratio of 0.7, the utilization factor is .79 where the spacing-mounting height ratio is 1.0 (luminaires as far apart as they are above the work-plane). With the same room conditions but with luminaires spaced .4 as far apart as they are above the work-plane, the utilization factor is only .65.

(Courtesy: General Electric Company, Nela Park, Cleveland, OH.)

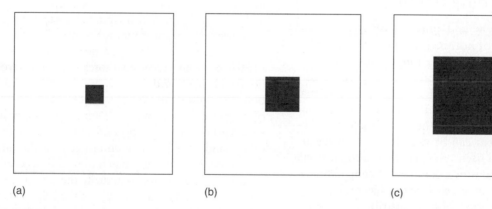

| (a) | (b) | (c) |

■ **FIGURE 17–4**
Lighting expressions of lighting designs with a square fixture at the center of a square room: (a) neutral; (b) expressive; (c) dominant.

class, whereas others are difficult to judge. A layout considered as one type may be totally different for another space occupancy. For example, a random layout as in Figure 17–7 (l) is considered "disorganized" in a classroom but will be "expressive" in a retail store. Exercise your judgment by filling in the blank spaces provided at the bottom of each figure, and discuss your choices with your instructor, fellow students, or colleagues.

17.2.3 Perspective

Although a geometric pattern of lighting fixtures on the ceiling is the starting point in most lighting design considerations, the designer must realize that lighting patterns will not be viewed in a two-dimensional plane. In fact, all architectural elements in a space, including lighting elements, are viewed in perspective. Often, a logical or attractive geometric pattern on the reflected

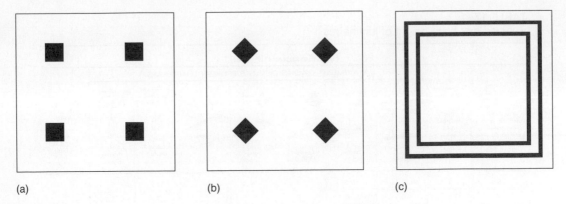

■ **FIGURE 17–5**
Lighting expression of sample lighting designs: (a) neutral; (b) confused; (c) dominant.

ceiling plane may create a confused expression in a space. Figure 17–9 (p. 538) illustrates this problem.

The designer should analyze the more telling three-dimensional perspective view of unusual lighting geometries to understand their impact on the interior architecture. With the more widespread use of computer-aided design, generating these perspectives is becoming faster and easier; however, just a simple thumbnail sketch should be sufficient for basic insight on the lighting configuration.

17.2.4 Architectural Lighting

The architectural form and purpose of a space will, whether consciously or not, be influenced by the lighting system and the light distribution it produces.

A well-integrated design might incorporate lighting into an architectural detail in order to provide visual reinforcement or highlight. An architectural cove or wall slot is a good example of how a completely integrated lighting element can be used to accent a key architectural feature. Often, the lighting designer will simply use a manufacturer's standard accent lights or wall washers to accentuate a surface or object selectively. This specific manipulation of light distribution and intensity allows the designer to enhance preferred characteristics of the architecture while directing the attention and focus of the occupants. Figure 17–10 (p. 539) is an example of how effective lighting can be in this role.

Generally, direct light brings out surface textures and more dramatically models three-dimensional objects. Soft, diffuse light will suppress shadows and tend to flatten architectural form.

Decorative luminaires, appropriately stylized or themed for the environment, can also provide a strong architectural statement, as shown in Figure 17–11 (p. 539).

Successful architectural lighting is most always a collaborate effort between the lighting specifier and the architect. Thoughtful, integrated design can be a by-product if this effort begins early enough in the design process to allow for joint development of the architectural lighting details and features as shown in Figure 17–12 (p. 539).

17.2.5 Lighting Reinforcement of Impressions

The importance of light as a way of reinforcing a desired impression of a room in the occupant's mind should not be overlooked. The intensity, direction, distribution, and color of light all contribute to a viewer's subjective response to an environment.

For example, guests at a fine restaurant illuminated to 50 fc by ceiling-recessed fluorescent troffers would have no trouble reading the menu, but the intensity and distribution of light would certainly not enhance an impression of elegance and romance. Dimmer, nonuniform light would provide a more appropriate reinforcement. On the other hand, too dim an illuminance over the dining area would require lighting a match to read the menu or to see what one is eating; this is actually the case in some high-priced restaurants—an example of the other extreme.

The *IESNA Lighting Handbook* includes a comprehensive review of light's influence in shaping human mood and impression. The designer should consult this source for further insight on using the lighting system to full advantage in creating pleasant, psychologically appropriate environments.

17.3 LIGHTING DESIGN DOCUMENTATION

Comprehensive documentation of a lighting design is required if it is to be successfully incorporated into the

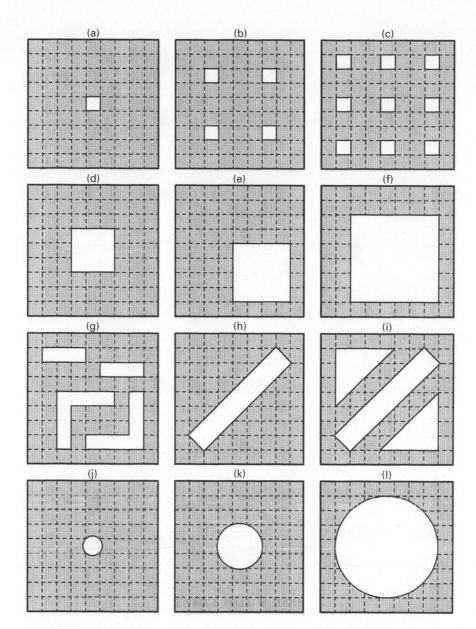

■ **FIGURE 17–6**

Exercise in evaluating lighting expressions in terms of form, pattern, orientation, and locations.

(a) _____   (b) _____   (c) _____
(d) _____   (e) _____   (f) _____
(g) _____   (h) _____   (i) _____
(j) _____   (k) _____   (l) _____

project's overall design. The goal of this phase is clear, concise, and unambiguous communication of the lighting concept to those who will construct it. These construction documents must not only translate the lighting design but also present a coordinated effort with the project's other disciplines. Although the architect or engineering project manager usually takes the lead in this effort, it is essential that the lighting designer provide feedback to the process.

17.3.1 Construction Drawings

Scaled construction drawings are used to describe the lighting design and its various systems and configurations. Typically, the luminaires are indicated on both an electrical lighting plan and an architectural reflected ceiling plan. The electrical drawings will show luminaire quantities and their approximate location and will designate each with a reference tag or "type." This type

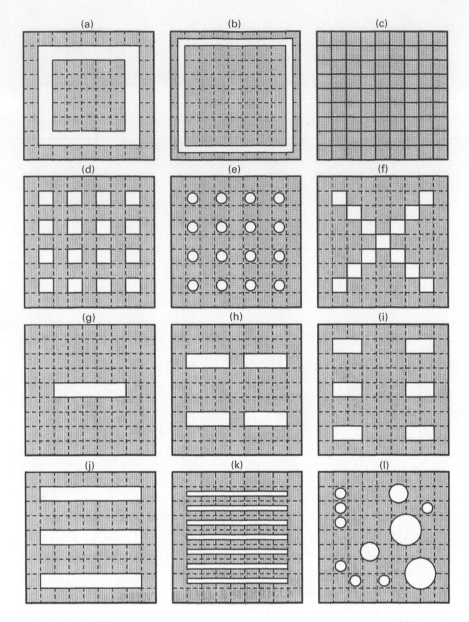

■ **FIGURE 17–7**

Exercise in evaluating lighting expressions in terms of form, pattern, orientation, and locations.

(a) _____   (b) _____   (c) _____

(d) _____   (e) _____   (f) _____

(g) _____   (h) _____   (i) _____

(j) _____   (k) _____   (l) _____

is cross referenced to a luminaire schedule and specification. The electrical plan will also include the circuiting and control wiring for the lighting equipment. Figure 17–13 (p. 540) is a sample area from a typical electrical lighting plan for a shopping mall.

The architectural reflected ceiling plan shows the luminaire layouts more accurately, including key dimensions and notes to assist the contractor in installation. Figure 17–14 (p. 541) shows a sample area from an architectural reflected ceiling and lighting plan for a typical office lobby. It is sometimes necessary to indicate certain luminaires on other architectural drawings, such as sconces on interior elevations, chandeliers on building sections, and ground-mounted accent lights on landscape plans. This clarifies their location and mounting height.

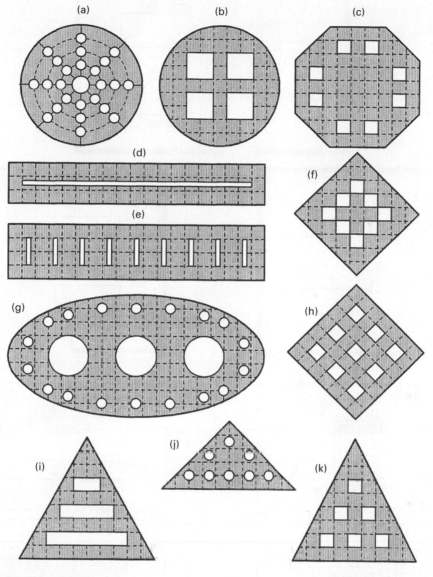

■ FIGURE 17–8

Exercise in evaluating lighting expressions in terms of form, pattern, orientation, and locations.

(a) _____   (b) _____   (c) _____
(d) _____   (e) _____   (f) _____
(g) _____   (h) _____   (i) _____
(j) _____   (k) _____

17.3.2 Specifications and Luminaire Schedule

These documents communicate technical construction issues of the lighting system. They outline the lighting system in detail, with explicit instructions concerning product, quality, execution, and final commissioning.

The luminaire schedule (also known as a lighting fixture schedule) provides the specific product information for every specified luminaire. It includes the designated manufacturer or manufacturers, catalog numbers, description, lamping, mounting, and special electrical notes. The specified equipment is cross-referenced to the electrical lighting plan using a luminaire "type" designation. A typical luminaire schedule for a commercial project is shown in Figure 13–9.

17.4 DAYLIGHT

Electric lighting systems are a relatively recent development in the history of the built environment. In early

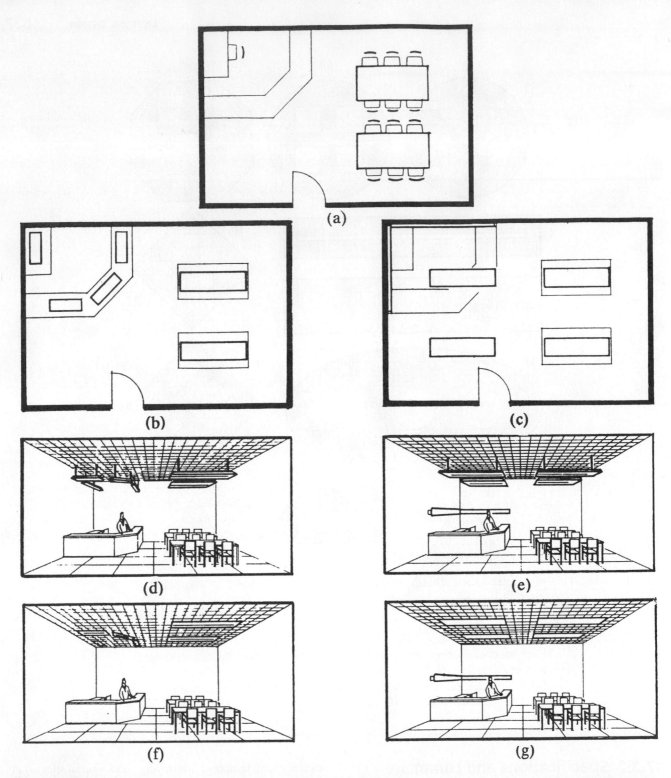

■ FIGURE 17–9

Evaluation of alternative lighting design in perspectives. (a) An architectural floor plan of a small reading room showing two reading tables, checkout counter, and librarian's desk against the wall. (b) Design 1: Lighting layout follows furniture plan. (c) Design 2: Lighting layout follows room shape. (d) Perspective view of design 1 with suspension-mounted luminaires. The ceiling appears to be confused, cluttered (disorganized). (e) Perspective view of design 2 with suspended luminaires. The ceiling appears to be orderly, and too regimented (neutral); however, a bracket-mounted wall luminaire against the wall over the librarian's desk provides a luminous relief and a practical solution to illumination need. (f) Perspective view of design 1 with ceiling recess-mounted luminaires softening ceiling clutter somewhat (still disorganized). (g) Perspective view of design 2 with ceiling recess-mounted luminaires and bracket-mounted luminaires. This appears pleasant, neutral, and expressive (a better solution). There are, of course, many other solutions to be considered.

■ **FIGURE 17–10**
Uplighting of a ceiling can both enhance the architecture and make it appear more spacious. (Courtesy: Randy Burkett Lighting Design, St. Louis, MO.)

■ **FIGURE 17–12**
Lighting in this lobby area uses decorative downlights to complement the architectural interior and avoid direct glare. (Courtesy of WTA.)

■ **FIGURE 17–11**
Custom light fixtures were used along with ductwork in this elementary school cafeteria to reinforce theme of movement and vitality. (Courtesy of WTA.)

design, buildings were generally long and narrow, with their primary axis oriented east and west, and very large windows at both the north and the south walls to take maximum advantage of natural light. The invention of electric lighting and mechanical systems freed designers from previous building forms, and daylight became less favored as electrical lighting grew more flexible and quality illumination could be provided to harmonize

with a building's design, however, with energy becoming more expensive and its use placing more of a strain on the natural environment, daylight has emerged again as an effective means for reducing energy consumption in buildings.

Daylight increases the awareness of the occupants of a building by allowing a visual connection between the interior and the exterior environment. The visual connection may have positive or negative effects on an occupant's anxiety and productivity, depending on the external environment.

17.4.1 Comfort and Energy

Daylighting saves energy only if lights are turned off or dimmed when natural lighting is suitable. Turning off or dimming can be manual or automatic. Dusk-to-dawn switching of outdoor lighting is an example of the concept that has been commonplace for many years. Using daylight sensors in buildings is becoming more commonplace as technology improves.

Daylight can be admitted by windows or skylights. In either case, the system must be carefully designed to prevent direct sunlight from striking occupied spaces, which would cause thermal discomfort and glare. Figure 17–15 shows simple diffusion screens used to block direct sun and direct light upward for deeper

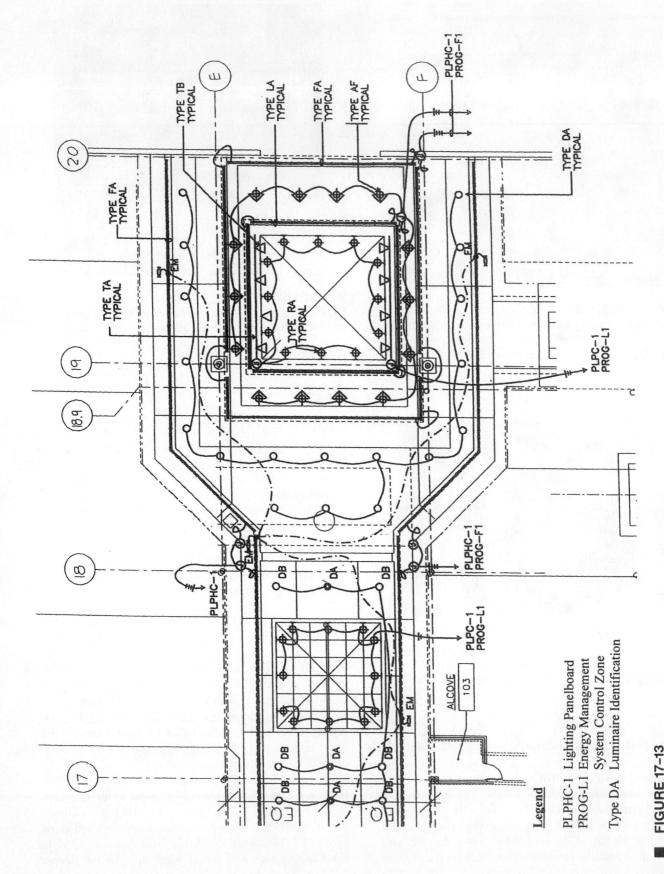

■ **FIGURE 17-13**

Sample area from a typical lighting plan for a shopping mall. (Courtesy: Randy Burkett Lighting Design, St. Louis, MO.)

<u>Legend</u>

PLPHC-1   Lighting Panelboard
PROG-L1   Energy Management
            System Control Zone

Type DA   Luminaire Identification

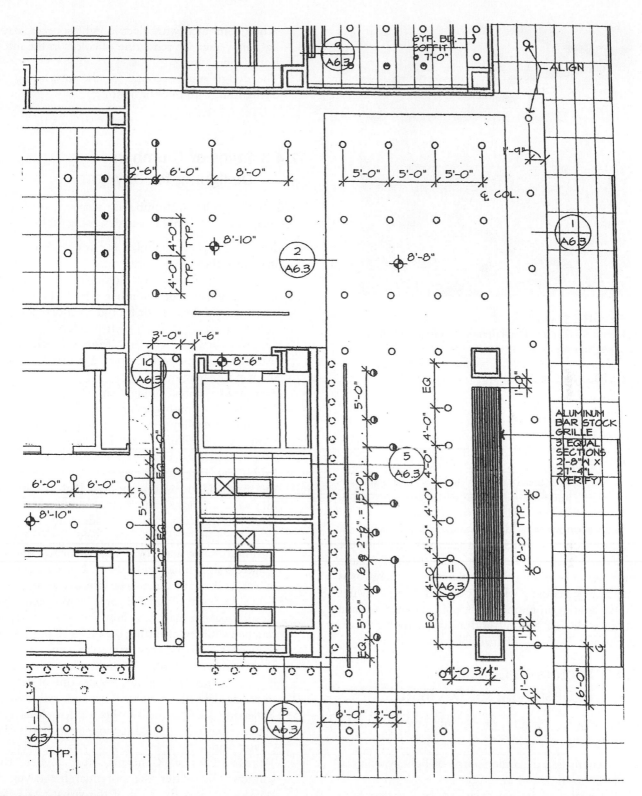

■ FIGURE 17-14

Sample area from an architectural reflected ceiling and lighting plan for a typical office lobby. (Courtesy: Randy Burkett Lighting Design, St. Louis, MO.)

◼ FIGURE 17–15
Lighting for this professional design office is coordinated with clerestory windows, structure and ductwork to produce shadow-less lighting, supplemented by day lighting. (Courtesy of WTA.)

penetration into the space. Light shelves, inside or outside the window, and other specially designed devices perform the same function, often with higher precision, usually at higher cost and with less occupant control.

The economics of daylighting design depends on climate and on how much extra glazing is used for the sake of daylighting, and on the orientation of the glazing. If the amount of extra glazing is significant, energy used for heating and cooling will likely exceed lighting savings from daylighting in temperate or cold climates. In mild climates with low heating and air-conditioning loads, daylighting is likely to enhance energy performance.

17.4.2 Factors to Consider in Designing for Daylight

In designing for daylight, one should take note of the following factors:

- Daylight is a dynamic source of light, varying in both position and intensity.
- External shading due to landscaping, the configuration of the building, and nearby structures must be considered.
- Proper interior and exterior controls of daylight should be provided.
- Attention must be paid to glare from windows and skylights.

- When combined with electric light sources, daylight may alter the rendering of colors in the interior of a building.
- The design must account for the interfacing of daylighting with interior lighting.

17.4.3 Rules of Thumb in Designing for Daylight

The following rules of thumb with respect to the features mentioned are applicable in designing for daylight:

- *Room geometry*   Minimize the room depth and maximize the ceiling height.
- *Sidelighting*   Orient workstations in offices so that daylight will be on either side of the worker (ideally, from the left side for right-handed persons and from the right side for left-handers).
- *Effective distance*   Daylight from sidelighting is usually assumed effective up to a depth 2.5 times the window height.
- *Solar gains*   Direct sunlight should be avoided.
- *Orientation*   South glazing provides the greatest amount of daylight. North glazing provides the most consistent illumination. East and west glazing have the greatest effect on the building cooling load. Large-area glazing may create overheating on the south-side rooms in modern high-rise buildings when windows cannot be opened.
- *Overhead lighting*   Daylight admitted through roof skylights can be effectively redistributed throughout the space by deflectors or translucent reflectors suspended below them. Translucent glass or acrylic skylights can also be effective at diffusing light entering through the roof. Newer technologies for glazing materials allow the view out to be retained while transmitting only 5 to 10 percent of the visible light and very little solar gain.
- *Architectural elements*   Consider the use of overhangs, screens, louvers, and other control devices to manipulate the amount of daylight reflectance. Reflecting surfaces below the window level can increase daylight by redirecting light across the ceiling plane.
- *Window location*   Generally, the higher the window, the better the potential for daylight. Windows below the level of the visual task offer little illumination value.
- *Room finish*   Lighter room finishes are more effective; however, highly reflective white surfaces may cause discomfort due to glare.
- *Space planning*   Open spaces allow daylight to penetrate deeper into the space.

- *Glazing area* Glazed areas need not be greater than 25 percent of the floor area. Excessive glazing adds to heating and cooling loads and rarely improves visual performance or enhances visual impressions.

17.5 EXTERIOR LIGHTING DESIGN

The lighting of a building's exterior and surrounding site can be an important factor in the success of a project. Exterior lighting possibilities other than code-required safety and security illumination are often overlooked by a building owner, despite the fact that a well-designed exterior building and site lighting system can significantly increase a property's actual or perceived value. Figure 17–16 illustrates this power of light.

Possible reasons for including exterior and site lighting in a building's design are building or corporate identification, increased vehicular and pedestrian safety, civic and community pride, historical or architectural significance, and marketing, among others. Some buildings and sites will be more naturally appealing for a lighting design strategy than others, simply because of their use or design characteristics. It is up to the project's lighting designer to discuss these possibilities with the client or owner, explaining both the strengths and the weaknesses of various approaches.

17.5.1 Design Considerations

Many factors go into the design of exterior building and site lighting, all of which should be reviewed and analyzed before proceeding into design.

- *Ambient conditions* The existing ambient lighting conditions around the building and adjacent sites should be studied to understand what visual "competition" is presented to the development of a lighting design.
- *Viewing conditions* Key views to the building and site should be examined to determine the impact of a lighting strategy. Distant drive-by, vehicular entry, and local pedestrian views could all be important.
- *Adjacent properties* It is important to keep in mind the possible impact of a lighting system on neighboring properties. Excessive spill light or perceived nighttime "sky glow" is to be avoided in most areas. This is an especially sensitive issue in residential communities.
- *Local codes or ordinances* Some regions and locales have ordinances in force that limit the extent of exterior lighting for anything beyond safety and security. The designer must research these before starting work.

17.5.2 Lighting System Selection

A wide variety of light sources and equipment are available for exterior environments. Design concerns for building and site lighting will govern the selection of the appropriate system.

Selecting the light source should include consideration of life, efficiency, size, cost, and color. Chapter 15 details the important attributes of each type of source. In illuminating building facades, color may be the decisive factor. The color and texture of the facade materials should be rendered as realistically and attractively as possible. Whenever practical, the designer should conduct on-site or lab test mockups to idealize the match between the source and the material being lighted. Landscape should also be illuminated with a flattering source. Usually metal halide or halogen is the lamp of choice, since high-pressure sodium distorts the plant material's natural appearance. Compact fluorescent is also a good option in warm climates.

Building floodlights and accent lights are available in a variety of distributions and intensities. The precise control of light may be critical if only selected areas of a building or building features are to be illuminated, as is shown in Figure 17–17. Equipment size and mounting may dictate what styles can be used. The daytime appearance of the luminaire could be important if it is outwardly visible. It should be reviewed with both the architect and the owner.

Landscape lighting equipment should be as discreet as is reasonably possible, since it is normally in plain view of the pedestrian. Glare to passersby should be minimized through the use of louvers or shields.

■ **FIGURE 17–16**
Exterior lighting can be effective in establishing a project's visual identity at night. Illuminating the landscape and special features of an adjacent site can heighten the impact. (Courtesy: Randy Burkett Lighting Design, Inc.)

■ FIGURE 17–17
Precise control of floodlight distributions can be important when the design demands varying the gradient of light across a surface or the minimization of spill light onto adjacent facades. (Courtesy: HOK and Robert Pettus, St. Louis, MO.)

Luminaire selection should place a great deal of emphasis on long-term sustainability as well as the appropriateness of the design. Equipment of poor quality and workmanship will quickly fail in the harsh outdoor environment. Accessibility for relamping and routine maintenance is of prime concern. Safeguards against vehicular and pedestrian damage may be necessary in some situations.

17.6 DESIGN PRACTICE AND ALTERNATIVE SOLUTIONS

Although every designer should arrive at a singular solution to a design problem, there is always more than one alternative. The following design exercise is intended to make this point.

Suppose you have been commissioned to design the lighting for a small architectural or engineering office, as shown in the floor plan below. The office contains seven workstations, one reception desk, and one conference area (see Figure 17–18). The room data and design criteria are as follows:

- Room data:
  Dimensions 48′ × 26′
  Ceiling 9′–6″ height, 2′ × 2′ lay-in grid

Ceiling reflectance 80 percent (base)
Wall reflectance 50 percent (base = effective)
Floor reflectance 20 percent (effective)
- Luminaire to be used: 2′ × 2′ parabolic, 3–20 watt (1500 lm), type 25 (see Figure 16–7).
- Design criteria: Maintained illuminance—50 fc minimum, LLF = 0.75

Refer to Figure 17–18.

1. Using the average illuminance calculation form, calculate the number of luminaires required for uniform illumination over the workstation area only (36″ × 26″). The balance of the room is to be designed separately.
2. If the calculated number of luminaires includes a fraction, use the appropriate integral number of luminaires to best suit the workstation layout and the interior aesthetics. Luminaires at each workstation should be in front of or above the drafting table whenever practical.
3. The central corridor space of the room is approximately one-third of the room width. Suppose the illuminance at the central corridor is not a concern, and make a modified layout to concentrate the luminaires over the workstations by using slightly more than two-thirds of the number of luminaires calculated previously to fit the workstation pattern.
4. Make your own design for the reception and conference area. Select any luminaire or luminaires from Figure 16–4. Describe the design, and make a perspective of the area viewed from the entrance door toward the conference table, as a visitor would see it.

Although this design exercise is for only a small architectural or engineering studio, it covers most aspects of the design practices. The exercise must address the following:

- Calculations of illumination for uniform lighting, nonuniform lighting, and lighting effectiveness
- Decisions on the layout of luminaires
- Coordination with ceiling grids
- Lighting aesthetics
- Lighting reinforcement of users' impressions

The owning and operating costs of the design are important in the final selection of a system. These factors are too complex to be included here; nevertheless, all designers must be keenly aware of cost considerations in every design decision.

1. *Calculations.* Using the average illuminance calculation form that follows, we find that for the zonal cavity method of calculation, the number of luminaires required is 21.

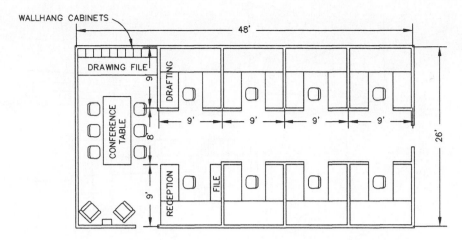

■ FIGURE 17–18
Floor plan of an architectural or engineering office.

2. *Uniform layout.* Given the space configuration, the luminaires may be arranged in three rows of seven each. This configuration is unsatisfactory for the workstation layout. Trials indicate that four rows of five each are most appropriate, even if this configuration is one luminaire short of the number required. Thus, we have:

- *Scheme A.* For uniformity within the entire area, the four rows of luminaires are more or less evenly spaced on the ceiling in a north–south direction and at the front of each workstation in an east–west direction. This scheme provides a neutral expression and satisfactory illumination.

3. *Nonuniform layouts.* To achieve maximum lighting of task areas, illuminance levels at the central corridor may be reduced. Several schemes are possible:

- *Scheme B.* The luminaires are moved within the workstation partition lines, with two luminaires in the front of each workstation and no luminaires in the corridor.
- *Scheme C.* In lieu of locating the luminaires as in Scheme A, the second row of luminaires is moved closer to the first row, and the third row closer to the fourth row.
- *Scheme D.* One $2' \times 4'$ luminaire is used to substitute for two $2' \times 2'$ luminaires within the workstation areas, and three single $2' \times 2'$ luminaires are used in the corridor area.
- *Scheme E.* If the workstation partitions are higher than the $5'$ level within the $9'$–$6''$ ceiling space, or if the partitions become full height to the ceiling, luminaires must be moved within each workstation room.

- *Scheme F.* The luminaires in each workstation room in Scheme E are irregular in location and aesthetically unpleasant. This is because of the continuation of the grid in the ceiling. If the ceiling grids are installed on a room-by-room basis, the lighting within the individual rooms can be centered within each room.

4. *Lighting plan for the conference area.* There are even more alternatives for illuminating the conference area than for the workstation area. With scheme C as the base for the workstation area, several schemes are possible for the conference area: These are just a few of the alternatives compatible with the basic lighting design (scheme C) of the space:

- *Scheme C1.* Continue the same $2' \times 2'$ luminaires for the conference table area.
- *Scheme C2.* Use the same $2' \times 2'$ luminaires, except that the unit over the conference table is composed of six $2' \times 2'$ luminaires suspended 3 ft below the ceiling. The composite $4' \times 6'$ unit may be selected for direct, indirect, or direct–indirect distribution. The file storage and seating areas are illuminated with incandescent or compact fluorescent down lights, or wall washers.
- *Scheme C3.* As a totally different design, two decorative luminaires using either incandescent or circular fluorescent lamps are suspended over the conference table. The suspended luminaires are for direct distribution only. A separate linear fluorescent unit is mounted over the board for vertical illumination.

Two solutions are shown in perspective (Figure 17–19).

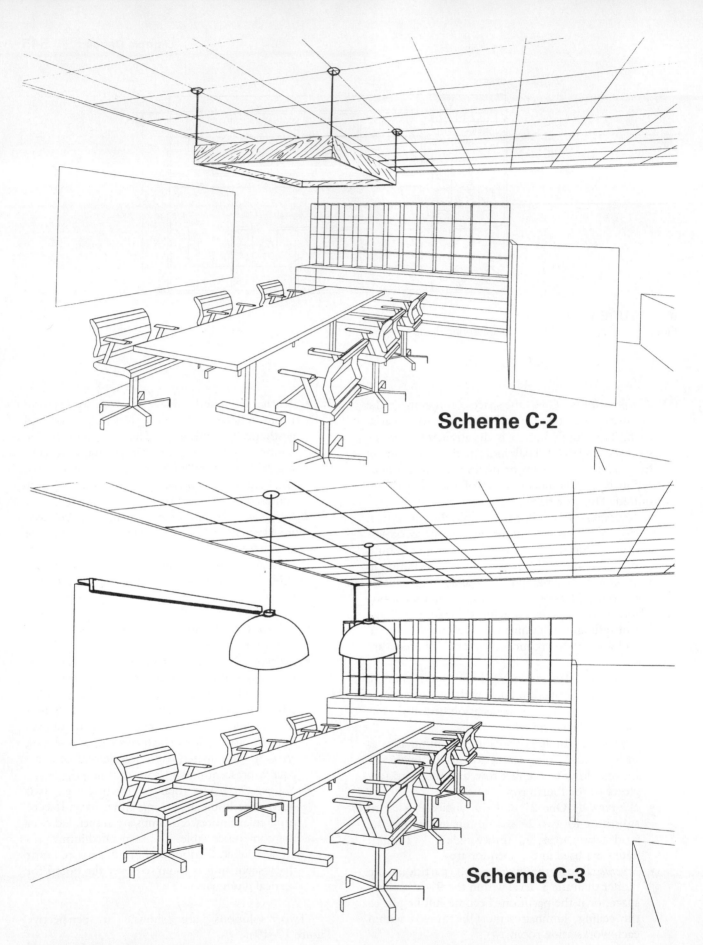

Scheme C-2

Scheme C-3

■ **FIGURE 17–19**
Perspective views of two of the many design solutions.

AVERAGE ILLUMINANCE CALCULATION FORM

FOR ROOM *Arch/Engr. office*

Visual Interpolation of CU
* Pw of 40% is determined by the weighted average of N,E & S walls @ 50% and W wall @ 0%, or no reflectance at all.
\# CU is found by interpolating between Pw @40% and RCR @ 2.28.

ILLUMINANCE CRITERIA	IES ILLUMINANCE CATEGORY					**E**
	MAINTAINED ILLUMINANCE, FC, (LUX)					**50**

FIXTURE DATA	MFR/MODEL	*Type 25, chapter 16*	
	TYPE DISTRIBUTION	*Direct*	
	NO. OF LAMPS PER FIXTURE		**3**
	RATED LAMP LUMEN & WATTS/LAMP	**1500/**	**20W**
	LUMENS PER FIXTURE (LPF)		**4500**

ROOM DIMENSIONS	h	**9.5**	WIDTH(W)	**26**	LENGTH(L)	**36**
ROOM CHARACTERS	h_{cc}	**0**	R_c	**.8**	R_{w1}	**.5**
	h_{rc}	**7**	R_w	**.5**	R_{w2}	**.5**
	h_{fc}	**2.5**	R_f	**.2**	R_{w3}	**.5**

RC RW	80				
	70	50	30	10	* $P_w = 40$
1	80	77	75	70	
2	74	70	66	63	\# $CU = 66$
3	69	63	58	54	@ RCR
4	64	57	51	47	= 2.28
5	59	51	45	41	
6	54	45	40	36	
7	50	41	35	31	
8	46	37	31	27	
9	42	33	27	23	
10	39	30	24	21	

P	PERIMETER, FT(M): *2×36 + 2×26*		**124**
A	AREA, SF(SM): *36 × 26*		**936**
PAR	PERIMETER/AREA RATIO (P ÷ A)		**.13**
CCR	2.5 × PAR × h_{cc}		**0**
RCR	2.5 × PAR × h_{rc} (#)		**2.28**
FCR	2.5 × PAR × h_{fc}		**.01**
P_{cc}	FROM R_c & R_{w1} & CCR		**.80**
P_w	SAME AS R_w OR R_{w2} (*)		**.40**
P_{fc}	FROM R_f & R_{w3} & FCR		**.20**
CU	FROM CU TABLE OF FIXTURE MFGR. INTERPOLATING BETWEEN RCR AND P_{cc}, P_w, P_{fc}		**.66**
LOF	BF – BALLAST FACTOR	**.95**	
	VF – VOLTAGE FACTOR	**.98**	**.93**
	OTHER	–	
LLF	LLD–LAMP LUMEN DEPREC.	**.85**	
	LDD–LUMINAIRE DIRT DEPREC.	**.95**	**.80**
	OTHER	–	

FLOOR OR CEILING PLAN

10' 36'

Considering there is an imaginary transparent wall separating the work area from the conference area.

Working Area

Floor Plan

9'6"

50"
26"

Elevation

(USE SEPARATE DRAWINGS FOR ADDITIONAL LAYOUTS)

CALCULATIONS

MAINTAINED ILLUMINANCE
$$E = \frac{N \times (LPF \times LOF) \times CU \times LLF}{A}$$

INITIAL ILLUMINANCE
$$E_i = E \div LLF$$

CALCULATION & REMARKS:
$$E = 50 = \frac{N \times (4500 \times .93) \times 0.8 \times 0.66}{936}$$
$$N = 21$$
$$E_i = E / LLF = 50/0.8 = 62.5 \, fc$$

* N – NUMBER OF LUMINAIRES

■ SCHEME A

For uniformity within the entire area, the four rows of luminaires are more or less spaced evenly on the ceiling in a N–S direction and at the front of each workstation in an E–W direction. Note that some of the luminaires are located directly above the 5-ft-high partitions, which is acceptable. This scheme provides a neutral expression and satisfactory illumination.

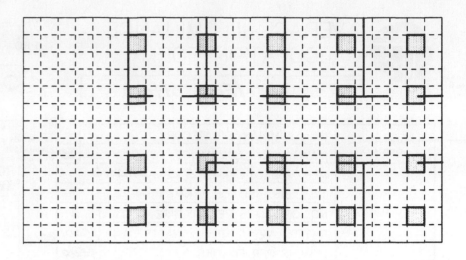

■ SCHEME B

With this scheme, the luminaires are moved within the workstation partition lines, having two luminaires in the front of each workstation and no luminaires in the corridor. A total of 16 luminaires are used, resulting in a 20 percent reduction in equipment over scheme A; however, lighting in the workstations and the overall space is spotty. This schedule is not a desirable solution.

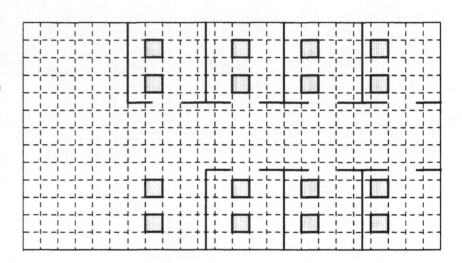

■ SCHEME C

In lieu of locating the luminaires shown in scheme A, the second row of luminaires is moved closer to the first row, and the third row closer to the fourth row. This reduces the illuminance level in the corridor and increases the illuminance level within the workstations to about 70 fc, which is a more desirable level for drafting tasks. This scheme is preferred.

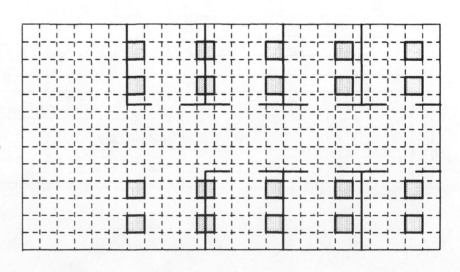

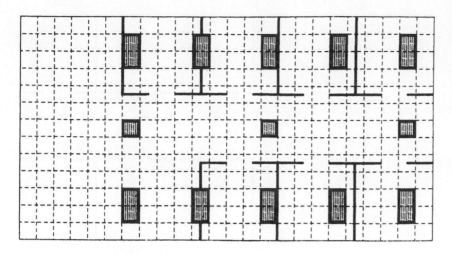

■ SCHEME D

With this scheme, (one) 2′ × 4′ luminaire is used to substitute for (two) 2′ × 2′ luminaires within the workstation areas, and (three) single 2′ × 2′ luminaires are used in the corridor area. Although this scheme uses more luminaires than scheme B, it is lower in cost, since 2′ × 4′ luminaires are the most popular type used commercially, and thus more competitively priced.

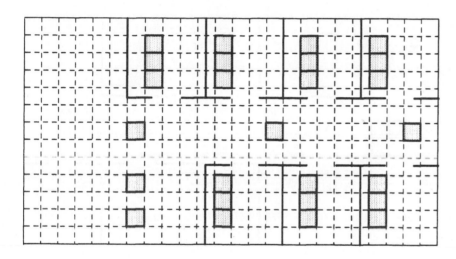

■ SCHEME E

If the workstation partitions are higher than the 5′ level within the 9′ 6″ ceiling space, or if the partitions become full height to the ceiling, lighting luminaires must be moved within each workstation room. In this case, the room cavity ratio of each workstation room will be greatly increased and the CU values decreased, resulting in a need to add another luminaire within each room.

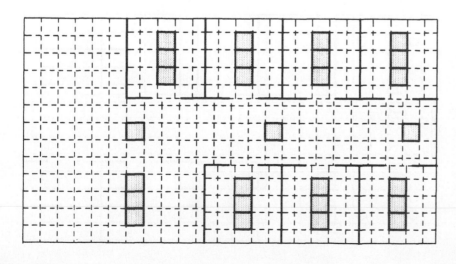

■ SCHEME F

The lighting fixture locations within each workstation room shown in scheme E are irregular in location and aesthetically unpleasant. This is because of the continuation of the ceiling grid in the ceiling. If the ceiling grids are installed on a room-by-room basis, lighting within the individual rooms can be centered within each room. This is a much better lighting design, although it will add to the cost of the ceiling.

■ SCHEME C1

Continuing the same 2′ × 2′ luminaire for the conference table area. This scheme provides a consistent neutral expression in spatial relations. There is no visual focal point.

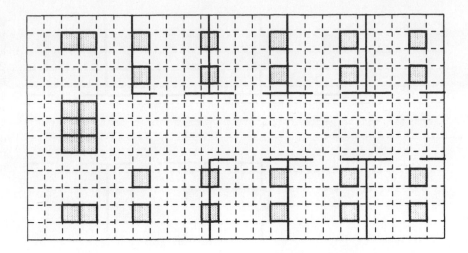

■ SCHEME C2

Using the same 2′ × 2′ luminaires, except that the unit over the conference table is composed of (6) 2′ × 2′ luminaires suspended 3′ below the ceiling. The composite 4′ × 6′ unit may be selected for direct, indirect, or direct–indirect distribution. The file storage and seating areas are illuminated with incandescent or compact fluorescent down lights or wall washers. The design provides good task illumination over the conference table, a relaxed atmosphere at the seating area, and a balanced spatial relationship at the storage area.

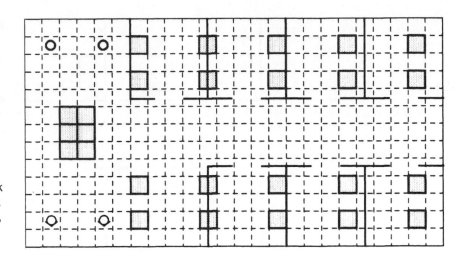

■ SCHEME C3

As a totally different design, (two) decorative luminaires using either incandescent or compact fluorescent lamps are suspended over the conference table. The suspended luminaires are for direct distribution only, which will not provide sufficient illumination on the wall-mounted illustration board. A separate linear fluorescent unit is mounted over the board for vertical illumination. This scheme provides an expressive to dominant visual atmosphere in the space.

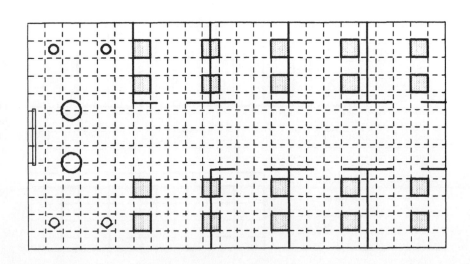

QUESTIONS

17.1 A lighting system is an integral part of the architectural expression of the space, which may be described as _____, _____, _____, _____.

17.2 Illumination must be designed for uniformity on a horizontal plane. (True) (False)

17.3 For uniformity in horizontal illuminance, luminaires must be mounted so as not to exceed the recommended spacing-to-mounting height ratio. (True) (False)

17.4 An ideal lighting system is a totally luminous ceiling. (True) (False)

17.5 To obtain the best visual modeling of a three-dimensional object, light flux should originate from one direction. (True) (False)

17.6 Lay out, on a reflected ceiling plan for uniform lighting, six round luminaires with general diffuse distribution in a 15-ft × 45-ft rectangular space.

17.7 If the ceiling height is 10 ft and the work plane is 3 ft above the floor, do you think the preceding layout will provide uniform lighting? If not, what would you suggest? Show your design in a reflected ceiling plan.

17.8 As an exercise in lighting design practice, lay out on a reflected ceiling plan, five 4-ft × 4-ft square luminaires in a 20 ft diameter round room (see figure below). The ceiling grid for this round room shown on the plan is 2 ft × 2 ft. The space is a casual living space where uniform lighting is neither mandatory nor desired.

From a lighting expression point of view, make an alternative layout, with any size or shape of luminaires that may be more compatible with this space and ceiling grid combination.

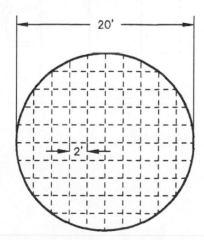

17.9 Daylight is effective up to a depth about   times the window height.

17.10 North glazing will provide the most consistent daylighting. (True) (False)

17.11 West glazing will have the greatest effect on a building's cooling load in the Northern Hemisphere. (True) (False)

17.12 South glazing may create overheating in spaces with southern exposure, even in winter. (True) (False)

17.13 Glazing should be extended to the floor level to gain the most benefits from daylight. (True) (False)

17.14 The glazed area need not be greater than _____ percent of the floor area for maximum daylight.

17.15 A Comprehensive Lighting Design Problem
This problem is intended as practice with lighting layout, controls, and wiring. The room to be designed is a lecture space with auxiliary areas. The architectural plan of the space, including the furniture layout and the lay-in ceiling grid, is shown in Figure 17–20.

From separate calculations, it was decided that the following luminaires would be installed:
- Main lecture area:
  - (10) type-A, semi-indirect fixtures. Each fixture is 1 ft × 4 ft, using two 32-W, T8 fluorescent lamps.
  - (6) type-B, semi-indirect luminaires to be installed in front of the chalkboard. Each luminaire is 1 ft × 4 ft using two 32-W, T8 fluorescent lamps. One of the two lamps provides downward asymmetrical distribution on the chalkboard, which should be separately switched. See photometric illustration.
- Lab area: (2) type-C, direct distribution luminaires. Each luminaire contains (3) 32-W, T8 fluorescent lamps. It is desirable to switch the center lamps from one switch and the outer two lamps from another switch.
- Coat room: (1) type-D, 2 ft × 2 ft direct distribution fluorescent luminaire using (4) 20-W, T8 fluorescent lamps.
- Closet: (1) type-E, surface-mounted fluorescent luminaires using (2) 40-W, T12 lamps.
  Using the calculated number of luminaires, develop a lighting layout in these rooms. Keep in mind that:
- The ceiling grid is already determined.
- There are VDT workstations along the west wall. (*Note:* There are additional computers on students' desks; however, no special lighting provisions will be required.)
  And as practice for electrical wiring, complete the branch circuit wiring for all lights in these rooms. Indicate the number of wires and the lamps or fixtures to be switched.

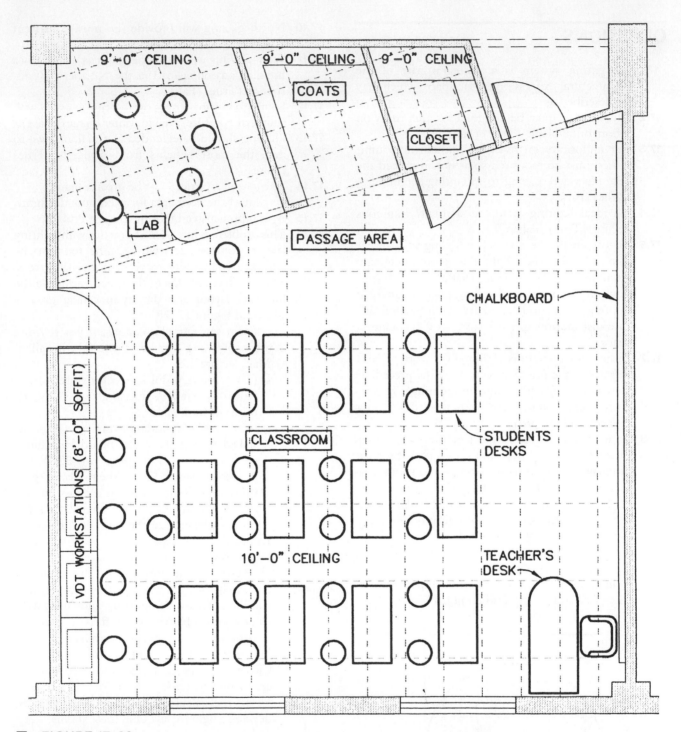

■ FIGURE 17–20

Furniture layout of the comprehensive lighting design problem for a typical junior high school.

17.16 What characteristics of light distribution intensity within a space will enhance an occupant's impression of relaxation?

17.17 Illuminating the walls of a small room with bright, uniform light intensity will help make the room appear more spacious. (True) (False)

17.18 Who should typically be involved in the selection process for a project's decorative luminaires?

17.19 A relatively symmetrical arrangement of ceiling-recessed lighting equipment which results in visible "dark-spots" on the task surface between luminaires likely means that the selected luminaires ( ) have been exceeded.

17.20 Key dimensions for locating luminaires accurately during installation are normally shown on the (electrical lighting plan) (architectural reflected ceiling plan) (circuiting and control plan).

17.21 Which of the following are not normally shown in a luminaire or lighting fixture schedule? (lamp code) (manufacturer and catalog number) (equipment cost) (voltage)

17.22 Lighting fixture shop drawing review is an important part of the project's early design development. (True) (False)

17.23 What are the principal considerations in developing an exterior lighting design strategy?

17.24 What is "sky glow" and when does it concern a lighting designer?

NOISE AND VIBRATIONS IN MECHANICAL AND ELECTRICAL SYSTEMS*

18

NOISE WAS DESCRIBED BY AMBROSE BIERCE (1911) AS unwanted sound, a stench in the ear, undomesticated music, and the chief product and authenticating sign of civilization.

Technologically, sound and vibration can be very complex. The intent here is to avoid technical jargon as much as possible while providing insights and guidelines to help avoid some of the egregious acoustical pitfalls in the practice of engineering and architecture design. It has been estimated that controlling noise and vibration in the initial design adds 2 to 3 percent to the cost, but retrofitting for the same results can increase the cost of control measures alone by a factor of 5 to 10.

18.1 RETROSPECTION

In ancient time there were open-air venues where the only interfering noises were the wind, waves, birds, and animals. With urbanization came manmade noises—hoofbeats and wagon wheels on cobblestone streets, for example. Buildings had evolved into massive stone edifices like the great castles of Europe. These effectively shut out urban noise and, for better or worse, provided the reverberation that was good for music but a disaster for speech.

Beginning in the nineteenth century, as steel evolved as a structural element, the race was on to package more and more usable space in buildings with less and less structural mass. Noise and vibration that were previously manageable if for no other reason than mass, became more and more an inherent part of the building. Since the end of World War II, there has been an exponential increase in the use of energy, manifested largely in HVAC systems. Concurrently building mass has decreased exponentially. Now we must design not

only for the noise and vibration generated within building equipment—fans, air conditioners, chillers, pumps, boilers, business machines of all kinds, and noises generated by throngs of people as they walk, talk, scuffle, and work—but for also the urban noises generated outside: buses, trucks, aircraft, construction equipment, jackhammers, and so on.

18.2 NOISE CONTROL: AN OVERVIEW

The control of noise is classically characterized as source-path-receiver. *Source* means the source of the noise—compressor, pump, air handling unit, horn, baby crying, etc. *Path* is the means by which the sound generated by the source reaches the receiver. *Receiver* is where the sound ends up, be it a person, a window, a property line, or whatever. Generally the path is airborne sound, structure-borne sound, or a combination of the two. Noise control consists of treating one or a combination of these three factors sufficiently to eliminate concerns about noise.

Source treatment usually involves changes in design and/or materials that lessen the generation of the sound. In designing buildings, other than specifying quieter equipment, we don't have much control over the generation of noise by our equipment.

Treatment of the path involves attenuating (lessening) noise along the path in some fashion without directly involving the source or the receiver. This might involve placing a wall between the source and the receiver or placing a resilient "break" in a concrete floor. Treating the path is the option most usually available.

Treating the receiver is a little more nebulous. It may involve moving people out of the noise field or having them wear ear protection. Architecturally, there is the possibility of locating mechanical equipment rooms away from sensitive spaces.

*J. T. Weissenburger is the contributing author of this chapter. See Acknowledgments.

18.3 BUILDING SPACES WHERE ACOUSTICAL CONCERNS MAY ARISE

The first step is to recognize and anticipate problem areas. Following is a brief, partial look at noise and vibration in buildings.

18.3.1 Public Areas

In some areas, public activities such as footfalls or conversation generate sound that not only may be objectionable within the area itself but may also intrude on adjacent areas. For example, fountains may be a source of pleasing sound in atriums but an irritating noise in adjacent spaces. Sound intrusion from mechanical and electrical systems or exterior sources can add to the cacophony, but architectural factors are responsible for the reverberant nature of the space itself. Examples include entrances, lobbies, atriums, reception areas, waiting rooms, toilets, and corridors.

18.3.2 General Offices

In spaces where private conversations may be held by small groups of people, intelligibility of speech is required, but confidential privacy is not. Surrounding walls are not necessarily required to be "soundproof." Interior acoustical treatment is recommended to control reverberation. Reasonable control of background noise, such as that from mechanical and electrical system components, and from exterior sources, is important. Examples: private offices, small conference rooms, open-plan offices, library reading rooms.

18.3.3 Executive Offices

These are spaces where private conversations may be held by small to medium-size groups. Intelligibility of speech is required, and confidential privacy is imperative. Enclosing walls must be "soundproof," and the interior acoustical treatment must be appropriate. "Cross talk" paths via HVAC ducts, pipe penetrations, conduits, etc., must be avoided. Background noise such as that from mechanical and electrical system components and from exterior sources can degrade speech intelligibility and should be controlled. Examples: private offices, medical consultation offices, personnel offices, conference rooms, boardrooms, and teleconferencing and videoconferencing rooms.

18.3.4 Assembly Areas

These are areas where the focus is on a central stage or podium and where the primary purpose is audible, intelligible communication from a speaker or speakers to an assembled audience. Interior acoustics for speech intelligibility are of primary importance. Speech may require electronic reinforcement. Recording and playback may be part of the electronic presentation equipment. Audiovisual equipment is usually present. Control of background sound from mechanical and electrical and exterior sources is important for good intelligibility. Special theater-type lighting is often required. Examples: large conference rooms, boardrooms, churches, auditoriums, theaters, recital halls, training rooms, lecture halls, classrooms.

18.3.5 Special-Purpose Areas—Quiet

These are areas where quiet is required for special technical reasons. Interior acoustic treatment is a prerequisite, as is the elimination of HVAC or other background noise and any exterior intrusions. Examples: TV studios, recording studios, broadcast studios, some laboratory situations, medical examination rooms.

18.3.6 Special-Purpose Areas—Noisy

These are areas where the intended activity is inherently noisy. Interior acoustical treatment may be indicated to minimize the noise within the area; however, of more general importance is protecting surrounding areas from that noise. Examples: mechanical/electrical equipment rooms, maintenance shops, exhibition spaces, computer rooms, kitchens, dining rooms, dishwashing areas, copy centers, projection rooms, vending machine areas, lobbies, health and fitness centers.

18.3.7 Assembly and Public Spaces Requiring Special Consideration

Many spaces require acoustical and other special considerations over and above those of public office-type buildings. These include theaters and auditoriums, churches, classrooms, lecture halls, restaurants, nightclubs, museums, display spaces, stores and retail spaces, concourses, open-plan offices, music practice rooms, recital halls, conference centers, boardrooms, and—a fairly recent arrival—audio and video teleconferencing rooms. These spaces demand acoustical treatment within as well as control of noise from the mechanical and electrical system.

18.3.8 Churches, Theaters, and Performance Spaces

Theaters, concert halls, auditoriums, churches, arenas, large lecture halls, and some radio and television studios are examples of live performance spaces. What they have in common is that they all accommodate performers and audiences in the same space at the same time. These are distinct, functionally and acoustically, from performance spaces that replace the binaural listener with one or more microphones and where the sound from the performers or recorded material is transmitted directly or indirectly (such as a radio or television studio) to an audience in a remote, physically unconnected listening space (a motion picture theater, a living room, etc.). Some performance spaces serve both functions. These spaces require an experienced acoustical consultant.

There is much common ground in the acoustical design of auditoriums, theaters, and arenas even though the space–volume characteristics may be vastly different. In all of these, intelligibility of the spoken word, often through a sound reinforcement system, is of paramount importance. The acoustics of the auditorium must be designed for speech as well as music. Control of background sound and reverberation is of primary importance in speech intelligibility.

18.4 BASIC CONCEPTS OF SOUND[1]

Sound in air consists of small pressure fluctuations. The human ear can often sense these fluctuations from as frequently as 20 times each second (20 hertz, Hz) to as frequently as 15,000 times per second (15 kHz) or more. The frequency of a sound is known as its *pitch*.

The strength of the pressure pulses is the loudness. The human ear can sometimes hear sound pressure as low as 2.9×10^{-9} pound per square inch, psi (20 micropascals, µPa), the threshold of hearing at 1000 Hz. It can tolerate sound pressure as great as 0.0029 psi (20 Pa) or greater before physically sensing pain.

18.4.1 The Decibel

Because of the wide range of pressure experienced by our ears, the concept of the decibel was introduced[2] as

a means of making the range of numbers more manageable. The decibel notation for sound pressure level[3] is expressed as

$$SPL = 10 \log \left(\frac{p}{p_{\text{ref}}} \right)^2 = 20 \log \frac{p}{p_{\text{ref}}} \qquad (18\text{--}1)$$

where p is the pressure fluctuation above atmospheric pressure in psi or µPa. A decibel quantity is always expressed in terms of a ratio with a reference quantity, p_{ref} in the preceding example. Above, p_{ref} is taken as the threshold of hearing: 2.9×10^{-9} psi in English units or 20 µPa in metric (SI) units. Thus, the range of hearing from the threshold of audibility to the threshold of pain becomes 0 dB to 120 dB.

18.4.2 Frequency

High pitch is high frequency. Low pitch is low frequency. The range of hearing can be from as low as 15 Hz to as high as 15 kHz (15,000 Hz). The range of frequency most important to speech intelligibility is from 300 to 3000 Hz.

Frequency of sound is often segmented into octave bands, although it is not uncommon to use one-third octave and narrower frequency bands in acoustical analyses. Octave bands have the same relationship in engineering acoustics as in music (i.e., the frequency of the upper limit of the octave is twice that of the lower limit). Engineering and music octave bands have the same relative mathematical description; however, engineering acousticians have chosen to describe the relative frequency range of octave bands somewhat differently from music octave bands. Scientific octave bands, called *preferred octave bands*, are related to 1000 Hz. The center frequencies most common in engineering are 62.5, 125, 250, 500, 1000, 2000, 4000, and 8000 Hz. The lower and upper limits of an octave band are $1/\sqrt{2}$, or 0.707, and $\sqrt{2}$, or 1.414, of its center frequency respectively. Two frequencies are an octave apart if one is twice the other.

18.4.3 Typical Sounds

Figure 18–1 shows the range in level and frequency of conversation and of orchestral music. Also shown, for reference, is the range of frequencies of the standard piano.

---

[1]See Reference 7 for a more detailed discussion of acoustical fundamentals.

[2]Introduced by Alexander Graham Bell, hence the name decibel. This is the reason for the capital B in its abbreviation, dB.

$$1 \text{ dB} = 10 \log (x)$$
$$1 \text{ Bel} = \log (x)$$
$$1 \text{ dB} = 10 \text{ Bel}$$

---

[3]Sound pressure level is usually designated SPL or L_{a}.

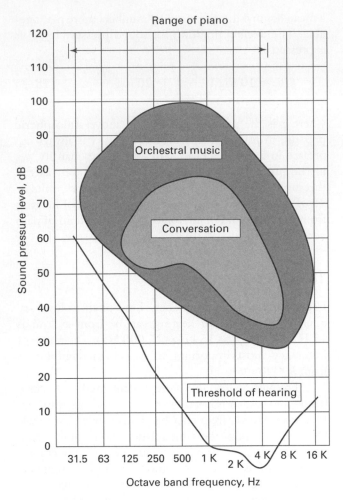

Range of piano

■ FIGURE 18–1
Levels and frequency of music and conversation.

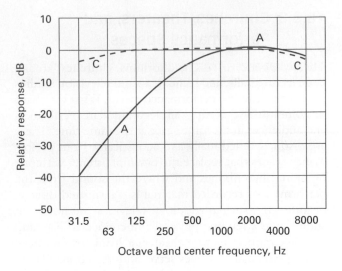

Octave band center frequency, Hz

■ FIGURE 18–2
A- and C-weighting: relative attenuation.

adjusts sound measurement to approximate the response of the ear at modest sound levels. C-weighting adjusts sound measurement to somewhat resemble the response of the ear at much higher levels. The relative weightings for A and C are shown in Figure 18–2. Linear makes no adjustments to the sound levels.

A-weighting has become the most prevalent measure. It is always used for OSHA-type sound level measurements and is generally used for environmental measurements.

C-weighting is not commonly used as an end in itself, however, it is a better measure of sound than A-weighting where low frequency is present (for example, in noise generated by fans and turbulence). It is not uncommon in practice to make both A- and C-weighted measurements where some diagnostic information is desired. The greater the difference between C-weighted and A-weighted levels, the greater is the low-frequency content of the sound being measured. When A-weighted and C-weighted levels are almost the same, most of the sound energy is above 500 Hz; however, A- and C-weighting are no substitute for octave band measurements.

18.4.4 A, C, and Linear Weighting in Acoustical Measurements

Overall sound levels, which measure a composite of all sounds generally over the range of human hearing, are commonly reported as A-weighted (usually designated dBA) and linear (usually designated dBL or just dB). C-weighted (usually designated dBC) is another weighting scheme.

In general, human hearing is most acute at frequencies around 1000 Hz. At other frequencies, particularly frequencies lower than 1000 Hz, greater sound pressure levels are required for a person to perceive the sound as being as "loud" as a sound at 1000 Hz. It has been rationalized that because the ear discriminates frequency of sound in this manner, there is justification for adjusting the measurement of sound to discriminate against low frequency, much as the ear does in hearing. A-weighting

18.4.5 General Perceptions of Sound

Subjective response to sound differs as much among individuals as the individuals themselves differ; however, there are some generally accepted perceptions of sound that may help in relating subjective experiences to quantified noise levels. For example:

■ A 10-dB increase in sound pressure level is generally perceived as twice as loud. A 10-dB decrease in

TABLE 18–1
Perceptions of sound

Perception	Nominal Level, dBA
Very quiet	10
Quiet	30
Moderate	50
Loud	70
Very loud	90
Deafening	120

TABLE 18–2
Some common sounds from household appliances, dBA measured at 3 ft

Appliance	Low	Medium	High
Freezer	38	41	45
Refrigerator	35	42	52
Electric heater		47	
Hair clipper		50	
Humidifier	41	43	65
Fan	37	56	69
Dehumidifier	52	57	63
Clothes dryer	51	57	66
Air conditioner	50	58	67
Electric shaver	47	60	69
Water faucet		61	
Hair dryer	59	61	65
Clothes washer	47	62	72
Water closet	46	63	76
Dishwasher	54	65	73
Electric can opener	54	66	76
Food mixer	49	67	79
Electric knife	65	71	75
Electric knife sharpener		72	
Sewing machine	70	72	74
Oral lavage	70	72	74
Vacuum cleaner	62	72	85
Food blender	62	75	88
Coffee mill	75	77	79
Food waste disposal	66	78	93
Edger/trimmer		81	
Home shop tools	63	83	97
Hedge trimmer		81	
Electric lawn mower	81	85	89

sound pressure level is generally perceived as half as loud.

- A 5-dB change in sound pressure level is perceptible.
- A 3-dB change in sound pressure level is hardly perceivable, even though this is a doubling or halving of the acoustical power.
- Low-frequency sound, below approximately 500 Hz, is perceived as less loud than high-frequency sound when both are experienced at the same level.
- 60 dBA is the long-term average voice level for speech communication in a normal voice, absent significant background sound.
- At 85-dBA background sound, conversation in a reasonable normal voice beyond a distance of 2 ft becomes strained.
- 120 dBA approaches the threshold of physical pain.
- Some subjective perceptions of loudness are shown in Table 18–1.

18.4.6 Common Sounds

To help develop a correlation between experience and sound level in decibels, Figure 18–1 and Table 18–1 and 18–2 show and list some typical sounds. Table 18–2 illustrates the range of some sounds that are often encountered around the home. As we shall see later, a background level above 60 dBA is the level at which it begins to be difficult to hold a face-to-face conversation.

18.5 USEFUL DESIGN CRITERIA

Before we can begin acoustical design, we need to have some idea of what noise levels are acceptable. In Section 18.3 we noted some generic considerations for building spaces. There are many different noise criteria for many different purposes. Of primary interest to the building team are those relating to noise within the building, but those relating to environmental noise should be of equal concern. Following are some of the more frequently encountered criteria.

18.5.1 Noise Criteria (NC)

The most common descriptor of sound in building spaces is *noise criteria* (NC). The purpose is to quantify acceptable background noise spectra from mechanical and other sources in unoccupied rooms.

Noise criteria and *noise criteria curves* were developed by L. L. Beranek as a means of expressing octave band spectral data in the range from 63 Hz through 8000 Hz (eight octaves) as a single-number rating. To do this, a family of curves was developed having a shape somewhat akin to the way the human ear hears frequency and sound level. This family of curves is shown in Figure 18–3.

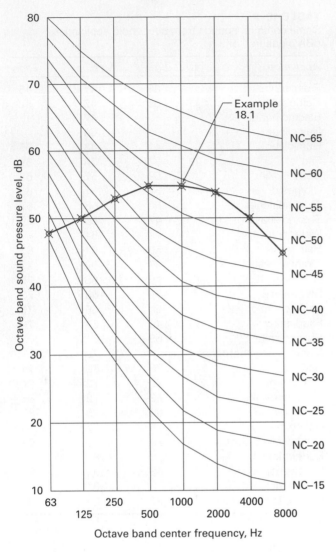

■ FIGURE 18–3
Noise criteria curves.

To determine NC level, octave band sound pressure levels (SPL) are plotted on the chart for each of the eight octaves. The highest NC curve of the family that is intersected, but not exceeded, by one of the octave SPLs is defined as the NC level.

Example 18.1   Determine the NC level for a sound spectrum in the following table:

Octave center frequency, Hz	Sound level, dB
63	48
125	50
250	53
500	55
1000	55
2000	54
4000	50
8000	45

See the plot of this spectrum in Figure 18–3. The spectrum curve touches the NC-55 contour in the 2000-Hz octave band. Therefore, the sound is said to be NC-55. When the data fall between two curves, the NC values can be interpolated, but are usually expressed as the nearest multiple of five.

Criteria using NC contours are given in Table 18–3. These are design guidelines relating to the intended use of the space. Note that the maximum A-weighted sound level associated with any NC curve is approximately 5 dB greater than the NC number.

18.5.2 Room Criteria (RC)

Although the NC method remains the most used, other methods have been proposed for specification and evaluation of noise in buildings. One alternative set of criteria is room criteria (RC). It is shown in the *Application Volumes of the ASHRAE Handbook* for 1995 and later. The method extends the frequency range down into the 16- and 32-Hz octave bands. This makes it valuable for evaluating an existing space where one has the luxury of appropriate noise measurements; but it is of little value in design, since the available data and theory do not adequately extend to these lower frequencies.

18.5.3 Guidelines for Speech Interferencem

Another useful criterion is the voice level required for face-to-face communication in the presence of background noise. Figure 18–4 provides some guidelines for nominal speech interference levels that can be used to judge suitability for speech communication. For example, as mentioned previously, at a background level of 60 dBA, communication in a normal voice can take place up to a distance of approximately 6 ft; with a 70-dBA background and at a distance of 4 ft, persons would have to speak with a raised voice to be understood.

For speech intelligibility in spaces where this is a primary requirement—churches, auditoriums, etc.—the background sound level should be at least 15 dB below the voice level.

18.5.4 Outdoor Noise Criteria

Outdoor sounds from rooftop units, cooling towers, chillers, emergency generators, and any other exterior equipment are potentially a matter of concern. More and more, urban areas are adopting ordinances to regulate emission of sound to adjoining property. Typical outdoor noise ordinances are drafted in terms of sound measured at the property line of the receiving property

TABLE 18–3
Criteria for background noise levels in unoccupied spaces

Occupancy Use	Range of A-Weighted Levels	Range of NC Criteria
Private residences	30–35	25–30
Apartments	35–40	30–35
Hotels/motels:		
Individual rooms or suites	35–40	30–35
Meeting/banquet rooms	35–40	30–35
Halls/corridors/lobbies	40–45	35–40
Service/support areas	45–50	40–45
Offices:		
Executive	30–35	25–30
Conference rooms	30–35	25–30
Private	35–40	30–35
Open-plan areas	40–45	35–40
Business machines/computers	45–50	40–45
Public circulation	45–50	40–45
Hospitals/clinics:		
Private rooms	30–35	25–30
Wards	35–40	30–35
Operating rooms	30–35	25–30
Laboratories	40–50	35–40
Corridors	35–40	30–35
Public area	40–45	35–40
Churches	35–40	30–35
Schools:		
Lecture rooms and classrooms	30–35	25–30
Open-plan classrooms	40–45	35–40
Libraries	40–45	35–40
Courtrooms	40–45	35–40
Legitimate theaters	25–30	30–35
Movie theaters	35–40	30–35
Restaurants	45–50	40–45
Concert and recital halls	20–25	15–20
Recording studios	20–25	15–20
TV studios	25–30	20–25

Source: 1991 *ASHRAE Handbook.*

and are based on the use, not the zoning, of the receiving property. There are usually different criteria for different receiving land uses and for time of day. Table 18–4 shows typical, but not universal, levels for continuous noise emitted to different categories of land use. The most stringent of these is residential land use. The consensus is that daytime (7:00 A.M. to 10:00 P.M.) levels should not exceed 55 dBA, nighttime (10:00 P.M. to 7:00 A.M.) levels should not exceed 50 dBA, and both should preferably be much lower. Other categories of land use have different levels. Some criteria provide for higher levels for shorter periods of time so long as the total acoustical energy for any 1-hour period does not exceed that for a constant criterion level for the same period of time.

18.6 ACOUSTICAL DESIGN CONSIDERATIONS IN HVAC SYSTEMS

Sound generated by and transferred through HVAC systems can be the most pervasive and unwanted noise in a building. A good HVAC system design incorporates proper control of HVAC-generated sound and vibration and does not permit them to intrude into occupied spaces. Sound produced by the HVAC equipment should be an unobtrusive part of the background and should not interfere with communication or other activities.

To control sound, the system designer must make estimates of sound from the sources (such as a fan or

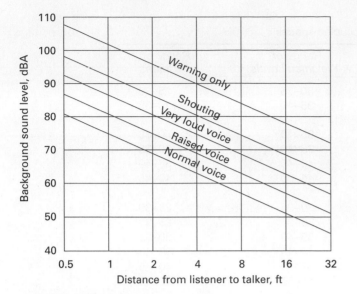

■ FIGURE 18–4
Speech interference levels.

TABLE 18–4
Typical urban noise criteria

Land Use Receiving Property	Continuous Sound Level, dBA	
	Daytime Limits 7:00 A.M.–10:00 P.M.	Nighttime Limits 10:00 P.M.–7:00 A.M.
Residential	55	50
Commercial	65	60
Light industrial	70	60
Heavy industrial	80	60

chiller) and its attenuation through various paths as it travels to the occupied spaces. If the projected noises are potentially objectionable, the design needs to be modified to provide the appropriate acoustical environment. Design changes are not always, or even necessarily, just to the HVAC system design. Optimum control of HVAC noise—or of any noise, for that matter—should not rest with the HVAC designer alone. The most cost-effective way to control HVAC noise is through a review of architectural and HVAC designs to identify the optimum control strategy. Wall and floor construction, location of mechanical facilities, and assigned uses of space are legitimately as much a part of noise control as are duct silencers and other in-system devices.

Some of the more common noise and vibration problems encountered in HVAC systems, and their treatment, are:

■ Failure to seal around duct and other penetrations through a wall.

Treatment: Seal large gaps with expand-in-place foam and a surface coating of acoustical

sealant. Small gaps can be sealed with acoustical sealant.

■ Breakout of noise from within the duct to occupied space via the ceiling plenum.

Treatment: Treat duct upstream with acoustical lining or duct silencers to eliminate the noise. Enclose duct in a drywall chase.

■ Rumble caused by bad flow conditions in duct system elements.

Treatment: Use good duct design practices; avoid sharp, abrupt edges at takeoffs; keep the velocity well below 1500 fpm. (See Chapter 6.)

■ Noise generated by flow through grilles.

Treatment: Reduce flow velocity and/or size grilles properly, or use different grilles.

■ Fan or flow-generated noise transmitted to occupied space via grilles.

Treatment: Attenuate duct noise upstream with acoustical lining or duct silencers; relocate the grille farther from the main duct.

■ Noise that enters ceiling plenum or occupied space through return openings in the shaft wall.

Treatment: Install silencers or other attenuation devices in the shaft wall opening.

■ Noise transferred to occupied space from ceiling plenum via return air grilles.

Treatment: Reduce noise in ceiling plenum by methods discussed above. Use sound traps on top of return air grilles.

■ Pipes hung from primary building structure resulting in structure-borne noise.

Treatment: Use properly sized resilient hangers.

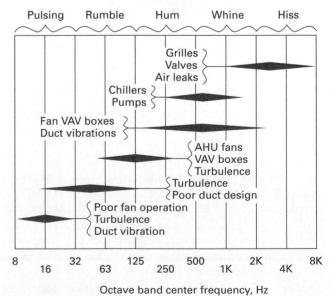

■ FIGURE 18–5
Principal frequency ranges of some HVAC noises.

- Machinery fastened solid to structural floor, resulting in structure-borne noise.

    Treatment: Use properly sized isolation mounts and machinery bases.

- Failure to use flexible coupling in pump, chiller, or other equipment, including electrical conduit, resulting in structure-borne noise.

    Treatment: Judicious use of flexible couplings.

- Terminal units that are too noisy or located improperly for the amount of noise they generate.

    Treatment: Use quieter units, possibly a size larger, or relocate.

- Through-the-wall cabinet units that are inherently noisy and improperly located in noise-sensitive areas.

    Treatment: Substitute equipment with acceptable noise generation for the space.

Figure 18–5 illustrates the range of the predominant frequencies generated by various sources.

In using manufacturers' sound data, make certain you understand the significance of the data. For example, manufacturers of grilles, terminal boxes, and similar equipment usually include in their data an assumed attenuation for the acoustical absorption of the room (room effect). Although this is generally reasonable, it can also be misleading.

18.7 MECHANICAL EQUIPMENT ROOMS (MERS)

Mechanical equipment rooms (MERs), by their nature, are generally very noisy areas. The noise in the interior of the MER should be held to a minimum by acoustical treatment within the room, but of more importance is the transmission of noise to adjacent areas. If at all possible, the MER should be distant from noise-sensitive areas. Figure 18–6 illustrates some of the paths by which MER noise can be transmitted to adjacent areas.

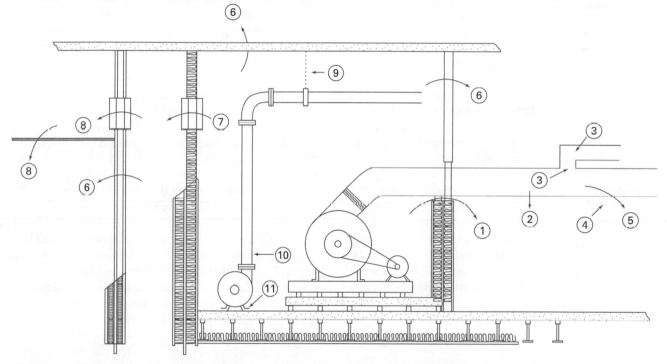

(1) Flanking through any penetrations of floor or wall. This includes not only ducts as shown, but also piping, conduits, etc.
(2) Breakout of noise through duct walls.
(3) Turbulence generated noise in bullhead tees and other abrupt changes.
(4) Noise generated by flow-through grilles.
(5) Ductborne noise via grilles in noisy trunk ducts.
(6) Direct transmission through floor slab. Often a dense, resiliently isolated ceiling with 12 in. of space between the slab and ceiling is required to adequately inhibit noise transmission through a floor slab. Direct transmission through walls.
(7) Air vents in walls leading to adjacent spaces.
(8) Noise from MER to open return ceiling plenums and through lay-in ceiling to occupied spaces.
(9) Pipes not on resilient hangers.
(10) No flexible connections at discharge of pumps.
(11) No vibration isolation.

■ **FIGURE 18–6**
Paths of noise transmission from mechanical equipment rooms (MERs).

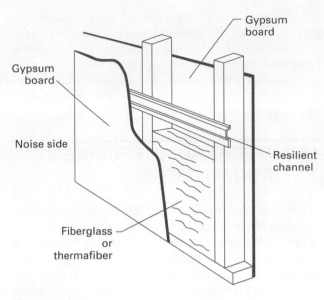

Gypsum board

Gypsum board

Noise side

Resilient channel

Fiberglass or thermafiber

■ **FIGURE 18–7**
Resilient channel on noise side of wall to help reduce low-frequency noise.

Adequate wall and ceiling acoustical sound transmission loss is required. The effectiveness of a wall or ceiling assembly as a noise barrier is limited by any air path, also called a *flanking path*. Treating the walls and ceiling of a mechanical equipment room with acoustically absorbent material-fiberglass duct liner, for example—can help by lowering the reverberant sound in the mechanical equipment room; however, a 5-dBA reduction in interior sound is the best that can be expected under the best of conditions. Acoustically absorbent materials have poor absorptive qualities in the lower frequencies. Low frequencies are generally the dominant sound in mechanical equipment rooms. Double layers of gypsum board on each side of steel or wood studs are often required. Dense concrete block is often used, but concrete block is a poor noise barrier if it is not painted, preferably with epoxy paint, to fill in the air paths through the block. Using resilient channels (Figure 18–7) on the noise side of a wall can help to reduce low-frequency transmission.

18.8 ROOFTOP UNITS (RTUS)

Rooftop units (RTUs; see Figure 18–8) have unique, significant noise problems. Not the least of these is a result of lightweight roof construction. Large roof openings are usually required for supply and return air duct connections. These ducts run directly from noise-generating rooftop air handlers to the building interior. There is insufficient space between the roof-mounted equipment and the closest occupied spaces below the roof to make

use of adequate methods of sound control. Rooftop units should be located above spaces that are not acoustically sensitive and should be placed as far as possible from the nearest occupied space. This measure can reduce the amount of treatment necessary to achieve an acoustically acceptable installation. Elevating the RTU above the roof so as to decouple it from the roof is effective in eliminating vibrations and reducing casing breakout into the ceiling plenum. Although the return air side is usually the larger problem, there are other potential problems that can be as bad. Some of the potential problems are illustrated in Figure 18–8.

18.9 NOISE IN AIR SUPPLY SYSTEMS

18.9.1 General Considerations

The air distribution system is a system of arteries that reaches throughout a building. It carries not only conditioned air but also noise. The general approach to analyzing ductborne noise is to begin with the supply fan and work through the duct system element by element through the terminus (grille) to the occupied space of interest. In systems where flow-generated noise is not a concern, usually it is necessary to analyze only the shortest duct path to an occupied space. If sound to this nearest space is within the criterion, it is usually not necessary to analyze rooms farther down the duct run, however, if flow velocities are high enough to generate significant noise, a more detailed analysis is in order. Terminal units can be noisy in low-velocity systems.

Figure 18–9 illustrates the evaluation of sound in a duct element. The attenuation of entering sound is reduced by the attenuation of the element. The generated sound is determined, and the two are combined as the sound entering the next element in line. This procedure is continued conveniently in tabular form or in a spreadsheet through the terminus of the duct run, and is usually done by octave band at least from 125 Hz through 4000 Hz, and preferably includes the 63-Hz octave band. Analysis is often limited to the 125-Hz octave band at the low end because of a lack of good sound data to work with.

18.9.2 Fans

General Characteristics of Fans

Fans are a major source of noise. In order of increasing primary frequency content, the common types of fans are listed below. (See Section 6.5 in Chapter 6 for fan design features and configurations.)

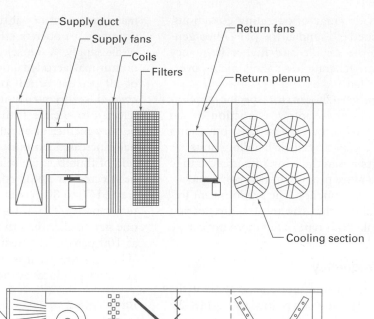

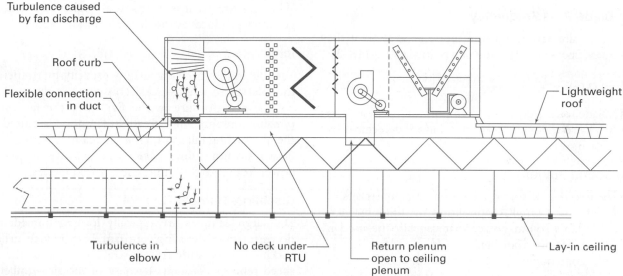

■ **FIGURE 18–8**

Top and section view of generic rooftop unit and sources of potential noise problems.

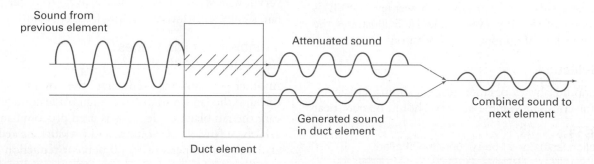

■ **FIGURE 18–9**

Analysis of sound in a duct element.

- *Propeller* Used in applications requiring movement of large quantities of air at low pressure drop, such as cooling coils and cooling towers.
- *Centrifugal* The workhorse of the HVAC system. Suitable for a wide range of applications and

operating conditions. Different blade types provide different capabilities and different noise-generation characteristics. Forward curved (FC) fan noise is generally in the low frequency (32- and 63-Hz octaves) characterized as rumble and can occur over a wide range of operating conditions. Backward inclined (BI) and airfoil (AF) often generate noise in the middle and higher frequency ranges, but they generate less sound energy overall than the FC fan.

- *Plug and plenum* Chosen for specific applications. Sound power is largely a function of fan housing. Often used for exhaust fans, but found more and more in packaged air-handling units.
- *Vaneaxial* Noise is in the higher frequency range and often has a more prominent narrow frequency due to blade pass frequency. The predominant frequency is usually within the most acute hearing range. That makes this type of fan more noticeable than others.

Fan Blade Pass Frequency

Fans generally have a signature noise generation at the blade pass frequency (Bf). It is determined by Eq. (18–2):

$$B_f(\text{Hz}) = \frac{\text{RPM} \times \text{number of blades}}{60} \qquad (18\text{–}2)$$

Blade pass frequency is important because it is nearly a pure tone, and thus much more detectable by the human ear.

Fan Sound Power

The best source of sound data is the manufacturers of the equipment. ASHRAE publishes methods for estimating a fan's sound power that can also be used if sound data are not available.

It is good design, both economically and in terms of controlling noise, for fans to operate at the point of optimum or peak efficiency. (See Section 6.5 in Chapter 6 for discussions of fans and fan curves.) Operation of the peak point can increase noise generation significantly. Bear in mind that a 3-dB increase in sound power level (PWL) is a doubling of sound power. Table 18–5 illustrates the noise "penalty" of nonoptimum operation.

Fan Selection

Fans should always operate near maximum efficiency. Fans for constant-volume systems should operate near

TABLE 18–5
Correction factor for off-peak operation of a fan

% of Peak Static Efficiency	dB Correction Factor, C
90–100	+0
85–89	+3
75–84	+6
65–74	+9
55–64	+12
50–54	+15

maximum efficiency at the fan design airflow rate. VAV systems must operate efficiently and stably over a range of air supply. A fan selected for peak efficiency at full output may aerodynamically stall at an operating point of 50 percent of full output, resulting in greatly increased low-frequency noise. On the other hand, a fan selected to operate at the 50 percent output point may be very inefficient at full output, resulting in substantially increased fan noise at all frequencies. As a general rule of thumb, a fan selected for a VAV system should be selected for a peak efficiency at an operating point of around 70 to 80 percent of the maximum required system capacity. This usually means selecting a fan that is one size smaller than that required for a peak efficiency at 100 percent of maximum required system capacity. When a smaller fan is operated at higher capacities, it generally produces as much as 5 dB more noise.

Inlet Vanes

Variable inlet vanes vary airflow capacity by restricting the inlet air to a fan wheel. This controls the total air volume and pressure at the fan while the fan speed remains constant. An increase in fan noise is caused by the obstructing inlet vanes that increase turbulence and flow distortions. Inlet vanes can increase fan sound power from 2 to 8 dB in the lower frequencies.

Discharge Damper Control

Discharge dampers are typically located immediately downstream of the supply air fan. They reduce airflow and increase pressure drop across the fan while the fan speed remains constant. Because of the air turbulence and flow turbulence created by the high-pressure drop across discharge dampers, duct rumble usually results near the damper. When the dampers are throttled to a very low flow, a stall condition is likely to occur at the fan, significantly increasing low-frequency noise.

Variable-Pitch Fan Blades for Capacity Control

Another method of reducing the overall airflow through the fan is variable-pitch fan blade controls that vary the fan blade angle. This is used predominantly in axial-type fans. As air volume and pressure are reduced at the fan, a 40 percent to 80 percent reduction in air volume achieves a noise reduction on the order of 2 to 5 dB in the 125- through 4000-Hz octave bands.

Example 18.2 A six-blade fan is operating at 1800 RPM. In which octave band should the 3 to 5 dB be added to account for the blade pass frequency?

$$B_f = \frac{1800.6}{60} = 180 \text{ Hz}$$

Answer From Section 18.4.2, the lower limit of the 250-Hz octave band is $0.707 \times 250 = 177$ Hz, and the upper limit is $1.414 \times 250 = 350$ Hz. Therefore, the fan noise due to blade pass frequency should be added to the 250-Hz octave band although it will equally contribute to the 125-Hz band. In most fan systems, the 250-Hz octave band is where blade pass frequency occurs most often.

Electronic Variable-Speed Controlled Fans

The primary acoustic advantage of variable-speed fans is the reduction of fan speed, which translates into reduced noise. Typically, the reduction in sound power can be estimated by Eq. 18–3:

$$\text{Sound power (dB)} = 50 \log \left(\frac{\text{lower speed}}{\text{higher speed}} \right) \quad (18\text{–}3)$$

Answer Because this reduction in speed generally follows the fan system curve, a fan initially selected at optimum efficiency (lowest noise) will not lose efficiency as the speed is reduced.

Example 18.3 (a) A fan operating at 1800 RPM has its speed reduced 10 percent. By how much is its acoustical power reduced?

$$\text{dB reduction} = 50 \ \log \left(\frac{0.9 \times 1800}{1800} \right) = 2.3 \text{ dB}$$

(b) What is the reduction in acoustical power if the speed is reduced 50 percent?

$$\text{dB reduction} = 50 \ \log \left(\frac{0.5 \times 1800}{1800} \right) = -15 \text{ dB}$$

18.10 SOUND IN DUCTS

18.10.1 Flow-Generated Sound

Although fans are a major source of sound in HVAC systems, they are not the only source. Flow-generated sound is generated by ducts, duct elbows, dampers, branch takeoffs, air modulation units, sound attenuators, and other duct elements. The sound power levels in each octave frequency band are a function of the geometry of duct elements and the turbulence and velocity of the airflow near an element. Duct-related aerodynamic noise problems can be avoided by:

- Sizing ductwork or duct configurations so that air velocities are low.
- Avoiding abrupt changes in the cross-sectional area of the duct.

- Providing for smooth transitions at duct branches, takeoffs, and bends.
- Attenuating sound generated at duct fittings with sufficient sound attenuation elements between a fitting and a corresponding air-terminal device.

18.10.2 Air Velocities in Ducts

The amplitude of aerodynamically generated sound in ducts is generally proportional to between the fifth and sixth power of the air velocity in the vicinity of a duct fitting. The velocity of sound in air is 1100 ft/sec or 66,000 ft/min; by comparison, the flow of air in ducts is 2000 ft/min or less for a low-velocity system, and 3000 to 5000 ft/min for medium- to high-velocity systems. Thus, air propagates upstream or downstream with equal ease. If there is undesirable flow-generated noise in a duct element, it may manifest itself upstream as well as downstream, as often happens with sharp takeoffs.

18.10.3 Characteristics of Ducts

Unlined Rectangular Sheet Metal Ducts

Straight, unlined rectangular sheet metal ducts are the ducts most commonly used in HVAC systems. They are the least costly and make efficient use of space, however, they provide very little sound attenuation. Attenuation is greater at low frequencies and tends to decrease as frequency increases. Also, attenuation lessens as the size of the duct increases. Maximum attenuation occurs at frequencies whose half-wavelength is equal to the small cross-sectional dimension of the duct. For standard wall thicknesses, low-frequency breakout noise can be a problem that requires additional treatment. This is illustrated in Example 18.4.

Example 18.4 A 24-in. $\times$ 24-in. cross-section duct is made of 22–gauge metal and is 20 ft long. The sound power that "breaks out" through the duct wall is approximated by the equation

$$\text{PWL}_{\text{out}} = \text{PWL}_{\text{in}} + 10 \ \log \left(\frac{S}{A} \right) - \text{TL}_{\text{out}}$$

where $\text{PWL}_{\text{out}} =$ sound power radiated from the outside surface of the duct

$\text{PWL}_{\text{in}} =$ sound power entering the duct

$S =$ area of outside radiating surface of duct (in.^{2})

$A =$ cross-sectional area of inside of duct (in.^{2})

$\text{TL}_{\text{out}} =$ normalized duct breakout transmission loss (dB)

The sound power entering the duct and the TL_{out} of the duct is as follows:

Octave center frequency, Hz	Entering sound power	TL_{out}	PWL_{out}
63	87	20	83
125	87	23	80
250	88	26	78
500	81	29	68
1000	76	32	60
2000	70	37	49
4000	66	43	39
8000	61	45	32

Breakout sound power is given by

$$A = 24 \times 24 = 576 \text{ in.}^2; \quad S = 4 \times 24 \times 240$$

$$= 23,040 \text{ in.}^2$$

$$10 \log (576/23040) = 10 \log (40) = 16$$

$$PWL_{out} = PWL_{in} + 16 - TL_{out}$$

The results are shown in the last column of the table.

Note that duct TL is different from wall-type TL.

Straight, unlined rectangular sheet metal ducts that are externally lagged tend to have slightly greater attenuation at low frequencies. Duct lagging is a specially formulated tarlike material that is applied to the outside of a duct to reduce breakout.

Circular and oval ducts generally exhibit greater TL characteristics than rectangular ducts because of the additional stiffness of the duct walls as a result of the curved surfaces.

Acoustically Lined Rectangular Sheet Metal Ducts

Rectangular sheet metal ducts with internal fiberglass or a similar lining are effective in absorbing high-frequency ductborne sound. Low-frequency attenuation is only slightly greater than that of unlined duct. Typical lining thicknesses range from 0.5 in. to 2 in. The greater the thickness, the greater the attenuation. A minimum thickness of 1 in. is recommended. Attenuation in lined rectangular ducts (unlike unlined rectangular ducts) is greater at high frequencies than at low frequencies. The addition of internal lining will increase the airflow velocity unless the duct size is increased to account for the lining. The density of the fiberglass lining used in lined rectangular sheet metal ducts usually varies between 1.5 and 3 lb/ft³. Fiberglass duct liner has a shellac-type coating on the air side to prevent erosion by the airflow. Breakout sound can be a problem with lined as well as unlined rectangular ducts.

Unlined Circular Sheet Metal Ducts

Unlined circular ducts provide very little sound attenuation. This should be considered when designing a duct system. Circular ducts are much more rigid than rectangular ducts. Therefore the walls do not vibrate as much or absorb as much sound energy as rectangular ducts. Circular ducts provide only about one-tenth the sound attenuation of rectangular ducts at low frequencies.

Acoustically Lined Circular Sheet Metal Ducts

Most available data for attenuation in acoustically lined circular sheet metal ducts come from the manufacturers. The ducts generally have double walls. The inner wall is perforated with an open area on the order of 23 percent. There is fiberglass between the inner and outer walls. Because of the stiffness of the duct walls, acoustically lined circular sheet metal ducts generally do not transmit breakout sound as readily as single-wall ducts.

Nonmetallic Insulated Flexible Ducts

Nonmetallic insulated flexible ducts can significantly attenuate ductborne noise. Attenuation varies with diameter and length of duct. Duct lengths should normally be limited to 5 to 8 ft. Care should be taken to keep flexible ducts straight, and any bends should have as long a radius as possible. An abrupt bend may provide some additional attenuation, but the flow-generated noise may be unacceptably high. Breakout noise from flexible ducts may be a problem, particularly in ceiling plenums above noise-sensitive areas. Flexible ducts are subject to damage, which can increase noise as well as reduce efficiency.

Rectangular Sheet Metal Duct Elbows

Attenuation of elbows is best where the half-wavelength of the frequency is near the dimension of the elbow in the plane of the bend. Larger elbows tend to attenuate more at low frequencies, but not necessarily more than smaller elbows. Square (mitered) elbows provide much better attenuation lined than unlined. To be most effective, the duct lining must extend at least two duct widths before and after the elbow. The thicker the lining, the better the attenuation. Elbows can be significant sources of flow-generated noise. Turning vanes in the elbow may help, but they may also induce more turbulence in the airstream.

18.11 DUCT SILENCERS

Duct silencers, as the name implies, are used to attenuate sound that is transmitted through HVAC duct systems, however, they serve many other sound attenuation objectives, as at intake and discharge louvers. Like all duct elements, silencers come with associated pressure losses. To minimize pressure losses, the face area of the silencer must be larger than the face area of the attached ducts. Face area is the area of the face of the silencers as opposed to the free area within the silencer.

There are three main types of in-duct silencers: absorptive or dissipative, reactive, and active. These are discussed below.

Silencers come in various shapes and sizes to fit the duct configuration. Straight silencers with both rectangular and circular cross sections are available with or without center splitters and pods. Elbow silencers are available as both absorptive and reactive silencers for use where there is insufficient space for a straight silencer.

Care should be taken in applying test data to actual installations. Adverse system effects can have a significant impact on the performance of all standard silencers. Standard silencers should be located at least three duct diameters from a fan, coil, elbow, or branch takeoff, or any other duct element. Locating a standard silencer closer than three duct diameters can result in a significant increase in both the pressure loss across the silencer and the generated noise. Active silencers may be located in this region without pressure loss, but the acoustic attenuation may be limited by turbulence.

18.11.1 Absorptive Silencers

Absorptive silencers are most commonly used. Rectangular silencers generally have aerodynamically shaped vanes or splitters that have perforated metal surfaces covering acoustic-grade fiberglass to absorb sound over a broad range of frequencies. Round silencers have an outer shell similar to double wall ducts. Most have an absorptive "bullet" in the center. Figure 18–10 shows examples of typical duct silencers. Airflow does not significantly affect the attenuation if duct approach velocities are under 2000 fpm.

Silencers add to the pressure drop (drag) in a system. Pressure drop is usually given as a function of face velocity of airflow by the manufacturer. Face velocity is the volumetric flow in cfm divided by the face area of the silencer, not the free area. Rectangular silencers come in stock sizes. Often, more than one silencer is required in parallel to provide the required face area, so several standard silencers are assembled into a bank of silencers.

Example 18.5 A 24-in. × 24-in. rectangular duct branches into a 24-in. × 6-in. duct and a 24-in. × 18-in. duct. What is the sound power split between the two branches?

$$24'' \times 6'' \text{ branch: Attenuation} = 10 \log\left(\frac{24'' \times 6''}{24'' \times 24''}\right)$$
$$= -6 \text{ dB}$$

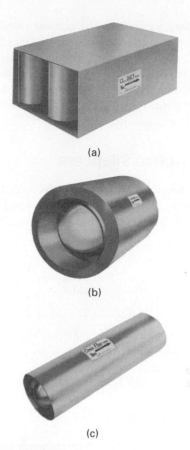

■ FIGURE 18–10

Typical duct silencers. (a) Rectangular configuration, (b) short tubular-type silencer, (c) long, tubular duct-size silencer. (Courtesy: Industrial Acoustics, Bronx, NY.)

$$24'' \times 18'' \text{ branch: Attenuation} = 10 \log\left(\frac{24'' \times 18''}{24'' \times 24''}\right)$$
$$= -1.2 \text{ dB}$$

Generally, the flow-generated noise is much less than the entering ductborne noise and does not contribute to the silenced noise level on the quiet side of the silencer, however, flow-generated noise should be evaluated if static pressure drop across the silencer exceeds 0.35 in. water gauge.

18.11.2 Reactive Silencers

Reactive silencers are the type used on motor vehicles. There are also applications for this kind of silencer in HVAC systems, but they are somewhat specialized.

A reactive silencer (unlike an absorptive silencer) usually contains no fiberglass or other absorptive media. This avoids the problem of entrapping contaminants from the airstream. Attenuation is achieved by a series of tuned cells that resonate at certain frequencies and in doing so absorb energy.[4] The outside physical appearance of reactive silencers is similar to that of absorptive silencers. Because of tuning, broadband attenuation is more difficult to achieve with reactive silencers than with absorptive silencers. Greater lengths may be required to achieve similar attenuation. Airflow generally increases the attenuation of reactive silencers.

18.11.3 Active Silencers

Active duct silencers are very effective in reducing noise at lower frequencies by producing sound waves that cancel the unwanted sound waves. Figure 18–11 shows

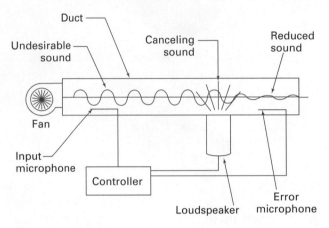

■ FIGURE 18–11
Schematic of an active noise control silencer showing cancellation of unwanted sound.

■ FIGURE 18–12
Typical performance of absorptive and active duct silencers.

a schematic of an active silencer. An input microphone measures the noise in the duct and feeds it to the controller. The controller is a digital computer that changes the phase of the sound so that it is the opposite of the input sound and plays it back into the duct. The level and phase of the canceling sound is adjusted by the controller so that it just cancels the undesirable sound. Because the components of the system are mounted outside the duct, there is no pressure loss or generated noise. Performance is limited, however, by the presence of excessive turbulence in the airflow detected by the microphones.

Active noise control has great potential, though it has not yet become commercially attractive. Installations where it has been properly used in HVAC systems for low-frequency noise control have shown it to be quite effective.

Figure 18–12 shows typical attenuation data for several different lengths of absorptive silencers and compares these with active silencers.

18.11.4 Other Types and Uses of Silencers

There are also elbow silencers that effectively attenuate the noise with splitters, which turn the air aerodynamically to minimize system pressure drop. There are special fan inlet and fan discharge silencers, including cone silencers and inlet box silencers, that minimize aerodynamic system effects and reduce noise at the source.

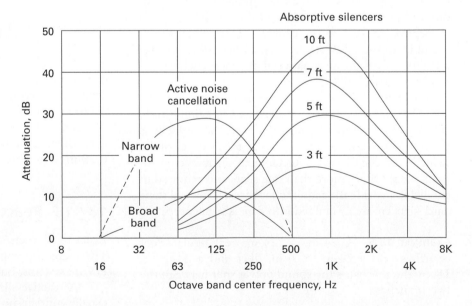

---

[4]These are known as Helmholtz resonators.

18.12 PLENUM CHAMBERS

Plenum chambers are often used to smooth out turbulent airflow associated with air as it leaves the outlet section of a fan and before it enters the ducted air distribution system of a building. The plenum chamber is usually placed between the discharge section of a fan and the main duct of the air distribution system. These chambers are usually lined with acoustically absorbent material to reduce fan noise and other types of noise. Plenum chambers are usually large rectangular enclosures with an inlet and one or more outlet sections. The main disadvantage of plenum chambers is that they require a large volume to be effective. On the other hand, they can be designed to fit in odd spaces.

18.13 SOUND POWER DIVISION IN DUCT BRANCHES

Where a duct branches into two or more separate ducts, the sound power contained in the incident sound waves in the main feeder duct is distributed between the branches associated with the junction. Many physical and aerodynamic factors can influence how sound power is distributed among the branch ducts. Absent any information that would permit assessing these factors, the best estimate of sound power division is to divide the main trunk sound power to each branch by the ratio of the area of that branch to the area of all branches (Equation 18–4). This division of sound power is referred to as *branch sound power division*. Duct branches can be big noise generators.

$$\text{Branch attenuation} = 10 \log\left(\frac{\text{branch area}}{\text{total area of all branches}}\right) \quad (18\text{--}4)$$

18.14 DUCT END REFLECTION LOSS

When low-frequency plane sound waves interact with openings into a large room, a significant amount of the sound energy incident at the opening is reflected back into the duct. This is similar to the way an organ pipe amplifies sound at specific frequencies associated with the pipe length. End reflection loss is greatest for small cross sections and low frequencies and decreases as pipes become larger and frequencies higher. Diffusers that terminate in a suspended lay-in acoustic ceiling can be treated as terminating in free space, but the grilles have

the effect of reducing the end loss somewhat. They usually have a restriction associated with them—a damper, guide vanes to direct airflow, a perforated metal facing, or a combination of these elements. Currently, no hard data are generally available to quantify the extent to which these elements effect the end loss attenuation.

18.15 RETURN AIR SYSTEMS

The plenum return air system makes use of the plenum space between the lay-in ceiling and the structural floor above. A ducted return air system is essentially the same as the ducted supply system and ducts the air from the room to the intake side of the AHU. When only total sound power of a fan is known, it is generally assumed that one-half is emitted to the supply side of the fan. Thus, the sound power to the supply side and return side of the fan are each 3 dB lower than the total sound power generated by the fan. The open or unducted return system to the ceiling plenum, however, is quite different.

The reentry of air from the ceiling plenum into the mechanical equipment room is often via a direct opening to the ceiling plenum. It may also be via a ducted opening directly from a corridor or lobby. In either case, there is usually little or no inherent attenuation between the mechanical equipment room and the ceiling plenum. Unless attenuated, high sound levels will be transmitted directly to the ceiling plenum, lobby, or corridor. Occupied areas in proximity to a mechanical equipment room can experience noise intrusion from the ceiling plenum as a result of noise emitted to the ceiling plenum.

Sound in ceiling plenums comes not only from direct connection to the mechanical equipment room but also from casing-radiated (breakout) noise from supply ducts in the ceiling plenum. The return air system should be designed so that the sound level in occupied rooms is less than that of the supply system. If the return system sound to a room is 5 dB less than the supply sound, the combination of the supply and return sound is only about 1 dB greater than the supply system alone.

Keep in mind that any air path is also a noise path. The effectiveness of a lay-in ceiling assembly as a noise barrier is often limited by these flanking paths, no matter how seemingly insignificant.

18.16 ROOM SOUND CORRECTION

The sound level throughout a room is a function of the sound power transmitted to the room by the HVAC system and other sound sources such as grilles, registers, and diffusers; air-valve and fan-powered air terminal units; fan-coil units located in ceiling plenums; and return air openings. It is also a function of the acoustical absorption properties of the room (room effect). The ceiling is usually the largest unencumbered surface in a room and as such is an important element in absorbing sound. Sound is also absorbed by wall treatment, furnishings, and people.

According to free-field theory, the sound pressure level decreases at the rate of 6 dB per doubling of distance from a point sound source. For example, if sound is measured to be 60 dB at a distance of 10 ft from a point sound source, it can be expected to measure 54 dB at a distance of 20 ft.

In a room, there is theoretically a direct sound that comes from the sound source to a receiver without reflecting from any surfaces. The sound field that is composed of all the reflections off of all surfaces is the reverberant field. It is one in which theoretically the sound travels in every direction at the same level with equal probability. Diffuse-field theory applies only to empty rooms in which there is no furniture or other objects that can scatter sound. There is theoretically a distance from the sound source at which the sound level in the room becomes constant, however, in the real world, there is scattering by furniture and other objects. Experience has shown that in real rooms, the sound pressure levels decrease at a rate more like 3 dB to 3.5 dB per doubling of distance from the sound source because of reflections and scattering of sound. Generally, a reverberant sound field does not exist in small rooms (room volume less than 15,000 ft³). In much larger rooms, reverberant fields may exist, but usually at large distances from the sound source.

18.17 TRANSMISSION OF SOUND THROUGH WALLS AND CEILINGS

18.17.1 Sound Transmission through Walls

When sound power is transmitted from one room to another, the sound level experienced is called the noise reduction (NR). Noise reduction is the difference (in decibels) between the sound pressure level in the source room and that in the receiving room. The reverberant sound field in the receiving room is a function of SPL_{src}, the transmission loss (TL) of the wall, the area of the common wall between the two rooms, and the acoustical absorption in the receiving room (A). The greater the common area (S_c), the more sound power is transmitted. The more sound absorbed in the receiving

room (A)[*], the lower the sound level experienced in the room. Remember that NR is an effective attenuation and that NR is reduced by the area and increased by the absorption. Noise reduction can be expressed by Eqs. 18–5 and 18–6:

$$NR = SPL_{src} - SPL_{rec} \qquad (18\text{–}5)$$

$$NR = TL - 10 \log (S_c) + 10 \log (A) \qquad (18\text{–}6)$$

Example 18.6 Two rooms have a common wall 9.5 ft × 20 ft. The wall is metal studs with drywall screwed to both sides of the studs; all joints are taped and finished. The TL of the wall is 37 dB in the 500-Hz octave. The sound level in the source room is 85 dB at 500 Hz. The acoustical absorption in the receiving room at 500 Hz is 487 sabins. What is the noise reduction at 500m Hz?

$$NR = TL - 10 \log (S_c) + 10 \log (A)$$

$$NR = 37 - 10 \log (20 \times 9.5) + 10 \log (487)$$

$$NR = 41$$

Transmission loss (TL) is measured in a laboratory setting by first measuring NR and correcting for S_c and A. Transmission loss is a property of the wall, ceiling, floor, etc., but NR is the meaningful measure of sound transmission from room to room. The TL is a laboratory measurement in which all of the sound energy is transmitted through the wall. Similar measurements are made in the field (between rooms in a house or office, for example) and include all the flanking paths as well as the wall itself. These are sometimes referred to as field transmission loss (FTL). Similarly, FNR is field noise reduction.

Transmission loss is measured in 16⅓-octave bands, 125 through 4000 Hz and, following a specific procedure, reduced to a single-number rating for wall acoustical insulation called *sound transmission class* (STC). When a similar test procedure is followed for field measurements, the result is referred to as *field sound transmission class* (FSTC). The FSTC can easily be 5 to 10 dB lower than the STC. When published STC data are used, it is recommended that they be reduced accordingly for design purposes.

When NR is based on FTL, it should be denoted as FNR, and the single-number rating should be field noise isolation class (FNIC).

An example of TL versus FTL for the wall construction in Figure 18–13(a) is shown in Figure 18–13(b),

---

[*]Acoustical absorption (A) is designated as sabins, named after one of the great contributors to understanding the role of absorption.

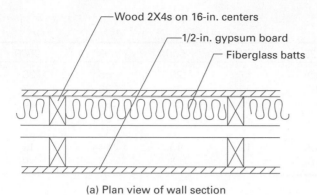

(a) Plan view of wall section

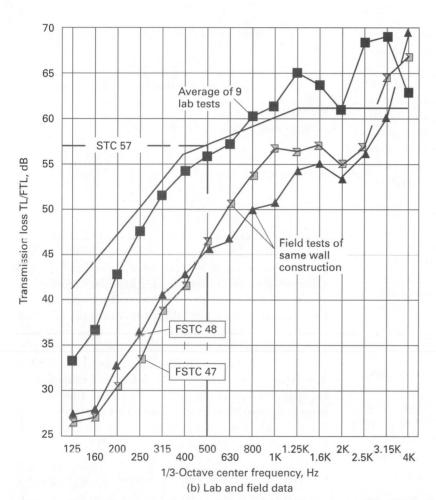

(b) Lab and field data

which records an average of nine laboratory tests and two different field tests of walls of the same design. The field tests were both in condominiums, in back-to-back bedrooms. The 8- to 10-dB difference in this case is the difference between privacy and the sense that one can hear and be heard through the wall.

In general, if the difference between TL and FTL is greater than 3 dB, most of the sound through the partition assembly is coming through flanking paths, not the wall per se. In the preceding example, this suggests poor workmanship in building the wall.

18.17.2 Sound Transmission through Ceiling Systems

When terminal units, fan-coil units, air-handling units, ducts, or return air openings to mechanical equipment rooms are located in a ceiling plenum above an occupied room, sound transmission through the ceiling system can be high enough to cause excessive noise levels in that room. One partial cure for this may be a better ceiling. Table 18–6 shows transmission loss for some ceiling assemblies.

TABLE 18-6
Typical transmission loss of ceiling assemblies, dB

Octave Center Frequency	63	125	250	500	1000	2000	4000
³⁄₈″ gypsum board	6	10	17	20	25	28	23
½″ gypsum board	9	14	20	23	26	27	25
⁵⁄₈″ gypsum board	10	14	20	25	27	27	26
1-½″ gypsum board	13	18	25	27	27	27	28
Typical ⁵⁄₈″ mineral fiber lay-in	4	5	6	8	10	12	14
Above w/ concealed spline	6	12	12	13	14	15	16
Fiberglass lay-in ceiling	2	3	4	5	7	9	10

Manufacturers of ceiling products rarely publish data that can be used in calculations. They do publish a ceiling attenuation class (CAC) rating that is similar to sound transmission class (STC). The CAC is different, however, in that it consists of a test in which there is a room of a specific size with a high-transmission-loss partition dividing it. Above the partition is the manufacturer's ceiling assembly, touching the top of the partition as in a real building. The specifics of the ceiling plenum are defined. This provides some comparison of different ceiling assemblies.

This problem is further complicated by the fact that ceilings usually have light fixtures, diffusers, grilles, speakers, sprinklers, etc. These elements significantly reduce the acoustical transmission loss of the ceiling by providing flanking paths.

Estimating the noise transmitted through a ceiling to a room is a guess, at best, using equal parts of available data and experience. Experience indicates that it is best to derate ceiling transmission loss data by 5 to 10 dB for design use. To estimate the sound levels in a room associated with sound transmission through the ceiling, the sound power levels in the ceiling plenum must be reduced by the transmission loss of the ceiling system before converting from sound power levels to corresponding sound pressure levels in the room. In the absence of a recognized test standard, the transmission loss values in Table 18–6 may be used.

18.18 ADDING DECIBEL QUANTITIES

A decibel, abbreviated dB, is 10 times the common logarithm of a quantity:

$$dB = 10 \log X \qquad (18\text{-}7)$$

where X is a quantity we wish to express as a decibel. For example, $10 \log 10^6 = 60$ dB. More generally, $10 \log 10^a = 10a$.

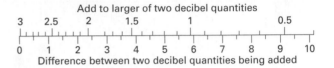

FIGURE 18–14
Addition of two decibel quantities.

If $A = 10 \log X$ and $B = 10 \log Y$, how do we add X and Y and express the result as a decibel, $10 \log (X + Y)$? Figure 18–14 shows a simple method. If X is greater than or equal to Y, take the difference $(X - Y)$ and enter the chart in Figure 18–14 on the bottom scale at this value; then, read on the top scale the increment to be added to the larger of the two quantities (X). For example, $X = 60$ dB and $Y = 58$ dB. Difference is 2. Read above and get 2.1; therefore $60 + 2.1 = 62.1$ dB.

If $Y = X$, then $X + Y = 2X$; $10 \log 2X = 10 \log 2 + 10 \log X = 10 \log X + 3$. Adding two equal decibel quantities results in a decibel quantity that is 3 dB greater than either of the two being added. To generalize this, adding n equal decibels X gives $10 \log X + 10 \log n$. For quick reference,

$10 \log 1 = 0$	$10 \log 2 = 3$	$10 \log 3 = 4.8$
$10 \log 4 = 6$	$10 \log 5 = 7$	$10 \log 6 = 7.8$
$10 \log 7 = 8.5$	$10 \log 8 = 9$	$10 \log 10 = 1$

Note that $10 \log 4 = 10 \log 2 + 3$; $10 \log 6 = 10 \log 3 + 3$; $10 \log 10 = 10 \log 5 + 3$.

Example 18.7 Determine the overall dB level from the octave band spectrum shown on the NC curve (Figure 18–3). Use the chart in Figure 18–14, taking two levels at a time. This step is illustrated in Figure 18–15. As a matter of convenience, the octave levels are arranged in descending order, however, it is not necessary to add the decibel values in any particular order.

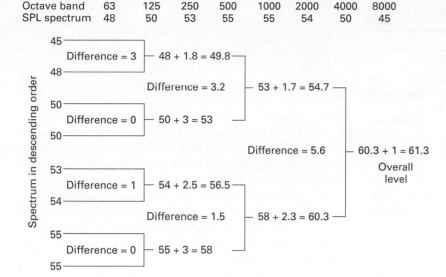

Octave band	63	125	250	500	1000	2000	4000	8000
SPL spectrum	48	50	53	55	55	54	50	45

■ **FIGURE 18–15**
Example from Figure 18–3 determining overall level by two-by-two decibel addition.

18.19 SOUND PRESSURE, SOUND POWER, AND SOUND INTENSITY LEVEL

Sound pressure, sound power, and sound intensity level are the basic quantities of acoustics, so it is best to introduce them at this time.

18.19.1 Definition of Quantities

Sound intensity level (IL), sound power level (PWL), and sound pressure level (*SPL*) are decibel quantities, all of which relate to acoustical power. Sound intensity (*I*) is a function of mean square of the sound pressure (p^2), the density of air (p), and the speed of sound (c).

$$I = \frac{p^2}{\rho c} \qquad (18\text{–}8)$$

Intensity level (IL) is defined as

$$IL = 10 \log \frac{I}{I_{ref}} \qquad (18\text{–}9)$$

If there is a point sound source emitting sound in all directions equally as illustrated (in two dimensions) in Figure 18–16, then the total sound power is the sum of the sound intensity times the area. The sound power level (PWL) in decibels is

$$PWL = 10 \log \frac{W}{W_{ref}} = 10 \log \frac{IS}{W_{ref}} \qquad (18\text{–}10)$$

where S is the surface area.

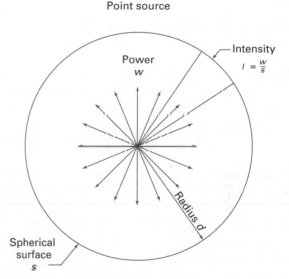

■ **FIGURE 18–16**
Point sound source radiation in free field.

The sound pressure level (SPL) is a special case of intensity level:

$$SPL = 10 \log \frac{p^2}{p_{ref}^2} = 20 \log \frac{p}{p_{ref}} \qquad (18\text{–}11)$$

Note that all the foregoing levels are a ratio of the acoustical power, intensity, and pressure to a reference quantity rather than the quantity per se. These reference quantities are given in Table 18–7.

At standard atmospheric conditions, IL and PWL are, for all practical purposes, numerically equal in English units.

TABLE 18-7
Standard accoustical reference quantities

Measure	English	Metric
Power (W_{ref})	10^{-12} W	10^{-12} W
Intensity (I_{ref})	9.289×10^{-14} W/ft^2	10^{-12} W/m^2
Pressure (P_{ref})	2.901×10^{-9} psi	2×10^{-5} newtons/m^2
	4.177×10^{-7} psf	20 µN/m^2

18.19.2 Propagation of Sound

Because sound pressure level is a special case of sound intensity level, the propagation of sound from a point source radiating in a spherical pattern can be generalized by Eqs. 18-12(a) and (b).

$$SPL = PWL + 10 \log\left(\frac{1}{4\pi d^2}\right) + 10.3 \text{ (English)}$$
$$(18\text{-}12\text{a})$$

$$SPL = PWL + 10 \log\left(\frac{1}{4\pi d^2}\right) + 0.3. \text{ (SI)} \quad (18\text{-}12\text{b})$$

Equations 18-12(a) and (b) are for propagation of sound in a free field. (See Example 18.8.)

Example 18.8   Consider the mythical point sound source in Figure 18-16 emitting acoustical power (W) of 10^{-3} W. The power level (PWL) is

$$\left(10 \log \frac{10^{-3}}{10^{-12}}\right) = 90 \text{ dB}$$

At a distance (radius) of 10 ft, the surface area of the sphere is $4\pi d^2$ or 1256 ft^2. The sound intensity (I) is

$$\left(\frac{10^{-3}}{1256}\right) = 7.96 \times 10^{-7} \text{W/sq ft}$$

The intensity level (IL) is

$$10 \log \left(\frac{7.96 \times 10^{-7}}{9.289 \times 10^{-14}}\right) = 69.3 \text{ dB}$$

At a distance of 20 ft, the surface area would be increased by a factor of 4, the intensity decreased by a factor of 4, and the intensity level decreased by 6 dB:

$$10 \log \frac{1}{4} = -6$$

This illustrates the fact that intensity level decreases 6 dB for each doubling of distance from a sound source.

18.19.3 Sound in Rooms

Sound in rooms was discussed briefly in Section 18.16. As we have all experienced many times, the sound in a room with all hard surfaces is louder than the same sound in a room with heavy draperies, furniture, and carpet—i.e., an acoustically "soft" room. The sound that is absorbed at the walls of the room is proportional to the intensity of the sound impinging on the surface. The acoustical absorption coefficient, usually designated as alpha (α), is the ratio of the sound absorbed by the surface to the total sound incident on the surface.

When heat is supplied in a room, the temperature rises until the rate at which heat flows into the room is equal to the rate at which it flows out. Similarly, when sound power is injected into a room, the sound pressure rises to a level where the rate of sound energy being absorbed by the walls is equal to the rate at which sound energy is introduced into the room, or power in equals power out. The theoretical sound pressure level of the reverberant field is

$$SPL_{rev} = PWL + 10 \log \left(\frac{4}{R}\right) + 10.3 \text{ (English)}$$
$$(18\text{-}13\text{a})$$

$$SPL_{rev} = PWL + 10 \log \left(\frac{4}{R}\right) + 0.3 \text{ (SI)}$$
$$(18\text{-}13\text{b})$$

Combining the direct SPL and the reverberant SPL, we obtain the following:

$$SPL = PWL + 10 \log\left(\frac{1}{4\pi d^2} + \frac{4}{R}\right) + 10.3 \text{ (English)}$$
$$(18\text{-}14\text{a})$$

$$SPL = PWL + 10 \log\left(\frac{1}{4\pi d^2} + \frac{4}{R}\right) + 0.3 \text{ (SI)}$$
$$(18\text{-}14\text{b})$$

Note that the first term in the parentheses in Eq. 18-14(a) and (b) is the same as in Eqs. 18-12(a) and (b).

18.19.4 Sound Fields

Sound fields in which sound travels from a source to a receiver are categorized as near field, far field, free field, and reverberant field. These are illustrated in Figure 18-17. These sound fields are not mutually exclusive.

Free field is the space around a source in which the sound travels directly from the source to the receiver without reflections such as would likely occur in an enclosed space. The sound pressure level decreases 6 dB for each doubling of distance. (See Example 18.8.) Near a finite-size sound source, such as the wall of a rooftop unit or noise enclosure, the sound may not decay at

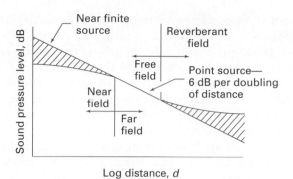

■ FIGURE 18–17
Definitions of sound fields.

6 dB per doubling of distance as it would for a point source. The space near a finite source in a free field where the sound does not follow the rule 6 dB per doubling of distance is the near-field sound of the source.

Reverberant field is a space in which the sound reflects from many surfaces many times. In a reverberant room, the sound is considered to be in the direct or free field until it first strikes a surface, after which it becomes part of the reverberant field. The second term in the parentheses in Eqs. 18–14(a) and (b) describes the reverberant field. In a reverberant field the sound decay does not follow the rule 6 dB per doubling of distance because of the many reflections.

Far field is the space in which the source can be considered a point source and would follow the rule 6 dB per doubling of distance were it not for the reverberant sound. Far field and reverberant field can overlap, as shown in Figure 18–17.

18.20 ACOUSTICAL ABSORPTION

When a sound wave falls on a surface, some of the energy is absorbed in and beyond the surface, and the remainder is reflected back into the surrounding medium. The absorptive properties of a material are defined by the coefficient of absorption α, which is in turn defined as the ratio of the energy absorbed by a surface to the total incident energy falling on the surface. The value of α may vary from 0.01 for a plate glass window (virtually all energy is reflected) to almost 1.0 for an open window (virtually all energy is absorbed).

The absorption characteristic of a particular material is a function of the frequency of the impinging sound. Tables of experimentally determined values of absorption for various materials are available in numerous publications. Most commonly, α is given by octave band, usually for octave bands of 125 Hz through 4000 Hz,

occasionally for the 63-Hz and 8000-Hz octaves, and rarely for the 32-Hz octave band.

As a general guideline, most materials offer very little absorption at low frequencies and increasing absorption as frequency increases. To balance this out, we frequently use thin panels such as thin plywood to absorb low frequency. Single-layer gypsum walls can absorb significant low-frequency sound. In these latter cases, sound energy is not strictly absorbed as with porous materials, but rather by forcing energy-absorbing vibrations of the paneled surface.

The noise reduction coefficient (NRC) is defined as the average of coefficients at 250, 500, 1000, and 2000 Hz.

Absorption is also a function of the incidence angle of the impinging sound wave, however, absorption coefficients as typically reported are for random incidence. Measurements of the coefficients are made in conditions that approximate a diffuse field as closely as is practical.

For a surface composed of many materials, the average absorption coefficient is given by

$$\bar{\alpha} = \frac{\alpha_1 S_1 + \alpha_2 S_2 + \alpha_3 S_3 + \cdots + \alpha_n S_n}{S_1 + S_2 + S_3 + \cdots + S_n} \quad (18–15)$$

where S_i is the area associated with material having absorption coefficient α_i.

18.21 SOUND TRANSMISSION LOSS

In Section 18.17, transmission of sound through walls was mentioned briefly. We will now fill in some additional details.

18.21.1 Sound Transmission Class (STC)

Sound transmission class (STC) is a method of rating the airborne transmission of sound through a wall or other structure at different frequencies by means of a single number. Sound is transmitted through walls and other dividing building elements differently at different frequencies. Generally, sound is transmitted more easily at low frequencies than at higher frequencies (i.e., sound attenuation is less at low frequencies than at higher frequencies).

The STC rating method is based on the laboratory test procedure specified in ASTM Recommended Practice E 90, in which the sound transmission loss of a test specimen is measured at 16 frequencies in 1/3-octave intervals covering the range from 125 to 4000 Hz. To determine the STC of a given specimen, its measured transmission loss values are plotted against frequency and compared with a reference curve (STC contour), shown in Figure 18–13(b).

The STC rating is then determined by adjusting a standard contour. The STC contour is shifted vertically relative to the test data curve to as high a position as possible while fulfilling the following conditions:

1. The maximum deviation of the test curve below the contour at any single test frequency shall not exceed 8 dB.
2. The sum of the deviations at all 16 frequencies of the test curve below the contour shall not exceed 32 dB. This is an average deviation of 2 dB.

This condition is illustrated in Figure 18–13(b). When the STC contour is thus adjusted (in integral decibels), the STC value is read from the vertical scale of the test curve as the TL value corresponding to the intersection of the STC contour and the 500-Hz frequency line. In the example, the STC value (57) is governed by the 32-dB total deviation below the contour.

18.21.2 Field Sound Transmission Class (FSTC)

The FSTC is determined exactly the same as STC, using the same reference contour and criteria as above, with the exception that STC is performed in a laboratory and FSTC is performed in the field on an actual wall installation. The FSTC is to be performed in accordance with ASTM E336.

18.21.3 Noise Isolation Class (NIC)

Noise isolation class (NIC) differs from sound transmission class (STC) in that, as explained in Section 18.17, NIC also considers the acoustical absorption in the receiving room and the area of wall common to the two rooms. The greater the acoustical absorption (A) in the receiving room, the lower the transmitted sound level in the room. The greater the wall area (S_c) common to the two rooms, the more sound power is transmitted.

18.21.4 Laboratory versus Field Transmission Loss

Another facet of NIC and STC is field versus laboratory measures and what this comparison tells us about the amount of acoustical energy that is transmitted through a partition relative to the amount that is transmitted via other paths (i.e., flanking paths). In the laboratory, STC tests are very careful to seal every flanking path so that the acoustical energy that is transferred to the receiving room is through the partition and not via flanking paths.

If one assumes that the integrity of the partition itself is the same in the field as in the laboratory, a field transmission loss (FTL) that is 3 dB less than the laboratory TL means that as much energy is reaching the receiving room by flanking paths as through the partition itself. It is not unusual to find FTL, and hence FSTC, measures that are 5 dB or more less than laboratory results for the same partition design. (*Note:* Do not specify STC in specifications if FSTC is intended.)

The example in Figure 18–13, while not a typical one, illustrates the axiom that effective acoustical isolation in the field is at least as much a matter of workmanship as design. Although partitions that a designer may draw on paper are perfect for the occasion, workmanship is ultimately the deciding factor.

A second observation from considerable experience with acoustical transmission loss and STC and NIC is that use of STC and NIC alone can often be misleading. For the types of partitions in question here, low-frequency transmission of sound is often the limiting factor in determining STC and NIC, as is illustrated in Figure 18–13(b). Low-frequency sound isolation (attenuation) usually requires special treatment.

18.21.5 Composite Walls

Where the transmitting wall is composed of sections with different TL characteristics, the nominal TL of the composite wall is given by

$$TL = 10 \log\left(\frac{S_1 + S_2 + S_3 + \cdots + S_n}{\tau_1 S_1 + \tau_2 S_2 + \tau_3 S_3 + \cdots + \tau_n S_n}\right)$$
$$(18-16)$$

where S_i is the area of wall sections of different composition and τ_i is the transmission coefficient of the wall section, defined as

$$\tau_i = 10^{-\frac{TL}{10}} \qquad (18-17)$$

and TL_i is the corresponding transmission loss.

18.22 ISOLATION OF MECHANICAL VIBRATION

In building structure design the usual concerns are for dead load and live load. These are both gravitational forces. Dead load is the weight of the structure alone. Live load, a misnomer, is the "dead" weight of machinery, people, bookcases, etc., that sits on the structure. A third loading condition is dynamic loading, resulting in dynamic motion of a building. Dynamic loading may be from an external force, such as wind or an earthquake. It may also be largely localized motion, as in the vicinity of a mechanical equipment room.

In buildings there are two primary concerns regarding isolation of vibration. The first is isolating equipment so that vibrations generated by machinery and other dynamic loads are not transmitted to the supporting structure. The second is isolation of equipment (instruments, surgical microscopes, etc.) from ambient vibrations that exist in the supporting structure. Such isolation consists primarily of permitting motion in one element (machine) so that forces transmitted to the base (floor) will be minimized. A third type of motion is a major concern—seismic-induced motion. Seismic restraining involves limiting large motions caused by earthquakes while permitting small motions required for isolation.

We live in a three-dimensional world. There are six degrees of freedom associated with the motion of a rigid body—one vertical, two horizontal, and three rotational. Analyzing dynamic behavior in more than one degree of freedom becomes quite complex.

In isolating equipment in buildings so that transmission of vibration to the supporting structure is minimized, the focus of attention is primarily on the vertical component of the vibration. This is because the primary flexibility of the supporting structure, the floor or roof, is generally vertical, however, it is not uncommon to have equipment fastened to a wall or building column. If the equipment is attached rigidly to the floor, the dynamic force is transmitted directly to the base. A floor is usually quite flexible in the vertical direction as compared with its flexibility in the horizontal directions.

It is convenient to be concerned with only vertical vibration isolation, assuming the floor to be rigid, because a system with a single degree of freedom is easily analyzed and provides insight for vibration isolation. But this is not always the case. Rocking motion, for example, is a significant possibility, particularly where the machinery has a center of mass that is high above the base.

18.22.1 Undamped Vertical Vibration with a Single Degree of Freedom

Figure 18–18 shows three examples of vibration often treated as simple models of one degree of freedom. A degree of freedom means that the body is free to move in that direction, however, motions may be "coupled" together in such a way that the vertical motion, for example, cannot exist alone, but rather is coupled to one or more motions in other principal directions. For instance, pushing down at the center of gravity in Figure 18–18(a)

may result in not only vertical displacement but also rotation, as shown in Figure 18–18(b).

The idealization of the single degree of freedom, undamped, rigid base illustrated in Figure 18–18(a) is shown in Figure 18–19(a) and (b). The machine to be isolated is idealized as a mass (m) sitting on a single spring. The spring is characterized by its spring constant, usually designated by k and having units of pounds per inch (lb/in.). The dynamic force is characterized as sinusoidal in the vertical direction with an amplitude of F_0. The natural frequency of the system ω_n is given by

$$\omega_n = \sqrt{\frac{k}{m}} \qquad (18\text{--}18)$$

Stiffness (k), mass (m), and natural frequency (ω_n) are all properties of the spring–mass, but only two are independent. Choosing any two immediately defines the third.

The static deflection of the system (δ) is the deflection the spring would have under the static weight of mass m (i.e., mg):

$$\delta = \frac{w}{k} = \frac{mg}{k} \qquad (18\text{--}19)$$

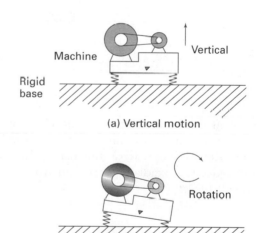

(a) Vertical motion

Machine

Vertical

Rigid base

(b) Rocking motion

Rotation

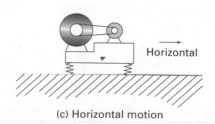

(c) Horizontal motion

Horizontal

■ **FIGURE 18–18**
Three types of motion with a single degree of freedom.

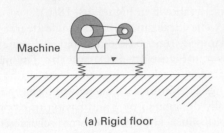

(a) Rigid floor

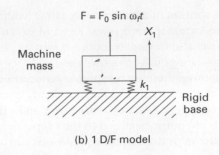

(b) 1 D/F model

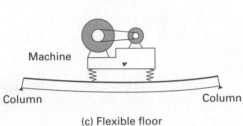

(c) Flexible floor

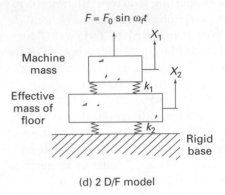

(d) 2 D/F model

■ **FIGURE 18–19**
Models of one and two degrees of freedom (D/F).

so that

$$\omega_n = \sqrt{\frac{g}{\delta}} \qquad (18\text{–}20)$$

where g is the gravitational constant (386 in./sec²) and δ is in inches. Normally we think of frequency in hertz (cycles per second, Hz) whereas the mathematics above gives the frequency in radians per second. To convert to hertz, divide by 2π:

$$f(\text{Hz}) = \frac{\omega}{2\pi} \qquad (18\text{–}21)$$

Frequency in hertz is usually denoted by f.

The dynamic force transmitted to the base (F_t) is the deflection (x) of the spring times the spring constant (k). The ratio of the peak transmitted force to the peak applied force is

$$\frac{F_t}{F_0} = \frac{1}{1 - \left(\dfrac{\omega_f}{\omega_n}\right)^2} \qquad (18\text{–}22)$$

This function, Eq. 18–22, is shown in Figure 18–20. If we think of beginning by shaking the mass at zero frequency and slowly increasing it, at the start the frequency ratio $\omega_f/\omega_n \approx 0$ and the system is essentially one

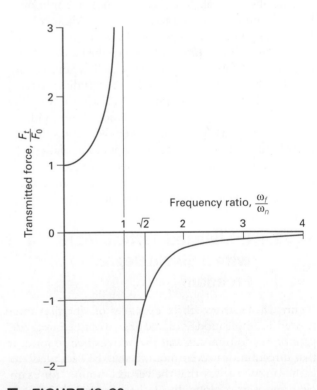

■ **FIGURE 18–20**
Transmitted force versus frequency for a system with a single degree of freedom.

in which the displacement is the equivalent static deflection equal to the amplitude of applied force. As the forcing frequency increases, the transmitted force and the deflection increase. The force transmitted is larger than the applied force. As the frequency ratio approaches one (1.0), the excitation frequency approaches the natural frequency of the system and, for the undamped case, the force ratio theoretically becomes infinite because the denominator in Eq. 18–22 becomes zero. Up to this point, the mass and the force are in phase (i.e., they move in the same direction). When the frequency ratio is greater than one, the force and the mass are out of phase (i.e., they move in opposite directions). But we note that as the frequency ratio increases above one, the force transmitted becomes smaller and smaller. When the frequency ratio is equal to $\sqrt{2}$, the force ratio is again negative one (-1.0). Above this frequency ratio the force ratio continues to decrease. The objective of vibration isolation is to operate at a frequency ratio much greater than $\sqrt{2}$.

The isolation efficiency, at forcing frequencies greater than the natural frequency, is the percent of the exciting force that is prevented from being transmitted to the supporting structure expressed as a positive number:

$$\text{Isolation efficiency (\%)} = 100\left(1 - \frac{1}{\left(\frac{\omega_f}{\omega_n}\right)^2 - 1}\right)$$

$$(18\text{--}23)$$

A frequency ratio of 3 to 1 eliminates about 88 percent of the exciting force from transmitting to the base. This is a good rule of thumb to use as a minimum in selecting vibration isolation.

A chart for determining isolation efficiency for the simple spring–mass Eq. (18–23), is shown in Figure 18–21. Since the natural frequency can be expressed by the static deflection, the static deflection is also shown in this figure. Frequencies are also shown in revolutions per minute (rpm) because machine speeds are usually expressed in rpm.

Example 18.9 Refer to Figure 18–21.

(a) We wish to provide 90 percent isolation efficiency for a motor running at 1800 rpm (revolutions or cycles per minute). Enter the chart at the bottom at 1800, read up to the 90 percent line, read across to the static deflection scale, and read approximately 0.12 in. An isolator needs to have only about 1/8-in. static deflection for 90 percent isolation efficiency.

(b) If we need to provide 99 percent isolation efficiency, we read up to the 99 percent line. We see that we would need about 1.3-in. static deflection.

18.22.2 Inertia Base

An inertia base involves adding mass to the machine being isolated, usually by adding a block of concrete between the machine and the isolation springs. It is important to note that if there are no changes other than adding mass, the dynamic deflection is decreased by the ratio of the masses, the static deflection of the spring is increased, and the isolation is improved. If the spring constant is adjusted so that the static deflection, and thus the natural frequency, of the base–machine configuration remains unchanged with the addition of the base, the isolation is unchanged. This does not change the isolation efficiency, because the frequency ratio is unchanged, as can be seen from Eq. 18–23. All this does is add to the dead weight on the floor.

An inertia base can add significant stability to the system and change the coupling between vertical and rocking motion, particularly where the center of gravity is high above the base of the machine.

For an inertia base to be effective, a good rule of thumb is that the weight of the inertia block should be at least twice the weight of the machinery. This is not often feasible for lightweight floor or roof construction, so one must look to the springs for isolation.

18.22.3 Coupled Vertical and Rocking Motion

In its simplest form, coupled vertical and rocking vibration can be modeled as a system with two degrees of freedom, as shown in Figure 18–22. Coupled, as discussed above, means that we cannot lift vertically (the dynamic load) without also causing a rocking motion. When the geometry is ideal, the vertical and rocking are uncoupled. In practice, there is usually some degree of coupling of the vertical and rocking degrees of freedom. The strength of the coupling depends on several factors that are beyond the scope of this discussion. To minimize coupling, the springs should be symmetrical about the center of gravity (CG), and the CG should be as low as possible. Using an inertia base will lower the CG of the system and make it more stable.

18.22.4 Undamped Vertical Vibration with Flexible Base

Machinery located on basement floors that are concrete and poured directly on earth or on large concrete pads in the earth behave much more like the case of a single degree of freedom, however, equipment located on flexible bases such as a floor may exhibit significantly different behavior, particularly when the floor is "springy."

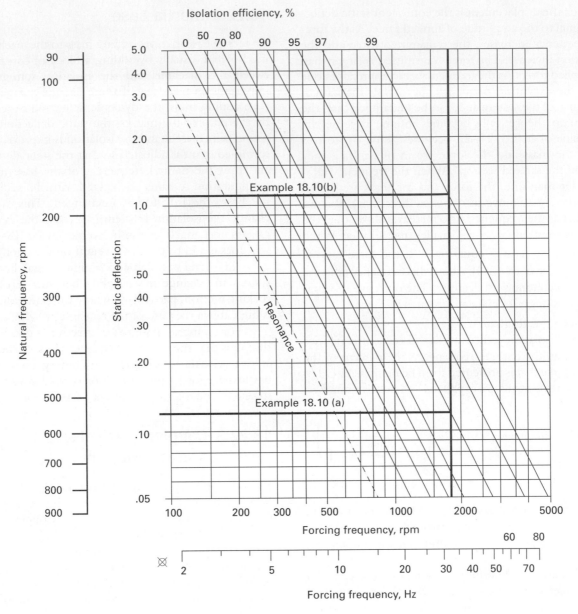

■ FIGURE 18–21

Determining isolation efficiency for a system with a single degree of freedom.

A simple model of the system is represented in Figure 18–19(c) and (d). This model has two degrees of freedom—motion of the machinery and motion of the base. For the simple case the floor is modeled as a simple spring and an effective mass of the floor, as shown in Figure 18–19(d). Analytical description of a system with two degrees of freedom, or any system with more than one degree of freedom, is beyond the scope of this chapter. In isolating a machine from a soft base such as a lightweight floor, the isolation springs must have a lesser spring constant than the base (floor).

A roof or floor structure has many modes and natural frequencies of vibration. Adding a mass or spring-isolated mass to a structure with more than one degree of freedom changes the natural mode shapes and frequencies of the structure. For most situations encountered in practice, a single degree of freedom is adequate to specify vibration isolation, however, there are cases that can inadvertently lead to trouble. These often involve very flexible roof or floor structures. An experienced structural dynamics engineer is recommended for such cases.

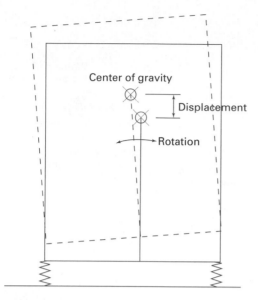

■ FIGURE 18–22
Two degrees of freedom with vertical displacement and rocking.

Some generalizations can be offered for complex systems, but like any generalizations, they must be applied judiciously.

- As long as the natural frequency of the isolation system is considerably lower than the natural frequency of the floor, one degree of freedom will generally be adequate (i.e., the forcing frequency is at least three times the natural frequency for a single degree of freedom system).
- Avoid natural frequency ratios (i.e., natural frequency with a single degree of freedom to floor natural frequency) that are close to one. If the floor is "soft," the isolation system must be "softer."
- If very soft floors are involved, and if the installed equipment is motion-sensitive, get help from an expert. It is possible, with lightweight construction, that the dynamic behavior of the entire building may be involved.

18.23 VIBRATION ISOLATORS

Vibration isolators come in all sizes and shapes. Figure 18–23 illustrates some of the common types.

Figure 18–23(a) is a typical neoprene waffle pad. This is used in noncritical applications where small static deflections are required. It is most effective in isolation of structure-borne sound. The example shows two pads in series with a steel shim between to get more deflection.

Figure 18–23(b) shows a double-deflection neoprene mount. It functions much like the waffle pad, but it is configured to provide greater deflection.

Figure 18–23(c) is a typical unhoused freestanding spring. The acoustical pad on the bottom is to minimize structure-borne noise as well as to keep the mount from slipping. Springs of this type generally require sideways restraint to avoid horizontal motion such as is shown in Figure 18–18(c).

Figure 18–23(d) is a restrained spring mount. Vertical motion is sufficient to provide the necessary isolation, but it is limited so that large motion is not possible. Horizontal motion is also restrained. Such a mount usually has a neoprene pad to isolate structure-borne noise.

Figure 18–23(e) is a housed spring that functions much like the restrained mount in providing for isolation motion, but not motion beyond this in any direction.

Note that all the mounts in Figure 18–23 provide a means of attaching to the equipment to be isolated. Most also provide a means of adjustment so that the load can be balanced out over several mounts.

Figure 18–24 illustrates two isolation hangers functionally typical of a wide variety of isolation hangers. These are most often used in isolating ceiling systems.

Figure 18–24(a) is a typical neoprene isolation hanger. Its primary function is to minimize transmission of structure-borne sound. Figure 18–24(b) is typical of a wide variety of spring hangers. It has a neoprene pad for structure-borne sound. Similar hangers are available with only a spring. These hangers are used for isolating ceiling systems in critical situations and for isolating ducts, pipe, conduits, hanging air-handling units, VAV boxes, and virtually anything that must be hung. They are available with different spring constants for a wide range of loads. The isolation efficiency is the same as for bottom mounts.

18.24 SEISMIC VIBRATION CONTROL AND RESTRAINT

The intent of this section is not to lead the reader to seismic design but to introduce some of the considerations in building planning. The objective of seismic restraint is to limit the motions of equipment that can occur during a seismic event so that dynamic loads and motions that would occur in unrestrained vibration-isolated equipment cannot happen. At the same time, seismic restraint must allow motions that are necessary for effective vibration isolation under normal nonseismic operating conditions. Because seismic isolation is important for life safety, seismic design has been codified by structural engineers.

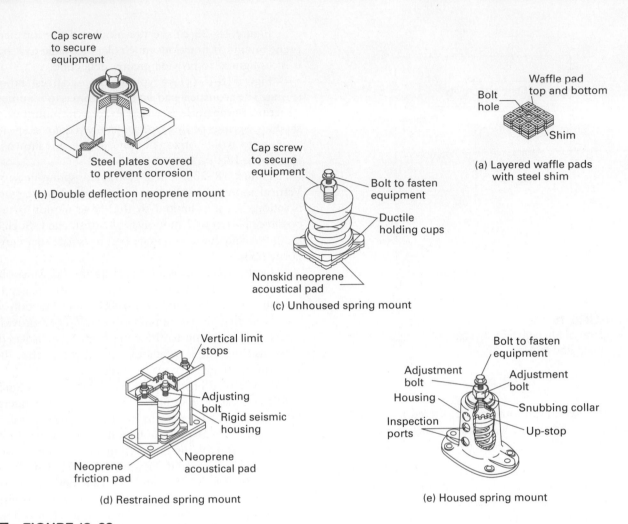

Cap screw
to secure
equipment

Steel plates covered
to prevent corrosion

(b) Double deflection neoprene mount

Cap screw
to secure
equipment

Bolt to fasten
equipment

Ductile
holding cups

Nonskid neoprene
acoustical pad

(c) Unhoused spring mount

Waffle pad
top and bottom

Bolt
hole

Shim

(a) Layered waffle pads
with steel shim

Vertical limit
stops

Adjusting
bolt

Rigid seismic
housing

Neoprene
friction pad

Neoprene
acoustical pad

(d) Restrained spring mount

Bolt to fasten
equipment

Adjustment
bolt

Adjustment
bolt

Housing

Snubbing collar

Inspection
ports

Up-stop

(e) Housed spring mount

■ **FIGURE 18–23**
Examples of spring isolators. (Courtesy: Mason Industries, Hauppauge, NY.)

Earthquake activity since 1900 on the west coast of the continent is shown in Figure 18–25. In 1971 there was a major earthquake in San Fernando, California, that measured 6.0 on the Richter scale with horizontal ground accelerations recorded at 0.2 g. More recently, there have been even more severe earthquakes in Turkey, Mexico, and Taiwan, with heavy loss of life. There was also tremendous damage to buildings and interior equipment. Most of the damage to mechanical equipment was a result of the "softness" of the vibration isolation, which permitted large motions of the equipment in response to seismic loading. In many cases, equipment jumped off the vibration isolation mounts, severely damaging not only the equipment itself but also the mounts and connections such as piping and conduit.

An example of the damage is shown in Figure 18–26. The machinery was mounted on unrestrained freestanding springs with inadequate thrust

restraint. The dynamic response of the system tore up the concrete housekeeping pad and broke the restraints. The photo tells an all too often repeated story. All-direction seismic snubbers, as shown in Figure 18–27, could have prevented this destruction. (Reference 16 contains many excellent photos of earthquake events.)

In other events, pipes, ducts, and conduits broke loose. Suspended HVAC equipment plunged to the floor. Equipment was rendered inoperative. It would have been better, in terms of preventing seismic damage, if the equipment had been attached rigidly to the supporting structure.

Far-ranging and severe damage can be inflicted in a short period of time. As a result of experience, particularly during the second half of the twentieth century, codes and specifications for seismic protection have evolved. Most local building codes have adopted seismic codes by reference. These codes provide empirical

Eye bolts
top and bottom for
wire or horizontal bolts

Double deflection
neoprene element
with a projecting
bushing to prevent steel
to steel contact

Precompression plate
hangers are precompressed
to 60% of rated load.
Plate open when
full load is applied.

(a) Double deflection neoprene hanger

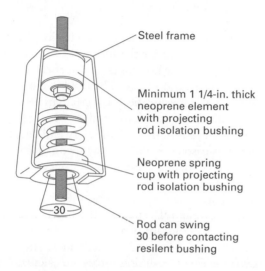

Steel frame

Minimum 1 1/4-in. thick
neoprene element
with projecting
rod isolation bushing

Neoprene spring
cup with projecting
rod isolation bushing

Rod can swing
30 before contacting
resilent bushing

(b) Spring and neoprene hanger

■ **FIGURE 18–24**

Examples of isolation hangers. (Courtesy: Mason Industries, Hauppauge, NY.)

equations for developing a maximum lateral force, and a percentage of the lateral force is used to determine the vertical force. A typical equation is $F_p = ZIC_p W_p$, where F is the equivalent static lateral design force, Z is based on seismic zone, I is an importance factor relating to the type of facility and occupancy, C_p is the type of nonstructural component, and W_p is the weight of the equipment. Check applicable codes for values for and application of seismic calculations.

To avoid the cost of overdesigning and the potential life safety penalties of underdesigning, it is important to work with these exact values when dealing with major structural components such as the steel framing. Vibration isolation and seismic restraints are a minuscule percentage of a building's construction cost and inexpensive insurance against cataclysmic damage should an earthquake occur.

18.25 THE RICHTER SCALE

In 1935 C. F. Richter defined what has come to be known as the Richter scale. Richter magnitude is given by the formula

$$M = \log_{10}\left(\frac{A}{A_0}\right) \qquad (18\text{–}25)$$

where A is the maximum amplitude recorded by a Wood–Anderson seismograph at a distance of 100 km from the center of the disturbance and A_0 is a reference amplitude of one-thousandth of a millimeter. Note that every tenfold increase in A is an increase in M of one. (The Richter scale is in Bels as opposed to decibels.) Table 18–8 shows the maximum acceleration and duration likely to be associated with an earthquake.

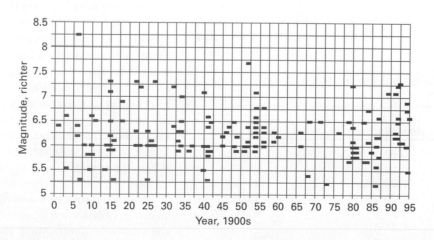

■ **FIGURE 18–25**

Earthquake history, 1900–1995: California, Nevada, Baja California. (Source: USGS earthquake catalogs web page.)

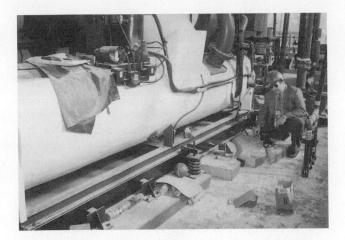

■ FIGURE 18–26
Seismic damage during San Fernando earthquake of 1971. (Courtesy: Mason Industries, Hauppauge, NY.)

■ FIGURE 18–27
Unhoused spring isolators with all-direction seismic snubbers. (Courtesy: Mason Industries, Hauppauge, NY.)

TABLE 18–8
Maximum ground acceleration and duration of strong phase of earthquakes

Magnitude, Richter	Maximum Acceleration	Duration, sec
5	0.09 g	2
5.5	0.15 g	6
6	0.22 g	12
6.5	0.29 g	18
7	0.37 g	24
7.5	0.45 g	30
8	0.50 g	34
8.5	0.50 g	37

18.26 GUIDELINES FOR SEISMIC DESIGN

Vibration isolation and seismic restraint are two different but related design disciplines. Vibration isolation, as has been noted, minimizes the transfer of dynamic forces caused by machinery to the building structure or, in the case of sensitive instruments, motion of the building to the instruments. Seismic restraints give the vibration isolators freedom to move in limited but sufficiently large deflections to provide the required isolation while preventing larger motions or forces in directions beyond the capacity of the isolators.

- Use the appropriate building codes for the model and seismic zone for the location of the building. These specify the horizontal and vertical maximum accelerations and forces to be used in the design.
- Simplify the specifications by specifying the higher level of forces as developed by the applicable codes, but err on the side of safety and protection.
- Never use isolation rails. Always use a one-piece reinforced concrete base or a one-piece structural steel frame.
- Spring static deflection has no meaning in seismic restraint design. Do not use the horizontal or vertical spring constant of an unhoused spring to calculate resistive forces. Although an earthquake is a dynamic event, resulting forces are calculated statically on the basis of codes.
- Wherever possible, limit the motion of spring-mounted bases using either separate double-acting seismic snubbers with neoprene cushioned interfaces, or steel springs within self-snubbing housings. Restraining housings may be either steel, ductile iron, or cast steel. Gray iron castings are not acceptable, because the lack of ductility can result in shattering when subjected to shock.
- Provide anchor bolts or drill-in anchors that are seismically approved and properly selected on the basis of design calculations. Anchor bolts must be embedded and spaced in accordance with ICBO or other recognized standards.
- All housekeeping pads must be structurally doweled or bolted to the structure and adequately reinforced to resist the seismic forces on anchor bolts.
- When steel frames or concrete piers are used to support equipment, they must be properly anchored to the structure and rigidly cross-braced.
- Install double-arched flexible rubber connectors at the interfaces of equipment and piping where rubber is acceptable for the service. Use braided stainless steel hoses or stainless steel expansion joints in

only those applications where rubber expansion joints are not suitable.

- All suspended equipment, including piping and ductwork, whether isolated or not, must be braced against sway and axial motion. Cable braces are recommended for isolated equipment and either cable or solid braces for nonisolated equipment. Suspension rods may require bracing to prevent them from buckling under compression stress.

- Wherever possible, use OSHPD or other government preapproved seismic devices with preapproved ratings. When such devices are not available, ratings based on tests are more reliable than ratings based on calculations. When testing is impractical, calculation should be made by a professional engineer with a minimum of 5 years' experience in seismic design.

18.26.1 Applicable Codes and Standards

- BOCA National Building Code, Building Officials and Code Administrators International, Inc.

- Standard Building Code (SBCCI), Southern Building Code Congress International, Inc.
- Uniform Building Code (UBC), International Conference of Building Officials
- State and local codes that have adopted one of the above codes by reference and have incorporated local amendments.
- National Fire Protection Association (NFPA).

18.26.2 Seismic Zones

The world has been classified into seismic zones that reflect the relative potential severity of earthquakes in these zones. They are based on past seismic activity and the estimated probability of future occurrences. The seismic zones for the United States are shown in Figure 18–28. Governing design codes vary with each jurisdiction. For design purposes, peak dynamic forces have been translated into equivalent static forces and are based on the applicable seismic zone. These static forces are prescribed by codes for use in calculations for design of seismic restraints. The methodology of analysis using these seismic forces is beyond the scope of this chapter. (See Reference 2.)

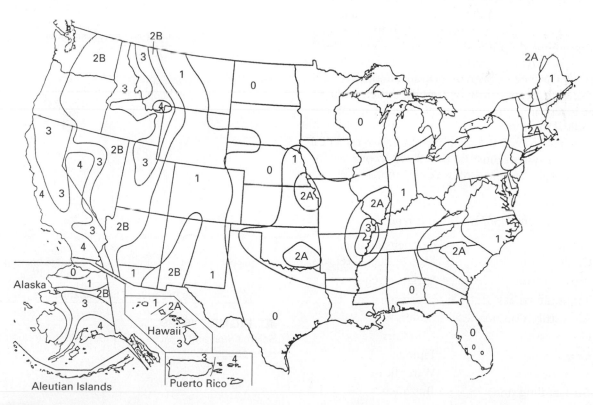

■ **FIGURE 18–28**
Seismic zones in the United States.

18.26.3 Commercially Available Seismic Restraints

Many types and sizes of seismic restraints are available. Figure 18–29 illustrates some typical restraint methods. Figure 18–29(a) is an anchor bolt going directly into a machine base that is otherwise vibration isolated. An installation using this type of snubber is shown in Figure 18–27. A bridge-quality neoprene bushing allows some motion but is attached to the floor by the restraining angle. Figure 18–29(b) shows another all-direction seismic snubber. The through-bolt passes through a bridge-quality neoprene unit and the upper bracket. The neoprene is attached to the bottom bracket. Any path of force from the restrained unit has to pass through the neoprene insert. Figure 18–29(c) illustrates a method for resiliently restraining structural elements.

Figure 18–30 shows a large-capacity restrained spring assembly that is designed to allow freedom sufficient for vibration isolation but restrain larger seismic displacements.

Restraint of vertical piping is important. Figure 18–31 illustrates a method for this. Again, any path of force from the restrained unit has to pass through the neoprene insert. This type of restraint is also good for preventing structure-borne noise from transferring from waste pipes to floor structures.

It is very important that ducts, pipes, conduits, and other hanging objects be restrained from becoming wild pendulums during a seismic event. Figure 18–32 illustrates one type of restraint for lateral bracing of pipe. A similar type of restraint is available that replaces the cable with an angle iron or a similar structural shape so that it restrains not only lateral movement but also longitudinal motion.

Just about everything in a building's mechanical and electrical system must be tied down in one way or another to resist seismic forces. For example:

All mechanical and electrical systems

AC units	Condensing units
Chillers	Motor control centers
Fans (all types)	Variable frequency
Tanks (all types)	drives
Air distribution boxes	Boilers
Compressors	Conduits
Generators	Piping
Transformers	Water heaters
Air-handling units	Bus ducts
Computer room units	Cooling towers
Heat exchangers	Pumps (all types)

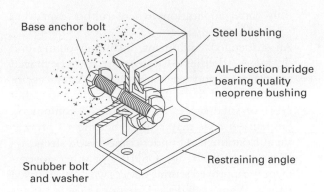

(a) All-direction seismic snubber

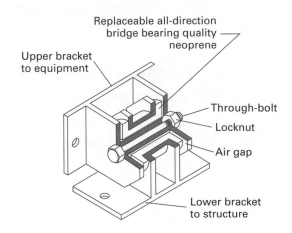

(b) All-direction seismic snubber

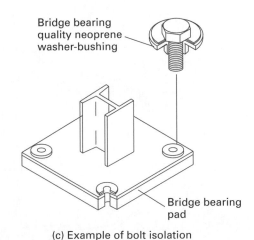

(c) Example of bolt isolation

■ **FIGURE 18–29**
Seismic restraint devices. (Courtesy: Mason Industries, Hauppauge, NY.)

■ FIGURE 18–30
Restrained spring assembly installed. (Courtesy: Mason Industries, Hauppauge, NY.)

Unit heaters	Cabinet heaters
Air separators	Ductwork
Condensers	Rooftop units
Luminaires	Cable trays
Unit substations	Electrical panels
Battery racks	Switching gear

Equipment buried underground is excluded, but entry of services through the foundation wall is included.

18.26.4 Life Safety Systems

Life safety systems are those systems necessary for the continued functioning of a facility for its intended use during and after a seismic event while ensuring continued safety for occupants. All life safety equipment should be so noted clearly on drawings and equipment schedules. Examples:

- All systems involved for fire protection, including sprinkler piping, fire pumps, jockey pumps, fire pump control panels, service water supply piping, water tanks, fire dampers, and smoke exhaust systems.
- All systems involved with or connected to the emergency power supply, including all generators, transfer switches, and transformers and all flow paths to fire protection and emergency lighting systems.
- All medical and life-support systems.
- Fresh air relief systems on emergency control sequence, including air handlers, conduits, ducts, dampers, etc.

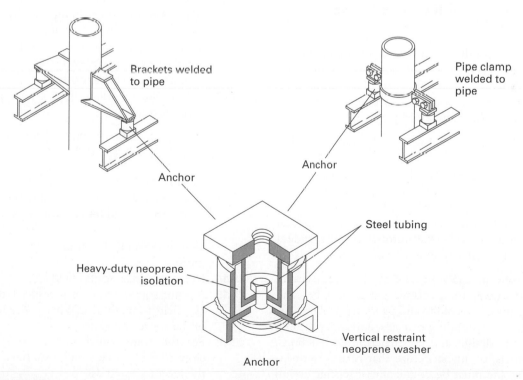

■ FIGURE 18–31
All-direction anchor. (Courtesy: Mason Industries, Hauppauge, NY.)

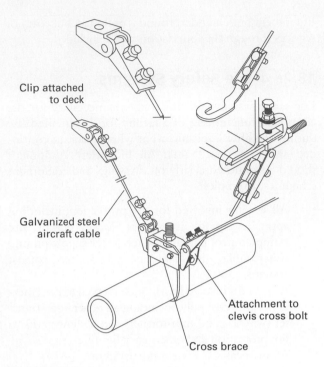

Clip attached
to deck

Galvanized steel
aircraft cable

Attachment to
clevis cross bolt

Cross brace

■ **FIGURE 18–32**
Piping cable restraint. (Courtesy: Mason Industries,
Hauppauge, NY.)

18.26.5 Submittal Requirements

Throughout the design process, manufacturers of equipment and products to be incorporated in the building are required to submit various data, information, and calculations. Manufacturers of vibration isolation and seismic restraints should be required to provide submittals for products, including:

- Descriptive data—catalog cuts, detailed schedules, etc.
- Shop drawings—fabrication details, suspension and support for ceiling-hung equipment, specific details of seismic restraints and anchors.
- Seismic certification and analysis—seismic restraint calculations, preapproval number or independent test data for restraining devices.

As we saw in 1999 in Turkey, Greece, Taiwan, Mexico, and other areas, seismic events can be cataclysmic in terms of loss of life and property damage. Architects, engineers, building officials, and everyone else involved in the design and construction of buildings, bridges, tunnels, or any structure that relates to or involves human life must be cognizant of seismic design criteria and see that they are implemented in new construction and in retrofit of existing buildings.

QUESTIONS

18.1 What are the five senses?

18.2 From now until the next class, listen carefully to the sounds around you. Keep a log of these sounds. Classify them as Pleasant, Neutral, Obtrusive, or Warning. What is the origin of the sounds you have identified? Aside from the sounds you have identified, what are some other sounds that fit each of the above five categories?

18.3 On a scale of 1 to 10, how would you rate your classroom acoustically for your learning experience?

18.4 What is the speed of sound at (a) 50 Hz, (b) 500 Hz, (c) 5000 Hz?

18.5 What is the wavelength of a (a) 50-Hz sound wave, (b) 100-Hz sound wave, (c) 1000-Hz sound wave?

18.6 What is the rationale for measuring sound with A-weighting?

18.7 In the *Devils Dictionary* (1911) Ambrose Bierce defined noise thus: "Noise, n, A stench in the ear. Undomesticated music. The chief product and authenticating sign of civilization." Do you agree? Why?

18.8 How would you define noise as opposed to sound?

18.9 Noise has grown more intense over the passing years. How do you account for this?

18.10 Identify some spaces in buildings where a quiet environment is of primary importance.

18.11 What are two primary components of sound, and what are their characteristics?

18.12 A ⅝-in. mineral ceiling tile has absorption coefficients as follows:

125	250	500	1000	2000	4000	8000
0.31	0.29	0.51	0.7	0.71	0.71	0.58

What is the NRC for the material?

18.13 Four sound sources individually produce sounds of 55, 55, 58 and 64 dB at a measurement location. What will be the sound level at the measurement location if all sounds are active at the same time?

18.14 Define a diffuse sound field.

18.15 A point sound source is measured at 300 ft in a free field to be 60 dB. What is the sound power of the source in watts?

18.16 For the point source in Question 18.15, how close to the source would one have to move for the level to be 70 dB?

18.17 For the distance in Question 18.16, what is the sound at the closer location?

18.18 The decrement (D) by which sound level diminishes is given by the general equation

$$\Delta = A \log\left(\frac{d_2}{d_1}\right)$$

If

$$A = 20 \text{ and } \frac{d_2}{d_1} = \frac{1}{2}$$

then

$\Delta = -6$ (i.e., 6 db per doubling of distance (dB/dd))

If $\Delta = -3$ dB/dd, what is the coefficient A?

18.19 If in Question 18.18 the decrement $\Delta = -4.5$ dB/dd, what is the coefficient A?

18.20 If one was measuring sound emitted from a large source—say, a large air-handling unit—how would one be able to measure approximately where the near field gave way to the direct field (i.e., the distance from the unit where the sound field began to act as if it were coming from a point source)?

18.21 A centrifugal fan operating at 38,000 cfm, 1800 rpm, 8 blades, 2-in. water generates the total sound power levels in each octave band given in the following table. These data do not contain a correction for blade passage frequency.

Center frequency	63	125	250	500	1k	2k	4k	8k
Fan PWL	92	92	91	86	81	75	71	69

What is the blade passage frequency?

18.22 The fan in Question 18.22 sits in a mechanical equipment room. The supply is ducted from the room. The HVAC system is an open return. The inlet to the fan is open to the room. The room absorption is given in the following table. What is the nominal reverberant SPL in the room? Assume that half of the total sound power is emitted to the room and that no direct sound reaches the opening.

Center frequency	63	125	250	500	1k	2k	4k	8k
Absorption, Sabins	98	168	100	160	250	580	1400	1750

18.23 For Question 18.23, assume that the reverberant sound in the MER is as given in the following table and the return opening in the wall is 2 ft wide × 2-ft high. What is the sound power level emitted through the opening?

Center frequency	63	125	250	500	1k	2k	4k	8k
Reverberant SPL	85	83	87	77	70	71	53	50
Incident SPL	79	77	81	71	64	54	47	44

18.24 For good vibration isolation, what should be the minimum ratio of the disturbing force frequency ω_f to the resonant (natural) frequency ω_n?

18.25 What is the isolation frequency for a system with a single degree of freedom whose frequency ratio is:

$$\frac{\omega_f}{\omega_n} = 2$$

18.26 Hawaii is famous for its volcanoes and earthquakes. Are there any areas in the United States that have a worse potential for earthquakes?

REFERENCES

Heating, Ventilating, and Air-Conditioning Systems and Applications, chap. 52, "Sound and Vibration Control," American Society of Heating, Refrigeration, and Air Conditioning Engineers, 1987. (Updated every third year.)

Seismic Restraint Guidelines, 2nd ed., Mason Industries, Inc., Hauppauge, NY, 1999.

Apfel, R. E., *Deaf Architects and Blind Acousticians? A Guide to the Principles of Sound Design*, Apple Enterprises Press, New Haven, CT, 1998.

Beranek, L. L., ed., *Noise and Vibration Control*, McGraw Hill Book Company, New York, 1971. Revised edition published by Institute of Noise Control, 1988.

Beranek, L. L., "Applications of NCB and RC Noise Criterion Curves for Specification and Evaluation of Noise in Buildings," *Noise Control Engineering Journal*, 45 (5): Sept.–Oct. 1997.

Cavanaugh, W. J., and J. A. Wilkes, *Architectural Acoustics*, John Wiley and Sons, New York, 1999.

Egan, M. D., *Concepts in Architectural Acoustics*, McGraw-Hill Book Company, New York, 1972; 2nd ed., 1988.

Ghering, W. L., *Reference Data for Noise Control*, Ann Arbor Science Publishers, Inc., Ann Arbor, Michigan, 1978.

Harris, C. M., and C. E. Crede, eds., *Shock and Vibration Handbook*, McGraw-Hill Book Company, New York, 1976.

Harris, C. M., ed. *Handbook of Noise Control*, McGraw-Hill Book Company, New York, 1979.

Lubman, D., and E. A. Wetherill, eds., *Acoustics of Worship Spaces*, Acoustical Society of America, Melville, NY, 1985.

Rettinger, M., *Studio Acoustics*, Chemical Publishing Co., New York, NY, 1981.

Sabine, W. C., *Collected Papers on Acoustics*, Harvard University Press, Cambridge, Mass., 1922. Reprinted by Dover Publications, New York, NY, 1964.

Thomson, W. T., *Theory of Vibration with Applications*, 2nd ed., Englewood Cliffs, NJ, Prentice-Hall, 1981.

Weissenburger, J. T., "The Significance of Laboratory versus Field Sound Transmission Loss," *Sound and Vibration*, October 1994.

Wiegel, R. L., *Earthquake Engineering*, Englewood Cliffs, NJ, Prentice-Hall, 1970.

American Society of Testing Materials

ASTM E90, Standard Practice for Laboratory Measurement of Airborne Sound Transmission Loss of Building Partitions.

ASTM E-413, Standard Classification for Determination of Sound Transmission Class.

ASTM E226, Standard Test Method for Measurement of Airborne Sound Insulation in Buildings.

ASTM E1414, Standard Test Method for Airborne Sound Attenuation between Rooms Sharing a Common Ceiling Plenum.

ASTM C423, Sound Absorption of Acoustical Materials in Reverberation Rooms.

American National Standards Institute

ANSI S1.1, Acoustical Terminology (Including Mechanical Shock and Vibration).

ANSI S1.2, Method for the Physical Measurement of Sound.

ANSI S1.4, Specifications for Sound Level Meters.

ANSI S1.8, Preferred Reference Quantities for Acoustical Levels.

ANSI S1.11, Specifications for Octave, Half-Octave, and Third-Octave Band Filter Sets.

ANSI S1.13, Methods for Measurement of Sound Pressure Levels.

U.S. Department of Health, Education, and Welfare

Jensen, P., C. R. Jokel, and L. N. Miller, *Industrial Noise Control Manual*, rev. ed., DHEW (NIOSH) Publication No. 79–117, U.S. Department of Health, Education, and Welfare, 1978.

Salmon, V., J. S. Mills, and A. C. Petersen, *Industrial Noise Control Manual*, HEW Publication No. (NIOSH) 75–183, Cincinnati, OH.

Published Proceedings

Proceedings of Noise-Con, various years, published by Noise Control Foundation, New York.

Proceedings International Conference on Noise Control (Inter-Noise), various years, published by Noise Control Foundation, New York.

ARCHITECTURAL ACCOMMODATION AND COORDINATION OF MECHANICAL AND ELECTRICAL SYSTEMS

19

PRECEDING CHAPTERS COVERED THE TECHNICAL aspects of individual mechanical and electrical systems and equipment and acoustical engineering for buildings. This chapter concentrates on integrating all of these systems into buildings through proper planning, space allocation, and location of equipment and services. There are many alternatives for placement and accommodation of systems and equipment. The solutions discussed in this chapter should be considered as illustrations of principles rather than strict guidelines.

Topics are selected for their potential to become problematic if not addressed and resolved early in the design. The materials are derived from experience working with owners, design teams, and contractors on issues that affect space, space planning, form, appearance, cost, and utility of buildings.

19.1 SYSTEMS TO BE INTEGRATED

This chapter will identify the types of spaces needed for various systems serving typical building types, along with concepts for locating these spaces and guidelines for their size. At the outset the design team should identify and list the mechanical and electrical systems that will be needed. Tables 19–1 and 19–2 will serve as checklists for systems and major planning considerations.

19.2 SPACE ALLOWANCES FOR MECHANICAL AND ELECTRICAL SYSTEMS

Many design professionals use the rule of thumb to allocate mechanical and electrical space in programming and schematic design. Commonly, 3 to 8 percent of the gross area is allocated based on judgment. Once the program or schematic design is set, the architect and engineer are bound to make the numbers work, sometimes without regard for design quality or value to the owner. These values may be greatly in error if the building requires higher levels of mechanical service than average. For instance, data centers often require 2 to 3 times as much mechanical and electrical space as is used for computers and operations staff. Laboratory buildings and healthcare facilities also require much more space than, say, classrooms and offices. On the other hand, a simple office building with pad-mounted transformers, recessed circuit panels, and rooftop units might require less than 1 percent mechanical and electrical space. Typical values are shown in Table 19–3.

Postponing realistic consideration of mechanical and electrical space can result in shoehorning these functions into inadequate space in inappropriate locations. The best way to plan mechanical and electrical space is to consider the spaces required as program elements the same as any other building functions such as the cafeteria or the conference room. Once the required spaces are identified, an initial estimate can be made of areas required, and the preferred location or cautions

TABLE 19–1
Checklist of mechanical systems

*Systems/functions marked with an asterisk (*) should be carefully evaluated during the preliminary design phase
so that the design process may be orderly and effective*

System/Function	Major Planning Considerations
▪ Energy source (*) for all M/E systems	Gas, oil, electrical power, coal, central (steam, hot water, chilled water), alternative energy, etc.
▪ Environmental issues (*)	Control zones (humidity, temperature), natural ventilation, daylighting, solar shading, integrated and socially responsive design (green building), etc.
▪ Heat rejection (*)	Cooling towers or condensers (water, air-cooled), and location, etc.
▪ Heating/cooling distribution (*)	Central or unitary systems, etc.
▪ Central plant (*)	Capacities and locations, etc.
▪ Ventilation and exhaust (*)	Outside air (minimum, 50%, or 100%); general, food preparation, toxic gas, and special exhaust, etc.
▪ Automation	BAS or BMS, etc.
▪ Water source (potable)	Pressure, capacity, size, location, etc.
▪ Water source (gray) (*)	Pressure, capacity, size, usage (sanitary, irrigation), etc.
▪ Hot water	Generators or heat exchangers, etc.
▪ Sewage disposal (*)	Storm, sanitary, sewage treatment (on site or in building), public sewers, etc.
▪ Storm water	Roof, ground, discharge locations, sewer, etc.
▪ Subsoil drainage	Drainpipes, sumps, pumps, discharge, location, etc.
▪ Sanitary facilities	Plumbing fixtures, water, waste, soil, and venting, etc.
▪ F.P. water storage (*)	Lake, pond, tanks, location, and capacities
▪ Fire and smoke detection	Thermal and smoke detectors, etc.
▪ Fire and smoke containment (*)	Fire shutters, building compartmentation (by zone or by floor) Smoke exhaust, floor pressure controls, stair pressurization, etc.
▪ Fire annunciation	Alarm, public address, fire department, etc.
▪ Fire extinguishing (*)	Portable, automatic sprinklers and types
▪ Firefighting	Fire hose and standpipe, Siamese, etc.
▪ Fire pumps (*)	Single or cascading, energy (gas, diesel, or electric), etc.
▪ Lightning protection	Air terminals, grounding, etc.

for location can be identified. If mechanical and electrical spaces are considered as part of the program, installation cost will be lower, performance will be enhanced, and maintenance will be easier and more economical.

19.3 UTILITY SERVICE CONNECTIONS

Early in planning, the architect, civil engineer, and building services engineers must work together to determine appropriate locations for utility service entries, which connect to surrounding site infrastructure. Most utilities are connected, although some services can also enter above grade. Services include the following. Those services that can enter either above or below ground are noted.

> Water, sanitary
> Water, fire (if separate from domestic)
> Sanitary sewer
> Storm sewer

Power*
Natural gas
Oil from on-site tanks for heating or generators
District (or campus) steam
District (or campus) chilled water
Telecom*
Cable TV*

*Can enter above or below ground

In addition to physically connected services, many buildings include antenna reception provisions for data and communications. Each of these services needs special design consideration for code compliance and maintainability.

19.4 HVAC DECISIONS AND COORDINATION

HVAC systems generally account for 15 to 35 percent of overall building cost and can be the most space

TABLE 19–2
Checklist of electrical systems

Systems/functions marked with an (*) should be carefully evaluated during the preliminary design phase so that the design process may be orderly and effective

System/Function	Major Planning Considerations
▪ Normal power source (*)	Utility or on-site power (capacity, phase, and voltage) service entrance, substations, vaults, etc.
▪ Emergency power source (*)	Separate service, on-site generation, etc.
▪ Power distribution	Primary or secondary voltages, panels, and substation locations
▪ On-floor distribution (*)	Underfloor ducts, cellular floors, raised floors, ceiling conduit network, poke-through, etc.
▪ Emergency power distribution	Critical equipment load, emergency lighting, etc. Critical building loads, power source (batteries, UPS, etc.)
▪ Power for building equipment	Mechanical, food service, process, transportation (vertical, escalators), etc.
▪ Major lighting systems (*)	Light sources and method of mounting (surface, lay-in, pendent), etc.
▪ Lighting design and layout	Light sources, fixture selections, layout, and controls, etc.
▪ Emergency lighting	Exit, exitway, critical, and emergency, etc.
▪ Feature lighting	Architectural expression and building features, etc.
▪ Daylighting (*)	Fenestration, skylights, controls, etc.
▪ Exterior lighting	Site, landscape, building facade, security, etc.
▪ Telephone communications	Type, lines, stations, switchboard, features, facsimile, modem, etc.
▪ Data distribution (*)	Cables, wire closets, LAN, etc.
▪ Public address (PA)	Intercom, paging, and music systems, etc.
▪ Audiovideo (A/V)	Radio, TV, and signal distribution systems, etc.
▪ Satellite dishes (*)	Number, diameter, orientation, and location, etc.
▪ A/V transmission towers (*)	Radio, TV, microwave, etc.
▪ Time and signal	Clock and program systems, etc.
▪ Fire alarm	Interface with FP, HVAC, BAS, etc.
▪ Security systems (*)	CCTV monitoring, detecting, alarming, etc.
▪ Automatic controls	Interface with HVAC, elevators, fire protection, lighting, security, etc.
▪ Specialty systems (*)	Numerous specialty systems

TABLE 19–3
Range of M/E floor area required for buildings

Type of Occupancy	Percent of Gross Building Area		
	Low	Medium	High
Computer centers	10	20	30
Department stores	3	5	7
Hospitals	5	10	15
Hotels	4	7	10
Offices	2	4	6
Research laboratories	5	10	15
Residential, single-occupancy	1	2	3
Residential, high-rise	1	3	5
Retail, individual stores	1	2	3
Schools, elementary	2	3	4
Schools, secondary	2	4	6
Universities and colleges*	4	6	8

*Buildings other than those used for classrooms follow the space required for specialty buildings such as laboratories, computer centers, and residences.

consuming of all mechanical and electrical systems. As described in Chapters 3 through 7, HVAC systems consist of air handling and distribution, cooling production, heating production, and distribution piping. Generally, small buildings will have package equipment such as through wall units, rooftop units, or split systems combining furnaces with air conditioners as found in most residences. Larger buildings may simply have a large number of these types of systems or may have more complex systems consisting of air-handling units, chillers, cooling towers, boilers, and pumps. The decision among systems is made on a number of factors including cost (initial and ongoing), performance, and feasibility of installation. Once the decision is made, the architect or builder must work with the engineer or subcontractor to accommodate systems with adequate space in the right location.

19.5 SELECTING THE ENERGY SOURCE FOR HEATING

The heating energy source must be decided very early because the choice impacts space requirements and many building details. There are four options that might be considered:

Generally, electric resistance heat will result in a higher operating cost for energy, lower initial cost for equipment, and less space required. Options include baseboard, heaters in terminal boxes, heaters in air-handling units, or any combination of these systems. Depending on climate, electric heat may or may not have a significant effect on the size of the electric service. In mild climates, the electric used for winter heating may be equal to or less than the power needed for summer cooling, in which case, there will be no effect on service. In cold climates, heating will likely increase service size and cost, but effect on space for equipment would not be significant.

Electric hydronic heat (uses one or more electric boilers to produce hot water) is generally higher in utility cost, higher in initial cost, and uses valuable space for boilers and pumps. Electric hydronic heat would be considered only in special cases where electric is very cheap in comparison with other energy sources.

Gas heat is generally lower in operating cost for energy, but higher in initial cost for equipment, and may require more space than electric if boilers are used. Options include furnaces in air-handling units, which require no additional space, or single or multiple boilers, which require flue stack provisions and additional mechanical space.

Oil heating provisions are similar to gas, except that on-site fuel storage will be required as discussed later in this chapter. Oil is often used in combination with gas for dual fuel boilers to enhance system reliability, primarily used for hospitals.

District or campus steam or campus hot water may be higher in operating cost for energy if steam is purchased from an outside utility, but will require less mechanical space than on-site boilers, reduce equipment cost at the building, and be lower in maintenance.

Example 19.1   The following energy costs are applicable for an upcoming 250,000-sq-ft project:

Electric $0.08 per kWh
Gas $1.20 per therm
Oil $2.15 per gallon
District steam $18.00 per million Btu

If the building is projected to use 30,000 Btu per sq ft per year, what will be the annual cost of energy for each source?

Answer At 30,000 Btu per sq ft per year, a 250,000-sq-ft building will consume . . .

At 30,000 Btu per sq ft per year, a 250,000 sq ft building will consume 7,500 million Btu. Requirements for various energy sources will be

Electric
Energy   (7,500 million Btu)/(3,413 Btu/ kWh) = 2.2 million kWh
Cost   2.2 million kWh @ $0.08 = $176,000

Gas (assume 80% efficiency)
Energy   (7,500 million Btu)/(100,000 Btu/ therm $\times$ 0.8) = 94,000 therms
Cost   94,000 therms @ $1.20 per therm = $113,000

Oil (assume 80% efficiency)
Energy   (7,500 million Btu)/(140,000 Btu/ gallon $\times$ 0.8) = 67,000 gallons
Cost   67,000 therms @ $2.15 per therm = $144,000

District steam
Energy   7,500 million Btu
Cost   7,500 million Btu @ $18.00 per million Btu = $135,000

19.5.1 Gas Service Provisions

Natural gas service generally enters in the basement, but can enter in other locations. Gas is distributed from the utility main at a higher pressure than the utilization pressure in the building, so a pressure-reducing station will be part of the service. The service entrance will also include a volume flow meter. For commercial and institutional buildings these components are normally located outside the building, as shown in Figure 19–1. The gas service needs to have protection from damage and, in some locations, needs protection against tampering and vandalism.

19.5.2 Oil Storage for Heating or Generators

Fuel oil is the most commonly used on-site fuel source. Oil is stored in one or more tanks either above ground or below ground. In either case federal, state, and local regulations should be followed regarding fire safety, and leak and spill containment. The volume of oil stored on-site will depend on the capacity of the equipment using the oil and on the length of time that the equipment is expected to be on standby fuel.

(a)

(b)

(c)

■ **FIGURE 19–1**
(a) Gas service provisions are generally outdoors for safety and (b and c) may require bollards or other protection from damage and tampering. (Courtesy of WTA.)

■ **FIGURE 19–2**
Aboveground fuel tank (belowground
tank shown in Chapter 5). (Courtesy
of Highland Tank & Manufacturing
Company, Inc.)

Once the required minimum storage volume is estimated, vendors can be consulted regarding the number, size, shape, and site requirements for standard tank configurations. Aboveground tanks are generally less expensive and easier to access for inspection and maintenance. Their appearance, however, must be considered when locating them; see Figure 19–2. Their location must also consider ease of truck access and proximity to the equipment using the oil. The fuel supplier should be contacted regarding the type of truck used for delivery so that access can be planned accordingly. Remote fill locations are often piped to the actual tank site when truck access is problematic.

Tank sizing includes allowances for oil tank volume over and above usable tank capacity. With this included, the following guideline can be used to estimate tank size at the beginning of the project:

Generators: 250 gallons per 100 kW per day
Boilers: 250 gallons per 1000 mBh per day

The mechanical and the electrical engineers can easily estimate the size of oil-using equipment through knowledge of climate and basic building program. Owner-furnished criteria or building code, especially in the case of healthcare facilities, will define the minimum time of standby fuel utilization.

Example 19.2   A data center is programmed as follows:

Computer room 10,000 sq ft
Average power density 150 watts/sq ft
Support power estimate 40% of computer load

The owner wishes to power the facility with standby generators for at least 48 hours. What will be the minimum storage volume of diesel fuel for the site?

Power requirement = (10,000 sq ft × 150 watts/sq ft × 40% / 1000 kW/watt) = 2100 kW Fuel storage = 250 gallons per 100 kW per day × 2100/100 kW ×

48/24 days = 21,000 gallons. Use this data to work with supplier on tank selection and site provisions.

19.5.3 Heating Systems and Equipment Coordination

HVAC Boilers

Boilers should not be placed in the same room as chillers due to possible refrigerant leaks, which will cause severe corrosion in operating boilers. Boiler rooms are similar in layout to chiller rooms, with ample clearance for maintenance, equipment removal and replacement, and auxiliary equipment such as pumps, expansion tanks, and water treatment. Boiler rooms need enough height for breeching and flue stacks over the equipment. Flues can be a problem in tall buildings if boilers are located in the basement and flues must be taken all the way to the roof. Expense of the flue stack and space lost in the core can be avoided if the boiler room is placed on the top floor, which is a common solution in high-rise construction.

If height is not available for large boilers, a number of smaller, modular boilers might be considered as shown in Figure 19–3.

District (or Campus) Steam

District steam may be available in dense urban areas and campuses. The advantage is centralization of maintenance at the steam plant and eliminating the need architecturally to accommodate boilers in the building. The cost of district steam may appear high in comparison with the energy cost of producing heat on-site, but there are cost avoidances in equipment, space, and maintenance. The comparison shown in Figure 19–4 shows that considerable space can be saved by using district steam. Initial cost reduction will include space savings and avoidance of boilers.

(a)

(b)

(c)

■ FIGURE 19.3

(a) Multiple small, modular boilers can be installed in equipment space with modest ceiling height. (b) Larger boilers will require taller spaces. (c) For large boiler installations, a separate building may be required. (Courtesy of WTA.)

Example 19.3   An office building is being planned for an urban area with district steam available. From the following facts, compare the initial cost and annual expenses for on-site gas-fired boilers and district steam.

Project Data

Estimated cost for boilers, flue, and fuel piping	$225,000
Estimated cost for heat exchangers and steam piping	$50,000
Space requirements	per Figure 19–4
Cost of space	$100 per sq. ft.
Annual cost of boiler maintenance	$3,000 per year
Annual cost of heat exchanger maintenance	$500 per year
Cost of gas for boilers, annual	$25,000
Cost of steam, annual	$50,000

Initial Cost Comparison

Boilers		District Steam	
Equipment	$225,000	Equipment	$50,000
Space (1,000 sq ft)	$100,000	Space (225 sq. ft.)	$22,500
Total	**$325,000**	**Total**	**$72,500**

Operating Cost Comparison

Boilers		District Steam	
Maintenance	$3,000	Maintenance	$500
Fuel	$25,000	Steam	$50,000
Total	**$28,000**	**Total**	**$50,500**

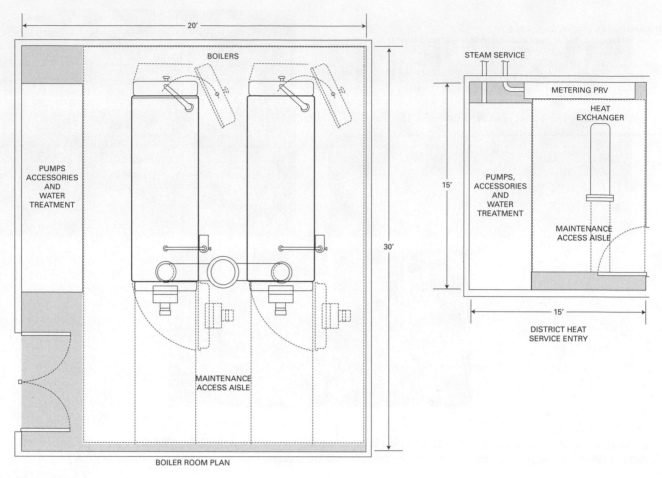

■ FIGURE 19–4
Space comparison, boilers versus district steam. (Courtesy of WTA.)

Operating cost savings by using boilers = $50,500 − $28,000 = $22,500

Initial cost penalty by using boilers = $325,000 − $72,500 = $252,500

In this instance, installing boilers will save operating cost and the initial investment will have approximately 10-year simple payback.

District steam is generally direct-buried insulated steel piping in a corrosion-resistant conduit, which may also contain condensate return piping. The conduit will enter the building at basement level or in a pit located in a mechanical room at grade. As with all pipe entries below grade, care must be taken to prevent leakage of groundwater into the building. A shutoff valve and steam metering equipment will be located near the service entry. Since the steam is provided at a higher pressure than needed by the heating system, the mechanical room at the service entry will need a pressure-reducing station.

19.6 COOLING EQUIPMENT AND SYSTEMS COORDINATION

19.6.1 Heat Rejection Equipment

Air-conditioning heat rejection equipment by the nature of its function needs to be located outdoors. There are several types: cooling towers, dry coolers and condensers, condensing units, air-cooled chillers, and the condensing sections of direct expansion (DX) rooftop units. Each has different characteristics that need to be considered. Their function and operation is discussed in detail in Chapter 4.

Condensers and Condensing Units

Condensers and condensing units are noisy, and most communities have ordinances that limit sound levels at

TABLE 19–4
Typical Values, Allowable Sound Levels at Property Line

Adjacent land use	Time	dBA
Residential	day	55
Residential	night	60
Commercial	continuous, 60 min.	65
Light industrial	continuous, 60 min.	70

property lines (see Table 19–4). Sound screens or noise abatement accessories might be necessary if units are placed too close to the line or other noise-sensitive locations. Typical condensing unit installations are shown in Figure 19–5.

DX rooftop units, condensing units, and air-cooled chillers contain compressors as well as fans and can produce more noise. They are also more prone to vibration transmission and they are heavier. These equipment items should not be placed over noise-sensitive areas such as executive offices or conference rooms. If located at grade, the site designer should take care to have someone on the design team estimate the noise level at the property line. Generally, local ordinances limit noise as described in Table 19–4 and, in any case, neighbors should be spared from excessive noise.

Cooling Towers

Cooling towers are used to reject heat from water chillers. Systems using cooling towers are generally more energy efficient than systems using air-cooled condensers, but maintenance is higher.

Cooling towers are also quieter than air-cooled condensers. Fan noise is generally not a problem if the tower is located on the roof, since most of the noise is directed upward. Towers located at grade can present noise problems to adjacent windows or occupied areas of the site. As discussed in Chapter 4, water drift can also be a problem with towers at grade.

Most cooling towers take air in from two sides and discharge it through fan(s) vertically at the top. Ideally, cooling towers are located unobstructed on the roof or at grade. Towers are often screened from view, and the screen needs to allow for adequate free airflow into the side of the tower. A rule of thumb is that the screen should be located at a distance at least the height of the tower. Towers can also be located simply on the roof, behind screens, or in roof wells as shown in Figure 19–6. Towers can also be located at grade adjacent to the chiller room (Figure 19–7).

19.6.2 HVAC Chillers

Water chillers are used to produce cooling on large projects. Chiller rooms must be designed with ample clearance for maintenance, equipment removal and replacement, and auxiliary equipment such as pumps, expansion tanks, and water treatment. Figure 19–8 shows a large chiller room with provisions for maintenance and auxiliary equipment. In order to route piping over equipment, chiller rooms generally need space with higher clearances than the rest of the building.

Chillers are heavy and prone to noise and vibration. Accordingly, a basement or slab on grade room might be most economical. Cooling towers should be located at a higher elevation than the chillers they serve, and as close as practical since condensing water piping is generally the largest, heaviest, and most expensive piping in the building. For this reason, chillers for a high-rise building will generally be located in the basement with the tower at grade or on the podium level if practical. If towers need to be located at the roof, chillers should be located in an upper level or penthouse equipment room.

Chiller location may also affect the schedule of building completion. If the main mechanical room is located at the upper level, this work cannot start until the building is topped out. Locating major mechanical at lower levels will allow early start of this time-consuming work.

19.6.3 District (or Campus) Chilled Water

Where available, district or campus chilled water is an alternative to on-site chillers. The advantages are low cost, reduced maintenance, and less space required in individual buildings. Similar to district steam, chilled water enters at the basement wall or in a pit if the mechanical room is at grade. Typically, the service will have main shutoff valves and metering equipment. Flow or flow and temperature difference (between supply and return) is metered. Using flow and temperature allows the calculation of actual cooling drawn at the building, but flow alone is sometimes used. Building chilled-water pumps are located at the service entrance. A typical room layout is shown in Figure 19–9 for a large campus building along with a chilled-water plant providing similar capacity. Note the difference in area requirement.

19.7 HVAC AIR HANDLING AND DELIVERY

Chapter 3 described the basic options for HVAC air handling and delivery, which include PTAC, fan coils, central air handling from fan rooms, and rooftop units. More than one of these options might be used in a single building.

(a)

(b)

(c)

■ **FIGURE 19–5**
Condensers and condensing units
can be roof mounted (a and b) with
attention to vibration and
appearance if desired, or (c) at
grade if properly protected.

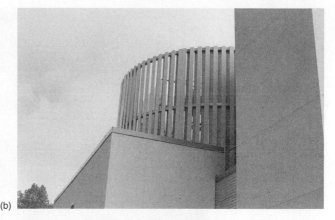

■ FIGURE 19–6

(a) Cooling towers can be exposed on roof or (b) concealed with screens or (c) in roof wells. (Courtesy of WTA.)

19.7.1 Small Units in Occupied Space

PTACs (package terminal air conditioners) are mounted in windows or through walls. Window units are generally used only in small utility buildings or where budget is critical. Through-wall units, however, are commonly used in hotels and apartment buildings. Unit ventilators and sometimes fan coils may also require an opening in the wall for ventilation. Architectural considerations include the details of wall openings. Wall openings can be problematic in historic buildings, and cause a freeze hazard in cold climates. Having air-conditioning units in the occupied space can also be noisy, which is especially objectionable for sleeping quarters and classrooms. Wall units do not require mechanical space, but they might create breeze and noise close to the units, reducing the amount of usable space. Fan coils are often located above ceilings to mitigate or avoid these problems.

19.7.2 Central Air-Handling Systems

Other systems use central air-handling equipment. Where roof area is available, the architectural planning of HVAC systems starts with deciding whether air handling will be from rooftop units or by air-handling units located in fan rooms.

(a)

(b)

■ **FIGURE 19–7**
Towers can also be located at grade (a), but may require elevation (b) if chiller plant is on level 1. (Courtesy of WTA.)

Commercial office buildings up to approximately five stories in height generally have adequate roof space to accommodate the required amount of equipment, and initial cost generally drives a decision to use rooftop units.

There are several disadvantages to rooftop equipment. For many buildings, there is not adequate roof space for rooftop units. Over five stories, roof space runs out, and air-handling equipment is best planned indoors on the floors served. This is due not only to roof space

■ FIGURE 19–8
Chiller room. Note the large aisle for maintenance and provisions for pumps and other auxiliaries.

limitations, but also to the extensive area required for air shafts if all the air serving the entire building starts at the roof.

Maintenance is less convenient on the roof. Parts such as motors and compressors need to be hoisted from the ground and transported across light roof structure. Personnel, tool, and materials access is often up a fixed ladder through a roof hatch. Equipment is exposed in inclement weather making repairs and maintenance difficult. These problems can be mitigated by good design to make rooftop units a more acceptable solution for a wider range of buildings.

Rooftop units are less expensive to purchase and install than indoor air-handling units in fan rooms. Indoor units do, however, have some advantages that can justify the higher cost. First, rooftop units, being located outdoors, are subject to weather and corrosion. The expected life of rooftop equipment is approximately 15 years in most climates. Obviously, in hot, dry climates rooftop units will last longer; and in moist temperate climates, life will be shorter. Indoor units will last indefinitely if properly maintained. Another advantage is maintenance. Outdoor units require outdoor maintenance. If continuous, reliable performance is critical, indoor units will be preferred.

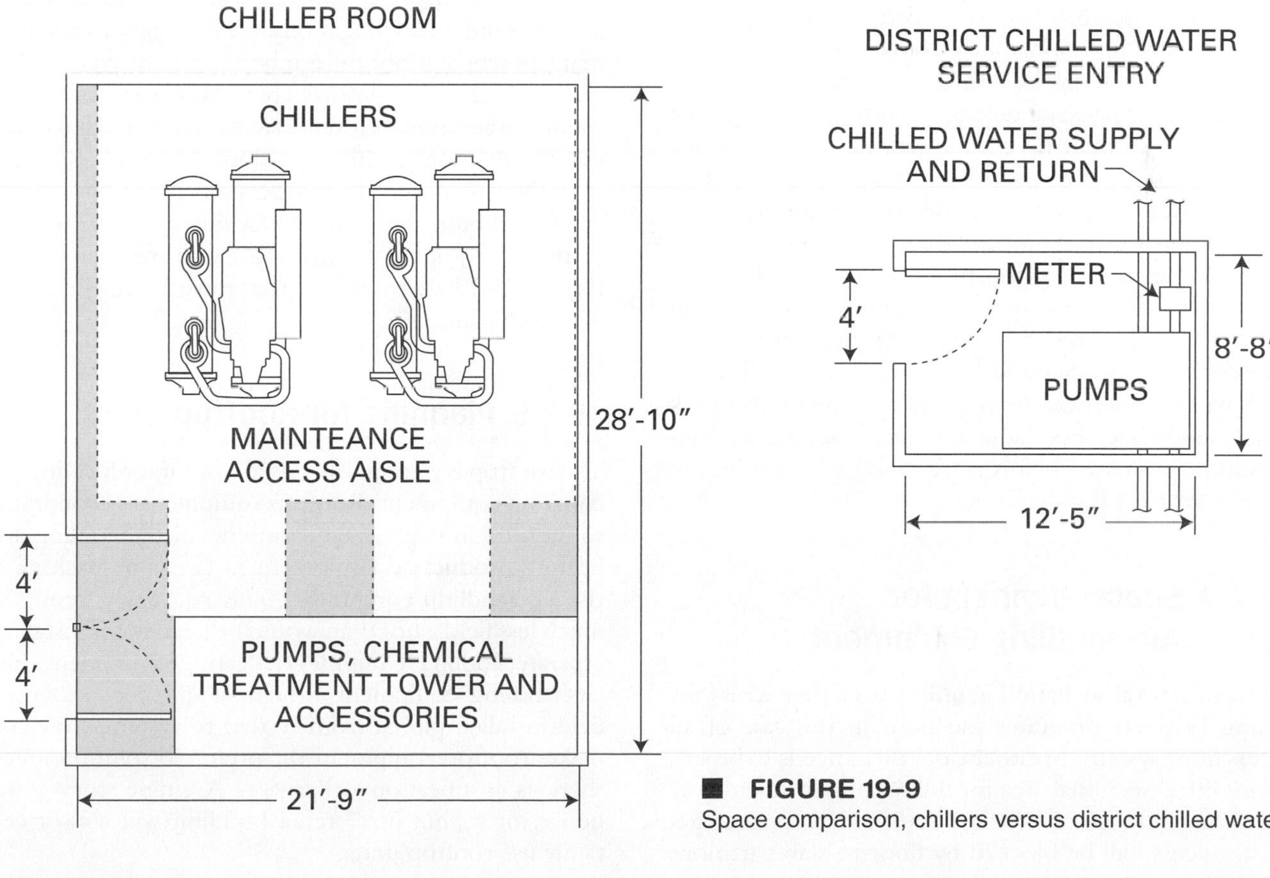

CHILLER ROOM

CHILLERS

MAINTEANCE
ACCESS AISLE

28'-10"

4'

4'

PUMPS, CHEMICAL
TREATMENT TOWER AND
ACCESSORIES

21'-9"

**DISTRICT CHILLED WATER
SERVICE ENTRY**

CHILLED WATER SUPPLY
AND RETURN

METER

4'

8'-8"

PUMPS

12'-5"

■ FIGURE 19–9
Space comparison, chillers versus district chilled water.

■ **FIGURE 19–10**
Unit ventilators as shown in Figure 19–10 are an example of small HVAC units in the occupied space. Care must be taken regarding noise and effects on furniture placement.

19.7.3 Unit Size and Area Served by Individual Units

Planning of central air-handling systems must consider how the total building will be divided for service by individual units. Many options are possible; see Figure 19–12. For instance, a high-rise office building might have a small, modular package unit on each floor, or might be served by large, built-up systems serving multiple floors. Similarly, rooftop units serving a medium-rise building might be designed to serve single floors, multiple floors, stacked portions of multiple floors, or other options. Using larger units serving more area will reduce the number of units and may reduce the cost of installation unless the units become so large that they must be custom construction. Serving larger areas from individual units will also increase the size and expense of ductwork and chase space. For large buildings, several options should be evaluated before proceeding with a solution; see Figure 19–11.

19.7.4 Space Planning for Air-Handling Equipment

Air from central air-handling units runs either across the ceiling in ducts or under the floor in the case of an access floor system. In either case, there needs to be sufficient cross-sectional area for the supply air to get away from the fan room or air chase and the return air to get back. Access will be blocked by floor-to-slab partitions as for a stairwell, an electrical closet, elevator, or other common core elements. As a rule of thumb for early planning, approximately 1 sq ft of net opening is required for every 1000 sq ft of area served by the system.

Fan rooms are generally noisy. The basics of acoustics and noise mitigation are included in Chapter 18. Many potential problems can be avoided by proper planning to keep noise-sensitive areas away from mechanical rooms. In addition, vestibules are recommended to isolate noise from noise-sensitive corridors.

Double doors should be considered for ease of entry with bulky filters and to facilitate possible replacement of equipment without extensive demolition. Figure 19–13 shows factors that must be considered in fan room planning.

19.7.5 Planning for Rooftop Units

The rooftop is a commonly used option for locating air-handling equipment. Rooftop equipment is economical to install. Units generally combine all the cooling and heating production equipment in the same package as the air-handling equipment. This packaging results in much less field labor than would be needed for installing separate cooling equipment, heating equipment, and air-handling equipment, which would need extensive field-installed piping from system to system. Low cost makes rooftop equipment the preferred solution unless there is justification otherwise. A single-story warehouse for a "big box" retail building will almost certainly use rooftop units.

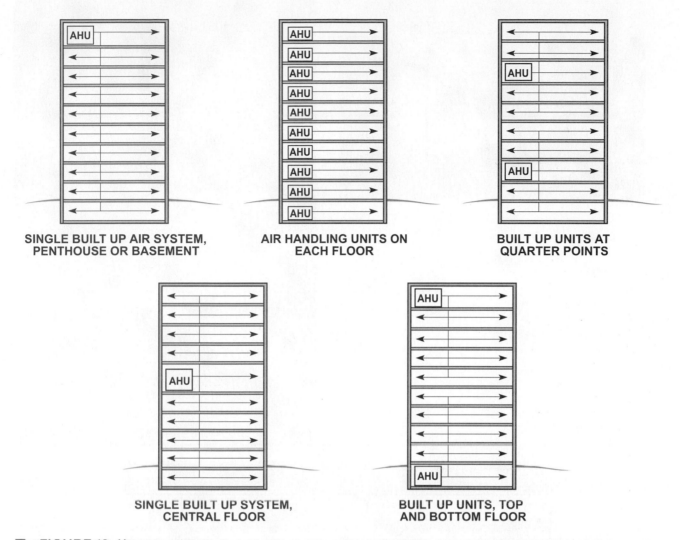

■ FIGURE 19–11
Options for fan rooms in a high-rise building; similar logic will apply to all large buildings regardless of height.

Screening can be used to conceal rooftop units from view. Designs can vary widely as shown in Figure 19–14. Screens will require support to the roof and must withstand wind loads. Some units are equipped with screens as an accessory that attaches to the unit. This is generally less expensive than providing a separately supported screen, but appearance may not be consistent with architectural intent.

As an alternative to exposed, screened equipment, a penthouse level can be constructed to keep equipment and maintenance staff out of the weather. Several options are shown in Figure 19–15. If used solely for mechanical equipment, lightweight construction with unfinished interiors can also be used for economy. As a variant, the penthouse enclosure can be prefabricated with skid-mounted equipment installed, piped, and wired in the factory prior to shipment. This type of construction is quick to install and greatly reduces the cost of field labor.

19.8 PLUMBING COORDINATION

19.8.1 Water Service Entrance

Water from the local provider may be provided at a pressure higher than is needed or wanted in a low-rise building. For this reason the service entry may contain a pressure-reducing station as well as a meter and backflow preventer. Space requirements are generally less than 100 sq ft, located in the basement as shown in Figure 19–16, or grade level if there is no basement. The space should be accessible to maintenance personnel on an outside wall at which the water service most conveniently enters. Ideally, the entry can be in a mechanical room used for other purposes. If this is not possible, a separate space of ample dimension will be needed.

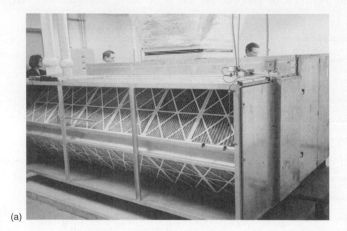

(a)

(c)

(b)

■ **FIGURE 19–12**

Many central air-handling unit configurations are possible. (a) Compact unit for high-rise building. (b) Modular package unit. (c) Large, built-up air-handling unit. (Courtesy of WTA.)

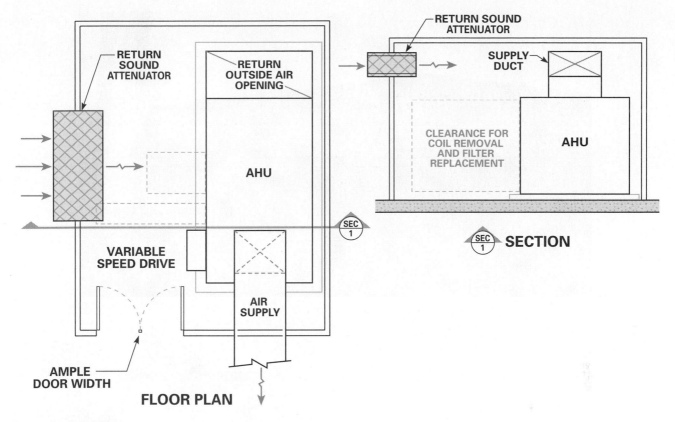

■ FIGURE 19–13
Planning for fan rooms must consider maintenance and clearance for ductwork.

Often, water heaters, possibly the hot-water storage tank, and fire protection service, including separate service, backflow preventer, and fire pump, if needed, are best placed in the same space. Buildings taller than about five stories will generally need booster pumps, and high-rise buildings will need extra provisions and space for pressure boosting and control.

19.8.2 Sanitary Sewer

Municipal sewers are generally located in the street at an elevation low enough to allow gravity drainage from the lowest point in the building, usually the basement floor. If the lowest point of the building is too high above the municipal sewer to allow gravity drainage, a lift pump station will be required as shown in Figure 19–17. The pump station will be designed to drain only the lower portion of the building. Higher levels will be drained by gravity.

There are no particular space requirements for sanitary sewers except for cleanouts on stacks, which need clearance for occasional access to clear blockages.

19.8.3 Storm Sewer

Storm sewers are located on-site, but must also collect roof drainage. Roof drains can consist of scuppers and downspouts, which need to be accommodated architecturally, but keep the system outside the building envelope. For large roofs or tall buildings or where exterior drains are architecturally unacceptable, drains will pass down through the roof into the buildings and be routed eventually outside to the storm sewer.

Plastic pipe is often used for storm drains. If the HVAC system uses the ceiling plenum for return air, plastic pipe will not be allowed due to noncompliant flammability and smoke-developed ratings. Water leak hazard is another serious coordination issue. Drains should not be located over high-value spaces such as computer rooms and museums. These spaces are not only high value, but also are generally humidified. Leaks can result indirectly from condensation if the piping is not insulated below the roof; snow melt, especially, will result in cold piping prone to condensation in humidified spaces. Many spaces

(a)

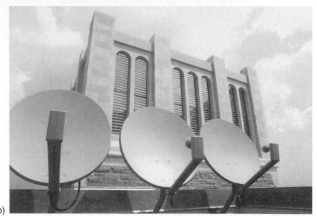

(b)

(a)

■ **FIGURE 19–14**

Rooftop units may set back far enough to be unobtrusive or may be screened by various designs to complement the building architecture. (a) A simple Mansard to hide roof clutter. (b) An elaborate architectural feature to screen equipment. (Courtesy of WTA.)

■ **FIGURE 19–15**

(a) Penthouse-level mechanical enclosures can be simple metal construction. (Courtesy of WTA.)

(b)

in healthcare facilities are also humidified in winter. Indoor storm drainage piping is generally combined under the roof to reduce the number of risers. Piping is routed most conveniently and economically to the building foundation wall and out to the site drainage system.

19.8.4 Water Heating

Options for water heating are similar to those discussed for HVAC heating production: fuel-fired units with flues, electric, or district steam heat exchangers. Some facilities, mainly hospitals, use large quantities of hot water, and systems are equipped with storage tank(s) for load leveling so that the heaters need not be sized for large instantaneous loads. Space for water heating can be as small as $6' \times 6'$ for a small heater to $10' \times 30'$ for a large hospital. These are not large requirements, but must be considered in overall planning.

19.8.5 Waste Piping

Plumbing chases are at toilet rooms, and the most economical arrangement is to place water closets and urinals back-to-back in adjacent rooms. Rooms should be stacked, as offsets in chases are expensive. Minimum wall thickness for plumbing is shown in Figure 19–18. Figure 19–19 shows a wider chase that allows maintenance access to back-of-house piping. This type of design might be

(c)

■ **FIGURE 19–15** *(Continued)*
(b) prefabricated with equipment inside or (c) an integral part of the building.

■ **FIGURE 19–16**

(a) Equipment at water service entry; note dual backflow preventer. (b) Booster pump system required when water pressure is not adequate. (Courtesy of WTA.)

(a)

(b)

considered for institutional owners committed to long-term occupancy.

In multistory and high-rise construction, toilets are generally located at the building's core along with elevators, exit stairs, electrical rooms, and HVAC provisions. When plumbing is needed at locations remote from the core, drain and vent piping must be routed under the floor below (drain) and the ceiling above (vent) to the core. This is sometimes impractical due to interferences with other services and inadequate height for required slopes. When routing to the core is impractical, a separate riser will be required, which can be very

FIGURE 19–17
The lift station for sanitary waste in the basement needs floor clearance for maintenance and head clearance to pull pumps. (Courtesy of WTA.)

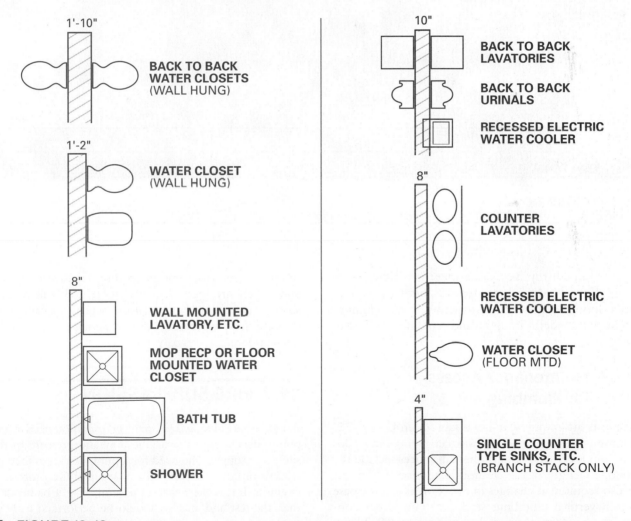

1'-10"
BACK TO BACK WATER CLOSETS (WALL HUNG)

1'-2"
WATER CLOSET (WALL HUNG)

8"
WALL MOUNTED LAVATORY, ETC.

MOP RECP OR FLOOR MOUNTED WATER CLOSET

BATH TUB

SHOWER

10"
BACK TO BACK LAVATORIES

BACK TO BACK URINALS

RECESSED ELECTRIC WATER COOLER

8"
COUNTER LAVATORIES

RECESSED ELECTRIC WATER COOLER

WATER CLOSET (FLOOR MTD)

4"
SINGLE COUNTER TYPE SINKS, ETC. (BRANCH STACK ONLY)

FIGURE 19–18
Minimum space requirements for plumbing chases.
All chase dimensions based on single 5/8" drywall and all back-to-back fixtures lined up. Structural elements in chase or wall will require additional depth.

■ **FIGURE 19–19**
Walk-in plumbing chase to facilitate inspection, repair, and replacement. (Courtesy of WTA.)

expensive and disruptive if done after the building is occupied. If the need for future remote plumbing is anticipated during design and construction, "wet columns" should be provided with drain and vent lines installed in advance.

19.8.6 Maintenance Access for Plumbing

Cleanouts are required in horizontal drain lines at 75′ minimum spacing for 4″ and smaller piping, and 100′ minimum for 5″ and larger. These distances include the developed length of the cleanout extension. Cleanouts are also required at changes in direction and at the base of each vertical plumbing stack and rain leader connected to a storm sewer, as shown in Figure 19–20.

Interceptors are used in plumbing systems to separate materials and prevent their entry into the public sewer system. Interceptors require access for service, and if they are located in the floor, they will require space in the ceiling below. Grease separators can be installed either at specific foodservice equipment or in the kitchen floor.

19.9 FIRE SUPPRESSION

A fire suppression system may or may not need a fire pump, depending on available flow and pressure of the site water supply. The need for a pump is determined by analysis of flow test results by a qualified fire protection engineer. If it is likely that a fire pump might be needed, then the test and analysis should be performed early so that the space and expense can be adequately planned. Figure 19–21 shows a typical fire pump installation to convey approximate size and nature of the required space.

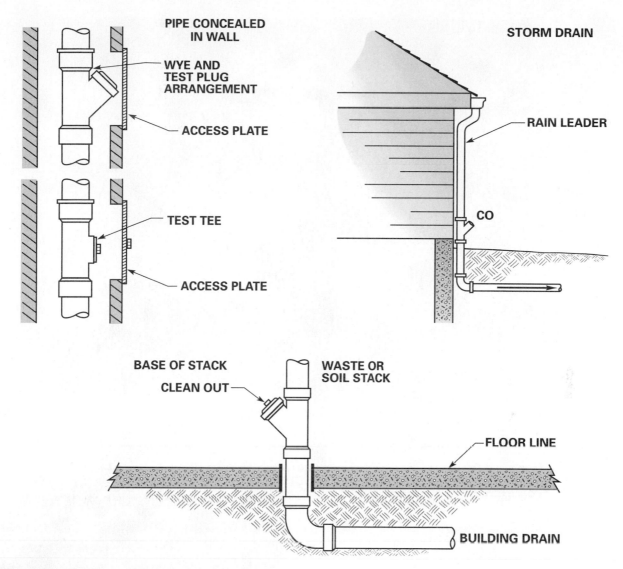

PIPE CONCEALED IN WALL

WYE AND TEST PLUG ARRANGEMENT

ACCESS PLATE

TEST TEE

ACCESS PLATE

STORM DRAIN

RAIN LEADER

CO

BASE OF STACK

CLEAN OUT

WASTE OR SOIL STACK

FLOOR LINE

BUILDING DRAIN

■ **FIGURE 19–20**
Plumbing cleanouts.

■ **FIGURE 19–21**
Typical fire pump installation. (Courtesy of WTA.)

19.10 ELECTRICAL DECISIONS AND COORDINATION

19.10.1 Power Service to Building

The prevailing trend is for underground power distribution, which is less prone to damage and less obtrusive visually than overhead lines. Typically power is distributed at higher voltage than that used in the building, and a transformer will need to be located on-site or in the building. For small buildings, mainly residential, a single transformer might serve multiple buildings. Figure 19–22 shows several outdoor transformers for various-size buildings. Utility feed to the transformer and the building feed from the transformer are undergound.

(a)

(b)

■ **FIGURE 19–22**

(a) Outdoor transformers may be pole mounted or (b) pad mounted, depending on available utilities. (Courtesy of WTA.)

Inside the building, a main panel and metering provisions will be located either in the basement or at grade level. Size and extent of the main panel will depend on building load. Large buildings will require a floor-mounted switchboard. The main panel must be readily accessible to shut off power in an emergency, should be centrally located for economy of distribution, and needs code clearance in the electrical space around the equipment. Very large buildings will be served at medium voltage (higher than 600 V). The higher voltage will require a substation to step down the voltage for distribution in the building. The substation and switchgear may be located inside or outside the building.

19.10.2 Electrical Service Voltage Implications on Planning

Small- and medium-sized buildings are served by underground feeders or overhead drops, and the service voltage is either 120/240 V single-phase or 120/208 V three-phase. In either case, there is no need for a transformer in the building since these are utilization voltages. The main distribution panel or switchboard will be located on an outside wall at the service entrance, either in the basement or on the first floor. There are strict code limits on the length of service feed inside the building before the main disconnect, which is located in the main distribution panel

(c)

■ **FIGURE 19–22** *(Continued)*
(c) Very large buildings or campuses may be served by medium-voltage transformers similar to utility substations.

or switchboard. The size of the electrical room or space is dictated by code clearances in front of the panel.

Large buildings are typically served at 480 volts. The 480-volt power is used directly for mechanical equipment and any other large loads. Transformers need to be accommodated in the building to produce 120/208-V power for appliances and lighting. There are two basic options for locating these transformers, and the decision is most often based on economics. The lower voltage 120/208-V power requires larger wire sizes. Long runs of large wire size will not be an unreasonable expense if the building is compact. For compact buildings, a single 120/208-V transformer can be used. But if the building is spread out horizontally or vertically, the use of multiple remote transformers served by 480-V feeders will be more economical as shown in Figure 19–24. The electrical engineer or contractor can generally decide very early in the design based on experience.

Very large buildings and campuses may distribute power at medium voltage (greater than 600 V). Medium-voltage switchgear may be indoor or outdoor and will require considerable space. An indoor installation is shown in Figure 19–23b. An outdoor, campus switchgear lineup is shown in Figure 19–25. Underground, medium-voltage feeders from this installation go to transformers inside or outside each building and are stepped down to low voltage for safe use at the building.

19.10.3 Electrical Panels and Receptacle Transformers

All buildings need electrical panels for branch circuits to receptacles and lighting. A panel bus is powered by a single feeder, and circuit breakers connected to the bus feed branch circuits (see details in Chapter 11). Generally,

(a)

(b)

■ **FIGURE 19–23**
(a) Electrical main service at a switchboard will contain the meter and breakers to feed various loads. (b) Higher voltage switchgear will contain a transformer. (Courtesy of WTA.)

branch circuit length is restricted to approximately 100 ft to avoid excessive voltage drop. This means that electrical panels need to be located within this or a shorter distance from any part of the building served by the panel. In small buildings, especially residential, panels can be located in the basement and serve multiple floors. In larger buildings, mostly commercial and institutional, panels should be located on the floor served by the panel.

Panels can be surface- or flush-mounted on walls (see Figure 19–26). For surface-mounted panels wiring into and out of the panel will be visible. If the panel is in a finished area, flush mounting is preferred for appearance. Panels can be located in occupied space if appearance is not objectionable and adequate, convenient clearance is available for occasional inspection, resetting tripped breakers, and modifying branch circuits.

FIGURE 19–24
Step-down transformer in electrical closet for 120/208-V receptacle power. (Courtesy of WTA.)

Required minimum clearance in front of electrical equipment depends on voltage of the panel and what is present across the maintenance space in front of the equipment. There are three relevant conditions. Condition 1 is defined as having live parts on one side and no live or grounded parts on the other side of the workspace. The code also contains qualifications and exceptions that might be applicable in special instances. Condition 2 is exposed live parts on one side and grounded parts on the other side. Building structure or metallic piping would constitute grounded parts as well as electrical components. Condition 3 is exposed live parts on both sides. Based on these conditions, minimum clearances by code will be as shown in Table 19–5.

Clearances for higher-voltage equipment are greater. Higher voltages are used at indoor substations and for distribution in very large buildings from the primary utility service entrance to remotely located substations. For large electrical equipment rooms or rooms that house high-voltage equipment, there are also special code requirements for entrances and considerations for installation and eventual replacement of components.

TABLE 19–5
Minimum working space for electrical equipment commonly used inside buildings

	Min. Clearance by Condition		
Voltage	Cond. 1	Cond. 2	Cond. 3
0–150	3 ft	3 ft	3 ft
151–600	3 ft	3 ½ ft	4 ft

Locating panels in electrical alcoves allows the use of doors to conceal them. This arrangement also allows the economy of exposed cables and conduit, but requires more space and the expense of doors. Still, panel alcoves require less space than electrical closets. Options are shown in Figure 19–27.

Electrical closets are preferred for ease of maintenance and flexibility of working on the system without disruption of activities in occupied spaces. Electrical closets also provide secure space for other electrical components such as transformers, wiring risers, and data/com system devices for building controls, fire alarm, etc. A typical

■ **FIGURE 19–25**
Outdoor medium-voltage switchgear.
(Courtesy of WTA.)

electrical closet is shown in Figure 19–28 with guidelines on clearances and room dimensions. If the building floor is too large to be served from a single closet, a satellite closet might be used, or a surface-mounted panel located at an appropriate remote location.

Higher voltage, generally 480 V is often used for distribution in large buildings to reduce the size and expense of conductors. In these systems transformers will be needed at intervals in the building to produce 120/240-V or 120/208-V power for receptacles and lighting. A single transformer can easily serve the panels in a number of electrical closets; therefore, not all closets will be equipped with transformers. In a high-rise building, for instance, a transformer might be located at every third or fourth floor in stacked electrical closets. Similar spacing concepts will apply to large horizontal buildings.

either shared space in electrical closets, or were located in separate, secure closets for administrative reasons. A central electronic switch was located at the service entry, and the switch required reliable power and continuous air conditioning. Today, virtually all telecom systems include rack-mounted electronics in every closet, requiring reliable power and continuous air conditioning. In addition, telecom rooms are much larger than was the case in older buildings. A sample room layout is shown in Figure 19–29 for a large office building floor. The owner's IT department should be consulted early in the design process to determine their space requirements and preferred layout. Also, power requirements and heat load from telecom equipment can vary greatly; therefore, the electrical and mechanical engineers need information for programming and design of support systems.

19.10.4 Telecom Wiring Closets and Server Rooms

In older buildings telecom closets consisted simply of punch-down blocks on wall-mounted plywood. These

19.10.5 Electrical Raceway System Coordination

Technical aspects of electrical raceway systems are described in Chapter 10. Outlets are needed for power,

■ **FIGURE 19–26**
(a) Surface-mounted electrical distribution in unfinished space.

(a)

data, and communications. Distributing the wiring to workstations involves coordination with the architecture, and selection among options will affect functionality, cost, and appearance of the building. Options include:

1. Conduit in walls or columns is the least-expensive option for running wire to outlets and is the most common solution for general-purpose housekeeping outlets. For outlets serving equipment at workstations (computers, task lighting, appliances, etc.), outlets in the wall or column will be satisfactory if there is a wall or column close to the point of use. Often, however, furniture and workstations are arranged in an open-plan environment, and walls are not contiguous.

2. Poke-through fittings are often used in multistory buildings and can be installed after the building shell is constructed once final location of outlets is known. Fittings can also be added as needs change. They are relatively inexpensive, but have the disadvantage of needing access to the floor below for installation, which is disruptive. Unlike wall outlets, they can be located anywhere on the floor.

3. Power poles are used to route wiring down from the ceiling plenum to workstations or cubicles. Like poke-through fittings, they can be located after the building shell is complete, and their location is flexible.

4. Underfloor duct systems route wiring through passageways, which are placed on the structural deck and covered by lightweight concrete fill. Knock-outs, called pre-sets, located at intervals along the ducts provide wire access for installation of outlets on the floor. Generally, ducts are located 5' on center, so there are limitations on locating outlets, which must be considered in furniture layout. Another problem with underfloor duct systems is their effect on the construction process. They must be layed after the structural slab is poured, then another pour of concrete is poured over the ducts. The extra step is time consuming and expensive.

5. Cellular deck is similar to the underfloor duct system, except that the wiring is routed through cells in the structural deck. This method is less expensive, easier to construct, and requires no additional building height.

■ **FIGURE 19–26** *(Continued)*
(b) Flush mounted electrical panels in a finished corridor. (Photos courtesy of WTA.)

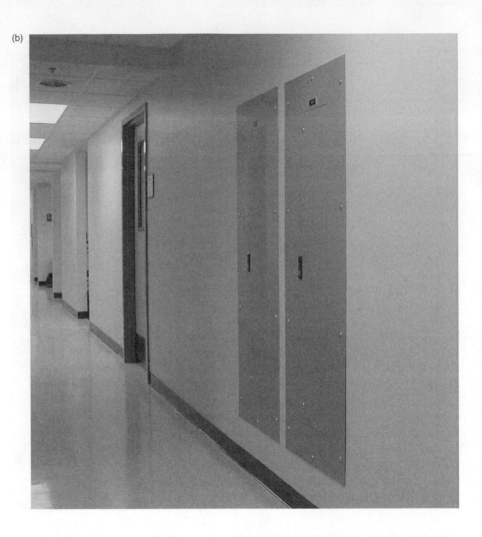

(b)

6. Flat wire is seldom used in new construction. It consists of bus (flat) conductors in an insulated flat cable, which is taped to the floor and covered with a steel protective cover. Flat wire must be used in conjunction with carpet tiles. Flat wire is expensive, but can be a flexible solution where other methods are not feasible, such as slab on grade, and power poles are deemed appropriate.

7. Access floor systems offer the highest flexibility for installing and reconfiguring wiring for offices. They are also the most expensive solution, but there are some trade-offs which mitigate the cost of the floor. If the floor is used for air distribution, ductwork in the ceiling can be avoided, and lights can be placed in cavities between structural elements. The resulting building can have shorter floor-to-floor height, which means less cost for skin and structure.

19.10.6 Options with Access Floors

Access floors can be used for airflow as well as for electrical wires, cables, and raceways. Raised floor systems can be classified in three categories. High raised floors, 12″ to 48″ standard are used mostly in computer rooms. For office applications, 6″ to 12″ floor heights can be used, and the system may or may not be part of the HVAC distribution. Low-profile systems are also available in heights as low as 2 ½″. Low-profile floors are used for wiring, but are too shallow for air distribution.

In buildings with raised floors, many spaces must still be equipped with conventional flooring at the structural slab elevation. These include toilet rooms, janitors closets, mechanical and electrical rooms and, sometimes, elevator lobbies. Access floors require offsets in floor heights, either ramps or steps down to floors at the structural floor elevation, or depressions in the structural slab if the finished floors are planned at the same elevation. Steps are adequate for controlled-access spaces, such as fan rooms and wiring closets, but toilet rooms require the same elevation or maximum 1″ in 12″ sloped ramps for accessibility.

For small floors air is simply blown into the floor plenum from the core shaft or air-handling unit. Large floors will require provisions for ducting to the far ends

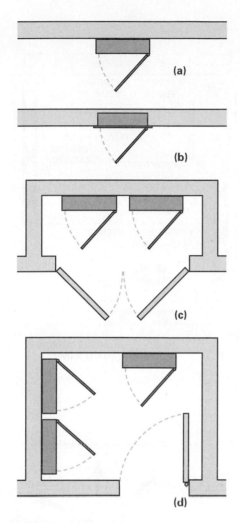

■ FIGURE 19–27
Panels serving branch circuits and lighting can be located exposed in occupied space (generally corridors) (a,b), concealed in alcoves (c), or in electrical closets (d).

of the floor. Ducting can be through actual ducts or through channels created by septum panels. Septum panels are less expensive, offer more cross-sectional area for their width, and are easier to cross with electrical wiring.

Raised floors are highly valued for their access to service components for maintenance and reconfiguration. Floor brands vary in their ease of tile and carpet removal and replacement. These features should be carefully checked before specifying raised-floor products.

19.11 GENERATORS

Engine-generators are furnished and installed as part of the electrical contract but require mechanical systems for fuel delivery, cooling and ventilation, and structural attention for support and vibration control, and

architectural care in placement, appearance, and sound control. Emergency and standby power systems are used in high-rise buildings, hospitals, data centers, and other facilities where safety or reliability are essential.

Systems most commonly used are diesel engine-driven electric generators. The engines are noisy and prone to vibration. They need large airflows for cooling the radiators and the engines themselves. Exhaust must be discharged outdoors away from windows, fresh air intakes, and pedestrians. Often, generators are placed outdoors in enclosures designed with provisions for ventilation and sound control. This is generally less expensive than providing accommodations in the building, but appearance on the site must be considered, and screening might be desirable. In addition, noise at the property line needs to be considered, both for engines in enclosures and for engines located in buildings or outbuildings (see Figure 11–7).

Generators with fuel storage are required for high-rise life safety systems, and placement can be problematic due to restricted sites. For vibration control, the ideal location will be at the lowest basement level. This will require either a large areaway for ventilation or remote radiators, perhaps located at the service area. An areaway might be useful not only for ventilation but also for equipment removal and replacement, not only of the generator but for other large items in the basement. Generators can be located at upper levels, but structure for support and vibration control will be substantial. Wherever large generators are located, a means for hoisting engine heads will be advisable to facilitate overhaul.

19.12 CEILINGS

19.12.1 Coordinating Multiple Services

Ceilings can be hard surface (drywall or plaster), exposed structure (concrete or steel), or suspended grid. In any case, ceiling coordination and reflected ceiling plans involve many disciplines including HVAC, fire protection, electrical, lighting, structural, and architectural. Most commercial and institutional buildings use suspended acoustical ceilings due to their finished appearance, accessibility to services, and low cost. The process of coordinating all the ceiling elements ideally goes through several iterations until all the services are laid out economically and to work well.

The architect starts the coordination process by deciding on the dimension and placement of the ceiling grid, considering the column spacing and window placement. This layout is followed by the lighting designer, who decides the dimension and spacing of light

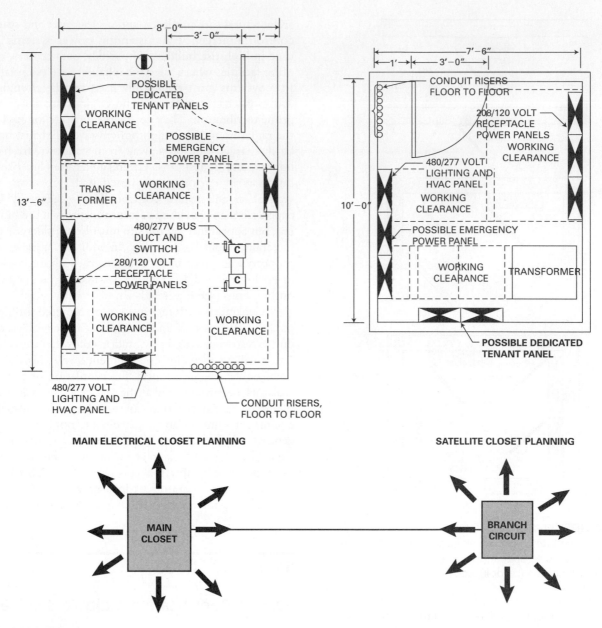

MAIN ELECTRICAL CLOSET PLANNING

SATELLITE CLOSET PLANNING

■ **FIGURE 19–28**
Electrical closet planning guidelines.

fixtures. Once these basic decisions are made, the ceiling elements with lesser visual impact (HVAC diffusers, sprinklers, speakers, smoke detectors) are coordinated in the overall plan. Ideally, the resulting pattern is regular and economical for each service. Often, however, the desire for regular spacing will result in more sprinkler heads or more diffusers than are actually needed. For instance, low-hazard sprinkler coverage is 225 sq ft, which corresponds ideally to 15-ft by 15-ft spacing. If the lighting spacing is 12 ft, then sprinklers might also be spaced at this frequency, resulting in more than the minimum number of heads required by code.

19.12.2 Exposed versus Concealed Ceilings

The cost, appearance, and functionality of a building's ceiling is very much affected by whether systems (e.g., structural, mechanical, electrical, fire suppression) are

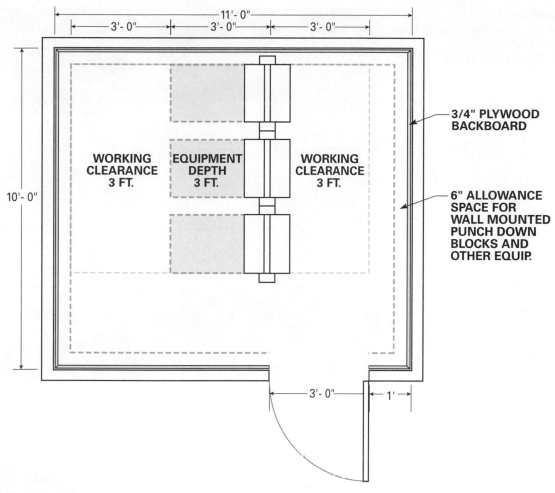

■ **FIGURE 19–29**

Telecom closet planning. The closet shown could handle a large office floor; smaller spaces could be served by a room as small as 6 ft × 10 ft.

exposed or concealed overhead. Exposed systems are generally more expensive due to the need for extensive coordination, higher finish materials, and better workmanship. The cost of a good-quality suspended ceiling at $1.50 to $3.00 per sq ft is virtually always a less-expensive option.

Exposed services can create an interesting architectural appearance, often intentionally linked to buildings housing technical programs as shown in Figure 19–30. Exposed services also give good access for inspection, emergency repairs, and modifications. This is especially advantageous in manufacturing, laboratories, data centers, and other technology centric facilities.

Housekeeping should be considered in deciding between exposed and concealed services. Accumulation of dust may be unacceptable in healthcare environments or clean spaces for research or manufacturing.

Acoustics is another factor in deciding whether to expose services. Suspended ceilings absorb sound within the room and block noise breaking out from ductwork. Generally, a space with exposed services will be 5 to 10 db noisier than a space with an acoustical ceiling.

Lighting should also be considered. A light-colored acoustical tile or hard ceiling can achieve a maintained reflectivity of up to 80 percent. This will result in more efficient utilization of lighting than a lower reflectivity ceiling cavity occupied by irregularly shaped and colored system components.

19.13 EQUIPMENT ACCESS ACCOMMODATIONS

Doorways and corridors to mechanical and electrical rooms should be large enough to remove equipment. Many equipment items, such as air-handling units, can be broken down into smaller pieces, but some items, such as chillers, are best removed and replaced in one

Exposed services are easier to access
and maintain in this laboratory
classroom. (Courtesy of WTA.)

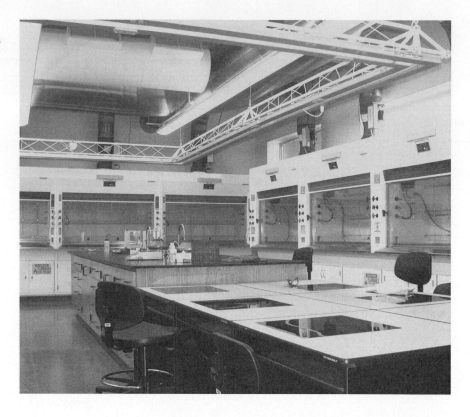

(a)

■ **FIGURE 19–31**
(a) Areaway adjacent to a major mechanical area in the
basement is used for equipment removal and replacement
and doubles as a source of ventilation air. (b) Large
windows at the upper level are designed to be removed
and a hoist beam extended to replace equipment.

(b)

626

piece. Placing rooms on outside walls will allow large openings constructed with removable panels or ventilation louvers. Areaways are desirable for removal of equipment from basements and can double as a source for ventilation air as shown in Figure 19–31. Removal and replacement of equipment in upper-level mechanical rooms can be accomplished by hoists or cranes through wall openings, or through roof hatches by crane or helicopter.

19.14 VERTICAL CHASES

Chases are needed for ducts, electrical raceways, plumbing risers, roof drains, and for distribution of HVAC-related piped services. In most instances, these services, once installed, will have no need for routine maintenance, but inspection panels are still good practice. For systems with run outs to the floors, such as for domestic water or heating water, isolation valves are recommended so that the piping on the floor can be opened without the need to drain the entire system. These valves need access, and may be located above ceilings or behind access panels in the chase. If the chase penetrates through floors with fire ratings, the shaft will also need to be rated. Any penetrations of the shaft will need to be sealed with appropriate fire-stopping materials and ducts and openings will need to be protected by fire dampers. Fire dampers need access panels in the duct for inspection and possibly to reset (open). Architecture and furniture arrangement must allow personnel to service the dampers at their access panels.

Chase planning for ductwork includes supply ducts, return ducts, outside air ducts, and exhaust ducts. If spaces are served from the roof or mechanical rooms on floors above or below, then supply and return air must pass vertically through a chase. The supply is always ducted. The return can be ducted, or it can pass through the chase area around the supply. Toilet and janitor closet exhaust ducts can share the same chase as supply and return, but other types of exhaust, such as kitchen hood exhaust and laboratory fume exhaust must be in separate chases.

Kitchen exhaust ducts serving grease hoods have special code requirements for fire safety and maintenance. Kitchen exhaust should be routed as direct to outdoors as practical. Ductwork is constructed of welded steel and must be equipped with access panels at every three stories vertically and every 20 ft horizontally for cleaning. The enclosure around the duct must be fire-rated construction, but penetrations are not required to have fire dampers.

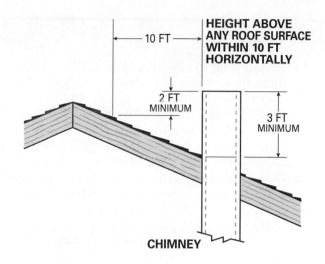

■ FIGURE 19–32
Vent stack clearance guidelines; see code for additional requirements.

19.15 ROOF ELEMENTS

19.15.1 Stacks and Exhaust Vents

Vents for most oil- and gas-fired boilers and furnaces need to be terminated not less than 2 ft above the roof and 2 ft higher than any portion of a building within 10 ft of the vent, as shown in Figure 19–32.

19.15.2 Exhaust Fans

Exhaust fans for toilets, janitors closets, and kitchens are typically located on the roof over the rooms they serve. The architect should be familiar with the types and appearance of these fans in the event that they will be visible from other portions of the building.

Kitchen exhaust discharge can be at a wall at least 10 ft above grade or at least 4 ft above a roof. The discharge must be at least 10 ft from an adjacent building or property line or 5 ft if directed away.

Research and healthcare buildings are usually equipped with fume hoods, which require exhaust at the roof. Hood exhaust is designed to discharge upward at high velocity so as to minimize backwash and downdraft, which could expose the building or its surroundings to hazardous fumes. Discharge of fume exhaust should be at least 7 ft from the surrounding walk surface to prevent exposure to personnel on the roof.

QUESTIONS

19.1 Recommend the location of air-handling equipment for the following design situations. State your reasons briefly.
Low-rise office building, inland Florida
Low-rise office building, Duluth, Minnesota
High-rise office building, Manhattan, New York
Suburban public school, Midwest

19.2 Design an electrical closet in a plan to house two 120/208-volt receptacle power panels (16″ wide × 8″ deep), one 277-volt lighting panel (16″ wide × 8″ deep), one BAS panel (24″ wide × 6″ deep), a bus duct passing through from floor-to-floor, and a 37-kVa step-down transformer (36″ × 36″).

19.3 Prepare a decision matrix to evaluate exposed services versus lay-in tile plenum ceiling for a laboratory building. Identify criteria you feel are relevant, weight them from the viewpoint of the owner who uses and operates the building.

19.4 Describe the difference in space requirements between a chilled water, hot-water system with central air handling, and a system with chilled water, central air handling, and electric heat.

19.5 A new low-rise campus on a sprawling "green field" rural site requires full standby power capability. What options, pros and cons, would you present for locating the generators? What options, pros and cons, would you present for a high-rise building in a congested urban site?

19.6 Your client is an investment developer with a high profit expectation. What method of on-floor electrical distribution will he likely favor to minimize his initial investment in construction for a rental office building? What would his choice be if he were hired to construct a building for a successful information technology company with considerable churn rate?

19.7 What are the pros and cons of power poles as a wiring solution for a new, corporate high-rise office building with a high churn rate? What other solutions might be more appropriate? State your reasons.

19.8 You are designing a total renovation (new M/E, finishes) of an existing low-rise building with low floor-to-floor height. The project has limited budget and will be occupied by tenants seeking inexpensive office space. Occupancy will be in phases as space is leased. What HVAC solutions would you consider? Why?

19.9 What wiring solutions might you consider for the building in Question 19.8?

19.10 A high-rise building is being planned. The initial massing concept takes the tower to the building line all around. Where would you place chillers, boilers, and cooling towers? Why? What problems can be anticipated with the locations you suggest?

19.11 What electrical spaces would you anticipate needing in the high-rise building described in Question 19.10? Where would you place them?

GLOSSARY OF TERMS, ACRONYMS, AND ABBREVIATIONS

THIS APPENDIX CONTAINS TERMS, ACRONYMS, AND abbreviations commonly used in the planning and design of mechanical, electrical, and illumination systems. For ease of understanding and quick reference, these terms are briefly explained rather than precisely defined. If a term is deemed essential for the basic understanding of the various systems, it will have been included in more detail in the text of the applicable chapters. If a term is good to know only as general or conversational knowledge, it may not be included in the text.

The terms are arranged alphabetically, with a letter in parentheses to identify their normal application in one or another of the various building systems, as follows:

(A)	Architectural
(AS)	Acoustics
(AX)	Auxiliary electrical systems
(DP)	Data processing/computer
(E)	Electrical power systems
(FP)	Fire protection systems
(G)	General, applicable to all or most mechanical and/or electrical systems
(H)	Heating and air-conditioning systems
(I)	Illumination systems
(PS)	Plumbing and sanitation
(S)	Structural systems
(TC)	Telecommunications

To find the exact location at which a term appears found in the text, see the index for page numbers.

Absolute filter (H) A high-efficiency mechanical filter built originally for removing radioactive particles and having an efficiency of 99.9% or higher.

Absolute humidity (H) Weight of water vapor per unit volume of air–steam mixture. Usually expressed in grains per cubic foot.

Absolute pressure (G, H) The pressure above an absolute vacuum; the sum of the pressure shown by a gage (or *gauge*) and that indicated by a barometer.

Absolute temperature (H) Temperature measured from absolute zero; on the Fahrenheit scale, the absolute temperature is approximately °F + 459.6° and on the Celsius scale °C + 273°.

Absorption (AS) The ability of a material to absorb acoustical energy. Measured in sabins. The product of area (*S*) and absorption coefficient (α). Frequency sensitive.

Absorption (H) An assimilation during which the absorbent undergoes a physical or chemical change (or both). Absorbers such as lithium chloride and calcium chloride are used in dehumidifying. The process is also used in absorption refrigeration.

Absorption coefficient (AS) The fraction of sound energy impinging on a surface that is absorbed by that surface, usually denoted by α. Frequency sensitive.

Absorption refrigeration (H) A process whereby a secondary fluid absorbs the refrigerant and, in doing so, gives up heat. Afterward, it releases the refrigerant, during which time it absorbs heat.

Acoustics (AS) The science of sound, including its generation, transmission, and effects.

Acrylonitrile-butadiene-styrene, ABS (PS) A thermoplastic compound from which fittings, pipe, and tubing are made.

Active sludge (PS) Sewage sediment, rich in destructive bacteria, that can be used to break down fresh sewage more quickly.

Adaptive reuse (G) Converting existing buildings to new functions to avoid demolishing them; generally considered a sustainable design by avoiding the need to dispose of existing building materials and conserving the use of new materials along with the resources needed to produce and erect them.

Adiabatic (H) A change at constant total heat; an action or process during which no heat is added or subtracted.

Aeration (PS) An artificial method in which water and air are brought into direct contact with each other to furnish oxygen to the water and to reduce obnoxious odors.

Aerobic (PS) Living or active only in the presence of free oxygen (said of certain bacteria).

Airborne sound (AS) Sound that is transmitted through air by a series of oscillating pressure fluctuations.

Air change (H) The quantity of infiltration or ventilation air, in cubic feet per hour or minute, divided by the volume of the room.

Airfoil fan (AS, H) A fan with an airfoil-shaped blade that moves the air in the general direction of the axis about which it rotates.

Air gap (PS) The unobstructed vertical separation between the lowest opening from a pipe or faucet conveying water or waste to a plumbing fixture receptor.

Alternating current, AC (E) Flow of electricity that cycles or alternates direction. The number of cycles per second is referred to as the *frequency*.

Ambient (AS, H, I) Encompassing on all sides. Thus ambient air is the surrounding air; ambient light is the lighting environment of a space; ambient noise is the surrounding sound in a space.

Ambient lighting (I) Lighting throughout an area that produces general illumination.

Ambient noise (AS) The sound associated with a given environment, usually a composite of many sources.

American Standard Code for Information Interchange, ASCII (DP) A system for referring to letters, numbers, and common symbols by code numbers. The numerical value assigned for each binary digit of an eight-digit byte is, from left to right, 128, 64, 32, 16, 8, 4, 2, and 1.

American Wire Gauge, AWG (E) The standard system for measuring the size of wires in the United States.

Ampacity (E) The current, in amperes, that a conductor can carry continuously under conditions of use without exceeding its temperature rating.

Analog (DP, TC) Varying continuously (e.g., sound waves). Analog signals have their frequency and bandwidth measured in hertz. See *Digital*.

Anechoic (AS) An acoustical environment free of any reflected sound; a free field.

Angstrom (I, TC) A unit of length, 0.1 nm or 10^{-10} m, often used to measure wavelength, but not part of the SI system of units. See *Nanometer*.

Apparatus dew point (H) Dew point temperature of the air leaving the conditioning apparatus.

Aspect ratio (H) In a duct, the depth of elbow along (parallel to) the axis of bend divided by the width in the plane of the bend.

Atmospheric pressure (G, H) Pressure indicated by a barometer. The standard atmospheric pressure is 760 mm of mercury (29.921 in. of mercury) or 14.7 lb per square inch (in.^{2}) at sea level.

Attenuation (AS) Lessening or reduction, e.g., from 80 dB to 70 dB.

Aural (AS) Pertaining to the ear or to the sense of hearing.

A-weighting (AS) Prescribed frequency response defined by ANSI Standard S1.4–1971. Used to obtain a single number representing the sound pressure level of a noise in a manner approximating the response of the ear, by deemphasizing the effects of the low and high frequencies. See dBA.

Axial flow fan (H) A fan with a disk- or airfoil-shaped blade that is mounted on a shaft and that moves the air in the general direction of the shaft axis.

Backflow (H, PS) In plumbing or HVAC systems, water flow into water-distributing pipes from any source other than that for which the system is designed.

Backflow connection (PS) In plumbing, any connection or arrangement which permits backflow.

Backflow preventer (PS) A device to prevent backflow into a water supply line from the connection at its outlet.

Background sound (AS) Noise from all sources in an environment, exclusive of a specific sound of interest.

Backup (DP) A copy of a program or document that can be used if the original is destroyed.

Ballast (I) A device used with fluorescent and HID lamps to provide the necessary starting voltage and to limit the current during operation of the lamp.

Bandwidth (DP, TC) The highest frequency that can be transmitted in an analog operation. Also (especially for digital systems), the information-carrying capacity of a system.

Baud (DP, TC) Strictly speaking, the number of signal-level transitions per second in digital data. For some common coding schemes, baud equals the number of bits per second, but this is not true for more complex coding, where the term is often misused. Telecommunication specialists prefer bits per second, which is less ambiguous.

Baud rate (DP) The number of bits per second. Baud rates are most commonly used as a measure of how fast data are transmitted by a modem.

Beam angle (I) The angle between the two directions for which the intensity is 50 percent of the maximum intensity, as measured in a plane through the nominal beam centerline.

Bell-and-spigot joint (PF) The joint commonly used in cast-iron pipe. Each piece is made with an enlarged diameter or bell at one end into which the plain or spigot end of another piece is inserted when the pipe is being laid. The joint is then made tight with cement, oakum, lead, or rubber caulked into the bell around the spigot.

Bellows (H, PS) An expansible metal device containing a fluid that will volatilize at some desired temperature, expand the device, and open or close an opening or a switch, as in controls and steam traps.

Bimetal (H) Two metals of different coefficients of expansion welded together so that the piece will bend in one direction when heated and in the other when cooled. Thus it can be used in opening and closing electrical circuits, as in thermostats.

Binary numbers (DP) The base–2 numbering system of almost all computers, as opposed to the base–10 (decimal) numbers in common use.

Binary system (DP) A numbering system using 2 as base, as opposed to 10 as base in the decimal system. For example, 1 in binary system is 1; 10 is 2; 11 is 3; 100 is 4; 101 is 5; 110 is 6; 111 is 7; 1000 is 8; 1001 is 9; and 1110 is 10. The binary system is used in almost all computers.

Bit (DP) Abbreviated from "binary digit," the smallest possible unit of information. A bit represents one of two things: yes or no, on or off, or, as expressed in the binary numbers used in computers, 0 or 1.

Blackbody (H, I) A body that absorbs all the radiant energy falling upon it.

Boiler (H)   A closed vessel in which fuel is burned to generate steam. (A vessel that produces hot water should be called a *hot-water heater.*)

Boiler horsepower (H)   The power required to evaporate 34.5 lb of water at 212°F per hour, or the equivalent of 33,475 Btu per hour (Btuh).

Booting (DP)   The starting up of a computer by loading an operating system into it. (The name *boot* comes from the idea that the operating system pulls itself up by its own bootstraps.)

Branch circuit (E)   The circuit conductors between the final overcurrent device protecting the circuit and the outlet(s).

Branch interval (PS)   In plumbing, a vertical length of waste pipe, usually one story high, to which the horizontal branches from that floor are connected.

Branch vent (PS)   A vent connecting one or more individual vents with a vent stack.

British thermal unit, Btu (G, H)   The amount of heat required to raise the temperature of 1 lb (0.45 kg) of water 1 degree Fahrenheit (0.565°C).

Broadband (TC)   In general, covering a wide range of frequencies. The broadband label is sometimes used for a network that carries many different services or for video transmission.

Buffer (DP)   An area of memory or a separate memory cache that holds information until it is needed. Buffers are used to speed up printing, redraw the screen, etc.

Bug (DP)   A mistake or unexpected occurrence in a piece of software (or, less commonly, in a piece of hardware).

Building drain (PS)   The section of the horizontal drainpipe connecting the drainage pipes inside the building with the building sewer outside the inner face of the building wall.

Building sewer (PS)   The pipe extending from the outer end of the building drain to the public sewer.

Building trap (PS)   A trap in the building drain to prevent air circulation between the building drainage system and the building sewer.

Bulletin board (bulletin board system), BBS (DP)   A computer dedicated to maintaining messages and software and making them available over phone lines at no charge. People upload (contribute) and download (gather) messages by calling the bulletin board from their own computers.

BX (E)   An electrical cable consisting of a flexible metallic covering inside of which are two or more insulated wires for carrying electricity.

Byte (DP)   A group of eight bits that the computer reads as a single letter, number, or symbol. For example, 01000101 is a typical byte of the binary system.

CAD/CAM (DP)   See *Computer-aided design* and *Computer-aided manufacturing.*

CAD/CAM Calorie (H)   In engineering, the large calorie is usually used; it is defined as one one-hundredth of the energy or heat required to raise the temperature of 1 kg of water from 0°C to 100°C. The small calorie is the heat required to raise the temperature of 1 g of water.

Candela, cd (I)   The unit of measurement of luminous intensity of a light source in a given direction.

Candlepower, Cp (I)   Luminous intensity of a light source in a specific direction.

Candlepower distribution curve (I)   A curve, generally polar, representing the variation in luminous intensity of a lamp or luminaire in a plane through the light center.

Cathode ray tube, CRT (DP, TC)   The display technology used on virtually all computer monitors and television sets.

Cathodic protection (PS)   The control of the electrolytic corrosion of an underground or underwater metallic structure by the application of an electric current.

Central processing unit, CPU (DP)   The central part of a computer. The CPU includes the circuitry built around the CPU chip and mounted on the motherboard that actually performs the computer's calculations.

Check valve (FP, H, PS)   A valve designed to allow a fluid to pass through in one direction only. A common type has a plate suspended so that the reverse flow aids gravity in forcing the plate against a seat, shutting off reverse flow.

Chip (DP)   A tiny piece of silicon or germanium impregnated with impurities in a pattern that creates different sorts of miniaturized computer circuits. A chip is the basic brain controlling all electronic or computer equipment and systems.

Circuit breaker (E)   A device designed to open and close a circuit manually or automatically on a predetermined overcurrent.

Circuit vent (PS)   A branch vent from two or more traps extending from in front of the last fixture connection of a horizontal branch to the plumbing vent stack.

Clean room (H)   A room in which the air is highly purified, particularly with regard to dust and other particulate matter. Clean-room techniques are used in the processing of delicate materials and products.

Clock rate (or clock speed) (DP)   The operations of a computer are synchronized with a quartz crystal that pulses millions of times each second. These pulses determine, for example, how often the screen redraws an image and how often the CPU accesses RAM or a hard disk. The frequency of the pulses—how often they occur—is measured in megahertz (millions of cycles per second) and is called the *clock rate* or *clock speed*.

Close nipple (H, FP, PS)   A nipple with twice the length of a standard pipe thread.

Closed cycle (H)   A system in which the fluid is used over and over without introduction of new fluid, as in a hot-water heating or mechanical refrigeration system.

Coefficient of performance, COP (H)   In a heat pump, the ratio of the effect produced to the electrical input in consistent units.

Coefficient of utilization, CU (I)   The ratio of the luminous flux (lumens) from a luminaire, calculated as received on the work plane, to the luminous flux emitted by the luminaire's lamps alone.

Color-rendering index, CRI (I)   Measure of the degree of color shift objects undergo when illuminated by a light source, as compared with a reference source, normally incandescent.

Color temperature (I)   The absolute temperature of a black-body radiator having a chromaticity equal to that of the light source.

Community antenna television, CATV (TC)   Cable television, a broadband transmission system generally using 75-ohm coaxial cable that simultaneously carries many frequency-divided TV channels.

Compact disk, CD (DP, TC)   A 4.5-in.-diameter disk containing digital information that can be read by a laser source.

Compact disk read-only memory, CD-ROM (DP, TC)   A 4.7-in.-diameter compact disk with read-only memory. It can store up to 660 megabytes of information.

Compiler (DP)   Software that implements a program by translating it all at once.

Computer-aided design, CAD (DP)   Refers to both hardware and software.

Computer-aided manufacturing, CAM (DP)   Computers and programs that run manufacturing machinery or even entire factories. (You seldom see the word CAM alone; it's usually combined with CAD.)

Condensate (G)   Water that has liquefied from steam or precipitated out from air.

Condensate (H, PS)   Liquid formed by condensation.

Condenser (H)   Apparatus used to liquefy a gas.

Condensing unit (H)   An assembly attached to one base and including a refrigerating compressor, motor, condenser, and receiver, and the necessary accessories.

Conductance (H)   The quantity of heat (usually BTU) transmitted per unit of time (usually 1 hour) from a unit surface of material under a unit temperature (usually 1°F) differential between the surfaces. Unit abbreviation is Btu/hr-ft²°F.

Conduction (H)   The transmission of heat from one part of a body to another part of the same body, or from one body to another in contact with it, without appreciable displacement of the particles of the body.

Conductivity (H)   The quantity of heat (usually Btu) transmitted per unit of time (usually 1 hour) from a unit surface (usually 1 sq ft) to an opposite unit of surface of one material per unit of thickness (usually 1 in., but occasionally 1 ft) under a unit temperature differential (usually 1°F) between the surfaces.

Continuous load (E)   A load such that the maximum current is expected to continue for 3 hours or more.

Contrast (I)   Difference in brightness between an object and its background.

Convection (H)   The transfer of heat from one point to another within a fluid (such as air or water) by the mixing of one portion of the fluid with another. If the motion is due to differences in density, from temperature differences, the convection is natural; if the motion is imparted mechanically, it is forced convection.

Convector (H)   A unit containing heating elements that allow air to be heated through natural convection without using external power.

Cooling tower (H)   A device for cooling water by evaporation in the outside air. A natural-draft cooling tower is a cooling tower in which the airflow through it is due to the natural chimney effect; a mechanical-draft tower uses fans.

Coprocessor (DP)   A chip that specializes in mathematics, graphics, or some other specific kind of computation. When the CPU is handed the kind of job the coprocessor specializes in, it hands the job off to the coprocessor.

Counterflow (H)   In a heat exchanger, a situation in which the fluid absorbing heat and the fluid losing heat are directed so that the lower and higher temperature of the one are adjacent to the lower and higher temperature of the other, respectively. Ordinarily, one fluid is flowing in the opposite direction from the other, hence the term.

Critical angle (I, TC)   The angle at which light undergoes total internal reflection.

Critical velocity (H)   The point above which streamline flow becomes turbulent.

dBA (AS) (TC)   Overall A-weighted sound pressure level expressed in decibels referenced to 20 micropascals. An A-weighting is a single number approximating the response of human hearing.

Dead room (AS)   A room whose boundaries and furnishings absorb a great amount of sound.

Debug (DP)   To search out bugs or defects in a piece of software and eliminate them.

Decay rate (AS)   The rate at which the sound pressure level (in dB) decreases when the source of the sound is eliminated.

Decibel, dB (AS, TC)   Ten times the logarithm to the base 10 of a quantity divided by a reference quantity; $dB = 10 \times \log_{10}(X/X\text{ref})$.

Degree day, DD (H)   The number of Fahrenheit degrees that the average outdoor temperature over a 24-hour period is less than 65°F.

Demand charge (E)   That part of a utility service charged for on the basis of the possible demand as distinguished from the energy actually consumed.

Demand factor, DF (E)   The ratio of the maximum demand to the total connected load of a system.

Demand load, DL (E, H, PF)   The load, in appropriate units, such as kW, Btuh, or gpm, that an electrical or mechanical system encounters.

Density (G)   The weight of a unit of volume—usually, pounds per cubic foot.

Developed length (H, PF)   The length along the centerline of a pipe and its fittings.

Dew point, DP (G, H)   The temperature of a gas or liquid at which condensation or evaporation occurs.

Dielectric fitting (H, PS)   A fitting having insulating parts or material that prohibits the flow of electric current.

Diffraction (AS)   Alteration of the direction of propagation of a sound wave in the vicinity of a boundary discontinuity on the edge of a reflecting or absorbing surface. Frequency sensitive.

Diffuse sound field (AS)   A sound field in which the intensity of the sound is independent of its direction; an area over which the average rate of sound energy flow is equal in all directions.

Diffusion (H)   The movement of individual molecules through narrow spaces by the molecular velocity and bouncing associated with individual gaseous or liquid molecules.

Digital (DP, TC)   Expressed in binary code to represent information. See *Analog*.

Diode (DP, TC)   An electronic device that lets current flow in one direction. Semiconductor diodes used in fiber optics contain a junction between regions of different doping. (Doping means adding impurity elements to pure semiconductors to improve their conductivity.) Diodes include light emitters (LEDs and laser diodes) and detectors (photodiodes).

Direct current, DC (E) Flow of electricity continuously in one direction from positive to negative.

Direct expansion (H) An arrangement wherein a refrigerant expands in an evaporator in the airstream as intended.

Direct-fired heater (H) A fuel-burning device in which the heat from the fuel is transferred through metal to air, which is then introduced to the space to be heated.

Direct glare (I) Glare resulting from high luminances or insufficiently shielded light sources in the field of view.

Direct sound field (AS) A sound field in which the energy arrives at the receiver in a direct path from the source, without any contribution from reflections. See *Reverberant sound field*.

Directivity index, DI (AS) A measure of the directionality of sound from a specific source, in decibels. The difference between the actual sound pressure level of the source and the sound pressure that would exist if the same source were a point source radiating spherically. The directivity index is sensitive to frequency.

Discomfort glare (I) Glare producing discomfort. This type of glare does not necessarily interfere with visual performance or visibility.

Discomfort index (H) Original name of temperature–humidity index (THI).

Disconnecting means (E) A device by which the conductors of a circuit can be disconnected from their source of supply.

Disk (DP) A round, coated platter used to store information magnetically for computer processing. The two main types are floppy disks, which are removable and flexible; and hard disks, which are nonremovable and rigid.

Disk operating system, DOS (DP) The operating system used on IBM personal computers and compatible machines.

District heating (H) The heating of a multiplicity of buildings from one central boiler plant, from which steam or hot water is piped to the buildings. District heating can be a private function, as in some housing projects, universities, hospitals, etc., or a public utility operation.

Diversity factor (E, H, PS) The ratio of the sum of the individual maximum loads during a period to the simultaneous maximum loads of all the same units during the same period. Always unity or more.

Dot-matrix printer (DP) A printer that forms characters out of a pattern of dots. Usually, each dot is made by a separate pin pushing an inked ribbon against the paper.

Dots per inch, dpi (DP) A measure of the resolution of a screen or printer; the number of dots in a line 1 in. long.

Downspout (PS) The rain leader (pipe) from the roof to the means of disposal.

Drainage fixture unit, DFU (PS) A measure of probable discharge into the drainage system by various types of plumbing fixtures. By convention, 1 DFU is equivalent to 1 CFM.

Dry-bulb temperature, DB (H) The temperature of air as measured by a thermometer.

Dynamic head (H) The pressure equivalent of the velocity of a fluid. Customary units of head are usually feet or inches of H_2O.

Dynamic random access memory, DRAM (DP) A memory chip that loses its memory when the computer is shut off.

Effluent (PS) Treated or partially treated sewage flowing out of sewage treatment equipment.

Electronic cleaner (H) An air cleaner in which matter in the airstream is electrically charged, then attracted to surfaces oppositely charged.

Electronic data interchange, EDI (TC) A series of standards that provide for the exchange of data between computers over phone lines.

Electronic mail, or E-mail (DP, TC) Messages sent from computer to computer over phone lines or over a local area network (LAN).

Emissivity (H, I) The ratio of radiant energy emitted by a body to that emitted by a perfect blackbody. A perfect blackbody has an emissivity of 1, a perfect reflector an emissivity of 0.

Energy management system, EMS (H, E) An automated control system designed to achieve higher energy efficiency or lower energy consumption in a building. An energy management system is not limited to the operation of HVAC, lighting, and other power systems.

Enthalpy (H) For most engineering purposes, heat content or total heat, above some base temperature. Specific enthalpy is the ratio of the total heat to the weight of a substance.

Entropy (H) The ratio of the heat added to a substance to the absolute temperature at which the heat is added. Specific entropy is the ratio of total heat to weight of the substance.

Equipment-grounding conductor (E) The conductor used to connect the non-current-carrying metal parts of equipment, raceways, and other enclosures to the system ground. (The equipment-grounding conductor is color-coded green.)

Equivalent length (H, FP, PS) The resistance to flow of a duct or pipe elbow, valve, damper, orifice, bend, or fitting, or some other obstruction, expressed in the number of feet of straight duct or pipe of the same diameter that would have the same resistance.

Equivalent sphere illumination, ESI (I) The level of spherical illumination that would produce a task visibility equivalent to that produced by a specific lighting environment.

Ethernet (DP) A relatively fast local area network (LAN) cabling system developed by Xerox. Ethernet components are also sold by other vendors.

Evaporative cooling (H) Cooling by the evaporation of water in air when the wet bulb is considerably lower than its dry bulb. This method is widely used in dry climates.

Expansion joint (H, PS) A joint designed to absorb longitudinal thermal expansion in the pipeline due to heat.

Expansion loop (H, PS) A bend of large radius in a pipeline designed to absorb longitudinal thermal expansion in the line due to heat.

Face velocity (H) The speed, in feet per minute, by which air leaves a register or a coil.

Fault (E) A short circuit—either line to line or line to ground.

FAX (TC) Facsimile—the transmitting of an image by means of telecommunication; also *fax* (lowercase), a form more commonly used today.

Feeder (E) Circuit conductors between the service equipment and the final branch-circuit overcurrent device.

Feet of head (H, PS)   Pressure loss in psi divided by the factor 0.433.

Fiber optics, FO (DP, I, TC)   A technology in which light is used to transmit information from one point to another. The transmitting medium is constructed of thin filaments (strands) of glass wires through which light beams are transmitted. For illumination purposes, the light may be any one visible spectral wavelength. For data transmission, light must be of a single wavelength to be totally reflected (refracted) within the fiber. Light sources are usually generated by laser or LED.

Field angle (I)   The angle between the two directions for which the intensity of light is 10 percent of the maximum intensity, as measured in a plane through the nominal centerline of the light beam.

File server (DP)   A computer on a network that everyone on the network can access and get applications and documents from.

Finned tube (H)   Tube or pipe containing fins used for heat transfer between water and air, usually by natural convection.

Fixture branch (PS)   In plumbing, the branch from the water-distributing pipe in the building to the fixture supply pipe.

Fixture carrier (PS)   A metal unit designed to support an off-the-floor plumbing fixture.

Fixture drain (PS)   The drain from the trap of a fixture to the point where it connects with another drainpipe.

Fixture unit, FU (PF)   An index of the relative rate of flow of water to a fixture (water supply fixture unit, WFU) or of sewage leaving a fixture (drainage fixture unit, DFU).

Flanking transmission (AS)   Transmission of sound from the source to the receiver by paths around barriers or other means intended to block the transmission of sound.

Floppy disk (DP)   A flexible, removable disk (although the case in which the actual magnetic medium is housed may be hard, as it is on 3½-in. floppies).

Floppy disk drive (DP)   A device for reading data from and writing data to floppy disks.

Flow rate (H, PF)   Cubic feet per minute (CFM) of air circulated in an air system or pounds of water per minute circulated through a hot-water heating system.

Flushometer valve (PS)   A device that discharges a predetermined quantity of water to fixtures for flushing purposes and is actuated by direct water pressure.

Font (DP, TC)   A collection of letters, numbers, punctuation marks, and symbols with an identifiable and consistent look.

Foot-candle, fc (I)   The illuminance on a surface 1 sq ft in area on which there is a uniformly distributed flux of 1 lm.

Foot-Lambert, fL (I)   A unit of luminance of a perfectly diffusing surface emitting or reflecting light at the rate of 1 lm per square foot.

Formatting (DP)   All the characteristics of text other than the actual characters that make it up. Formatting includes things such as italics, boldface, type size, margins, line spacing, and justification. Also another term for initializing a disk.

Free field (AS)   A field free from boundaries that would otherwise tend to reflect sound.

Frequency (AS, E)   Number of complete oscillation cycles per unit of time. A unit of frequency often used is the hertz.

Function keys (DP)   Special keys labeled F1, F2, etc., on some extended keyboards.

Furnace (H)   Either the combustion space in a fuel-burning device or a direct-fired air heater. In the latter case, not to be confused with a boiler.

Fuse (E)   An overcurrent protective device with a circuit-opening fusible part that is heated and severed by the passage of overcurrent through it.

Galvanizing (PS)   A process whereby a surface of iron or steel is covered with a layer of zinc.

Gigabyte, (DP)   A measure of computer memory, disk space, and the like, equal to 1024 megabytes (1,073,741,824 bytes), or about 179 million words. Sometimes a gigabyte is treated as an even billion bytes, but plainly, that is almost 74 million bytes short. Sometimes abbreviated gig (more often in speech than in writing).

Glare (I)   The sensation produced by luminance within the visual field that is sufficiently greater than the luminance to which the eyes are adapted. Glare may cause annoyance, discomfort, or loss of visual performance and visibility.

Ground (E)   A conducting connection, whether intentional or accidental, between electrical circuit or equipment and the earth.

Ground fault (circuit) interrupter, GFI or GFCI (E)   A device that senses ground faults and reacts by opening the circuit.

Grounded conductor (E)   A system or circuit conductor that is intentionally grounded.

Grounding conductor (E)   A conductor used to connect equipment or the grounded circuit of a wiring system to a grounding electrode or electrodes.

Halon (FP)   A bromtrifluoromethane gas that is effective in extinguishing fires.

Handshake (DP)   What computers do when communicating, in order to establish a connection and agree on protocols for the transmission of data.

Hard disk drive (DP)   A rigid, usually nonremovable disk or the disk drive that houses it. Hard disks store much more data and access the data much more quickly than floppy disks do.

Head end (TC)   The central facility where signals are combined and distributed in an airborne sound (AS) system.

Heat pump (H)   An all-electric heating/cooling device that takes energy for heating from outdoor air (or groundwater).

Hertz, Hz (AS, DP, E)   A measure of frequency (cycles per second).

High-density (or high-resolution) television, HDTV (DP)   Television with about double the resolution of present systems.

High-intensity discharge (HID) lamp (I)   A lamp whose light source is mercury, metal halide, or high-pressure sodium.

High-pressure sodium (HPS) lamp (I)   HID lamp in which light is produced by radiation from sodium vapor.

Home run (E)   The wiring run between the panel and the first outlet in the branch circuit. (Looking upstream, the wiring run between the last outlet and the panel.)

Horsepower (E, H, PF)   A unit of power that equals 746 W, or 1 hp = 3/4 kW.

Humidity (H)   Usually, water vapor mixed with dry air. Absolute humidity is the weight of water (or steam) per unit volume of the air–water mixture.

Hydrogen ion concentration, pH (H)   The logarithm of the reciprocal of the gram ionic hydrogen equivalents per liter. On the pH scale 7.0 is neutral; smaller values to 0 are acid, progressively so; and higher values to 14 are alkaline, progressively so.

Icon (DP)   A graphic computer symbol usually representing a file, folder, disk, or tool.

Illuminance (I)   The density of the luminous flux incident on a surface; the quotient of the luminous flux divided by the area of the surface when the latter is uniformly illuminated.

Illuminance (lux or foot-candle) meter (I)   An instrument for measuring illuminance on a plane. Instruments that respond accurately to more than one spectral distribution are color-corrected.

Incandescent lamp (I)   A lamp in which light is produced by a filament heated to incandescence by an electric current.

Index of refraction (I, TC)   The ratio of the speed of light in a vacuum to the speed of light in a material, usually abbreviated n.

Indirect lighting (I)   Lighting by luminaires distributing 90 to 100 percent of the emitted light upward.

Input/output, I/O (DP)   Signal fed into and transmitted out of a circuit.

Integrated circuit, IC (DP, TC)   An electronic device that contains hundreds or thousands of separate components, such as transistors, resistors, and switches. Encapsulated in a plastic enclosure, it is also called a *chip*.

Integrated services digital network, ISDN (TC)   A digital standard calling for 144-kbit/sec transmission, corresponding to two 64-kbit/sec digital voice channels and one 16-kbit/sec data channel.

Intensity, or I (AS)   The average rate of sound energy flow per unit of area in a direction perpendicular to the area, usually in W/m².

Intensity, or I (I)   The luminous flux per unit solid angle, expressed in lumens per steradian (lm/sr) or candela (cd).

Intensity level, IL (AS)   Ten times the logarithm of the ratio of the sound intensity to a reference sound intensity (I_{ref}) of 10^{-12} W/m² (9.29×10^{-14} W/ft²).

International Organization for Standardization, ISO (G)   An international organization that establishes standards on scientific and technology quantities.

Internet (DP)   A computer network that joins many government and private computers over phone lines. Internet was started in 1969 by the Defense Department, is managed by the National Science Foundation, and is now the most popular computer network in the world. Among the services is the World Wide Web (www).

Inverse square law (I)   The law stating that the illuminance at a point on a surface varies directly with the intensity of a point source and inversely as the square of the distance between the source and the point.

Invert (PS)   Lowest point on the interior of a horizontal pipe.

Isolux chart (I)   A series of lines, plotted on any appropriate set of coordinates, each of which connects all the points on a surface having the same illumination.

Kilobyte, K (DP)   A measure of computer memory, disk space, and the like, equal to 1024 characters, or about 170 words.

Laminar flow (H)   Flow occurring in fluid layers, in a conduit or space, with no eddy currents. Each particle of fluid flows directly from one end to the other or from top to bottom, or in some planned manner.

Lamp efficacy (I)   The ratio of lumens produced by a lamp to the watts consumed. Expressed as lumens per watt (lpw).

Lamp lumen depreciation, LLD (I)   Multiplicative factor in calculations of illumination for reduction in the light output of a lamp over a period of time.

LASER (TC)   An acronym for *light amplification by stimulated emission of radiation*. Laser light is directional, covers a narrow range of wavelengths, and is more coherent than ordinary light. Semiconductor diodes are the standard light sources in fiber-optic systems.

Laser diode (DP, TC)   A semiconductor diode in which the injection of current carriers produces laser light by amplifying photons generated when holes and electrons recombine at the junction between *p*- and *n*-doped regions.

Latent heat (H)   Inherent heat in the form of fluid without a phase change.

Life-cycle cost analysis (G)   Accounting of cash flows for installation, debt retirement, maintenance, utilities, and salvage value of a system or component over the life of the element being analyzed; generally done to evaluate alternative options.

Light-emitting diode, LED (AX, DP, TC)   Small lamp illuminated by the movement of electrons in a semiconductor.

Light loss factor, LLF (I)   A factor used in calculating illuminance after a given period and under given conditions. (Formerly called *maintenance factor*.)

Liquid crystal display, LCD (DP, TC)   A product that uses liquid crystals sealed in glass. The pixels contain individual transistors to generate alphanumeric and graphic images.

Live room (AS)   A room that has very little sound absorption from the boundaries and furnishings. Opposite of dead room.

Local area network, LAN (DP)   A network of computers and related devices that transmits data among many nodes in a relatively small area, such as one office or one building. See *Wide area network*.

Loop vent (PS)   A branch vent serving more than one fixture and looping back to connect with a stack vent instead of the vent stack.

Loudness (AS)   A subjective description of the level of sound. Typically, a 10-dB increase in sound pressure level is judged to be twice as loud as the level before the increase.

Low-pressure sodium (LPS) lamp, (I)   A discharge lamp in which a single wavelength of visible yellow light is produced by radiation of sodium vapor at a low pressure.

LP gas (H)   Liquefied petroleum gas, a gas of high heating content, stored under high pressure in liquid form. Delivered to consumers in containers where piped gas is not available, and used by gas utilities for peaking (when demand exceeds source of natural gas).

Lumen, lm (I)   The unit of luminous flux; the luminous flux emitted within a unit solid angle (one steradian) by a point source having a uniform luminous intensity of one candela (1 cd).

Luminaire (I)   A complete lighting unit consisting of a lamp or lamps together with the parts designed to distribute the light, to position and protect the lamps, and to connect the lamps to the power supply.

Luminaire dirt depreciation, LDD (I) Multiplicative factor used in calculations of illuminance for reduced illuminance due to dirt collecting on the luminaires.

Luminaire efficiency (I) The ratio of the luminous flux (lumens) emitted by a luminaire to that emitted by the lamp or lamps used therein.

Luminance (I) The amount of light reflected or transmitted by an object.

Lux, lx (I) The metric unit of illuminance. One lux (1 lx) is one lumen per square meter (1 lm/m²).

Macro (DP) A command that incorporates two or more other commands or actions. (The name comes from the idea that macro commands incorporate "micro" commands.)

Maintenance factor, MF (I) A factor used in calculating illuminance after a given period and under given conditions.

Makeup air (H) Air brought into a building from the outside to replace that exhausted by a ventilating system.

Makeup water (H, PS) Water supplied to replenish that lost by leaks, evaporation, etc.

Malleable iron (PS) Cast iron heat-treated to reduce its brittleness. The process enables the material to stretch to some extent and to withstand greater shock.

Masking (AS) The rendering undetectable of one sound of interest by other sounds.

Mbh (H) Thousands of British thermal units (Btu) per hour (h).

MCM (E) Thousand circular mil; used to describe large wire sizes.

Meg (DP) An abbreviation for *megabyte*.

Megabyte, MB (DP) A measure of computer memory, disk space, and the like, equal to 1024K (1,048,576 bytes). Sometimes people try to equate a megabyte with an even million characters.

Megahertz, MHz (DP, TC) A million cycles, occurrences, alterations, or pulses per second. Used to describe the speed of computers' clock rates.

Memory (DP) The retention of information electronically, on chips. There are two main types of memory: RAM, which is used for the short-term retention of information (that is, until the power is turned off), and ROM, which is used to store programs that are seldom, if ever, changed.

Menu (DP) A list of commands to operate a computer. There are many types of menus: pop-up menus, submenus, etc.

Mercury lamp (I) A high-intensity discharge (HID) lamp in which the major portion of the light is produced by radiation from mercury.

Metal halide (MH) lamp, (I) A high-intensity discharge (HID) lamp in which the major portion of the light is produced by radiation of metal halides and their products of dissociation, possibly in combination with metallic vapors such as mercury.

Modulator-demodulator, or modem (DP) A device that lets computers talk to each other over phone lines. Modems are used to send digital signals through telephone lines, to be converted back to analog signals at the receiving end.

Mollier chart (H) A form of graph, named for its inventor, covering heat properties of a fluid and having entropy as one coordinate and enthalpy as the other.

Motherboard (DP) The main board in a computer (or another computer device).

MS-DOS (DP) The original, and still the most popular, operating system used on IBM PCs and compatible computers. (The name stands for *Microsoft Disk Operating System*. Also called PC-DOS.)

Multimedia (DP, TC) The combination of multiple media in an integrated system, such as audio, video, text, graphics, FAX, and telephone.

Multiplex (DP, TC) To transmit two or more signals over a single channel.

Multiuser (DP) Said of software or hardware that supports use by more than one person at one time.

Nadir (I) Vertically downward directly below the luminaire or lamp; designated as 0°.

Nanometer (AX, G, I, TC) A unit of length, 10^{-9} m. It is part of the SI system and has largely replaced the non-SI unit angstrom (0.1 nm) in technical literature.

Nanosecond (DP, G, I, TC) A billionth of a second. Used to measure the speed of memory chips, among other things. Abbreviated ns.

National Television Standard, NTS (TC) The analog video broadcast standard used in North America and set by the National Television Standards Committee.

National Television Standards Committee, NTSC (TC) Committee that sets television transmission standards in North America at 30 frames per second and 525 horizontal lines per frame.

Near field (AS) The field in the vicinity of a real sound source wherein the sound level contours are different from those of a point source in a free field.

Needle valve (H) A valve provided with a long tapering point in place of the ordinary valve disk. The tapering point permits fine gradation of the opening.

Net present value (G) The theoretical inception cost or savings due to operation of an element or system including initial investment and cash flow from operations, generally discounting future cash flows to the present, thereby accounting for interest, inflation, and/or foregone profit.

Network (DP, E) Two or more computers (or other computer-related devices) connected to share information. Usually, the term refers to a local area network.

Network operating system, NOS (DP) A software program for a computer network that runs in a file server and controls access to files and other resources from multiple users.

Nipple (H, PS, FP) A tubular pipe fitting usually threaded on both ends and under 12 in. long. Pipe over 12 in. long is regarded as cut pipe.

Noise isolation class, NIS (AS) A single-number rating derived in the same manner as STC but based on NR (noise reduction) rather than TL (sound transmission loss). It includes the acoustical absorption in the receiving room.

Noise reduction (AS) The difference in decibels between the sound in one space and the sound in a second space attenuated by some intervening medium; e.g., from one room to another.

Noise reduction coefficient, NRC (AS) The average of the sound absorption coefficients in the 250-, 500-, 1000-, and 2000-Hz octave bands.

Octave (AS, AV) An interval between two frequencies having a ratio of two to one (2:1).

Octave band (AS, AV) A frequency band whose upper limit is twice the lower limit.

Off line (DP) Said of things done while one is not actively connected to a computer or a network. For example, you might work on a message off line, then log onto an electronic mail system to send it. Opposite of on line.

Ohm's law (E) The relationship between current and voltage in a circuit. The law states that current is proportional to voltage and inversely proportional to resistance. Algebraically, in DC circuits, $I = V/R$; in AC circuits, $I = V/Z$.

On line (DP) On, or actively connected to, a computer or computer network. For example, on-line documentation appears on the screen rather than in a manual. Opposite of off line.

One-pipe hot-water system (H) System that carries heated and cooled water to and from the radiators in the same "main" and in the same direction, with the hot water on top. This system is rare in modern-day applications.

One-pipe steam system (H) A steam heating system in which steam supplied to a radiator travels in the same pipe but in opposite directions as the cold water or condensate returns from the radiator. This system is rare in modern-day applications.

Open protocol (DP, TC) A set of standard procedures that are agreed upon by all manufacturers so that data can be transmitted without obstructions.

Operating system (DP) The basic software that controls a computer's operation.

Optic fiber cable, or OFC (DP, I, TC) An information transmission medium consisting of a core of glass or plastic surrounded by a refractive cladding, and a protective jacket. It transmits stranded and ribbon configurations. See *Fiber optics*.

Orifice (H, PS) An opening. The term is commonly applied to disks placed in pipelines or radiator valves to reduce the flow of a fluid to a desired amount.

Orsat (H) A device for measuring the percentage of carbon dioxide, oxygen, and carbon monoxide in flue gases.

Overcurrent (E) Any current in excess of the rated current of equipment or the ampacity of a conductor.

Overcurrent device (E) A device, such as a fuse or a circuit breaker, designed to protect a circuit against excessive current by opening the circuit.

Panel or panelboard (E) A box containing a group of overcurrent devices intended to supply branch circuits.

Parabolic reflector lamp, PAR (I) A lamp with internal reflector having the contour of a parabola to achieve beam control.

Partial pressure (H) That part of the total pressure of a mixture of gases contributed by one of the constituents.

Pascal, Pa (AS) Measure of pressure in the SI system ($= 1.45 \times 10^{-4}$ psi).

Payback period (G) Elapsed time after installation of an element or system for savings in utilities and maintenance to equal the additional cost of its installation compared with another competing option.

Percentage humidity (H) Ratio of weight of vapor in a mixture of air and water vapor to the weight of vapor when the vapor is saturated at the same temperature. Contrast with relative humidity.

Performance interaction (G) The effects that one or more systems or subsystems have on other systems or subsystems; e.g., better insulation reduces HVAC loads and electrical system capacity to support the HVAC.

Personal computer, PC (DP) IBM machine introduced in 1981; also, its various clones.

pH (H) See *Hydrogen ion concentration*.

Phase alternate line, PAL (TC) A television transmission standard used mostly in Europe; 25 frames per second and 625 lines per frame.

Picture element, or pixel (DP, TC) Any of the little dots of light that make up the picture on a computer (or TV) screen. The more pixels there are in a given area—that is, the smaller and closer together they are—the higher is the resolution of the screen. Sometimes pixels are simply called *dots*. The number of pixels on a screen is usually expressed by the number of horizontal and vertical rows, such as 640×480 or 1800×1280.

Pink noise (AS) Sound that has equal energy in each octave or suboctave band.

Pitch (AS) A subjective description of sound quality referring to the principal frequency content. The sound pressure level of a tone.

Point source (AS, I) A source of essentially zero dimensions that radiates sound or light uniformly in all directions.

Polarization (I) The process by which the transverse vibrations of light waves are oriented in a specific plane.

Polyvinyl chloride, PVC (P) An inert plastic material commonly used for pipes. It has high resistance to corrosion.

PostScript (DP) A page-description programming language developed by Adobe and designed specifically to handle text and graphics and their placement on a page. Used primarily in laser printers and image setters.

Pounds per square inch pressure, psi (H, PF) A unit measure of pressure, or force per unit area.

Power factor, PF (I) The ratio of the working power to the apparent power of an AC circuit.

Presbycusis (AS) A condition of deterioration of hearing acuity due primarily to aging.

Private automatic branch exchange, PABX (TC) See *Private branch exchange*.

Private branch exchange, PBX (AX, TC) A private telephone system that usually interconnects with public telephone systems to serve a business or an organization. It can also provide access to a computer from a data terminal. PBX and PABX are used interchangeably.

Programmable ROM, PROM (DP) A read-only memory chip that can be changed with a special device.

Propeller fan (H) A fan with airfoil blades that move the air in the general direction of the axis of the fan.

Protocol (DP, TC) A set of standard procedures that control how information is transmitted between computer systems.

Psychrometer (H) A device employing a wet-bulb and a dry-bulb thermometer to measure the humidity in the air.

Psychrometric chart (H) A graph used in air conditioning and showing the properties of airsteam mixtures.

Pure tone (AS) A sound in which the instantaneous sound pressure wave is a pure sinusoid.

Raceway (E)   An enclosed channel of metal or nonmetal designed for holding wires or cables.

Radiation (H)   The transfer of energy in wave form from a hot body to a (relatively) cold body, independently of matter between the two bodies.

Radio frequency, RF (TC)   Electromagnetic waves operating between 5 kHz and the MHz range. Although off-air (over-the-air) and satellite TV signals operate far above these frequencies (54 MHz and up to many GHz), these TV signals are frequently incorrectly referred to as RF signals in commercial applications.

RAM cache (DP)   An area of memory set aside to hold information recently read in from disk, so that if the information is needed again, it can be gotten from memory (which is much faster than getting it from disk).

RAM disk (DP)   A portion of memory set aside to act as a temporary disk.

Random access memory, RAM (DP)   The part of a computer's memory used for the short-term retention of information (in other words, until the power is turned off). Programs and documents are stored in RAM while you use them. Actually, just about all kinds of memory are accessed randomly. See *Read-only memory*.

Random noise (AS)   Sound whose magnitude cannot be predicted at any time.

Rapid-start (RS) fluorescent lamp (I)   A fluorescent lamp designed for operation with a ballast that starts the lamp light output within a split second.

Read-only memory, ROM (DP)   The part of a computer's memory used to store programs that are seldom or never changed. A user can read information from ROM but cannot write information to it. See *Random access memory*.

Reboot (DP)   To restart a computer from scratch.

Reflectance (I)   The ratio of the reflected light to the incident light falling on a surface.

Reflected glare (I)   Glare resulting from specular reflections of high luminances in polished or glossy surfaces in the field of view.

Reflector (R) lamp (I)   A lamp with an internal reflector to redirect the light output to a controlled direction.

Refraction (I, TC)   The process by which the direction of a ray of light changes as the ray passes obliquely from one medium to another in which its speed is different.

Refrigerant (H)   A substance that absorbs heat while vaporizing and whose boiling point and other properties make it useful as a medium for refrigeration.

Regenerative heating or cooling (H)   An arrangement whereby heat rejected in one portion of a cycle is utilized in another portion of the cycle.

Relative humidity, RH (H)   The ratio of the water vapor (by weight) in air to the water vapor saturated in air at the same pressure and temperature.

Relief vent (PS)   A vent designed to provide circulation of air between drainage and vent systems or to act as an auxiliary vent.

Residual pressure (FP)   Pressure remaining in a system while water is being discharged from outlets.

Resolution (DP)   The number of dots (or pixels) per square inch (or in any given area). The more dots there are, the higher is the resolution of the device.

Resonance (AS, G)   A state in which the forces of oscillation of a system occur at or near a natural frequency of the system.

Reverberant sound field (AS)   Sound that is reflected from the boundaries of and furnishings within an enclosed space. Excludes direct sound. (See *Direct sound field*.)

Reverberation (AS)   The persistence of sound in an enclosed space as a result of repeated reflection or scattering of the sound.

Reverberation time (AS)   The time required for the sound pressure level in a reverberant sound field to decay 60 dB after the source has been extinguished.

Riser diagram (E)   Electrical block-type diagram showing the connection of major items of equipment and components.

Romex (E)   One of several trade names for NEC-type nonmetallic sheathed flexible cable.

Romex (NM) cable (E)   A cable composed of flexible plastic sheathing inside of which are two or more insulated wires for carrying electricity.

Room cavity ratio, RCR (I)   A number indicating the proportions of a room cavity, calculated from the length, width, and height of the room.

R-value (H)   Resistance rating of thermal insulation.

Sabin (AS)   Unit of measure of acoustical absorption. Named after Wallace Clement Sabin, U.S. physicist.

Saturated air (H)   Air containing saturated water vapor, with both the air and the vapor at the same dry-bulb temperature.

Saturated pressure (H)   That pressure, for a given temperature, at which the vapor and the liquid phases of a substance can exist in stable equilibrium.

Saturated steam (H)   Steam at the boiling temperature corresponding to the pressure at which it exists. Dry saturated steam does not contain water particles in suspension. Wet saturated steam does.

Seasonal energy efficiency ratio, SEER (H)   Measure of the efficiency of HVAC equipment on a seasonal, rather than a design-load, basis.

Self-contained cooling unit (H)   A combination of apparatus for room cooling complete in one package. Usually consists of compressor, evaporator, condenser, fan motor, and air filter. Requires connection to electric line.

Semiconductor (DP, TC)   A material that has an electrical resistance somewhere between that of a conductor (e.g., metal) and that of an insulator (e.g., plastic). Silicon and germanium are the two most commonly used semiconductors. The flow of electric current in a semiconductor can be changed by light or by electric or magnetic fields.

Sensible cooling effect (H)   The difference between the total cooling effect and the dehumidifying effect.

Sensible heat (H)   Heat that raises the air temperature.

Septic tank (PS)   A tank designed to separate solid waste from liquid waste.

Serial port (DP)   The jacks on the back of a PC into which you can plug printers, modems, etc. (*Serial* refers to the fact that data are transmitted through these ports serially, one bit after another, rather than in parallel, several bits side by side.)

Service conductors (E)   The supply conductors that extend from the street main or from transformers to the service equipment of the premises supplied.

Shielded twisted pair, STP (AX, DP, TC) See *Twisted pair*.

Simplex (TC) Single element (e.g., a simplex connector is a single-fiber connector).

Small computer system interface, SCSI (DP) An industry-standard interface for hard disks and other devices that allows for very fast transfer of information.

Software (DP) The instructions that tell a computer what to do. Also called programs or, redundantly, software programs.

Sol-air temperature (H) That temperature of the outdoor air that, in contact with a completely shaded building surface, would give the same rate of heat entry into the outdoor surface of the outdoor air, the actual intensity of solar and sky radiation incident upon that surface, and the actual wind velocity.

Sound power (AS) The rate at which acoustic energy is radiated. Usually measured in watts.

Sound power level, PWL (AS) Ten times the logarithm of the ratio of sound power, in watts, to the reference level (W_{ref}) of 10^{-12} W.

Sound pressure (AS) The fluctuations of pressure about atmospheric pressure. Usually measured in micropascals (μPa).

Sound pressure level, SPL (AS) Ten times the logarithm of the ratio of the mean square pressure to the square of a reference pressure (P_{ref}) of 20 μPa.

Sound transmission class, STC (AS) A single-number rating system designed to provide a cursory estimate of the sound-insulating properties of a wall or partition.

Sound transmission loss, STL (AS) The difference in decibels of the sound pressure level on the receiver side of a partition or barrier from that on the source side, with the receiver side being free-field conditions.

Specific heat (H) The heat absorbed by a unit weight of a substance per unit temperature rise of the substance.

Specific humidity (H) Weight of water vapor, in grains or pounds, per pound of dry air.

Specific volume (H) The volume occupied by a unit of air. Usually given in cubic feet per pound.

Specular reflection (I) Mirrorlike reflection.

Split system (H) Historically, a combination of warm-air heating and radiator heating; the term is also used for other combinations such as hot water-steam and steam-warm air. It is also applicable to a DX cooling system when the evaporator (cooling) section is separated from the condensing (heat rejection) section.

Stack (PS) The vertical main of a system of soil, waste, or vent piping extending through one or more stories.

Stack vent (PS) The extension of a soil waste stack above the highest horizontal drain connected to the stack.

Standard air (H) Air at 70°F and 29.921 in. of mercury and weighing 0.07488 lb per cu ft.

Standing wave (AS) A sound wave that has a fixed distribution in space, the result of progressive waves of identical frequencies. It is characterized by the existence of nodes or partial nodes and antinodes at fixed points in space.

Static pressure (H) The pressure exerted by a fluid in all directions; the pressure which would tend to burst the container; the pressure exerted by the fluid if stationary.

Steam trap (H) An apparatus for allowing water or air to pass but preventing passage of steam.

Storm sewer (PS) A sewer used for conveying rain or surface or subsurface water.

Streamline flow (H) Also termed *viscous* or *laminar flow*. In a streamline flow, the fluid particles are moving in a straight line parallel to the axis of the pipe or duct. With streamline flow, the friction varies directly as the velocity.

Structure-borne sound (AS) Sound transmitted through solid material by means of vibration waves in the material.

Subcooling (H) Cooling of a liquid refrigerant below the condensing temperature at constant pressure.

Subsoil drain (PS) A drain that receives only subsurface or seepage water and conveys it to an approved place of disposal.

Sump pump (PS) A mechanical device for removing liquid waste from a sump.

Superheated steam (H) Steam at a temperature higher than the boiling temperature corresponding to the pressure at which it exists.

Supply fixture unit, SFU (PS) A measure of the probable hydraulic demand on the water supply by various types of plumbing fixtures.

Surcharge (PS) Rising of water levels above sewer inlets owing to instantaneous loads in excess of flow capacity, generally a temporary phenomenon.

Sustainability (G) Quality of an activity, policy, design, or design element to enhance functionality, improve habitability, conserve resources, and reduce harm to the environment.

Swing joint (H, PS) An arrangement of screwed fittings and pipe to provide for expansion in pipelines.

Switchboard (E) A large panel containing switches, overcurrent devices, buses, and, usually, instruments.

System disk (DP) Any disk containing the system software a PC needs to begin operation.

System software (DP) A catchall term for the basic programs that help computers work; system software includes operating systems, programming languages, and certain utilities.

Système International d'Unites (French), or International System of Units, SI (G) System of measurement based on the metric system (meter, kilogram, second); different from the traditional or British system (foot, pound, second). The United States agreed to convert to the SI system in Public Law 94–168, signed in 1975; however, the conversion has been slow.

Task cooling (H) Provision for local comfort conditioning, generally in a small area, where requirements are different from those of surrounding spaces.

Task lighting (I) Provision for local illumination in appropriate areas where requirements are different from those of surrounding spaces.

Tee (H, FP, PS) A pipe fitting that has one side outlet at right angles to the run.

Text file (DP) An ASCII file—just characters and no formatting.

Therm (H) A unit of heat equal to one hundred thousand (100,000) Btu.

Thermocouple (H) Two dissimilar metals joined together to produce an electromotive force that varies with the temperature. Used to measure temperature.

Timbre (AS) A characteristic of sound whereby the listener is capable of distinguishing between two sounds even though they are of equal level and pitch.

Tone (AS)   A sound having a pitch and capable of causing auditory sensation.

Tons of refrigeration (H)   A common commercial measure of refrigeration capacity, especially of the cooling capacity of air conditioning apparatus, the equivalent of the heat required to melt one ton (2000 lb) of ice (heat of fusion is 144 Btu per pound) in 24 hours, hence, 288,000 Btu per day or 12,000 Btu per hour.

Total heat (H)   The sum of sensible heat and latent heat in a substance or fluid above a base point, usually 32°F (0°C).

Total internal reflection (TC)   Total reflection of light back into a material when the light strikes the interface with another material having a lower refractive index at an angle below a critical value.

Total pressure (H)   The sum of the static pressure and velocity pressure of a fluid.

Transformer (E)   A device used to raise or lower electrical voltage by means of an electromagnetic core and windings.

Trap (PS)   A fitting or device designed to provide a liquid seal that will prevent a fluid from passing back to where it came from.

Tubeaxial fan (H)   An airfoil (propeller) or disc fan within a cylinder and including driving mechanism supports either for belt drive or direct connection.

Tungsten–halogen lamp (I)   A gas-filled tungsten incandescent lamp containing a certain proportion of halogens.

Turbulent flow (H)   Flow in which the fluid particles are moving in directions other than a straight line parallel to the axis of the pipe or duct.

Twisted pair, or TP (E, DP, TC)   A type of wire made of two insulated copper conductors twisted around each other to reduce induction (and thus interference) from one conductor to the other. The twist, or lay, varies in length to reduce the potential for interference from signals between pairs in a multipair cable. The lay usually varies between 2 and 12 in. Closer lay provides better attenuation between the conductors. TP cables are classified as unshielded (UTP) and shielded (STP). The latter includes a metal sheath surrounding the pairs within the protective jacket.

U coefficient (H)   Rate of heat transmission in Btuh/sq ft-°F.

Ultrasonic (AS)   Sound above the audible range.

Uninterruptible power supply, UPS (DP, E)   A power supply or system that provides a steady source of electrical power even when the normal (utility) power supply is interrupted. The system usually contains a storage battery floating on line with an inverter to convert battery power from DC to AC. Other UPSs may utilize a DC motor–driven AC generator set.

Unshielded twisted pair, UTP (E, TC)   See *Twisted pair.*

Vacuum breaker (PS)   Check valve open to the atmosphere when the pressure in piping drops to atmospheric pressure.

Vaneaxial fan (H)   An airfoil (propeller) or disk fan within a cylinder, equipped with air guide vanes either before or after the wheel and including driving mechanism supports for either belt drive or direct connection.

Vapor barrier (H)   A material intended to prevent the passage of water vapor through a building wall or a pipe so as to prevent condensation within the wall or within the insulating material.

Vapor pressure (H)   The equilibrium pressure of the vapor of a liquid in contact with the liquid.

Veiling reflections (I)   Regular reflections superimposed on diffuse reflections from an object that partially or totally obscures the details to be seen by reducing the contrast. Veiling reflections are sometimes called *reflected glare.*

Velocity pressure (H)   The pressure exerted by a moving fluid in the direction of its motion. It is the difference between the total pressure and the static pressure.

Ventilation (H)   The art or process of supplying outside (so-called fresh) air to, or removing air from, an enclosure.

Venturi (H)   A contraction in a pipeline or duct to accelerate the fluid and lower its static pressure. Used for metering and other purposes.

Virtual memory (DP)   A technique that lets a computer treat part of a hard disk as if it were RAM.

Virus (DP)   A program that functions on a computer without the user's consent. A benign virus may do nothing more than duplicate itself, but some viruses are meant to destroy data.

Viscosity (H)   The property of a fluid by which it resists an instantaneous change of shape or arrangement of its molecules. Broadly, resistance to flow.

Visual comfort probability, VCP (I)   The rating of a lighting system, expressed as a percent of people who, when viewing from a specified location and in a specified direction, will be expected to find the system acceptable in terms of discomfort glare.

Visual display terminal, VDT (DP, G)   A data terminal with a TV screen.

Voltage, V (E)   The electric pressure in an electric circuit, expressed in volts.

Voltage drop, VD (E)   The diminution of voltage around a circuit, including the wiring and loads. Must equal the supply voltage.

Voltage to ground (E)   For grounded circuits, the voltage between the ungrounded conductor and the ground; for ungrounded circuits, the greatest voltage between the given conductors.

Warming-up allowance (H)   An addition to the capacity of heating system (as calculated for heat loss) to provide for quick pickup in the morning.

Water hammer (PF)   Banging of pipes caused by the shock of closing faucets or other flow-control devices.

Water hammer arrester (PS)   A device, other than an air chamber, designed to provide protection against excessive surge pressure.

Wavelength (electromagnetic) (D, I, TC)   The distance between nodes of an electromagnetic wave, such as radio, TV, light, or radar; given by L (length, m) $= 3 \times 10^8/f$, where f is the frequency, in hertz.

Wavelength (sound) (AS)   The distance between nodes in a sound wave; given by L (length, ft) $= 1130/f$, where f is the frequency, in hertz.

Wet-bulb temperature, WB (H)   The temperature of the air as measured by a wet-bulb thermometer. Except when the air is saturated, the wet-bulb temperature is lower than the dry-bulb temperature in inverse proportion to the humidity.

White noise (AS)   Sound that has equal energy in every pass band of equal bandwidth.

Wide area network, WAN (DP)   A network of computers and related devices that transmit data among many nodes between buildings or cities. See *Local area network*.

Window (DP)   An enclosed area on the VDT or LCD screen that has a title bar (which one can use to drag the window around). Disks and folders open into windows, and documents appear in windows when recalled. Windows is also the name of an operating system.

Zonal cavity method (I)   A lighting design procedure used for predetermining the relation between the number and types of lamps or luminaires, the room characteristics, and the average illuminance on the work plane. The zonal cavity method takes into account both direct and reflected flux.

Appendix

GLOSSARY OF TECHNICAL ORGANIZATIONS

GLOSSARY OF TECHNICAL ORGANIZATIONS

Abbreviation	Full Name
ACEC	American Consulting Engineers Council
AGA	American Gas Association
AGC	Associated General Contractors of America
AIA	American Institute of Architects
AIID	American Institute of Interior Designers
AIPE	American Institute of Plant Engineers
AMCA	Air Movement and Control Association, Inc.
ANSI	American National Standards Institute
ARI	Air Conditioning and Refrigeration Institute
ASCII	American Standard Code for Information Interchange
ASHRAE	American Society for Heating, Refrigerating, and Air Conditioning Engineers
ASME	American Society of Mechanical Engineers
ASPE	American Society for Plumbing Engineers
ASSE	American Society of Sanitary Engineers
ASTM	American Society for Testing and Materials
AWWA	American Water Works Association
BICSI	Building Industry Consulting Service International
BOCA	Building Officials and Code Administration
BOMA	Building Owners and Managers Association, International
BRI	Building Research Institute
CSI	Construction Specification Institute
EJC	Engineers Joint Council
EPA	Environmental Protection Agency
FIA	Factory Insurance Association
FMS	Factory Mutual System
IBR	Institute of Boiler and Radiator Manufacturers
IEEE	Institute of Electrical and Electronics Engineers
IES	Illuminating Engineering Society
IESNA	Illuminating Engineering Society of North America
ISO	International Organization for Standards
NCAC	National Council of Acoustical Consultants
NEC	National Electrical Code
NECA	National Electrical Contractors Association
NEMA	National Electrical Manufacturers Association
NFPA	National Fire Protection Association
NPC	National Plumbing Code
NSPE	National Society of Professional Engineers
SMACNA	Sheet Metal and Air Conditioning Contractors National Association
TIA/EIA	Telecommunication Industry Association (Formerly EIA/TIA)
UBC	Uniform Building Code
UFC	Uniform Fire Code
UL	Underwriter Laboratories, Inc.

Appendix

C

Units and Conversion of Quantities

This Appendix Contains the Commonly Encountered units of quantities and derived quantities in conventional (imperial or British) and SI units. Although the United States will eventually phase out the imperial system in favor of the International Standard (Système International, or SI) system, most U.S. construction measures are still expressed in the conventional system. For example, building dimensions are still given in feet and inches; pipe and conduit sizes are expressed in inches. Thus, a conversion table between the two systems is needed until such time as only the SI system is in use.

This appendix provides an abbreviated list of units and conversion factors between conventional and SI units. For more complete information on units, the reader is referred to the following publications:

- AIA: *Metric Building and Construction Guide*
- ASTM: *Standard Practice for Use of the International System of Units*

For convenience, this appendix also includes information on the standard exponents and Greek letters.

TABLE C–1
Greek letters

A	α	Alpha	E	ϵ	Epsilon	I	ι	Iota	N	ν	Nu	P	ρ		Rho	Φ	ϕ	Phi
B	β	Beta	Z	ζ	Zeta	K	κ	Kappa	Ξ	ξ	Xi	Σ	$\sigma \varsigma$		Sigma	X	χ	Chi
Γ	γ	Gamma	H	η	Eta	Λ	λ	Lambda	O	o	Omicron	T	τ		Tau	Ψ	ψ	Psi
Δ	δ	Delta	Θ	$\vartheta\ \theta$	Theta	M	μ	Mu	Π	π	Pi	Y	υ		Upsilon	Ω	ω	Omega

TABLE C–2
Exponents

Prefix	Symbol	Factor by Which the Unit Is Multiplied
exa	E	$1{,}000{,}000{,}000{,}000{,}000{,}000 = 10^{18}$
peta	P	$1{,}000{,}000{,}000{,}000{,}000 = 10^{15}$
tera	T	$1{,}000{,}000{,}000{,}000 = 10^{12}$
giga	G	$1{,}000{,}000{,}000 = 10^{9}$
mega	M	$1{,}000{,}000 = 10^{6}$
kilo	k	$1000 = 10^{3}$
hecto	h	$100 = 10^{2}$
deka	da	$10 = 10^{1}$
deci	d	$0.1 = 10^{-1}$
centi	c	$0.01 = 10^{-2}$
milli	m	$0.001 = 10^{-3}$
micro	μ	$0.000{,}001 = 10^{-6}$
nano	n	$0.000{,}000{,}001 = 10^{-9}$
pico	p	$0.000{,}000{,}000{,}001 = 10^{-12}$
femto	f	$0.000{,}000{,}000{,}000{,}001 = 10^{-15}$
atto	a	$0.000{,}000{,}000{,}000{,}000{,}001 = 10^{-18}$

645

TABLE C–3
Partial list of units and abbreviations

Abbreviation	Unit	Abbreviation	Unit
A	ampere	kV	kilovolt
atm	atmosphere	kVA	kilovolt-ampere
bps	bits per second	kVAr	reactive kilovolt-ampere
cd	candela	kW	kilowatt
°C	degree Celsius	kWh	kilowatt-hour
cal	calorie	L	luminance
cgs	centimeter-gram-second (system)	lm	lumen
cm	centimeter	lx	lux
cp	candlepower	m	meter
CRI	color-rendering index	m^2	square meter
CU	coefficient of utilization	mA	milliampere
dB	decibel	MHz	megahertz
emf	electromotive force	min	minute (time)
°F	degree Fahrenheit	mm	millimeter
fc	foot-candle	mph	mile per hour
ft	foot	nm	nanometer
ft^2	square foot	ns	nanosecond
fL	foot-Lambert	R	reflectance factor
h	hour	rad	radian
hp	horsepower	sec or s	second
Hz	hertz	sq	square
in.	inch	sr	steradian
$in.^2$	square inch	V	volt
J	joule	VA	volt-ampere
K	kelvin	Var	reactive volt-ampere
kcal	kilocalorie	W	watt
kg	kilogram	μA	microampere
kHz	kilohertz	μV	microvolt
km	kilometer	μW	microwatt
km/s	kilometer per second		

TABLE C–4
Conversion of general, mechanical, and electrical units

Multiply	By	To Obtain
Angstrom units	3.939×10^{-9}	Inches
Acre	0.4047	Hectares
Acre	43,560	Square feet
Atmospheres	760.0	Millimeters of mercury (at 32° F)
Atmospheres	29.921	Inches of mercury (at 32° F)
Atmospheres	33.97	Feet of water (at 62° F)
Atmospheres	10,333.0	Kilograms per square meter
Atmospheres	14.697	Pounds per square inch (psi)
Barrels (oil)	42	Gallons
Boiler horsepower	33,475	Btu per hour (Btuh)
Btu	0.252	Calories (large)
Btu	252	Calories (small)
Btu	778	Foot pounds
Btu	1.055	Kilojoules (kj)
Btu	0.000293	Kilowatt-hour (kWh)
Btu per pound (Btu/lb)	2.326	Joules per gram (J/g)
Btu per hour (Btuh)	0.252	Calories (large) per hour
Btu per hour per square foot per F (Btu/h · ft² · F)	0.488	Calories (small) per hour per square centimeter per C (cal/h · cm² · c)
Btu-inch per square foot per hour per F (Btu-in/ft² · h · F)	1.24	Calorie (small)-centimeter per sq cm per hr per C (cal-cm/cm² · h · C)
Calories (large)	3.968	British thermal units (Btu)
Calories (large)	1.1619	Watt-hours (Wh)
Calories (large) per hour	3.968	Btu per hour
Centimeters	0.3937	Inches
Centimeters of mercury	136.0	Kilograms per square meter
Centimeters of mercury	0.1934	Pounds per square inch
Cubic centimeters	0.06102	Cubic inches
Cubic feet	0.028317	Cubic meters
Cubic feet	1728.0	Cubic inches
Cubic feet	7.48052	Gallons
Cubic feet	28.32	Liters
Cubic feet of water	62.37	Pounds (at 60° F)
Cubic feet per minute (CFM)	472.0	Cubic centimeters per second (cm³/s)
Cubic feet per minute	0.4720	Liters per second (L/s)
Cubic inches	16.39	Cubic centimeters
Cubic meters	35.3145	Cubic feet
Feet	30.48	Centimeters
Feet of water	62.37	Pounds per square foot
Feet of water	0.4335	Pounds per square inch
Feet per minute (fpm)	0.5080	Centimeters per second (cm/s)
Feet per minute	0.3048	Meters per minute
Feet per second (fps)	30.48	Centimeters per second (cm/s)
Feet per second (fps)	18.29	Meters per minute
Gallons (U.S.)	0.1337	Cubic feet
Gallons (U.S.)	231.0	Cubic inches
Gallons (U.S.)	3.7853	Liters
Gallons of water	8.3453	Pounds of water (at 60° F)
Gallons per minute (GPM)	0.06308	Liters per second
Grains	0.0648	Grams
Horsepower	33,000.0	Foot-pounds per minute
Horsepower	2546.0	British thermal units per hour
Horsepower	42.42	British thermal units per minute
Horsepower	0.7457	Kilowatts

(Continued)

TABLE C–4 (*Continued*)

Multiply	By	To Obtain
Horsepower (boiler)	33,475	British thermal units per hour (Btuh)
Inches	2.540	Centimeters
Inches of mercury (at 62° F)	13.57	Inches of water (at 62° F)
Inches of mercury (at 62° F)	0.4912	Pounds per square inch
Inches of water (at 62° F)	0.07355	Inches of mercury
Inches of water (at 62° F)	25.40	Kilograms per square meter
Inches of water (at 62° F)	0.03613	Pounds per square inch
Inches of water (at 62° F)	5.202	Pounds per square foot
Kilograms	2.20462	Pounds
Kilogram-calories	3.968	British thermal units
Kilograms per square centimeter	14.220	Pounds per square inch
Kilograms per square meter	0.2048	Pounds per square foot
Kilometers	0.62137	Miles
Kilowatts	1.341	Horsepower
Kilowatt-hours	3415.0	British thermal units
Kilowatt-hours	860.5	Kilogram-calories
Latent heat of ice	143.33	British thermal units per pound
Liters	0.03531	Cubic feet
Liters	61.02	Cubic inches
Liters	0.2642	Gallons
Meters	3.28083	Feet
Meters	39.37	Inches
Meters per minute	3.281	Feet per minute
Meters per minute	0.05468	Feet per second
Meters per second	3.281	Feet per second
Miles	1.60935	Kilometers
Miles per hour (mph)	1.61	Kilometers per hour (km/h)
Pounds	7000.0	Grains
Pounds	0.45359	Kilograms
Pounds of water (at 60° F)	27.68	Cubic inches
Pounds of water evaporated from and at 212° F	970.4	British thermal units
Pounds per square foot (psf)	4.883	Kilograms per square meter
Pounds per square inch (psi)	2.309	Feet of water (at 62° F)
Pounds per square inch	6.895	Kilo pascal (kpa)
Pounds per square inch	0.0703	Kilograms per square centimeter
Square feet	929.0	Square centimeters
Square inches	6.452	Square centimeters
Square meters	10.765	Square feet
Temperature (C) + 273	1	Absolute temperature (C)
Temperature (C) + 17.78	1.8	Temperature (F)
Temperature (F) + 460	1	Absolute temperature (F)
Temperature (F) − 32	5/9 (0.556)	Temperature (C)
Tons of refrigeration (Ton)	12,000.0	British thermal units per hour (Btuh)
Tons of refrigeration	200.0	British thermal units per minute (Btu/min)
Tons of refrigeration	50.4	Calories per minute (cal/min)
Watts	3.415	British thermal units per hour (Btuh)
Watts	0.01434	Kilogram-calories per min. (kcal/min)
Watt-hours (Wh)	3.415	British thermal units (Btu)
Watt-hours	0.8605	Kilogram-calories (kg-cal)

INDEX